The Cartographic Atlas of the Brain

Paulo Abdo do Seixo Kadri

The Cartographic Atlas of the Brain

Springer

Paulo Abdo do Seixo Kadri
Department of Neurosurgery
Brigham and Women's Hospital, Harvard School of Medicine
Boston, MA, USA

ISBN 978-3-031-38061-7 ISBN 978-3-031-38062-4 (eBook)
https://doi.org/10.1007/978-3-031-38062-4

This Springer imprint is published by the registered company Springer Nature Switzerland AG
The registered company address is: Gewerbestrasse 11, 6330 Cham, Switzerland

Foreword

This atlas presents breathtaking illustrations of the most exquisite preparation of the intrinsic anatomy of the brain. For several years, I witnessed Professor Kadri painstakingly carry out the most elegant dissections. Each time, he was not satisfied with an excellent one, he replaced it with a more perfect one.

Such a finely detailed and comprehensive atlas could not have come at any better time. With unprecedented advancements in all aspects of neuroscience, such a work becomes essential to any clinical practice or research related to the brain. In this atlas, the surgeon has the most precise map for safely approaching and dissecting intrinsic lesions. Neuronavigation has become routine and this atlas provides the authentic basis for such application. This work is the cornerstone for restoring brain function, as restoration depends on targeting functional centers and their connections. Neuroimaging, particularly with high-Tesla MRI, has opened the door for functional delineation and tractography. The preparations in this atlas show the actual anatomical structures for such advanced imaging, which can be used to verify and maintain the accuracy of the images. Advances in neuropharmacology restore, enhance, or modify brain function, all based on neurotransmitter and synapses from certain brain structures and its interconnectivity. Even rehabilitation is enhanced through knowledge of supplemental areas and the concept of brain plasticity. As traditional neurology builds on the anatomical location of symptoms and disease, advancements in understanding and treating the brain depend on the knowledge presented in this atlas. Simply stated, any student, clinician, or researcher of any aspect of brain function will find this work indispensable.

Notwithstanding its invaluable scientific contribution, this atlas is a striking work of art. The late Dr. Kenichiro Sugita wrote, "The living brain has dignity and beauty. We neurosurgeons are allowed to look at and touch it." Professor Kadri has put the beauty on display for all to see.

A classic is born.

Ossama Al-Mefty

Department of Neurosurgery
Brigham and Women's Hospital, Harvard School of Medicine,
Boston, MA, USA

Homage

Perhaps the first well-known mention of the concept of *gratitude* is found in Homer (8th century BCE): "Hera says to Sleep that if he will grant her request, she will always be grateful to him." Although expressed in this passage, gratitude is not defined. In the work of Theognis of Megara (6th century BCE), gratitude is given another dimension; it is no longer a simple obligation to return: "The favor cannot continue; the gratitude for it can." In his *Crito*, Plato attempts to answer the question "what is the nature of gratitude?" through Socrates' arguments on his duties to Athens, even though facing his own end. Aristotle placed gratitude among the virtues. Hesiod (active c. 750–650 BCE) described the Charites as personifying multiple virtues. These goddesses then became the Romans' Gratiae (The Graces). In his De Beneficiis ("On the Favors"), Seneca the Younger (4 BCE–65 AD) described the popular representation of the three goddesses' transfiguration as dancing hand-in-hand in an unbroken circle: A "benefit passing hand to hand nevertheless returns to the giver; the beauty of the whole is destroyed if the course is anywhere broken." For him, then, gratitude is definitely not a duty.

In his unfinished work, Summa Theologiae, Saint Thomas Aquinas (1225–1274) defined three rules that must be followed to accomplish gratitude: recognize the received favor, express one's appreciation and thanks, and repay the favor at a suitable place and time according to one's means.

To Thomas Hobbes (1588–1679), the law of nature regarding gratitude is important: "A man who receives benefit from another out of mere grace should try to bring it about that the giver of the benefit doesn't come to have reasonable cause to regret his goodwill."

The gratitude described by David Hume (1711–1776) is a virtue that "promotes the interest of our species and bestows happiness on human society." To Immanuel Kant (1724–1804), it is a debt that cannot be fully repaid.

Ever since I first met Professor Al-Mefty, in the year 2000, there has never been a single occasion when I should not have expressed my gratitude. Unfortunately (because Fortune cannot be blamed for my faults), I am humbly aware that I have not always succeeded at Aquinas' requirements. My coarseness often precluded me from recognizing the finesse of his acts. Chaperoned by timidity, I fear to become aware of the myriad occasions when I did not even whisper the acknowledgments, which I should have shouted. Beyond all that, the forthcoming days of my finitude won't be enough to repay the simplest of his favors should I dare to enumerate them. So, I am communing with Kant, in the hope of satisfying Hobbes and returning to humility using the virtue of Hume without breaking the circle of Seneca. For this, I have found my prayer in the divine Quran: "wa man shakara fai in nama yaskuro lenfseh" (Surat Loqman verse 12). Translated, this phrase means when you give gratitude to others, you are giving to yourself. I pray in the certainty that Professor Al-Mefty is among those who truly and sacredly deserve to be graced with gratitude.

For Ana, Renata, Gabriela, and myself, our GRATITUDE toward Professor Ossama and Mrs. Janice Al-Mefty is to be written in capital letters, always holding the hands of our capital Love.

Preface

> Anyone who brings out a new book in neuroanatomy at the present day should justify its existence on the first page.—Wendell J. S. Krieg, 1957

I am fortunate to have another opportunity to rediscover some of the best books I have read and, regarding neuroanatomy, to study some that I had only heard about. This specific opportunity seeded the present work. To be concise and be able to fulfill the requirement of the quote from Professor Krieg in his astonishing *Brain Mechanisms in Diachrome* from 1955, I have prepared this homage. I also want to express my gratitude to Professor Ossama Al-Mefty in the first paragraph of this piece. This present work would not even be dreamed of without him, and the ultimate application of the work is shown by his hands at the end of the introduction.

In response to Professor Krieg, a perusal of this atlas will supply the justification. It is a compilation of dissections that I have done over the past 20 years. For many of my peers, the graphic design appeared crowded, while other severe critics found it quite polluted. One quasi-reasonable explanation for this clutter stems from at least one of my own restrictions. My language challenges became obvious during this period in Boston, and there is a trend in the actual historical social context to abbreviate, to make it quick. In medicine and in neuroscience, this trend is expressed by the proliferation and widespread use of abbreviations. This trend, to me, creates an elite, distinct group that understands this language. In seeking an egalitarian view, I refrained from the use of abbreviations as much as I could. As it is the ambition of this author to reach as many people as are interested, selecting the anatomical structures to be labeled was also of concern, as that selection depends on the target audience. In this, maybe my genealogy has been the determinant. As a descendent of Phoenicians, through my Lebanese roots, and from the Portuguese, whose adventures through the unknown were a determinant of the history of the human being to unveil the universe, I always admired the complexity and details of cartographic maps. Such maps summon the knowledge of an era and are a key to the beyond. As the classical cartographers who shined the way for seekers of the wonders of vastness, it is my wish that this cartography of the brain candles the way for seekers of the wonders of the brain.

> The brain is a very big place, is the universe, in a very small space.—Carl Sagan, *Cosmos*

> As long as our brain is a mystery, the universe, the reflection of the structure of the brain, will also be a mystery.—Ramón y Cajal

Boston, MA, USA
May 2023

Paulo Abdo do Seixo Kadri

Acknowledgments

To Gabriela, the Love of my life.

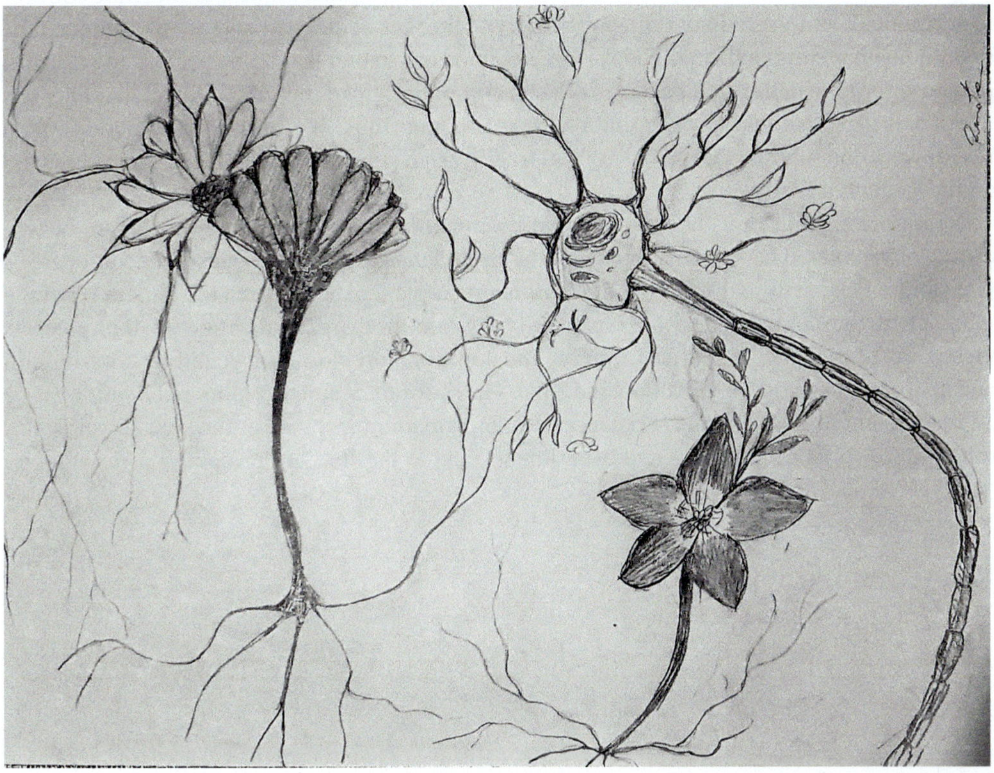

I am genuinely indebted to Professor Ossama Al-Mefty for opening the door through which I was blessed to encounter Professor M. Gazi Yaşargil. Professor Yaşargil instilled in me his passion for neuroanatomy and microneurosurgery. This same entrance brought me fortuitously to this special circle of scholars who commune with the same passion, among them my dear late Professor Evandro de Oliveira (1945–2021) and Professor Ugur Türe. With such a constellation of masters, my sole duty to learn was never an onus but a delight. I shall not say that climbing on their shoulders was an easy chore—indeed even now it is not—but surely a mere glimpse of their vision is magnificent. Through and with them, I have met countless colleagues, every one of them from whom I learned. An attempt to name them all would definitely be an invitation to unfairness. Prudently, as I still have so much to learn from them and because they know whom I am referring to, I humbly express my gratitude to all of them, in the name of my close friends Professors Luis Borba and Jean de Oliveira.

The *Cartographic Atlas of the Brain* is a collection of photographs of anatomical preparations done by me over 20 years and couldn't be completed if not for the generosity of multiple institutions throughout these years. Of special note are the M. Gazi and Dianne Yaşargil

Microsurgical Laboratory at the University of Arkansas for Medical Sciences, in Little Rock, Arkansas, and Ossama Al-Mefty's Microsurgery and Skull Base Laboratory at Brigham and Women's Hospital, Harvard School of Medicine, in Boston, Massachusetts. In these places, as well as in many others, I was accompanied by relentless altruistic colleagues and friends and, to honor them all, it seems to me fair to name Dr. Wenya Linda Bi. Her help, along with that of her colleagues, is easily seen as one observes and studies the demonstrations in this atlas. Any and all anatomical books must pay tribute to the donors of the specimens whose ultimate gift is what made possible this piece.

As a father, I am really proud of my two daughters, Renata and Ana. As a student, I am enormously helped by the illustrations to understand the complexity of neuroanatomy. As the author of this book, I am proudly indebted to my oldest daughter Renata Abdul Lucchesi Kadri, and her artistic skills. At a young age, with her dreamful eyes, captivating smile, and always helpful attitude, she diligently attended my unconventional request to illustrate some of the anatomical demonstrations. Not surprisingly to me, her dedication and talent produced the illustrations that enlightened this book. The depth of her thoughts is to be seen in the illustration chosen for the dedication page. A sole and complex-shaped individual form, that happens to be a cell that we call a neuron; embracing and connecting, in a blessed communing, with three flowers, the sublime face of beauty and fragility in Nature. She called it the three-flowered neuron. I called it family.

The ultimate goal of this book is to assist its audience, among them many who are or will become physicians. The supreme purpose of a physician is to serve the ones in need of care—the patients. The needs and suffering of every neurosurgical patient soften the harshness of the treatment of the coming ones, on every and all scale. The materialization of this piece is indebted to Emma and her parents, Sheryl and Jon Sokoloff, for their generous contribution that supported this work; and Sandra and Mike Gross for their support in its publication.

Finally, I am thankful for the assistance of Julie Yamamoto, who, over the past 20 years, has brought into words my thoughts and experiments, and to the staff at Springer.

Contents

1 Introduction

In the history of human knowledge, the brain, if it was not, undoubtedly now should be considered the most complex part of the human body. Assuredly it was, and still is for many, the most complex structure of creation.

The eight times that the hieratic symbols associated with the brain appear in the 3700-year-old Edwin Smith Papyrus are interpreted as descriptions of the meninges, cerebrospinal fluid, and the external surface of the brain. These descriptions, associated with an attempt to describe the diagnosis and prognosis of cephalic traumatic injuries, are the first known references to the brain on record [1].

In the fifth century BCE, Alcmaeon of Croton was the first writer to champion the brain as the seat of the senses and the central organ of intellect. He is credited with the first description, through dissection, of the optic nerves and Eustachian tubes and recognition of the arteries and veins as blood vessels. He also pioneered the approach of considering the internal causes of illness through the humoral theory. The same line of thought is seen in the Hippocratic Corpus' most famous treatise "On the Sacred Disease," probably written by Hippocrates of Kos (460–375 BCE) and his followers. However, the Hippocratic doctors were not registered as practitioners of dissection, and their knowledge of anatomy is reported as slight [2]. The essentially negative influences of Plato (427–347 BCE) and Aristotle (384–322 BCE) are surprising. Plato's seeking of the divine principle of pure reason rather than observing and experimenting on natural forms, advocating the study of ideas rather than the study of actual objects, was a strong and nefarious influence on the development of neuroscience. His concept of a tripartite soul (the immortal soul, responsible for reasoning and residing in the head; the superior mortal soul, responsible for executing reason and located in the heart; and the inferior mortal soul, responsible for controlling desire and emotions and residing in the liver) was an attempt to localize mental functions and would impact and reverberate in the field for about 2000 years [3].

Even more intriguing is the position of Aristotle. Why would the greatest biologist of antiquity, the founder of comparative anatomy, the first embryologist, the first evolutionist, and the first systematic student of animal behavior dismiss the brain as wet, cold, and devoid of sensation? Furthermore, why would one of the greatest minds favor the heart as the center of sensation, intellect, and movement to the detriment of the views on the hegemony of the brain held by his predecessors, such as Alcmaeon, his contemporaries, such as Hippocrates, and even his successors, such as Theophrastus of Eresus (372–287 BCE)?

Nevertheless, the scientific approach of Aristotle's Lyceum continued and expanded into Alexandria. Ptolemy I Soter (366–282 BCE) along with Alexander, a direct pupil of Aristotle, recruited Demetrius of Phalerum (350–280 BCE) and Strato of Lampsacus (335–269 BCE), themselves disciples of Theophrastus, and founded the Great Mouseion (Museum). In Alexandria, free of the restrictions that the Greek reverence for the body inflicted on the Athenians, where human dissection was not accepted, the seeking of knowledge encountered the quasi-scientific method of Egyptian embalmers, who had been practicing mummification for centuries, and the systematic study of the human body flourished. The Museum, with its more than one hundred professors, included an astronomic observatory, a zoo, botanical gardens, and the wonder of the ancient world, the Library [4]. Furthermore, it included dissection and operating rooms, thus creating the birthplace of neuroanatomy, which reached its peak with Herophilus of Chalcedon (335–280 BCE) and Erasistratus of Ceos (314–250 BCE). Herophilus distinguished and named the cerebrum and cerebellum, described the meninges and the sinuses, provided the first clear description of the ventricles, and distinguished the sensory and motor nerves and traced their origin to the brain. Erasistratus attributed the superior intelligence of humans to the greater number of cerebral convolutions and was perhaps the first to describe the cerebral aqueduct.

P. Kadri, *The Cartographic Atlas of the Brain*, https://doi.org/10.1007/978-3-031-38062-4_1

Nothing original was added to neuroanatomical knowledge over the next 300 years until the peak of the Roman Empire in the second century, when the greatest neuroanatomist of the classical era, Claudius (or Aelius) Galen of Pergamon (129–216), discovered and described more than 250 parts of the nervous system [5]. As a comparison, the next greatest discoverer, Andreas Vesalius (1514–1564) is to be credited with naming less than half of the parts. In fact, practically nothing was discovered for almost 1300 years after Galen's death, with the exception of the Middle Eastern philosophers of the Islamic Golden Age (the eighth to the thirteenth centuries), the Salerno School of Medicine (the late eleventh to the early thirteenth centuries), and the Bologna School of Medicine (the thirteenth to the fourteenth centuries). The vast contribution of the Islamic philosophers cannot be fully addressed here but is reverenced by citing the work of Ibn al-Haytham (Alhazen, 965–1040), who is credited with the first illustration of the optic nerves and optic chiasm [6]. The major contributions of the Salerno School were the revival of Galen's work and the identification of the olfactory bulb and the cauda equina described in the *Anatomia Magistri Nicolai Physici* (c.1150–1200). From the Bologna School, during the earliest stirrings of the Renaissance, the book by Mondino de Luzzi (1270–1326), *Anathomia Corporis Humani*, written in 1316, is considered the first example of a modern dissection manual and the first true anatomical text. It revived human cadaver dissection and described a few other anatomical structures, such as the mammillary bodies [5]. The influence of Galen would last till the Enlightenment, but we had to await the studies of Sir Charles Bell (1774–1842) and François Magendie (1783–1855) to glimpse the next advances in understanding spinal function after Galen's studies on the effects of spinal cord transection at different levels, which led him to conclude that the spine is an extension of the brain and the conduit of sensory and motor signals.

The Renaissance had a great impact with the introduction of illustration to systematic descriptions and teaching in anatomy. Attempting to explain the historical lack of illustrations of the brain, the French anatomist and physician Jacques Dubois, also known as Jacobus Sylvius (1478–1555), professor of Plato, Aristotle, and Galen at the University of Paris, taught that illustrations were crude and misleading representations of what could be seen in dissection. He even attempted to explain the motives that prevented their antecessors, such as Pythagoras of Samos (c.570–495 BCE) and Socrates (c. 470–399 BCE), from writing their thoughts because even the written words were too crude.

The few attempts to illustrate the brain preceding, the *Fabrica* of Vesalius, published in 1543, were merely crude, raw, stylized drawings. These included the 1490 edition of the *Philosophia Pauperum* of Albertus Magnus (c. 1200–1280) and those of Guido da Vigevano da Pavia (1280–1349). Plate XV of da Vigevano's *Anothomia Philippi Septimi* of 1345 is considered the first known illustration of the cortical surface of the brain. It was not until 1517 that a small and an unlabeled, but naturalistic, drawing of the brain was printed in *Feldtbüch der Wundartzney* ("Field Book of Surgery") by Hans von Gersdorff (c.1455–1529). Six years later, in 1523, the first realistic engraving of the brain was printed in the second edition of *Isagogae Breves* by Jacopo Berengario da Carpi (c. 1460–1530), who is considered the most influential anatomist prior to Vesalius.

The masterpiece of Renaissance anatomy, of course, is the 1543 *De Humani Corporis Fabrica Libri Septem* (*On the Fabric of the Human Body in Seven Books*), by the Flemish anatomist and physician Andreas Vesalius (1514–1564). Vesalius is considered the founder of modern anatomy and, with Nicolaus Copernicus (1473–1543), the father of the scientific revolution. The revolutionary series of the most astonishing and breathtaking drawings were produced under the direct supervision of Vesalius by several artists including the Flemish painter and student of Tiziano Vecellio (Titian) (c. 1488–1576), Jan van Calcar (c. 1499–1546). Regarding the nervous system, these illustrations were unmatched and undoubtedly remain in the pantheon of neurosciences. Vesalius was a discoverer surpassed only in the number of discoveries by Galen, and his illustrations were unrivaled for more than two centuries, until the significant advent of fixation methods during the Age of Enlightenment. In 1662, the English physician and anatomist, William Croone (1633–1684) presented his method of hardening and preventing putrefaction by the use of "spirits of wine" (alcohol) [7]. Although this method became well accepted among anatomists, it took more than another century for the paragon atlas of the French anatomist and physician Félix Vicq d'Azyr (1748–1794) to be published in 1786. During this interval, several works contributed enormously to the development of neuroanatomy: *Cerebri anatome: cui accessit nervorum descriptio et usus* (1664) by the English physician and anatomist Thomas Willis (1621–1675); *Discours sur l'anatomie du cerveau* (1669) by the Danish scientist, pioneer in anatomy and geology, and Catholic saint Nicolas Sténon (1638–1686); and *Neurographia Universalis* (1685) by the French anatomist Raymond Vieussens (c. 1641–1715). Shortly after the work of d'Azyr, the Scottish anatomist and surgeon Sir Charles Bell (1774–1842) presented his atlas with his own engravings, *Anatomy of the Brain*, in 1802. Mastering the use of alcohol to harden the specimens, the German physician, physiologist, anatomist, and psychiatrist Johann Christian Reil (1759–1813) presented his work in 1808–1809 [8]. Many other bastions of the classical descriptive era of neuroanatomy would present their astonishing, exquisite artworks. However, despite its splendor, the development of the microscope propitiated the advent of the cell theory by the German botanist Matthias Jacob Schleiden (1804–1881) in 1838 and by the German physician and physiologist Theodor Schwann (1810–1882) in 1839. Meanwhile, the development of physiology was advanced by the experiments and acrimonious debate between the Italian philosopher and physician Luigi Galvani (1737–1798) and the Italian chemist and pioneer of electricity Alessandro Volta (1745–1827), among others. These two factors might at least have contributed to appeasing the modesty of Vesalius on the potential of anatomy for understanding the brain, especially in its functions: "How the brain performs its functions in imagination, in reasoning, in thinking and in memory… I can form no opinion whatsoever. Nor do I think anything more will be found out by anatomy" [3].

In this way, although dissection was an established approach in the study of neuroscience, gradually the interest of neuroscientists shied away from macrodissection of the brain to emphasize the histological and physiological levels. Despite the work of the Dutch microbiologist and microscopist Antonie van Leeuwenhoek (1632–1723) on nerves and the retina [5], neuroscientists would have to wait for technological improvement of the microscopes (with the achromatic lenses) and techniques (such as the microtome and selective staining methods) to harvest the benefits of cell theory. The prevailing tenet to elucidate the nervous system was the reticular theory founded by the German anatomist Joseph von Gerlach (1820–1896) in 1871, even after demonstrations that nerve fibers come from nerve cells by the Swiss anatomist Rudolf Alfred von Kölliker (1817–1905) and the distinguishing of axons from dendrites by the German anatomist Otto Friedrich Karl Deiters (1834–1863) in 1863. Ironically, the method developed in 1873 by the most famous apologist of reticular theory, the Italian biologist, and the pathologist Camillo Golgi (1843–1926) was known as the black reaction. This step was the missing flame to unleash the geniality of the Spanish physician and neuroscientist Santiago Ramón y Cajal (1852–1934), whose revolutionary ideas would launch the foundation of the neuronal doctrine, the basis for our current understanding of the information process in the brain.

Even somehow eclipsed by these developments, nevertheless, the macroscopic study of the brain continued its evolution, absorbing the new technologies. New techniques of documentation and registration helped the practice of dissection survive concomitant with the explosion of new information derived from the microscopic approach to understanding the brain. The first lithograph of the brain was published by the German anatomist and embryologist Emil Huschke (1797–1858) in 1854. Huschke named the centralis anterior and posterior, and the fusiform and lingual gyri, and his lithograph was the first step toward the photography of the brain [9]. In 1896, Magnus Gustaf Retzius (1842–1919), a Swedish anatomist and anthropologist, used photographs to demonstrate brain dissections [8].

In 1935, the Swiss professor of anatomy Josef Klingler (1888–1963) improved the preparation method to realize dissections of the inner structures of the brain, which, in conjunction with exquisite dissections and high-quality, natural-sized black-and-white photographs, would produce one of the most beautiful atlases in the history of neuroscience, in 1956 [10]. Despite the uniqueness of this method, it was never widely used, being underestimated and therefore practically neglected. However, an offspring of neuroscience, neurosurgery, made it one of its pillars. Through the teaching of Klingler, M. Gazi Yaşargil "… gained an inimitable neuroanatomical perspective…" which helped to develop and sustain the advent of microneurosurgery [9]. Ugur Türe, a pupil of Yaşargil, also recognized the value of Klingler's brain dissections and, propelled by the growing interest in an anatomical substrate to validate modern neuroimaging techniques, once again appreciated the macroscopic studies of the brain, this time through the lens of the neurosurgical operative microscope. The importance of microanatomy was promptly recognized by the neurosurgical community, becoming a field of interest per se in the neurosurgical realm. The main exponent of this approach appeared in the figure of the American neurosurgeon Albert L. Rhoton, Jr. (1932–2016), especially regarding the anatomy of the skull base.

Among several other factors, the knowledge acquired through the laboratory studies of the skull base combined with the intrinsic anatomy of the brain is the state-of-the-art condition for each individual who dares to conceive of and master the current practices. A single, concise neurosurgical case, provided by Professor Ossama Al-Mefty, exemplifies this powerful concept (Fig. 1.1).

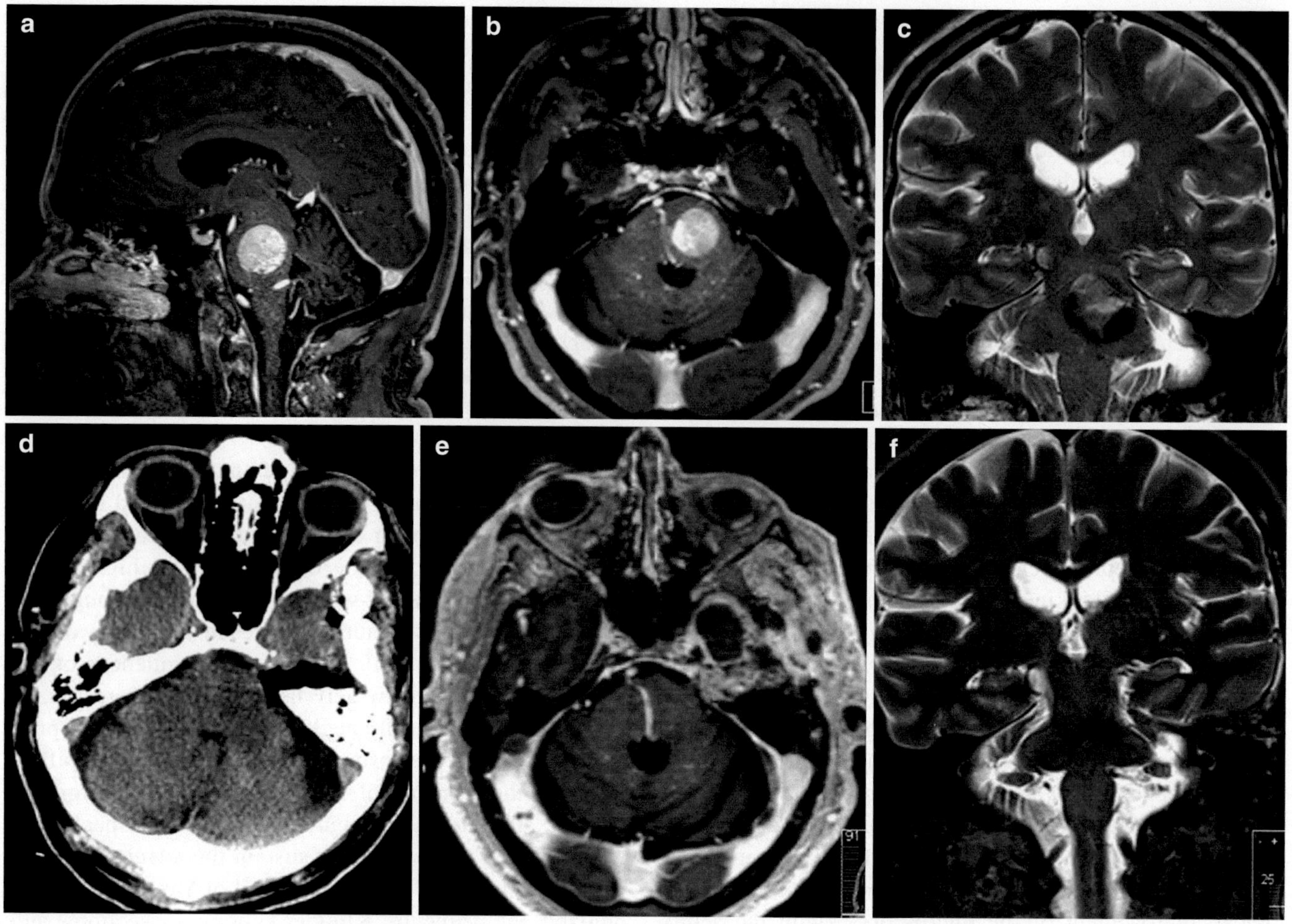

Fig. 1.1 Illustrative case (*courtesy of* Professor Ossama Al-Mefty) depicting the application of skull base approaches (anterior petrosal approach) combined with the knowledge of the intrinsic anatomy of the brainstem to safely access and remove a cavernoma of the base of the pons. (**a**) Preoperative sagittal contrast-enhanced T1-weighted magnetic resonance imaging (MRI) of the head. (**b**) Preoperative axial contrast-enhanced T1-weighted MRI of the head. (**c**) Preoperative coronal T2-weighted MRI of the head. (**d**) Postoperative axial computer tomography of the head depicting the extent of the bone removal on the petrous apex. (**e** and **f**) Postoperative axial contrast-enhanced T1-weighted MRI (**e**) and coronal T2-weighted MRI (**f**) of the head depicting the total removal of the cavernoma

References

1. Breasted JH. The Edwin smith surgical papyrus. Chicago, IL: University of Chicago Press; 1930.
2. Jones WHS. Hippocrates. London: Heinemann; 1923.
3. Gross CG. Early history of neurosciences. In: Adelman G, editor. Encyclopedia of neuroscience. Boston, MA: Birkhäuser; 1987. p. 843–7.
4. Bagnall RS. Alexandria: library of dreams. Proc Amer Philosoph Soc. 2002;146:348–62.
5. Swanson L. Historical trends in neuroanatomical terminology. In: Neuroanatomical terminology: a lexicon of classical origins and historical foundations. New York: Oxford University Press; 2014. p. 7–14.
6. Najjar J. From anesthetic sponge to nonsinking skull perforator, unitary work neurosurgery in the ancient Arabic and Islamic world. World Neurosurg. 2010;73:587–94. https://doi.org/10.1016/j.wneu.2010.01.029.
7. Cole FJ. A history of comparative anatomy: from Aristotle to the eighteenth century (1st Dover edition). New York: Dover Publications; 1975.
8. Türe U, Yaşargil MG, Friedman AH, Al-Mefty O, Yaşargil DG. Fiber dissection technique: lateral aspect of the brain. Neurosurgery. 2000;47:417–27. https://doi.org/10.1097/00006123-200008000-00028.
9. Yaşargil MG. Microneurosurgery, Vol. IVA. New York: Thieme; 1994. p. 2–114.
10. Ludwig E, Klingler J. Atlas cerebri humani. Basel: S. Karger; 1956.

2 Intrahemispheric Association Fibers of the Telencephalon

The connective fiber system of the telencephalon can be systematically divided according to the type of connection established by its neurons to both cortical systems—the neocortical and the limbic. Fibers that connect two neurons within the telencephalon are called *association fibers*. Fibers that connect a telencephalic neuron with a sub-telencephalic neuron (at the diencephalon, brainstem, or spinal cord) are called *projection fibers*. The association fibers can be divided into intrahemispheric and interhemispheric fibers. Intrahemispheric fibers connect cortical areas in the same hemisphere and are known as association fibers. Interhemispheric fibers cross the midline and connect both the hemispheres; they are also known as *commissural fibers* (Table 2.1). Projection fibers connect the telencephalon (the cortex and basal ganglia) with the subcortical areas of the diencephalon, brainstem, and spinal cord. The neocortex forms the corona radiata, the fibers of which converge medial to the lentiform nucleus and are called the internal capsule. The fibers of the paleoencephalon and basal ganglia project from different systems—the limbic and the extrapyramidal systems. Referred to simply as association fibers, intrahemispheric fibers connect the related regions within the same hemisphere and can be further classified as short and long fibers.

Table 2.1 Systematic representation of the association fibers of the telencephalon

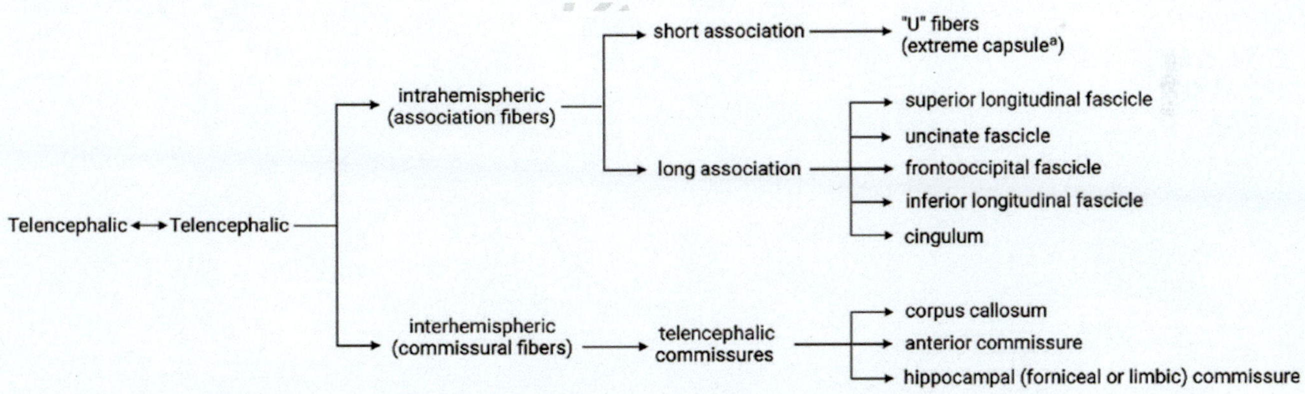

*The extreme capsule is composed of "U" fibers connecting the insular gyri and these gyri to the overlying opercula

P. Kadri, *The Cartographic Atlas of the Brain*, https://doi.org/10.1007/978-3-031-38062-4_2

Short Association Fibers

The short association fibers are also designated as "U" (of Meynert), arcuate (of Arnold), or intergyral fibers [1, 2]. These fibers pass through the cortical and subcortical layers connecting the adjacent gyri and constitute the most superficial lamina of the white matter. They originate from the cortex at the wall of the sulci and travel at the depth of the sulci [3, 4]. When removed, they resemble the letter "U," and hence, it is designated as U fibers (Fig. 2.1).

The "U" fibers are more prominent on the neocortical surface of the brain, especially on the lateral surface, and lie immediately superficial to the long association fibers (Fig. 2.2) As they gradually deepen to connect more distant areas, they are incorporated into the long association fibers (Fig. 2.3). Within the limbic lobe, the "U" fibers are poorly developed.

Extreme Capsule

One particular group of "U" fibers constitutes the extreme capsule. The extreme capsule corresponds to the lamella of white matter located between the cortical layer of the insula and the claustrum (Fig. 2.4) [5–7].

Herein is a short description of the insula as it relates to the extreme capsule: The insular lobe is hidden in the depth of the lateral fissure (Sylvian fissure) by the orbital, fronto-parietal, and temporal *opercula* (Fig. 2.5). It has a triangular shape and is delineated by the anterior, superior, and inferior

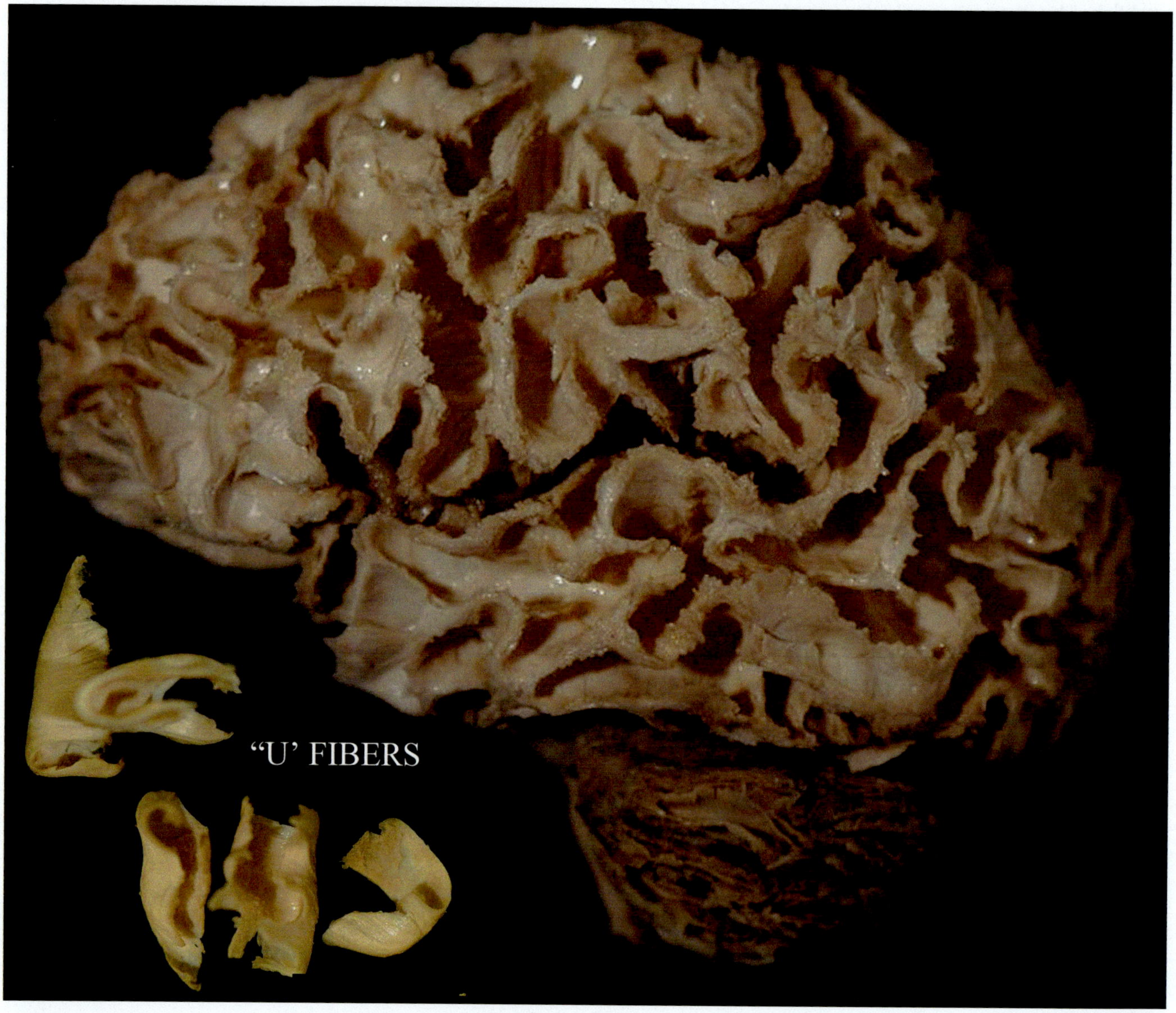

Fig. 2.1 Left-side preparation of the encephalon. After the cortex on the lateral neocortical aspect of the telencephalon is removed, the well-developed short association fiber system is unveiled. Inset: short association fibers resembling the letter "U." (©Kadri 2023. All Rights Reserved)

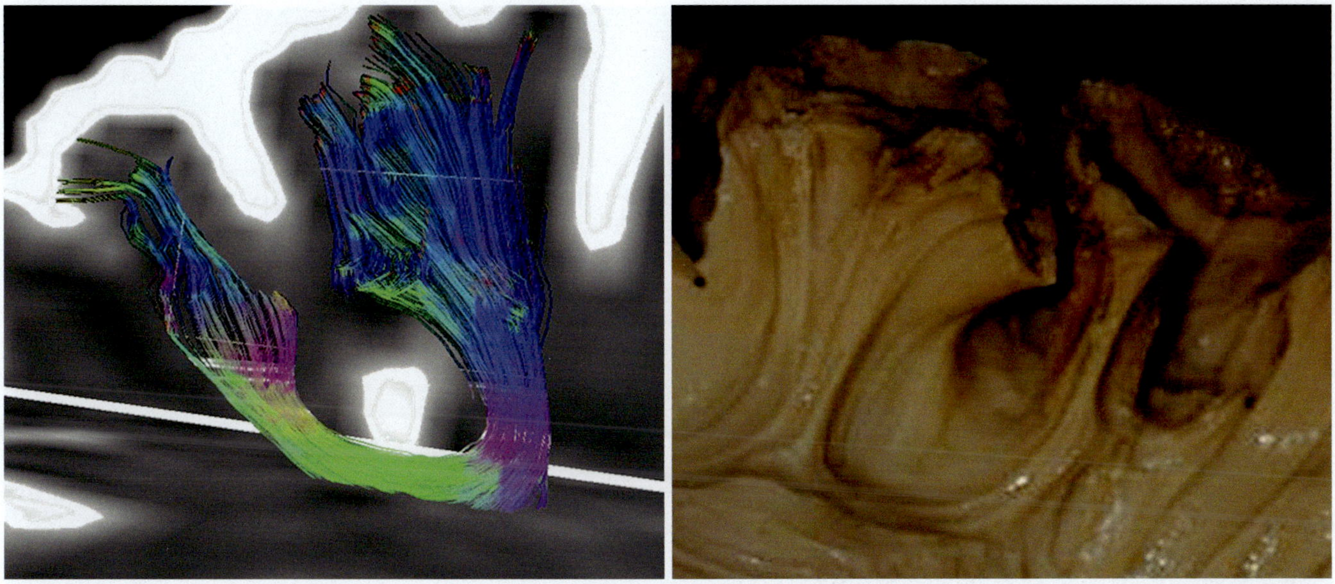

Fig. 2.2 Magnetic resonance tractography of the "U" fibers (left) and their counterpart in the anatomical preparation. (©Kadri 2023. All Rights Reserved)

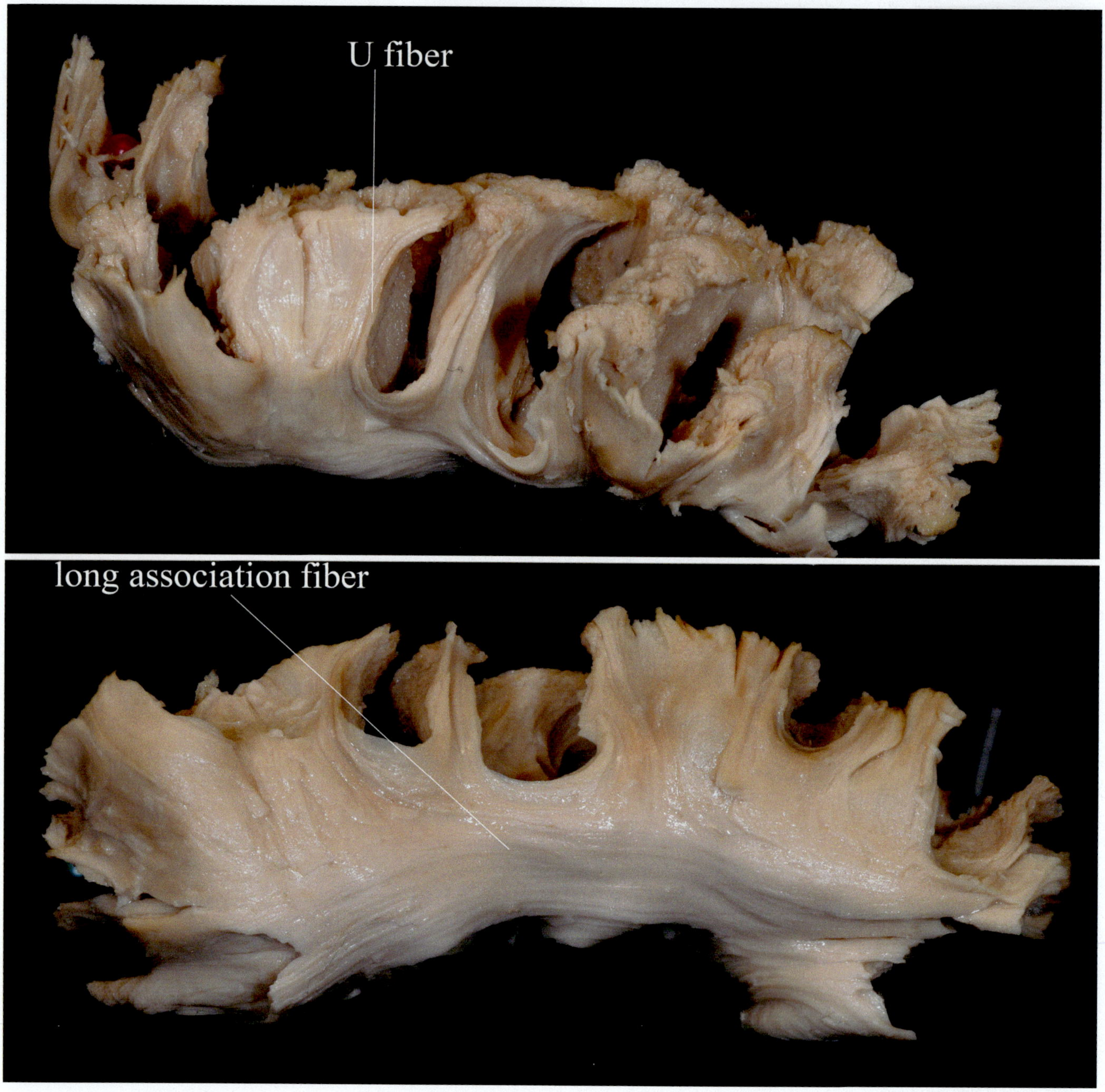

Fig. 2.3 Anatomical preparation of "U" fibers from the lateral (neocortical) surface. Notice the short connection between the adjacent gyri, resembling the letter "U," and its incorporation into a deeper layer of the long association fiber bundle that progressively interconnects more distant regions within the same hemisphere. (©Kadri 2023. All Rights Reserved)

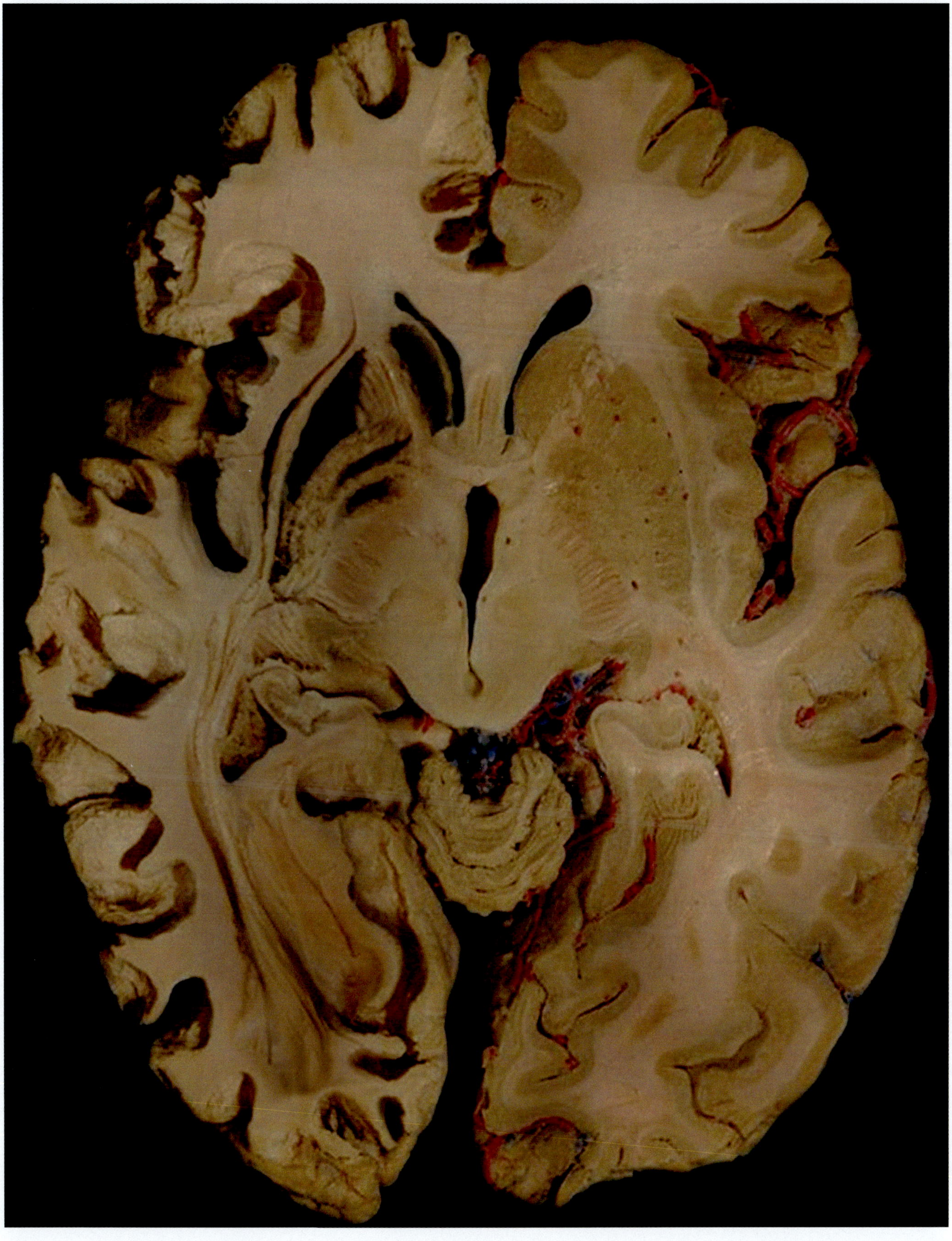

Fig. 2.4 Anatomical preparation and illustration of the axial cut of the brain at the level of the foramen of Monro. The gray matter on the left side was removed to unveil the fiber layers. (©Kadri 2023. All Rights Reserved)

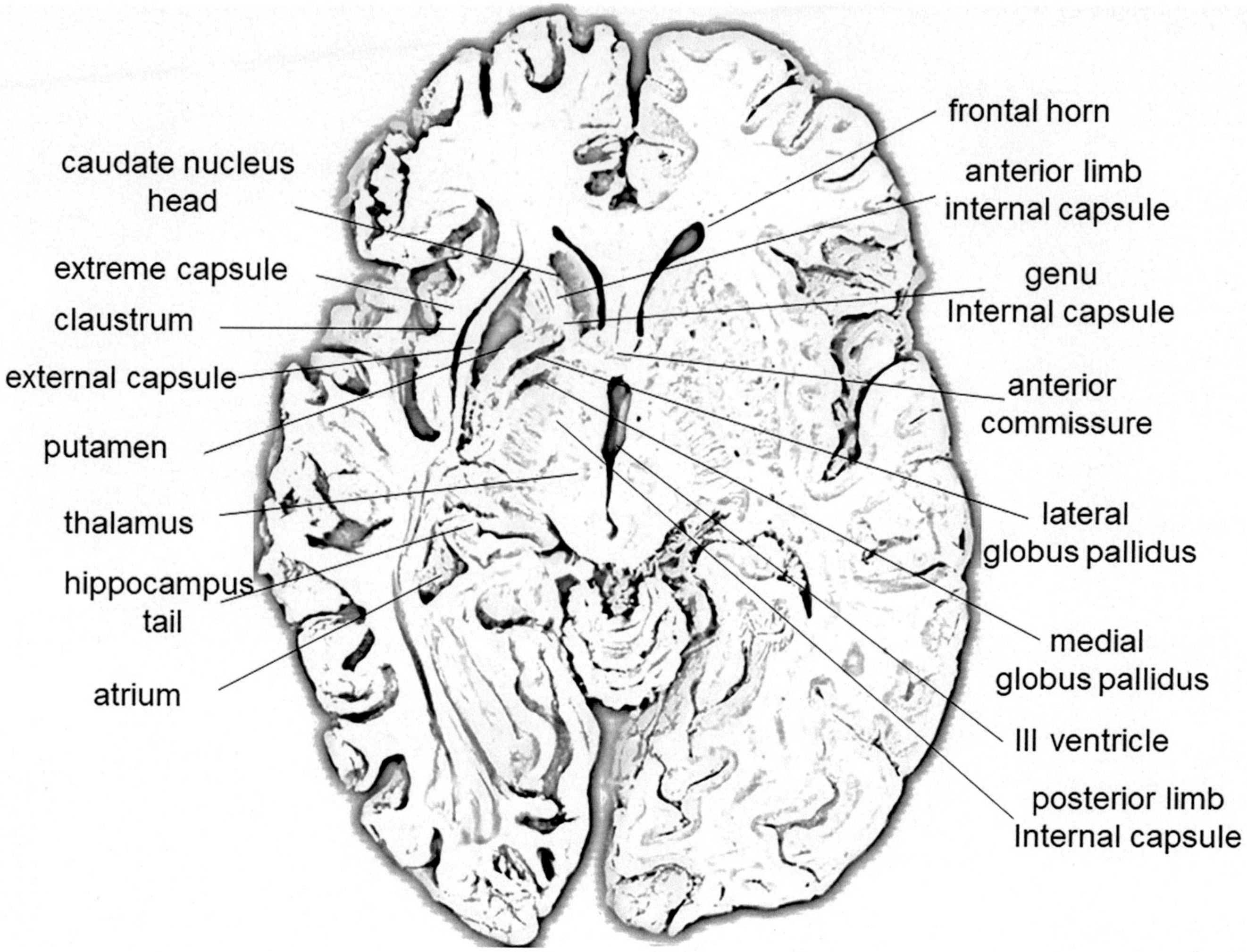

Fig. 2.4 (continued)

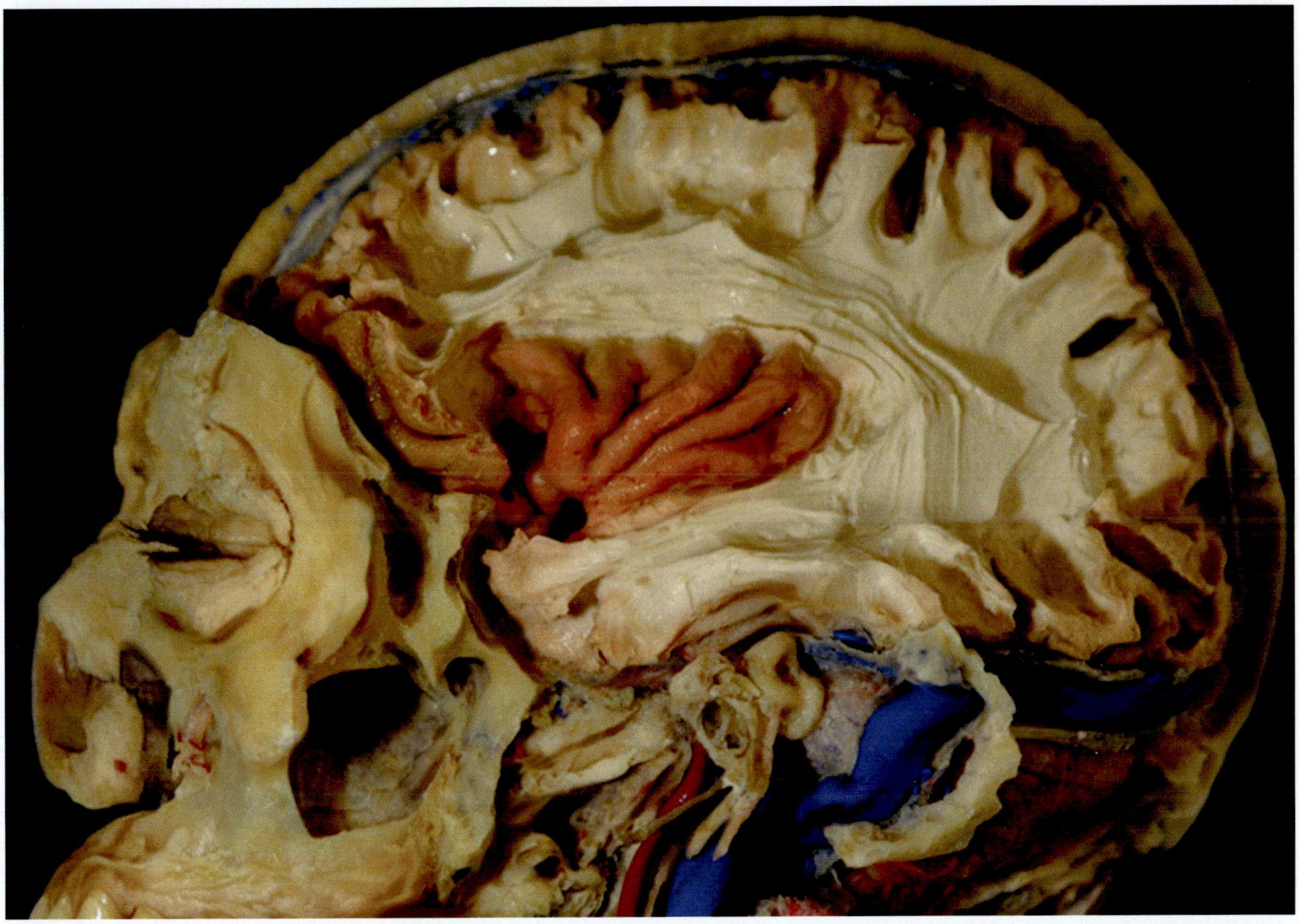

Fig. 2.5 Anatomical preparation and illustration of the left side of the head. Removing the opercula exposes the insula at the depth of the Sylvian fissure. (©Kadri 2023. All Rights Reserved)

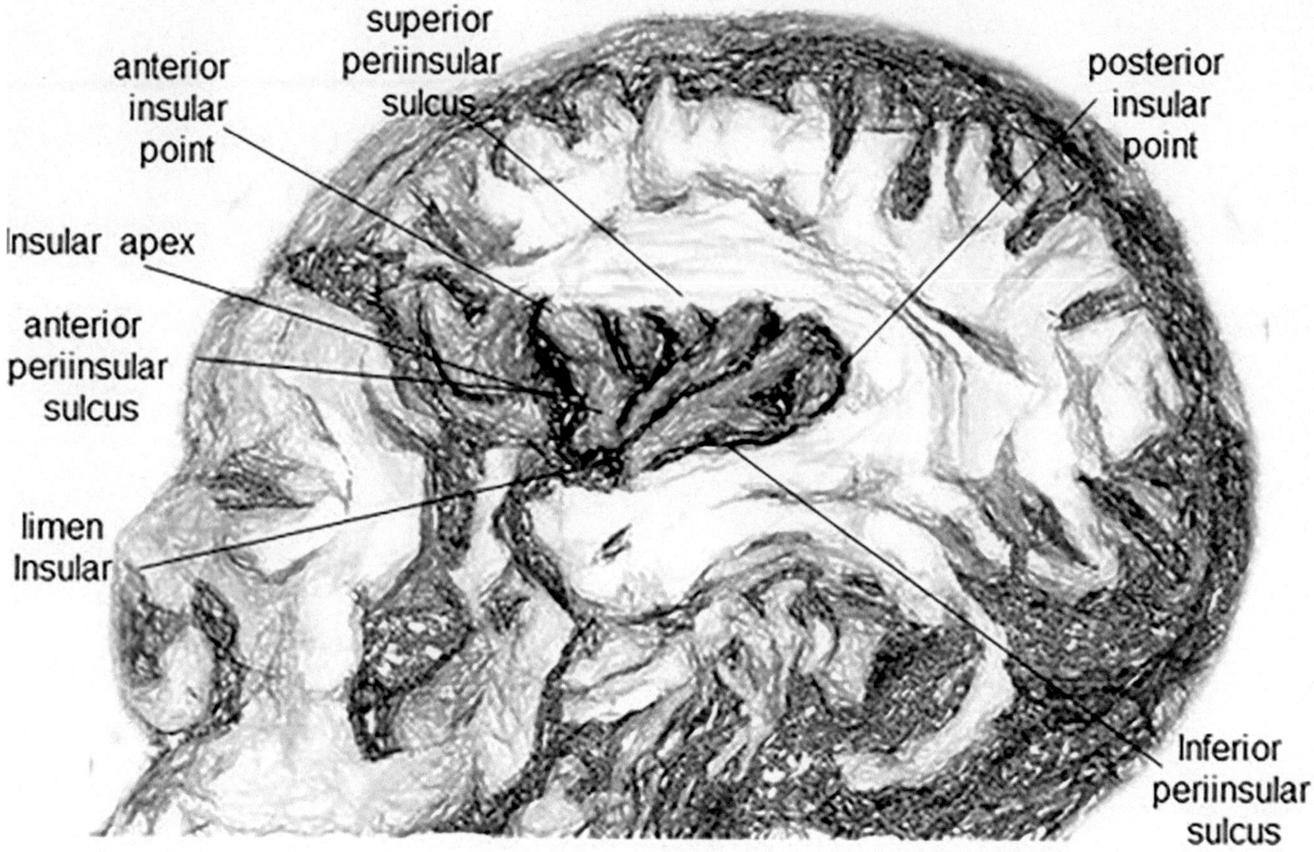

Fig. 2.5 (continued)

peri-insular sulci. The cortex of the insula is continuous with the subopercular cortex through the depth of the peri-insular sulci. This cortex has an anterior and a lateral surface. The anterior surface faces the orbital operculum, and the lateral surface is covered by the frontoparietal and temporal opercula. The insula is divided into anterior and posterior portions by the central sulcus of the insula, which reaches the limen insulae. The anterior insula harbors the short gyri of the insula, usually in threes: the anterior, medial, and posterior. The anterior short gyrus of the insula has an anterior surface that faces the sub-pars orbitalis area and may harbor the accessory and anterior transverse gyri of the insula. The anterior transverse gyrus of the insula traverses the anterior aspect of the limen insulae to the posteromedial region of the orbital surface of the frontal lobe. The posterior insula harbors the long gyri of the insula, usually in twos: the anterior and posterior. In general, the anterior gyrus is more prominent, and the posterior one tends to be smaller, or even rudimentary; however, two equal-size or even three gyri can appear. The limen insulae is the area of deflection of the lateral surface to the basal surface of the brain and is the limit between the insula laterally and the anterior perforated substance medially (Figs. 2.6 and 2.7) [2, 8, 9].

Two parts of the extreme capsule can be identified: the dorsal or superior posterior portion and the ventral or anteroinferior portion [6, 10] (Fig. 2.8). The thin dorsal portion corresponds to the short fibers that interconnect the insular gyri and connect these with the frontal, parietal, and temporal opercula. The ventral portion is located at the depth of the limen insulae. Superiorly, this portion connects the anterior surface of the insula to the sub-pars orbitalis and caudal orbitofrontal region. Anteriorly, it connects the

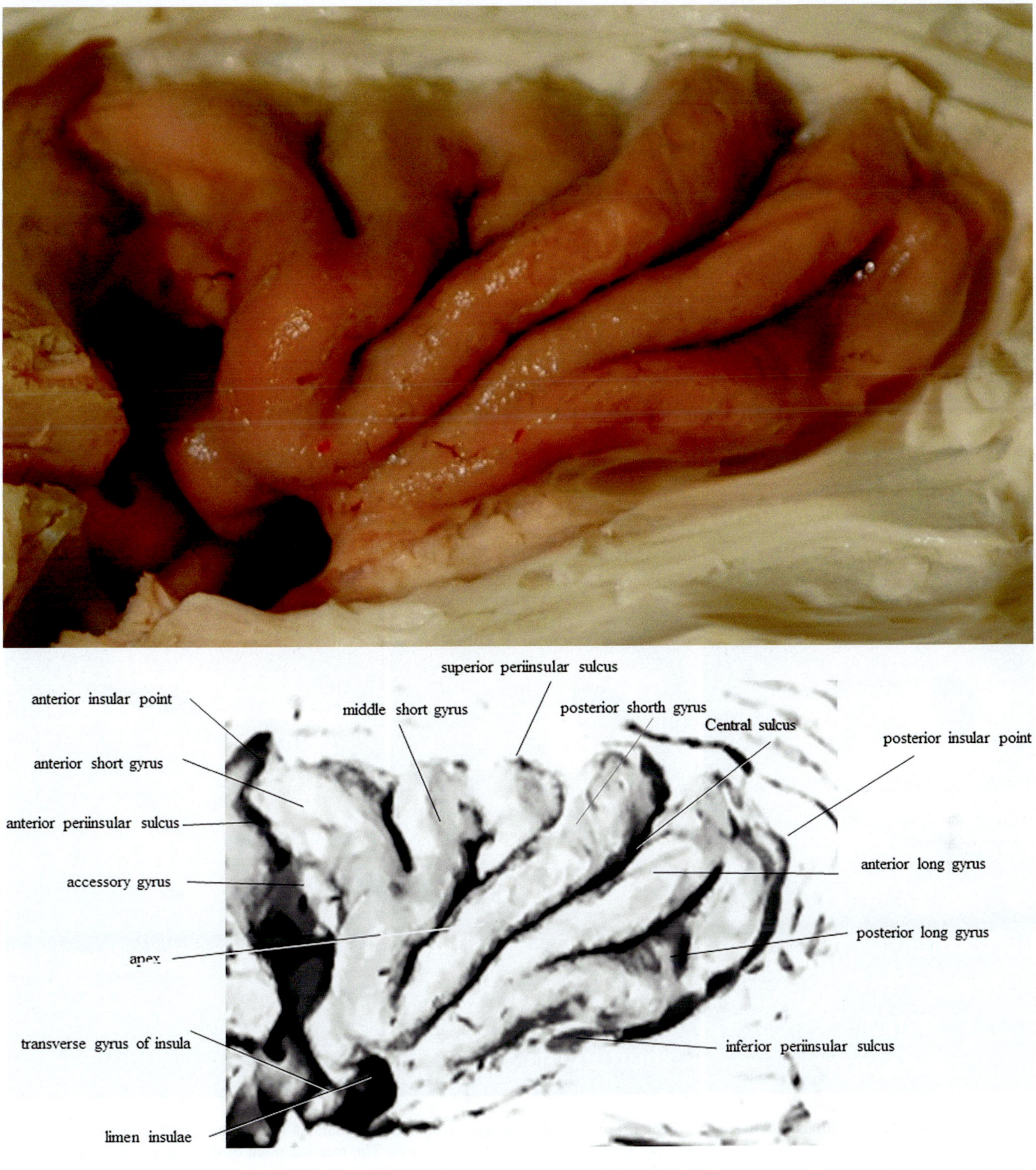

Fig. 2.6 Anatomical preparation and illustration of the left insula depicting the detailed anatomy of the cortical surface. (©Kadri 2023. All Rights Reserved)

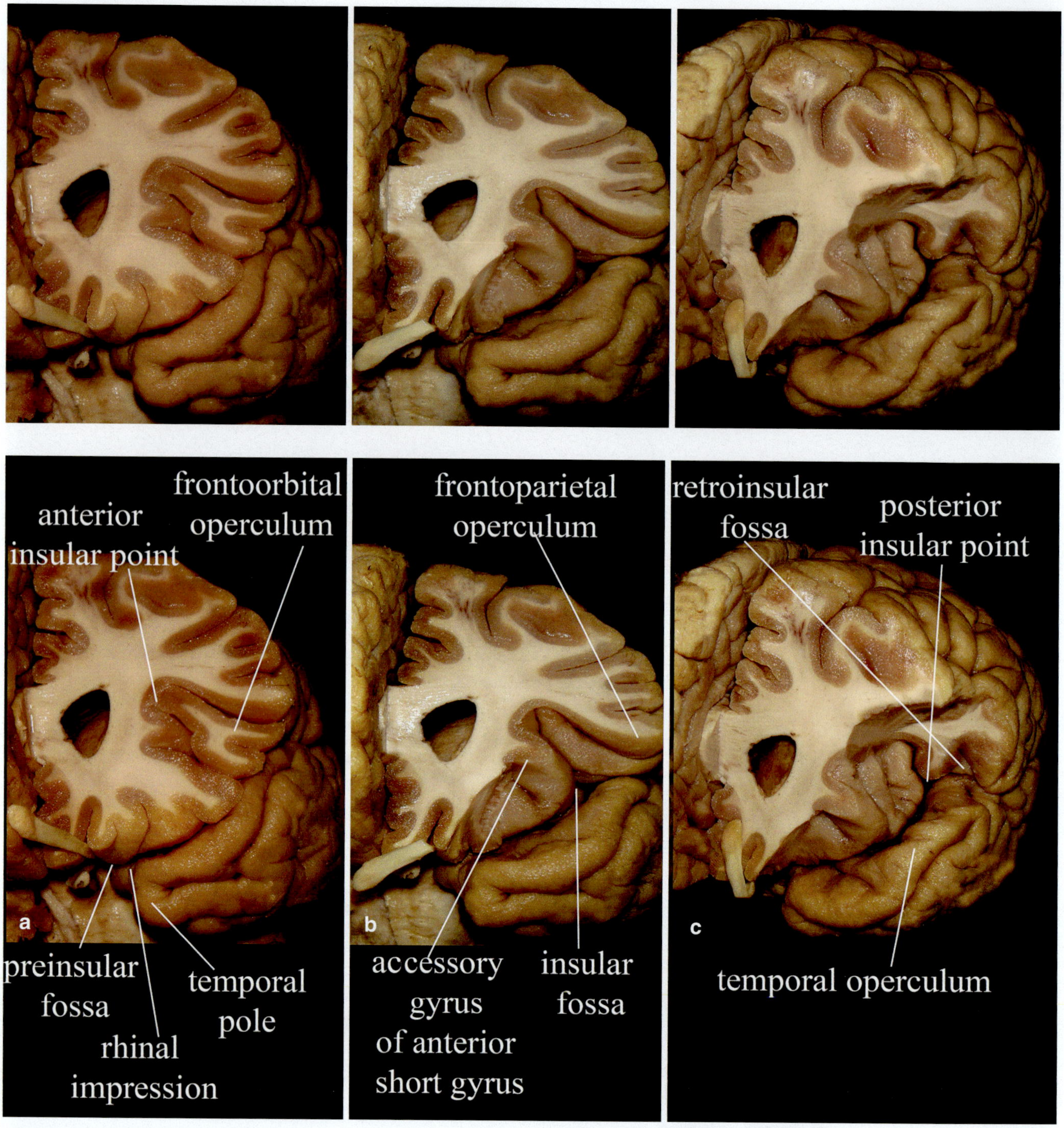

Fig. 2.7 Anatomical preparations and illustrations demonstrate the different compartments of the Sylvian fossa. (**a**) The pre-insular fossa, with the anterior surface of the insula covered by the fronto-orbital operculum. This corresponds to the sphenoidal (proximal, medial, anterior, trunk, or main stem) compartment of the Sylvian fissure, located proximal to the Sylvian point, where the main stem trifurcates at the lateral surface of the cerebral hemisphere. (**b**) The insular fossa, covered superiorly by the frontoparietal operculum and inferiorly by the temporal operculum, contains the insula, which is delineated by the peri-insular sulci. The meeting point of the anterior and superior peri-insular sulci is designated as the anterior insular point. The frontal recess of the frontal horn is located medial to it. (**c**) The retroinsular fossa, distal to the insula, is the deepest portion of the Sylvian fissure. The posterior insular point corresponds to the meeting point of the superior and inferior peri-insular sulci. The atrium of the lateral ventricle is located medial to it.

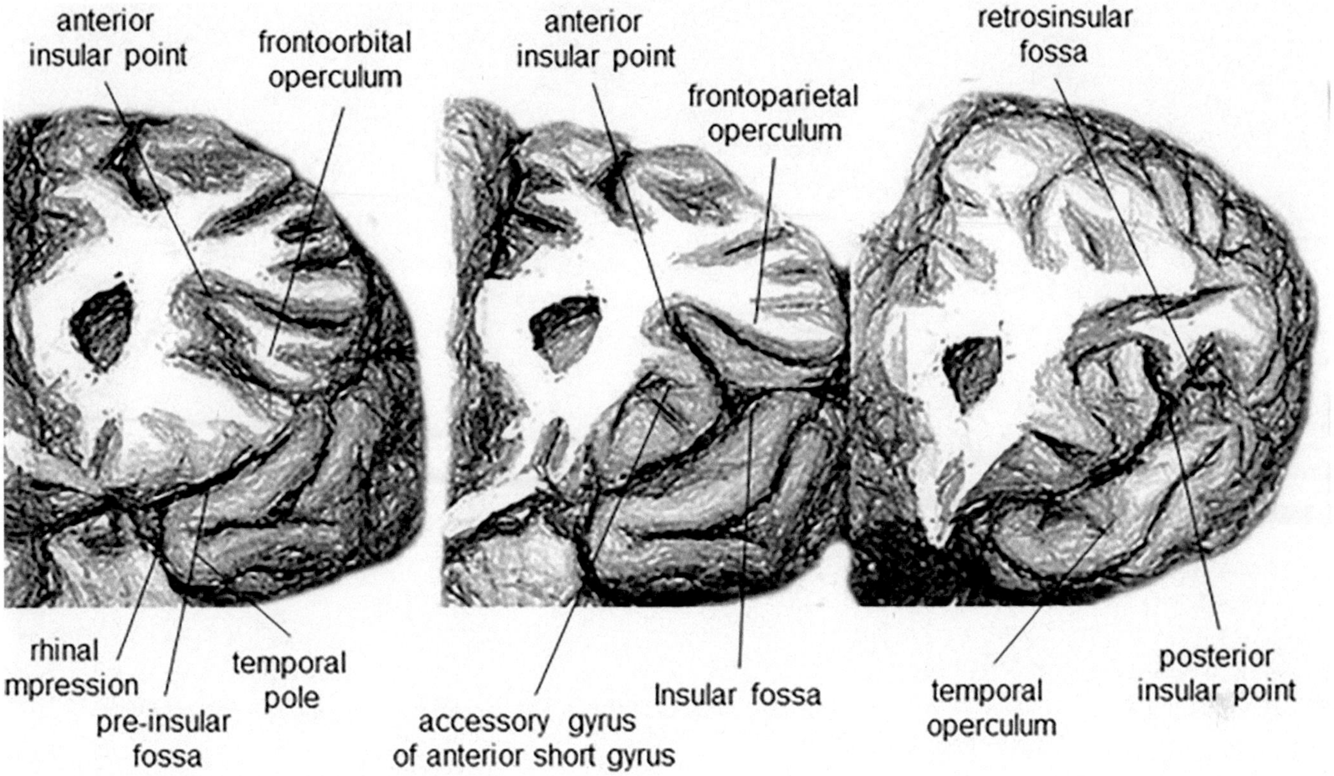

Fig. 2.7 (continued)

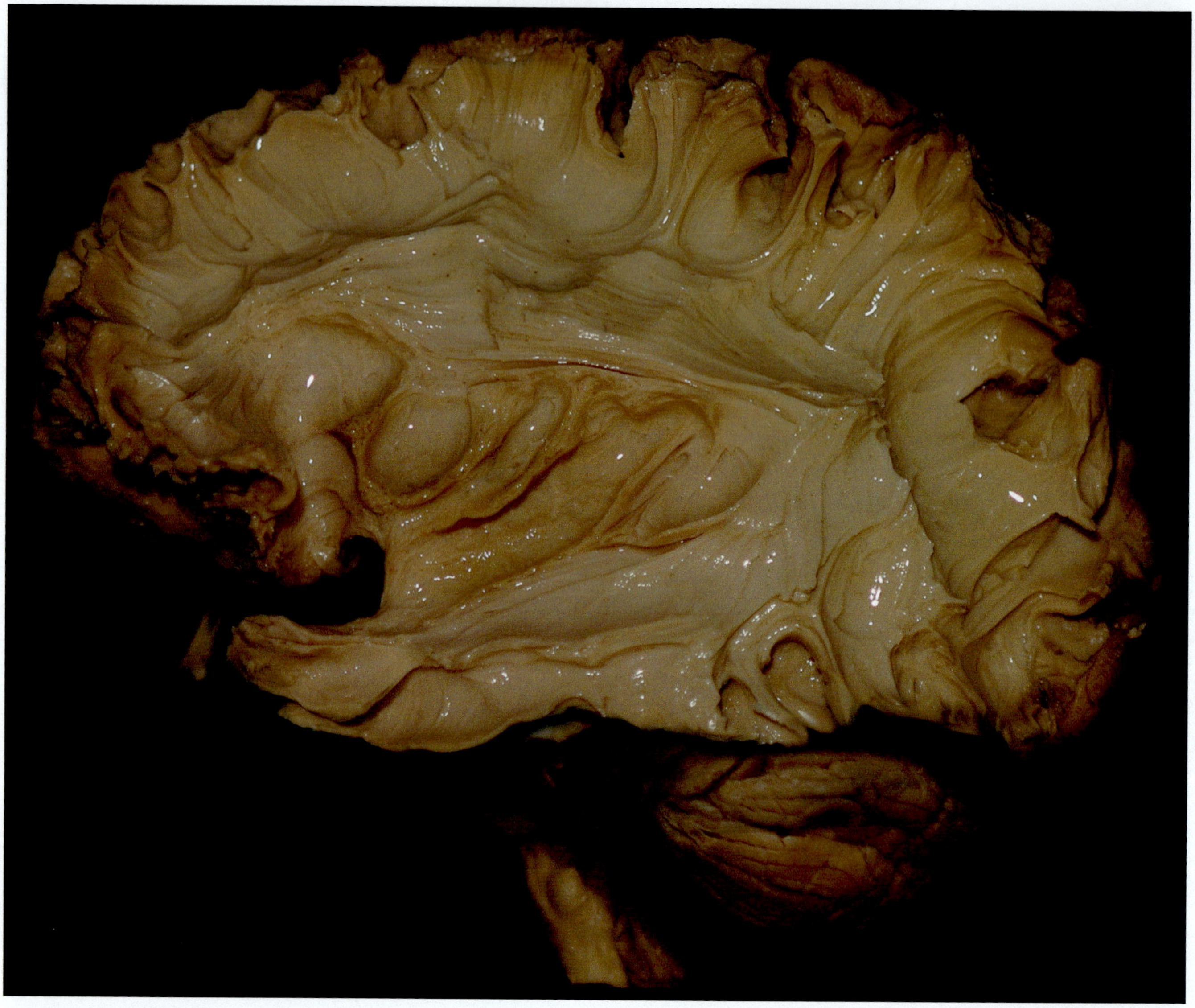

Fig. 2.8 Anatomical preparation of the left encephalon. After the "U" fibers and insular opercula are removed, the superior longitudinal fasciculus is seen at the perimeter of the peri-insular sulcus. Removing the cortical layer of the insula unveils the short association fibers interconnecting the insular gyri reciprocally and with the subopercular surface through the depth of the peri-insular sulcus. (©Kadri 2023. All Rights Reserved)

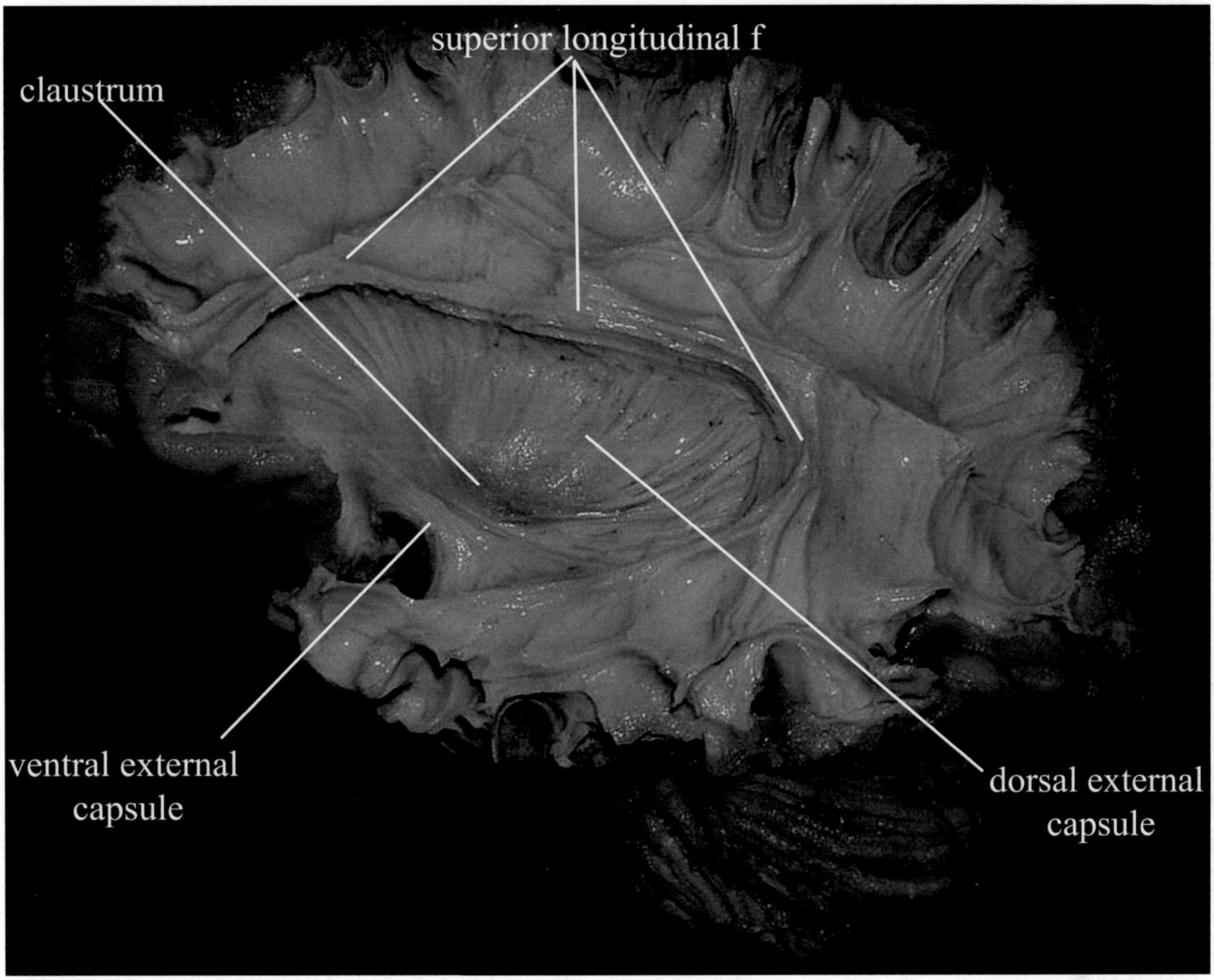

Fig. 2.9 (continued)

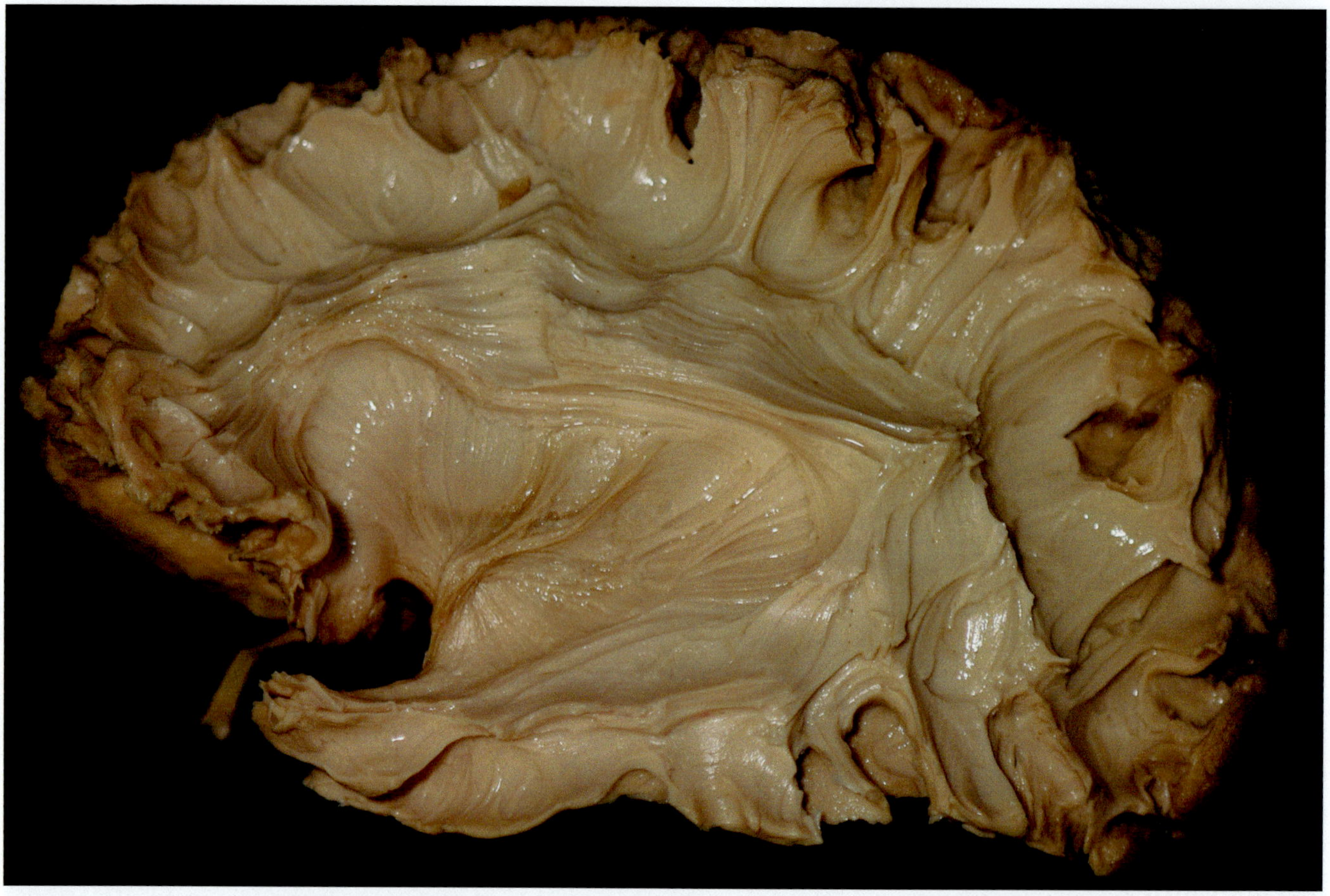

Fig. 2.10 Anatomical preparation of the left encephalon depicting the *insula amygdala reuniens* fasciculus, a connection to the limbic system. Further removal of the extreme capsule unveils the cortical connections of the basal ganglia with the neocortex through the external capsule and the claustrum. *f* fasciculus. (©Kadri 2023. All Rights Reserved)

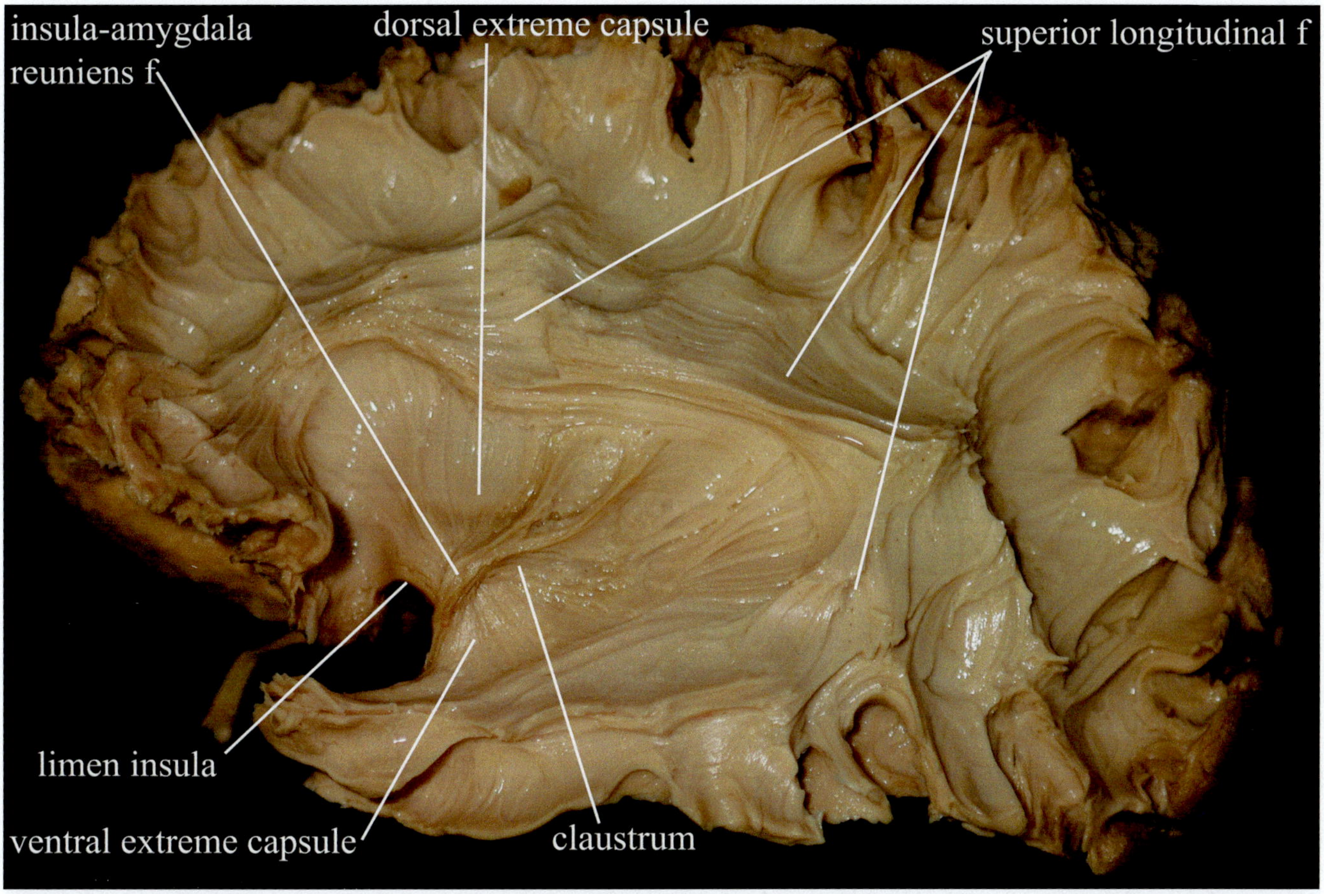

Fig. 2.10 (continued)

Broca, and hippocampal–septal connections through the fornix and the longitudinal stria (medial and lateral) are also considered long association fibers and are discussed with the limbic connections.

The Superior Longitudinal Fasciculus

The superior longitudinal fasciculus is considered the major fiber association system on the lateral surface of the brain, connecting the frontal, parietal, occipital, and temporal lobes. It resembles the letter C as it arches and wraps around the posterior limit of the Sylvian fissure (Fig. 2.11). This fasciculus is mainly located at the depth of the middle and inferior frontal gyri, inferior parietal lobule, and superior and middle temporal gyri [3, 5, 7, 11–14].

Several attempts to segment the superior longitudinal fasciculus have been described. These segments are the frontoparietal, temporoparietal, and frontotemporal, although no clear border or boundary between the proposed segments can be identified [6, 10]. The frontoparietal or horizontal segment connects the secondary and tertiary cortical motor-related areas (premotor and motor supplementary) of the frontal cortex, at the depth of the inferior and middle frontal gyri, to the superior and inferior parietal lobule. The temporoparietal or vertical segment connects the inferior parietal lobe to the posterior region of the superior and middle temporal gyri. The frontotemporal or arcuate segment is the deepest of the three parts and adjoins the posterior limit of the insula, connecting the fronto-orbital and frontal opercula and posterior superior temporal regions. The fibers of the vertical and arcuate segments inferior to the Sylvian fissure are also referred to as the inferior portion of the superior longitudinal fasciculus (Figs. 2.12, 2.13, 2.14, 2.15, and 2.16) [3, 15].

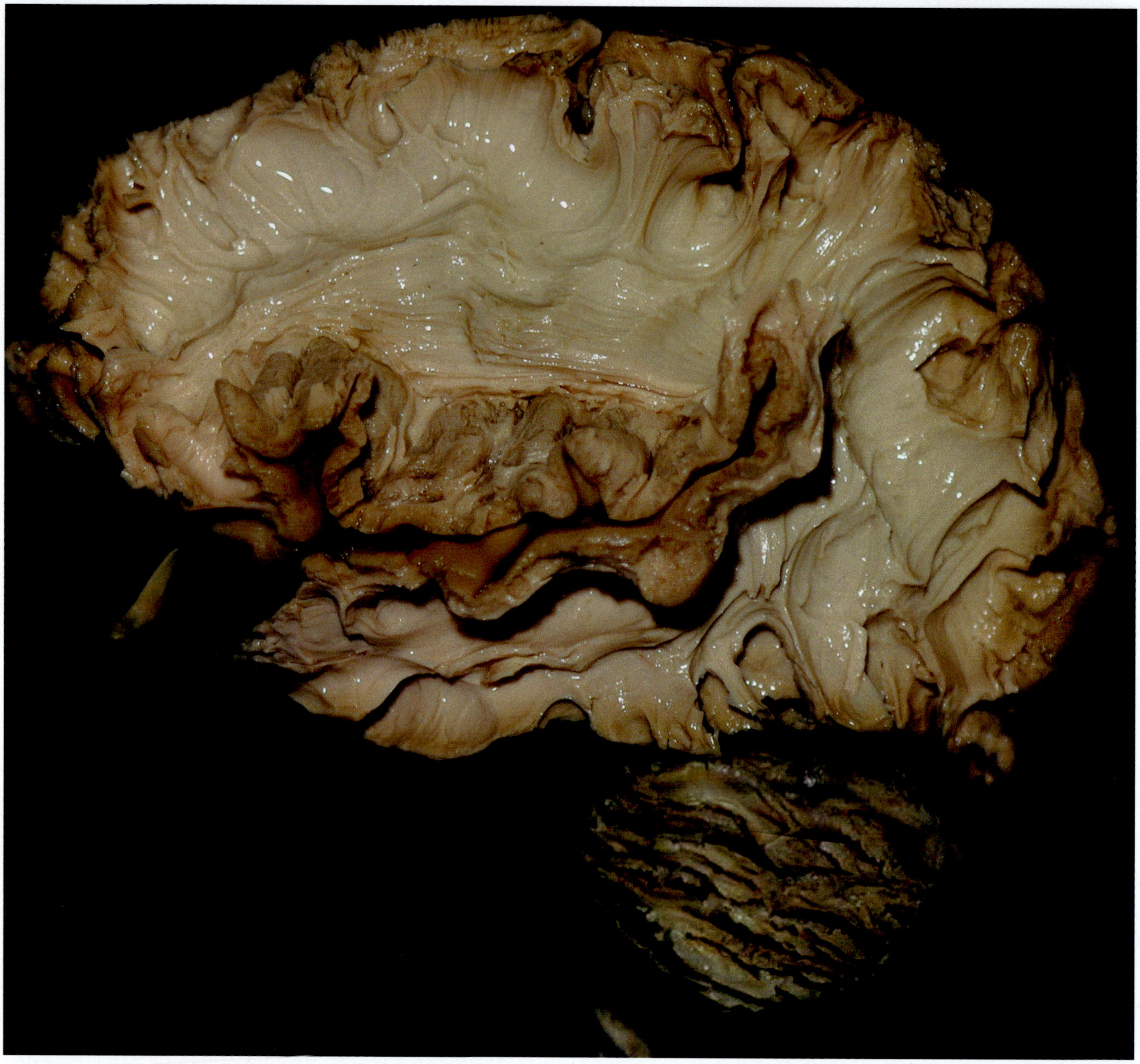

Fig. 2.11 Anatomical preparation and illustration of the left side of an encephalon after the removal of the "U" fibers around the Sylvian fissure. The fronto-orbital, frontoparietal, and temporal opercula are the superficial limits of the Sylvian fissure, and the insula lies at the depth of the Sylvian fissure. *ant* anterior, *g* gyrus. (©Kadri 2023. All Rights Reserved)

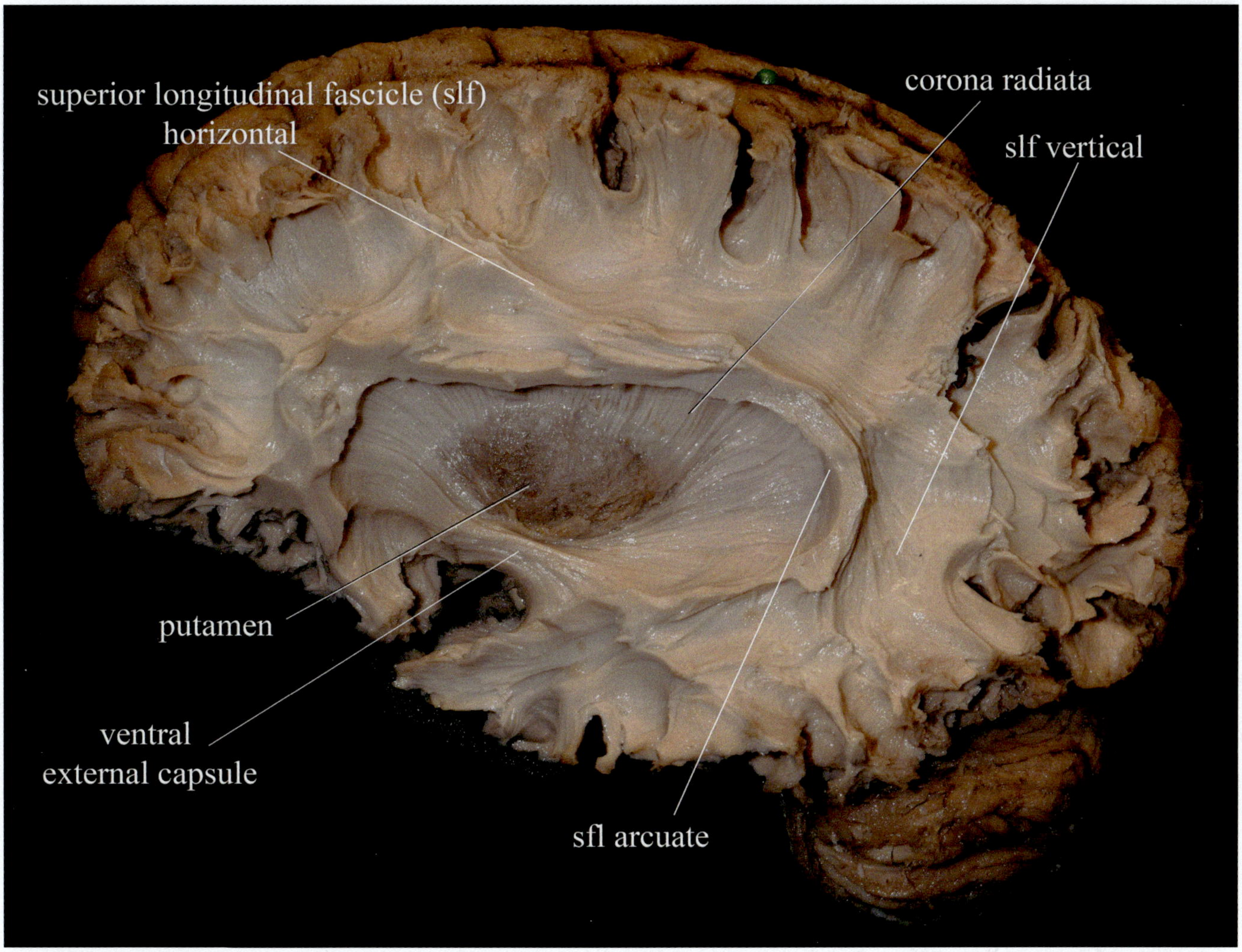

Fig. 2.13 (continued)

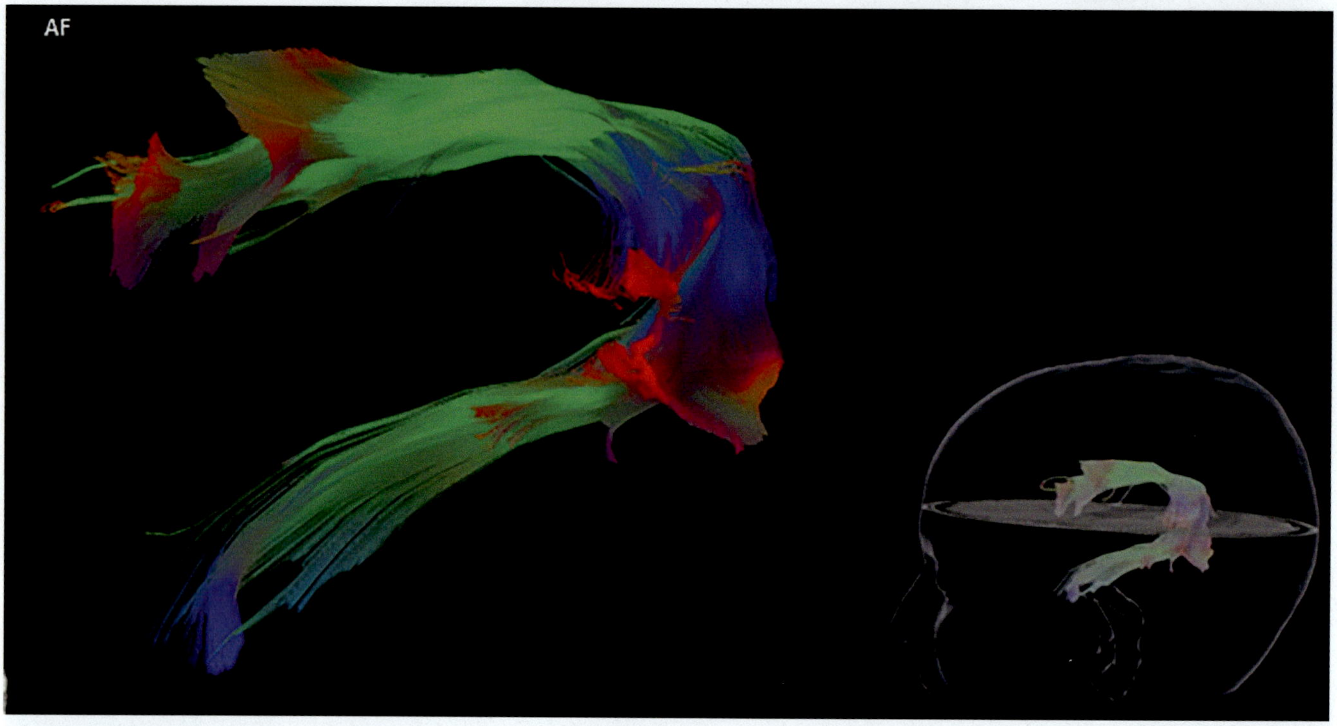

Fig. 2.14 Magnetic resonance tractography depicting the superior longitudinal fasciculus. (©Kadri 2023. All Rights Reserved)

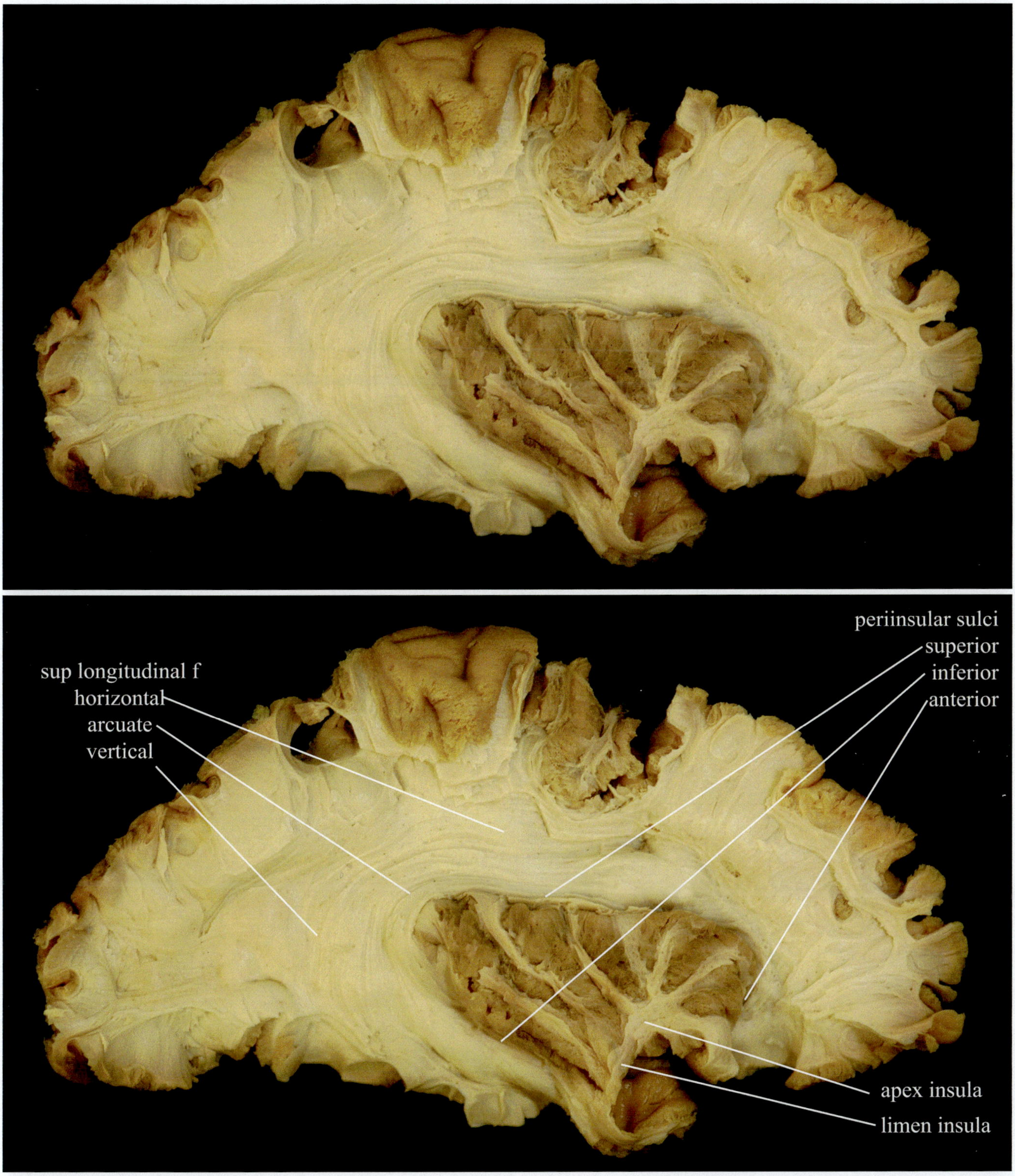

Fig. 2.15 Anatomical preparation through the stepwise removal of the multiple layers from medial to lateral directions. The cortex of the insular gyri is exposed, and the superior longitudinal fasciculus can be identified on the periphery of the insula. *f* fasciculus, *sup* superior. (©Kadri 2023. All Rights Reserved)

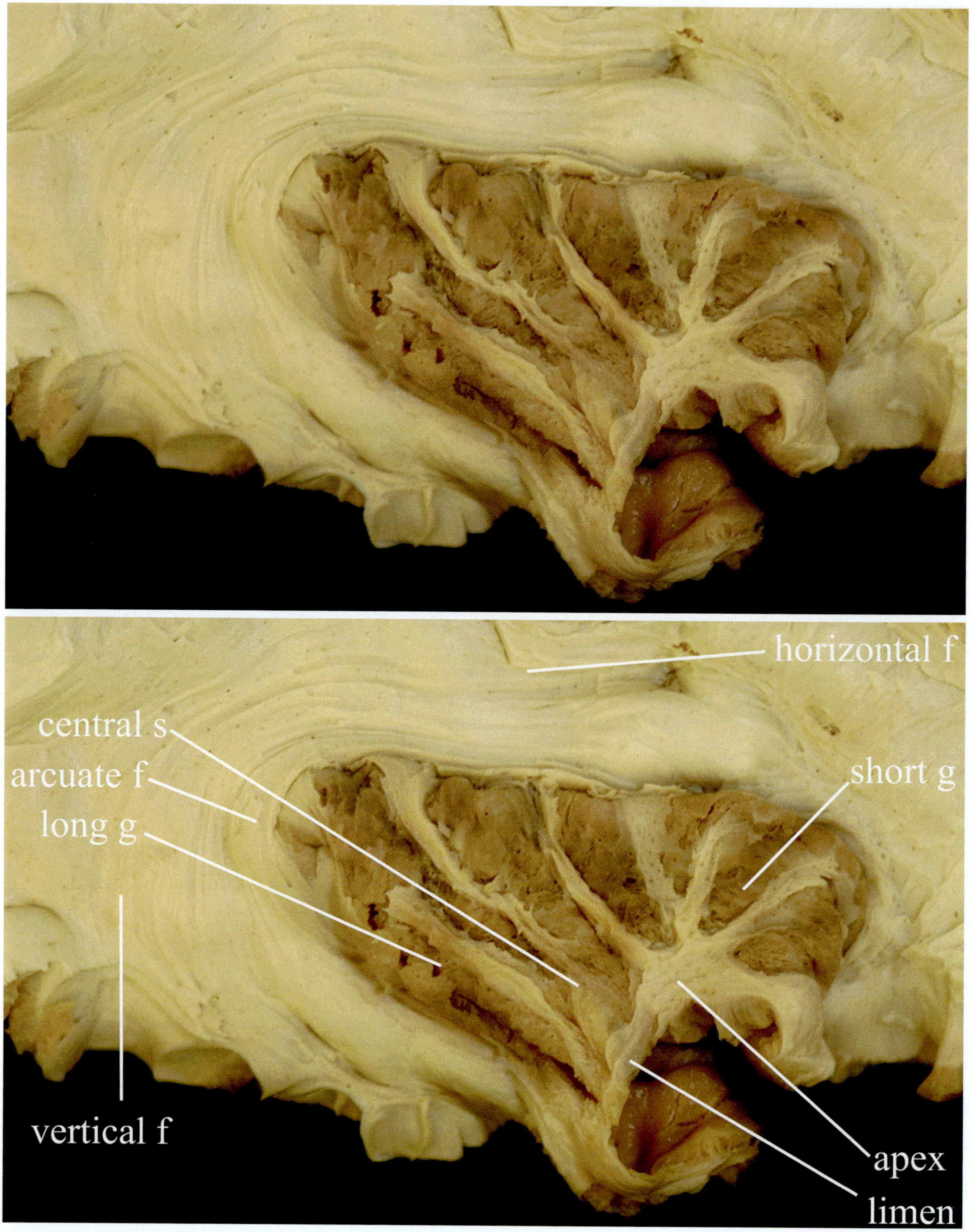

Fig. 2.16 Close-up view of the anatomical preparation demonstrating the inner aspect of the cortical insular surface. *f* fasciculus, *g* gyrus, *s* sulcus. (©Kadri 2023. All Rights Reserved)

The Occipitofrontal Fasciculus

The occipitofrontal fasciculus connects the frontal and occipital lobes through the insula and temporal lobes [3, 5, 7, 11–14, 16]. This fasciculus was previously known as the inferior fronto-occipital fasciculus, but once the so-called superior fronto-occipital fasciculus was misidentified by the horizontally oriented fibers of the anterior thalamic peduncle, the nomenclature of the occipitofrontal fasciculus was accepted [16]. Its fibers are located dorsal to the uncinate fasciculus at the level of the limen insulae, without a clear border [3]. This fasciculus extends from the prefrontal areas to the middle and posterior temporal regions, the occipital region, and the inferior parietal region. The occipitofrontal fasciculus is located medial to the inferior arm of the superior longitudinal fasciculus and superolateral to the fibers of the inferior longitudinal fasciculus and is part of the *sagittal stratum*. It is further divided into two components: (1) dorsal, more superficial, connecting the frontal lobe with the parietal and superior areas of the occipital lobes and (2) ventral, deeper, connecting the frontal lobe with the inferior occipital areas and the medial and posterior temporal areas (Fig. 2.17) [17].

The Uncinate Fasciculus

The uncinate fasciculus receives its name because of its hooked shape (Latin: *uncus = hook)* as its fibers travel through the limen insula interconnecting the frontal and temporal lobes [1]. It courses in a caudal direction from the white matter of the frontal lobe and is compacted at the limen insula. There, it has a ventral curve and fans out toward the anterior portions of the temporal lobe, mainly the inferior and middle temporal gyrus, and the amygdala [5, 7, 11–14, 18].

Three portions are described: frontal, insular, and temporal [18]. Two segments are also included: (1) the ventromedial, which connects the amygdala and piriform cortex to the subcallosal area and rectus gyrus; and (2) the dorsolateral, connecting the anterior portions of the middle and inferior temporal gyri to the orbitofrontal region (Figs. 2.16, 2.17, 2.18, and 2.19).

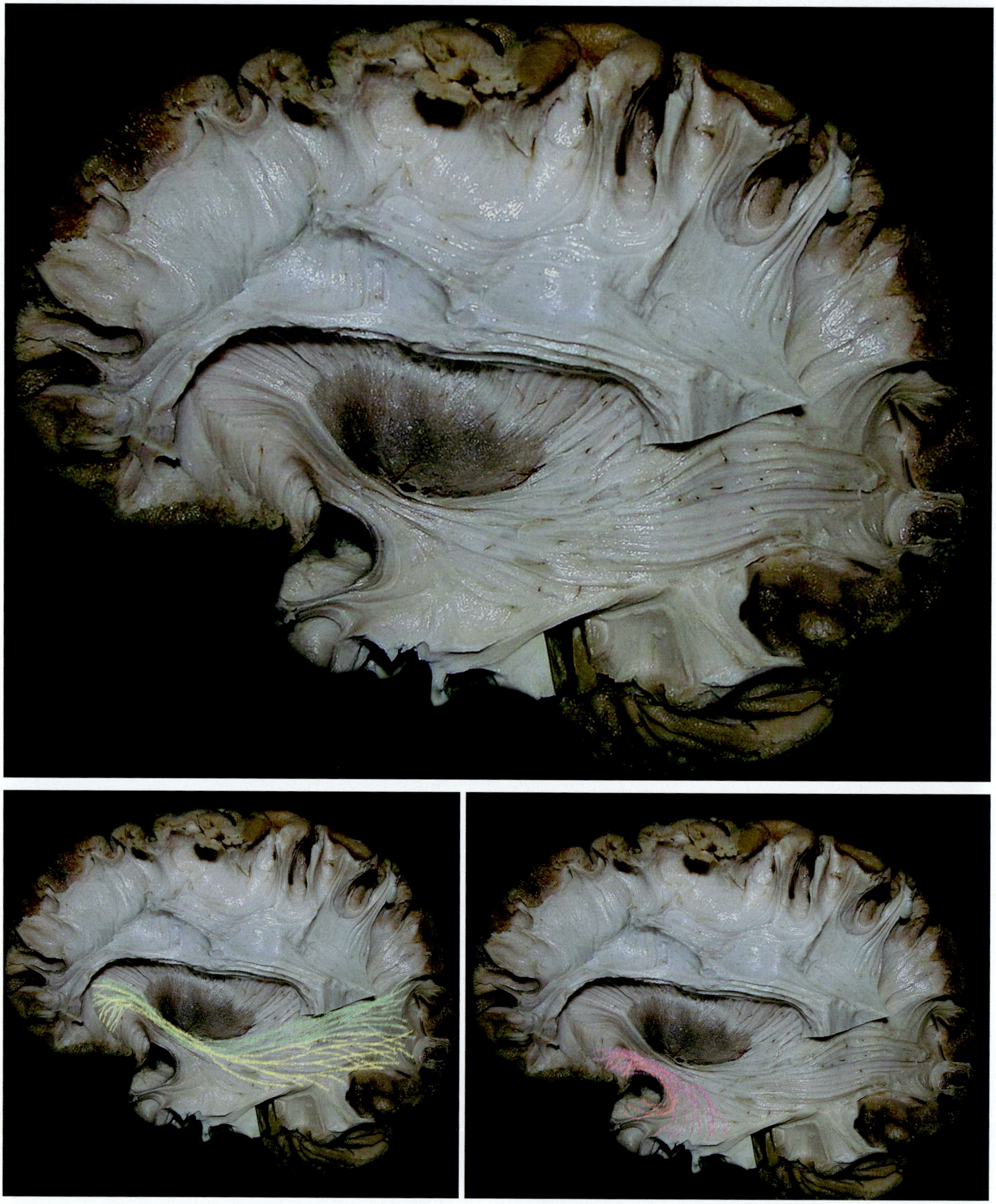

Fig. 2.17 Anatomical preparation and illustration of the left hemisphere depicting the ventral portion of the external capsule. This portion is composed of fibers of the uncinate fasciculus and occipitofrontal fasciculus. The claustrum and the dorsal portion of the external capsule have been removed, and the putamen is exposed. The uncinate fasciculus is further divided into ventral (red) and dorsal (purple) segments. The occipitofrontal fasciculus is divided into ventral (yellow) and dorsal (green) segments. (©Kadri 2023. All Rights Reserved)

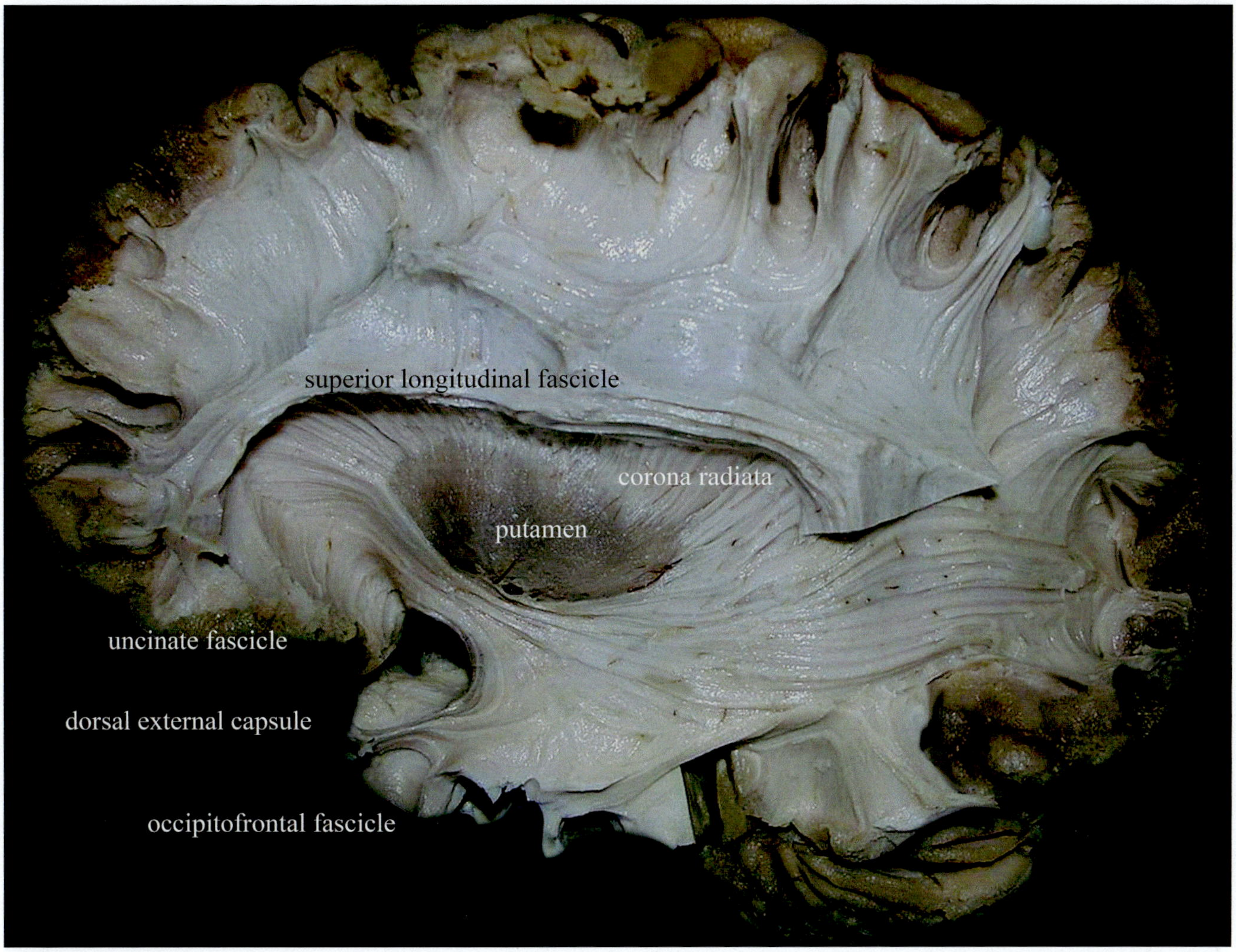

Fig. 2.17 (continued)

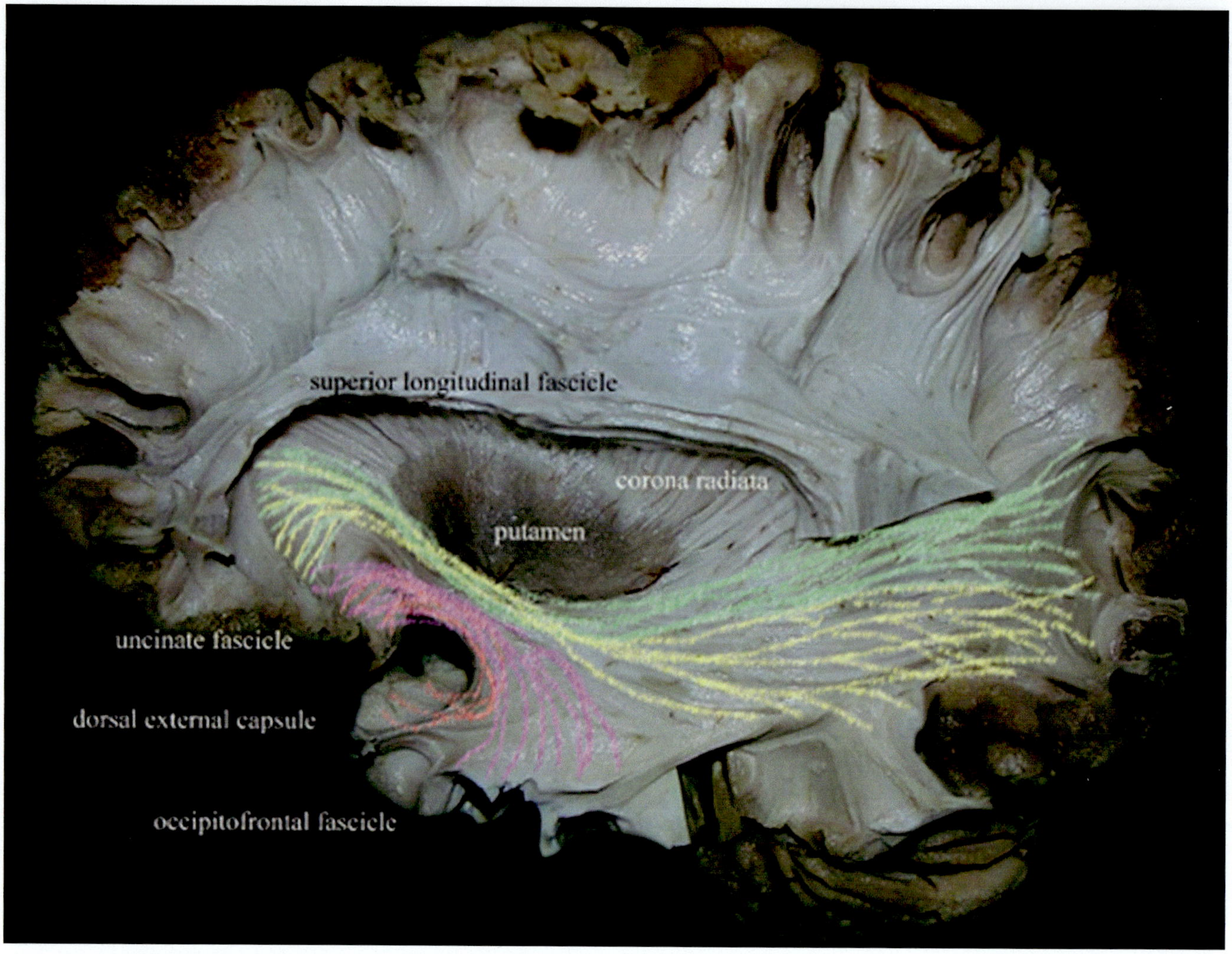

Fig. 2.17 (continued)

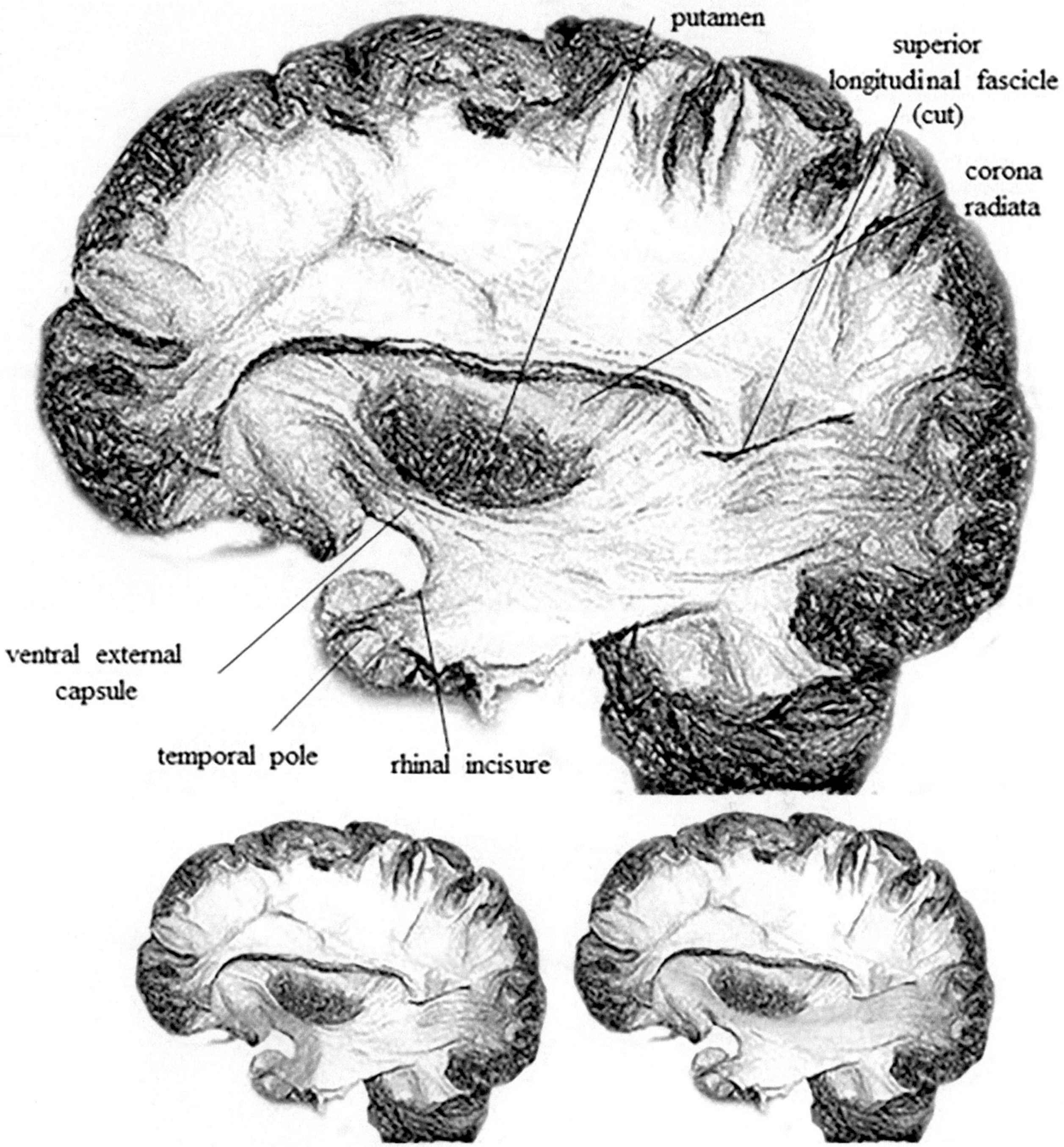

Fig. 2.17 (continued)

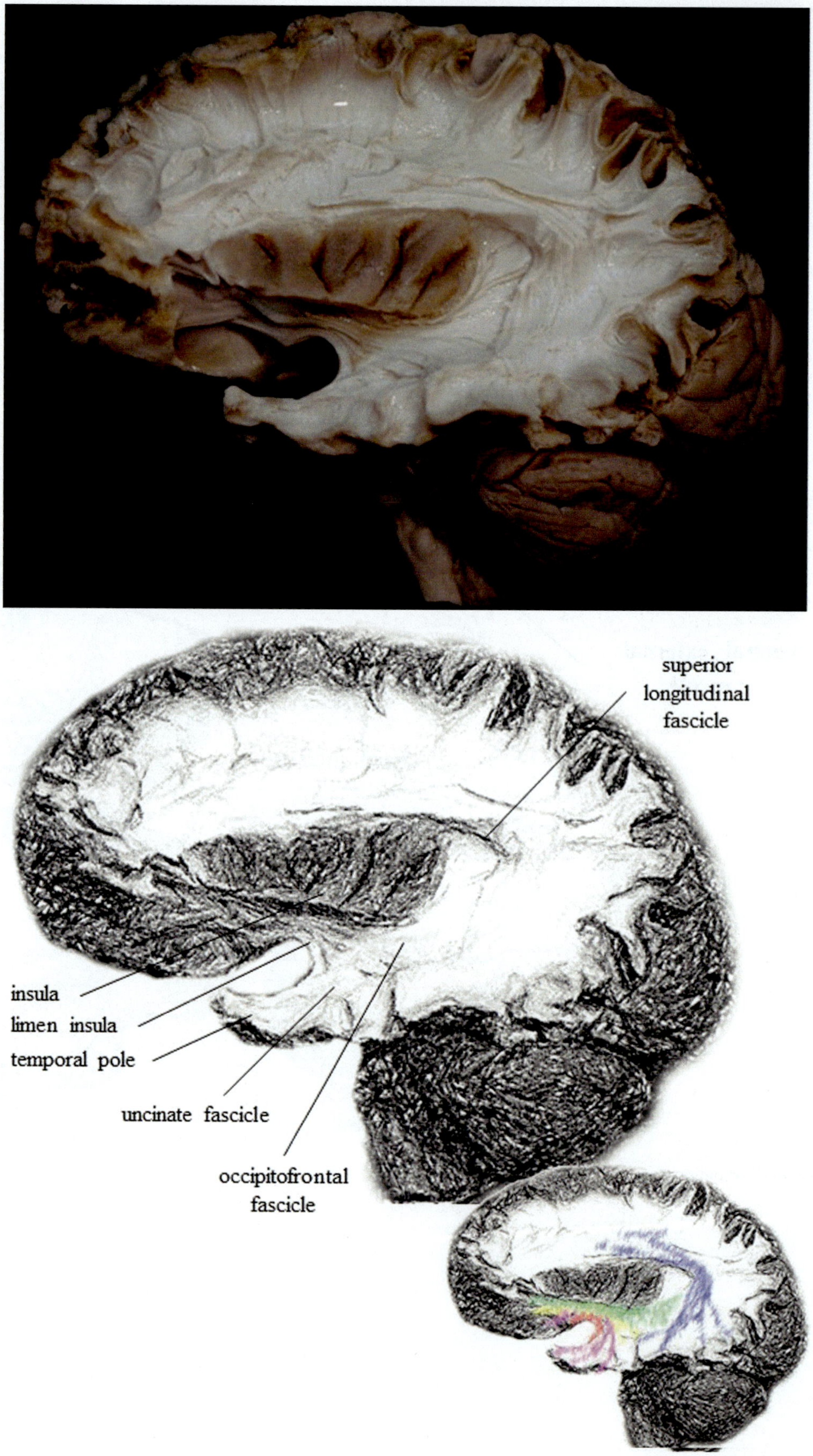

Fig. 2.18 Anatomical preparation and illustration of the left hemisphere. The arcuate fiber system on the lateral aspect of the brain has been removed, exposing the superior longitudinal fasciculus (blue). The cortex of the limen insula has also been removed, and the ventral portion of the external capsule is exposed. The ventral portion is composed of the ventral (red) and dorsal (purple) segments of the uncinate fasciculus and the ventral (yellow) and dorsal (green) segments of the occipitofrontal fasciculus. (©Kadri 2023. All Rights Reserved)

Fig. 2.19 Anatomical preparation of the left hemisphere. Removing the fibers of the occipitofrontal fasciculus delineates the uncinate fasciculus and exposes the lateral surface of the putamen ventrally. (©Kadri 2023. All Rights Reserved)

Fig. 2.19 (continued)

The External Capsule

The external capsule is a fine layer of white matter located lateral to the putamen and is separated from the more lateral extreme capsule by the claustrum. Its fibers join the fibers of the internal capsule and envelop the lentiform nucleus. At the limen insula, there is no exact limit between the fibers of the extreme and external capsules, nor is there a boundary between the fibers of the uncinate fasciculus and occipitofrontal fasciculus that compose the ventral segment of the external capsule [3].

The external capsule can be divided into ventral and dorsal segments. The ventral segment is composed of uncinate and occipitofrontal fasciculus fibers. The dorsal segment is the main cortico-striatal fiber system. The most superficial fibers of the dorsal segment project toward the claustrum (cortico-claustrum-cortical fibers). The deeper fibers connect the cortex with the putamen laterally (cortico-putaminal-cortical fibers) and the caudate nucleus medially (cortico-caudate-cortical fibers) (Figs. 2.20 and 2.21) [7].

The Inferior Longitudinal Fasciculus

The inferior longitudinal fasciculus is a subtle arrangement of fibers on the inferior surface of the temporal lobe that connects the anterior areas of the temporal lobe to the occipital lobe. It is at the depth of the inferior temporal and fusiform gyri, lateral to the collateral sulcus [5, 7, 11–14]. This fasciculus is formed by different and successive small bundles of association fibers, which have a macroscopic appearance of one fiber tract (Figs. 2.22 and 2.23) [19].

The Cingulum

The cingulum is the largest association fiber bundle of the limbic system and is located within the limbic gyrus. The limbic gyrus, also designated as the external ring of the limbic lobe, corresponds to the subcallosal area, the cingulate gyrus, the isthmus of the cingulate gyrus, and the parahip-

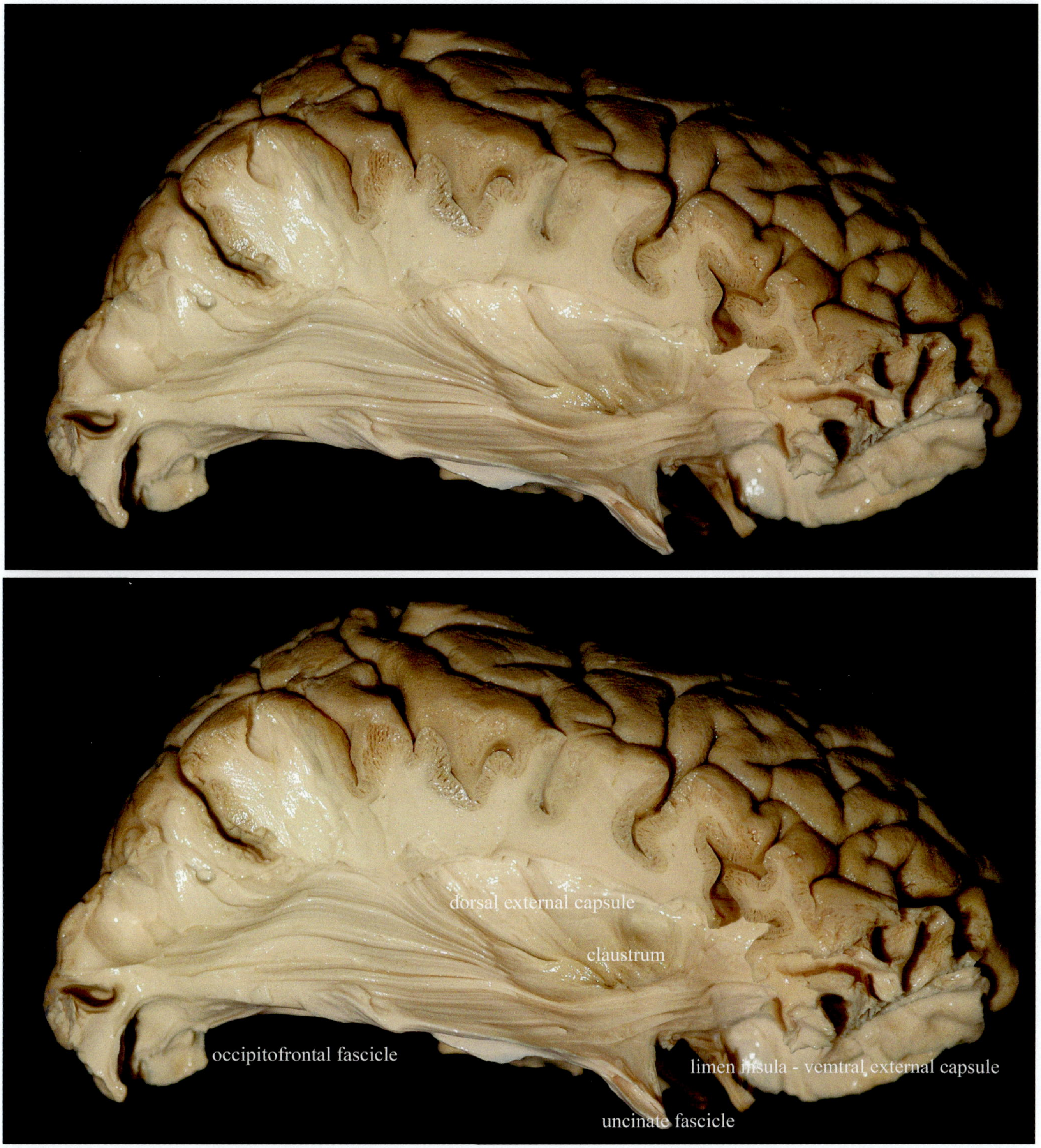

Fig. 2.20 Anatomical preparation of the right hemisphere. The "U" fiber system and the inferior arm of the superior longitudinal fasciculus are removed. The insular cortex and the extreme capsule have been removed to unveil the claustrum and the external capsule. The dorsal external capsule (blue) is composed of cortico-claustral-cortical fibers and cortico-striatal-cortical fibers. The ventral external capsule is located at the limen insula and comprises the ventral (red) and dorsal (purple) segments of the uncinate fasciculus and the ventral (yellow) and dorsal (green) segments of the occipitofrontal fasciculus. (©Kadri 2023. All Rights Reserved)

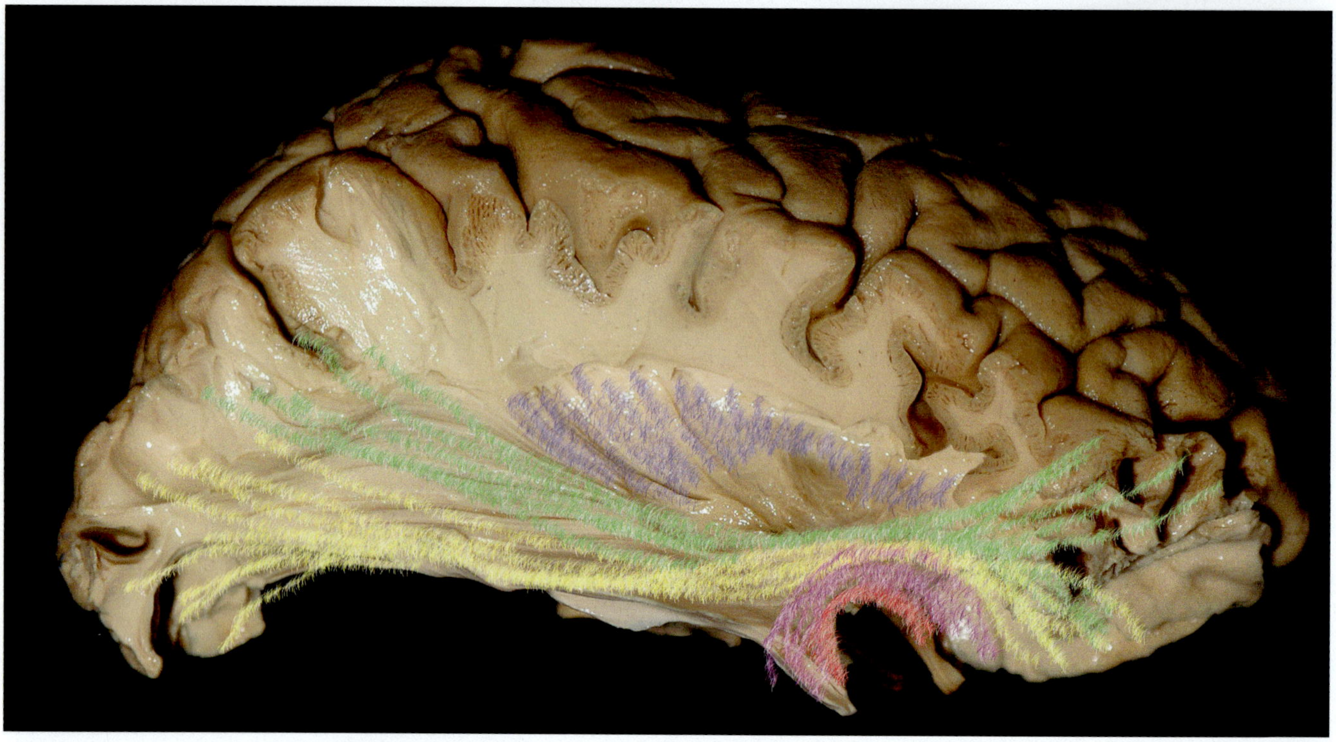

Fig. 2.20 (continued)

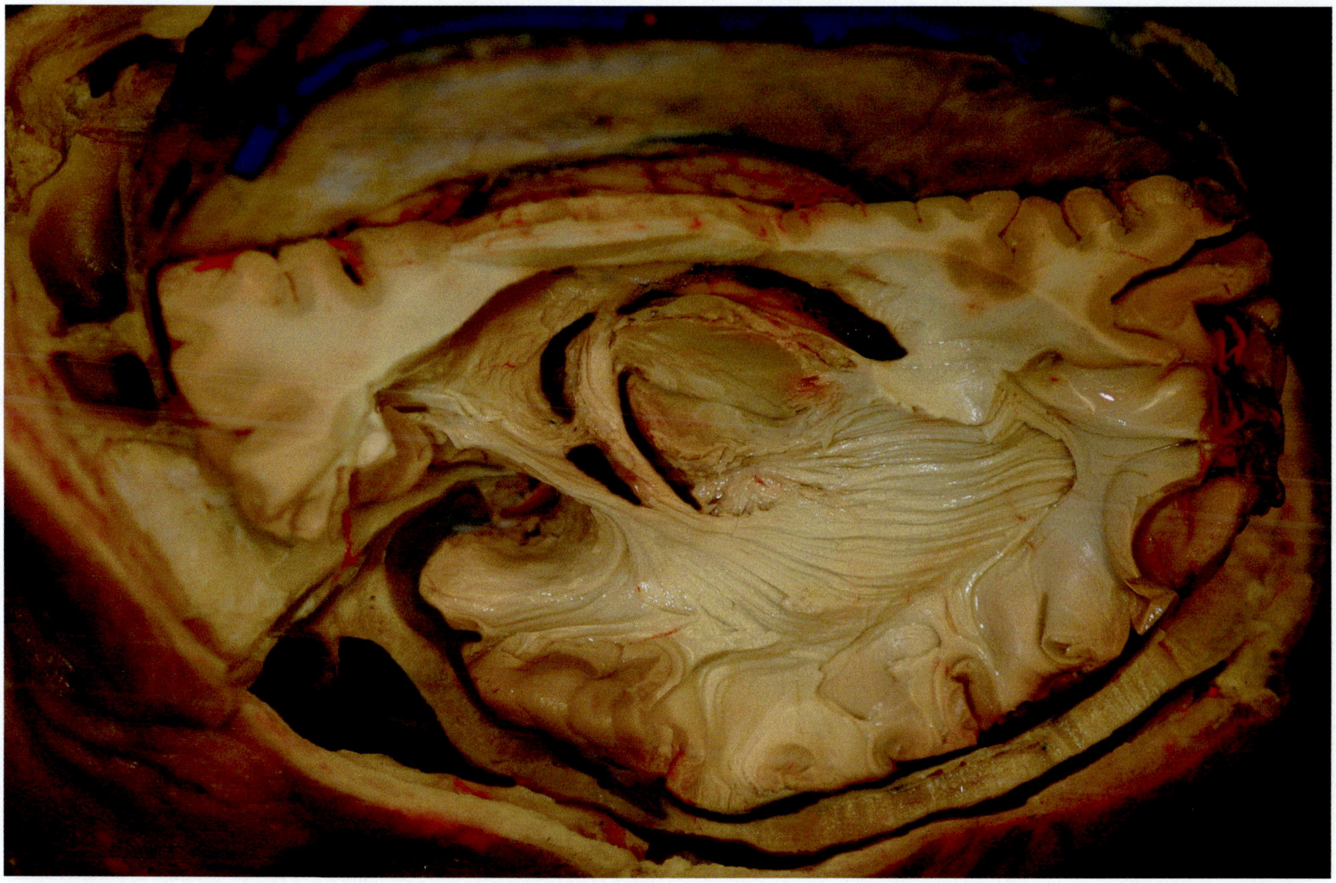

Fig. 2.21 Anatomical preparation of the left side of the head. The insula, extreme capsule, claustrum, dorsal external capsule, putamen, and ventral striatum have been removed. The ventral external capsule with the fibers of the ventral (red) and dorsal (purple) segments of the uncinate fasciculus and the ventral (yellow) and dorsal (green) segments of the occipitofrontal fasciculus are seen around the compact segment of the anterior commissure (blue).

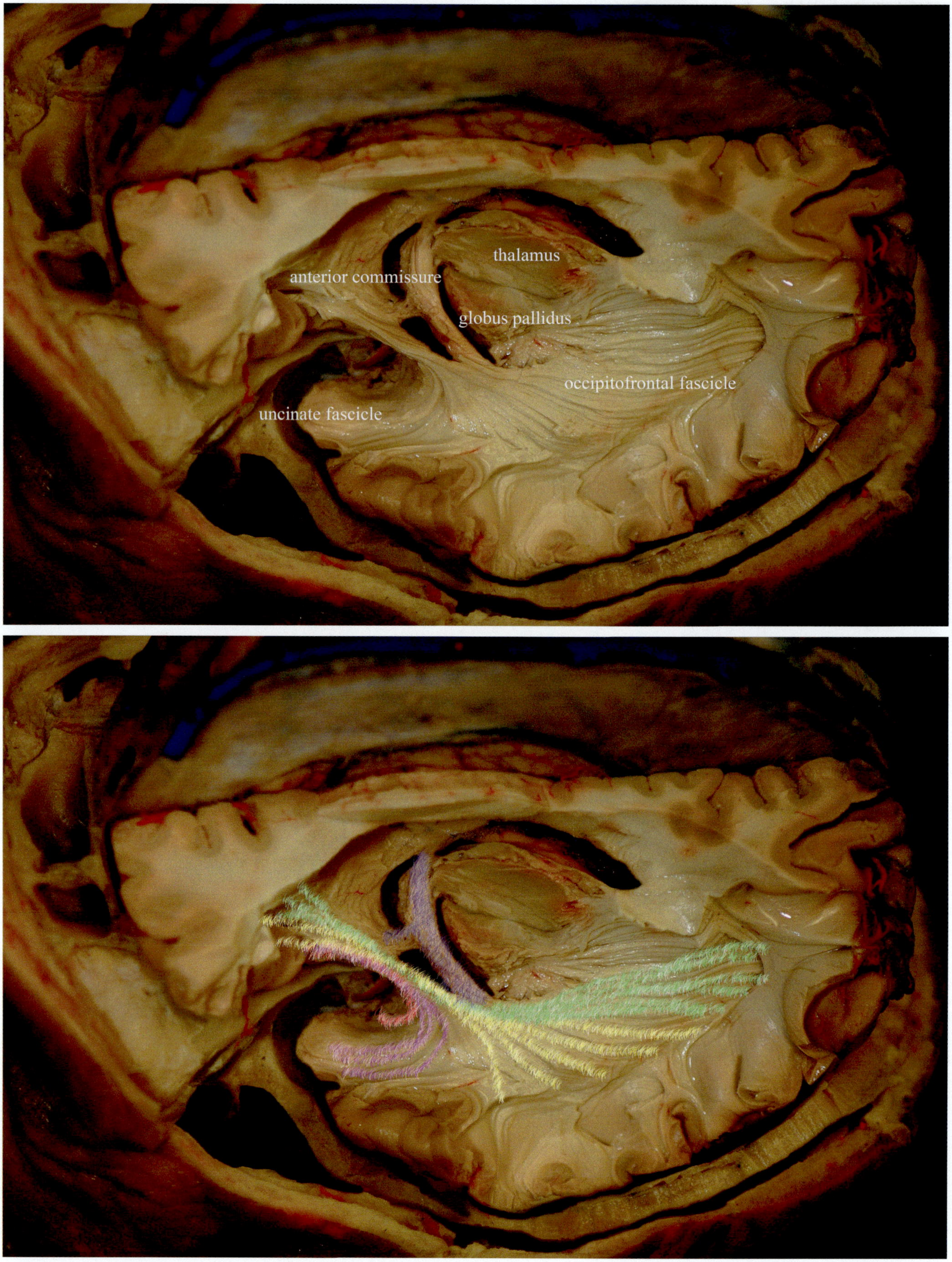

Fig. 2.21 (continued)

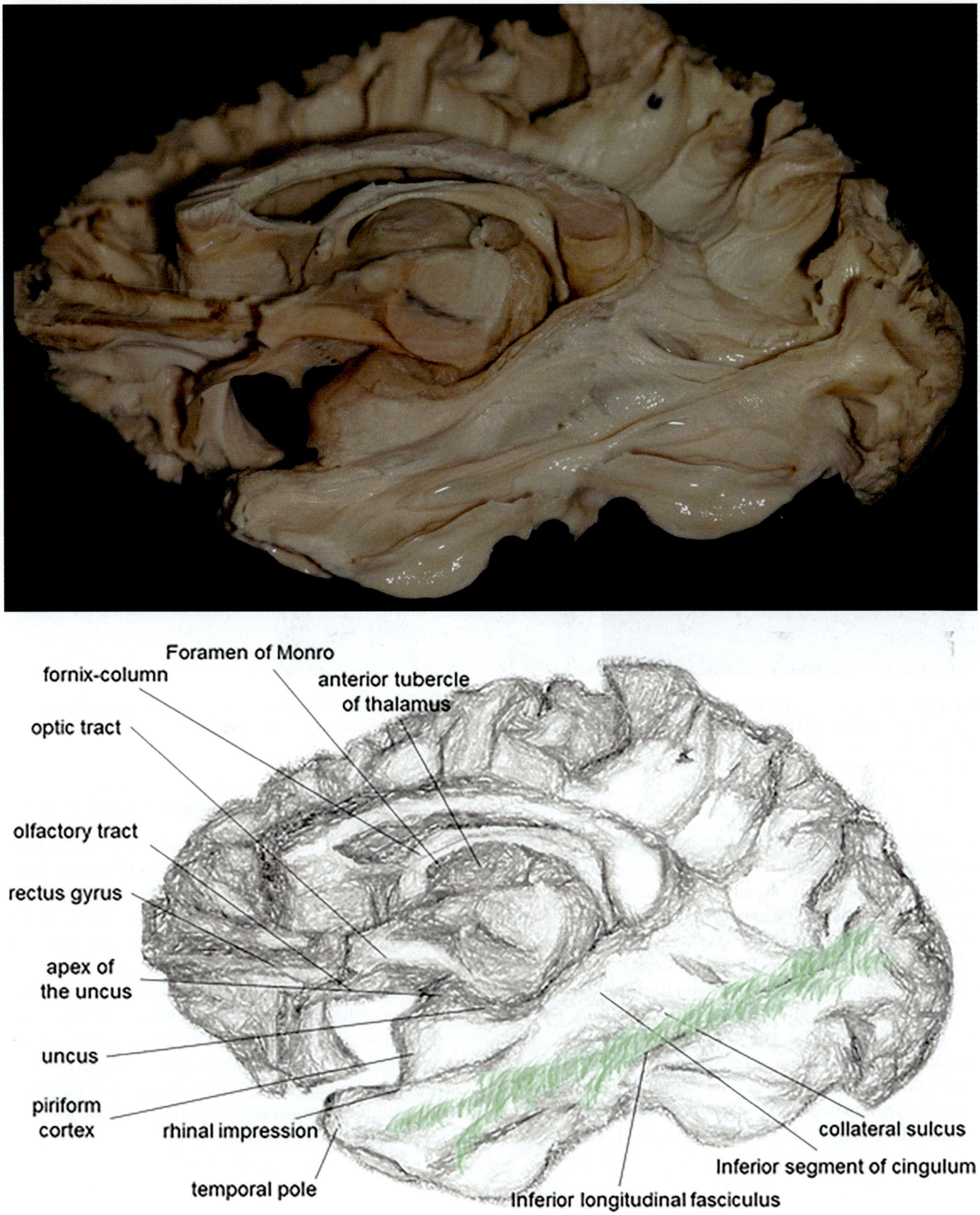

Fig. 2.22 Anatomical preparation and illustration of the left hemisphere. The cortex of the medial and basal aspects has been removed. The inferior segment of the cingulum was also partially removed. The ependyma of the collateral eminence at the floor of the temporal horn is visualized. The collateral eminence corresponds to the impression of the collateral sulcus in the ventricular cavity. The inferior longitudinal fasciculus is at the depth of the fusiform and inferior temporal gyri.

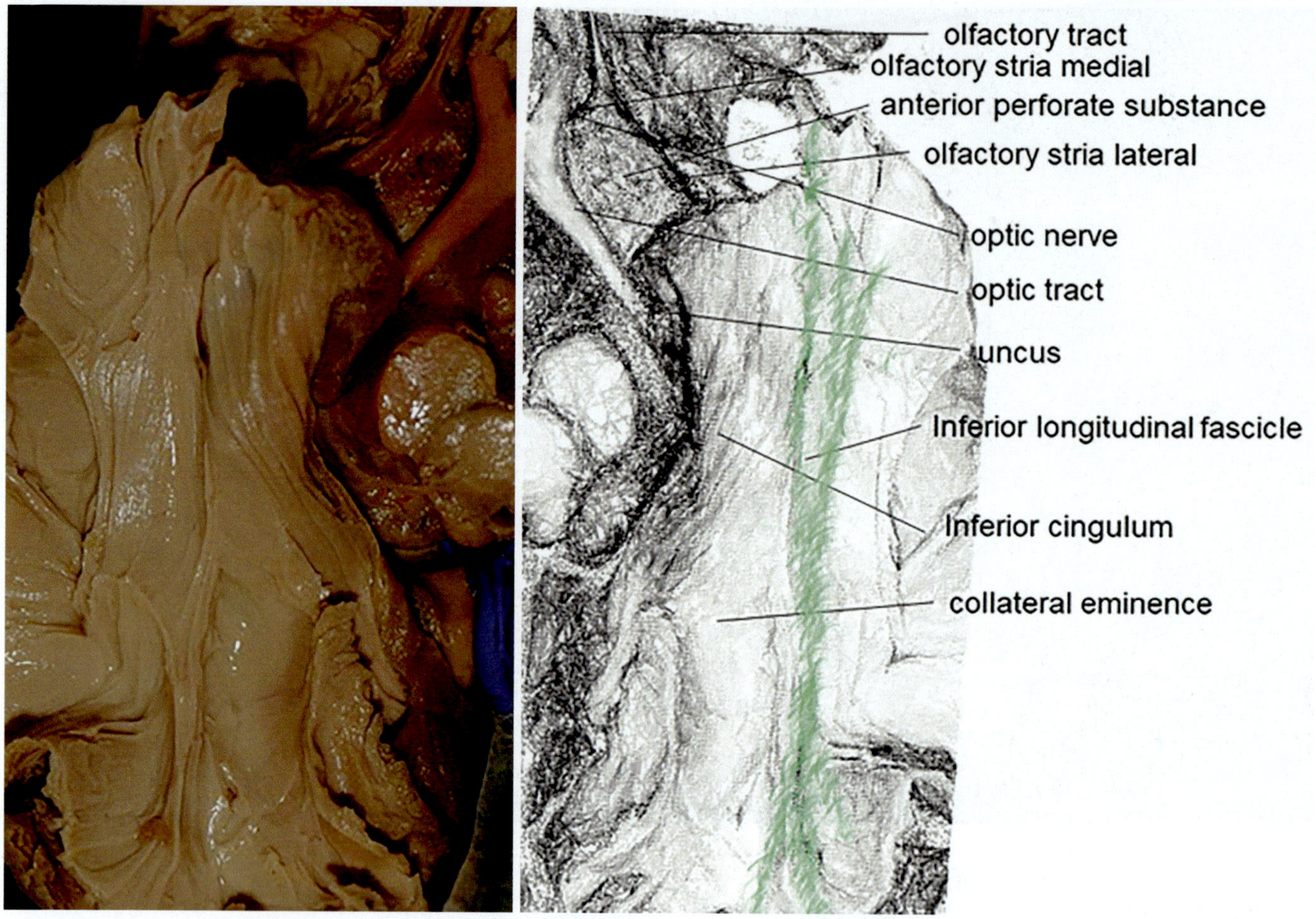

Fig. 2.23 Anatomical preparation and illustration of the right hemisphere, basal view. The cortex of the basal surface has been removed. The inferior segment of the cingulum is at the depth of the parahippocampal gyrus and extends toward the anterior portion of the parahippocampal gyrus and temporal pole. The inferior longitudinal fasciculus extends from the anterior temporal areas to the occipital pole. (©Kadri 2023. All Rights Reserved)

pocampal gyrus. It is separated from the neocortex by the limbic sulcus. This sulcus is complex, formed by the sequence of the cingulate sulcus, subparietal sulcus, the anterior portion of the calcarine sulcus, and the collateral sulcus, which may or may not be continuous with the rhinal sulcus. It marks the limits of the limbic lobe from the subcallosal area toward the rhinal impression (Fig. 2.24).

The cingulum extends from the subcallosal area, around the genu of the corpus callosum, toward the subparietal region. It then travels around the splenium of the corpus callosum along the isthmus of the cingulum and the parahippocampal gyrus to reach the most anterior portion of the temporal lobe (Figs. 2.25 and 2.26) [3, 5, 7, 11–14].

Two segments of the cingulum, the superior and the inferior, can be identified. The superior segment contains fibers connecting the subcallosal area mainly with the subparietal portion of the precuneus. The remaining fibers of the superior segment turn around the splenium of the corpus callosum and continue within the isthmus of the cingulate gyrus (Fig. 2.27). The inferior segment originates from the junction of fibers of the superior segment within the isthmus, and very few fibers originate from the posterior portion of the precuneus and the anterior lingual gyrus. Together, these three fiber bundles form the inferior segment of the cingulum in the core of the parahippocampal gyrus and reach the most anterior portion of this gyrus (Fig. 2.28).

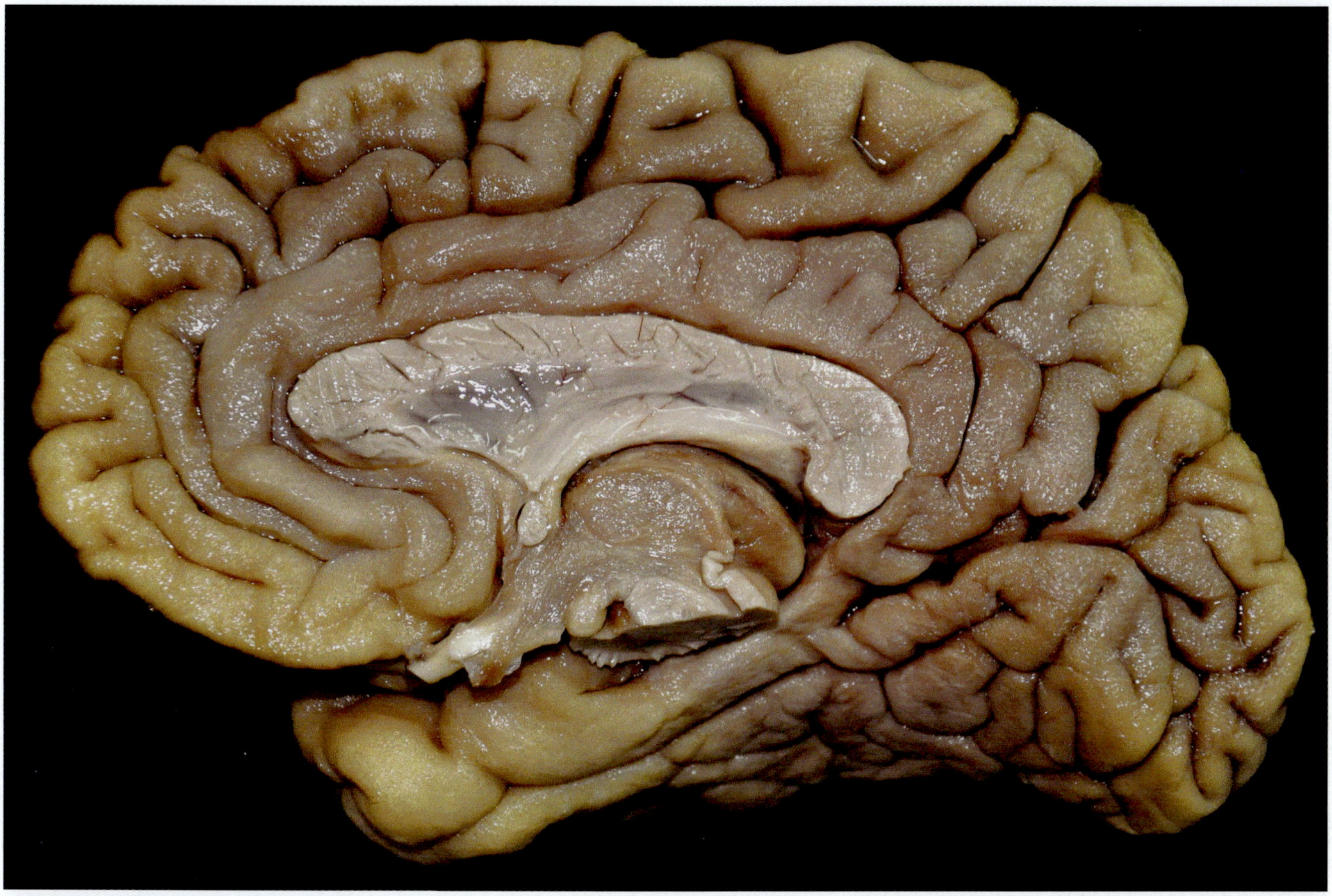

Fig. 2.24 Anatomical preparation of the right hemisphere, medial view. The limbic lobe (purple) is limited by the limbic sulcus (green).

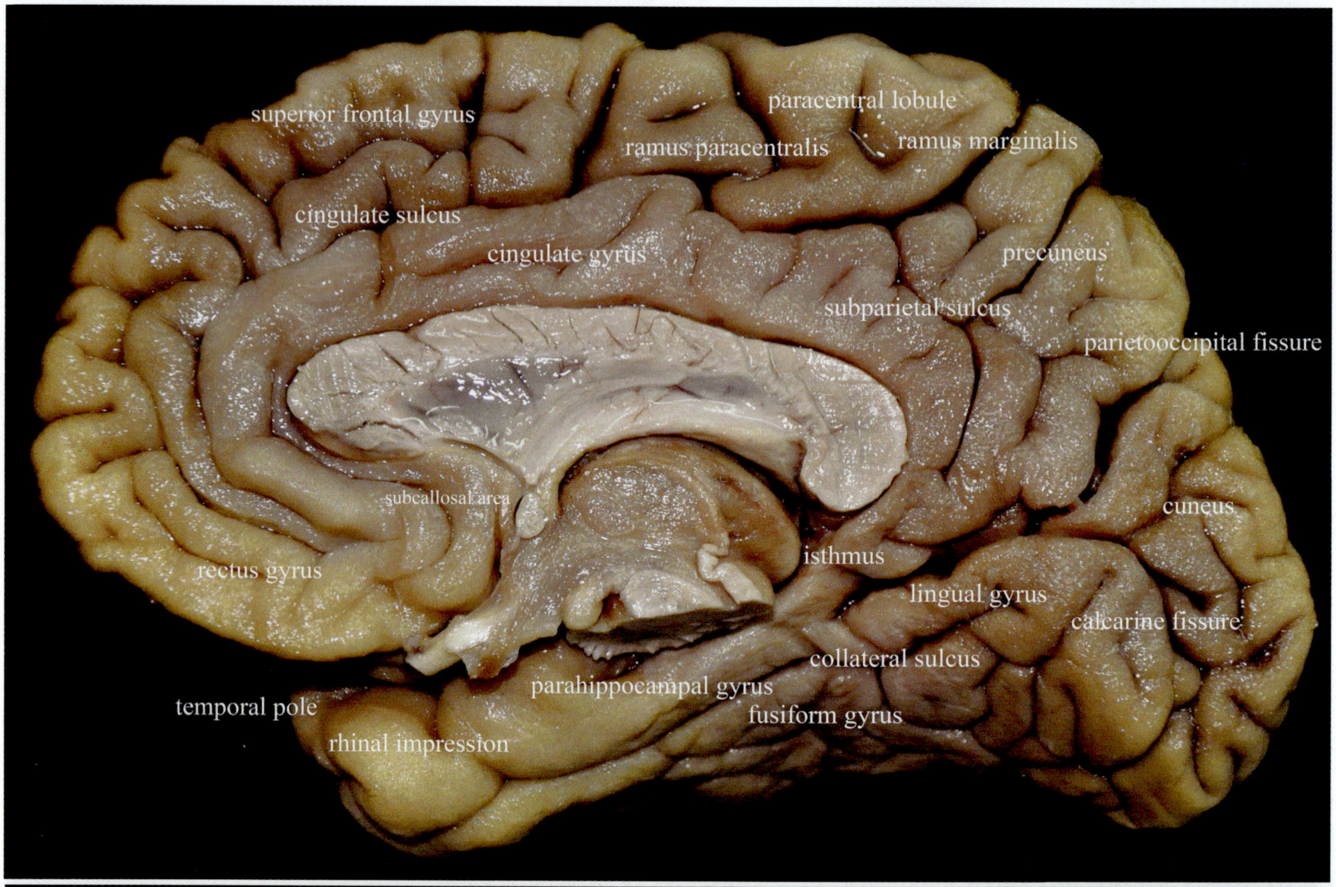

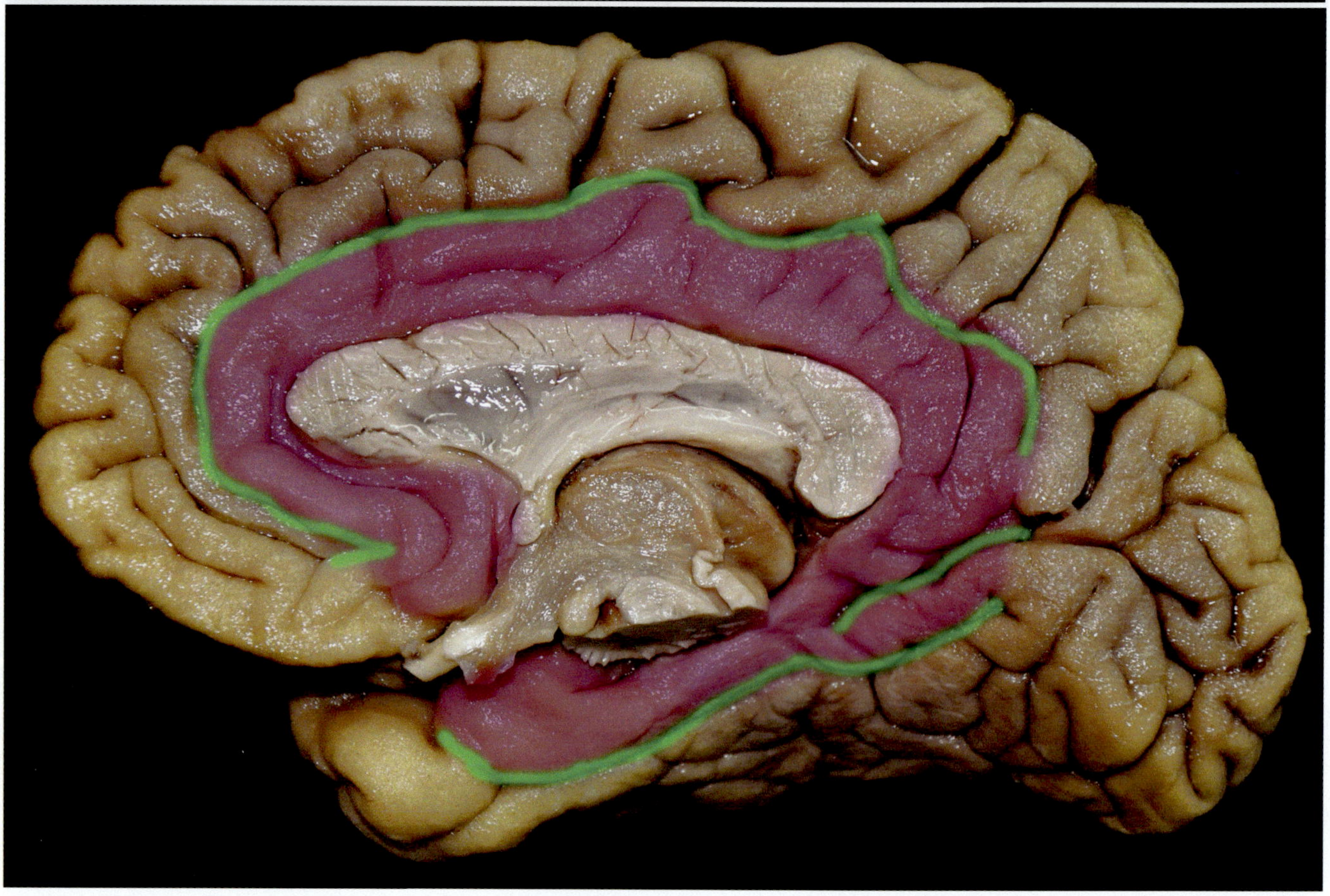

Fig. 2.24 (continued)

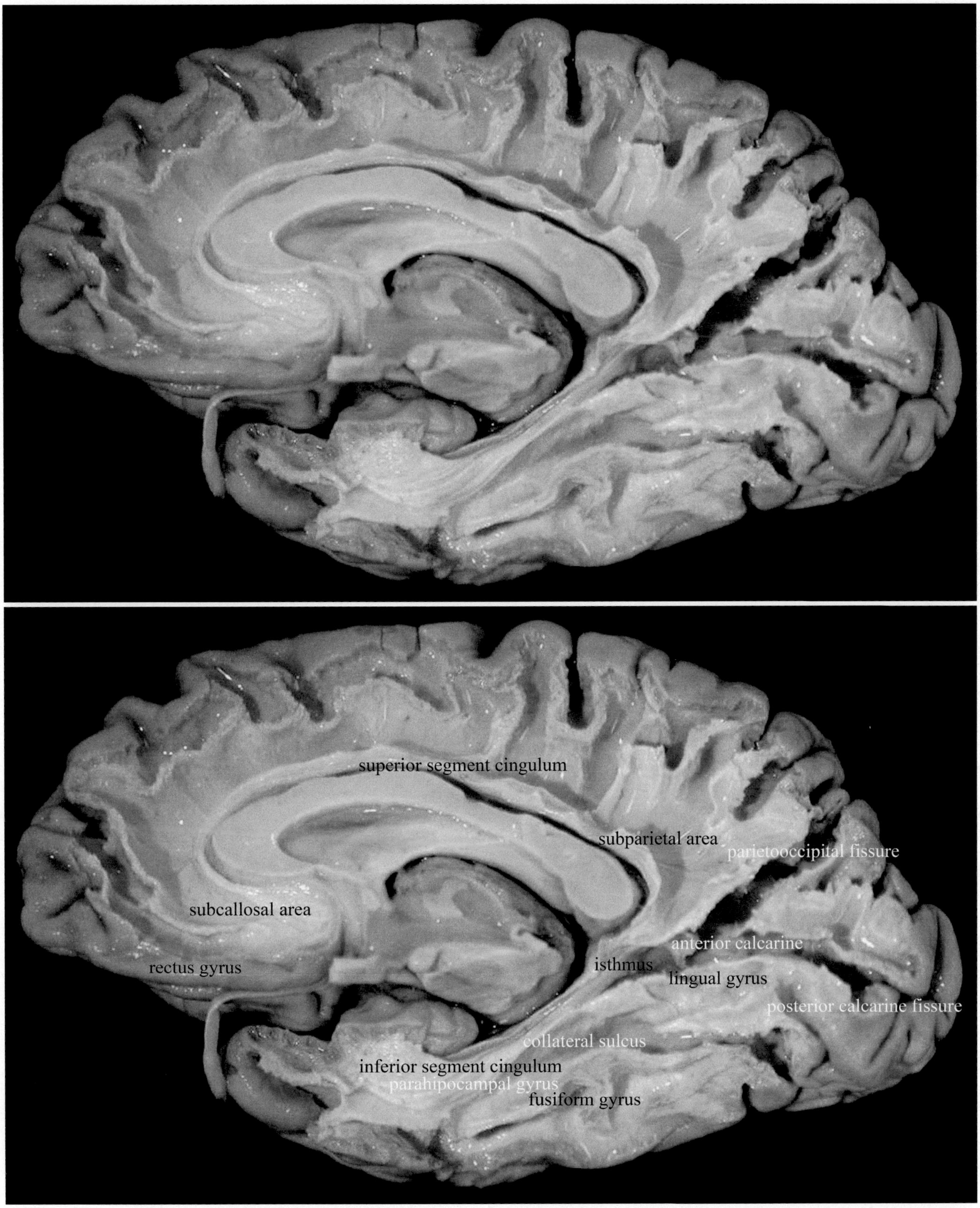

Fig. 2.25 Anatomical preparation of the right hemisphere, medial view. Removal of the cortex unveils the "U" fiber system, which is not prominent along the limbic lobe. Removing the few "U" fibers unveils the long association fiber of the limbic system: the cingulum. The cingulum has a superior segment (green) and an inferior segment (red).

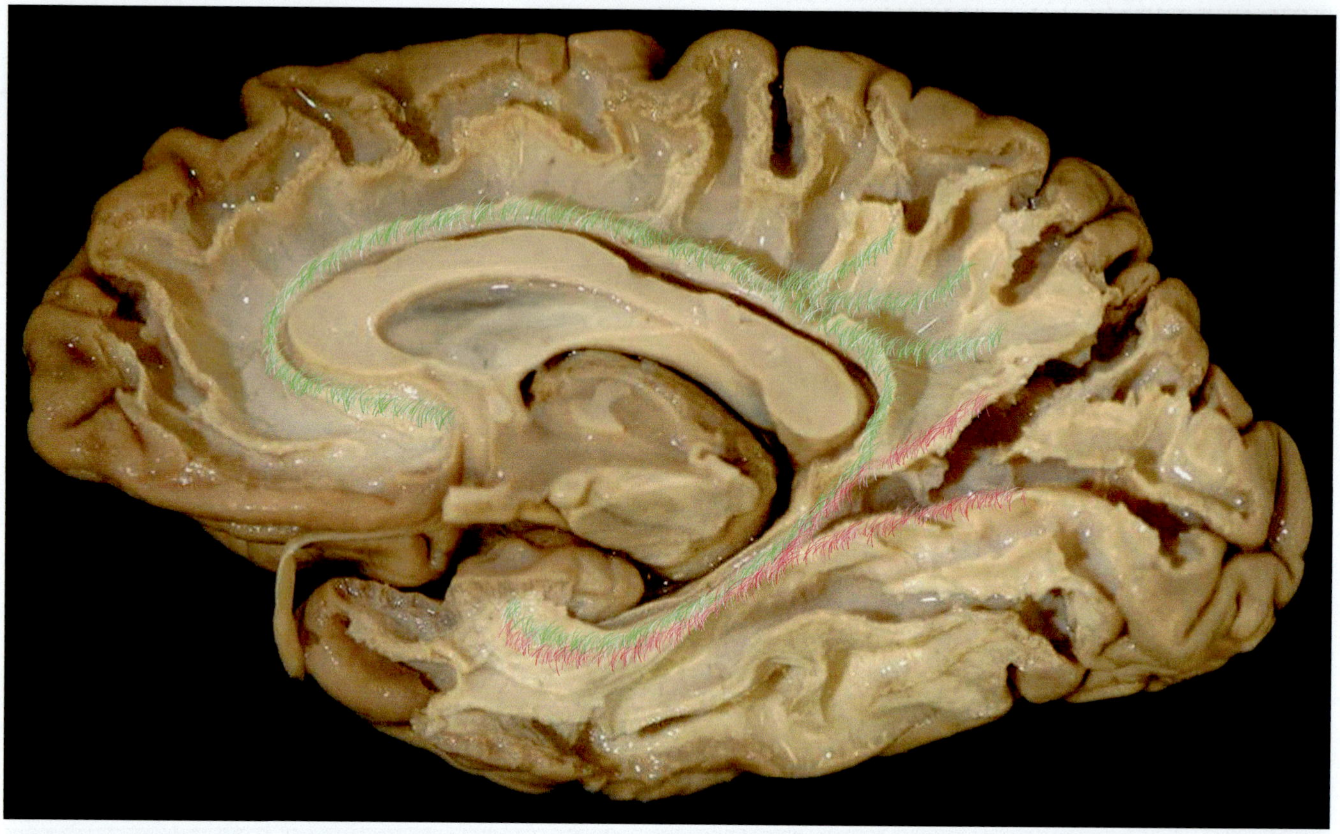

Fig. 2.25 (continued)

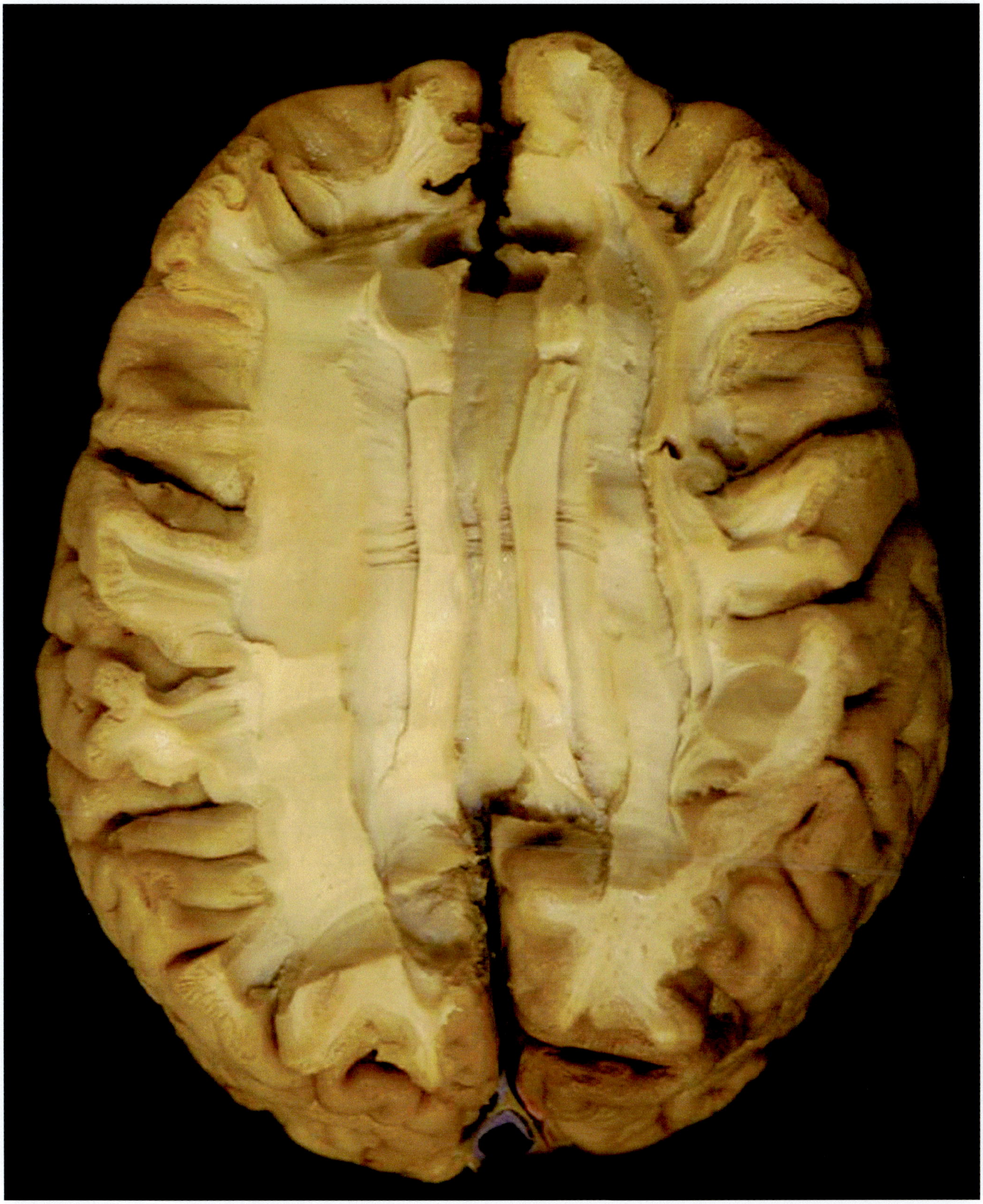

Fig. 2.26 Anatomical preparation of the brain from a superior view. The superior frontal gyrus, superior portions of the precentral and postcentral gyri, and the superior parietal lobe have been removed to expose the interhemispheric fissure. The cingulate gyrus is decorticated, and the cingulate bundle (green) is depicted lying in a horizontal anteroposterior orientation over the transverse interhemispheric fibers of the corpus callosum. (©Kadri 2023. All Rights Reserved)

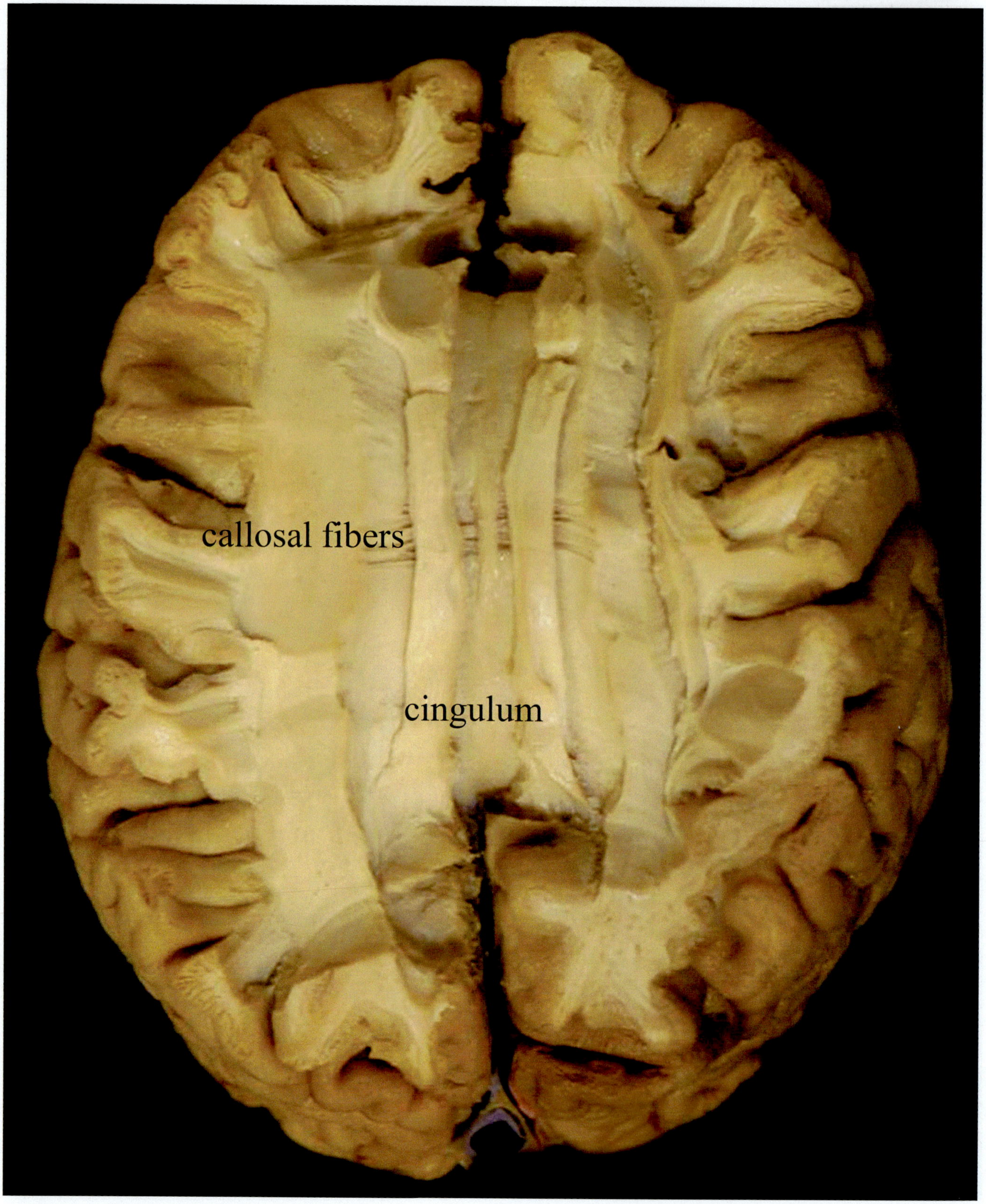

Fig. 2.26 (continued)

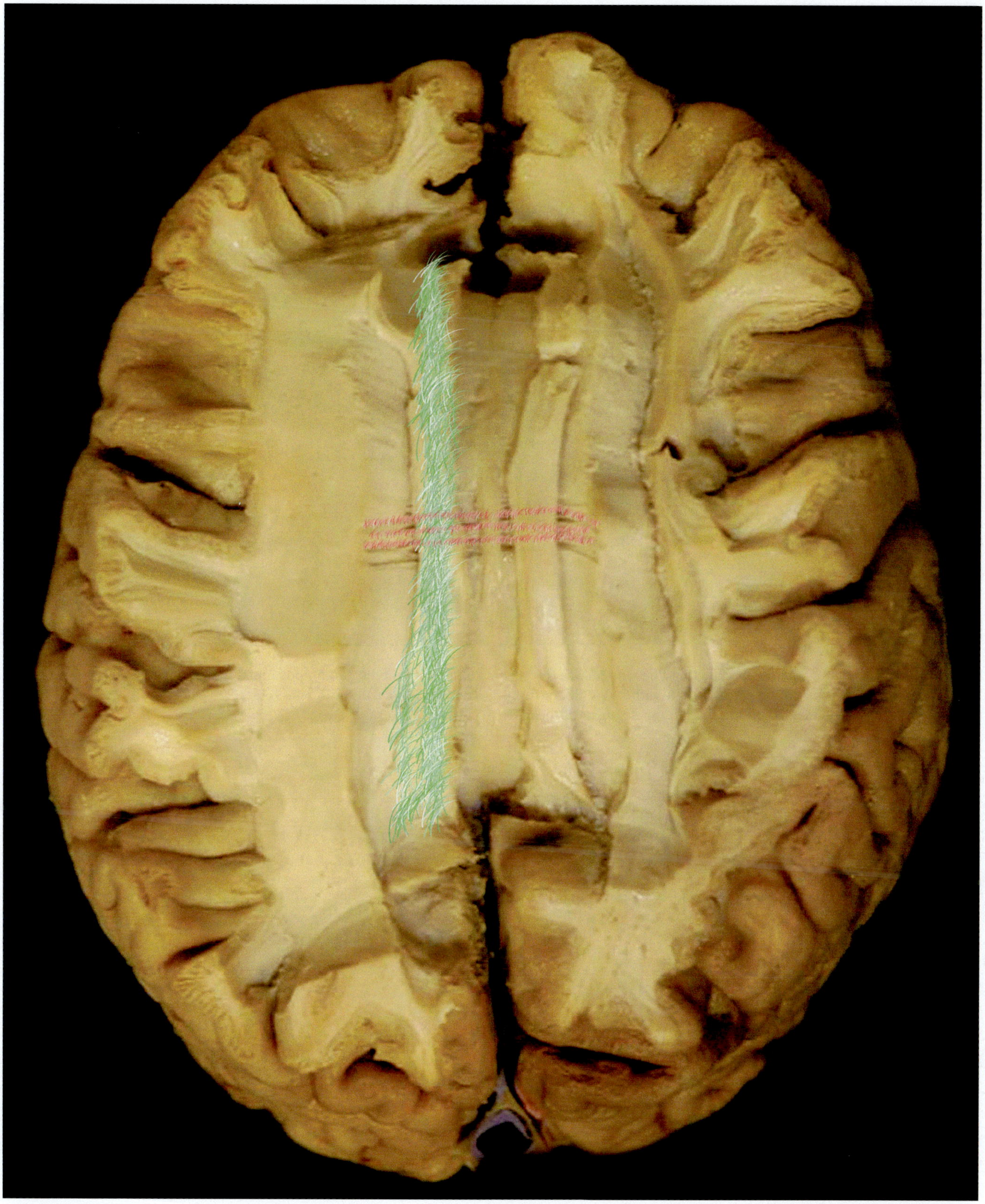

Fig. 2.26 (continued)

Fig. 2.27 Anatomical preparation depicting the confluence of the fibers of the superior segment of the cingulum with a few fibers from the subparietal and lingual gyri to form the inferior segment of the cingulum within tvhe parahippocampal gyrus. *g* gyrus.

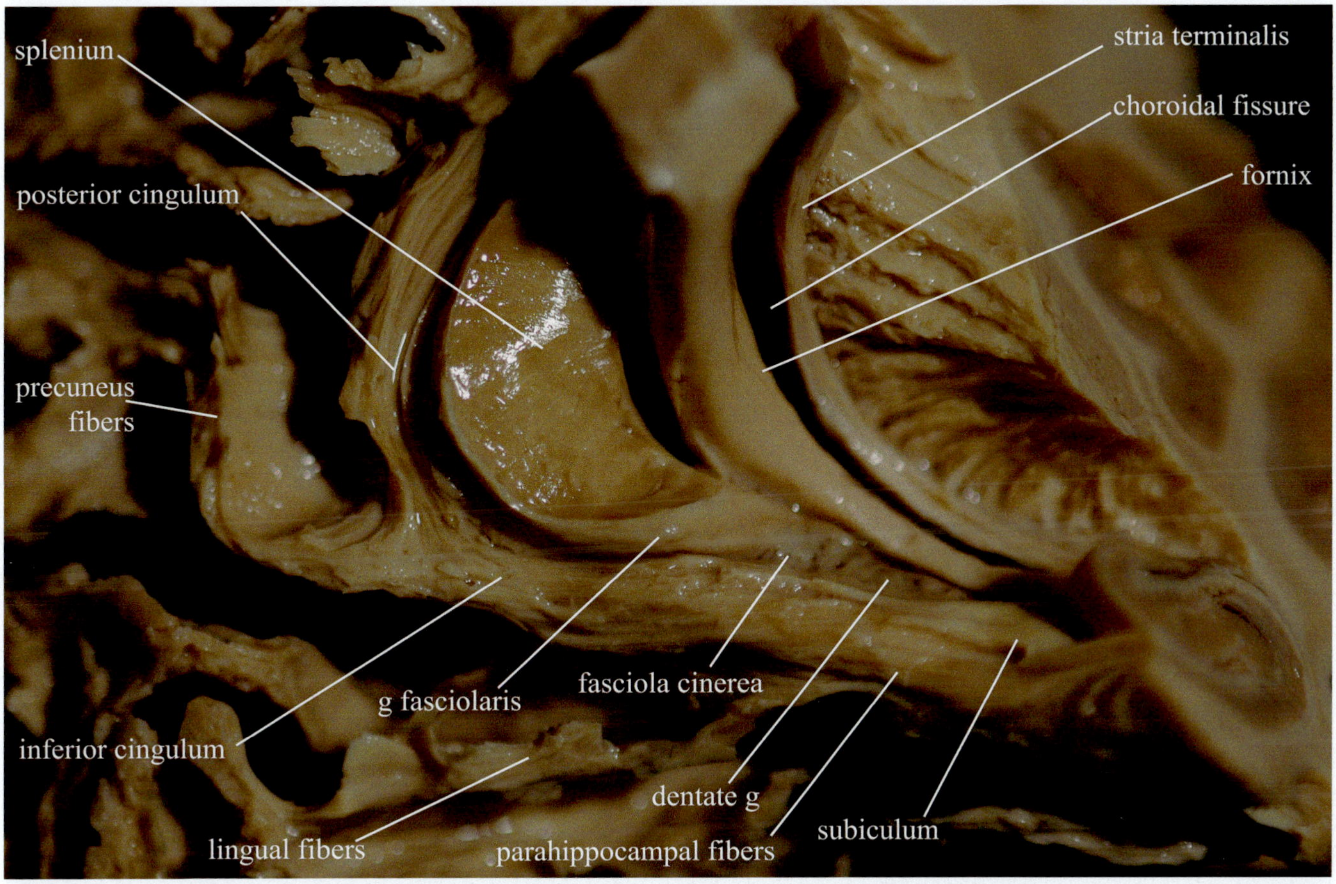

Fig. 2.27 (continued)

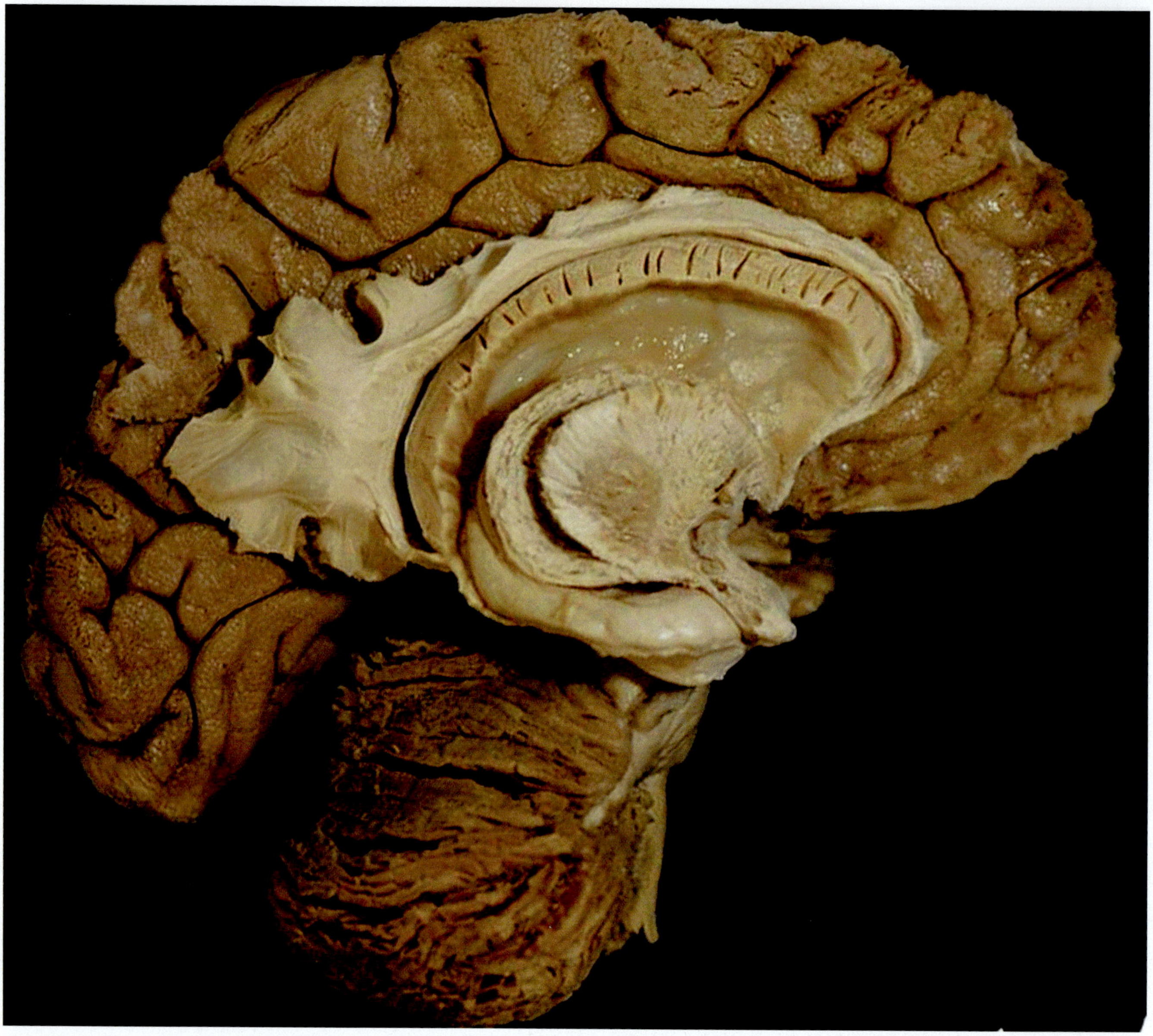

Fig. 2.28 Anatomical preparation of the right cingulum projecting over the medial aspect of the contralateral hemisphere with the superior segment (green) and inferior segment (red). (©Kadri 2023. All Rights Reserved)

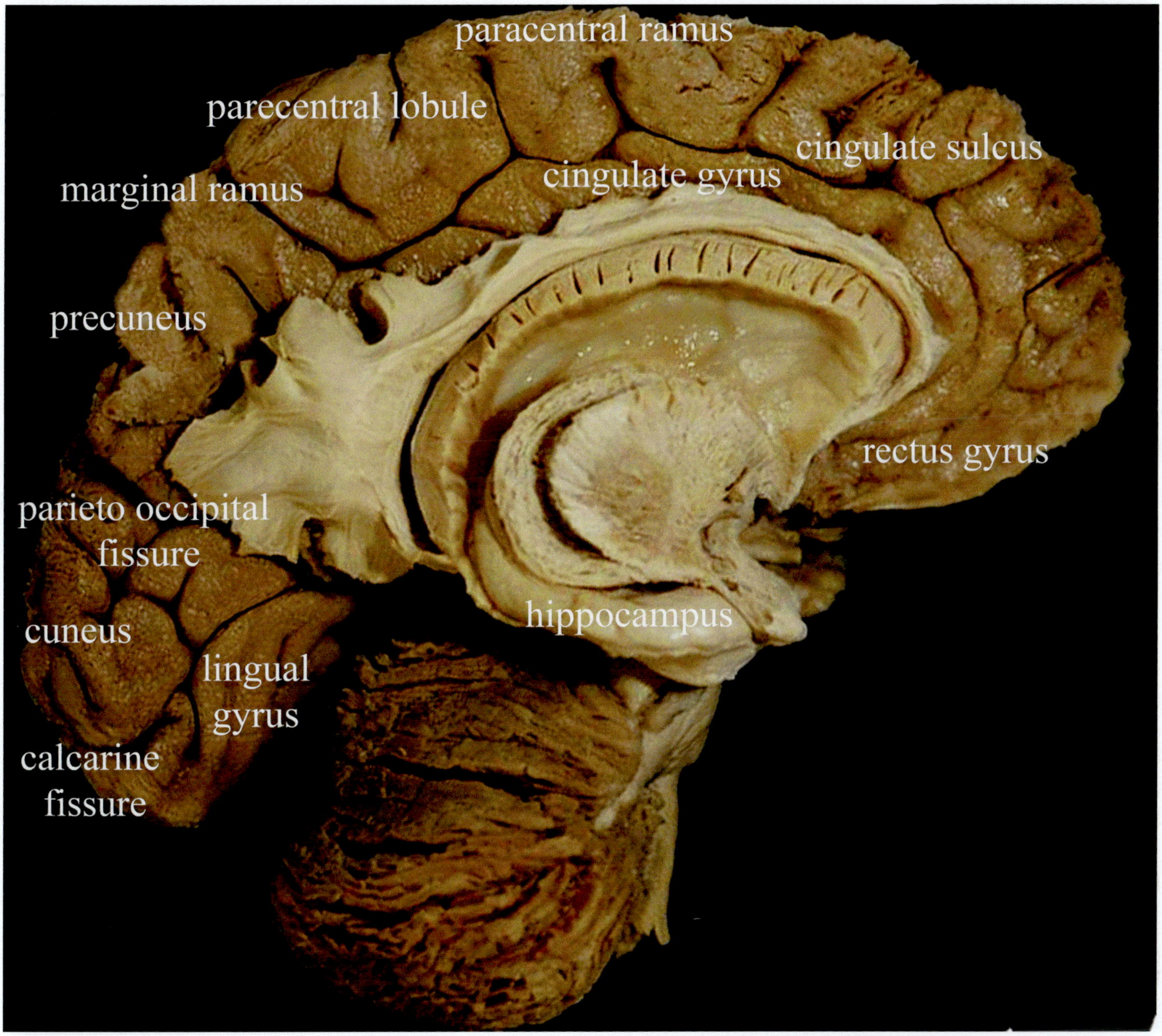

Fig. 2.28 (continued)

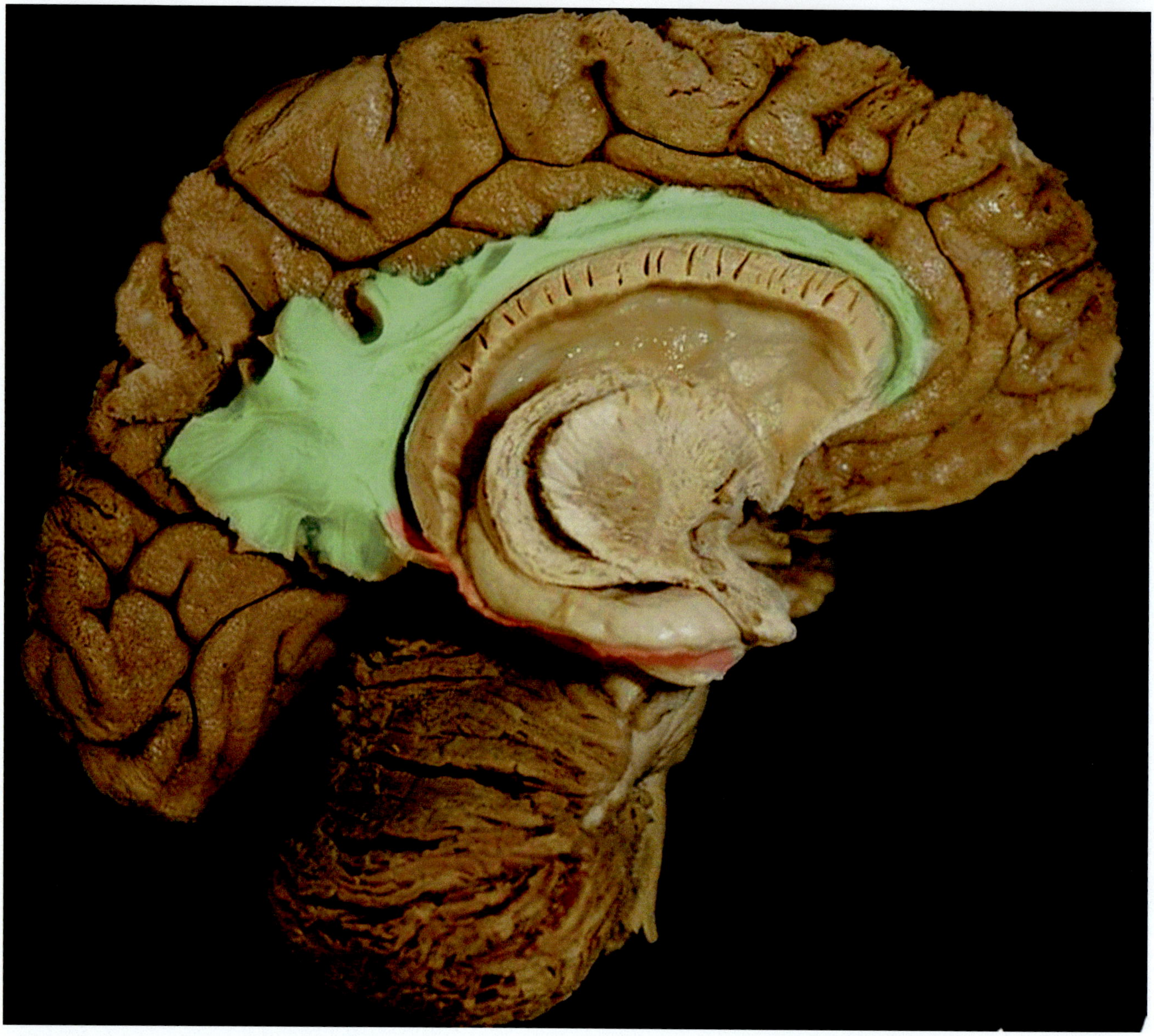

Fig. 2.28 (continued)

References

1. Testut L, Latarjet A. Tratado de anatomia humana, vol. 2. 9th ed. Barcelona: Salvat Editores; 1966.
2. Rhoton AL Jr. The cerebrum. Neurosurgery. 2002;51(4 Suppl):S1–51. https://doi.org/10.1097/00006123-200210001-00002.
3. Türe U, Yaşargil MG, Friedman AH, Al-Mefty O, Yaşargil DG. Fiber dissection technique: lateral aspect of the brain. Neurosurgery. 2000;47:417–27. https://doi.org/10.1097/00006123-200008000-00028.
4. Naidich TP, Krayenbühl N, Kollias S, Bou-Haidar P, Bluestone AY, Carpenter DM. White matter. In: Naidich TP, Castillo M, Cha S, Smirniotopoulos JG, editors. Imaging of the brain. Philadelphia, PA: Saunders; 2013. p. 205–44.
5. Carpenter MB. Core text of neuroanatomy. Subsequent ed. Baltimore, MD: Williams &Wilkins; 1991.
6. Standring S. Cerebral hemisphere. In: Gray's anatomy: the anatomical basis of clinical practice. 40th ed. Philadelphia, PA: Elsevier; 2008. p. 335–59.
7. Nieuwenhuys R, Voogd J, van Huijzen C. The human central nervous system: a synopsis and atlas. 4th ed. Berlin: Springer; 2008.
8. Türe U, Yaşargil DC, Al-Mefty O, Yaşargil MG. Topographic anatomy of the insular region. J Neurosurg. 1999;90:720–33. https://doi.org/10.3171/jns.1999.90.4.0720.
9. Ribas GC. The cerebral sulci and gyri. Neurosurg Focus. 2010;28:E2. https://doi.org/10.3171/2009.11.FOCUS09245.
10. Fernández-Miranda JC, Rhoton AL Jr, Álvarez-Linera J, Kakizawa Y, Choi C, De Oliveira EP. Three-dimensional microsurgical and tractographic anatomy of the white matter of the human brain. Neurosurgery. 2008;62(6 Suppl 3):989–1028. https://doi.org/10.1227/01.neu.0000333767.05328.49.
11. Heimer L. The human brain and spinal cord: functional neuroanatomy and dissection guide. 2nd ed. New York: Springer; 1994.
12. Klingler J, Gloor P. The connections of the amygdala and of the anterior temporal cortex in the human brain. J Comp Neurol. 1960;115:333–69. https://doi.org/10.1002/cne.901150305.
13. Ludwig E, Klingler J. Atlas cerebri humani. Basel: S. Karger; 1956.
14. Williams PL. Gray's anatomy: the anatomical basis of medicine and surgery. 38th ed. New York: Churchill Livingstone; 1995.
15. Kadri PAS, De Oliveira JG, Krayenbühl N, Türe U, De Oliveira EP, Al-Mefty O, et al. Surgical approaches to the temporal horn: an anatomic analysis of white matter tract interruption. Oper Neurosurg (Hagerstown). 2017;13:258–70. https://doi.org/10.1093/ons/opw011.
16. Türe U, Yaşargil MG, Pait TG. Is there a superior occipitofrontal fasciculus? A microsurgical anatomic study. Neurosurgery. 1997;40:1226–32. https://doi.org/10.1097/00006123-199706000-00022.
17. Martino J, Brogna C, Robles SG, Vergani F, Duffau H. Anatomic dissection of the inferior fronto-occipital fasciculus revisited in the lights of brain stimulation data. Cortex. 2010;46:691–9. https://doi.org/10.1016/j.cortex.2009.07.015.
18. Ebeling U, Von Cramon D. Topography of the uncinate fasciculus and adjacent temporal fiber tracts. Acta Neurochir. 1992;115(3–4):143–8. https://doi.org/10.1007/BF01406373.
19. Tusa RJ, Ungerleider LG. The inferior longitudinal fasciculus: a reexamination in humans and monkeys. Ann Neurol. 1985;18:583–91. https://doi.org/10.1002/ana.410180512.

3 Interhemispheric Association Fibers (Commissural Fibers)

The interhemispheric fibers cross the midline and connect the same regions (homotopic) and different regions (heterotopic) of the telencephalon reciprocally [1, 2]. Heterotopic fibers have a tendency to direct to contralateral areas that give rise to long intrahemispheric association fibers [2].

In humans, the three main telencephalic commissures are the corpus callosum, the anterior commissure, and the hippocampal commissure (also called the limbic or forniceal). Other less prominent commissures are the anterior hypothalamic (of Ganser), the dorsal supraoptic (of Meynert), the ventral supraoptic (of Gudden), the posterior, the habenular, and the inferior and superior colliculi commissures [3].

Corpus Callosum

The corpus callosum is the major interhemispheric commissure that connects most of the neocortical areas (Figs. 3.1, 3.2, 3.3, 3.4, 3.5, 3.6, and 3.7). It was named by Galen of Pergamon (129–216 AD) and described and illustrated for the first time in humans in 1543 by Andreas Vesalius (1514–1564), who recognized that it linked the two halves of the brain and was continuous with the white substance of the cerebral hemispheres.

The corpus callosum is composed of axons originating from neocortical regions of the brain that converge medially as a compact bundle arching at the bottom of the interhemispheric fissure [3]. From the midline, the corpus callosum spreads out laterally like a butterfly spreading its wings. Its fibers are described as homotopic and heterotopic, reciprocally connecting the cerebral hemispheres [1, 4, 5]. Four portions are described: the rostrum, genu, body, and splenium. A fifth portion, the isthmus, may be evident as a narrowing at the transition point of the body and splenium.

The rostrum, or *lamina rostralis,* extends from the anterior commissure to the genu of the corpus callosum. Its fibers seem to connect the frontobasal regions reciprocally. The genu is a thick portion of the corpus callosum, and its fibers connect the frontal lobes (orbitofrontal and prefrontal regions) and the anterior portions of the cingulate gyrus [3]. The body of the corpus callosum corresponds to the horizontal portion. Laterally, its fibers cover the roof of the lateral ventricle, intermingling with the fibers of the anterior and superior thalamic peduncles and connecting motor, premotor, and supplementary motor areas and adjacent regions of the insula and cingulate gyrus [3]. The isthmus, when evident, corresponds to the region where the fornix meets the inferior surface of the corpus callosum and carries commissural fibers from the precentral and postcentral gyri and from the anterior temporal transverse gyri (Heschl's gyrus) [3]. The splenium is the most posterior portion, and its radiating fibers form the forceps major and the tapetum [3]. The *forceps major* connects parieto-occipital areas, calcarine areas [6], and parahippocampal areas [5]. The *tapetum*, first described in 1892 by Bernard Sachs (1858–1944), corresponds to the inner fibers of the splenium that turn anterior to form the inner layer of white matter fibers on the roof and lateral wall of the atrium and temporal horn of the lateral ventricles [7]. The tapetum is located immediately under the ependymal layer, lateral to the tail of the caudate nucleus. There, it extends to the most anterior portions of the temporal horn and is medial to the fibers of the posterior and inferior thalamic peduncles [8].

P. Kadri, *The Cartographic Atlas of the Brain*, https://doi.org/10.1007/978-3-031-38062-4_3

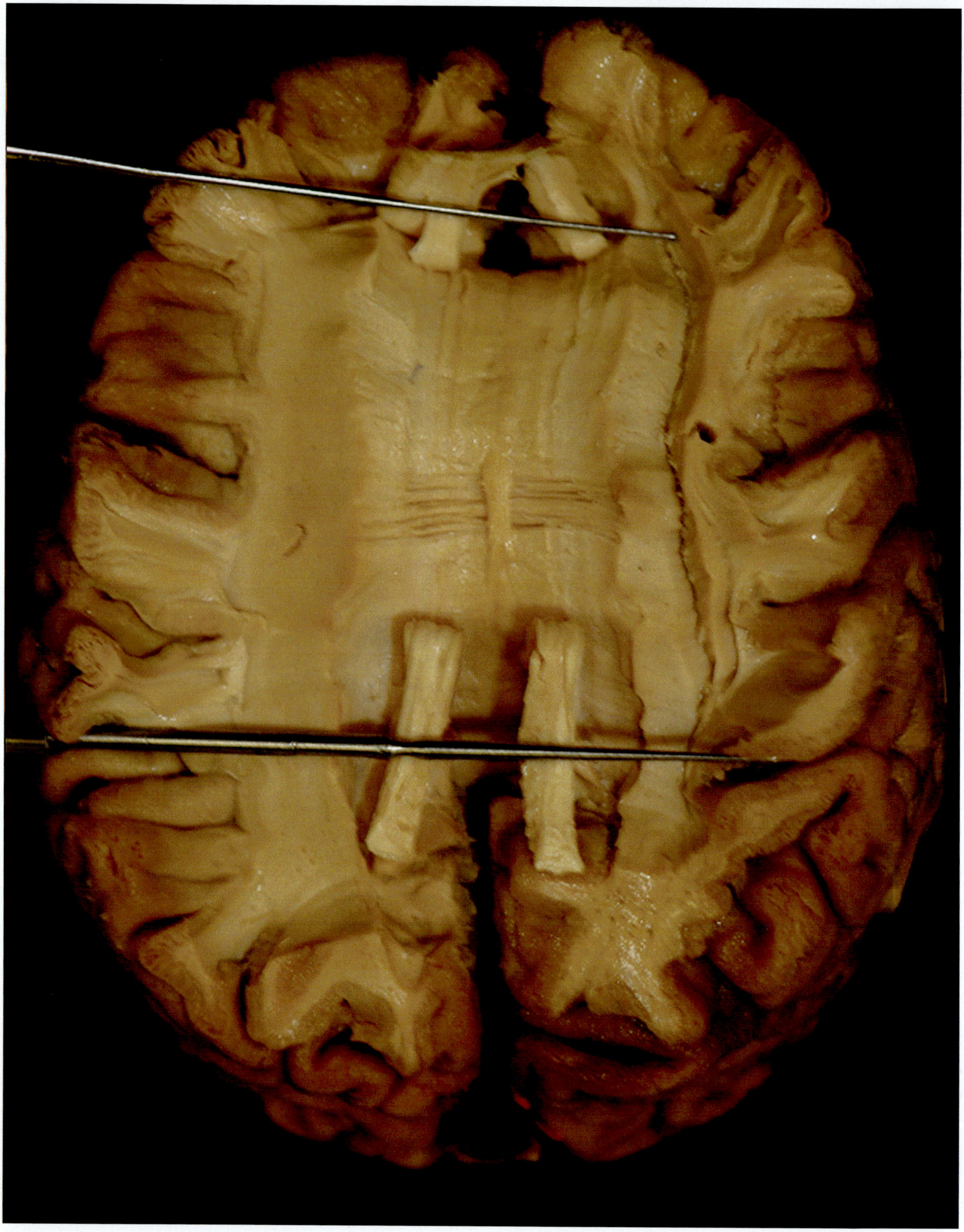

Fig. 3.1 Anatomical preparation and illustration of the corpus callosum. The superior frontal, precentral, and postcentral gyri and the superior parietal lobe have been removed. The cingulate gyrus was decorticated, and the cingulum (light green) was exposed, cut and reflected to expose the transverse fibers of the corpus callosum (red) at the depth of the interhemispheric fissure. The structures of the supracommissural hippocampus, the medial and lateral longitudinal striae (dark green), and the *indusium griseum* (light gray) are depicted.

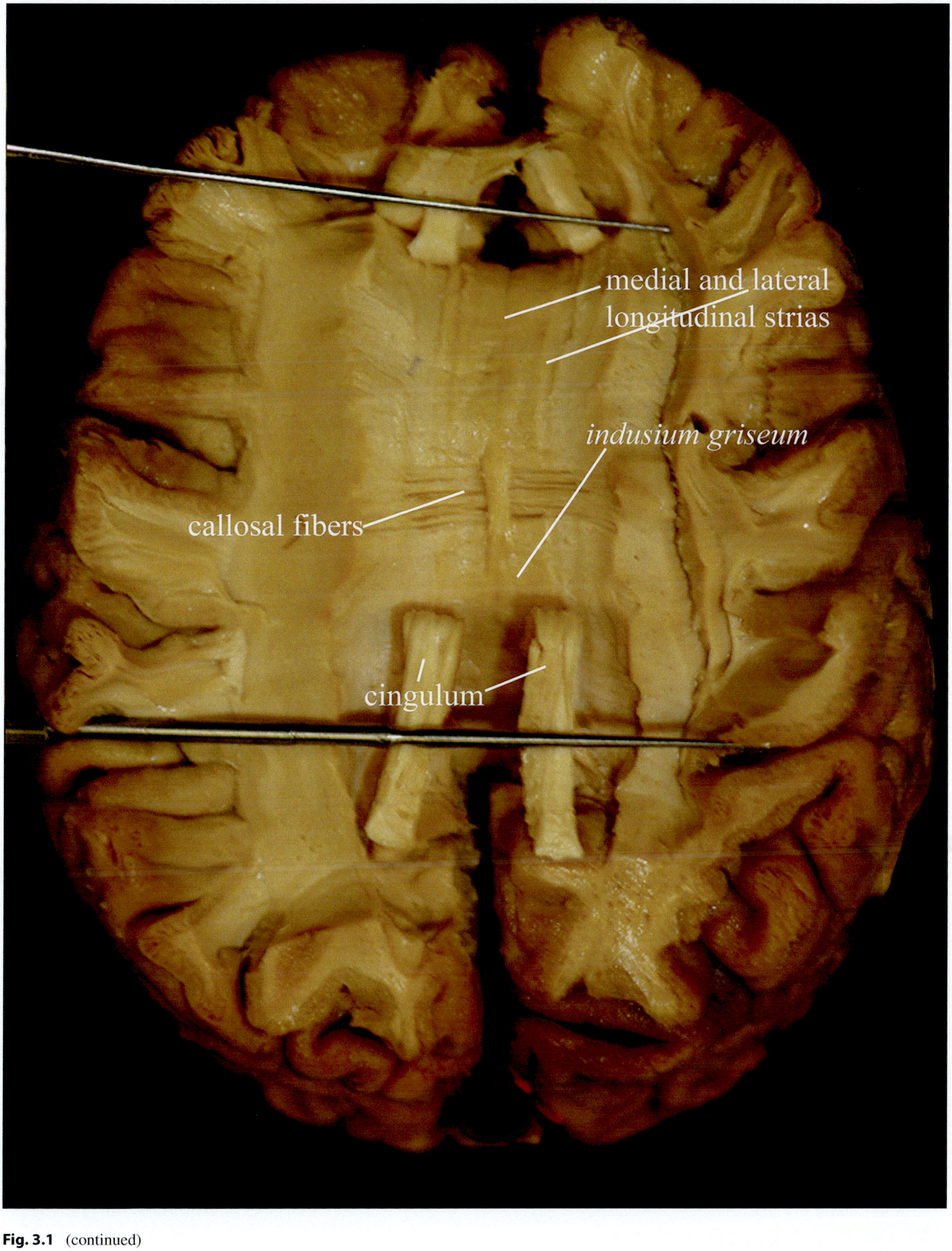

Fig. 3.1 (continued)

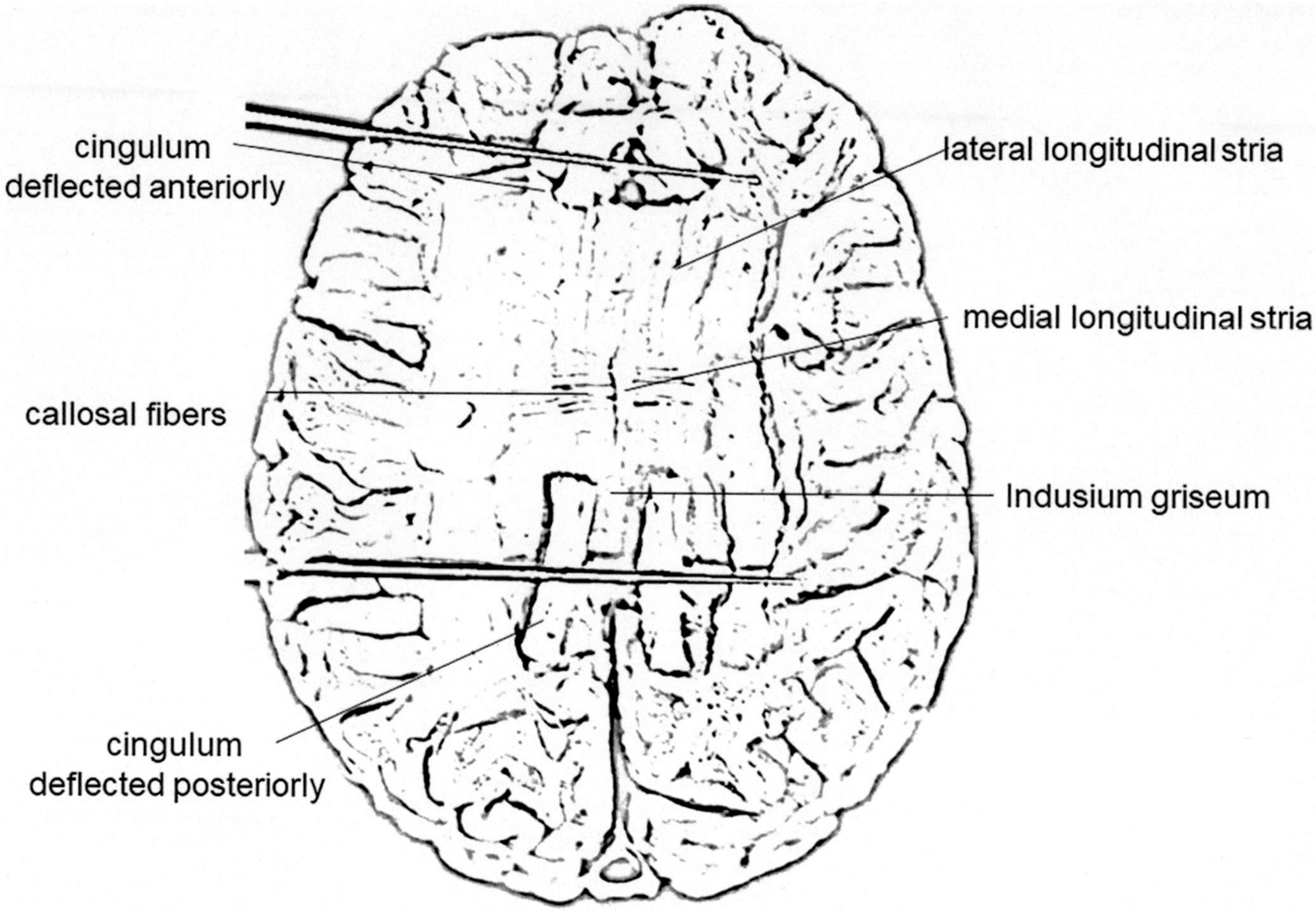

Fig. 3.1 (continued)

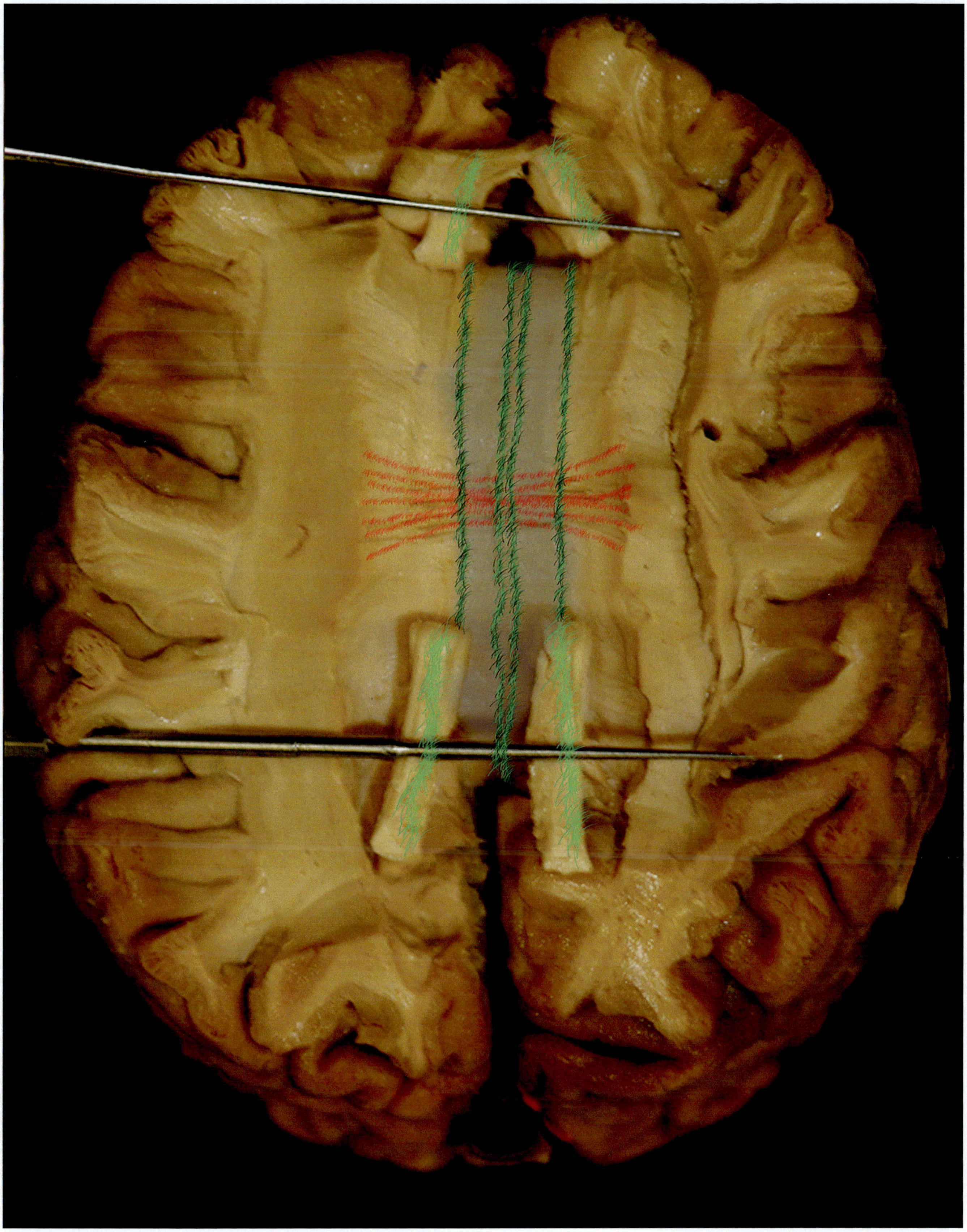

Fig. 3.1 (continued)

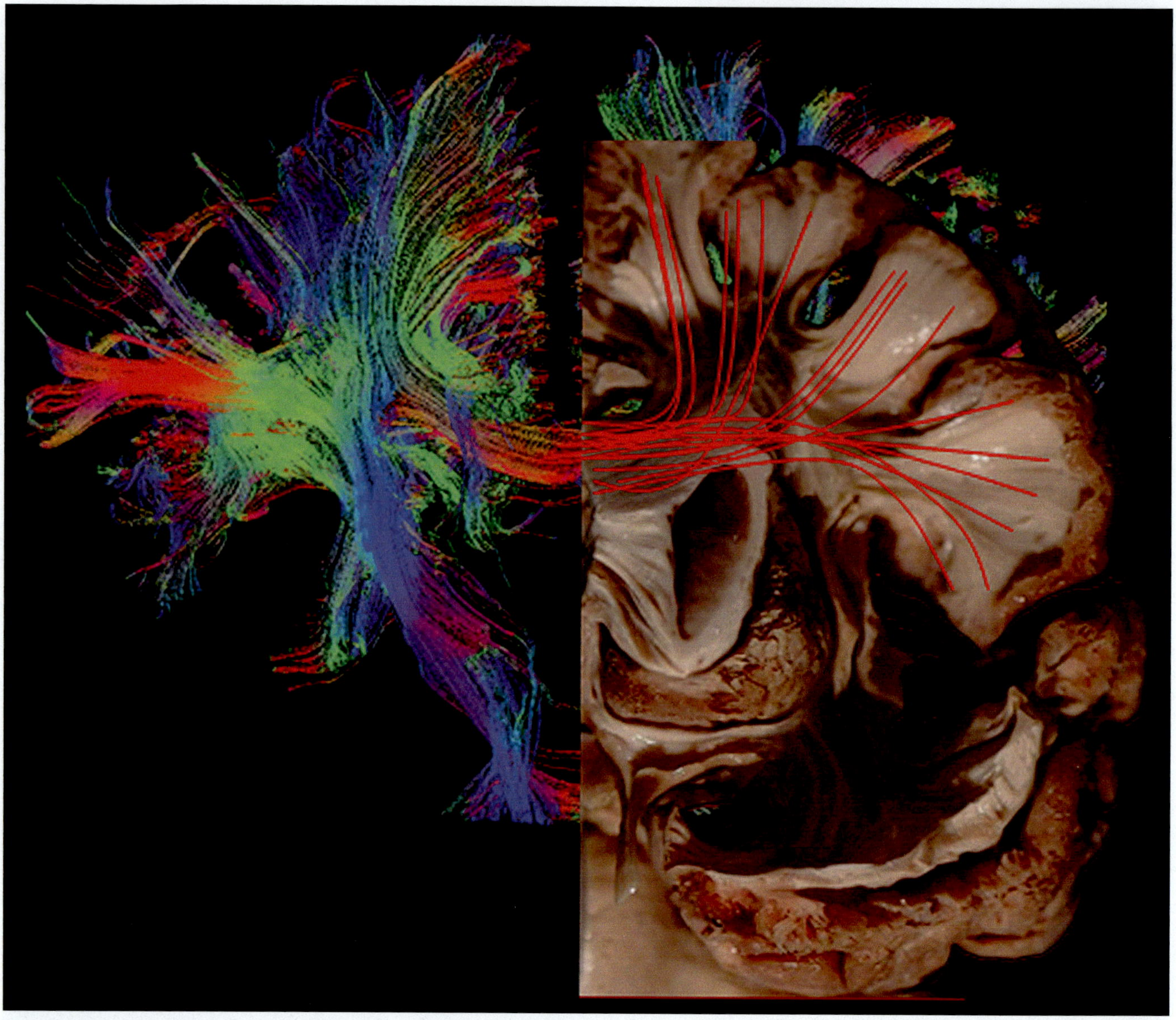

Fig. 3.2 A three-dimensional reconstruction of a tractography image overlying an anatomical preparation (coronal view) of the left hemisphere. This image depicts the radiating fibers of the corpus callosum. (©Kadri 2023. All Rights Reserved)

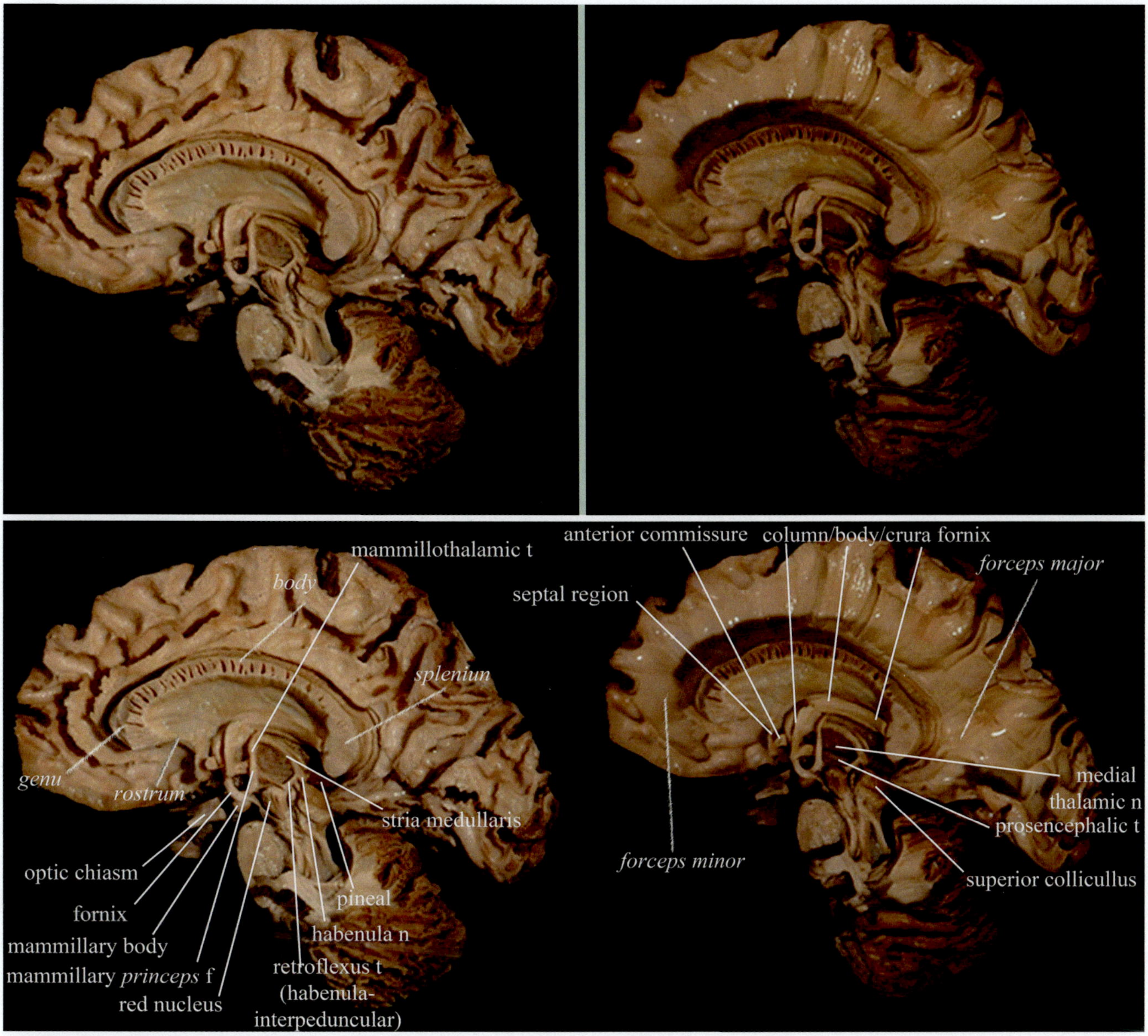

Fig. 3.3 Anatomical preparation of the left cerebral hemisphere, medial view. The rostrum, genu, body, and splenium of the corpus callosum are identified. On the left, the cortical surface has been removed, exposing the arcuate fiber system. On the right, the radiating fibers of the corpus callosum were followed laterally, after the removal of the cingulum, exposing the forceps minor radiating from the genu and the forceps major radiating from the splenium. *f* fasciculus, *n* nerve, *t* tract. (©Kadri 2023. All Rights Reserved)

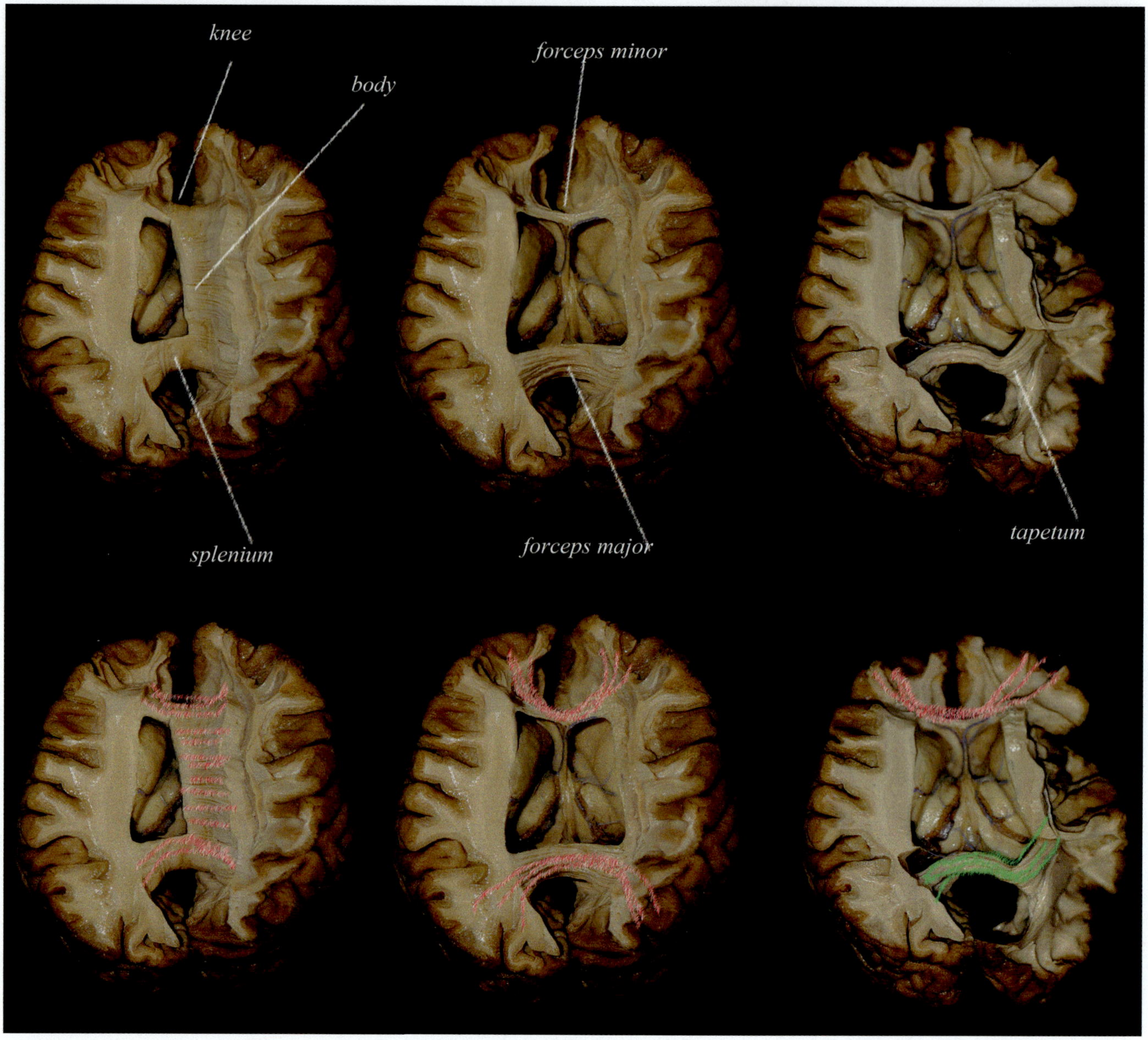

Fig. 3.4 Anatomical preparation of the serial dissections of the corpus callosum, superior view. The body of the corpus callosum is sequentially removed, and the radiating fibers of the genu and the splenium (colored areas) can be identified and followed. (©Kadri 2023. All Rights Reserved)

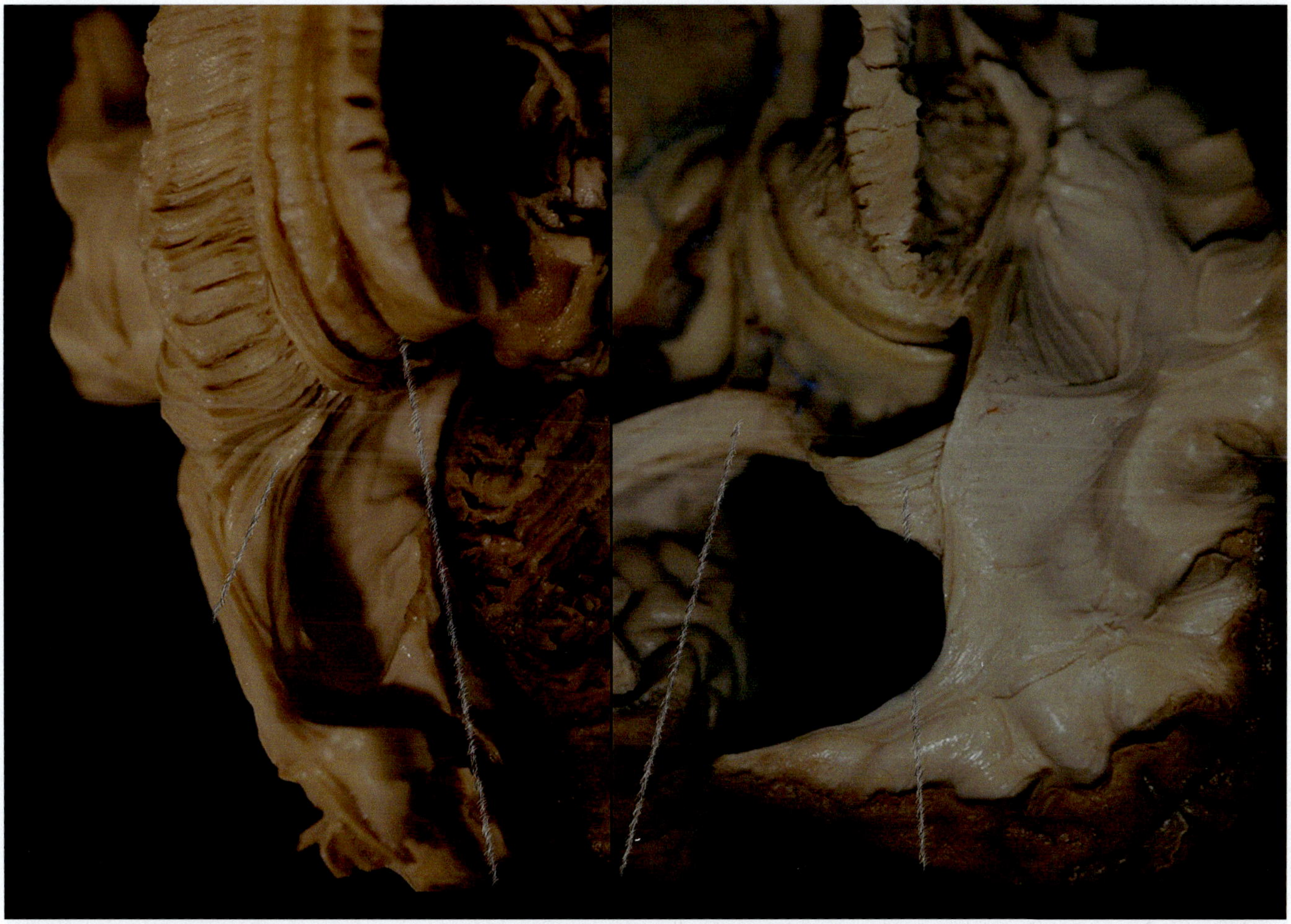

Fig. 3.5 Anatomical preparation of the splenium of the corpus callosum. On the left, the superficial fibers of the splenium radiate posteriorly and are designated as the forceps major. On the right, the inner fibers radiate anteriorly, on the roof and lateral walls of the atrium and temporal horn, and are designated as the tapetum. *n* nucleus, *pl* plexus, *st* striae. (©Kadri 2023. All Rights Reserved)

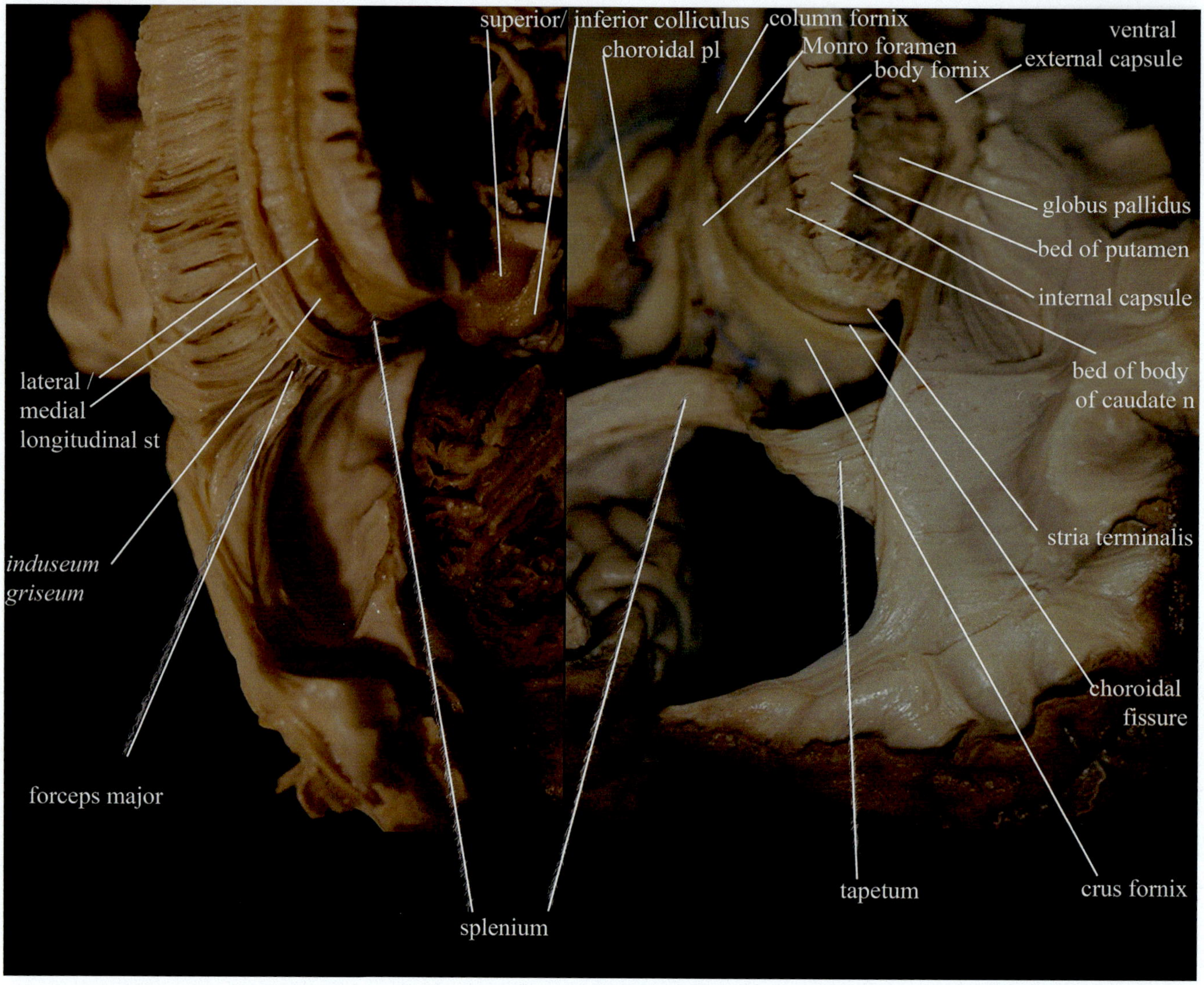

Fig. 3.5 (continued)

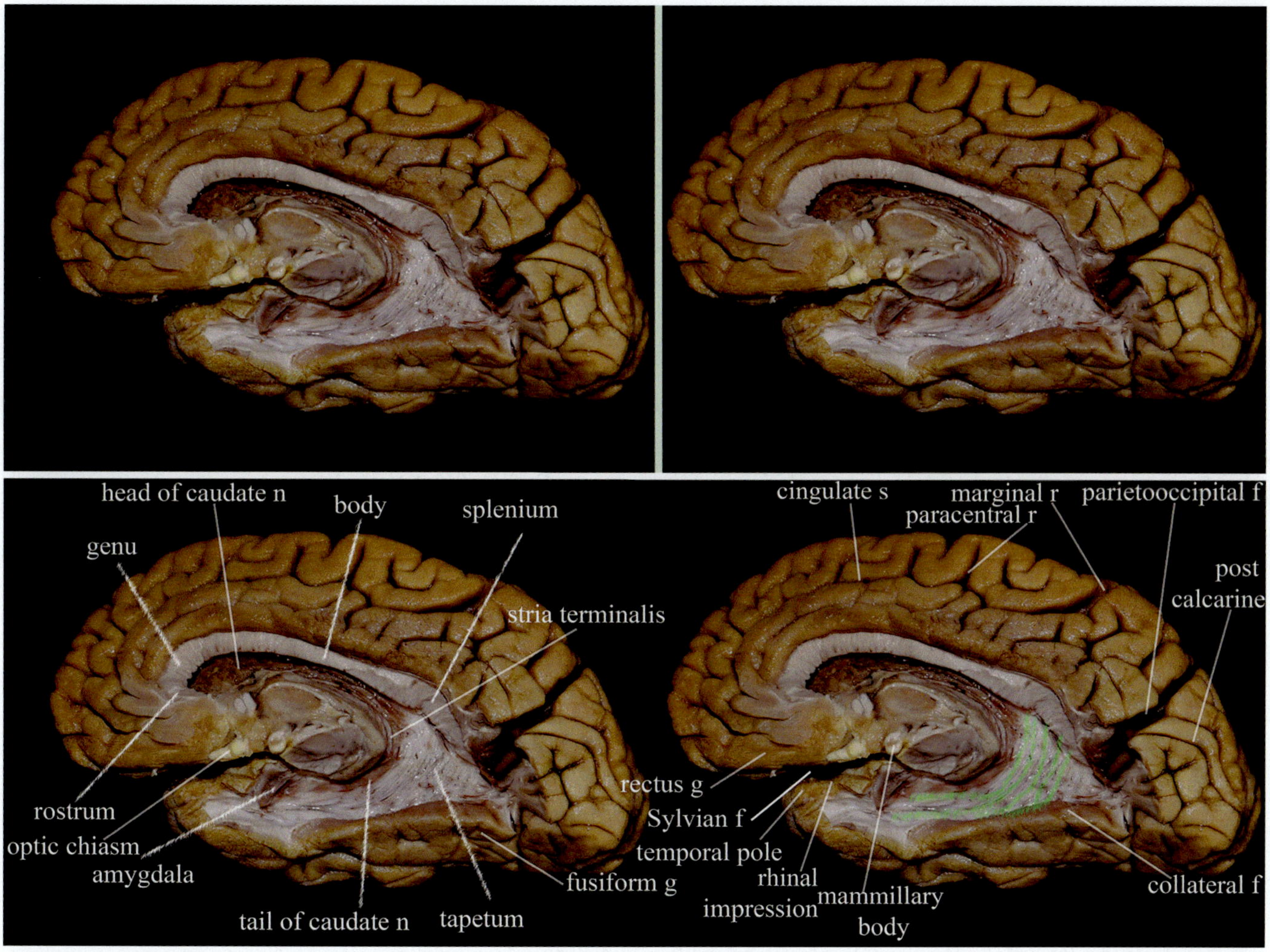

Fig. 3.6 Anatomical preparation of the right hemisphere, medial view. The fusiform and parahippocampal gyri have been removed. The ependyma on the lateral wall of the lateral ventricles has also been removed. The tapetum (green) can be seen radiating from the splenium of the corpus callosum and directed obliquely on the roof and lateral walls of the atrium and temporal horn. *f* fissure, *g* gyrus, *n* nucleus, *r* ramus, *s* sulcus. (©Kadri 2023. All Rights Reserved.)

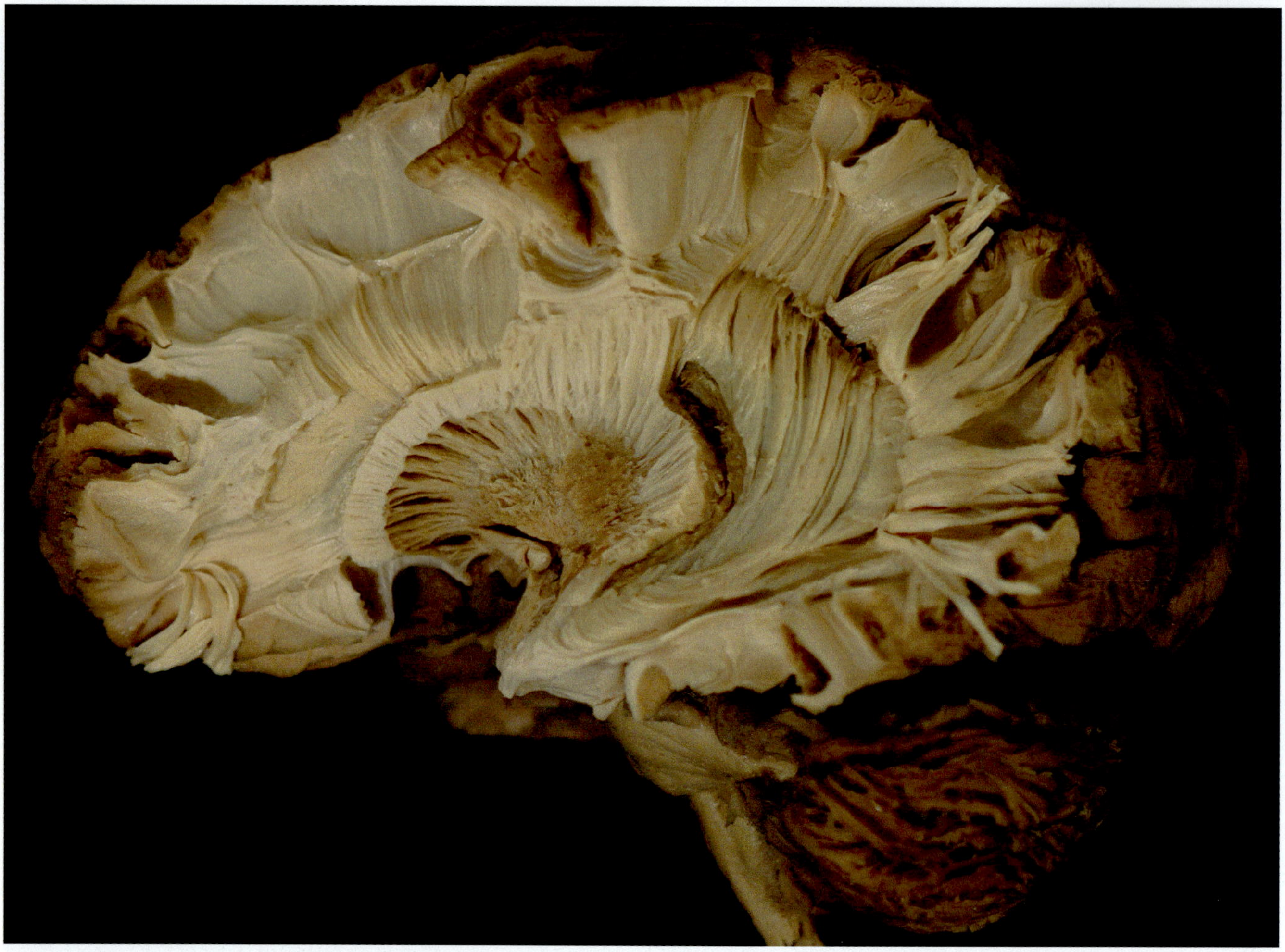

Fig. 3.7 Anatomical preparation of the left hemisphere. The short and long association fibers, the putamen, and the anterior commissure have been removed. The projection fibers of the corona radiata converge to form the internal capsule. The peculiar loop of the posterior thalamic peduncle (Meyer's loop), which contains the optic radiation, is partially preserved anteriorly. The horizontally oriented fibers of the sagittal stratum on the lateral wall of the temporal horn and atrium have been deflected posteriorly. The inner fibers of the tapetum are located just under the ependyma and on the outer limit of the tail of the caudate nucleus. *f* fasciculus, *n* nucleus, s sulcus. (©Kadri 2023. All Rights Reserved)

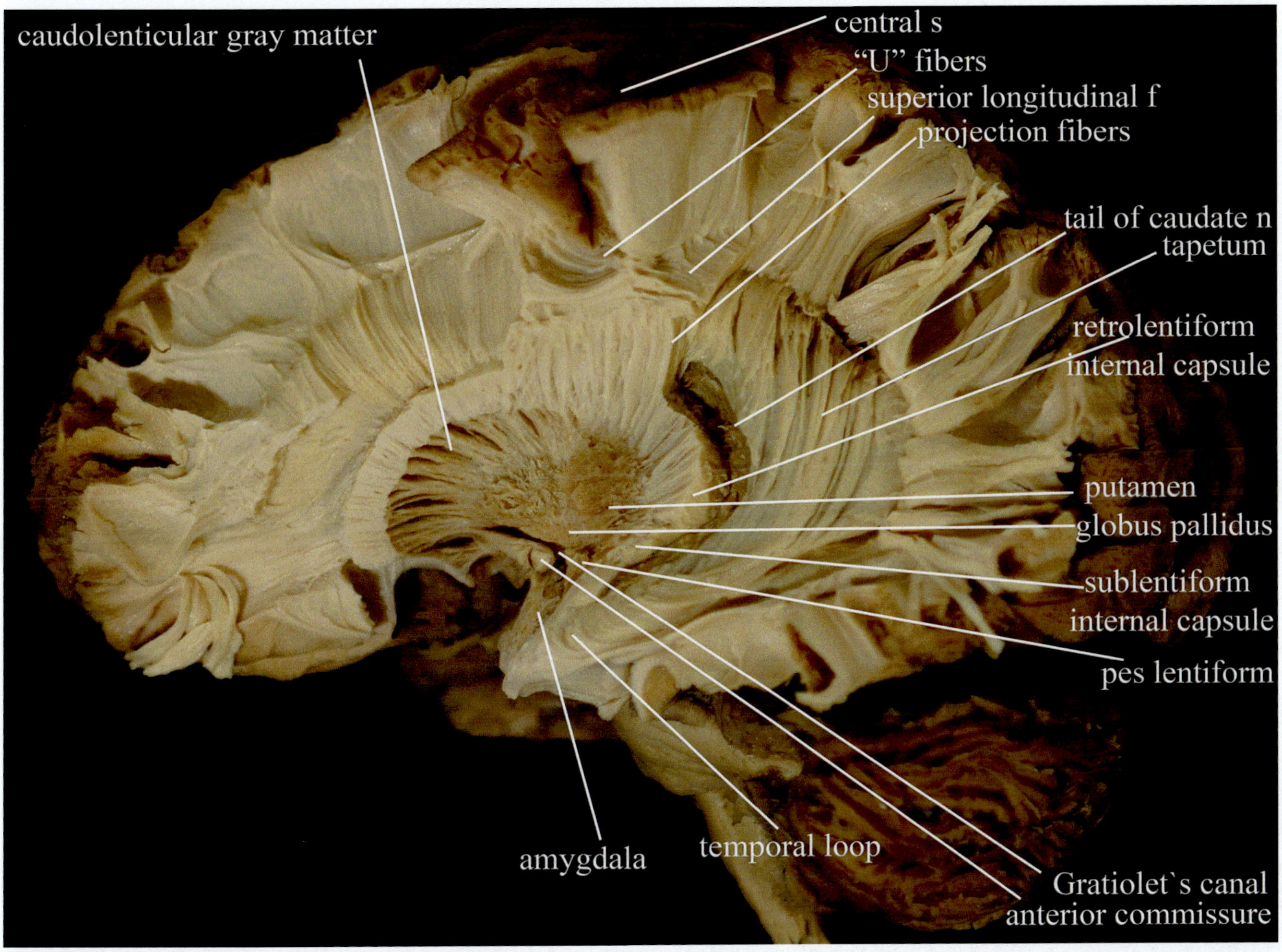

Fig. 3.7 (continued)

Anterior Commissure

The anterior commissure (Figs. 3.8, 3.9, 3.10, 3.11, 3.12, and 3.13) was named by Louis Pierre Gratiolet (1815–1865), a French anatomist and zoologist. This commissure is a compact, prominent bundle that crosses the midline and connects the caudal portion of the frontal lobe to the anterior portions of the temporal lobes. It can be divided into three segments: ventricular, compact or striatal, and lateral extensions.

The ventricular segment is located in the most anterosuperior portion of the third ventricle and is bordered anteriorly by a few fibers from the fornix (the precommissural forniceal fibers) and posteriorly by the retrocommissural fibers of the fornix. The compact, or striatal, component is a prominent fiber bundle perpendicular to the ventral aspect of the anterior limb of the internal capsule. It protrudes directly laterally and slightly caudally, indenting the ventral surface of the globus pallidus in what was termed by Joseph Dejerine (1949–1917)

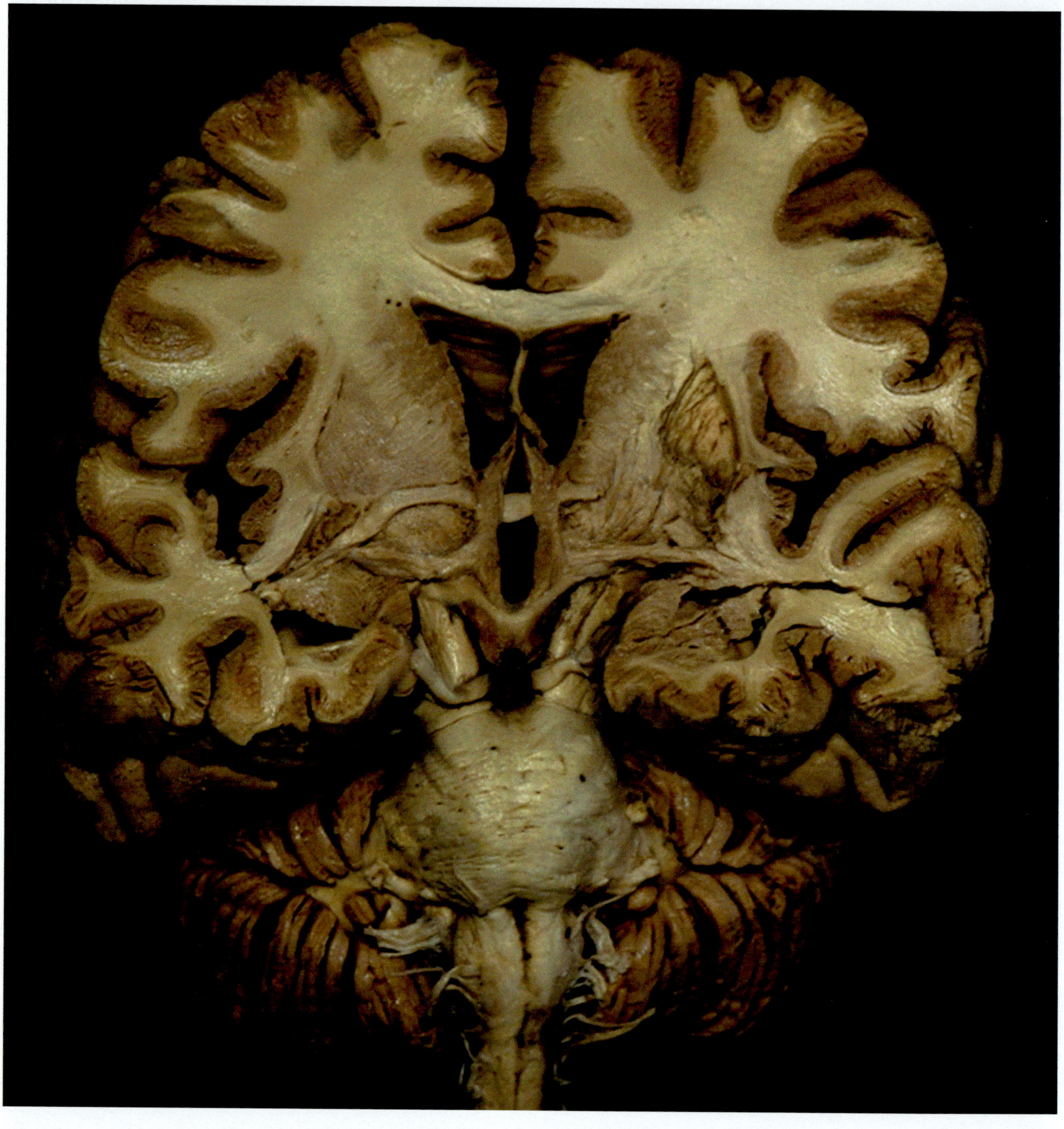

Fig. 3.8 Anatomical preparation and illustration, frontal view of a coronal plane through the anterior third ventricle. The anterior commissure segments are the ventricular segment, the compact segment, and the lateral extension. (©Kadri 2023. All Rights Reserved)

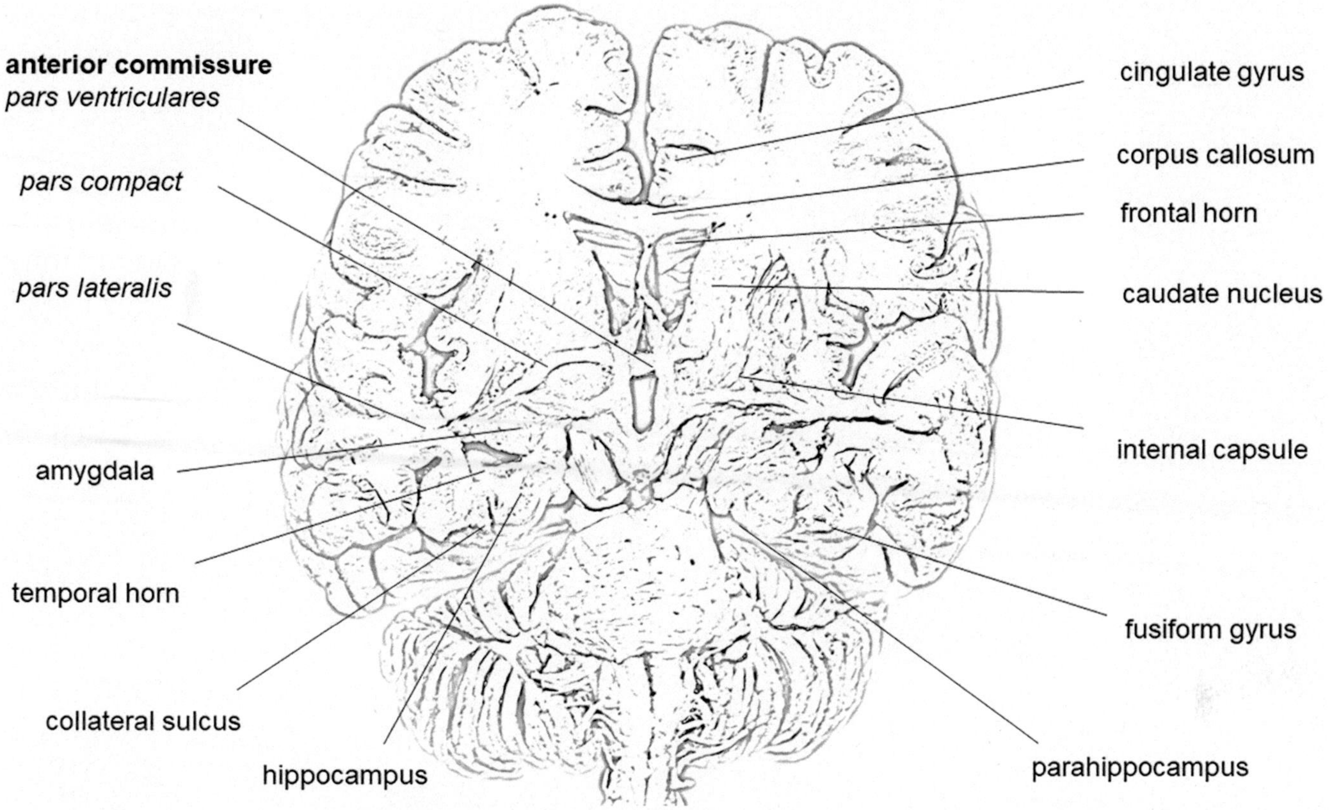

Fig. 3.8 (continued)

as the "Gratiolet canal." It extends laterally under the basal putamen, passing over the superior surface of the amygdala to fan out laterally, anteriorly, and posteriorly in the temporal lobe. A small component of fibers, which are not always present [9], interconnects the caudal orbitofrontal region and is referred to as the anterior portion, olfactory component, rhinencephalic (of Meynert) or basal telencephalic commissure. This structure connects the olfactory bulbs, the anterior olfactory nucleus, the olfactory cortex, and probably the amygdala and the nucleus accumbens septi [1, 2, 10, 11]. In the anterosuperior view, the anterior commissure resembles the handles of a bicycle [9]. The lateral extension extends toward the amygdala, temporal pole, inferior and middle temporal regions, and parahippocampal gyrus. Lateral and posterior portions of the lateral extension have been described [8, 10]. The anterior portion is medial, parallel, and in intimate relation with the fibers of the uncinate fasciculus [8, 10]. The posterior portion is parallel to the fibers of the occipitofrontal fasciculus and to horizontal fibers of the posterior thalamic peduncle from the designated *stratum sagittale* [8, 10].

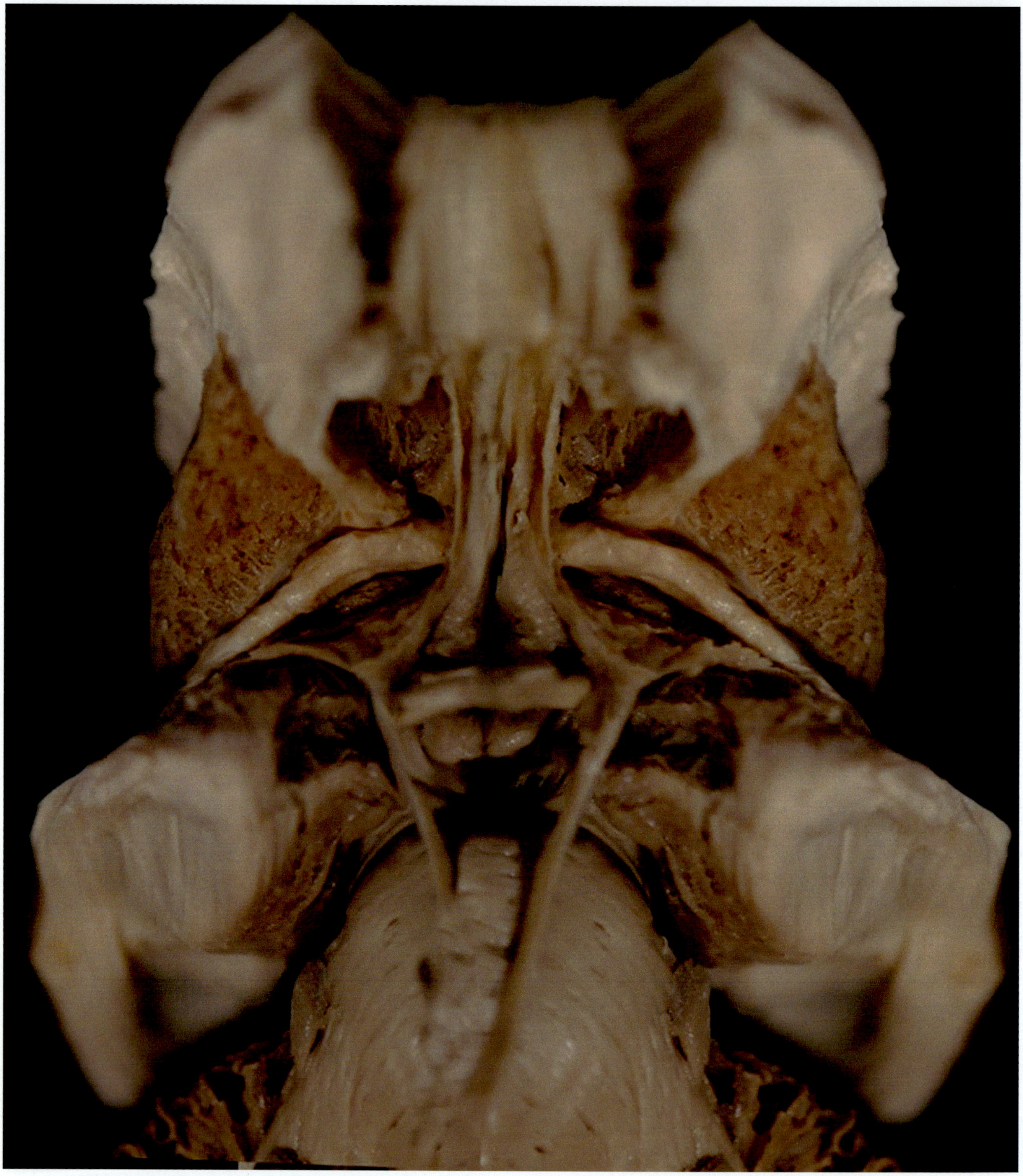

Fig. 3.9 Anatomical preparation and illustration of the brain to depict the anterior commissure. In this anterior view, the ventral striatum (ventral portions of the putamen, the head of the caudate, and the nucleus accumbens) has been removed. The medial and lateral olfactory striae and the anterior borders of the anterior perforated substance are reflected caudally. The medial and lateral longitudinal striae over the corpus callosum are depicted. The longitudinal striae and the medial olfactory stria converge in the paraterminal gyrus under the rostrum of the corpus callosum.

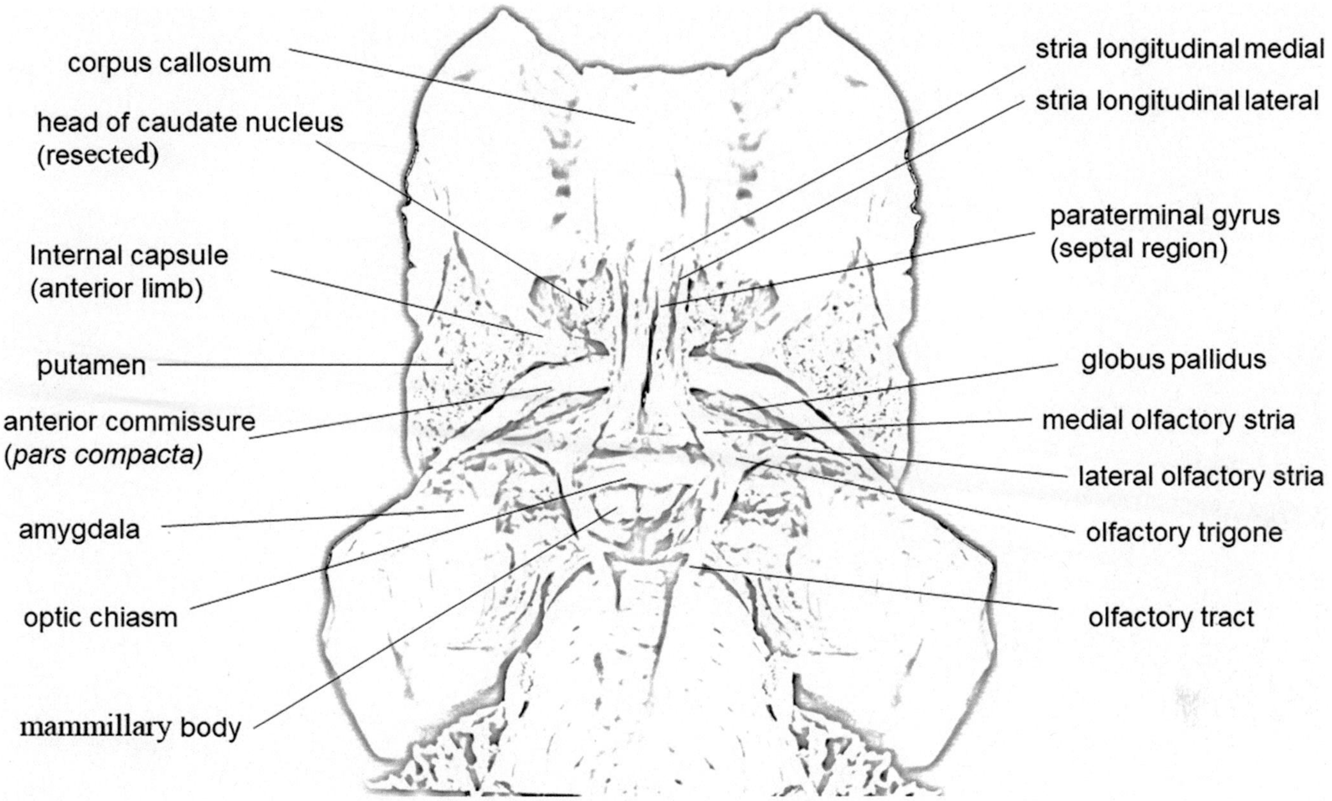

Fig. 3.9 (continued)

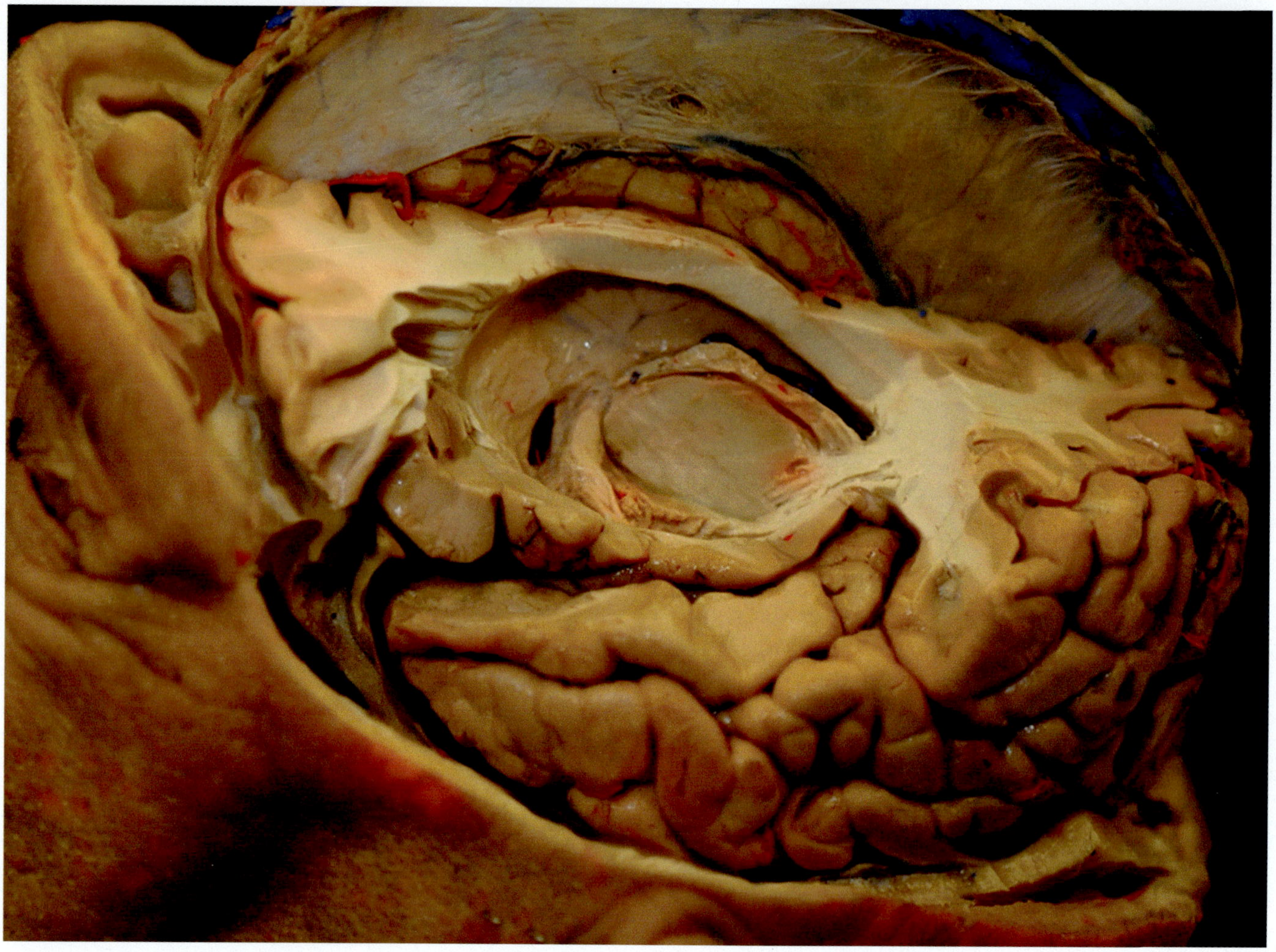

Fig. 3.10 Anatomical preparation and illustration of the left side of the head. The structures superior to the Sylvian fissure have been removed. The insular cortex was preserved to show the relation of the anterior commissure to the limen insulae. The putamen was removed, and the *pars compacta* of the anterior commissure is unveiled at Gratiolet's canal. (©Kadri 2023. All Rights Reserved)

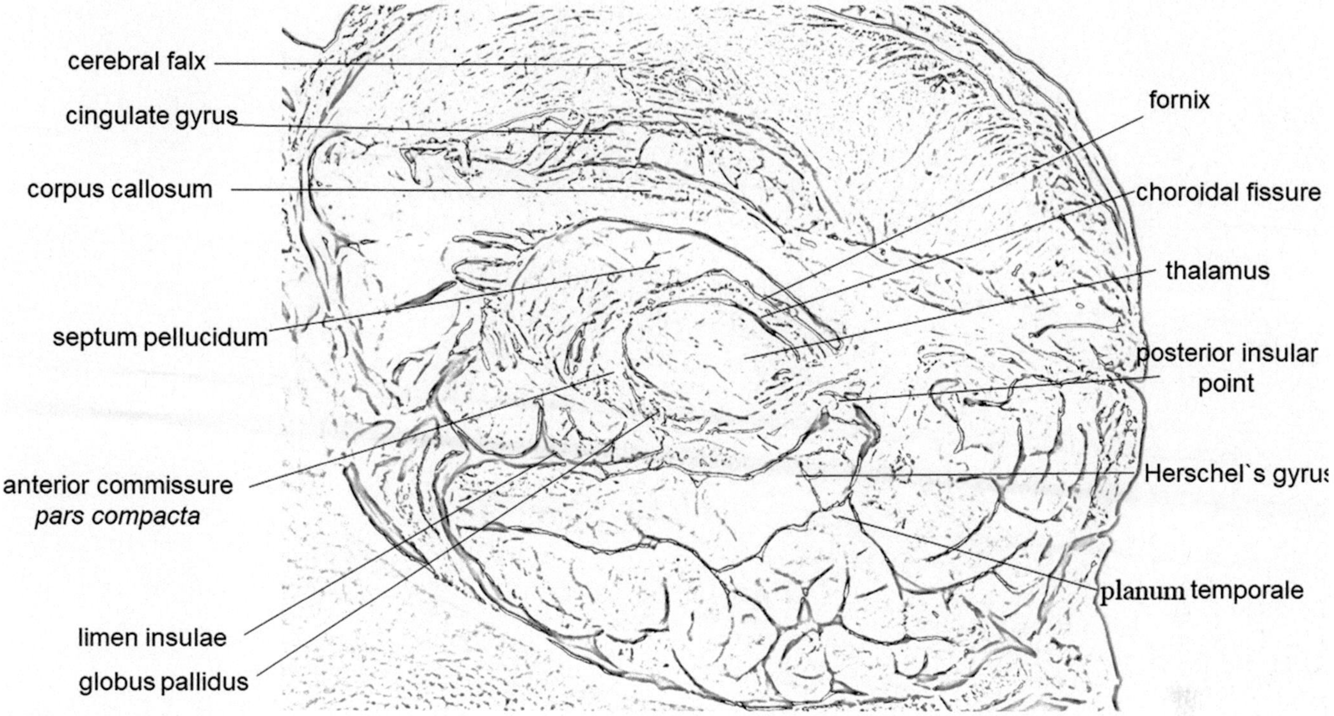

Fig. 3.10 (continued)

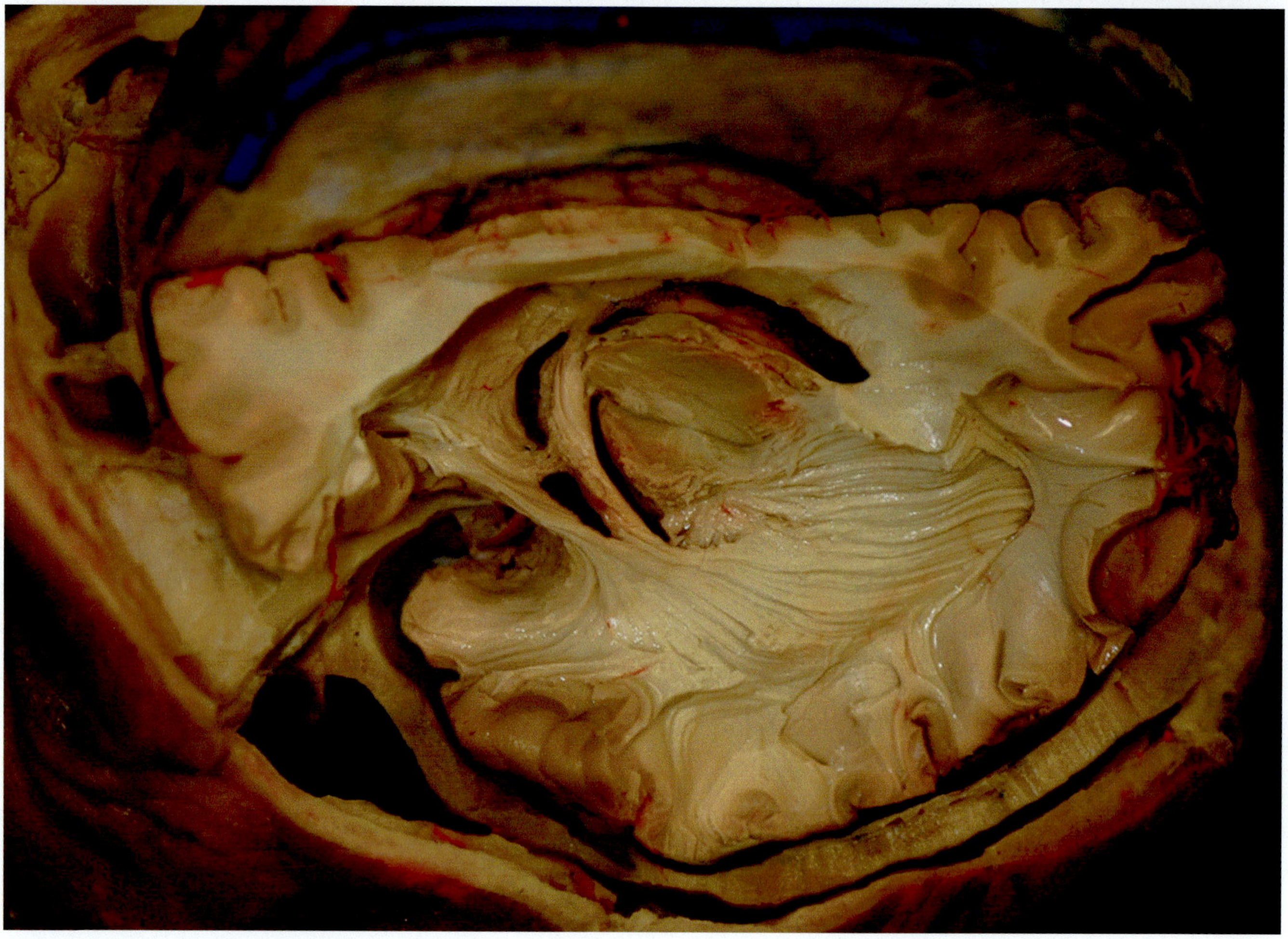

Fig. 3.11 Anatomical preparation and illustration depicting the olfactory component of the anterior commissure. (©Kadri 2023. All Rights Reserved)

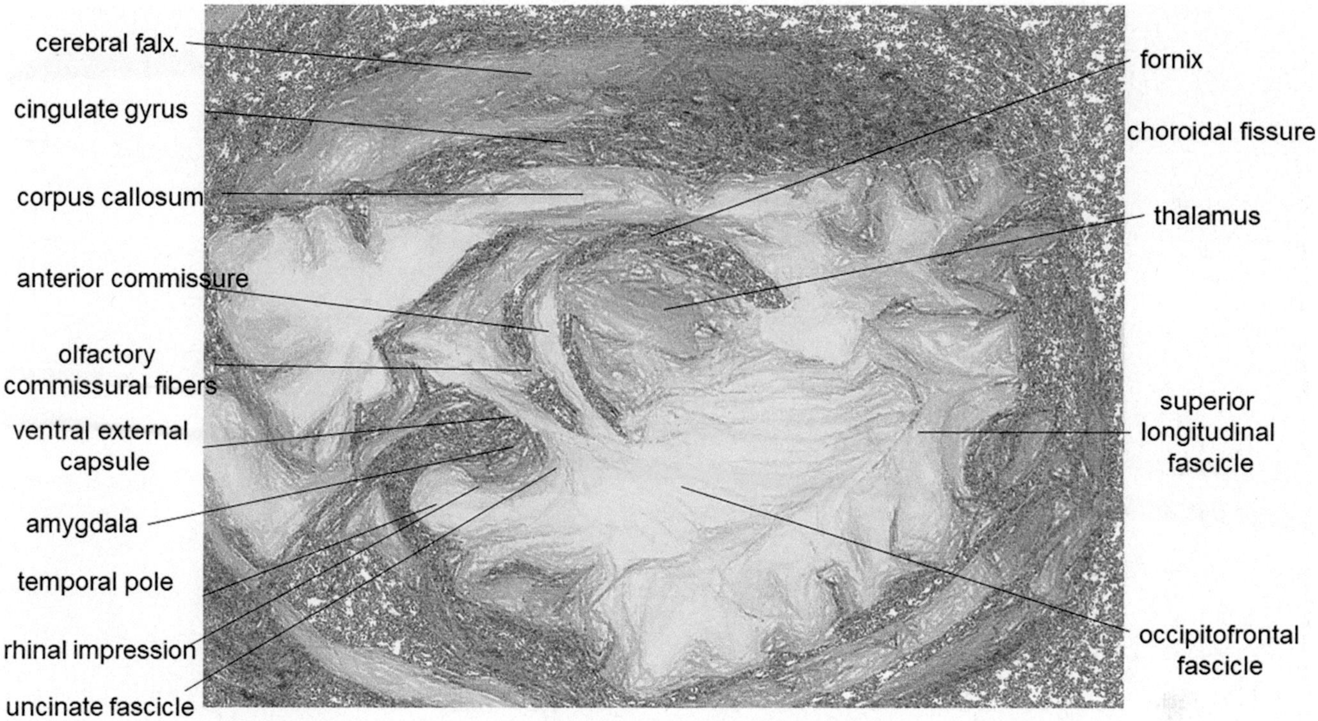

Fig. 3.11 (continued)

Fig. 3.12 Anatomical preparation and illustration of the left hemisphere. Removing the putamen exposes the continuation of the fibers of the corona radiata and the internal capsule. The caudolenticular gray matter is formed by bridges of the striatum through the holes of the internal capsule. The globus pallidus is a firm nucleus at the ventromedial portion of the lentiform nucleus and is impinged by the anterior commissure (*pars compacta*).

Fig. 3.12 (continued)

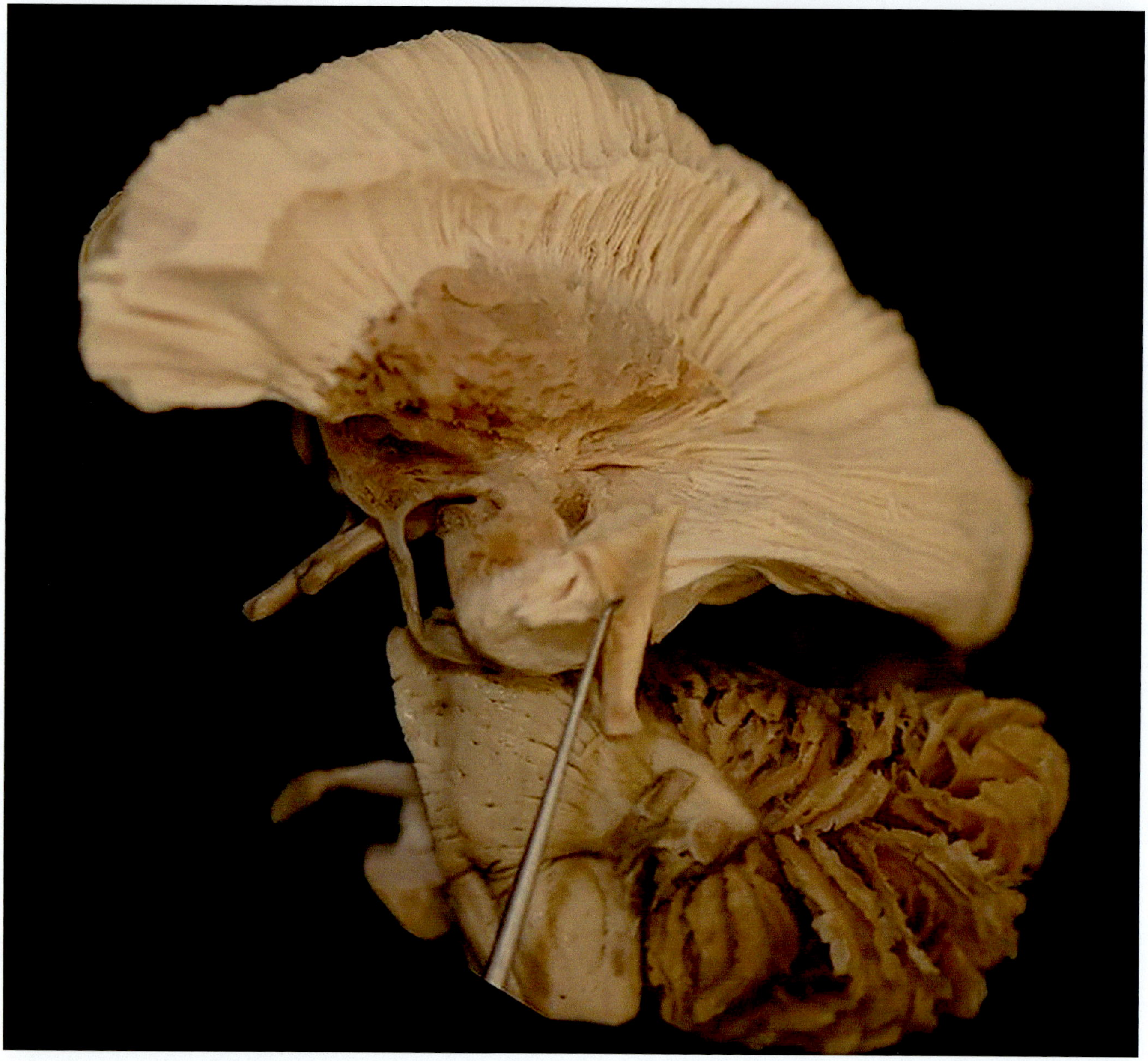

Fig. 3.13 Anatomical preparation and illustration of the left hemisphere. The anterior commissure is deflected from Gratiolet's canal. The relationships of the gray matter of the amygdala, putamen, globus pallidus, substantia innominata, and nucleus accumbens can be appreciated.

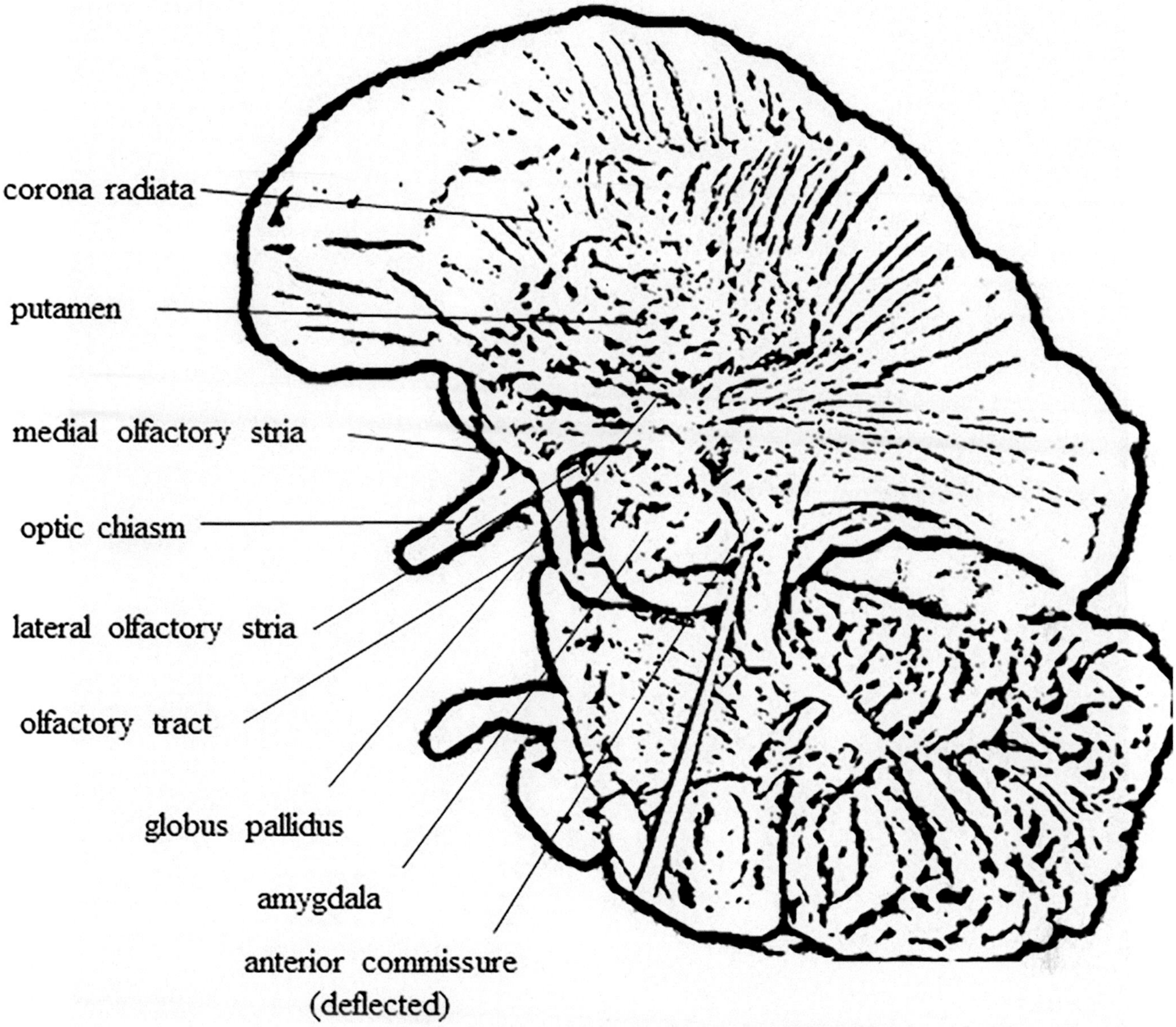

Fig. 3.13 (continued)

Hippocampal Commissure

Three limbic and paralimbic connections across the hemispheres have been described: (1) the ventral hippocampal commissure, (2) the dorsal hippocampal commissure, and (3) the hippocampal decussation [12]. The ventral commissure is at the transition of the body and column of the fornix, just posterior to the foramen of Monro. It contains fibers of the most anterior portions of the hippocampal formation. The dorsal hippocampal commissure is at the transition of the crura and the body of the fornix, where the fornix meets the inferior surface of the corpus callosum. Despite its name, it contains no actual fibers from the hippocampus, but rather from the presubiculum, entorhinal cortex, and posterior parahippocampal gyrus [13]. The hippocampal decussation is situated between both commissures and contains fibers from the hippocampal formation directed toward the contralateral septal nuclei.

The dorsal hippocampal commissure is visible in gross dissection (Figs. 3.14, 3.15, and 3.16). It is also designated as the limbic, forniceal, or ammonian commissure and connects the crura of the fornix [1, 8, 14–18]. Classical anatomists also designated it as the lyre of David (*salteryum, psalterium,* or *corpus psalloides*) [4, 13]. It is a tiny layer of fibers with a triangular shape between the crura of the fornix, situated dorsal to the posterior wall of the third ventricle and ventral to the splenium of the corpus callosum. Approximately, 20% of the fibers of the fornix cross the midline [3].

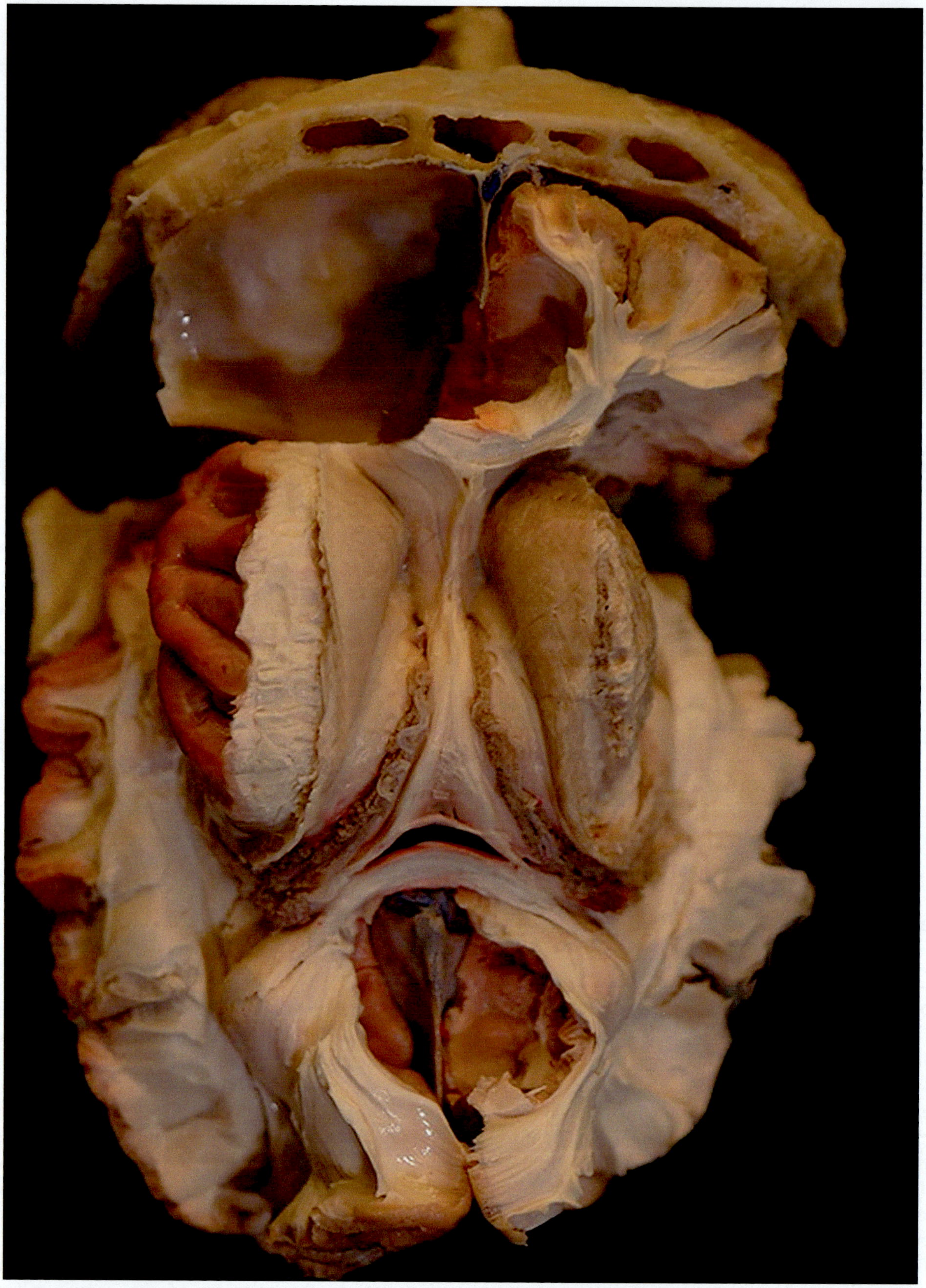

Fig. 3.14 Anatomical preparation and illustration of the dorsal hippocampal commissure. The inferior portion of the splenium with the radiating forceps major remains in place. The tiny component of fibers crossing at the level of the crura of the fornix can be seen.

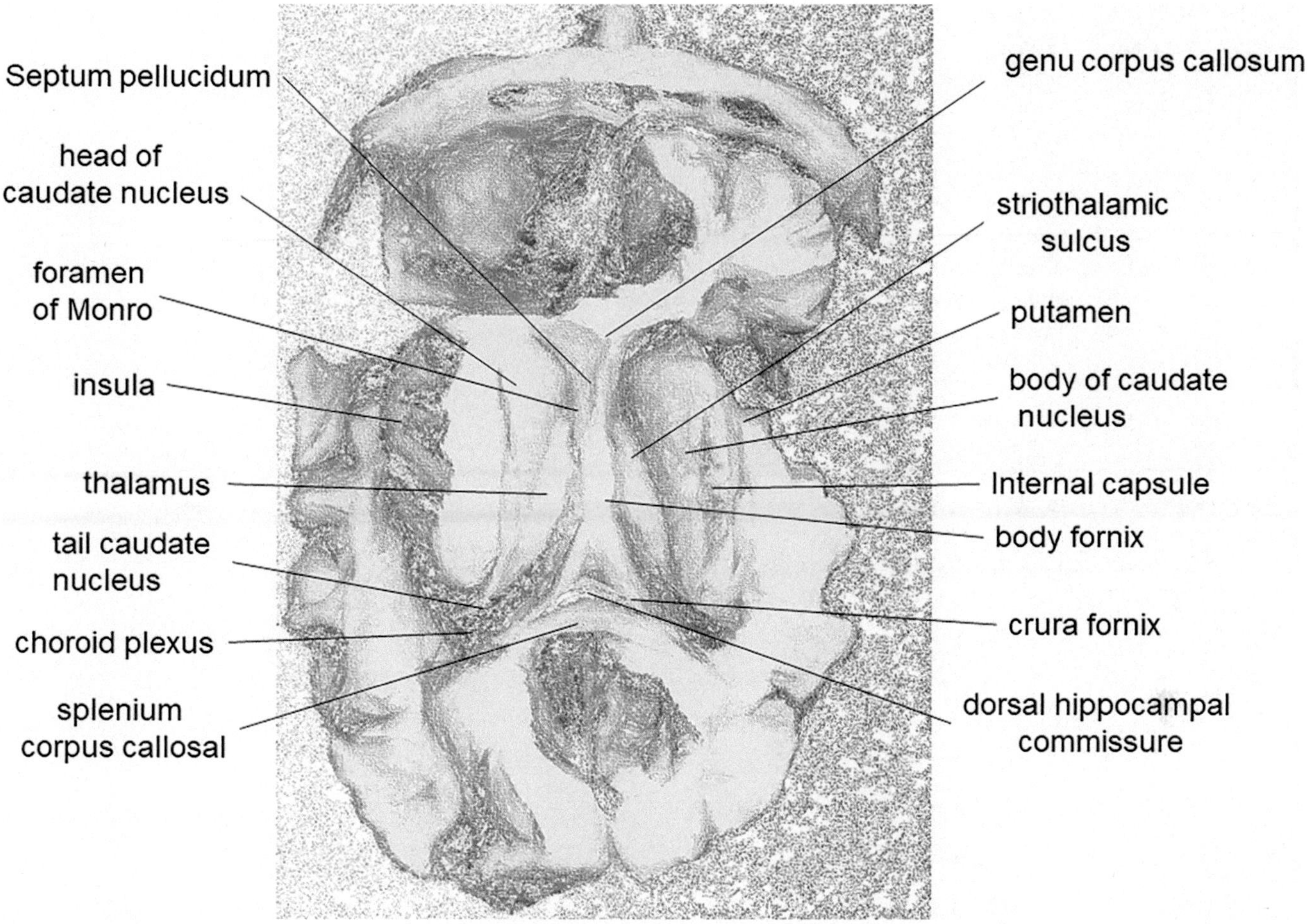

Fig. 3.14 (continued)

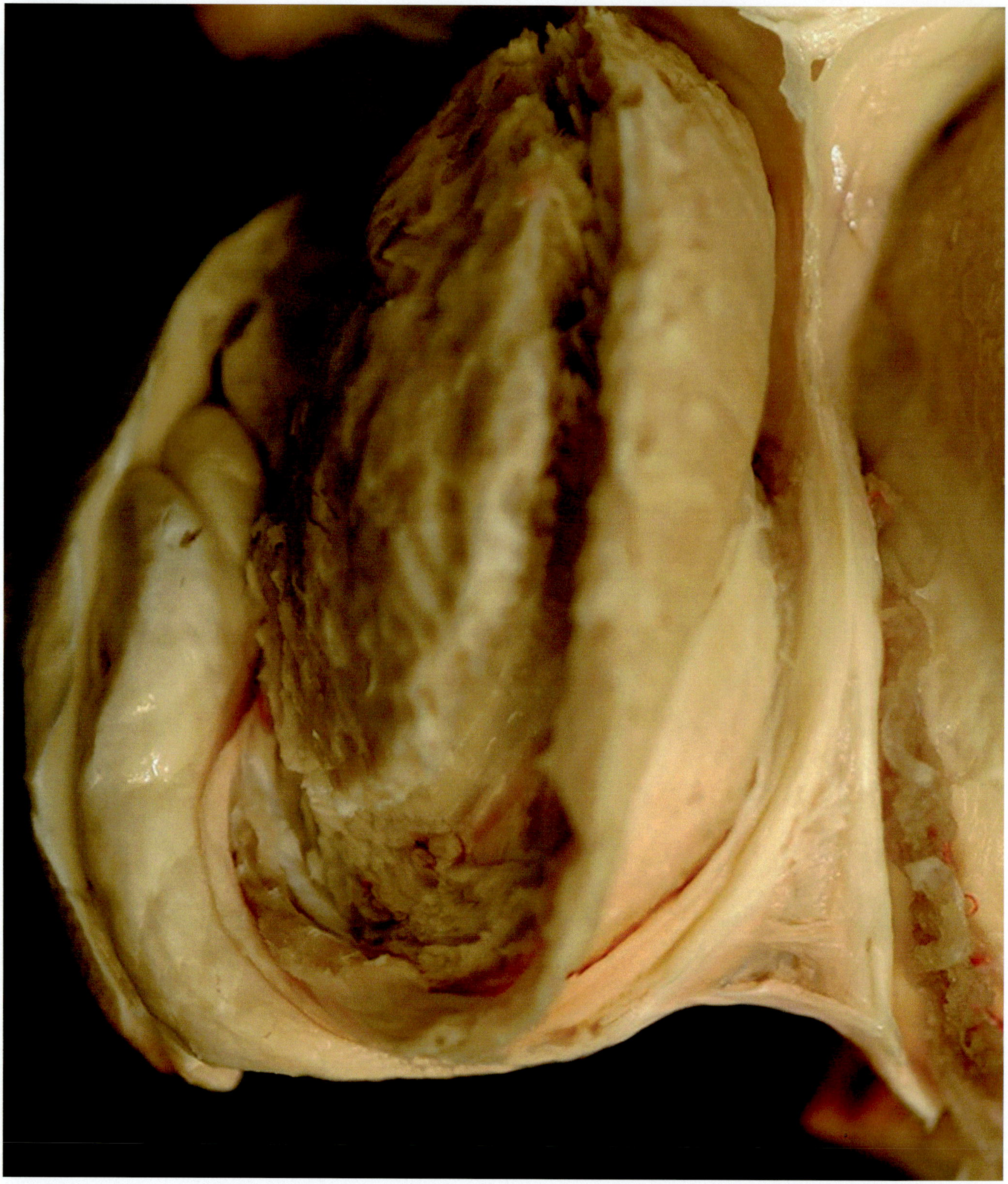

Fig. 3.15 Anatomical preparation and illustration of a close-up view of the dorsal hippocampal commissure. (©Kadri 2023. All Rights Reserved)

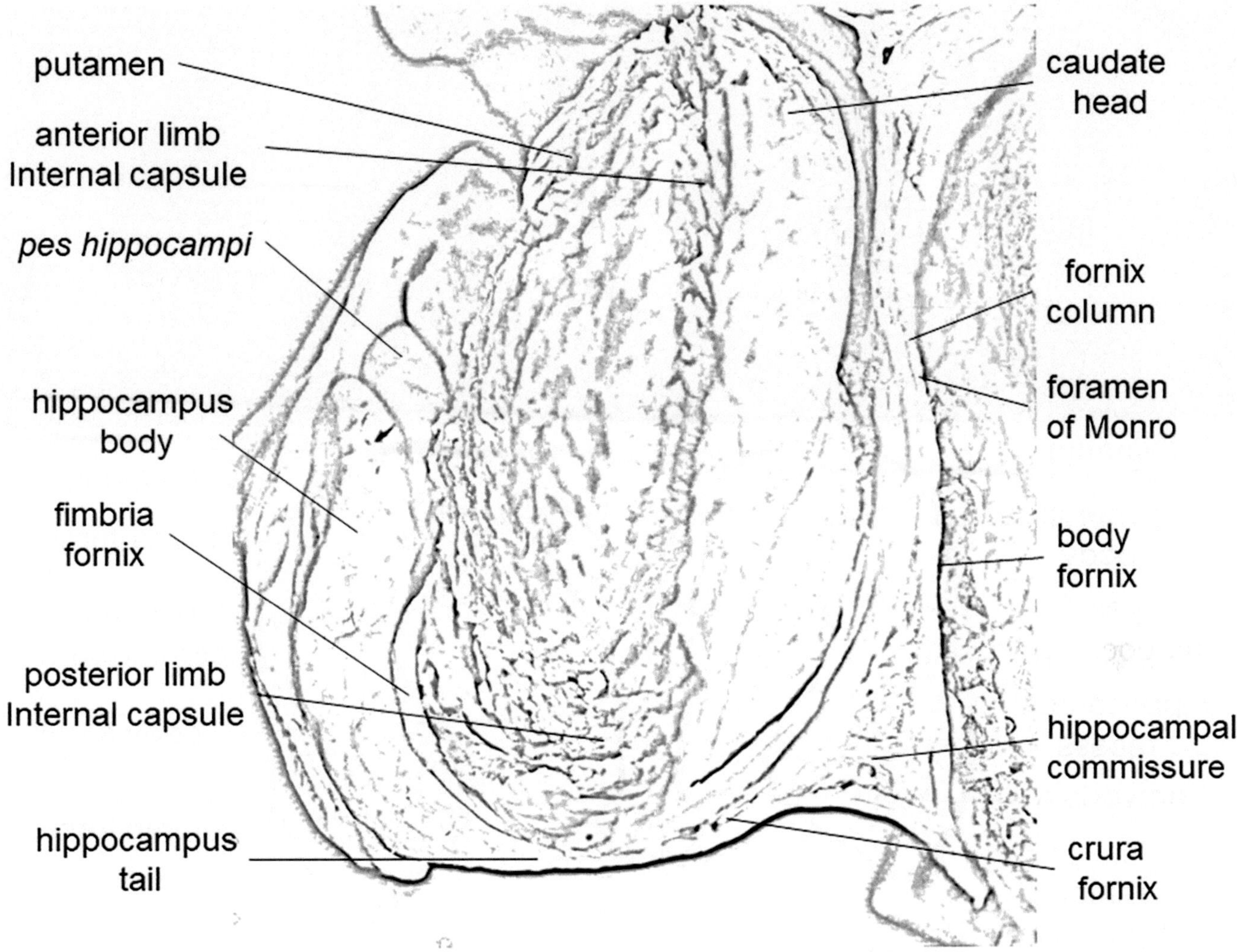

Fig. 3.15 (continued)

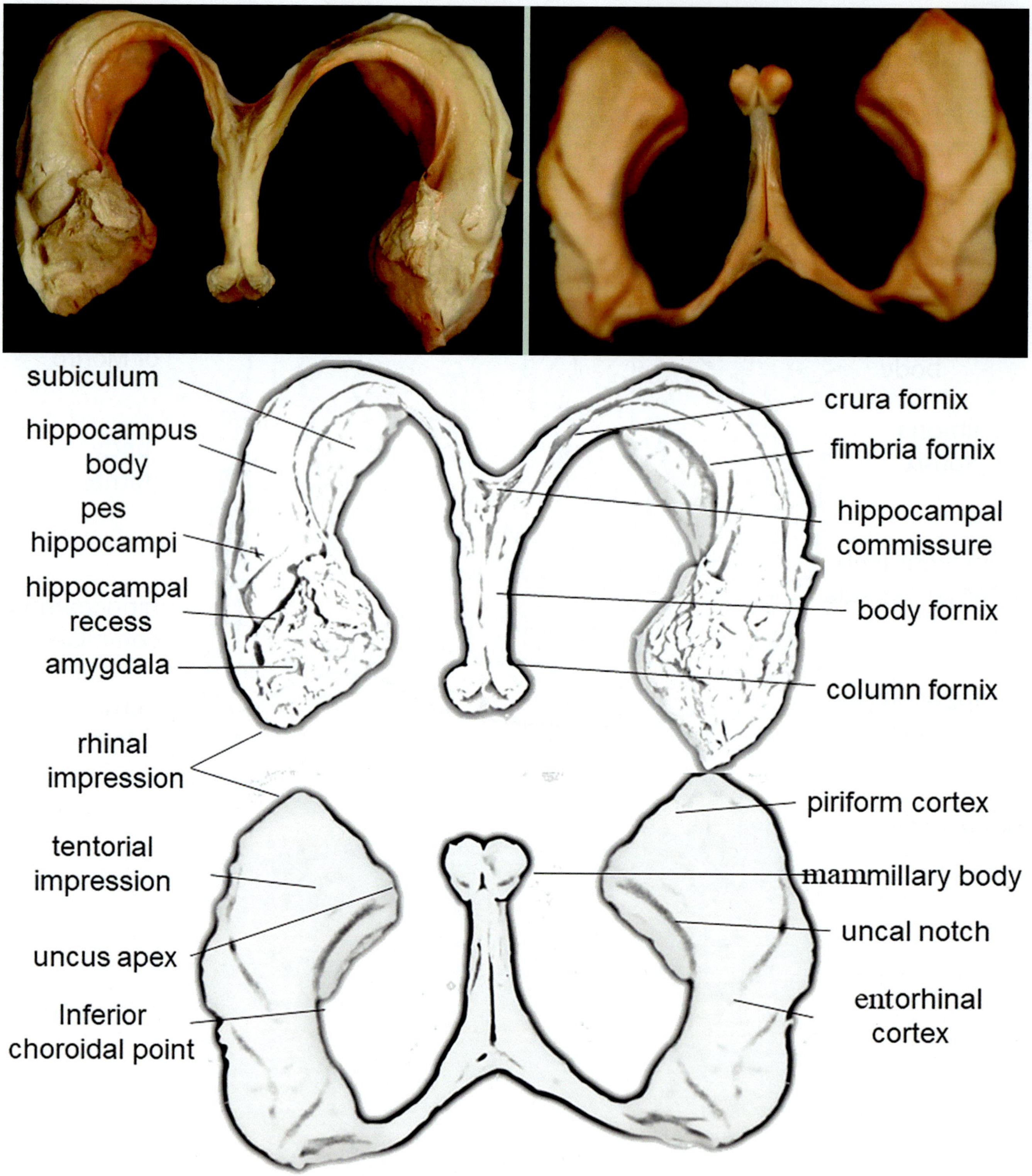

Fig. 3.16 Anatomical preparation and illustration of isolated bilateral parahippocampal gyri. The parahippocampal gyrus contains the two most important structures of the limbic system: the hippocampal formation and the amygdala. The fornix is the main efferent fiber system of the hippocampal formation and is segmented into the fimbria, crura, body, and column. The hippocampal commissures and decussations are within the crura and the body. (©Kadri 2023. All Rights Reserved)

References

1. Nieuwenhuys R, Voogd J, van Huijzen C. The human central nervous system: a synopsis and atlas. 4th ed. Berlin: Springer; 2008.
2. Standring S. Cerebral hemisphere. In: Gray's anatomy: the anatomical basis of clinical practice. 40th ed. Philadelphia, PA: Elsevier; 2008. p. 335–59.
3. Naidich TP, Krayenbühl N, Kollias S, Bou-Haidar P, Bluestone AY, Carpenter DM. White matter. In: Naidich TP, Castillo M, Cha S, Smirniotopoulos JG, editors. Imaging of the brain. Philadelphia, PA: Saunders; 2013. p. 205–44.
4. Testut L, Latarjet A. Tratado de anatomia humana, vol. 2. 9th ed. Barcelona: Salvat Editores; 1966.
5. Yaşargil MG. Microneurosurgery, Vol. IVA. New York: Thieme; 1994.
6. Fernández-Miranda JC, Rhoton AL Jr, Álvarez-Linera J, Kakizawa Y, Choi C, De Oliveira EP. Three-dimensional microsurgical and tractographic anatomy of the white matter of the human brain. Neurosurgery. 2008;62(6 Suppl 3):989–1028. https://doi.org/10.1227/01.neu.0000333767.05328.49.
7. Polyak SL. The vertebrate visual system; its origin, structure, and function and its manifestations in disease with an analysis of its role in the life of animals and in the origin of man, preceded by a historical review of investigations of the eye, and of the visual pathways and centers of the brain. Chicago, IL: University of Chicago Press; 1957.
8. Türe U, Yaşargil MG, Friedman AH, Al-Mefty O, Yaşargil DG. Fiber dissection technique: lateral aspect of the brain. Neurosurgery. 2000;47:417–27. https://doi.org/10.1097/00006123-200008000-00028.
9. Peltier J, Travers N, Destrieux C, Velut S. Optic radiations: a microsurgical anatomical study. J Neurosurg. 2006;105:294–300. https://doi.org/10.3171/jns.2006.105.2.294.
10. Raybaud C. The corpus callosum, the other great forebrain commissures, and the septum pellucidum: anatomy, development, and malformation. Neuroradiology. 2010;52:447–77. https://doi.org/10.1007/s00234-010-0696-3.
11. Catani M, Jones DK, Ffytche DH. Perisylvian language networks of the human brain. Ann Neurol. 2005;5:8–16. https://doi.org/10.1002/ana.20319.
12. Schmahmann JD, Pandya DN. Fiber pathways of the brain. New York: Oxford University Press; 2009.
13. Gloor P. The temporal lobe and limbic system. New York: Oxford University Press; 1997.
14. Carpenter MB. Core text of neuroanatomy. Subsequent ed. Baltimore, MD: Williams &Wilkins; 1991.
15. Heimer L. The human brain and spinal cord: functional neuroanatomy and dissection guide. 2nd ed. New York: Springer; 1994.
16. Klingler J, Gloor P. The connections of the amygdala and of the anterior temporal cortex in the human brain. J Comp Neurol. 1960;115:333–69. https://doi.org/10.1002/cne.901150305.
17. Ludwig E, Klingler J. Atlas cerebri humani. Basel: S. Karger; 1956.
18. Williams PL. Gray's anatomy: the anatomical basis of medicine and surgery. 38th ed. New York: Churchill Livingstone; 1995.

4 Projection Fibers

Projection fibers are fibers that connect telencephalic areas (cortical and basal ganglia) with more caudal areas within the central nervous system: the diencephalon, the brainstem, and the spinal cord [1–8]. Projection fiber bundles are often referred to as tracts, radiations, or peduncles (Figs. 4.1, 4.2, 4.3, 4.4, 4.5, 4.6, and 4.7).

There are different classifications of these fibers. According to Testut and Latarjet [2], they can be divided into two main systems: (1) fibers originating from olfactory areas, the septal area, and the limbic lobe (paleo- and archicortical) and (2) fibers originating from neocortical areas that converge to form the corona radiata. Yaşargil also divided these fibers into two main systems: (1) the cortico-thalamic-thalamic-cortical (optic, acoustic, vestibular, somatosensitive, gustative, and olfactory radiation) and cortico-estrio-palido-thalamo-cortical fibers and (2) the corticopontine, corticobulbar and corticospinal fibers [6].

The projection fibers constitute three divisions: a) the corona radiata, which converges to form the internal capsule, b) the hippocampal fornix, and c) the amygdalofugal fibers. The amygdalofugal fibers are divided into ventral (amygdalo-thalamic, amygdalo-hypothalamic, and amygdalo-septal) and dorsal fibers (stria terminalis) [7]. The fornix and the projection fibers of the amygdala are discussed in the chapter on the limbic system.

P. Kadri, *The Cartographic Atlas of the Brain*, https://doi.org/10.1007/978-3-031-38062-4_4

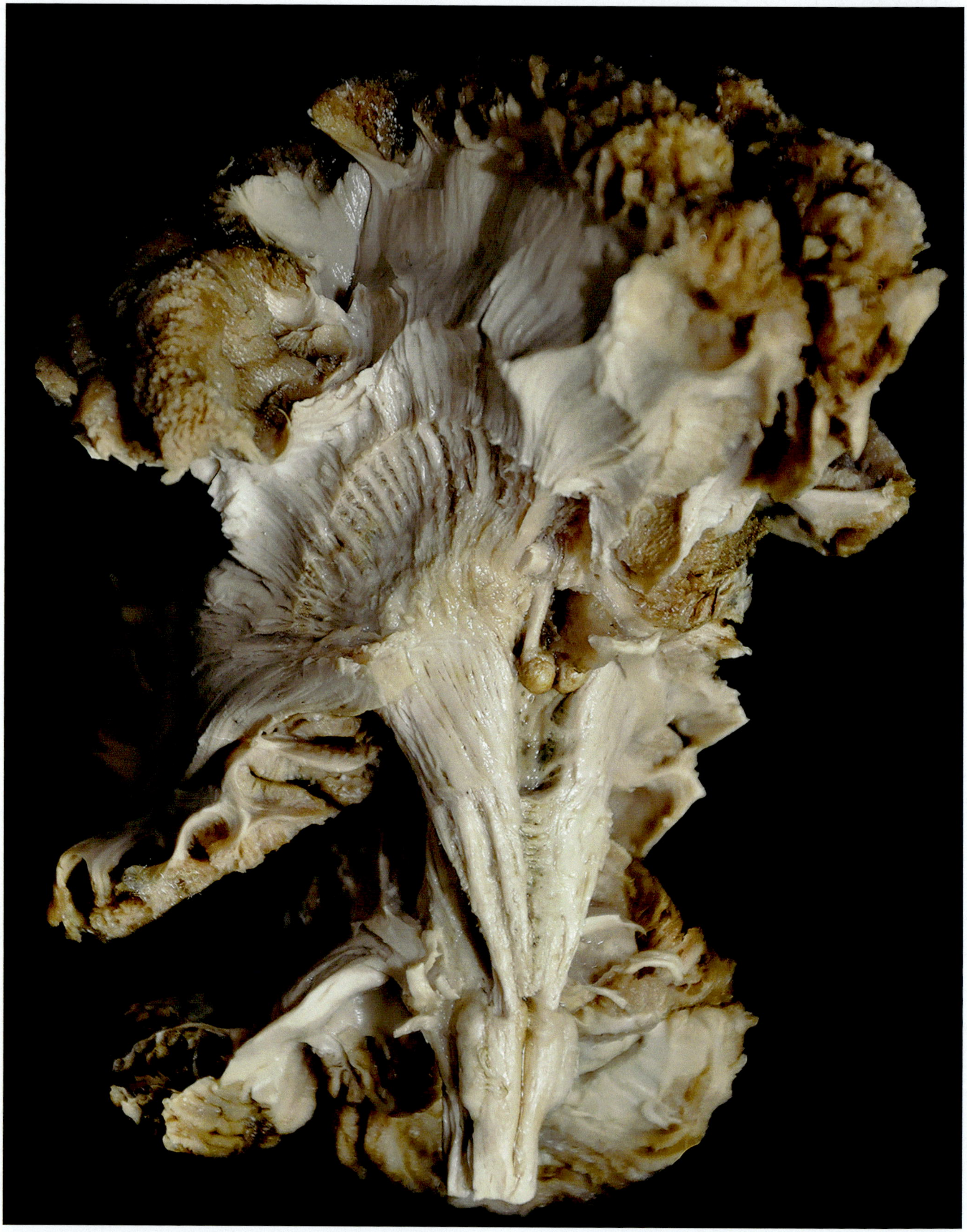

Fig. 4.1 Anatomical preparation of the projection fibers. The putamen and the globus pallidus form the lentiform nucleus. Removing the putamen exposes the globus pallidus. Removing part of the globus pallidus and the optic tract exposes the continuous projection fibers of the corticospinal tract, from the cortical area of the precentral gyrus toward the pyramidal decussation. The pyramidal decussation is the caudal limit of the medulla. *f* foramen, *g* gyrus, *n* nerve, *nn* nucleus, *s* sulcus. (©Kadri 2023. All Rights Reserved)

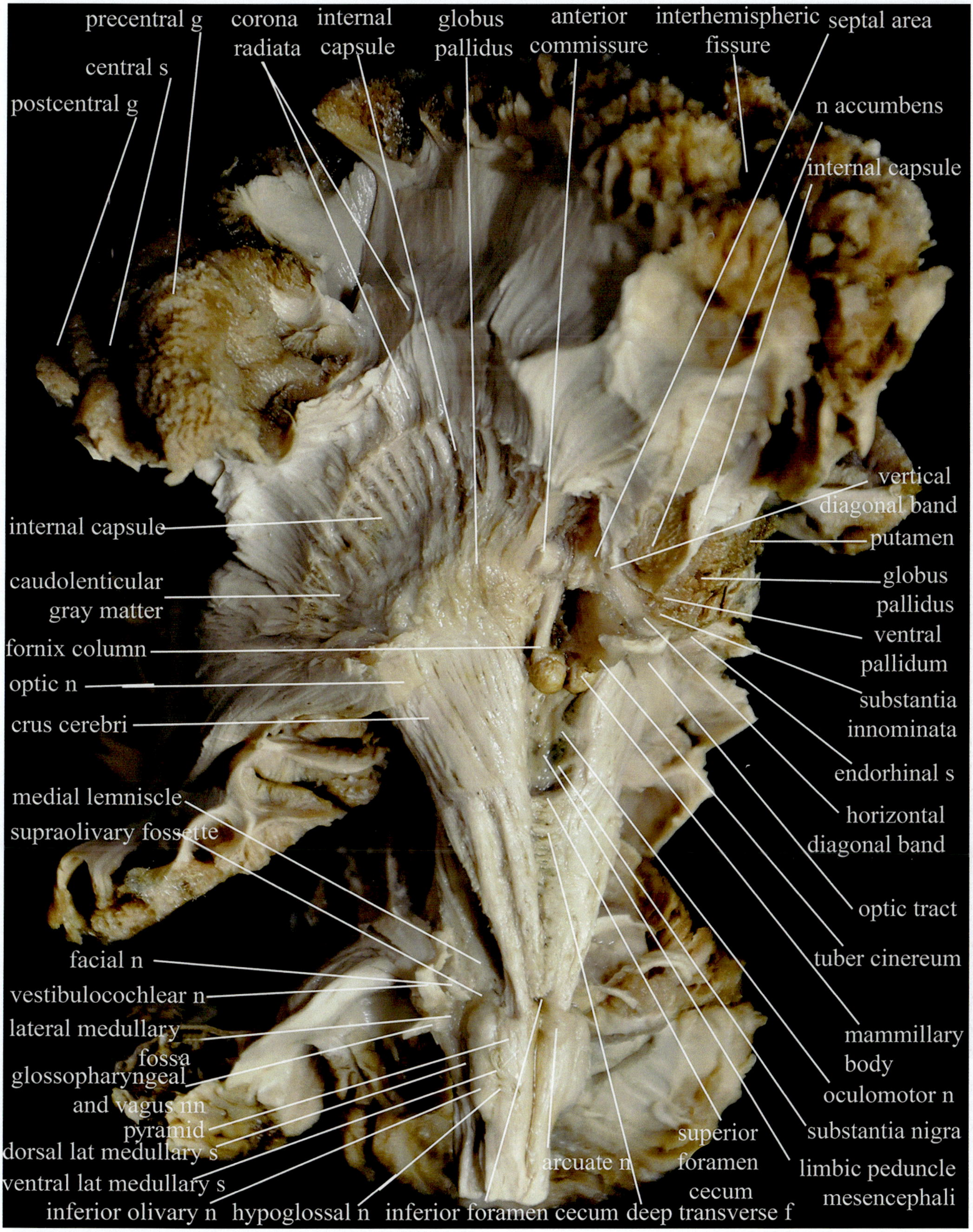

Fig. 4.1 (continued)

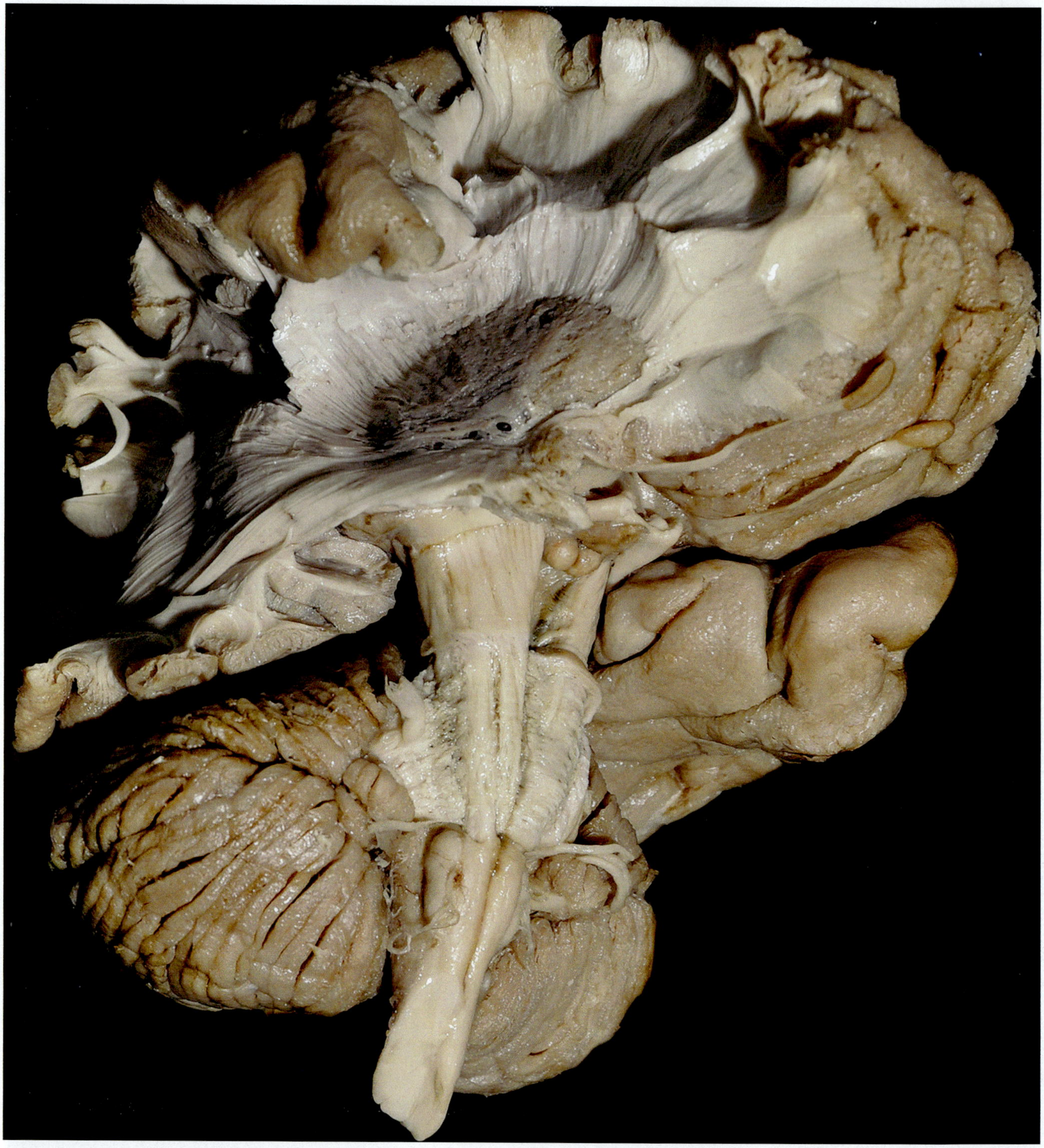

Fig. 4.2 Anatomical preparation of the projection fibers in the right hemisphere. The corona radiata is formed by the projection fibers that converge to compact into the internal capsule, medial to the putamen. The limit between the internal capsule and the crus cerebri, on the base of the mesencephalon, is the optic tract. The ventral limit between the mesencephalon and the pons is the pontomesencephalic sulcus. This sulcus is formed by the rostral superficial transverse fibers of the pons. The superficial transverse fibers are superficial to the descending projection fibers of the crus mesencephalic. These descending fibers are composed of the corticospinal, corticopontine, and corticonuclear tracts. The corticospinal tract at the pons is rearranged in several bundles separated by the intermediated transverse fibers of the pons. The bundles of the corticospinal tract are reunited at the ventral pontomedullary transition to reappear at the surface as the paired medullar pyramids. *f* fasciculus, *g* gyrus, *n* nerve, *s* sulcus, *t* tract.

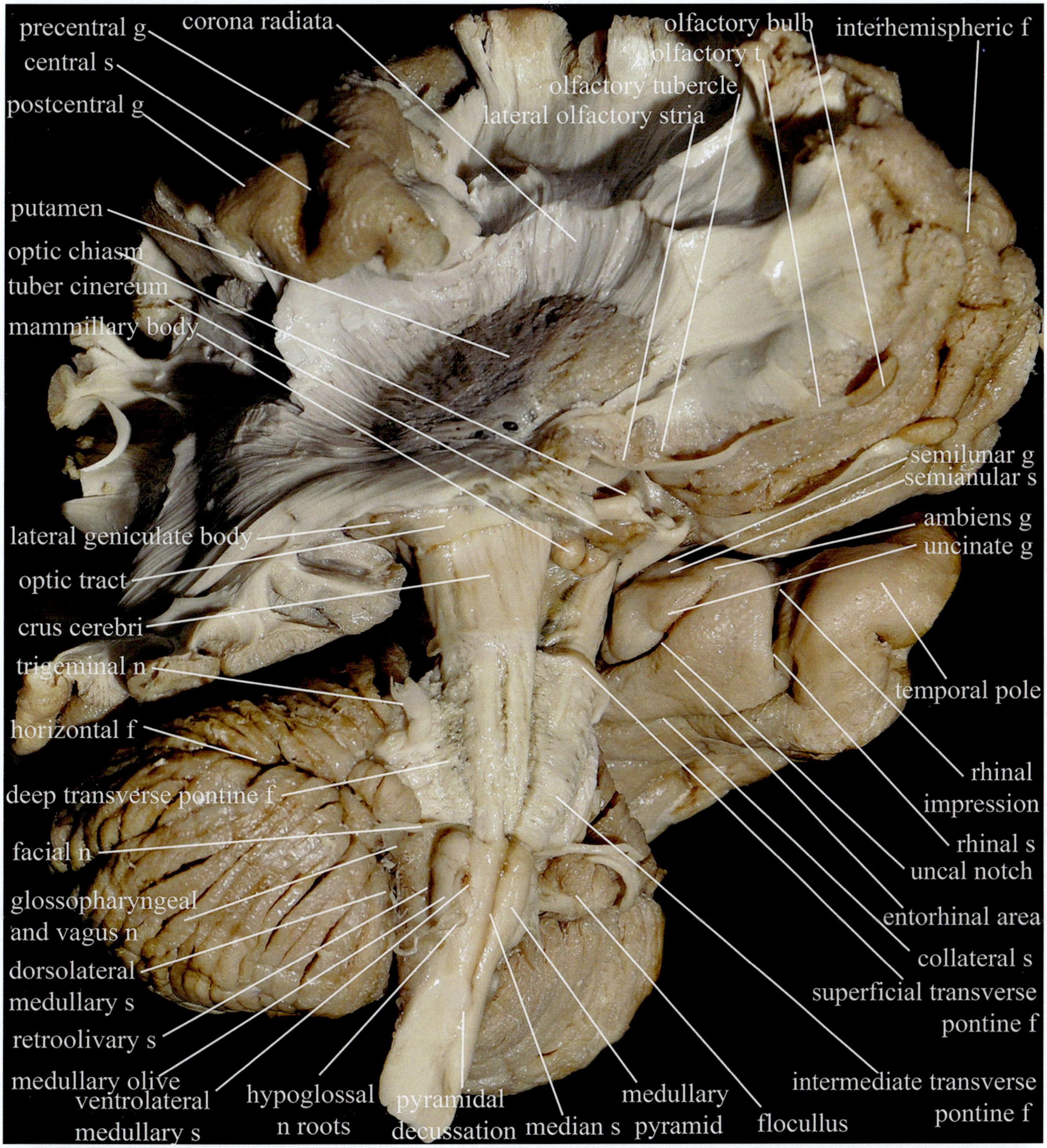

Fig. 4.2 (continued)

Fig. 4.3 Anatomical preparation and illustration of the projection fibers depicting the continuity of the corona radiata, internal capsule, and peduncle via the pons and pyramid. (©Kadri 2023. All Rights Reserved)

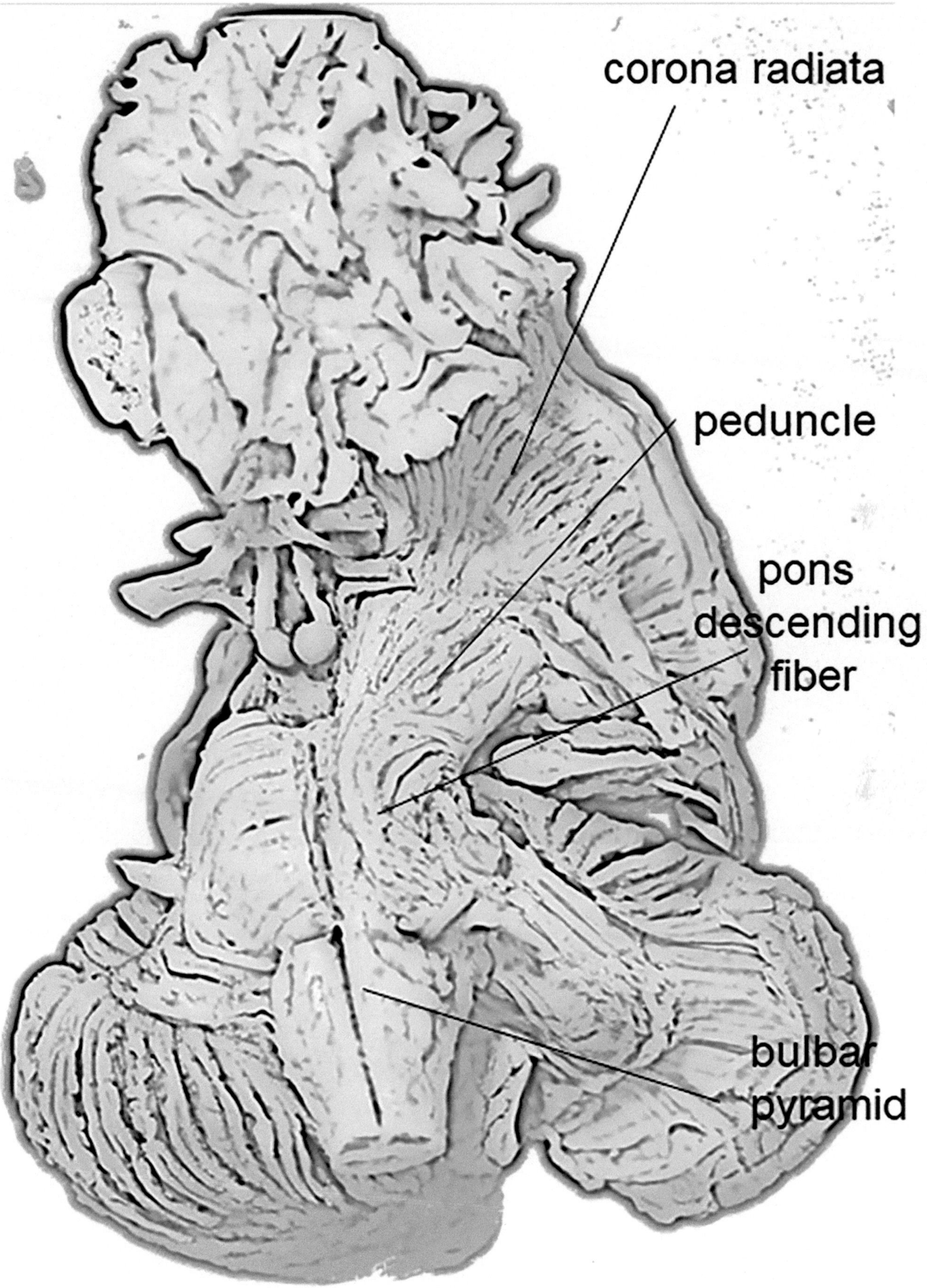

Fig. 4.3 (continued)

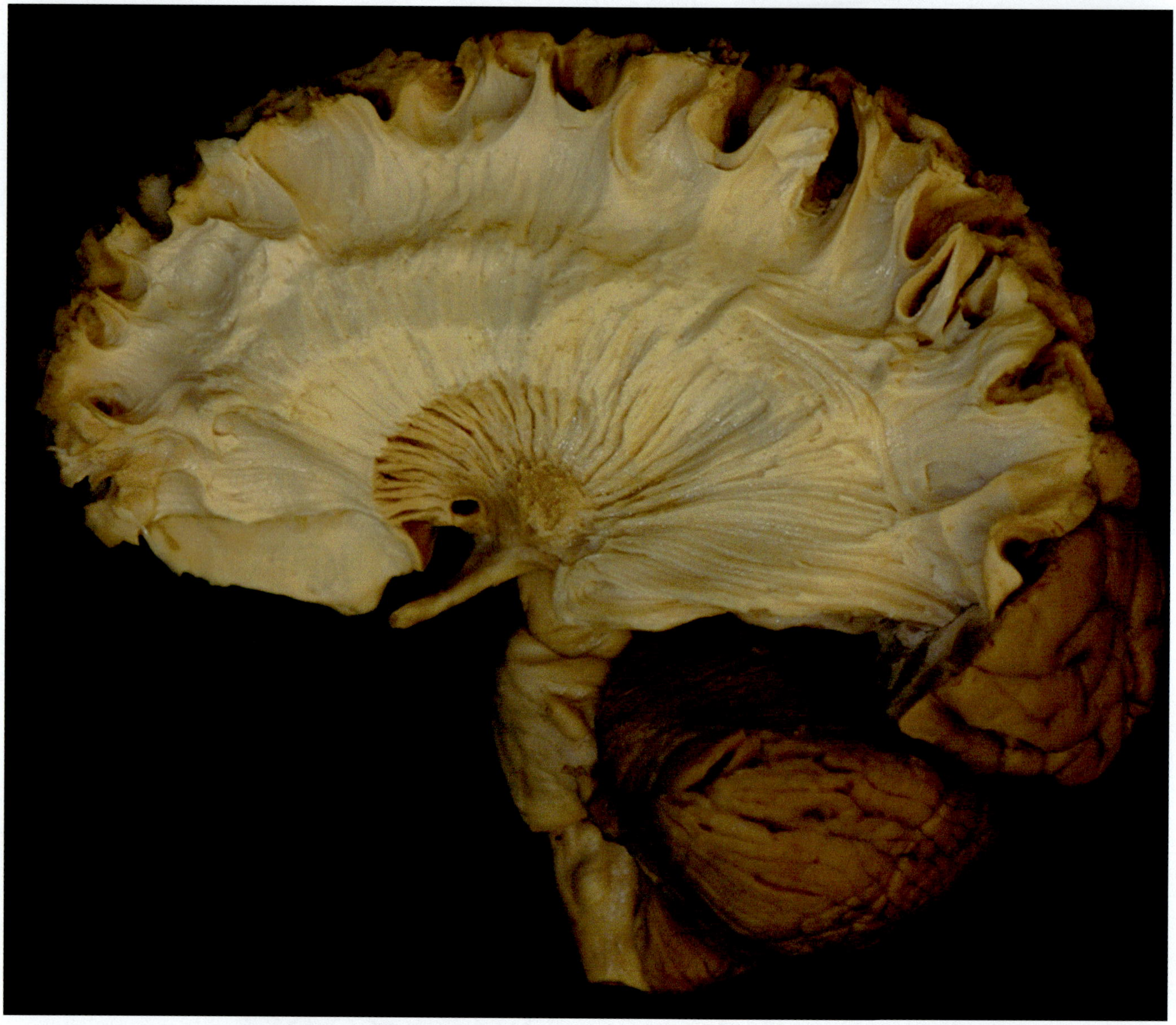

Fig. 4.4 Anatomical preparation and illustration of the left encephalon. The anterior commissure and the limbic structures of the temporal lobe have been removed, and the corona radiata and internal capsule are visualized. The tunnel of the anterior commissure is seen traversing the internal capsule to reach the anterior portion of the third ventricle. The inferior thalamic peduncle and part of the anterior loop of the posterior thalamic peduncle were removed, and the globus pallidus was kept as a reference point. The optic tract is the limit between the internal capsule and the mesencephalic peduncle. (©Kadri 2023. All Rights Reserved)

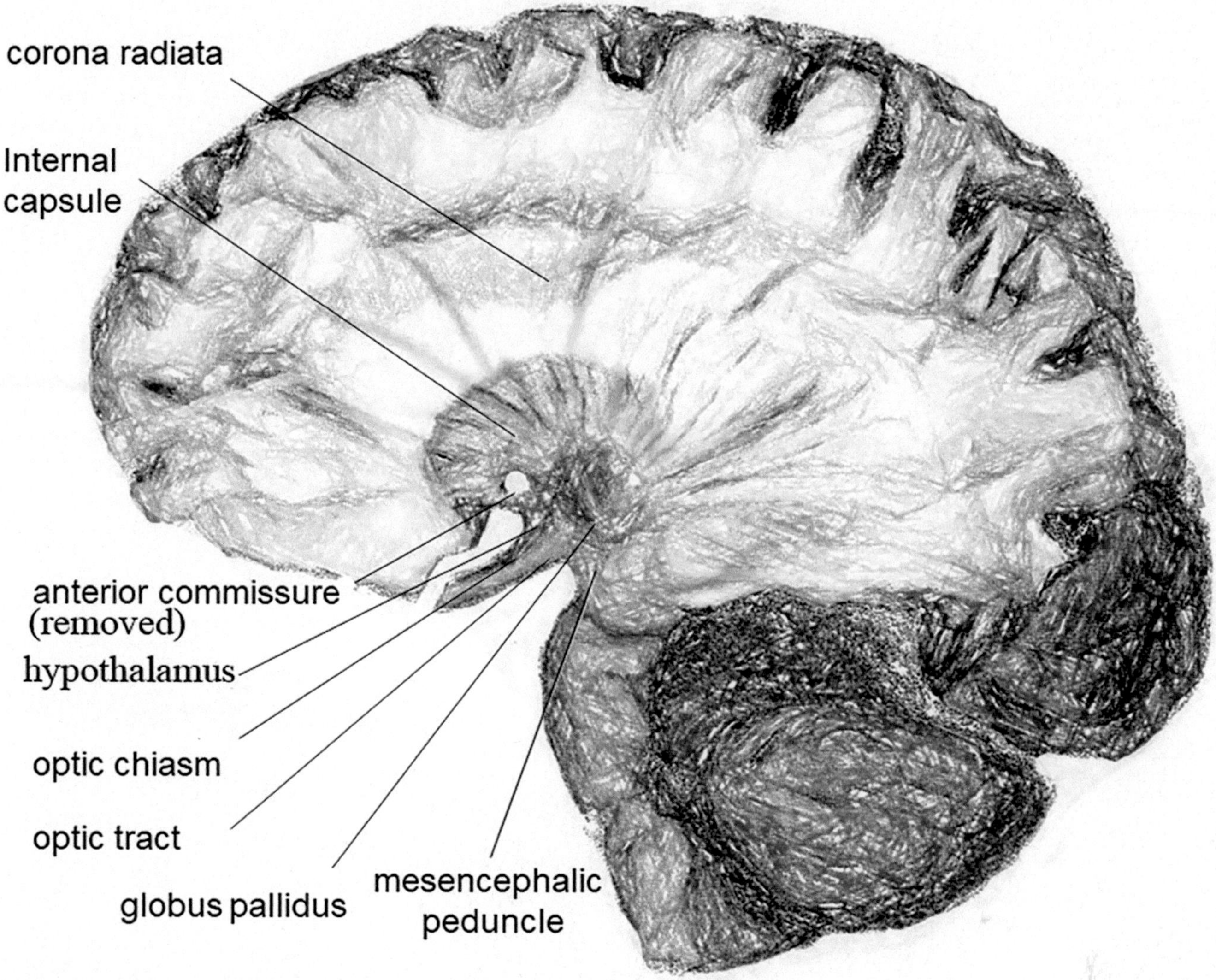

Fig. 4.4 (continued)

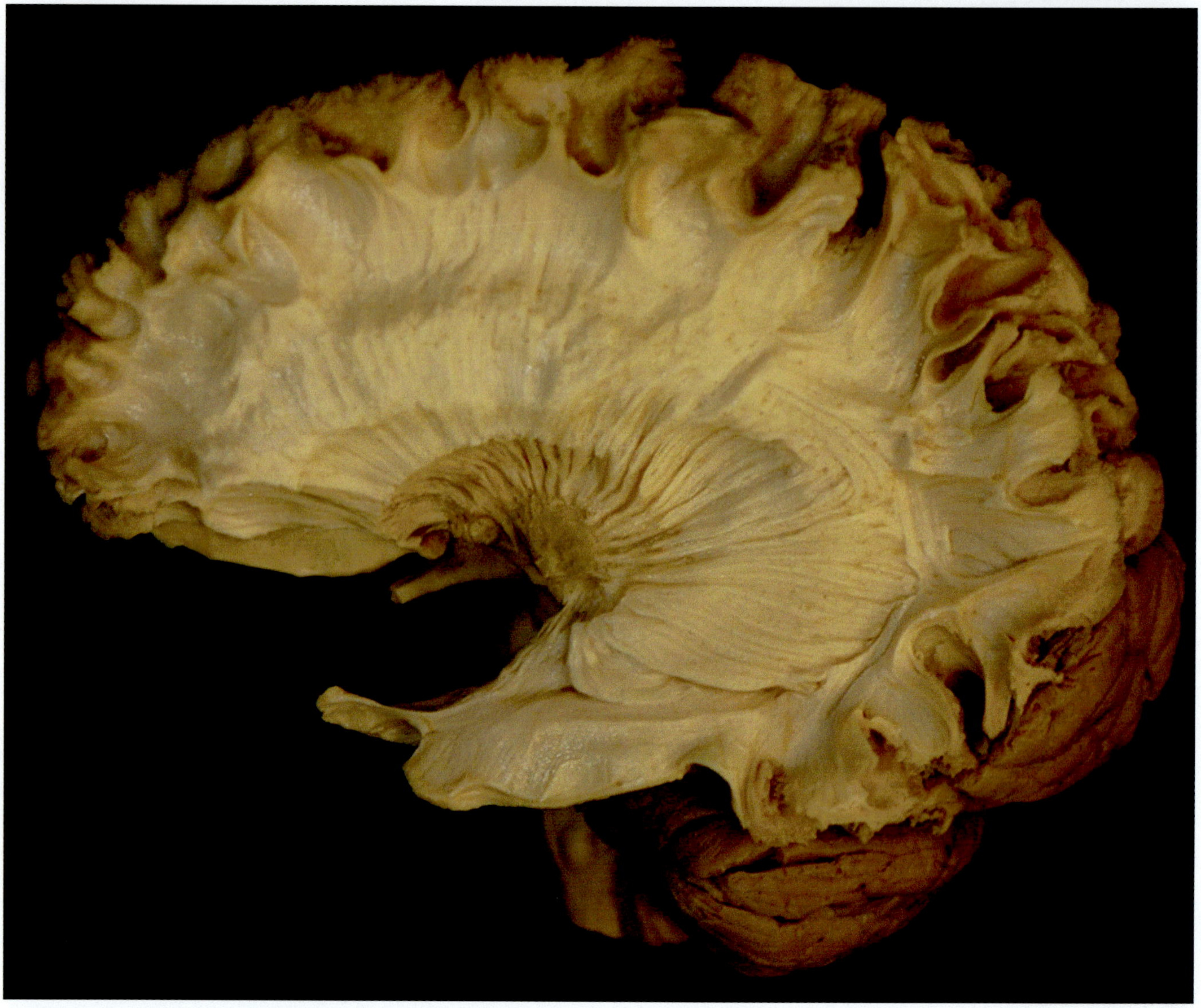

Fig. 4.5 Anatomical preparation and illustration of the left hemisphere, lateral view. The association fibers, putamen, and anterior commissure have been removed, and the corona radiata is visualized. The internal capsule corresponds to the compact corona radiata medial to the lentiform nucleus and is divided into five segments: anterior limb, genu, posterior limb, retrolentiform, and sublentiform.

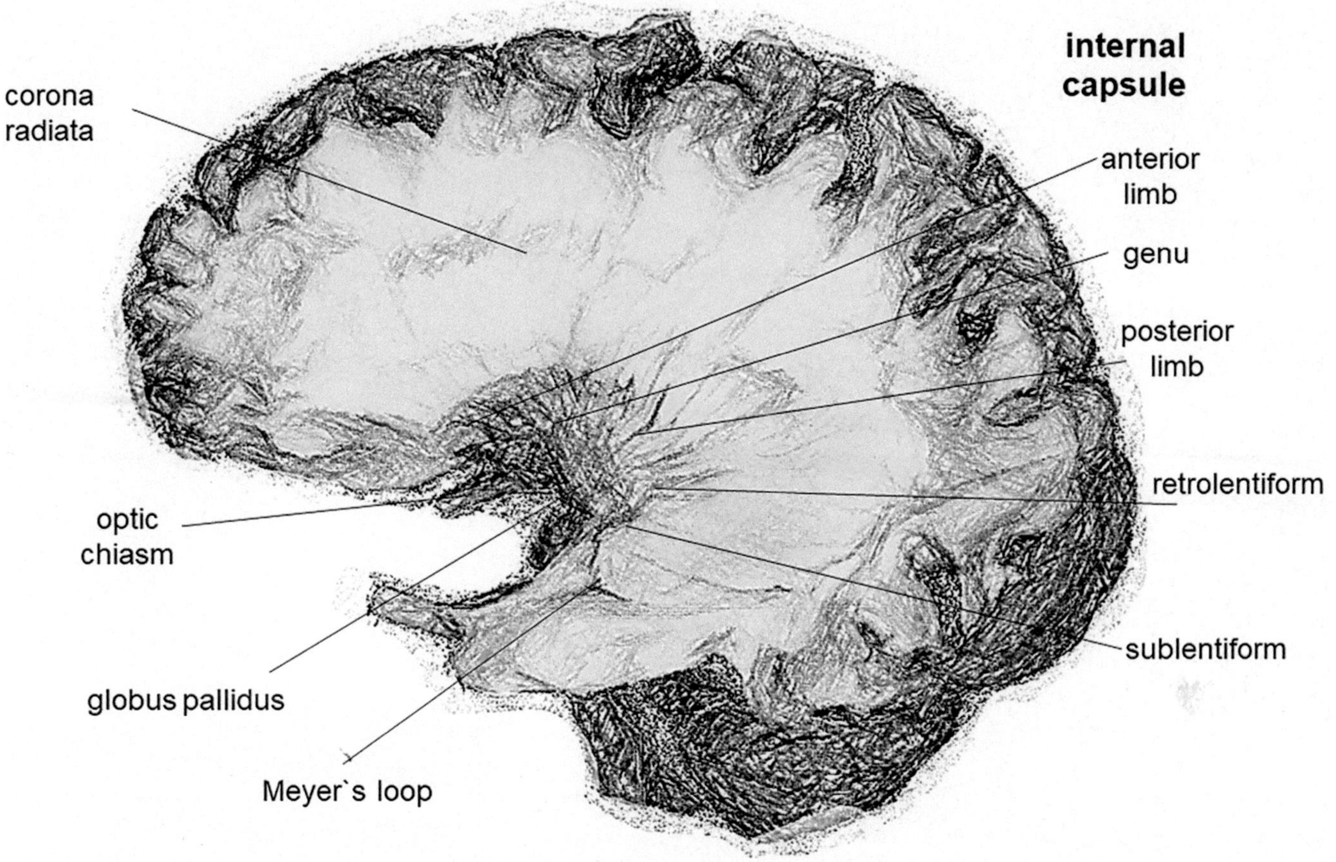

Fig. 4.5 (continued)

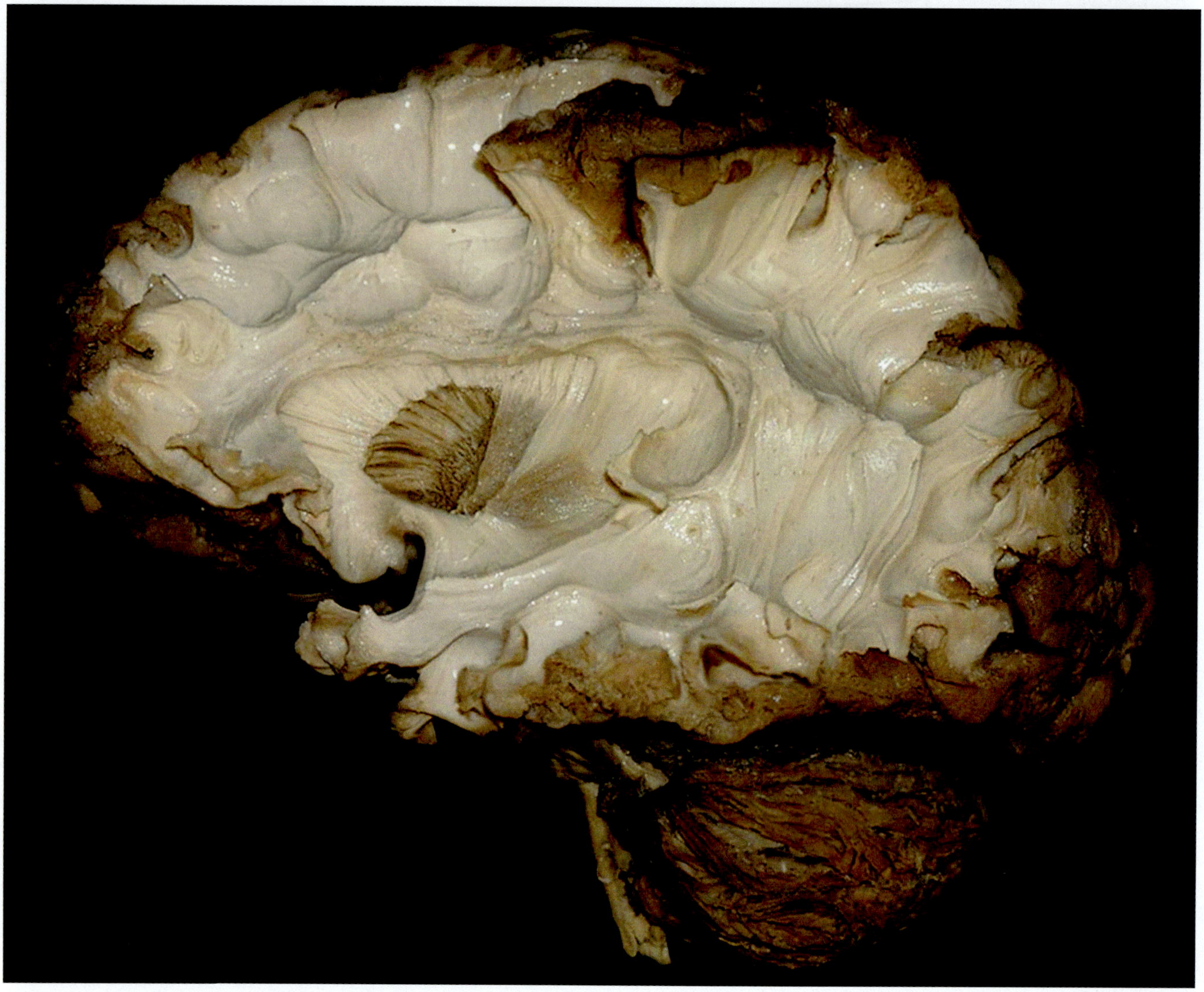

Fig. 4.6 Anatomical preparation and illustration of the main white matter fiber system on the lateral surface of the left hemisphere. (©Kadri 2023. All Rights Reserved)

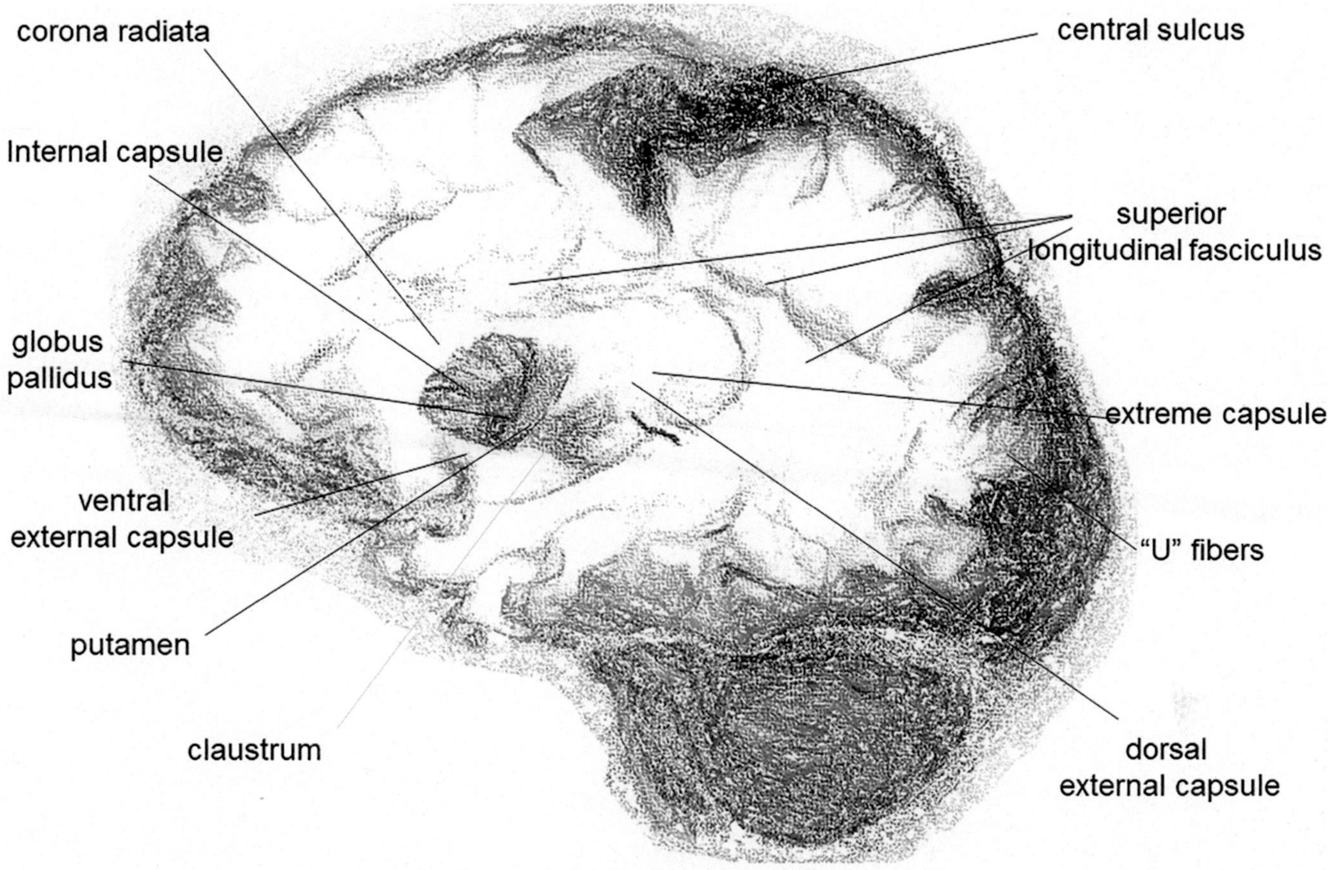

Fig. 4.6 (continued)

Fig. 4.7 Anatomical preparation and illustration, lateral view, of the left hemisphere depicting the transition between the internal capsule and the crus mesencephalic. The optic tract has been removed. The lateral mesencephalic sulcus anteriorly limits the descending fibers from the ascending fibers posterior to it. (©Kadri 2023. All Rights Reserved)

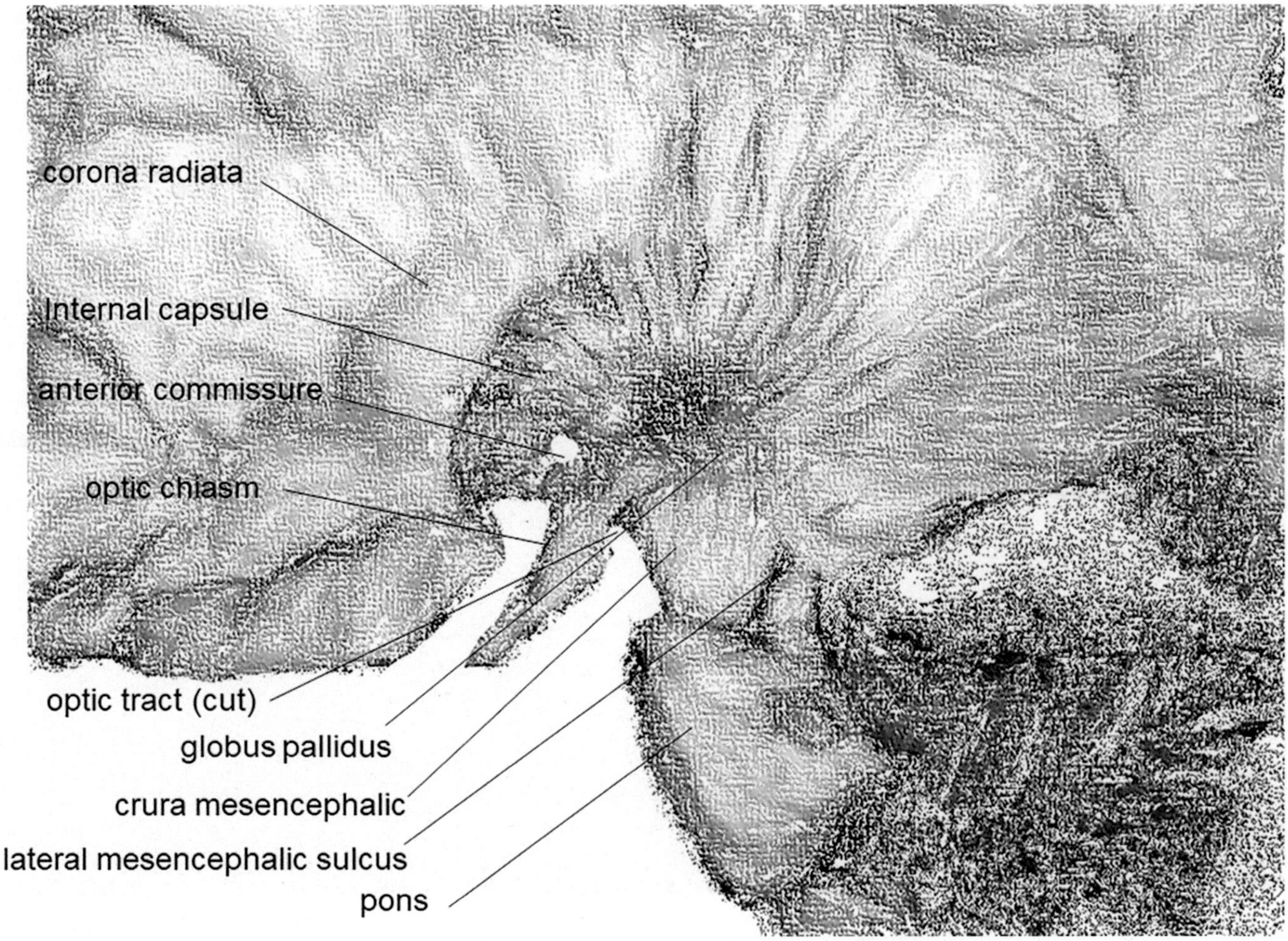

Fig. 4.7 (continued)

The Neocortical Projection Fiber System: Corona Radiata and Internal Capsule

Although first depicted by Vesalius in 1543, the continuity of the fibers of the corona radiata and internal capsule were probably first described and illustrated by the French anatomist Raymond Vieussens (1635–1715) in his *Neurographia Universalis* published in 1684. Vieussens also demonstrated the continuity of the fibers of the internal capsule through the cerebral peduncle and pyramidal tract in the pons and medulla [7]. In 1809, Johann Reil coined the name *corona radiata*.

The corona radiata does not correspond to a specific tract; rather, it is the descriptor of the large fanlike arrangement of the projection fibers of the cerebral hemisphere. It connects the cortical regions to the diencephalon, the brainstem, and the spinal cord and is composed of multiple different tracts [1, 3–5, 7–10]. Fibers of the corona radiata are intersected by commissural fibers of the corpus callosum and anterior commissure transversely [11] and by the long association fiber fasciculus sagittally (Figs. 4.5, 4.6, and 4.8).

Four segments of the corona radiata can be identified: (1) anterior—a tributary of the frontal lobe and superior segment of the limbic lobe, (2) superior—a tributary from the Rolandic circumvolution (precentral and postcentral gyri) and paracentral lobule, (3) posterior—a tributary of the parietal, occipital, and posterior temporal lobes, and (4) inferior—a tributary of the middle and anterior temporal lobes and the inferior segment of the limbic lobe (Figs. 4.9, 4.10, and 4.11) [2].

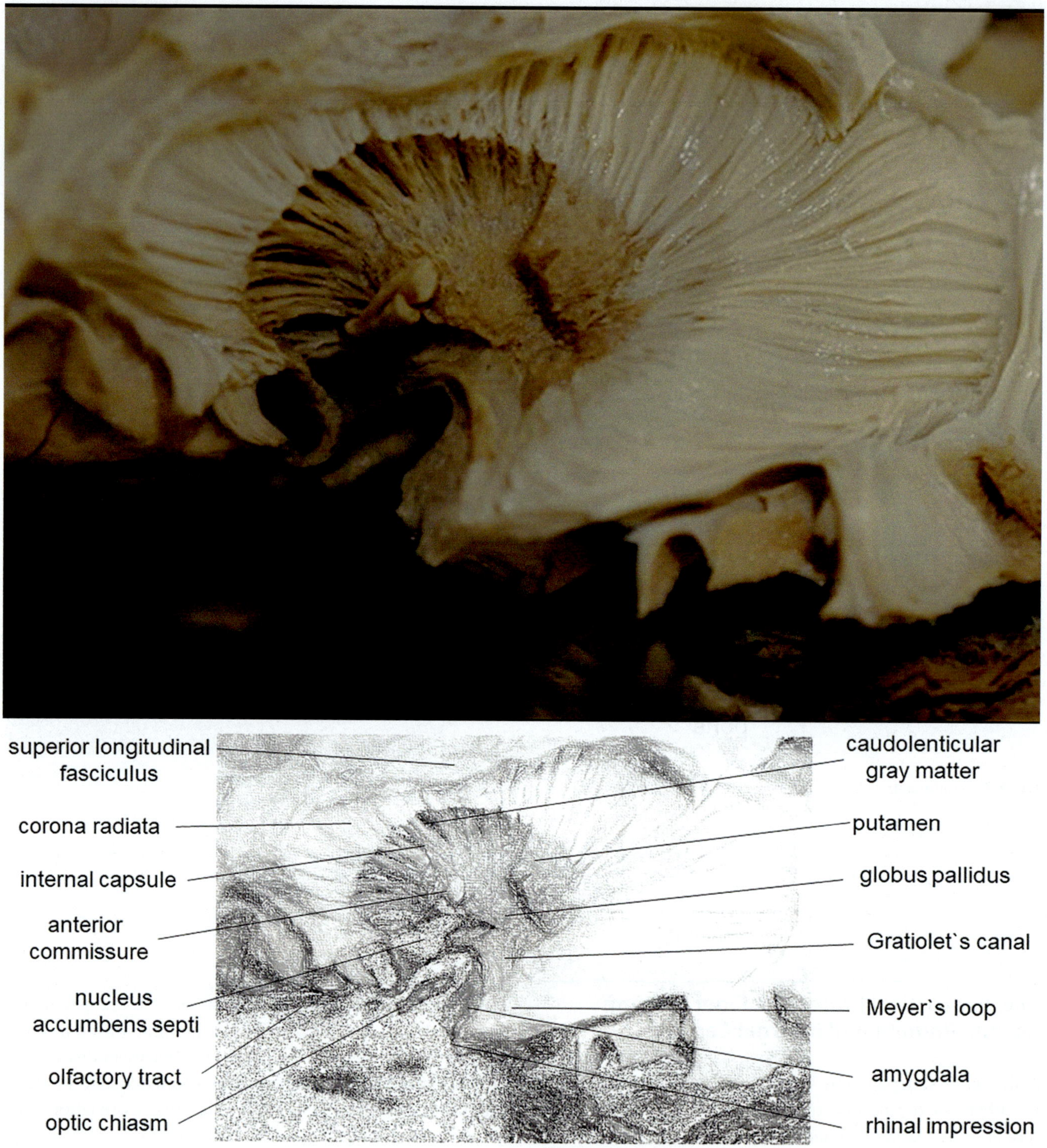

Fig. 4.8 Anatomical preparation and illustration, lateral view, of the left hemisphere depicting the internal capsule. The putamen has been partially removed to expose the globus pallidus. The anterior commissure was sectioned in its compact portion, and its lateral extension was removed. Gratiolet's canal and the superior surface of the amygdala are exposed. The projection fibers are visualized on the temporal lobe, with its peculiar Meyer's loop formed by the anterior loop of the posterior thalamic radiation. The optic radiation is part of the fiber system of the posterior thalamic peduncle. The amygdala is medial to the rhinal impression. (©Kadri 2023. All Rights Reserved)

Fig. 4.9 Anatomical preparation and illustration of the internal capsule, left side, lateral view. The external capsule has been removed and the putamen unveiled. The internal capsule corresponds to the fibers of the corona radiata that compact and pass through the nucleus striatum. The putamen is the lateral striatal nucleus, and the caudate nucleus is the medial one. The term *striatum* is derived from the striae (white lines) of the internal capsule in cuts through these nuclei. Two portions, the retrolentiform and sublentiform, are located outside the borders of the lentiform nucleus. (©Kadri 2023. All Rights Reserved)

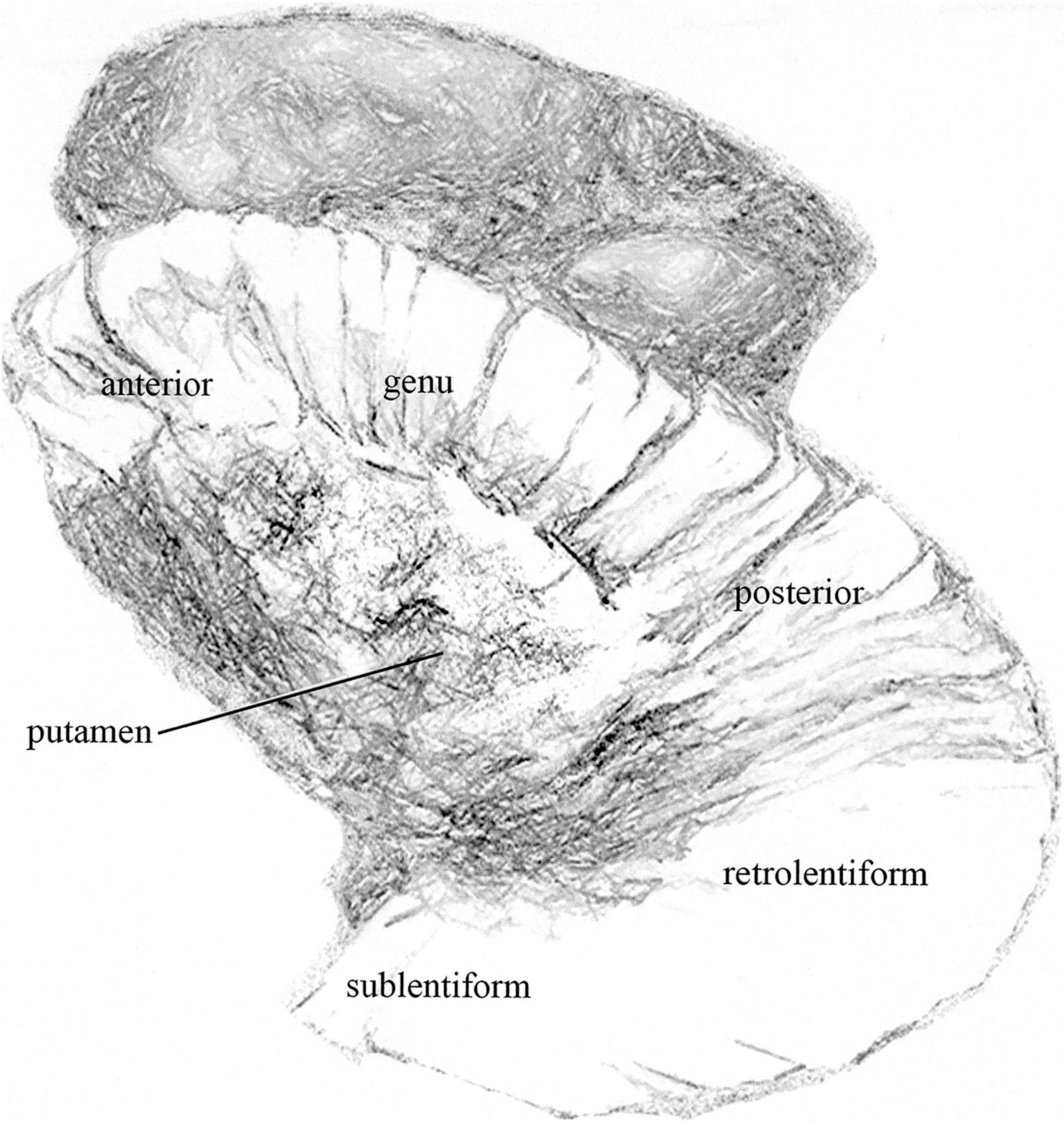

Fig. 4.9 (continued)

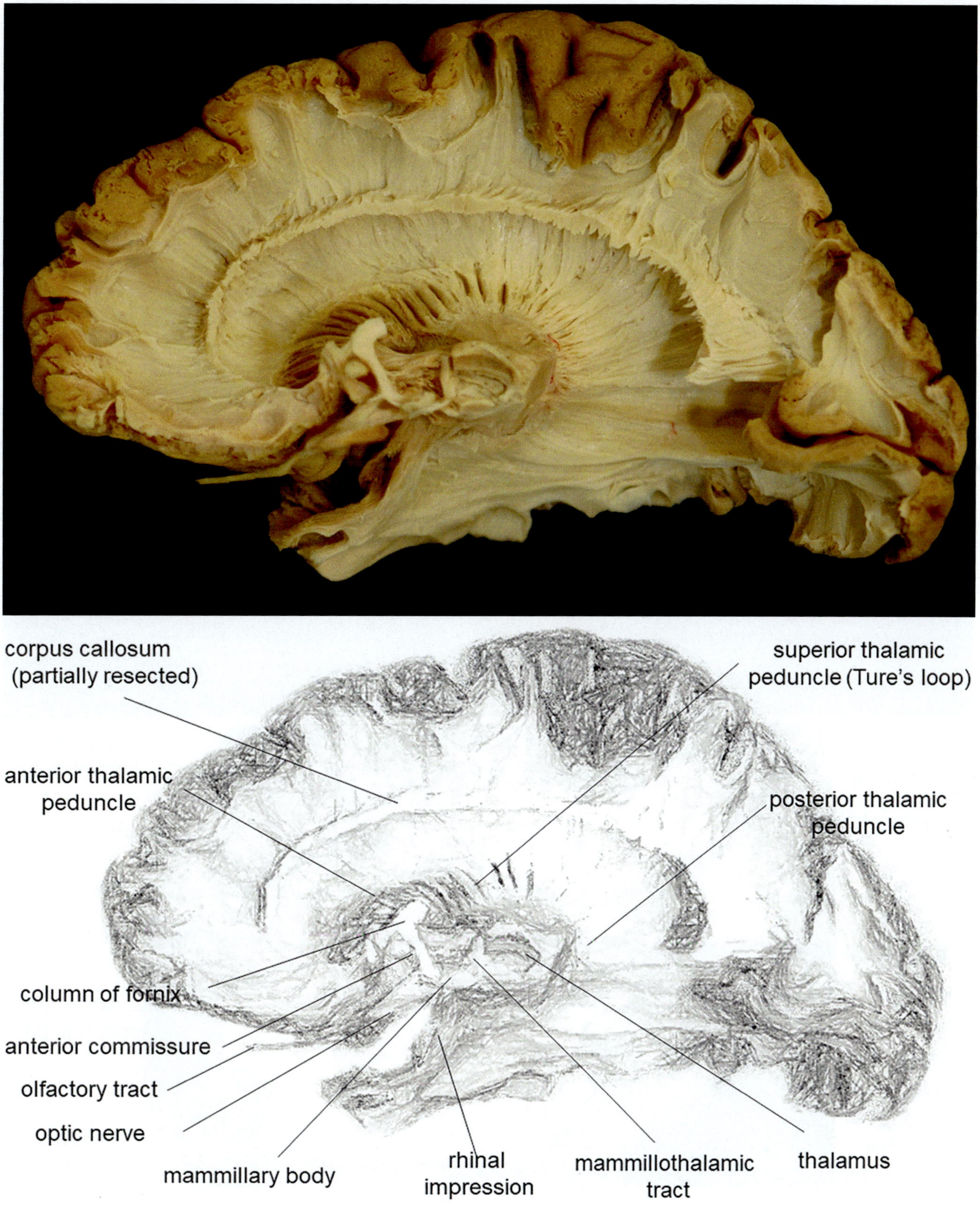

Fig. 4.13 Anatomical preparation and illustration, right hemisphere, viewed from the medial side. Removing the ependyma and caudate nucleus exposes the medial surface of the internal capsule. This surface contains the thalamic peduncles: anterior, superior, posterior, and inferior. (©Kadri 2023. All Rights Reserved)

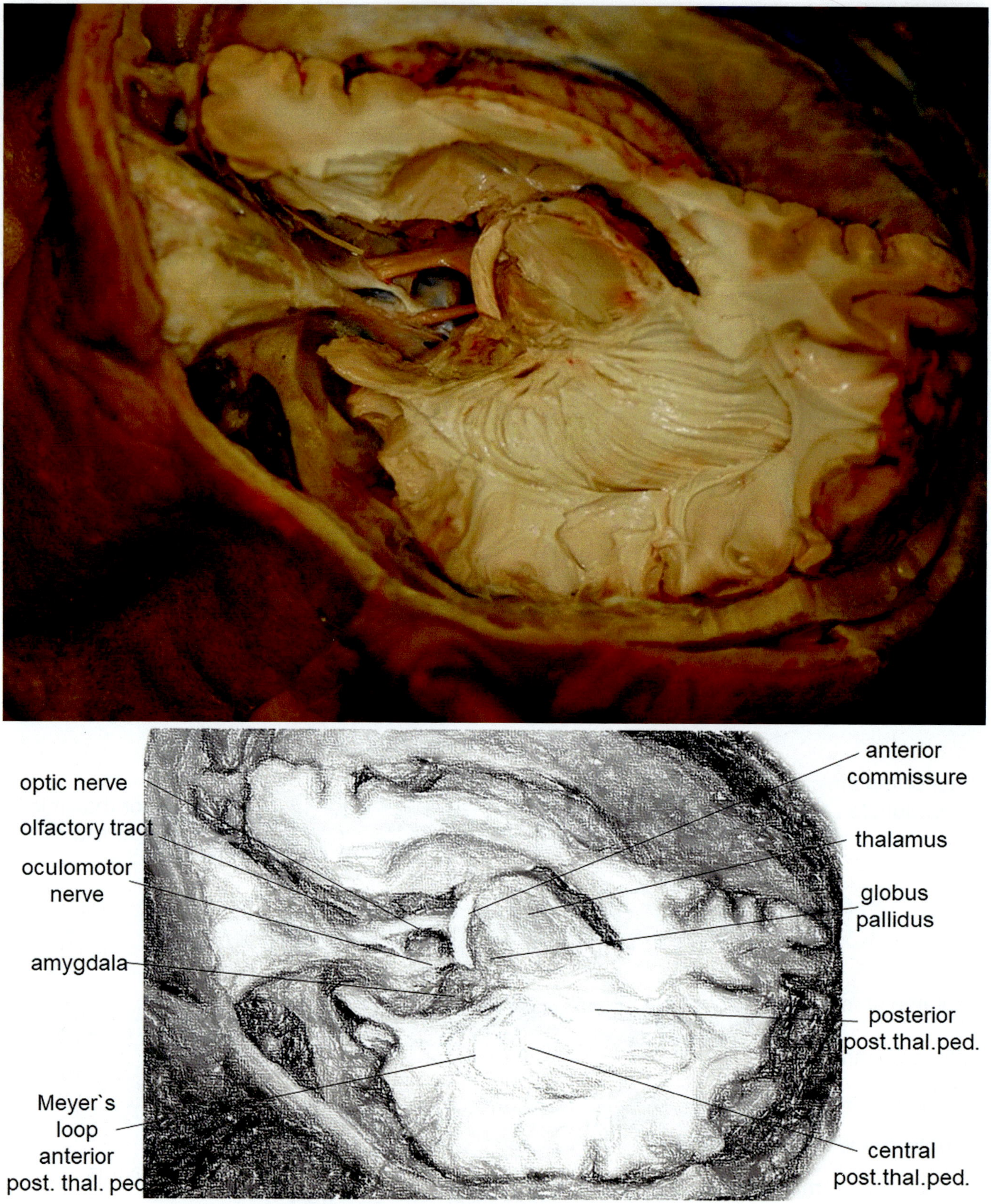

Fig. 4.14 Anatomical preparation and illustration depicting the optic radiation of the left side of the hemisphere. The association fibers and anterior commissure have been removed to expose the projection fibers within the temporal lobe. The characteristic anterior loop of the posterior thalamic peduncle can be seen. *post.thal.ped* posterior thalamic peduncle. (©Kadri 2023. All Rights Reserved)

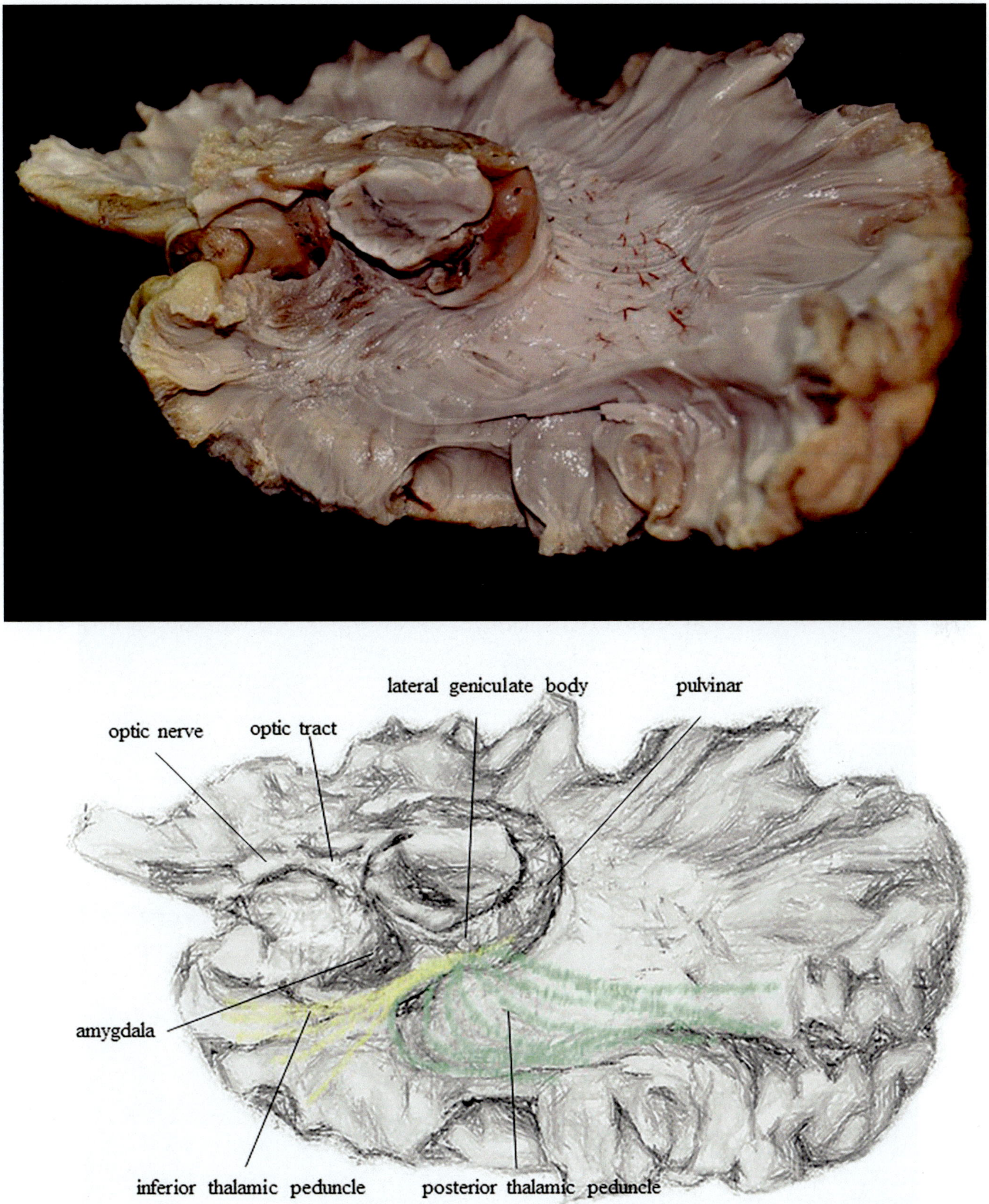

Fig. 4.15 Anatomical preparation and illustration depicting the posterior and inferior thalamic peduncles of the left side of the hemisphere, inferior view. The ependyma and the tapetum have been removed. The inferior thalamic peduncle connects with the temporal pole and anterior temporal regions. The posterior thalamic peduncle is depicted with its peculiar loop in the anterior portion.

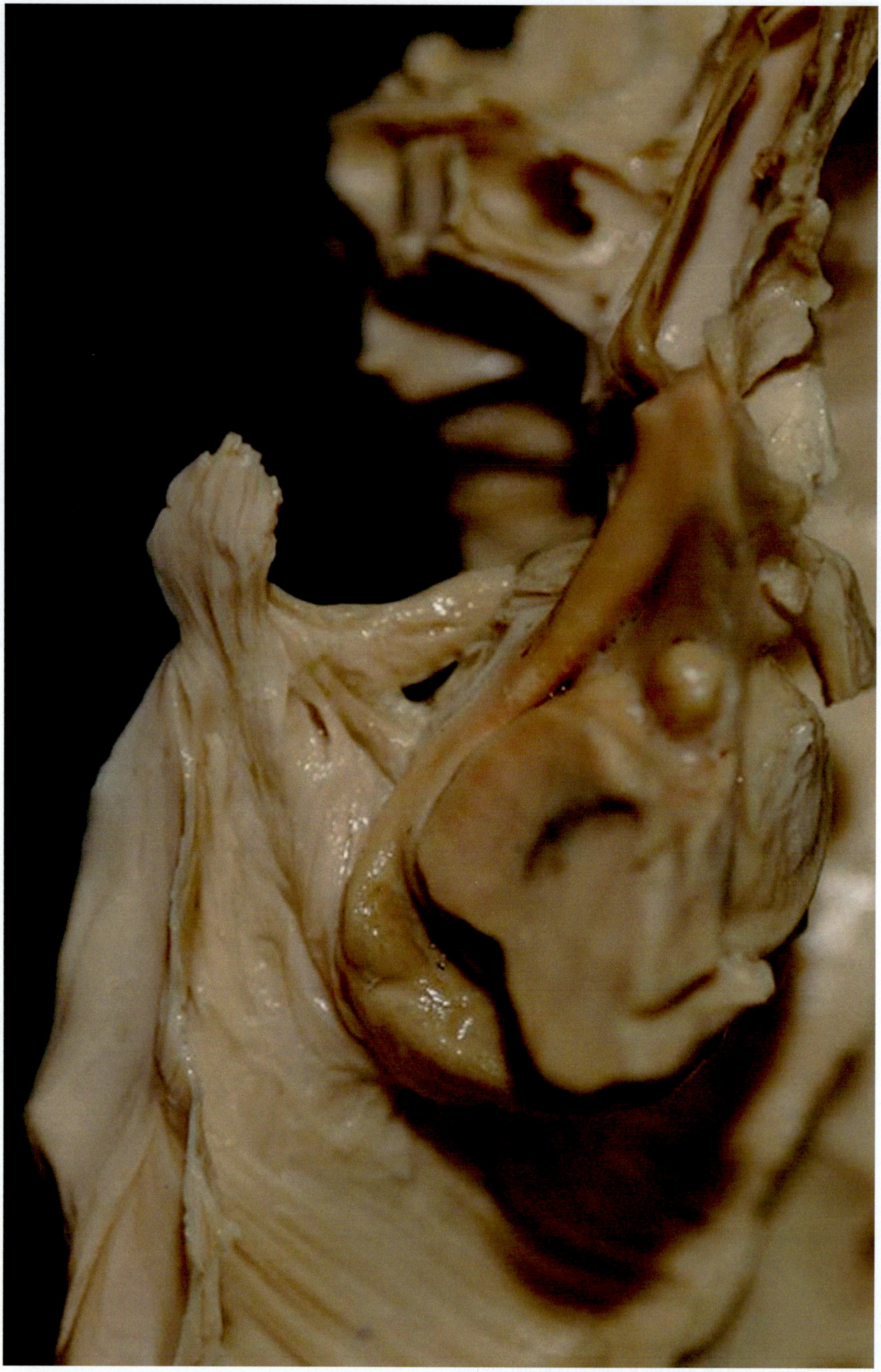

Fig. 4.16 Anatomical preparation and illustration of the right hemisphere, inferior view. The optic tract is depicted from the chiasm to the lateral geniculate body. The inferior and posterior thalamic peduncles are visualized. (©Kadri 2023. All Rights Reserved)

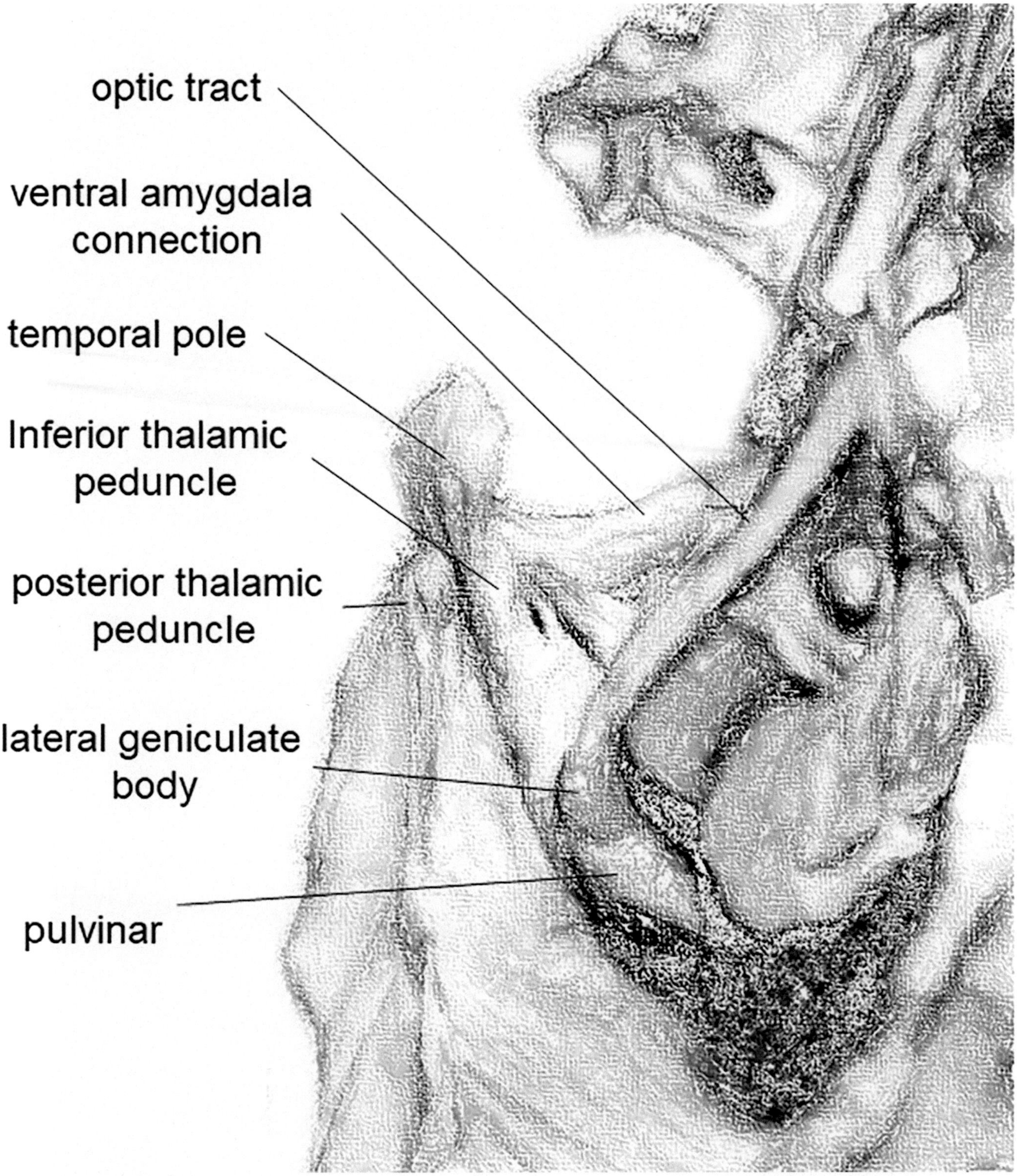

Fig. 4.16 (continued)

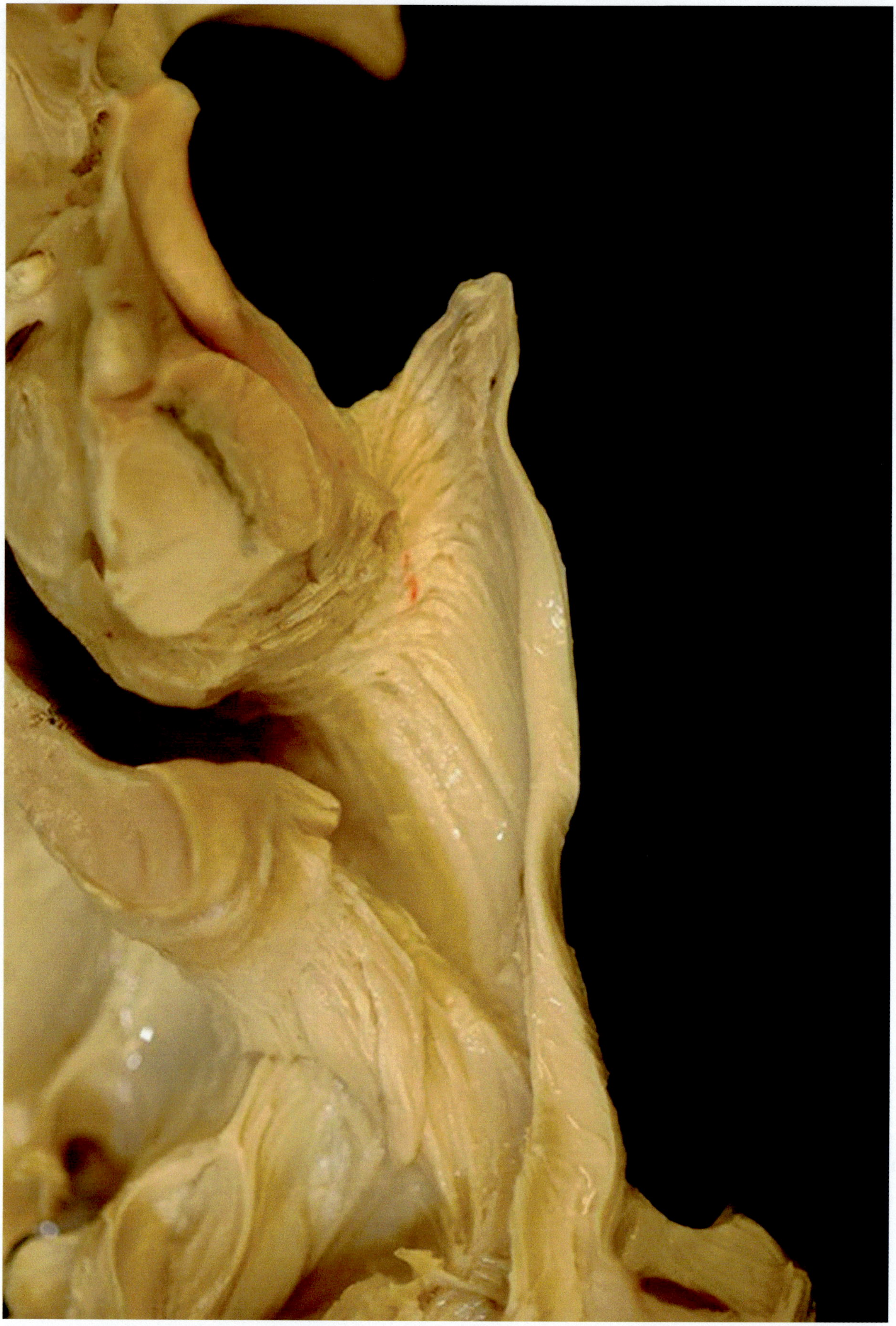

Fig. 4.17 Anatomical preparation and illustration of the left cerebral hemisphere, inferior view. The majority of fibers of the posterior thalamic peduncle emanate from the pulvinar of the thalamus. The geniculocalcarine tract (optic radiation) is one of the components of the posterior thalamic peduncle. (©Kadri 2023. All Rights Reserved)

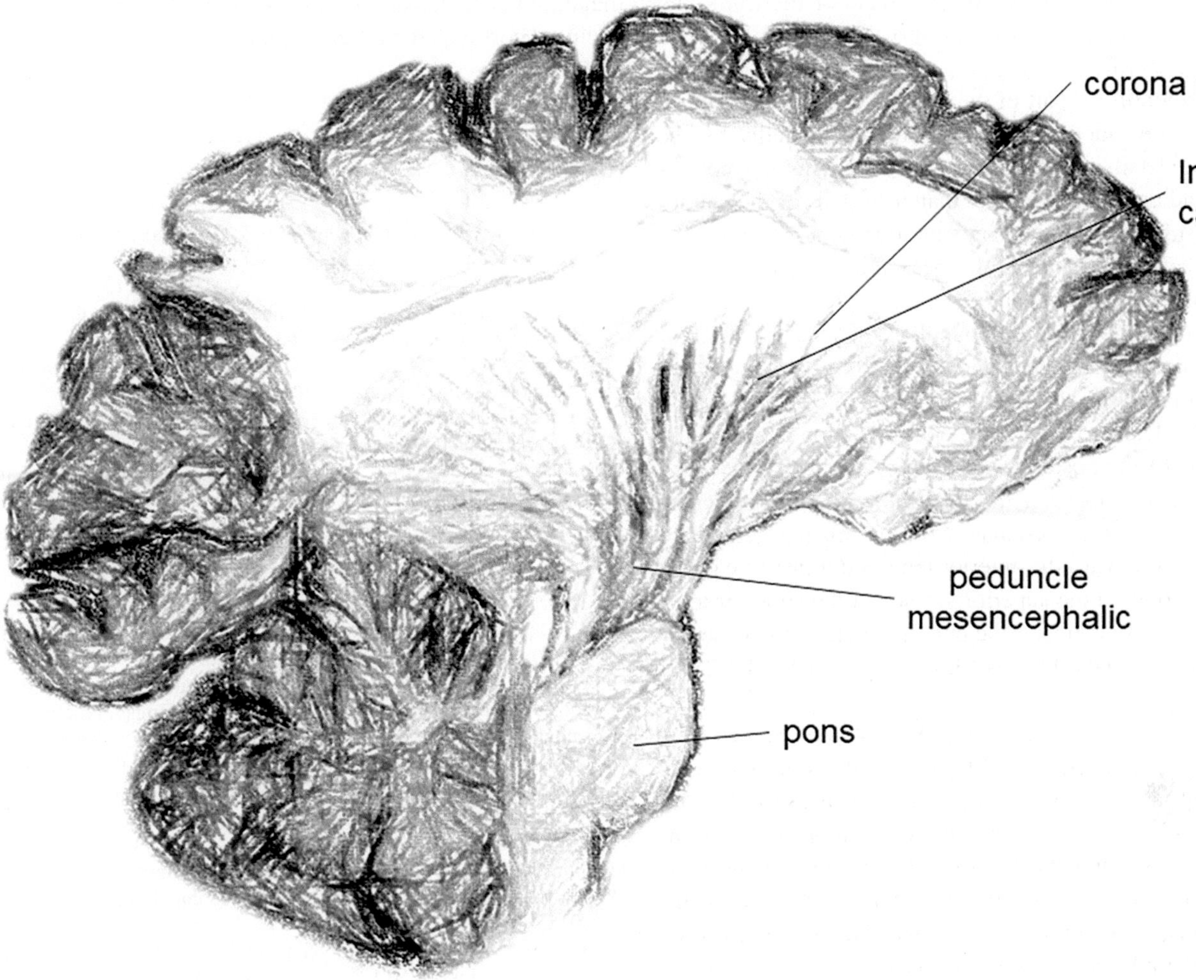

Fig. 4.19 (continued)

and arm anteriorly, followed by the fibers of the trunk and legs (Figs. 4.1, 4.2, 4.3, 4.4, 4.5, 4.6, 4.7, 4.9, 4.10, 4.12, 4.13, and 4.19) [6].

The retrolentiform part of the internal capsule is posterior to the lentiform nucleus but still bordered medially by the thalamus and tail of the caudate nucleus. This segment is compact and devoid of the gray matter bridges of the caudolenticular substance [2]. It contains ascending fibers of the posterior thalamic peduncle and descending parietopontine, occipitopontine, temporopontine, and occipitotectal tracts [5, 6, 11]. Its most important fiber bundle is the optic radiation of Gratiolet, also named the geniculocalcarine tract, in which fibers originate from the lateral geniculate body and extend toward the calcarine cortex. The optic radiation is one of the components of the posterior thalamic peduncle. This peduncle also contains fibers of the acoustic radiation (geniculotemporal tract), which originate from the medial geniculate body to extend toward the anterior temporal transverse gyrus (Heschl's gyrus) on the superior surface of the superior temporal gyrus. It also connects the pulvinar thalami to the occipital and parietal lobes (Figs. 4.13, 4.14, 4.15, 4.16, 4.17, and 4.18) [6].

The optic radiations are still a subject of debate since their description by Gratiolet, and Meyer's subdivision into anterior, central, and posterior bundles is still in use [15]. The anterior bundle extends anteriorly and turns posteriorly on the anterior roof and lateral wall of the temporal horn. It forms Meyer's loop and reaches the inferior calcarine fissure [15]. Initially, the central bundle extends laterally on the roof of the temporal horn but then turns posteriorly, following the lateral wall of the temporal horn and atrium. The posterior, or upper, bundle has a direct posterior trajectory on the upper lateral wall of the atrium and posterior horn to reach the superior calcarine fissure [15].

The sublenticular portion of the internal capsule is situated anterior and inferior to the retrolenticular segment and is at the roof of the temporal horn. It contains descending temporopontine fibers (Turck's fasciculus) and the temporothalamic fasciculus of Arnold (interconnecting the temporo-occipital cortex to the pulvinar thalami and lateral geniculate body) [2].

References

1. Klingler J, Gloor P. The connections of the amygdala and of the anterior temporal cortex in the human brain. J Comp Neurol. 1960;115:333–69. https://doi.org/10.1002/cne.901150305.
2. Testut L, Latarjet A. Tratado de anatomia humana, vol. 2. 9th ed. Barcelona: Salvat Editores; 1966.
3. Carpenter MB. Core text of neuroanatomy. Subsequent ed. Baltimore, MD: Williams &Wilkins; 1991.
4. Heimer L. The human brain and spinal cord: functional neuroanatomy and dissection guide. 2nd ed. New York: Springer; 1994.
5. Nieuwenhuys R, Voogd J, van Huijzen C. The human central nervous system: a synopsis and atlas. 4th ed. Berlin: Springer; 2008.
6. Yaşargil MG. Microneurosurgery, Vol. IVA. New York: Thieme; 1994.
7. Türe U, Yaşargil MG, Friedman AH, Al-Mefty O, Yaşargil DG. Fiber dissection technique: lateral aspect of the brain. Neurosurgery. 2000;47:417–27. https://doi.org/10.1097/00006123-200008000-00028.
8. Williams PL. Gray's anatomy: the anatomical basis of medicine and surgery. 38th ed. New York: Churchill Livingstone; 1995.
9. Ludwig E, Klingler J. Atlas cerebri humani. Basel: S. Karger; 1956.
10. Naidich TP, Krayenbühl N, Kollias S, Bou-Haidar P, Bluestone AY, Carpenter DM. White matter. In: Naidich TP, Castillo M, Cha S, Smirniotopoulos JG, editors. Imaging of the brain. Philadelphia, PA: Saunders; 2013. p. 205–44.
11. Standring S. Cerebral hemisphere. In: Gray's anatomy: the anatomical basis of clinical practice. 40th ed. Philadelphia, PA: Elsevier; 2008. p. 335–59.
12. Rhoton AL Jr. The cerebrum. Neurosurgery. 2002;51(4 Suppl):S1–S51. https://doi.org/10.1097/00006123-200210001-00002.
13. Türe U, Yaşargil MG, Pait TG. Is there a superior occipitofrontal fasciculus? A microsurgical anatomic study. Neurosurgery. 1997;40:1226–32. https://doi.org/10.1097/00006123-199706000-00022.
14. Ebeling U, von Cramon D. Topography of the uncinate fasciculus and adjacent temporal fiber tracts. Acta Neurochir. 1992;115:143–8. https://doi.org/10.1007/BF01406373.
15. Ebeling U, Reulen HJ. Neurosurgical topography of the optic radiation in the temporal lobe. Acta Neurochir. 1988;92:29–36. https://doi.org/10.1007/BF01401969.

The Limbic Connections of the Brain

5

The term *limbic* indicates a border or limit, and the limbic region in the brain refers to the limit (limbus) between the cortex and the white matter. The limbic lobe is different from the limbic system. The *limbic lobe* is a telencephalic area, the surface of which is represented by the cingulate and parahippocampal gyri. The *limbic system* contains the limbic lobe and subcortical, diencephalic, and brainstem structures.

Components of the Limbic System

There is no general agreement on the structures that comprise the limbic system [1–7]. This chapter describes the main cortical and subcortical elements and the limbic circuits.

Cortical Structures

The cortical structures are the following: (1) the cingulate gyrus, (2) the parahippocampal gyrus, and (3) the hippocampal formation.

The Cingulate Gyrus

This gyrus is visible in the medial aspect of the cerebral hemisphere concentrically surrounding the corpus callosum. It is limited by the sulcus of the cingulate gyrus, the subparietal sulcus, and the anterior calcarine fissure. The sulcus of the cingulate gyrus begins at the cingulate pole and separates the cingulate gyrus from the medial aspect of the rectus gyrus. As it turns around the genu of the corpus callosum and runs posteriorly, it sequentially separates the cingulate gyrus from the medial surface of the superior frontal gyrus and the paracentral lobule. The sulcus of the cingulate gyrus has two important ramifications, the paracentral and marginal rami, that limit the paracentral lobule on the medial surface. The marginal ramus separates the paracentral lobule from the precuneus and is an important landmark to identify the central sulcus of the brain. The central sulcus reaches the interhemispheric fissure just anterior to the supramarginal ramus. It extends from the subcallosal area, where it is continuous anteriorly with the gyrus rectus. The connection between these two gyri is known as the pole of the cingulate gyrus [7].

The subcallosal area is situated inferior to the rostrum of the corpus callosum. It contains three other diminutive gyri: the rudimentary *indusium griseum* of the precommissural hippocampus (see below), the medial paraolfactory gyrus, and the paraterminal gyrus. The medial paraolfactory gyrus is limited by the anterior and posterior paraolfactory sulci. The anterior sulcus separates the paraolfactory gyrus from the cingulate gyrus; the posterior sulcus separates it from the paraterminal gyrus. The paraterminal gyrus is a tiny whitish gyrus, just anterior to the anterior commissure (Figs. 5.1 and 5.2). It is the cisternal surface of the septal nuclei (see below). The posterior cingulate gyrus is separated from the medial aspect of the superior parietal lobe by the subparietal sulcus. Posterior to the splenium of the corpus callosum, the cingulate gyrus narrows and is separated from the superior aspect of the anterior lingual gyrus by the anterior segment of the calcarine fissure. This narrowing strip of the cingulate gyrus is named the isthmus and, as it joins the anterior lingual gyrus, constitutes the posterior limit of the parahippocampal gyrus [8].

P. Kadri, *The Cartographic Atlas of the Brain*, https://doi.org/10.1007/978-3-031-38062-4_5

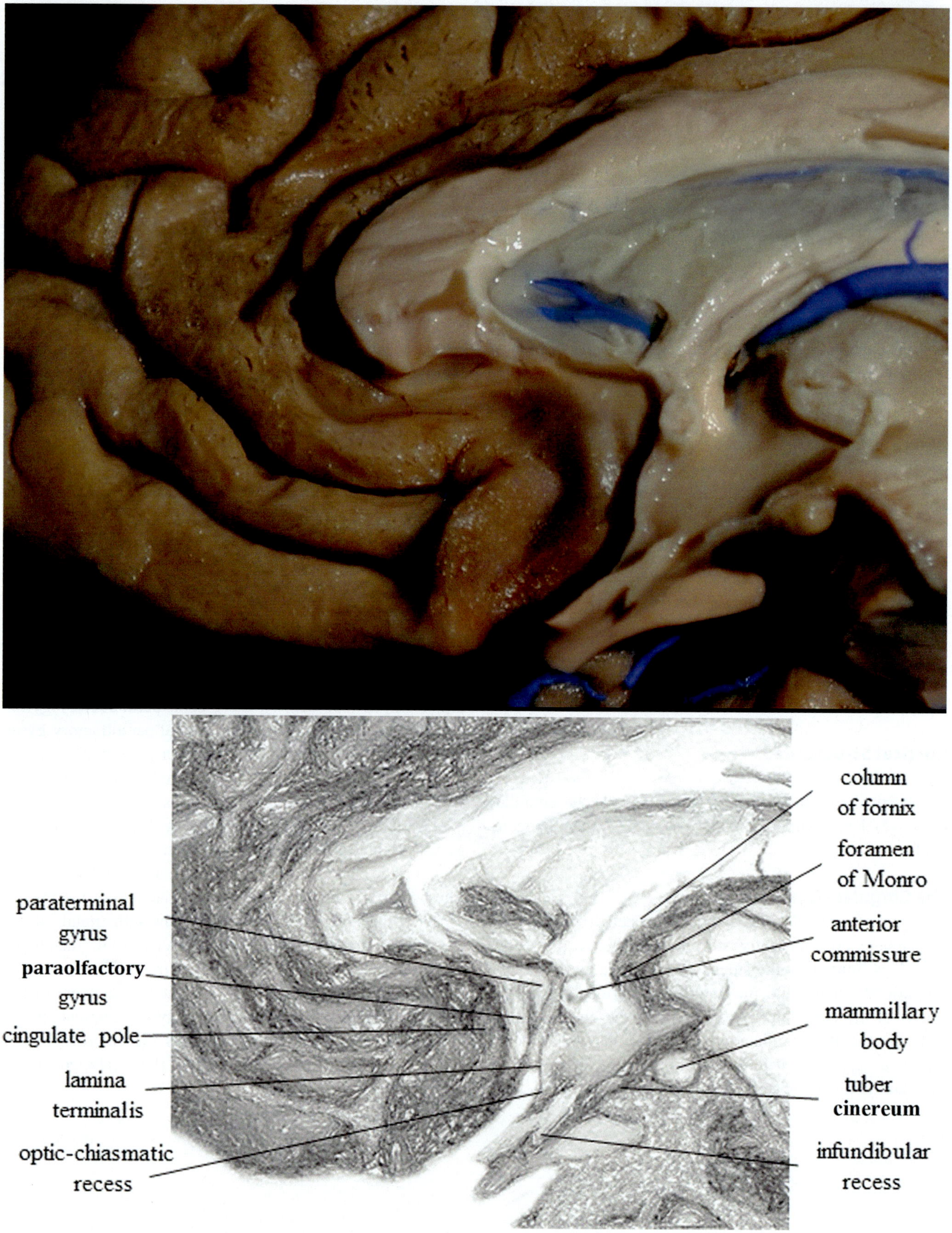

Fig. 5.1 Anatomical preparation and illustration of the right side of the cerebral hemisphere depicting the subcallosal area. (©Kadri 2023. All Rights Reserved)

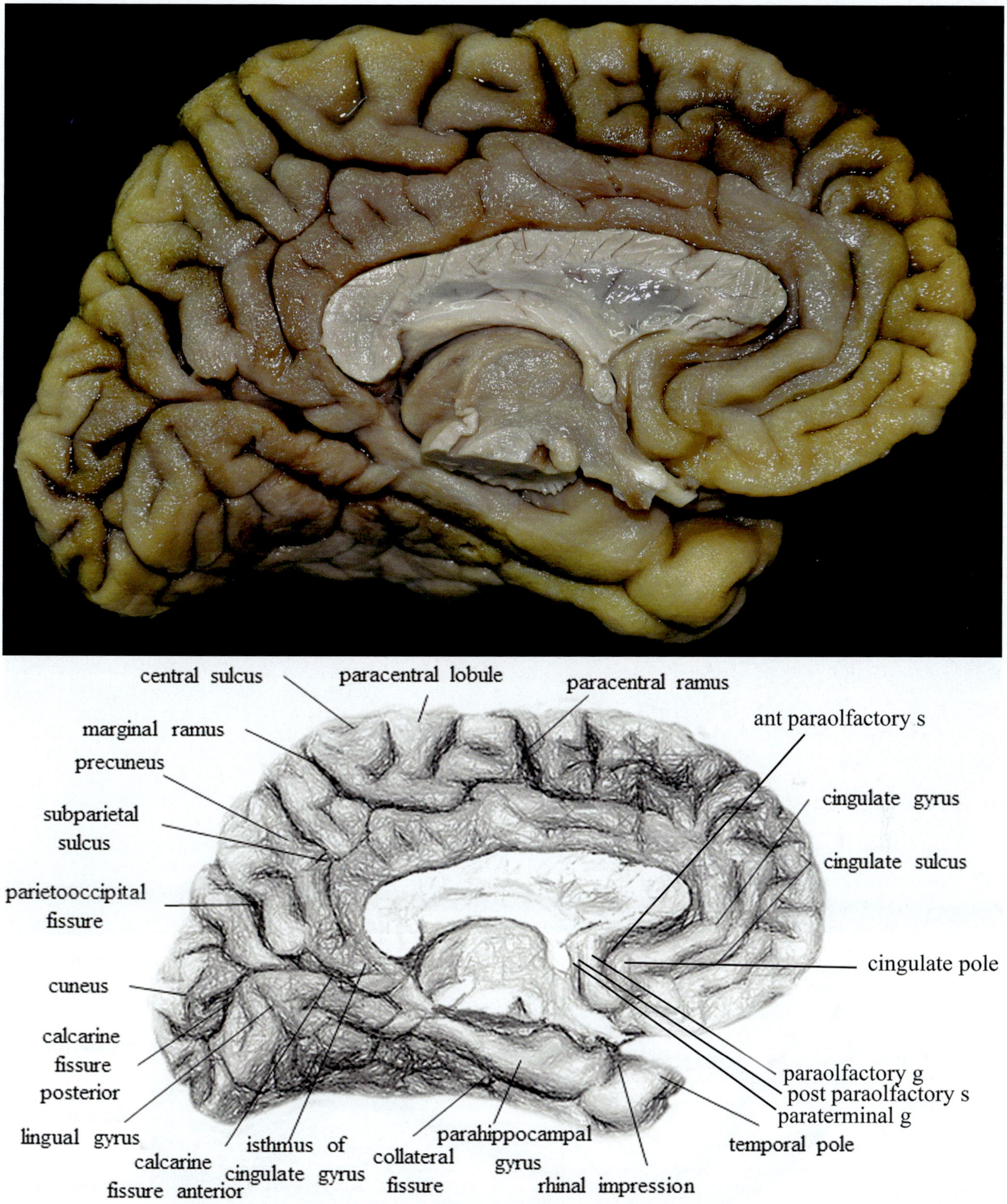

Fig. 5.2 Anatomical preparation and illustration of the left side of the cerebral hemisphere. On the surface, the limbic lobe is composed of the cingulate and parahippocampal gyri, limited by the limbic sulcus. The complex limbic sulcus is composed of the cingulate gyrus sulcus, the subparietal sulcus, the anterior portion of the calcarine fissure, the collateral sulcus, and the rhinal sulcus. *g* gyrus, *s* sulcus (©Kadri 2023. All Rights Reserved)

The Parahippocampal Gyrus

This gyrus is located on the inferomedial surface of the cerebral hemisphere and encircles the mesencephalon (Figs. 5.3, 5.4, 5.5, 5.6, 5.7, and 5.8). It is located inferior to the hippocampal formation and is separated from the dentate gyrus, one of the components of the hippocampal formation, by the hippocampal sulcus. It can be divided into anterior and posterior parts by an arbitrary line passing transversely to the posterior limit of the uncus. The anterior part is covered primarily by the entorhinal cortex, described and named area 28 by Korbinian Brodmann (1868–1918) [9]. It is therefore referred to as the entorhinal area. This area is connected to the major portions of the association areas of the cerebral cortex; primary olfactory areas; and the hippocampal formation, septal area, and amygdala [1, 2, 5]. It receives information from all sensory modalities. The rhinal sulcus is the anterolateral limit of the anterior part and is continuous with the collateral sulcus in about a third of cases [8]. The most anterior part of the region merges with the lateral orbitofrontal region and the insula at the level of the limen insula. It contains the lateral olfactory stria and the lateral paraolfactory gyrus. The lateral paraolfactory gyrus consists of the prepiriform cortex, also known as the piriform cortex. The anteromedial limit is the endorhinal sulcus, which separates the most anterior part of the parahippocampal gyrus from the anterior perforated substance. The anterior region contains two gyri, the semilunar and ambiens, separated by the semiannular

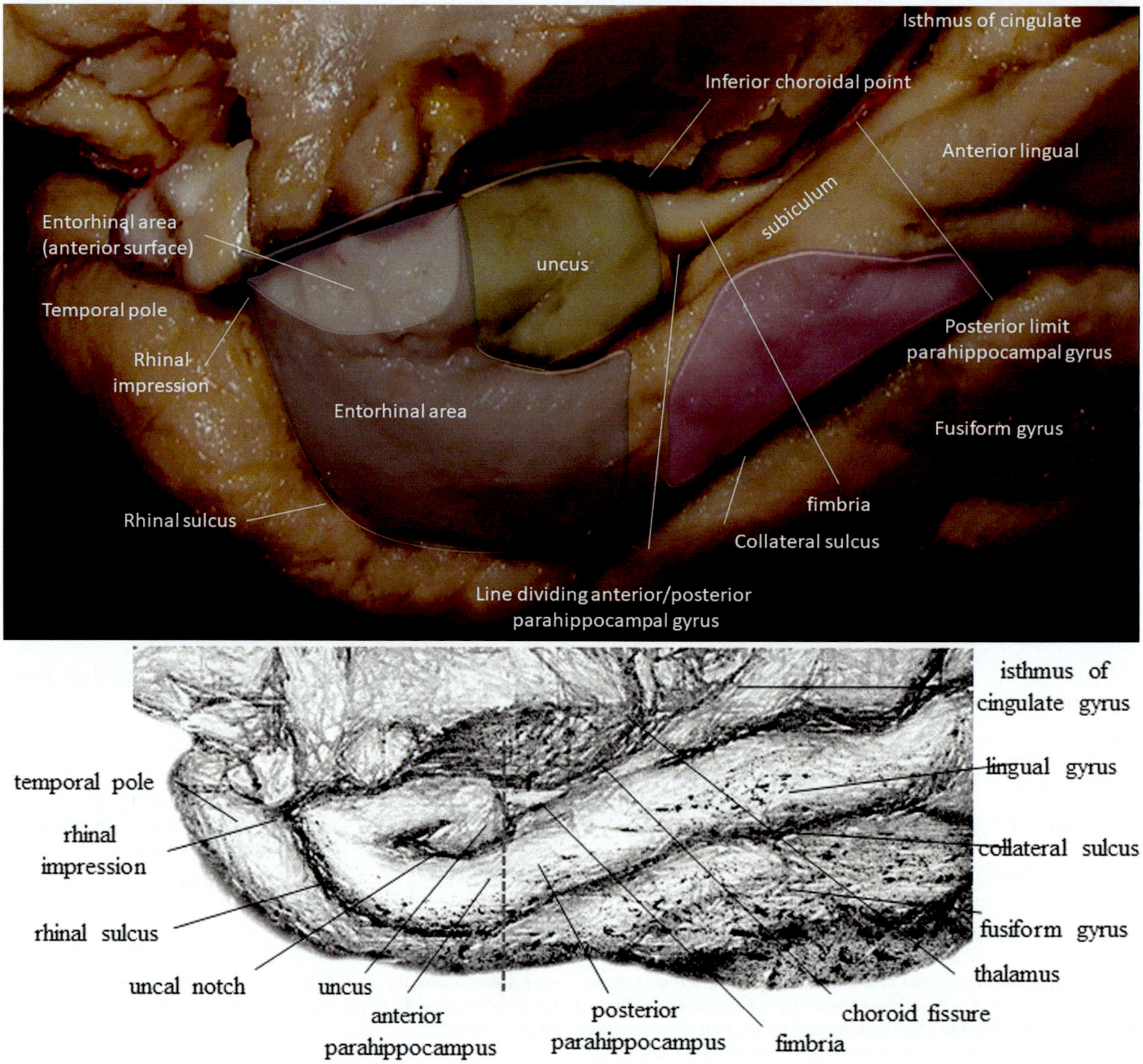

Fig. 5.3 Anatomical preparation and illustration of the right mediobasal surface of the cerebral hemisphere depicting the medial view of the parahippocampal gyrus. (©Kadri 2023. All Rights Reserved)

sulcus. The semilunar gyrus is related to the cortical nuclei of the amygdala and the ambiens gyrus to the lateral nuclei [3].

The ambiens gyrus is separated from the entorhinal area by the intrarhinal sulcus, which in most cases corresponds to the tentorial impression; nevertheless with carefully inspection, it can be identified in most of the specimens. This gyrus is separated from the uncinate gyrus of the uncus by the uncal sulcus. In most neurosurgical literature, the uncus is described in anterior (related to the semilunar and ambiens gyrus of the parahippocampal gyrus, i.e., the amygdala) and posterior regions (related to the extraventricular head of the hippocampal formation). Thereafter, it is described as part of the parahippocampal gyrus; however, the uncus is anatomically exclusively a region of the hippocampus (see below). The posterior portion extends from the posterior part of the uncus to the joining point of the isthmus of the cingulate gyrus and the anterior lingual gyrus. Laterally, it is limited by the collateral sulcus, the impression of which on the floor of the temporal horn is designated as the collateral eminence. It has a narrow and flat superior surface, the subiculum, which is separated from the hippocampus by the hippocampal sulcus [2, 3].

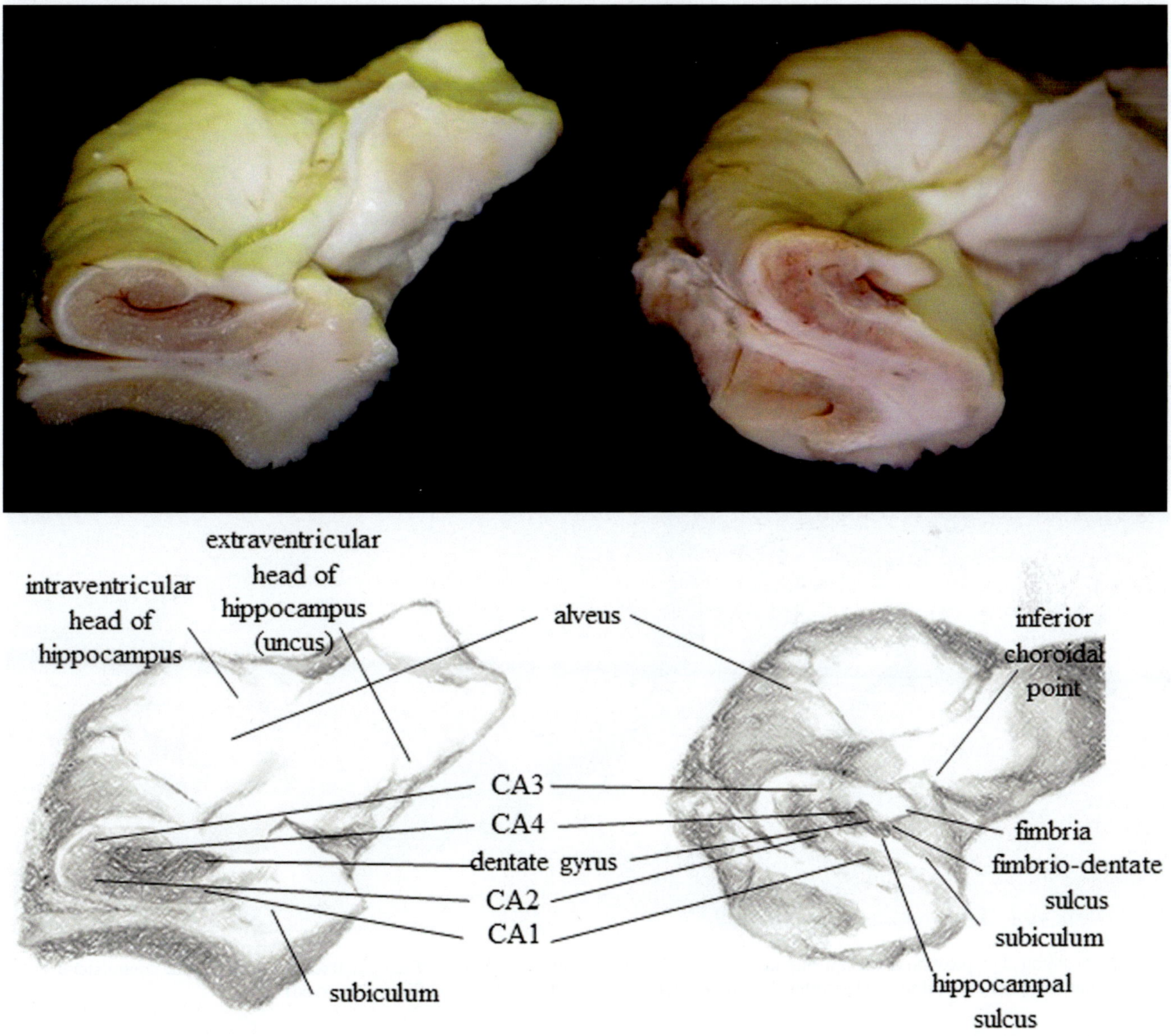

Fig. 5.4 Anatomical preparation and illustration of the isolated parahippocampal gyrus. The hippocampal eminence protrudes inside the ventricular cavity and is covered by a layer of white matter known as the *alveus*. The axons of the alveus converge to form the fimbria. The hippocampal eminence corresponds to the protrusion of the hippocampal sulcus, a virtual sulcus that separates the subiculum from the dentate gyrus. The subiculum is the flat superior portion of the parahippocampal gyrus. The subiculum, the cornu Ammonis, and the dentate gyrus form the hippocampal formation. The cornu Ammonis is subdivided into cytoarchitectonic categories of CA1–CA4. (©Kadri 2023. All Rights Reserved.)

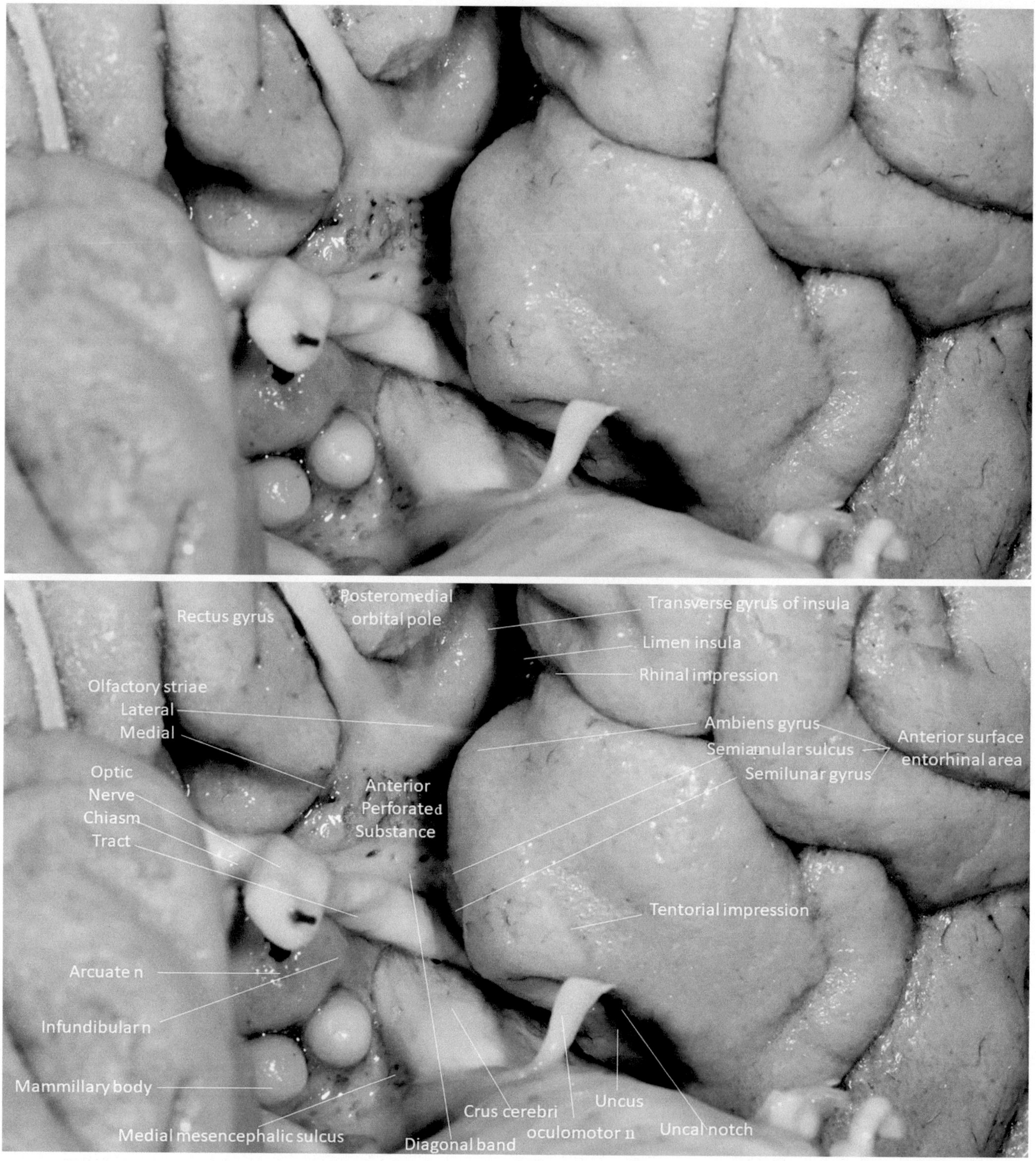

Fig. 5.5 Anatomical preparation of the left anterior perforated substance and its boundaries. The optic tract is reflected to the contralateral side. The landmarks of the piriform and entorhinal areas are seen. *n* nerve (©Kadri 2023. All Rights Reserved)

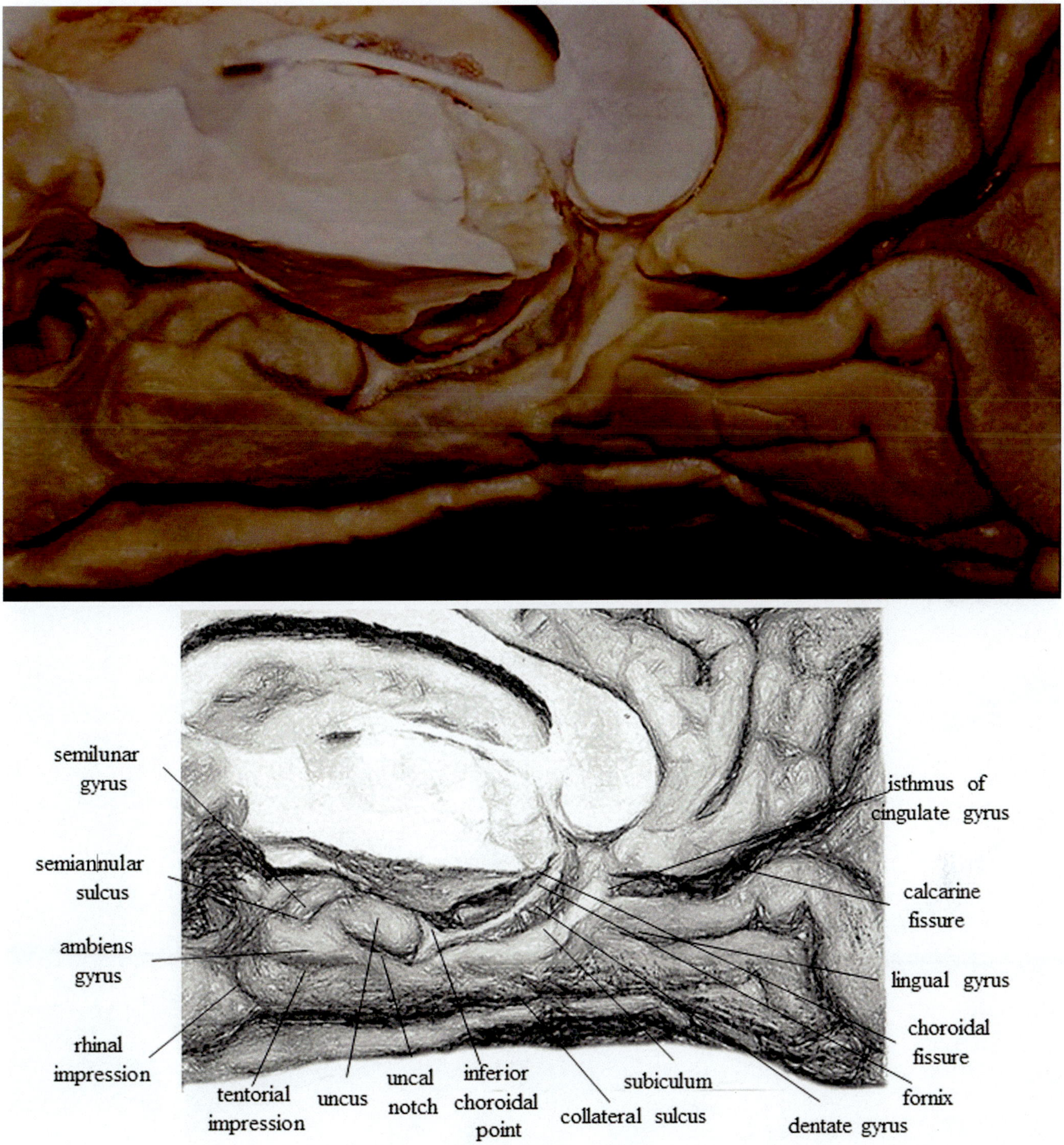

Fig. 5.6 Anatomical preparation and illustration of the left anterior perforated substance and its boundaries. (©Kadri 2023. All Rights Reserved)

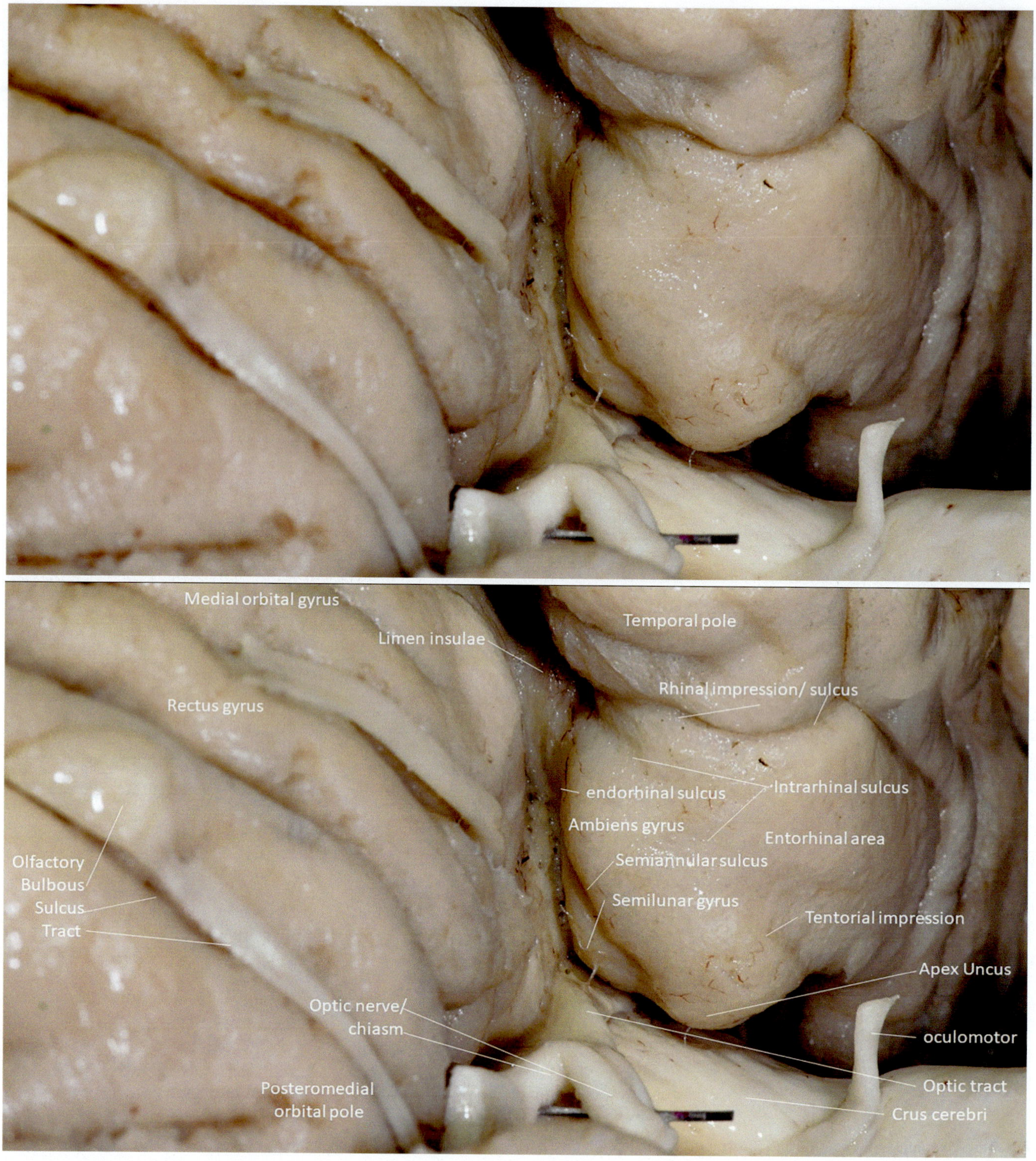

Fig. 5.7 Anatomical preparation of the left anterior perforated substance and the related landmarks. (©Kadri 2023. All Rights Reserved)

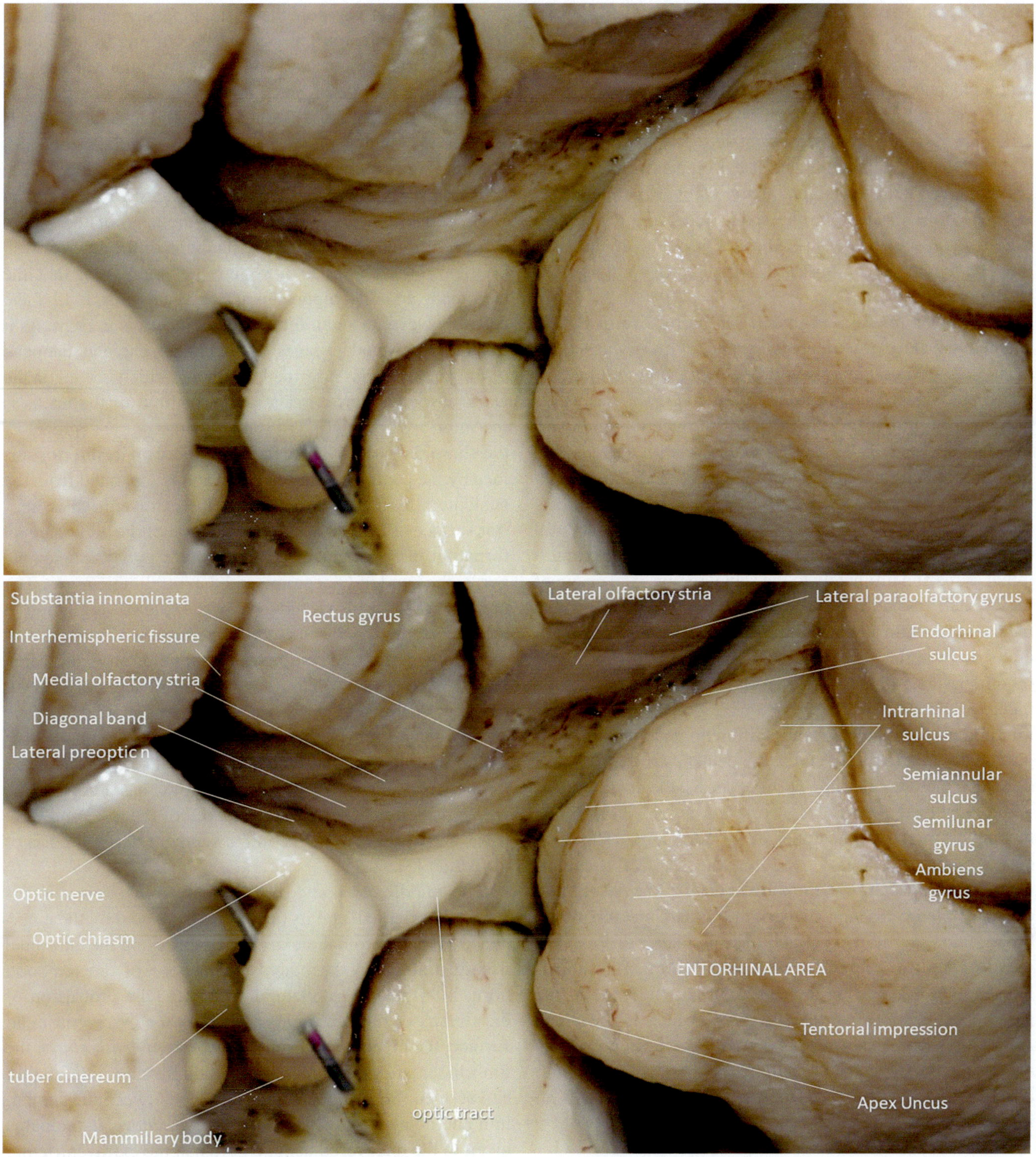

Fig. 5.8 Anatomical preparation of the right cerebral hemisphere, medial view, depicting the inferobasal aspect of the inferior segment of the limbic lobe. (©Kadri 2023. All Rights Reserved)

The Hippocampal Formation

The hippocampal formation (or simply the hippocampus) is a C-shaped structure on the mediobasal surface of the cerebral hemisphere (Figs. 5.8, 5.9, 5.10, 5.11, 5.12, 5.13, 5.14, and 5.15). Its limits are defined based on cytoarchitectonic arrangement and comprise the dentate gyrus, the hippocampus, and the subicular complex. It represents the archipallial cortical layer and is constituted by a relatively simple three-layered cortex throughout its extent. According to its position in relation to the corpus callosum, it can be divided into three segments: precommissural, supracommissural, and retrocommissural [3].

The precommissural segment is a narrow vertical strip situated just anterior to the paraterminal gyrus in the subcallosal area. It is continuous with the supracommissural hippocampus, which is constituted by the *indusium griseum*, a thin layer of gray matter that extends through the outer surface of the corpus callosum in continuity with the pre- and the retrocommissural hippocampus. The medial and longitudinal striae run sagittally within the supracommissural hippocampus. They represent a rudimentary supracallosal component of the fornix, and the fibers extend to the paraterminal gyrus and ultimately to the septal region. The origin of these small fasciculi in the hippocampus is uncertain. The medial stria might stem from the fasciola cinerea and the lateral stria from the gyrus fasciolaris [3] (see below). The retrocommissural hippocampal formation expands and is incorporated into the parahippocampal gyrus. The retrocommissural hippocampus is clearly differentiated into three longitudinally arranged structures: the dentate gyrus, Ammon's horn, and the subiculum. In corticogenesis, the dentate gyrus is the most medial part of the cortex. Ammon's horn forms the largest part of the hippocampal formation. The term cornu Ammonis (CA) is derived from the historical terminology of the hippocampal formation applied to the description of the histological arrangement of the cytoarchitecture of the hippocampal gyrus. This structure is divided into four segments or fields, CA1 to CA4, of Ammon's horn. The subiculum is the flat portion on the superior surface of the parahippocampal gyrus, the surface of which is medial to Ammon's horn. The retrocommissural segment arches around the mesencephalon, and the arch can be divided into anterior, middle, and posterior parts. All parts present an intraventricular and extraventricular surface.

The different components of the hippocampal formation have been ascribed different names according to their location, appearance, and intra- or extraventricular surface: the anterior, the middle or body, and the posterior or tail. The anterior portion, or head, is transversely oriented and harbors three or four bumps on its surface called the digitations hippocampi. These bumps give the hippocampus the appearance of a lion's paw, thus the denomination *pes hippocampi.* In some cases, the ependyma covering the head of the hippocampus may be fused with the ependyma of the anterior wall of the ventricle, which may give the impression that the amygdala and the hippocampus head are fused. In most cases, however, a small hippocampal recess is present. As it turns medially and assumes a hooklike appearance, the head of the hippocampus is termed the uncus, which is seen on the extraventricular surface. The posterior limit of the uncus is the inferior choroidal point, which is the initial segment of the choroidal fissure. The uncus is covered on its inferior surface by the parahippocampal gyrus and is separated from the subiculum by the uncal notch. On the extraventricular inferior surface of the uncus, the components of the hippocampal formation are recognized as three small gyri: the uncinate gyrus, the band of Giacomini, and the intralimbic gyrus. The uncinate gyrus is also referred to as the external digitation of the hippocampus and is mainly formed by the CA1 field. It is the most anterior and comprises two small protrusions separated by a shallow sulcus. The band of Giacomini is composed of the dentate gyrus but is devoid of its characteristic toothlike appearance. It is separated from the uncinate gyrus by the anterior portion of the hippocampal sulcus. The intralimbic gyrus, also referred to as the apex, has a cone shape with a posterior vertex. It is composed of the CA3 and CA4 fields and appears to be the initial segment of the fimbria of the fornix [3].

The middle, or body, of the hippocampal formation is sagittally oriented, and its beginning is marked by the emergence of the choroidal plexus, the fimbria, and the choroidal fissure. Lateral to the body, on the floor of the temporal horn, is located the collateral eminence, the protrusion of the fundus of the collateral sulcus. On the extraventricular surface, the hippocampal sulcus, the *margo denticularis*, the fimbriodentate sulcus, the fimbria, and the choroidal fissure can be seen. The hippocampal sulcus is between the subiculum and dentate gyrus. The *margo denticularis* is the name given to the uncovered dentate gyrus on the extraventricular

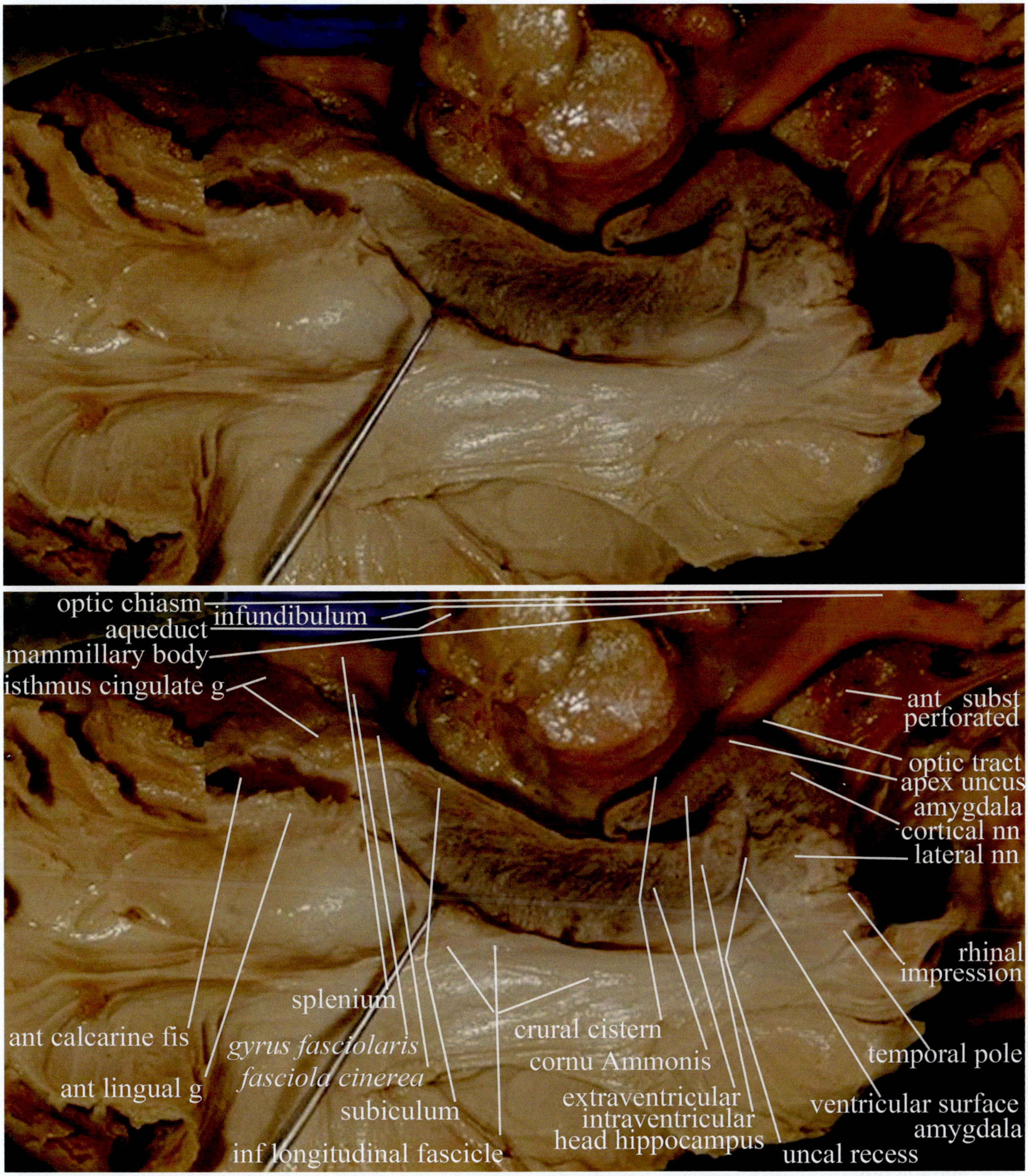

Fig. 5.9 Anatomical preparation of the left basal cerebral hemisphere. Removing the cortex of the inferior surface of the parahippocampal gyrus, followed by the sequential removal of the "U" fibers and the inferior segment of the cingulum, exposes the inferior surface of the hippocampal formation (retrocommissural) and the amygdala. *ant* anterior, *fis* fissure, *g* gyrus, *inf* inferior, *nn* nucleus, *subst* substance (©Kadri 2023. All Rights Reserved)

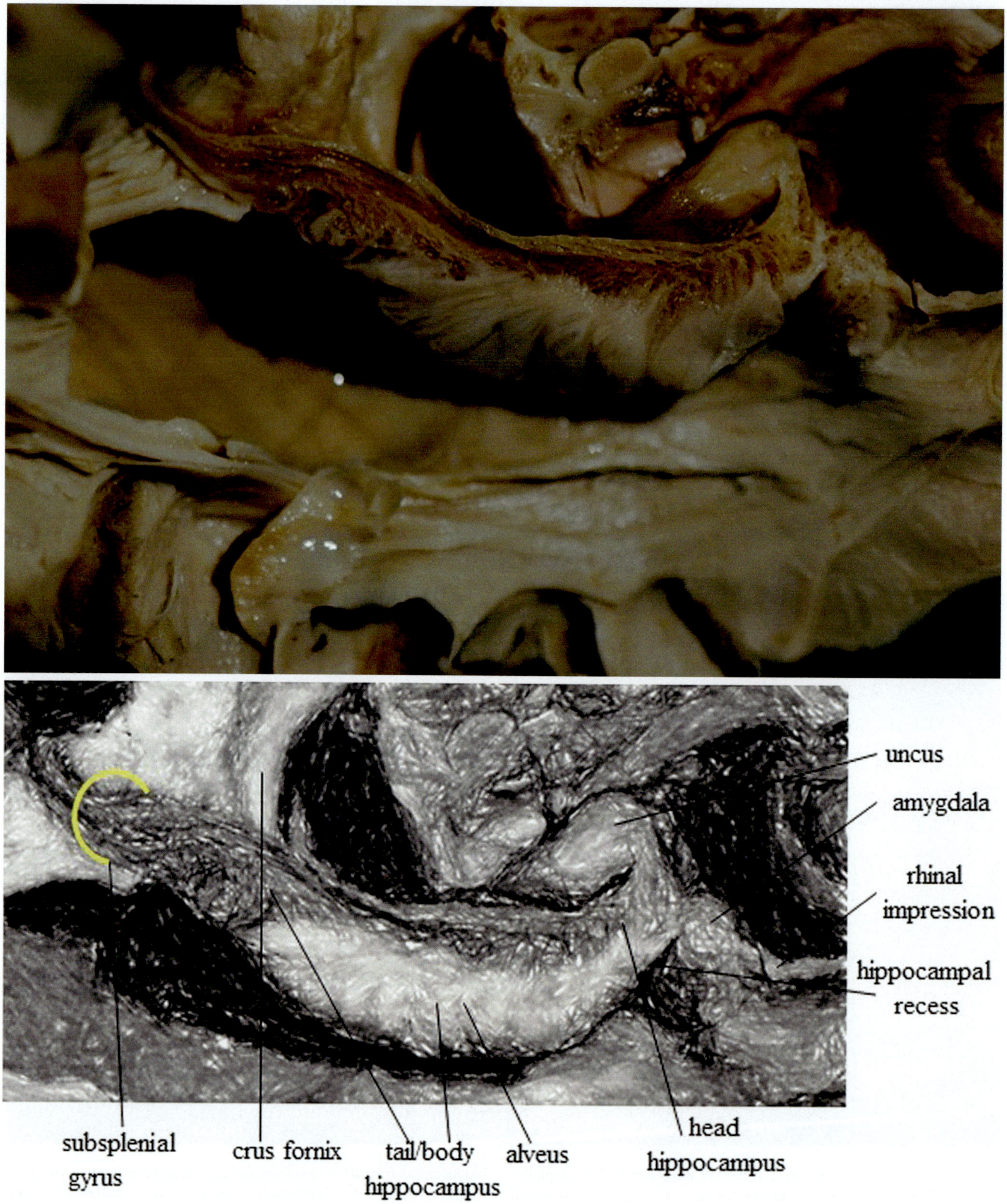

Fig. 5.10 Anatomical preparation and illustration of the left cerebral hemisphere, inferior view of the dissection of the parahippocampal gyrus. The cortex, "U" fibers, and the inferior segment of the cingulum have been removed. The hippocampal formation (retrocommissural) and the amygdala are exposed. (©Kadri 2023. All Rights Reserved)

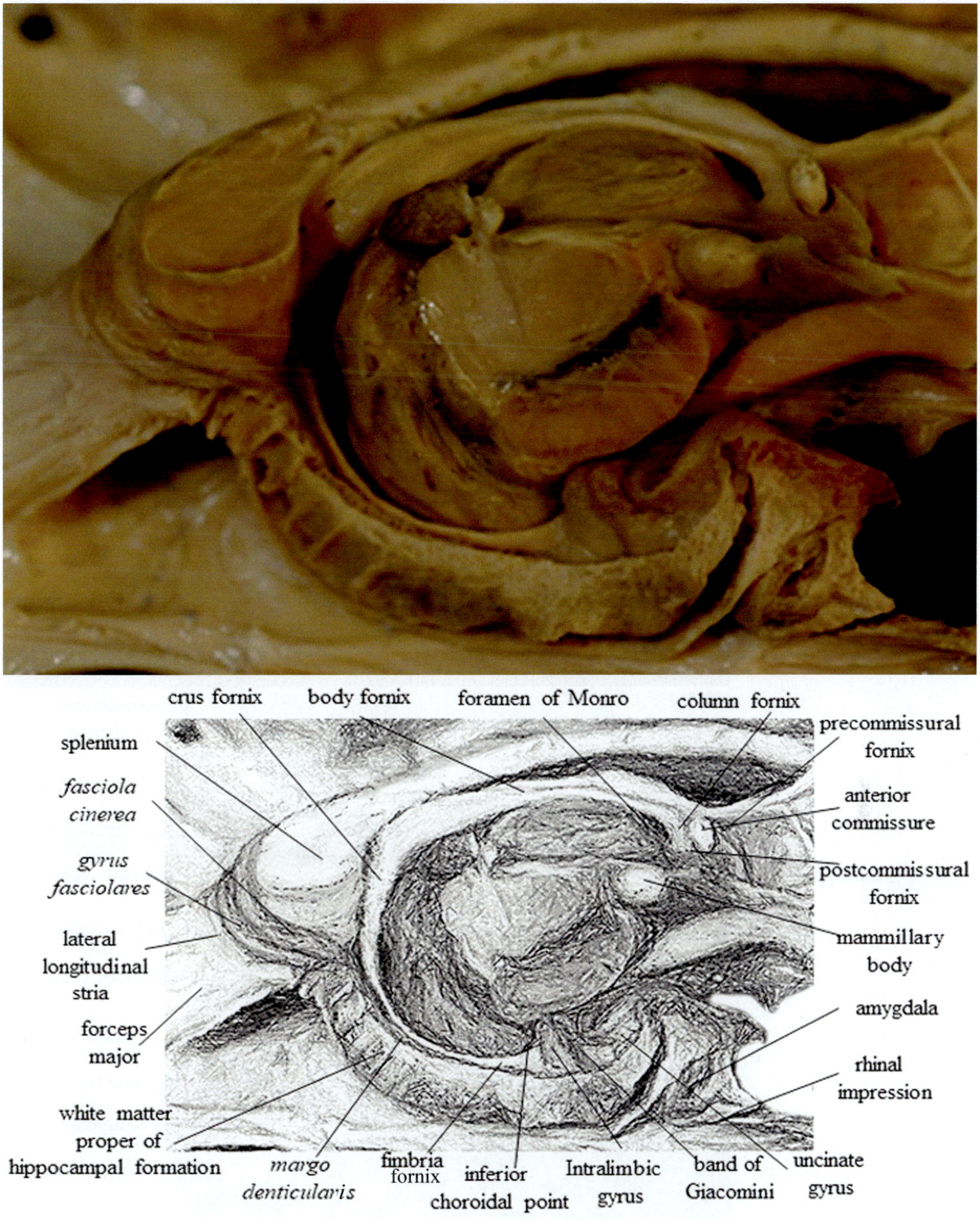

Fig. 5.11 Anatomical preparation and illustration of the left cerebral hemisphere, inferior view of the dissection of the hippocampal formation. Removing the cortex of the subiculum and inferior cornu Ammonis exposes the white matter of the hippocampus. The extraventricular extension of the head, body, and tail are unveiled. The amygdala at the anterior parahippocampal gyrus is exposed. (©Kadri 2023. All Rights Reserved)

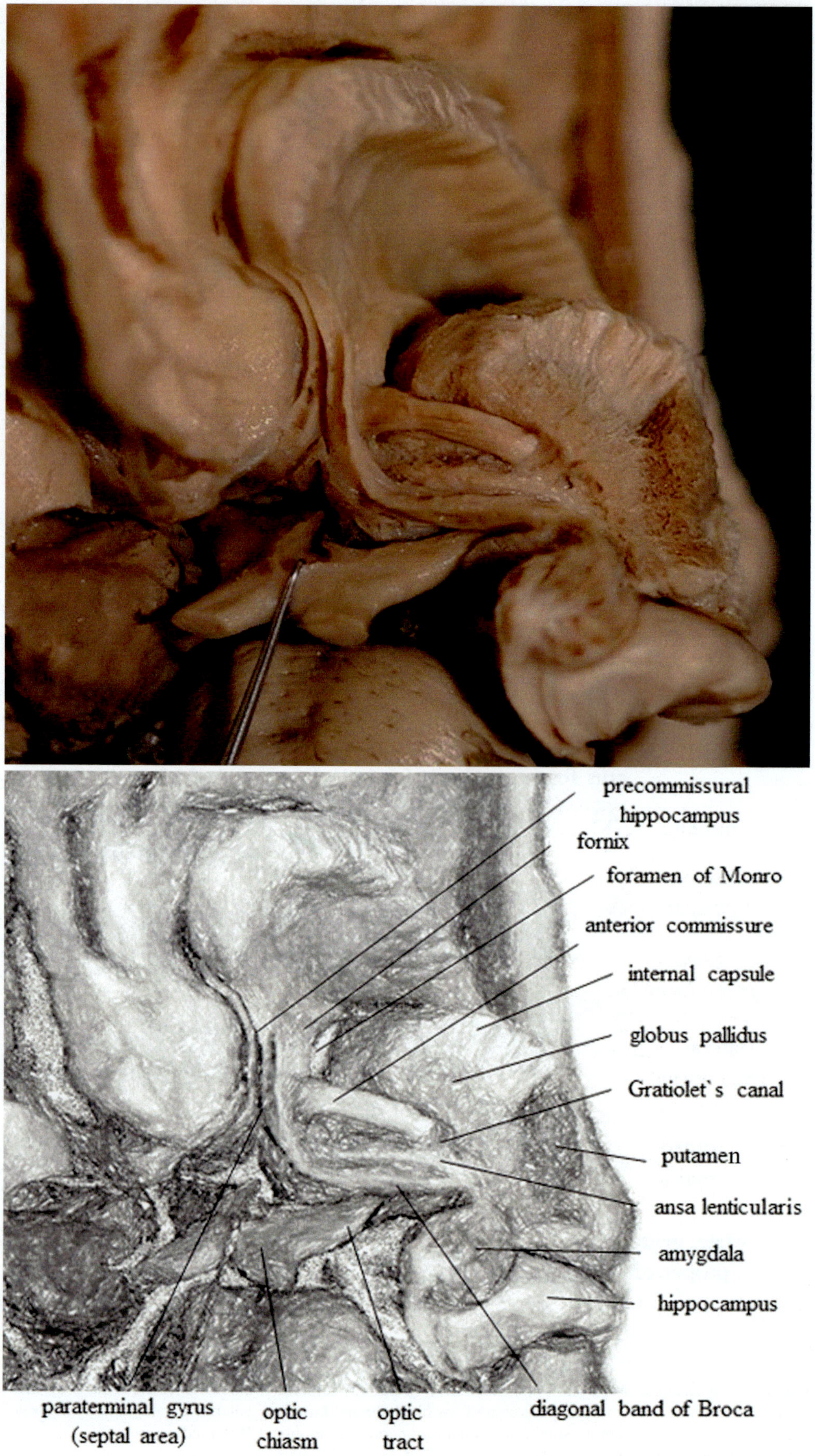

Fig. 5.12 Anatomical preparation and illustration of the precommissural hippocampus and amygdala. The anterior commissure has been severed at Gratiolet's canal. The globus pallidus, ansa lenticularis, and ventral amygdalofugal fibers are exposed. The amygdala septal fibers encompass the diagonal band of Broca. (©Kadri 2023. All Rights Reserved)

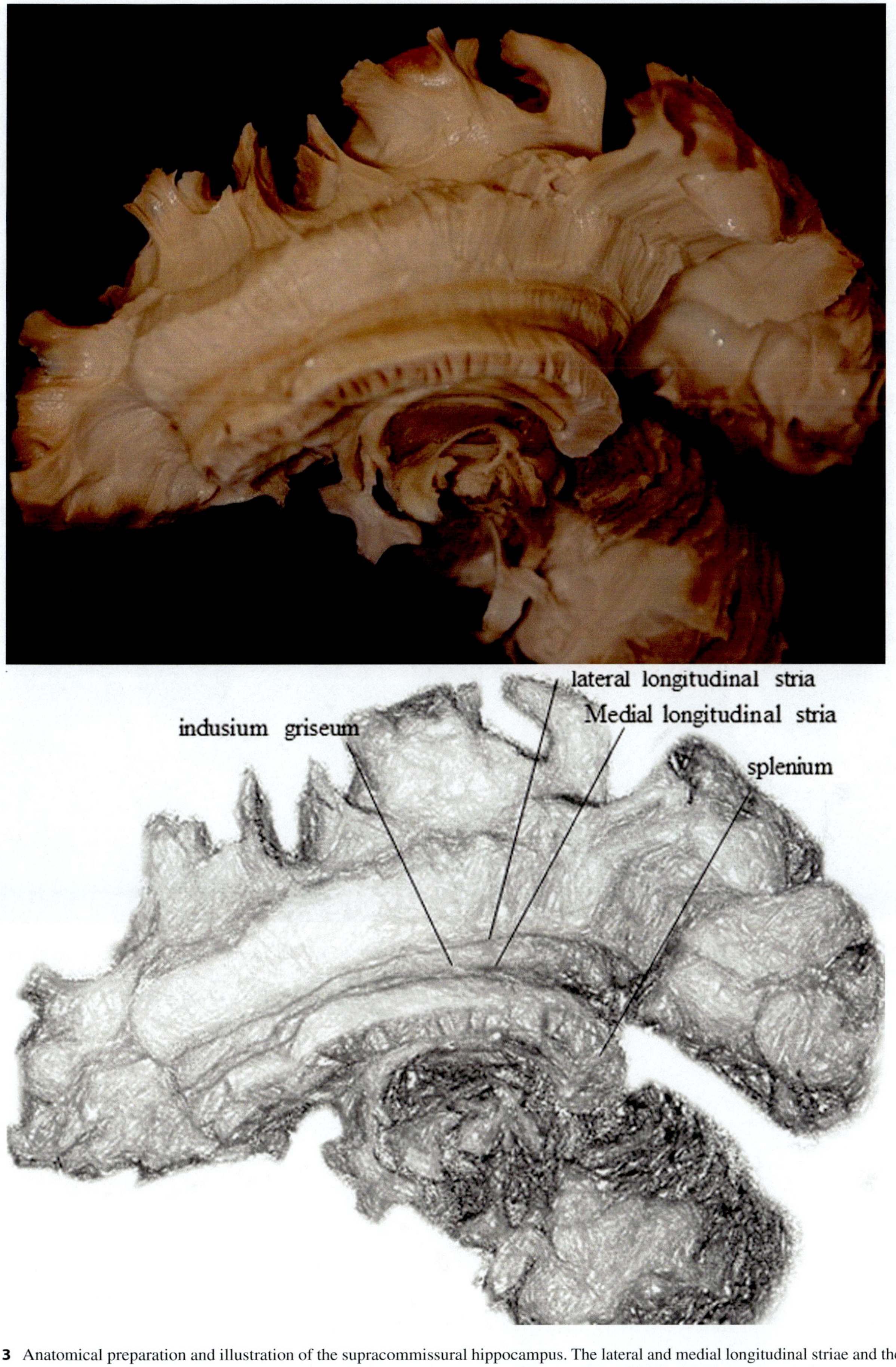

Fig. 5.13 Anatomical preparation and illustration of the supracommissural hippocampus. The lateral and medial longitudinal striae and the indusium griseum are on the external surface of the commissural fibers of the corpus callosum. (©Kadri 2023. All Rights Reserved)

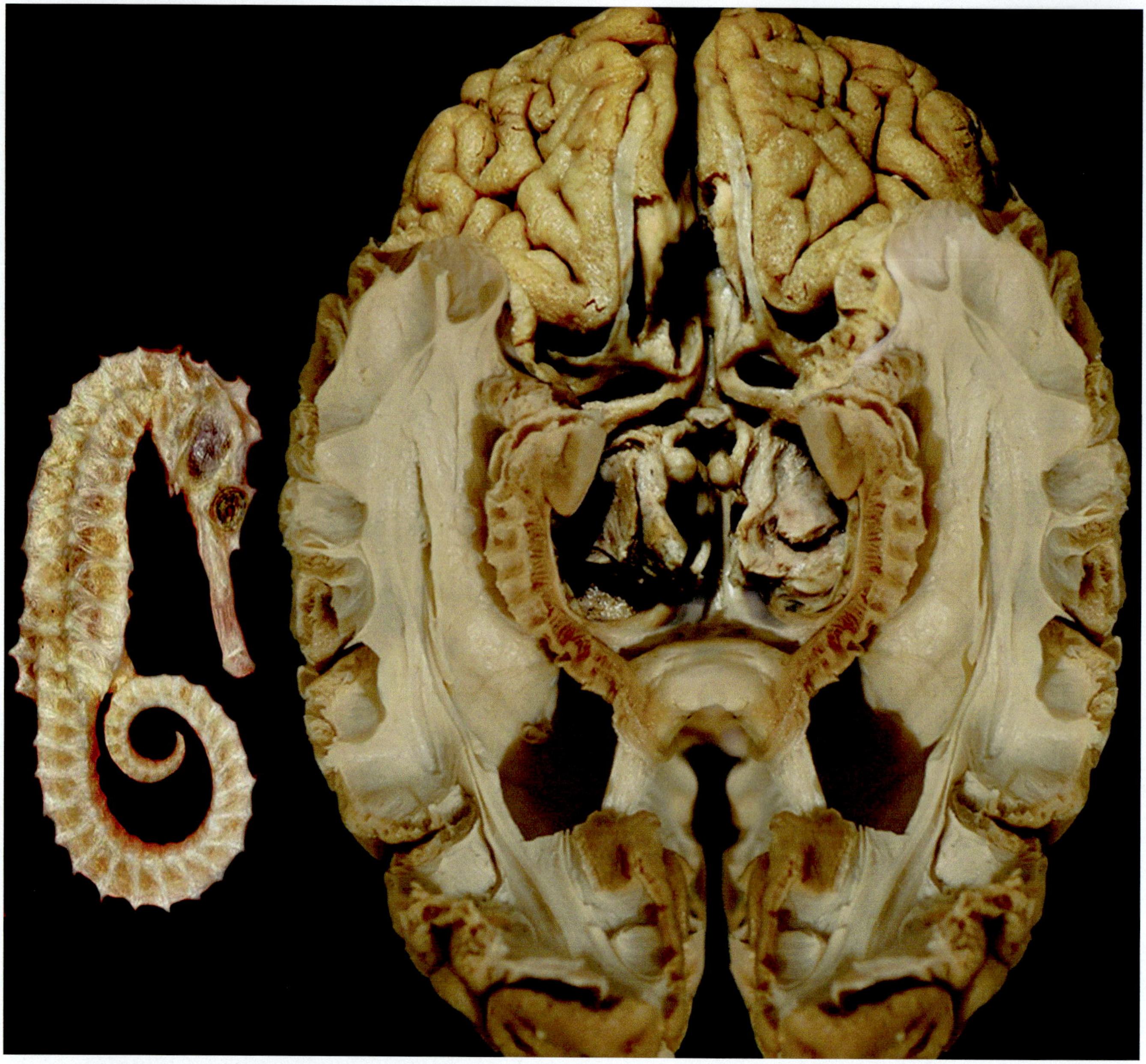

Fig. 5.14 Anatomical preparation of the hippocampal formation. The sagittally oriented structures of the retrocommissural hippocampus and its supracallosal and precallosal extensions can be identified. A comparison with a dry seahorse fossil is depicted to address the interesting comparative anatomical and mythological history of the terms involved in the naming of limbic structures. (©Kadri 2023. All Rights Reserved)

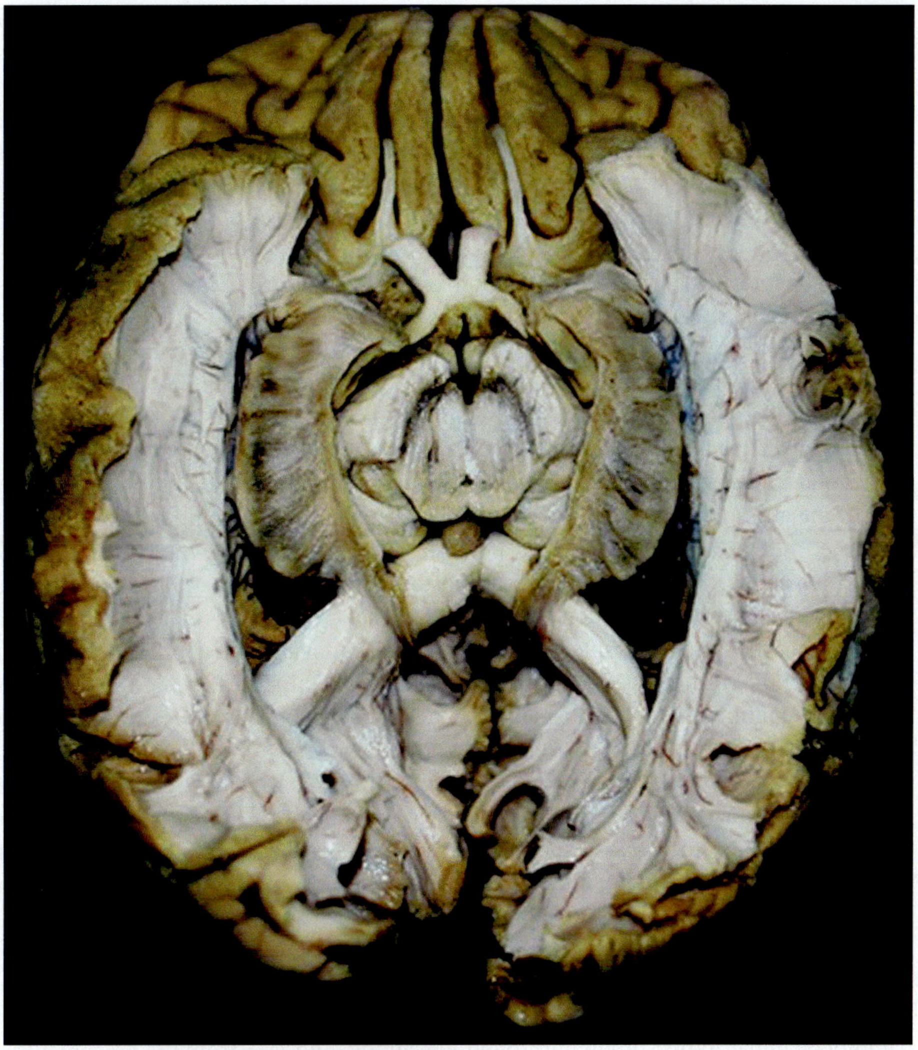

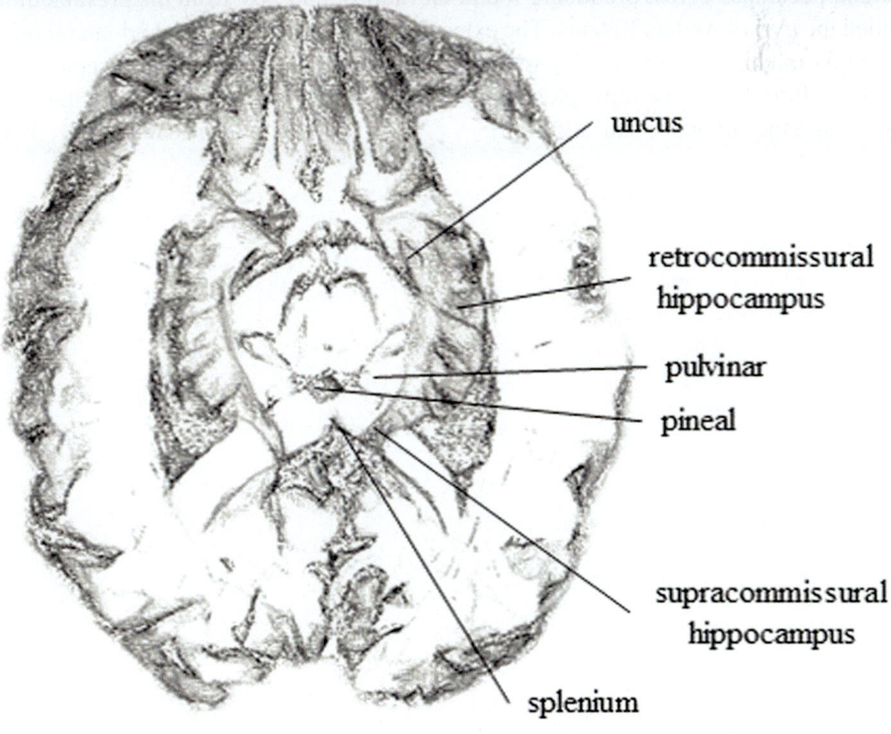

Fig. 5.15 Anatomical preparation and illustration of the inferior view of the hippocampal formation. The subiculum has been removed, unveiling the inferior surface of the cornu Ammonis and the CA1 and CA2 portions. The extensions of the gray and white matter around the splenium and their continuation on the superior surface of the corpus callosum can be observed. Also seen is the extension of the fornix on the undersurface of the splenium. (©Kadri 2023. All Rights Reserved)

surface. The fimbriodentate sulcus is between the fimbria and the dentate. The fimbria is the initial segment of the fornix (see below). The choroidal fissure is a cleft between the thalamus and the fornix, extending from the inferior choroidal point toward the interventricular foramen of Monro [10].

The posterior, or tail, of the hippocampal formation is also transversely oriented and turns medially. Its ventricular surface constitutes the anteromedial limit of the collateral trigone, which is the triangular shape of the floor of the atrium limited laterally by the collateral eminence, anteromedially by the hippocampal formation, and posteromedially by the calcarine eminence. The calcarine eminence is commonly referred to as the *calcar avis* and represents the ventricular impression of the calcarine fissure. At the medial wall of the atrium, the tail of the hippocampus seems to fuse with the *calcar avis*. On the extraventricular surface, the separation of the tail from the fimbria and the consequent enlargement of the fimbriodentate sulcus are seen. The fimbria continues under the splenium of the corpus callosum as the crus of the fornix (see below), while the tail of the hippocampus continues as the subsplenial gyrus. The visible components of the hippocampal formation are given different names as they turn around the corpus callosum to continue as the supracommissural segment. The enlargement of the fimbriodentate sulcus also exposes other segments of the hippocampus that were not visible. The CA3 field, covered by a small layer of the alveus, appears as a whitish strip and is called the *gyrus fasciolaris*. The *margo denticularis* loses its toothlike appearance here and is named the *fasciola cinerea*. In some cases, the CA1 field rises to the surface of the parahippocampal gyrus, producing round elevations, and is called the gyri of Anders Retzius. The extension of the retrocommissural hippocampus, as it passes around the splenium, is also referred to as the retrosplenial gyrus, the subsplenial gyrus, and the eminence subcallosa [3].

The Subcortical Structures

The subcortical structures are the following: (1) the amygdala, (2) the septal area, (3) the hypothalamus, (4) the limbic nuclei of the dorsal thalamus, and (5) the epithalamus (habenular nuclei).

The Amygdala

The name of the amygdala (*corpus amygdaloideum* or amygdaloid body) was coined by Karl Friedrich Burdach (1776–1847) because of its almond shape (Figs. 5.16, 5.17, 5.18, 5.19, and 5.20). The amygdala is a large nuclear complex at the anterior wall and anterior roof of the temporal horn. It is composed of six main nuclei: the more superficial cortical, the medial, and the more deeply located lateral, basal, accessory basal, and central [1, 3, 5]. The amygdala is diffusely connected. Its main inputs originate from the olfactory bulb and olfactory cortex; the septal region, *substantia innominata*, and hypothalamus; the thalamus; the brainstem; the hippocampus; and the neocortex. The lateral amygdaloid nucleus is the largest and is the main recipient of sensory information from the neocortex. The cortical and medial amygdaloid nuclei receive input from the lateral olfactory stria, while the medial and central amygdaloid nuclei receive input from the hypothalamus and septal region. The medial, central, basal, and accessory basal receive input from the thalamus. Several portions of the brainstem connect with the central nucleus of the amygdala via ventral amygdala pathways (see below). The hippocampal afferents are mainly from the presubiculum and CA1 and terminate in the central, basal, and lateral nuclei. The main output of the amygdala is toward the septo-preoptic-hypothalamic continuum, the dorsal thalamus, the brainstem, the corpus striatum, and several cortical areas [1, 3–5].

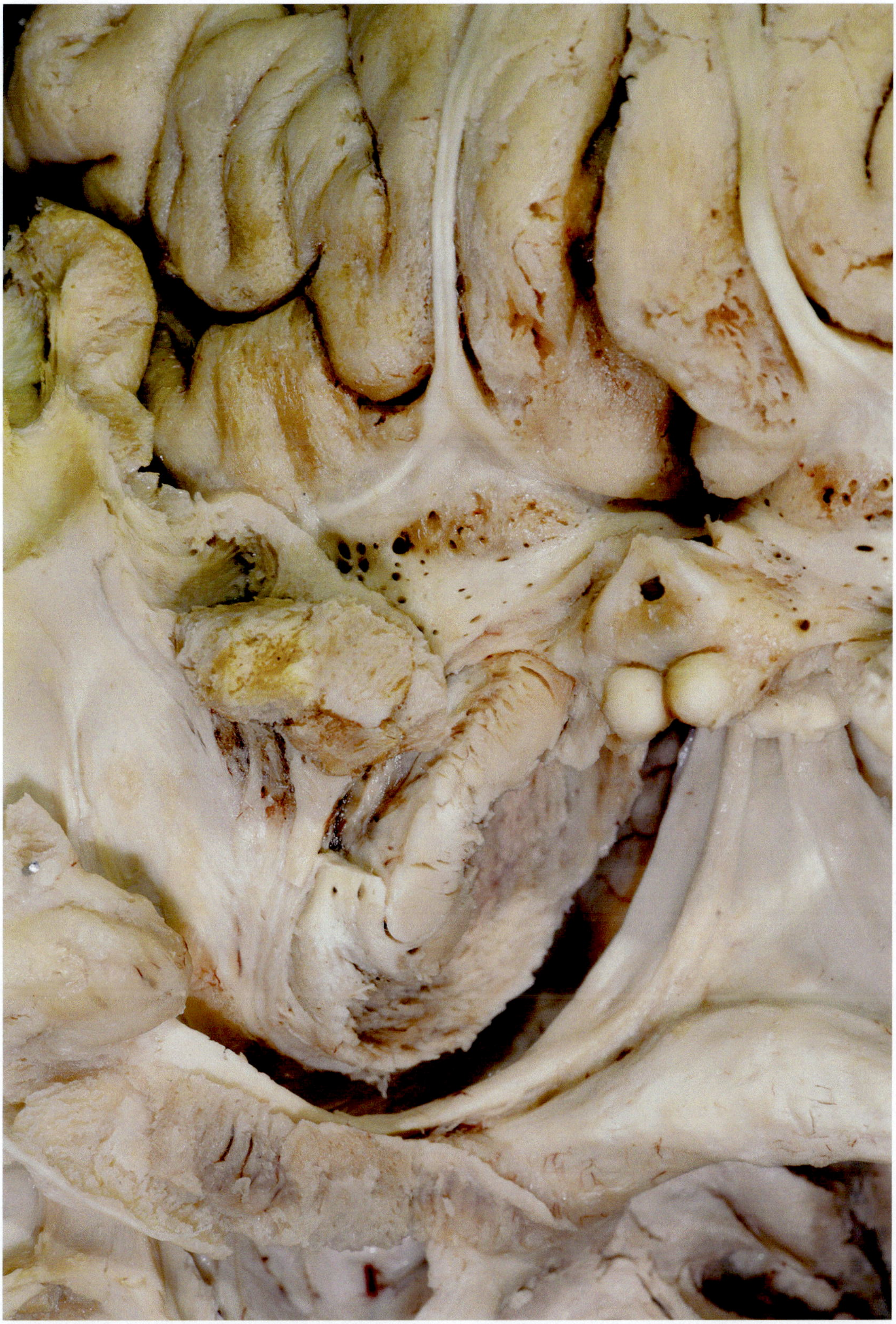

Fig. 5.16 Anatomical preparation of the amygdaloid complex, inferior view, after exposing the roof of the temporal horn of the lateral ventricles. The ependyma has been removed. The amygdaloid complex can be seen in the anterior wall and anterior region of the roof of the ventricle. The optic nerve, chiasm, and tract have been removed to expose the ventral connections of the amygdala and the hypothalamus. The ventral and dorsal connections of the amygdala are appreciated. *ant* anterior, *f* fasciculus, *g* gyrus, *n* nucleus, *s* sulcus, *st* stria, *subs* substance, *t* tract (©Kadri 2023. All Rights Reserved)

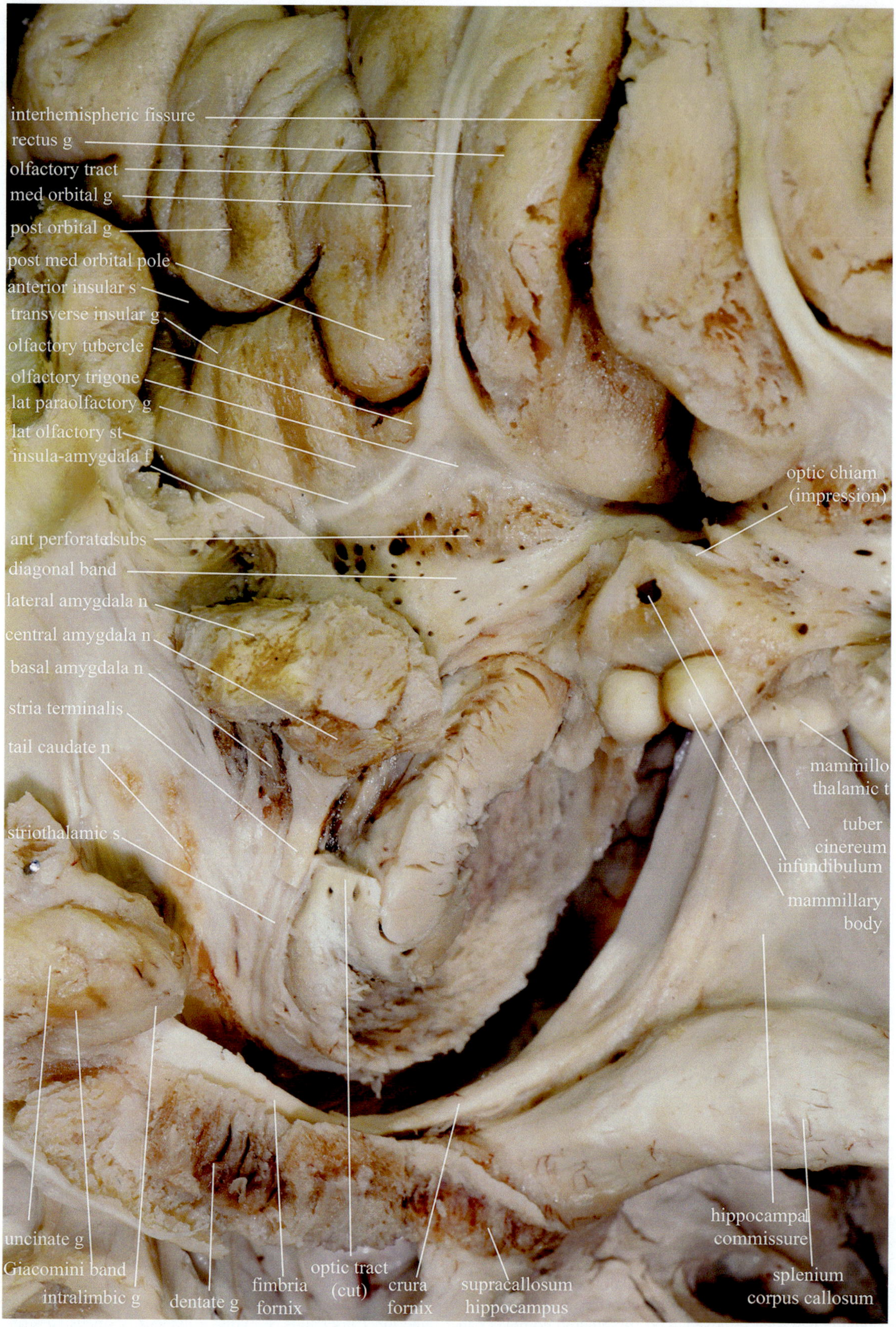

Fig. 5.16 (continued)

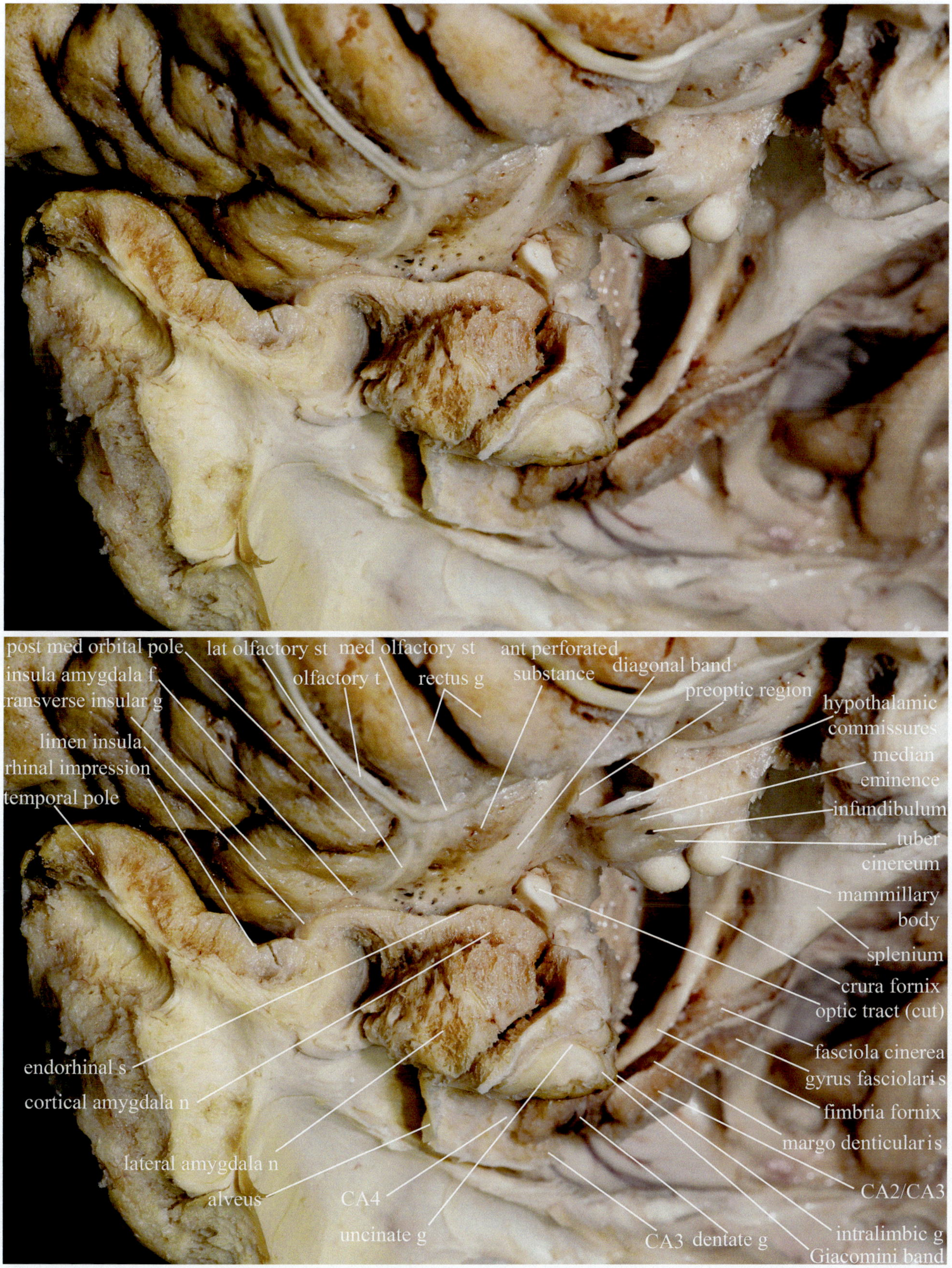

Fig. 5.17 Anatomical preparation of the amygdala, hippocampal formation, anterior perforated substance, and hypothalamus, inferior view. The optic apparatus has been removed. The contents of the floor (hippocampal formation) and anterior wall (amygdala) of the temporal horn are seen from below. The limits of the anterior perforated substance are visualized. *ant* anterior, *f* fasciculvus, *g* gyrus, *lat* lateral, *med* medial, *n* nucleus, *s* sulcus, *st* stria, *t* tract (©Kadri 2023. All Rights Reserved)

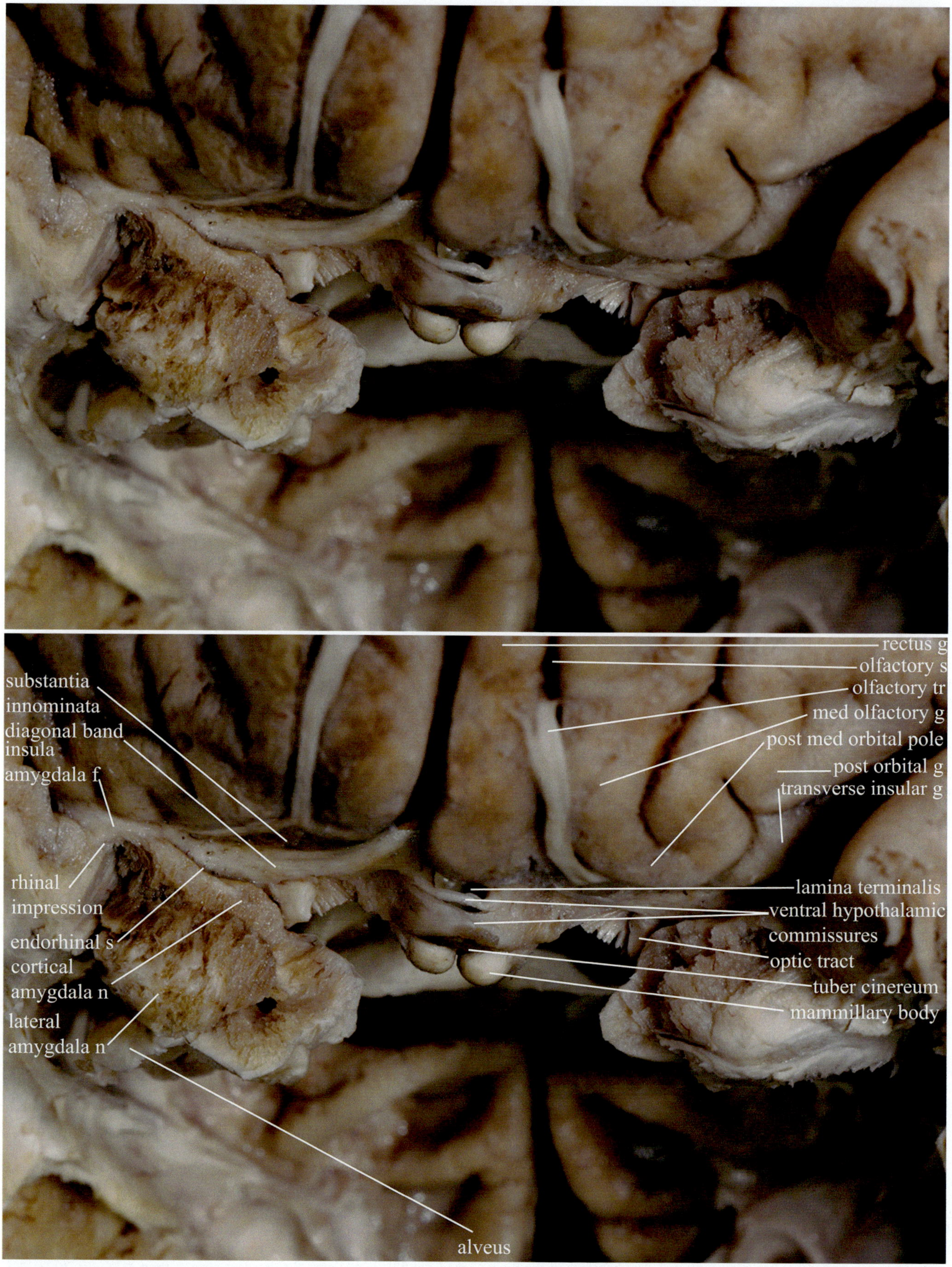

Fig. 5.18 Anatomical preparation of the amygdala, frontal view. Removing the optic apparatus exposes the ventral surface of the hypothalamus. *f* fasciculus, *g* gyrus, *med* medial, *n* nucleus, *post* posterior, *s* sulcus, *tr* tract (©Kadri 2023. All Rights Reserved)

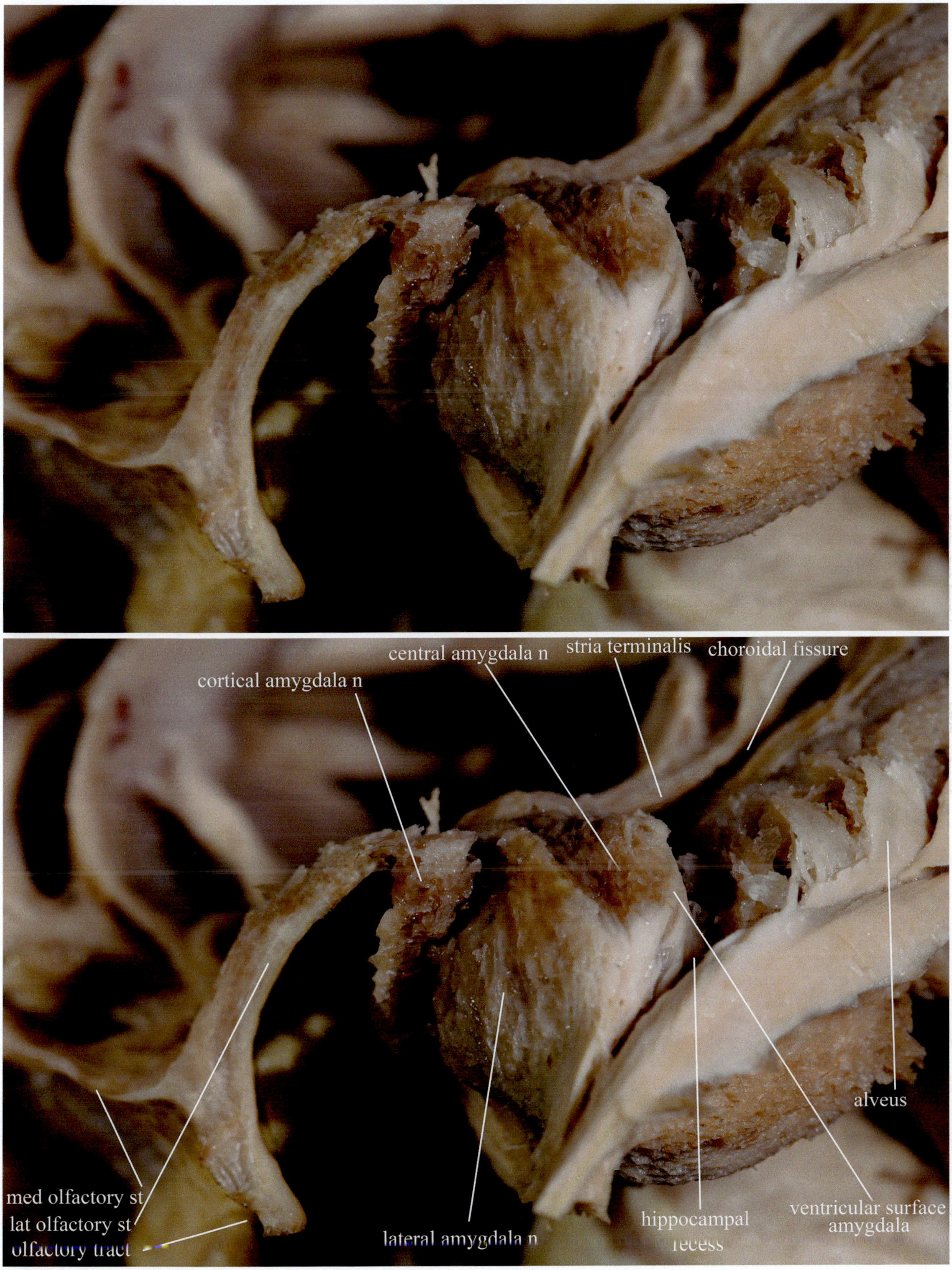

Fig. 5.19 Anatomical preparation of the amygdala and olfactory connections, lateral view, exposing the amygdala and hippocampal formation. The lateral olfactory stria extends to the cortical nucleus of the amygdala. *lat* lateral, *med* medial, *n* nucleus, *st* stria (©Kadri 2023. All Rights Reserved)

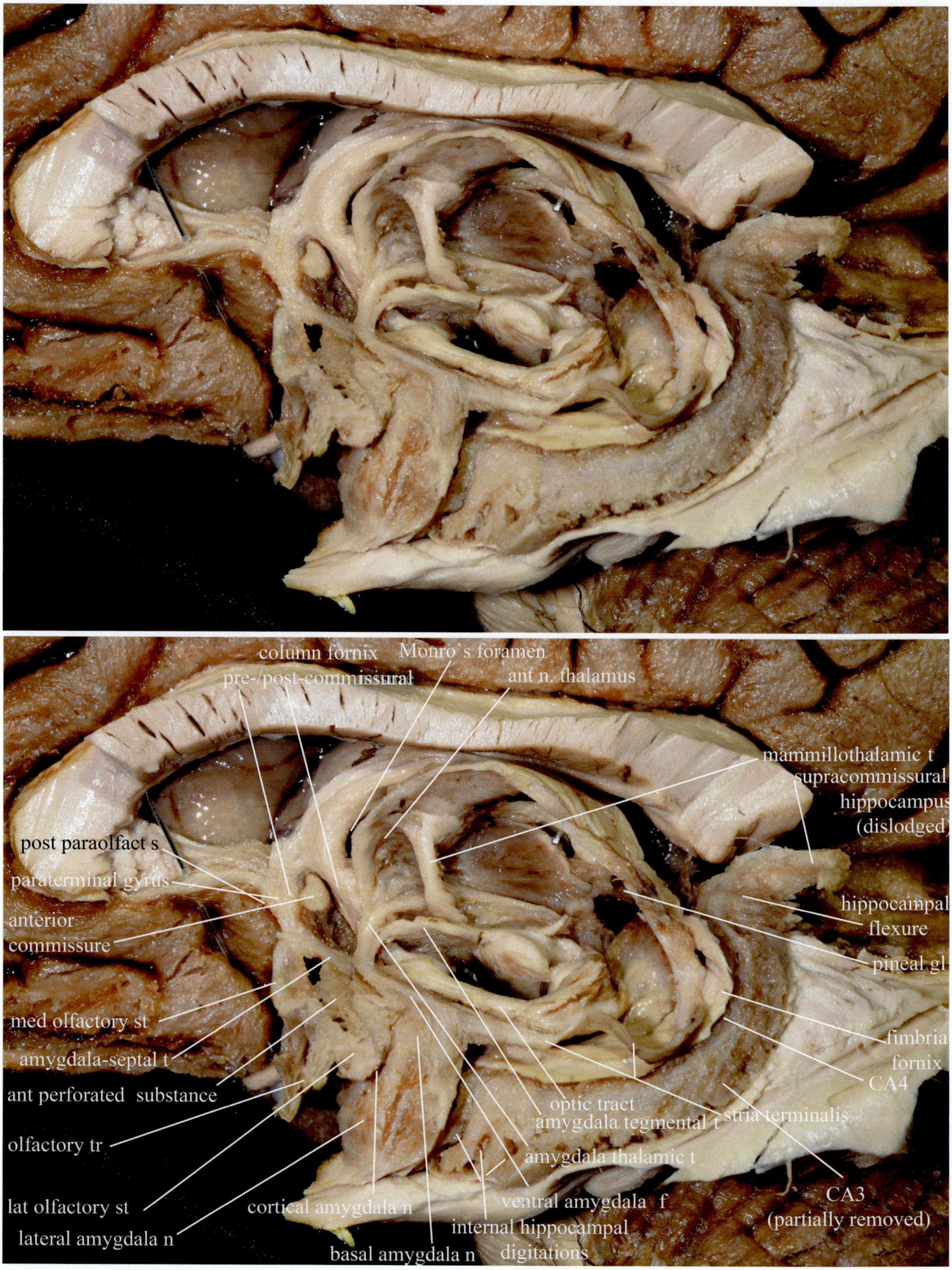

Fig. 5.20 Anatomical preparation of the limbic connections. The dorsal and ventral connections of the amygdala are exposed. *ant* anterior, *f* fasciculus, *gl* gland, *lat* lateral, *n* nucleus, *s* sulcus, *st* stria, *t* tract, *tr* tract (©Kadri 2023. All Rights Reserved)

The Septal Area

The septum is divided into dorsal and ventral areas. The dorsal area corresponds to the septum pellucidum. The ventral area is commonly referred to as the septal area or septum verum. On the interhemispheric surface, the septal area is represented by the paraterminal gyrus and contains two main nuclei, the lateral and medial septal nuclei. The septal nuclei form a relay between the hippocampal formation and the hypothalamus [1, 3–5] (Figs 5.21 and 5.22).

The Hypothalamus

The hippocampal formation connects diffusely with the hypothalamus via the fornix, but more abundantly with the mammillary bodies, which are prominent and characteristic eminences on the basal surface of the brain (Figs. 5.1,5.2, 5.16, 5.17, 5.18, and 5.23 and figures of the hypothalamus in Chap. 7). The mammillary bodies are located at the limit of the hypothalamic and mesencephalic surfaces of the floor of the third ventricle, posterior to the tuber cinereum. The mammillary body has medial, lateral, and posterior nuclei. These

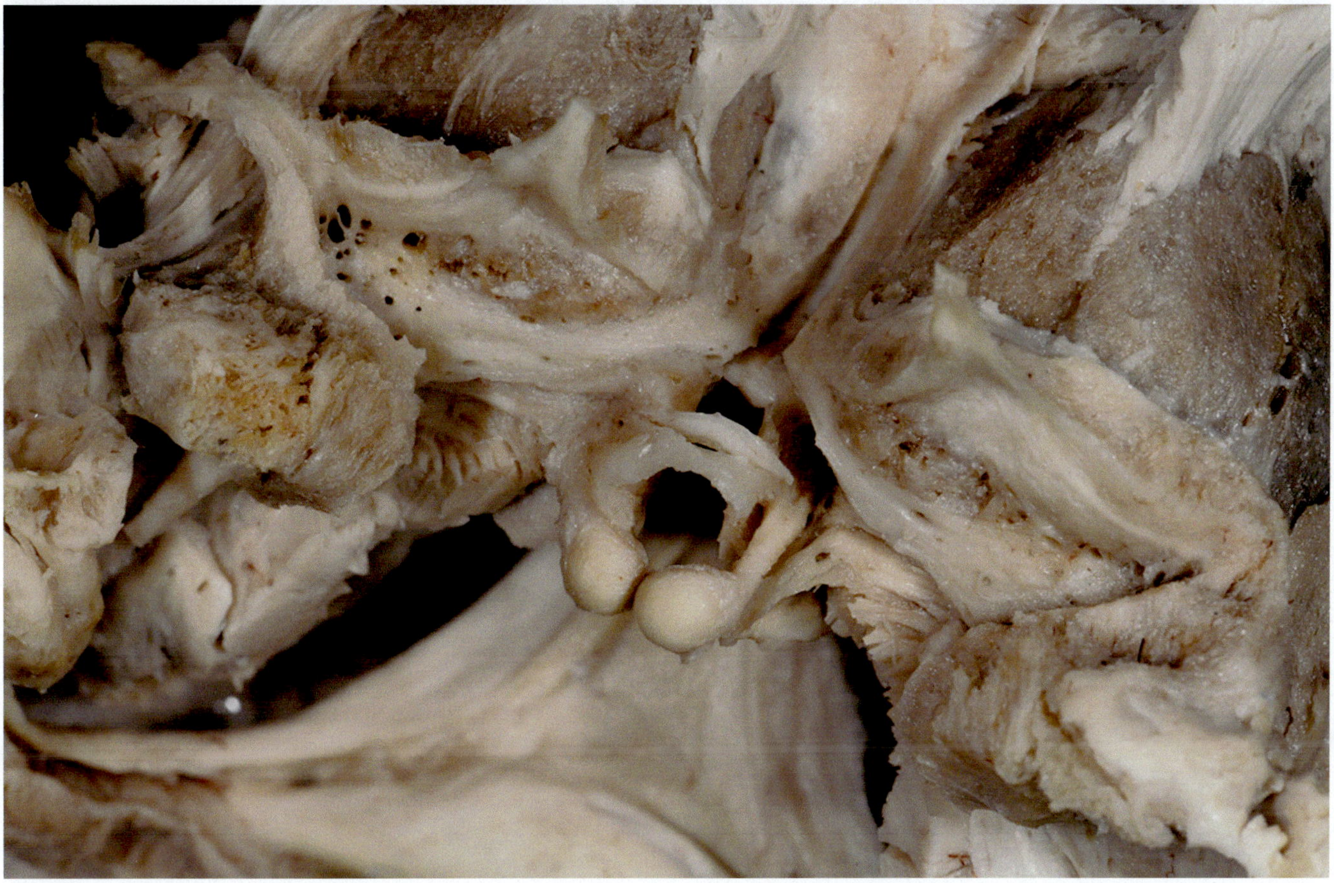

Fig. 5.21 Anatomical preparation of the anterior perforated substance. This structure is encircled by limbic structures. Its anterior limits are the olfactory striae. Its lateral limit is the limen insula. Its posterior lateral limit is the endorhinal sulcus. Its anteromedial limit is the interhemispheric fissure harboring the subcallosal area. Its posteromedial limit is the optic tract on the surface, but the structural arrangement extends to the border with the hypothalamus. *lat* lateral, *n* nucleus, *seg* segment, *str* stria (©Kadri 2023. All Rights Reserved)

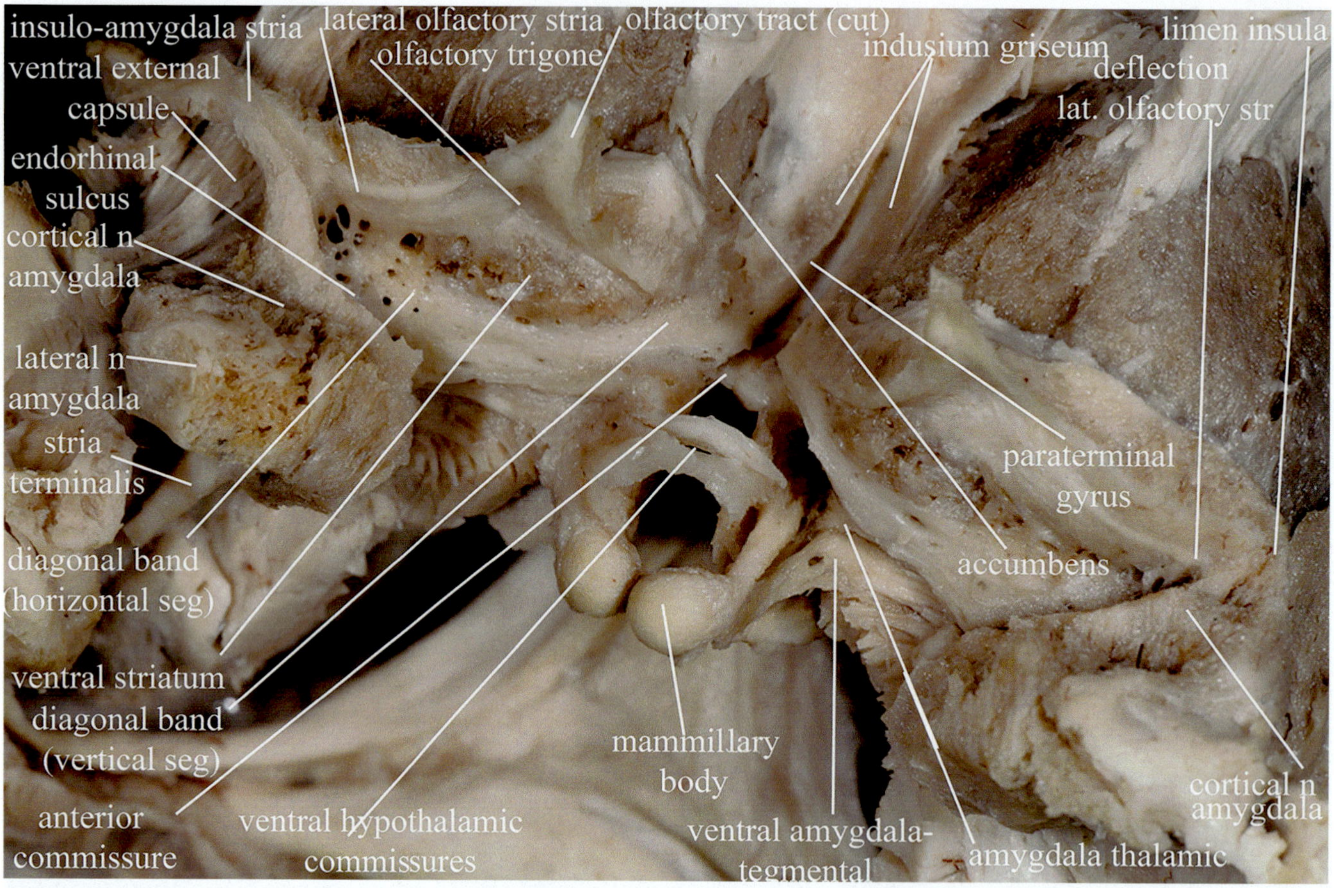

Fig. 5.21 (continued)

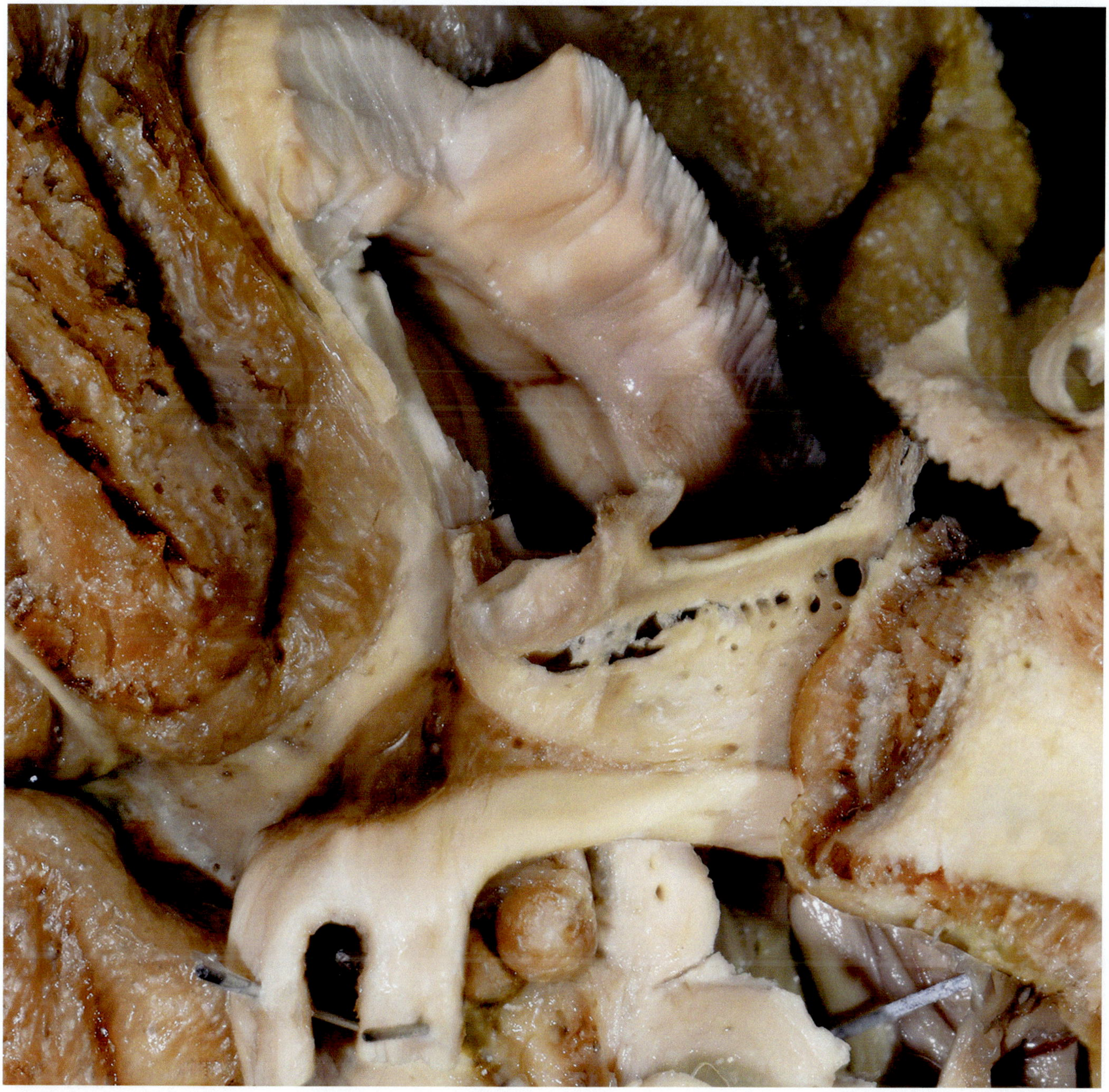

Fig. 5.22 Anatomical preparation of the septal region and anterior perforated substance. The anterior perforated substance is limited by the limen insula laterally, the lateral olfactory stria anterolaterally, the endorhinal sulcus posterolaterally, the optic tract posteromedially, and the anterior olfactory stria anteromedially. The vertical segment of the diagonal band connects to the paraterminal gyrus, which is white in color. The paraterminal gyrus is separated from the medial paraolfactory gyrus by the posterior paraolfactory sulcus. The medial paraolfactory gyrus is limited dorsally by the medial olfactory stria. The lateral paraolfactory gyrus is limited by the lateral olfactory stria. *g* gyrus, *n* nerve, *s* sulcus, *st* stria, *tr* tract (©Kadri 2023. All Rights Reserved)

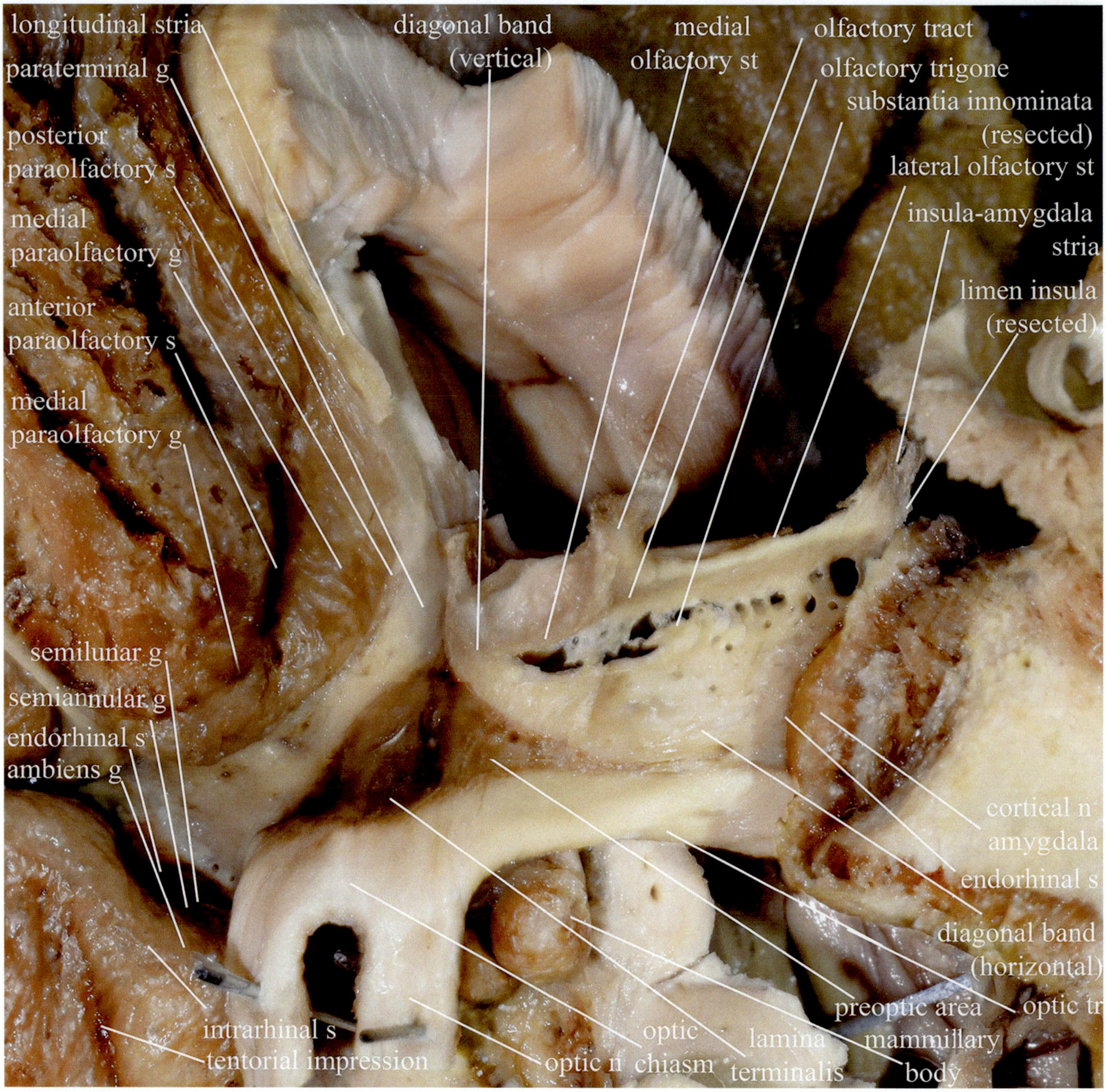

Fig. 5.22 (continued)

nuclei connect with the hippocampal formation via the fornix; with the amygdala via the ventral connections of the amygdala, known as amygdala-hypothalamic fibers; and with the tegmentum of the mesencephalon, ventrally via the mammillary peduncle and dorsally via the mammillotegmental fasciculus [1, 4, 5].

The Limbic Nuclei of the Thalamus

The limbic nuclei of the thalamus comprise the anterior nuclear complex and the medial dorsal nuclei (Figs. 5.20, 5.23, 5.24, 5.24, 5.26, and 5.27 and figures in Chap. 7). The anterior nuclear complex is one of the five major nuclear complexes of the thalamus, and its main connection is the mammillothalamic tract. It consists of the anteromedial, anterodorsal, and anteroventral nuclei. The anterior nuclei are related to the anterior tubercle of the thalamus, which is the most anterior limit of the thalamus and is seen as a protuberance on the posterior wall of the foramen of Monro [1, 4, 5].

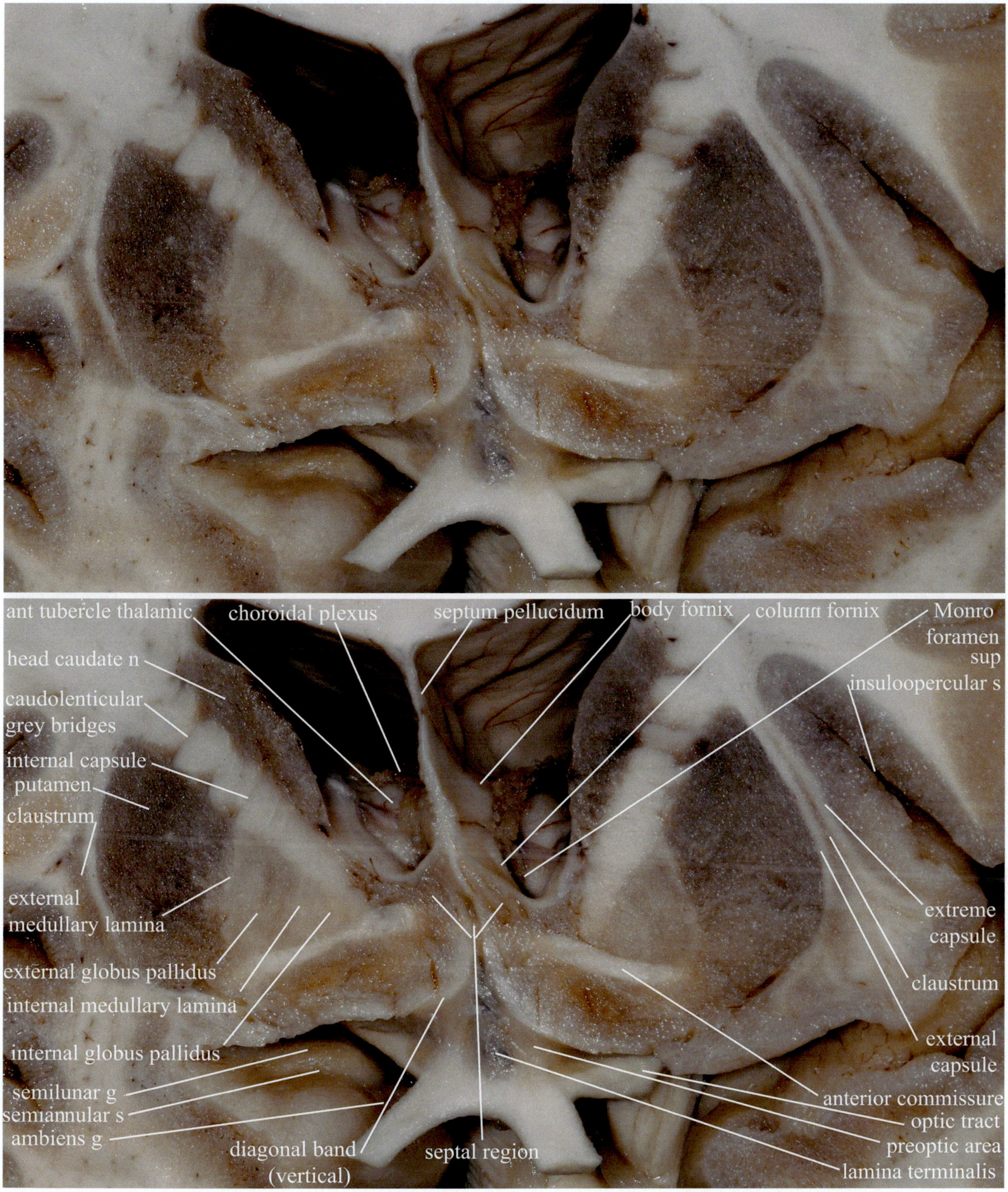

Fig. 5.23 Coronal cut at the level of the anterior commissure. On the surface of the interhemispheric fissure, the septal region is related to the paraterminal gyrus. In coronal cuts, the septal nuclei are located anterior and superior to the pars compacta of the anterior commissure. The septal nuclei are limited laterally by the ventral internal capsule and posteriorly by the column of the fornix on the medial side and the bed nucleus of the stria terminalis on the lateral side. The amygdalo-septal fibers embedded in the vertical segment of the diagonal band separate the paraterminal gyrus on the surface and the septal nuclei at a depth from the preoptic area. The septal-habenular and the septal-tegmental tracts separate the septal nuclei from the anterior hypothalamus. These fibers connect to the limbic midline continuum of the diencephalon and tegmentum of the brainstem. *ant* anterior, *g* gyrus, *n* nucleus, *s* sulcus, *sup* superior (©Kadri 2023. All Rights Reserved)

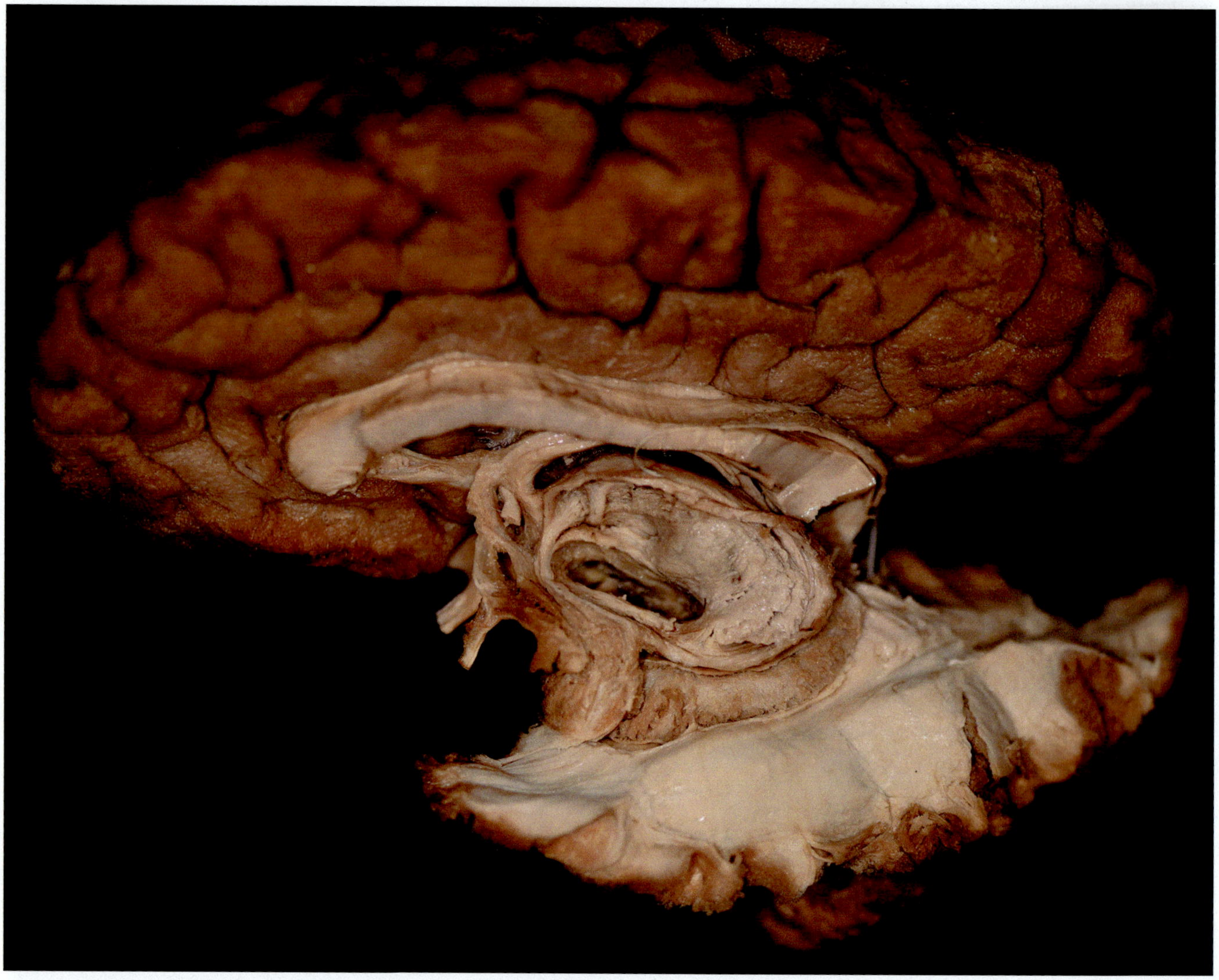

Fig. 5.24 Anatomical preparation of the telencephalic and diencephalic limbic connections. The projection fibers of the internal capsule and cerebral peduncle have been removed, and the capsular surface of the thalamus is exposed. The capsular surface is mainly related to the lateral nuclei, the ventral surface of the posterior nuclei, and the so-called metathalamus or geniculate bodies. The stria terminalis extends from the superior aspect of the central and lateral nuclei of the amygdala and inferior to the basal nucleus and runs in the striothalamic sulcus. The anterior commissure was cut, and the septal nuclei around it were removed. The white matter connection of the septal region is not exclusive but concentrates on the cortical surface and gives the characteristic whitish appearance of the paraterminal gyrus in the subcallosal area. In the paraterminal gyrus, the diagonal band, the precommissural fornix, and the stria terminalis seem to extend. The fibers of the stria terminalis that pass ventral to the anterior commissure, to the septal region, are lateral to the former; therefore, they extend to the lateral septal nuclei. The ventral fibers of the amygdala are directed to the septal region, hypothalamus, thalamus, and tegmentum of the brainstem. In the anterior perforated substance, these fibers are located on the posterolateral floor and appear on the surface as the horizontal segment of the diagonal band. The amygdalo-septal fibers seem to extend to the paraterminal gyrus and, on the surface, are represented by the vertical segment of the diagonal band. *ant* anterior, *f* fasciculus, *g* gyrus, *lat* lateral, *med* medial, *n* nucleus, *st* stria (©Kadri 2023. All Rights Reserved)

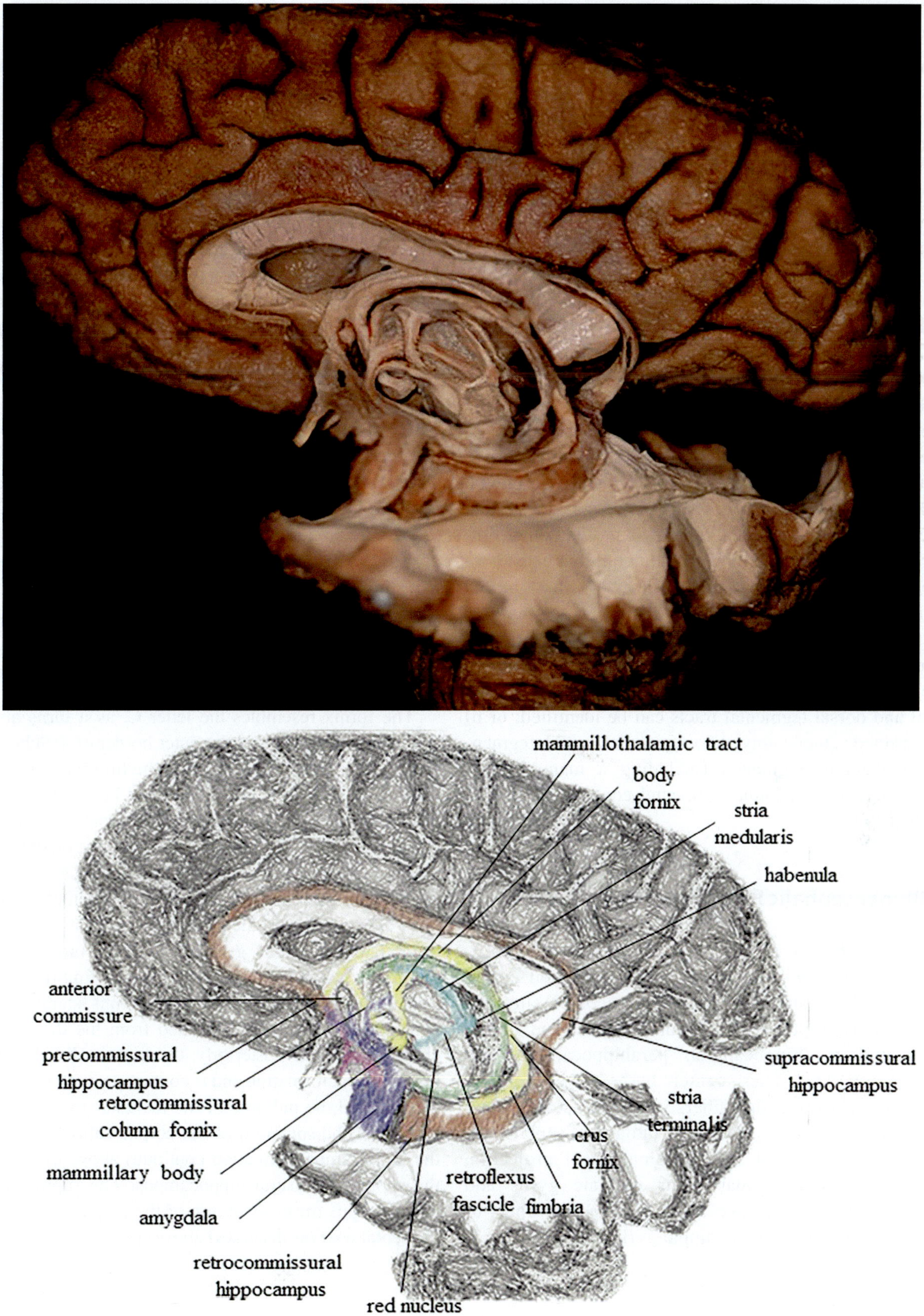

Fig. 5.27 Anatomical preparation and illustration of the left telencephalic and diencephalic portions of the limbic system. The multiple integrational circuits of the limbic system connect the cortical structures of the paleocortex (the olfactory areas among others), archicortex (the hippocampal formation and entorhinal area), and neocortex (the prefrontal cortex among others) directly or through a relay with the basal ganglia related to the limbic system. These are the amygdala and the septal region, with diencephalic structures and autonomic centers in the midline, along with the dorsal and ventral tegmental areas of the brainstem. (©Kadri 2023. All Rights Reserved)

The Habenula

The habenular complex includes medial and lateral nuclei and is an epithalamic structure located in the posterior third ventricle (Figs. 5.28, 5.29, 5.30, and 5.31). Its main afferent is from the septal area and hypothalamus via the stria medullaris. Its main efferent is the habenulo-interpeduncular tract (retroflexus fasciculus of Meynert), which travels on the medial surface of the red nucleus and terminates mainly in the interpeduncular nucleus.

The Fiber System of the Limbic System

The limbic system extends from the telencephalon to the spinal cord [1, 4, 5]. The telencephalic connections can be grouped into the olfactory connections and the hippocampal connections. The circuits encompass different nuclei of the amygdala and the septal region to connect the telencephalon and the hypothalamus. The circuits connecting the limbic telencephalic structures to the hypothalamus and other limbic areas of the thalamus form a continuum that extends to the tegmentum of the brainstem. Collectively, these limbic prosencephalic (telencephalic and diencephalic) connections are termed the prosencephalic tract. Segmentation of the prosencephalic tract can be based on two premises: a) the relation of its components to the brainstem tegmentum; ventral and dorsal tegmental tracts can be identified; or b) the connected centers; amygdalo-tegmental, septal–tegmental, hypothalamic–tegmental (including a mammillotegmental tract), and thalamic–tegmental connections can be identified.

The Rhinencephalic Projection Fibers

Sir William Turner (1832–1916), a British anatomist, coined the term *rhinencephalon* for the cerebral cortex areas related to olfactory function, encompassing the olfactory lobe, septal region, dentate gyrus, and limbic circumvolutions (hippocampal, parahippocampal, and cingulate gyri) [11]. These centers harbor several diffuse projections: (1) projection fibers of the olfactory lobule and septal region—the olfactory radiation, septal–thalamic connections, and stria terminalis; (2) projection fibers of the hippocampal gyrus (dentate gyrus and Ammon's horn)—the fornix; and (3) projection fibers of the cingulate and parahippocampal gyri [12].

The Olfactory Circuits

The olfactory tract connects the olfactory bulb and the olfactory tubercle (Figs. 5.5, 5.7, 5.8, 5.16, 5.17, 5.19, 5.20, 5.22, 5.24, 5.32, 5.33, and 5.34). There is no consensus on the extension of the olfactory tubercle in humans, but it seems to be continuous with the nucleus accumbens and the ventral striatum [5, 6]. The olfactory tract splits at the anterior border of the anterior perforated substance into medial and lateral olfactory striae. A small depression at the angle in the origin of the olfactory striae is termed the olfactory trigone. The medial olfactory stria extends toward the subcallosal area. The lateral olfactory stria extends into the semilunar gyrus, which consists mainly of the cortical nuclei of the amygdala [10, 13].

The Fornix

The fornix is the main efferent of the hippocampus and contains projections and commissural fibers (Figs. 5.20, 5.25, 5.26, 5.31, 5.35, 5.36, and 5.37) [3, 12]. It is also referred to as the ventricular dome of Winslow (Jacques-Bénigne Winslow, 1669–1760) [13]. It is located at the roof of the third ventricle and connects the hippocampal formation with the ipsilateral and contralateral septal nuclei, the anterior nuclei of the thalamus, the nuclei of the stria terminalis, and the mammillary bodies [12, 13].

The fornix resembles the letter C, as it turns around the thalamus and constitutes the outer border of the choroidal fissure. It is divided into four parts: the fimbria, crus, body, and column. At the floor of the temporal horn, the axons of the hippocampal formation cover its ventricular surface and are called the *alveus*. The alveus fibers converge on the superomedial border of the hippocampus and form the fimbria of the fornix. The fimbria is separated from the dentate gyrus by the fimbrodentate sulcus, the most anterior point of which is posterior to the intralimbic gyrus of the extraventricular surface of the hippocampal head. At this point, the fimbria is the upper limit of the inferior choroidal point. It runs lateral to the lateral geniculate body and is separated from the thalamus by the choroidal fissure. Posteriorly, the fimbria detaches from the hippocampal formation and is continuous with the crus, wrapping around the pulvinar of the thalamus to join the inferior surface of the splenium of the corpus callosum. Alternately, the hippocampal formation continues above the splenium as the supracommissural hippocampus. The supracommissural hippocampus traverses on top of the corpus callosum toward the septal area (as discussed above) [3, 13–16].

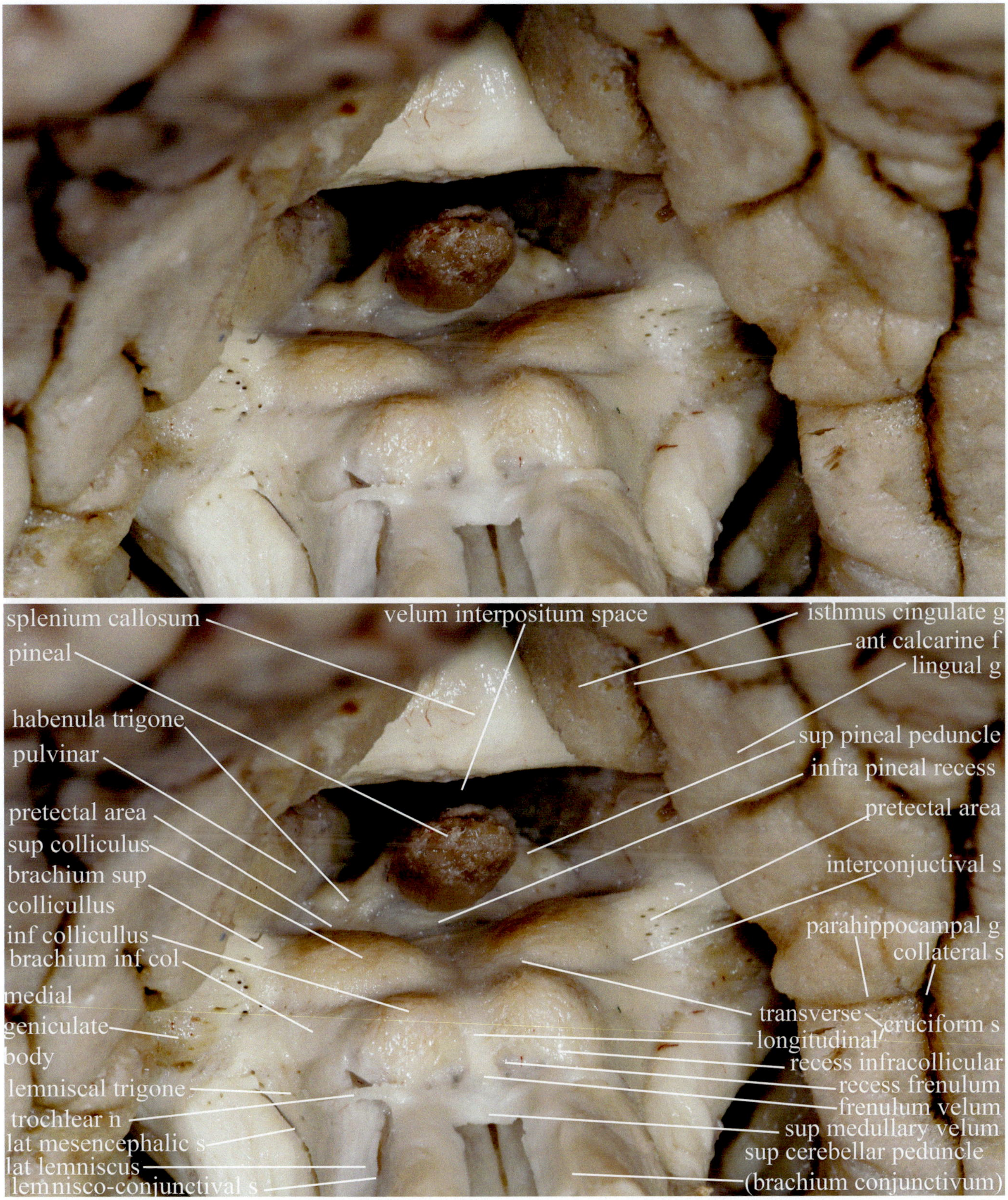

Fig. 5.28 Anatomical preparation of the epithalamus and quadrigeminal plate. The epithalamus is composed of the pineal gland and the habenular nuclei and its connections. The posterior commissure is the caudal limit of the epithalamus. The posterior commissure is composed of, in part, the interthalamic connections. The quadrigeminal plate is the posterior surface of the mesencephalic tectum. *col* colliculus, *f* fasciculus, *g* gyrus, *inf* inferior, *lat* lateral, *n* nerve, *s* sulcus, *sup* superior (©Kadri 2023. All Rights Reserved)

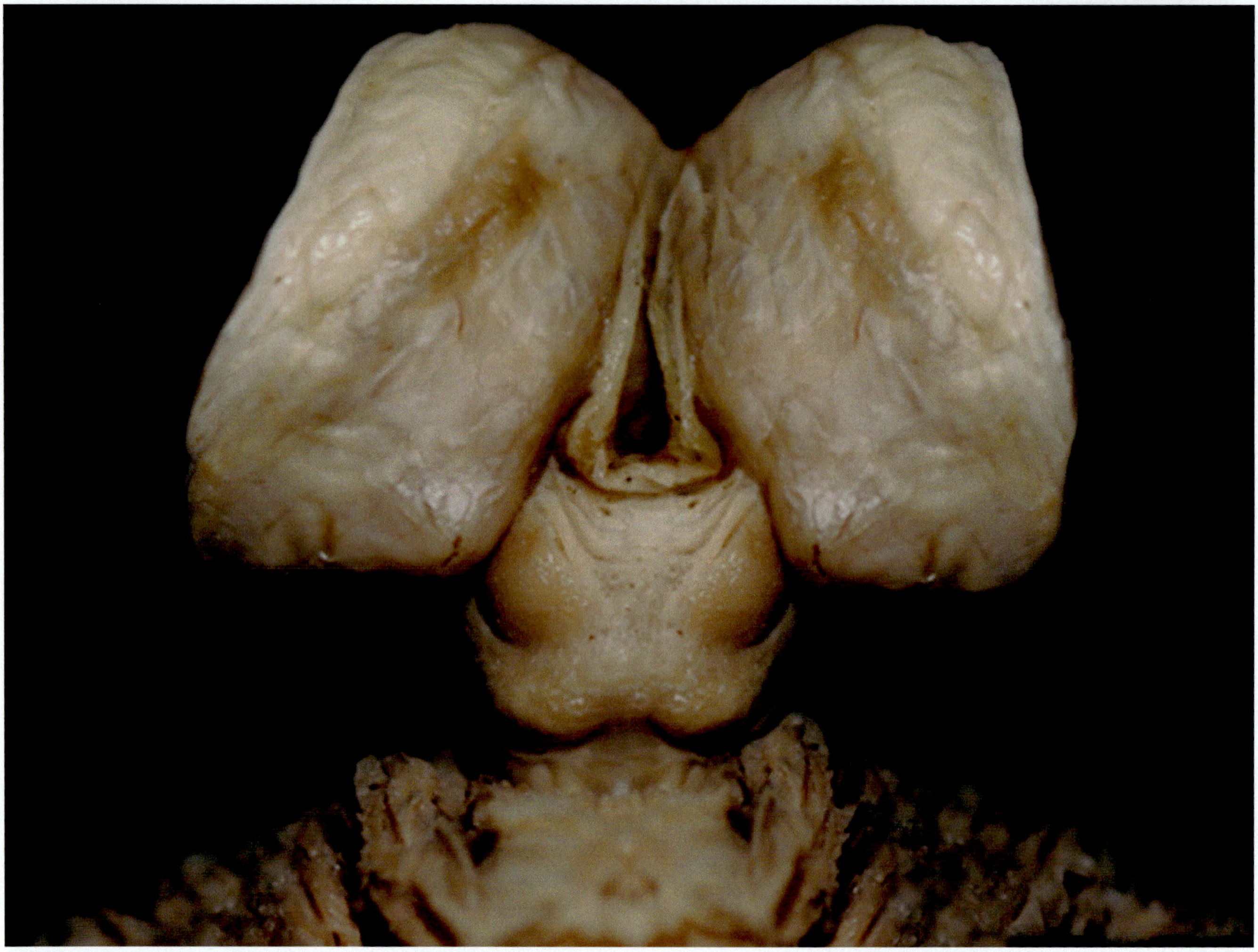

Fig. 5.29 Anatomical preparation of the cisternal surface of epithalamus and quadrigeminal plate. The pineal gland has been removed, and the posterior third ventricle, the stria medullaris, and the habenular commissure are exposed. The superior and inferior colliculi are demarcated by the longitudinal and transverse segments of the cruciform sulcus. *inf* inferior, *n* nucleus, *s* sulcus, *seg* segment (©Kadri 2023. All Rights Reserved)

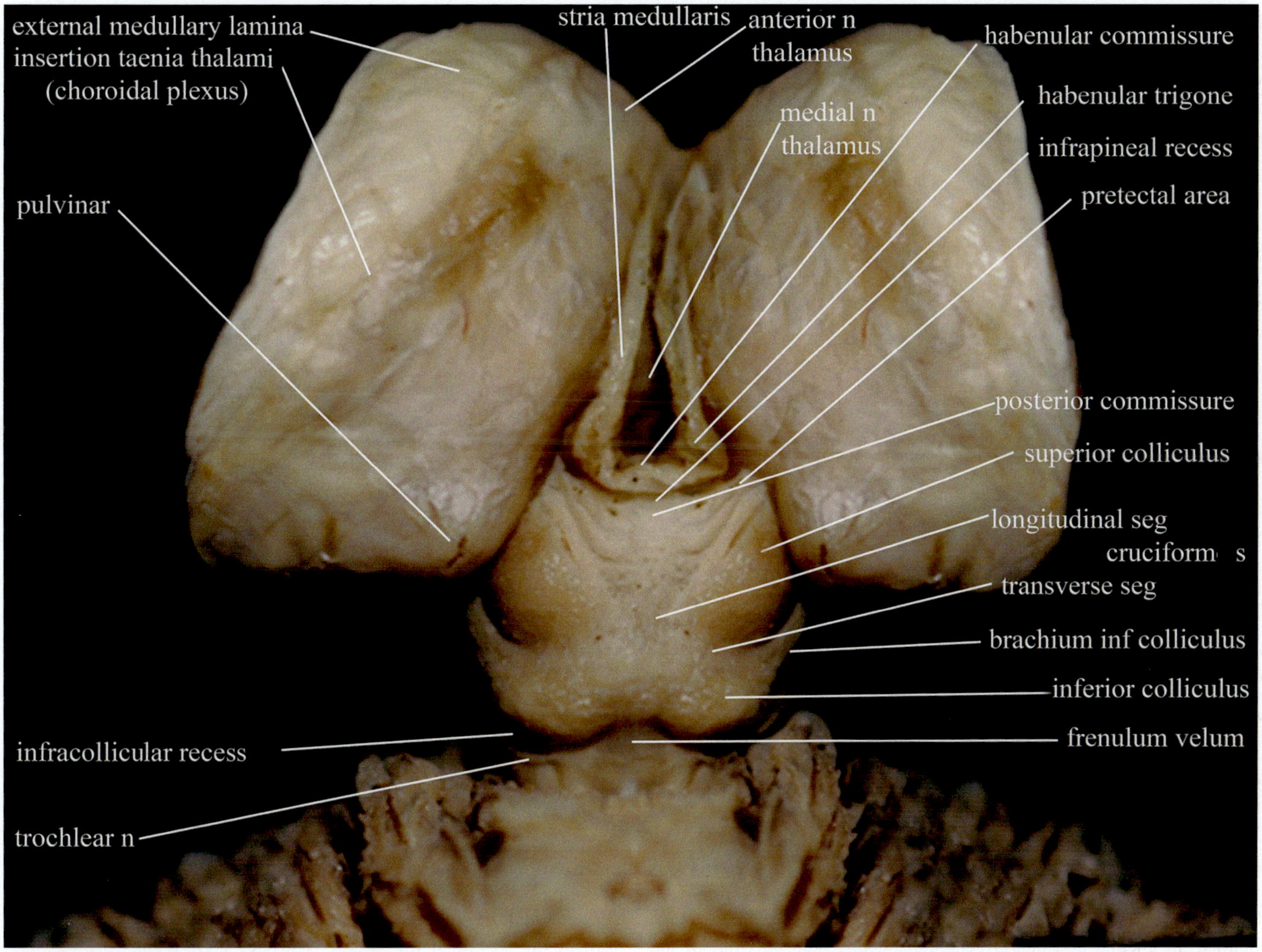

Fig. 5.29 (continued)

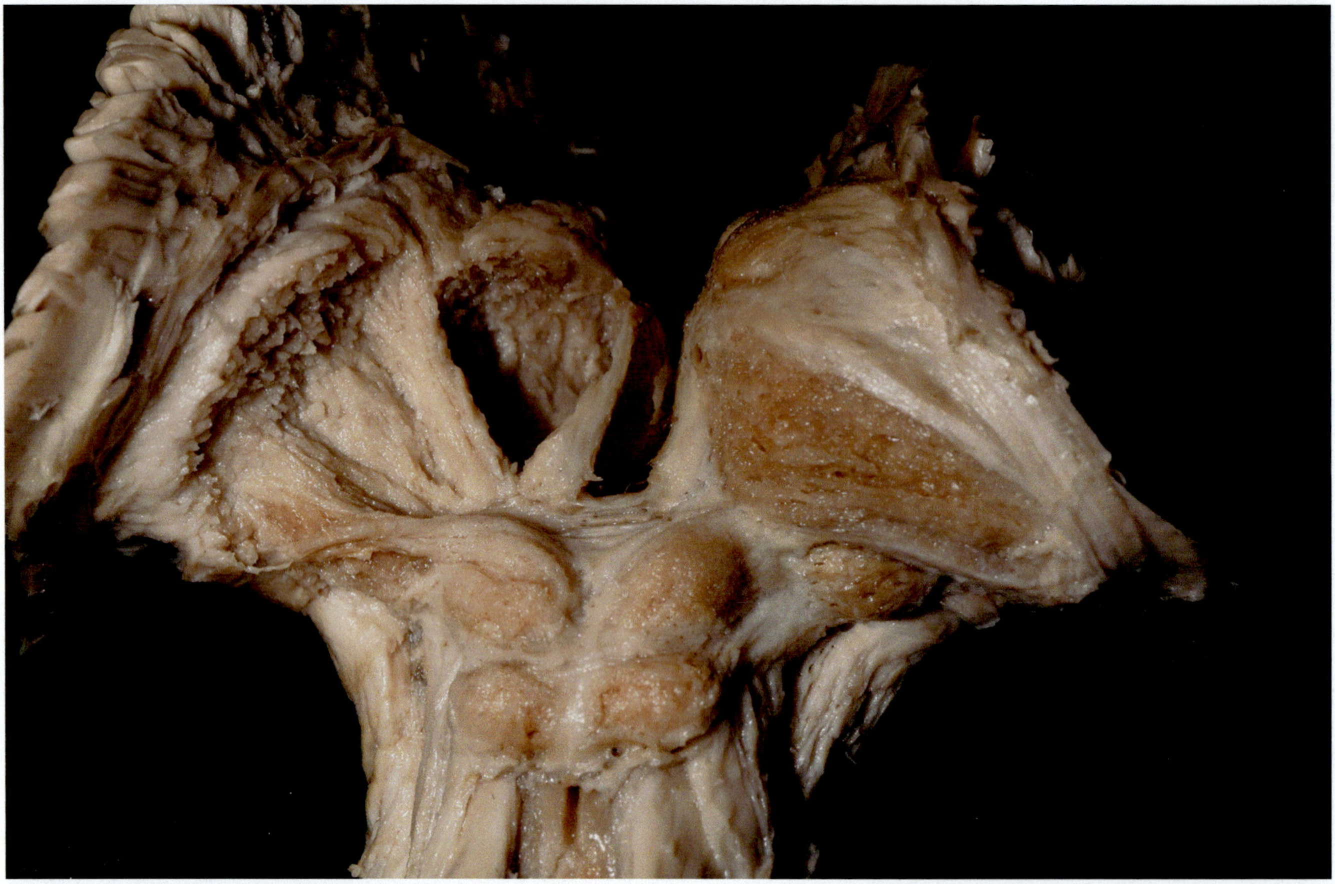

Fig. 5.30 Anatomical preparation of the posterior surface of the epithalamus and quadrigeminal plate. The pineal gland has been removed. The pial surface and, on the left side, the brachium of the inferior colliculus, the medial and lateral geniculate bodies, the pulvinar of the thalamus, and the medial dorsal thalamic nuclei have been removed. The internal medullary lamina of the thalamus is exposed. The stria medullaris is formed by connections of the habenular nuclei with the septal, preoptic, and anterior thalamic regions. The pulvinar is the posterior portion of the lateral dorsal thalamus. The pulvinar is also the anterior limit of the lateral extension of the quadrigeminal cistern (cisternal surface), the lateral inferior limit of the velum interpositum (velar surface), and the small part of the lateral floor of the body and anterior wall of the atrium of the lateral ventricles (ventricular surface). The brachium of the superior colliculus is hidden by the pulvinar. Removing the pulvinar exposes the superior brachium and its limits between the middle and lateral geniculate bodies. *ant* anterior, *b* body, *inf* inferior, *lat* lateral, *med* medial, *p* peduncle, *s* sulcus, *sup* superior, *tr* tract (©Kadri 2023. All Rights Reserved)

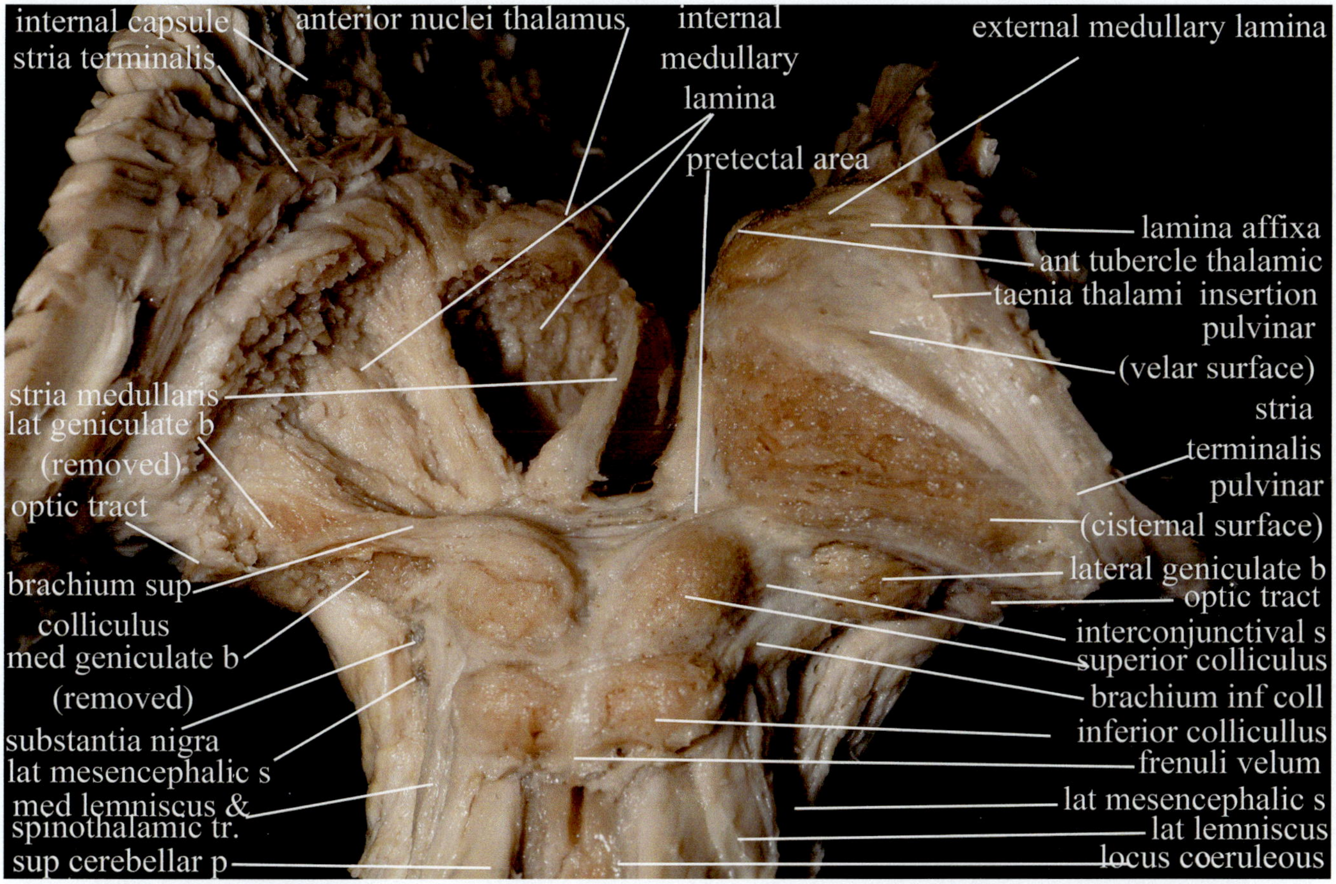

Fig. 5.30 (continued)

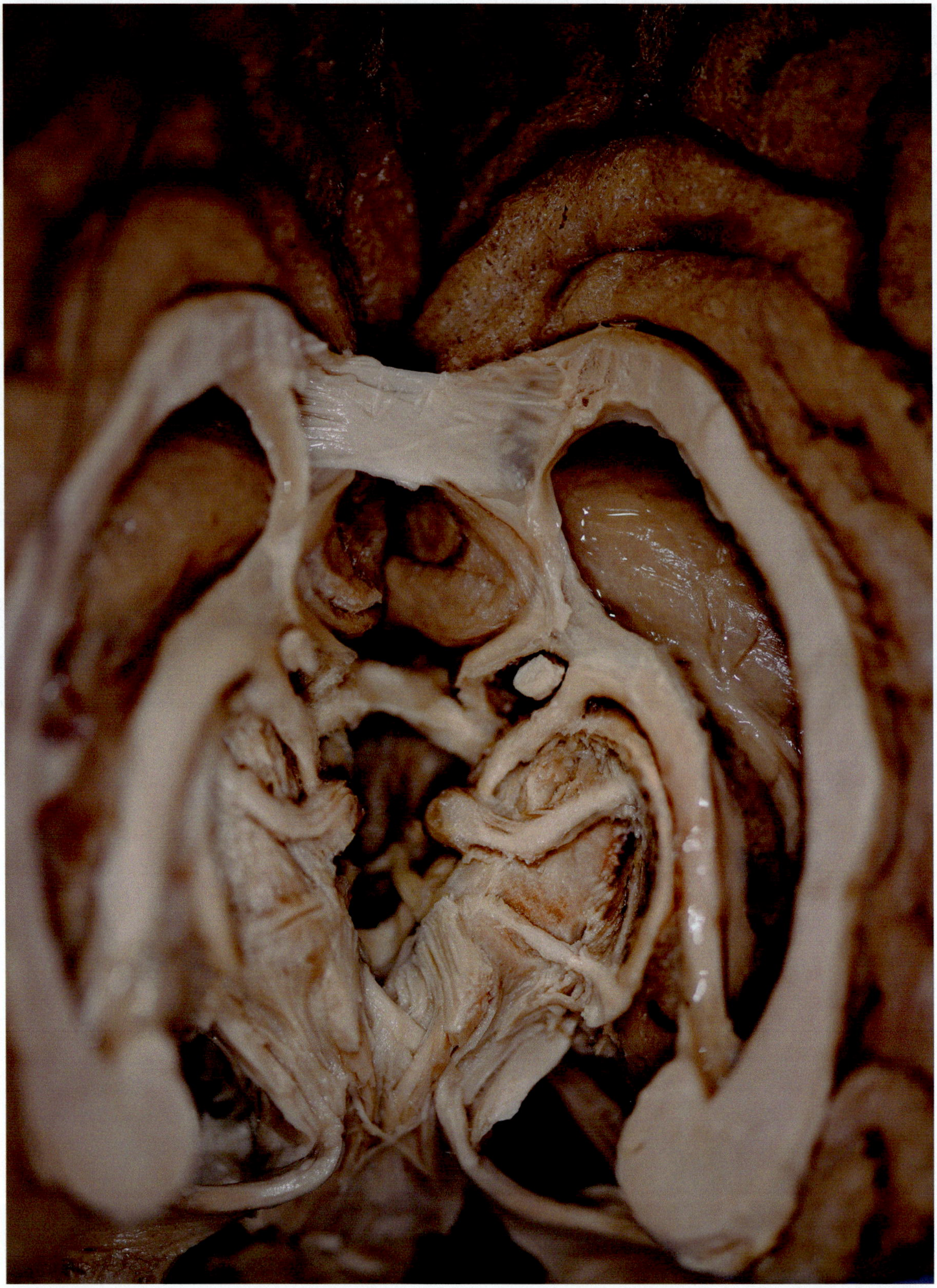

Fig. 5.31 Anatomical preparation from a dissection, superior view, of the subcortical diencephalic and mesencephalic limbic connections. The habenular nuclei are the main neuronal components of the epithalamus. The habenular nuclei connect to the interpeduncular nuclei via the retroflexus fasciculus. This fasciculus is also termed the habenulo-interpeduncular fasciculus and connects the habenular nuclei and the interpeduncular nucleus. The ventral connections of the habenular nuclei are mainly from the stria medullaris, but also connect with the amygdala, septal region, and hypothalamus. These fibers are part of the medial prosencephalic bundle, a limbic connection between the telencephalic and the diencephalic nuclei, and telencephalic and tegmental nuclei. The habenular connections via the medial prosencephalic bundle are the most lateral. *f* foramen, *g* gyrus, *n* nucleus, *pl* plexus, *s* sulcus (©Kadri 2023. All Rights Reserved)

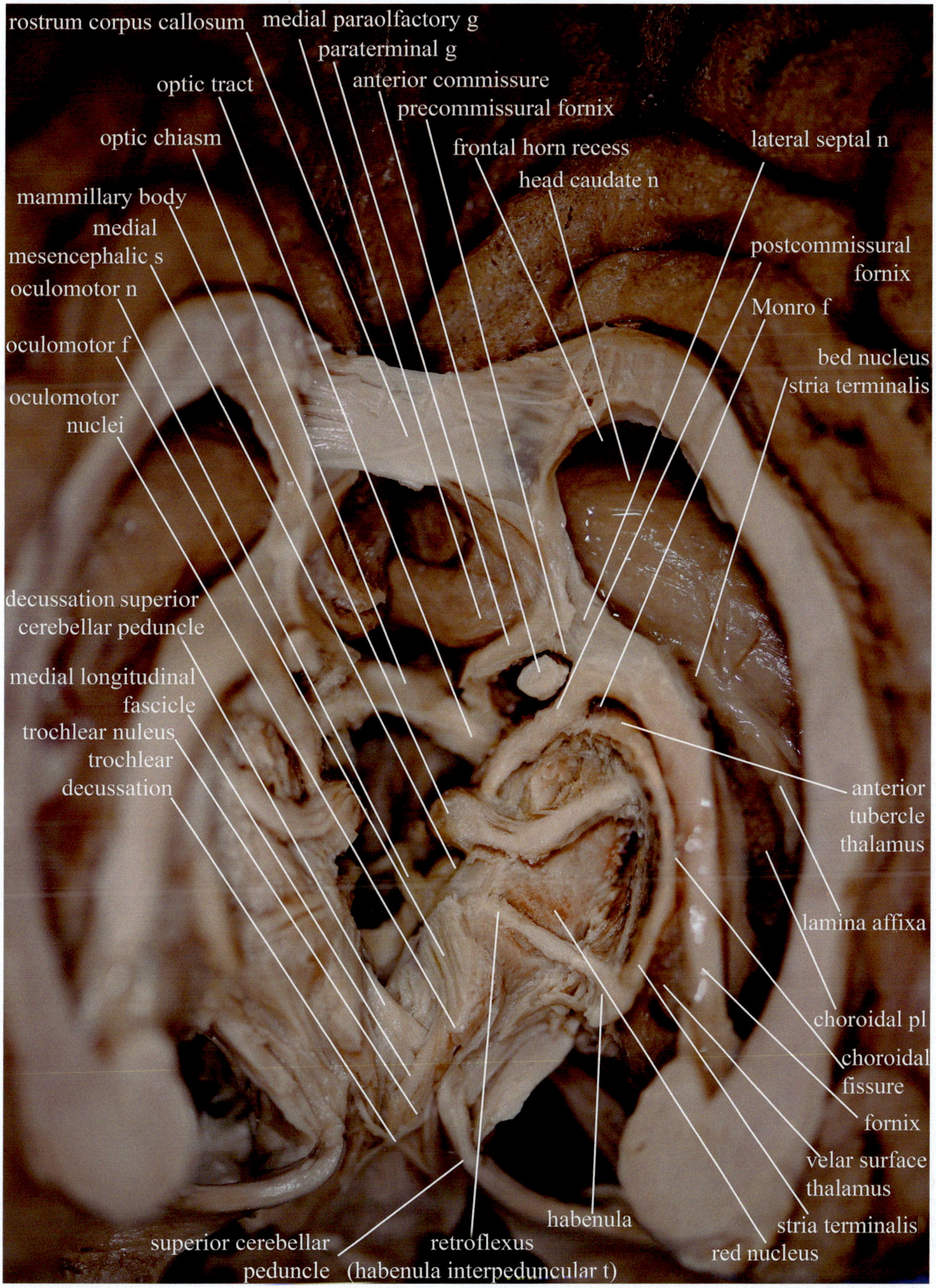

Fig. 5.31 (continued)

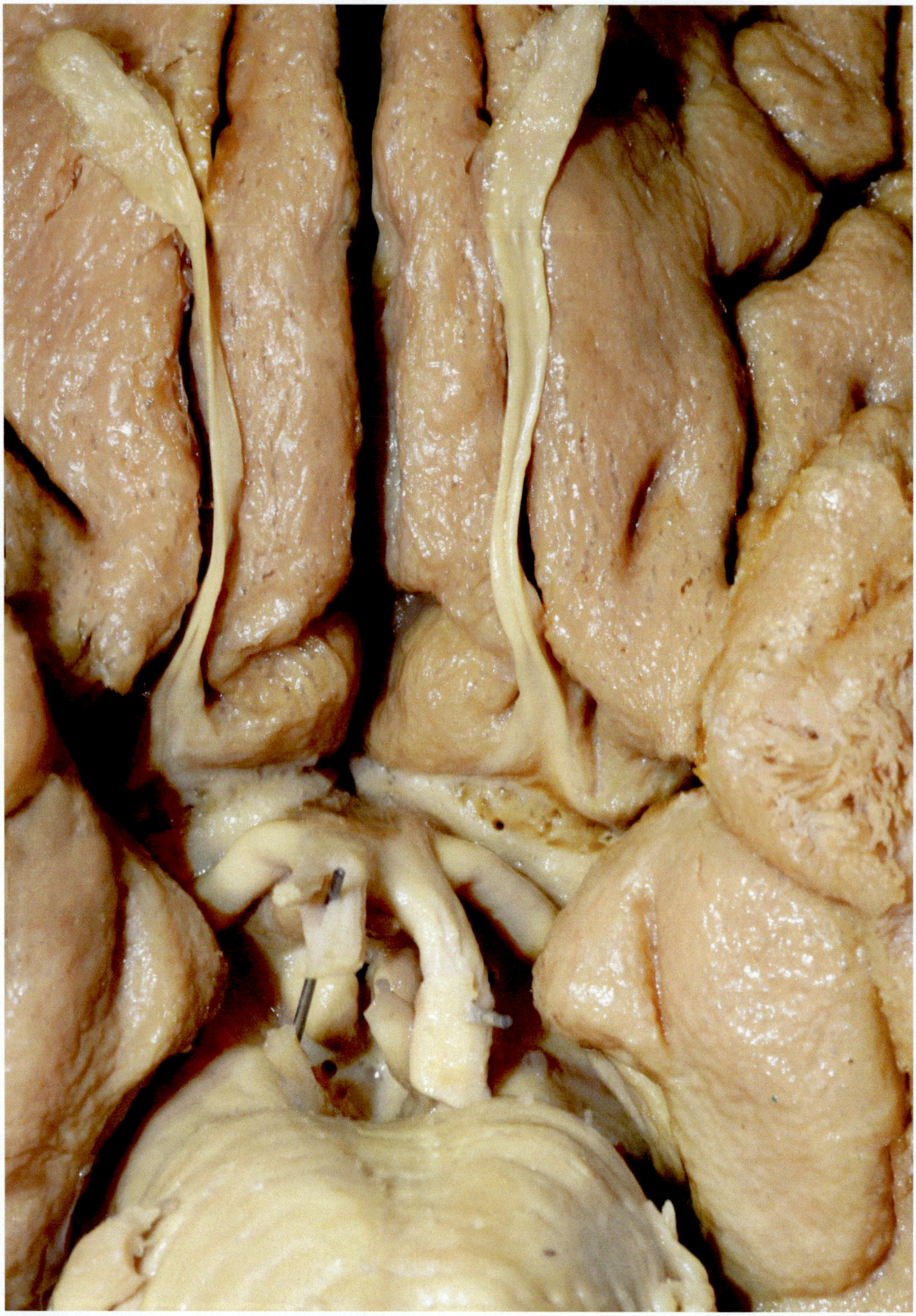

Fig. 5.32 Anatomical preparation of the olfactory region. The olfactory bulb receives input from the olfactory nerve fibers and projects through the olfactory tract that covers the olfactory sulcus. The olfactory sulcus is a deep sulcus that separates the inferior surface of the rectus gyrus from the medial orbital gyrus. At the rostral end of the olfactory sulcus, at the surface, the olfactory tract separates into medial and lateral strips—the medial and lateral olfactory striae. With the substantia innominata, these structures delineate a flat, sometimes depressed triangular area called the olfactory trigone. The medial olfactory stria is shorter and separates the medial paraolfactory gyrus from the substantia innominata. The lateral olfactory stria is longer and extends to the limen insula. It separates the lateral paraolfactory gyrus and the transverse gyrus of the insula from the substantia innominata of the anterior perforated substance. The lateral olfactory stria fuses with the lateral horizontal segment of the diagonal band and limits the anterior perforated substance and the limen insula. *g* gyrus, *lat* lateral, *med* medial, *post* posterior, *s* sulcus (©Kadri 2023. All Rights Reserved)

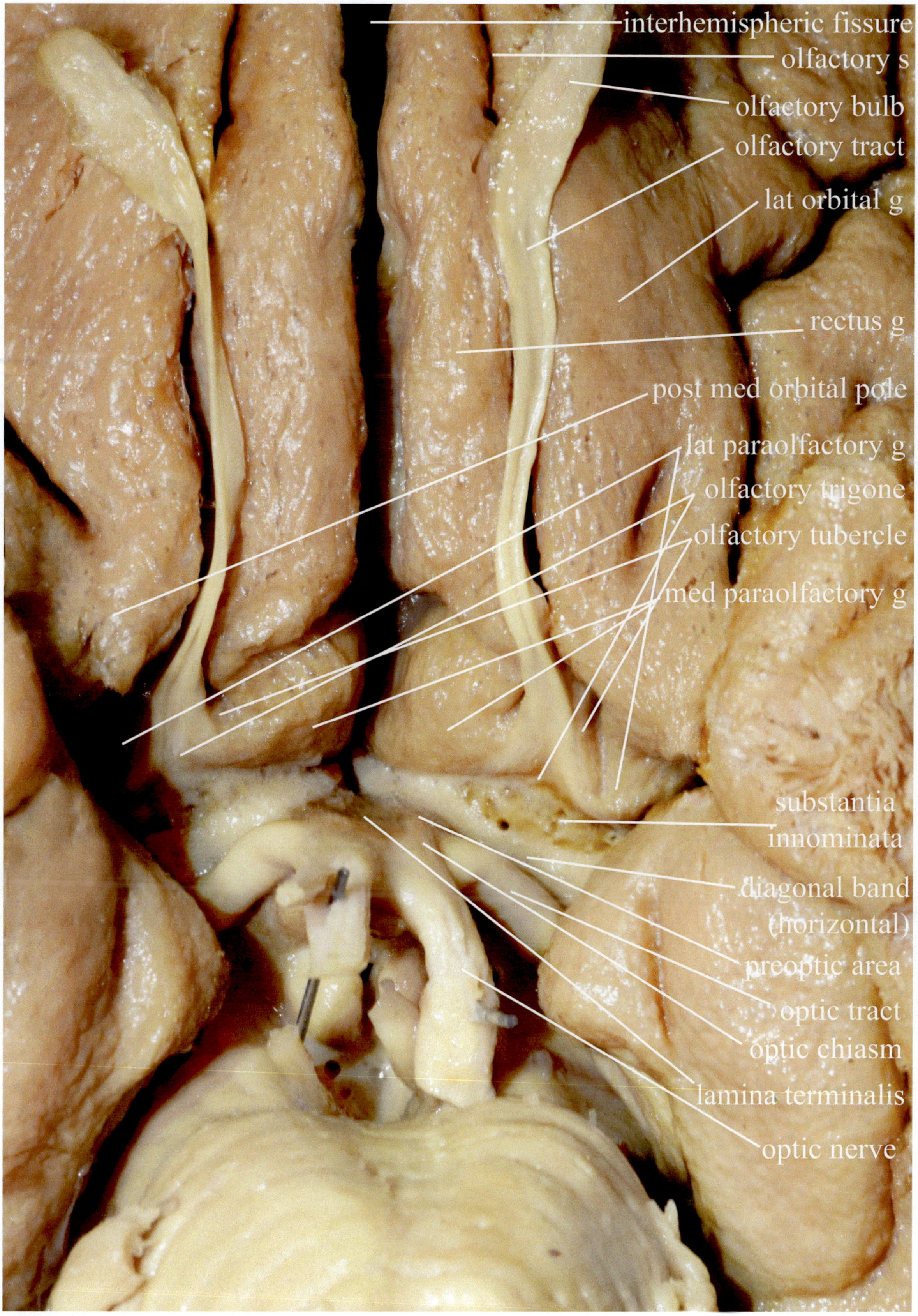

Fig. 5.32 (continued)

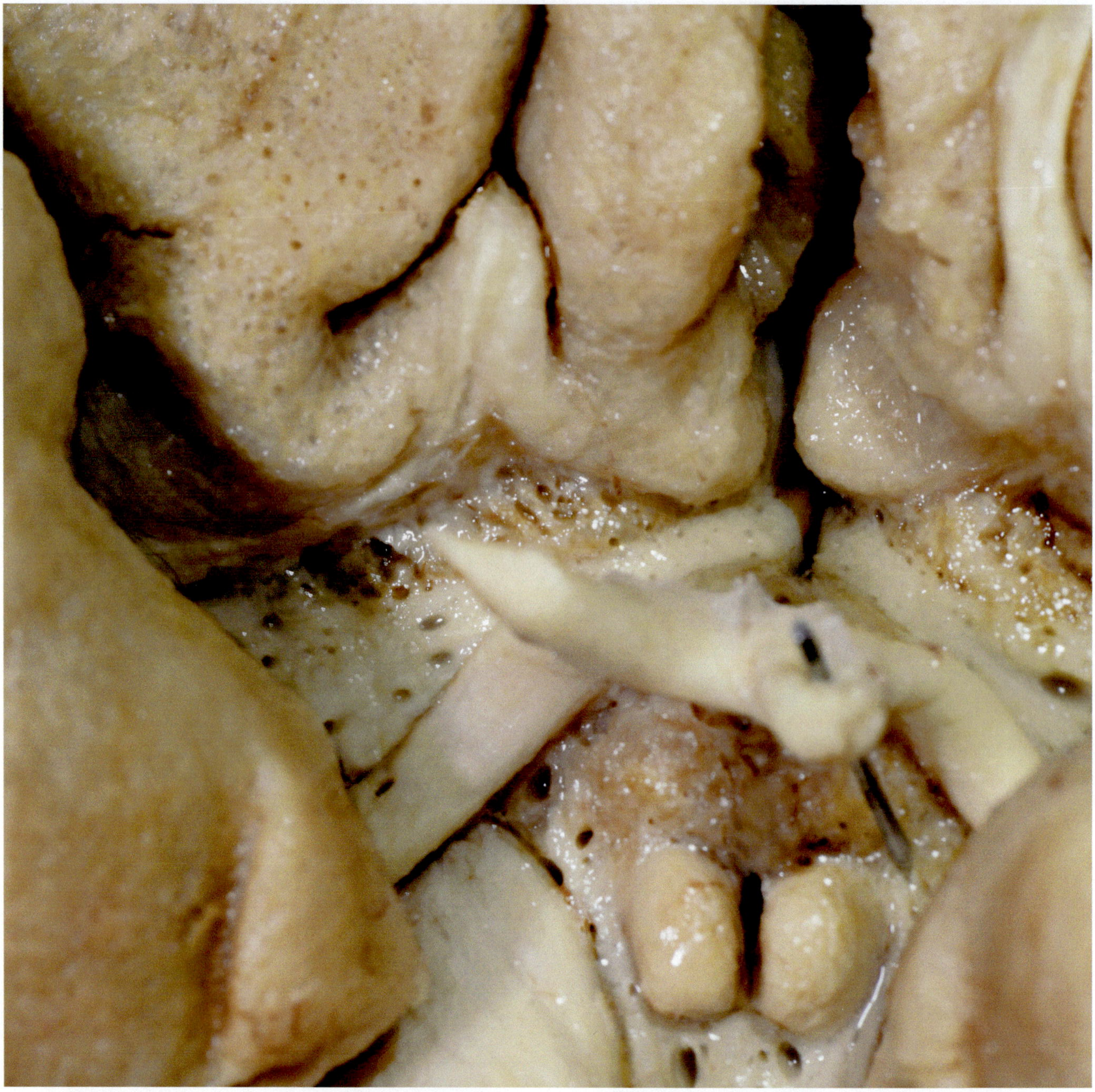

Fig. 5.33 Anatomical preparation of the olfactory tubercle. The olfactory tract has been removed on the right side to expose the rostral end of the olfactory sulcus. However, in the depth of its rostral end, the olfactory sulcus terminates in small lateral and medial rami that limit the pyramid-shaped olfactory tubercle. *g* gyrus, *lat* lateral, *med* medial, *s* sulcus, *st* stria (©Kadri 2023. All Rights Reserved)

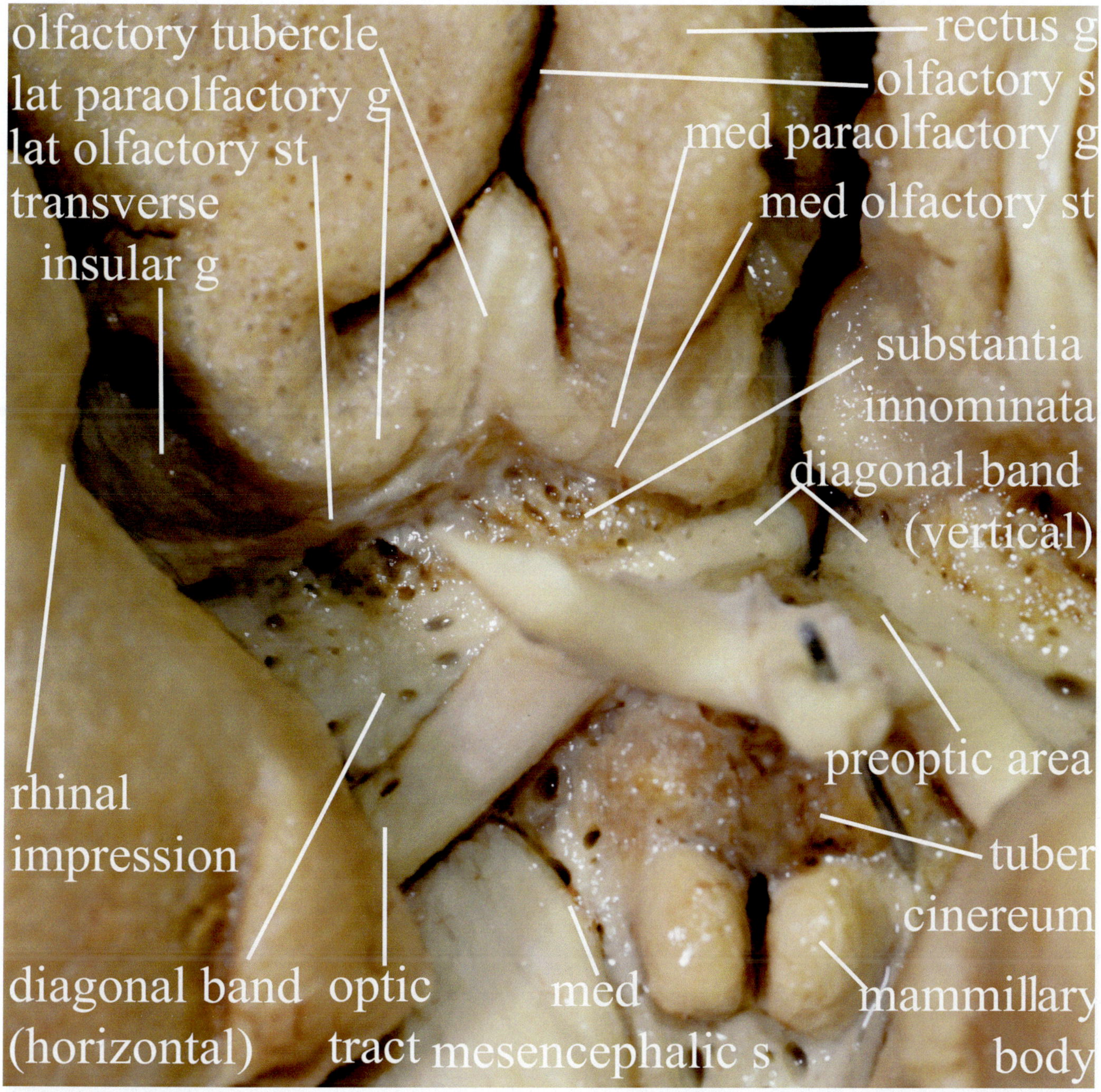

Fig. 5.33 (continued)

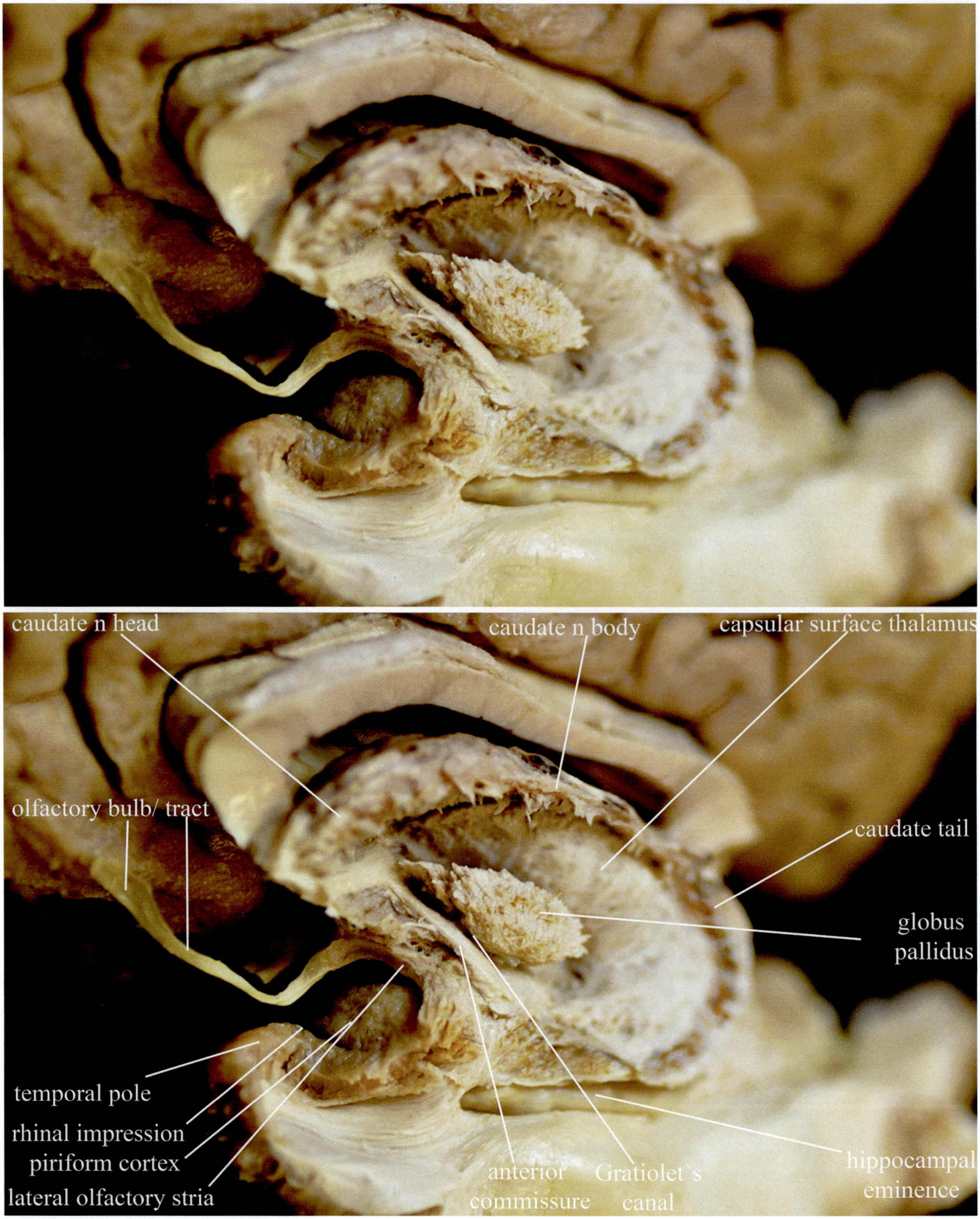

Fig. 5.34 Anatomical preparation of the limbic connections and olfactory centers. The amygdala, globus pallidus, and ventral striatum form a continuum on the base of the telencephalon. *n* nucleus (©Kadri 2023. All Rights Reserved)

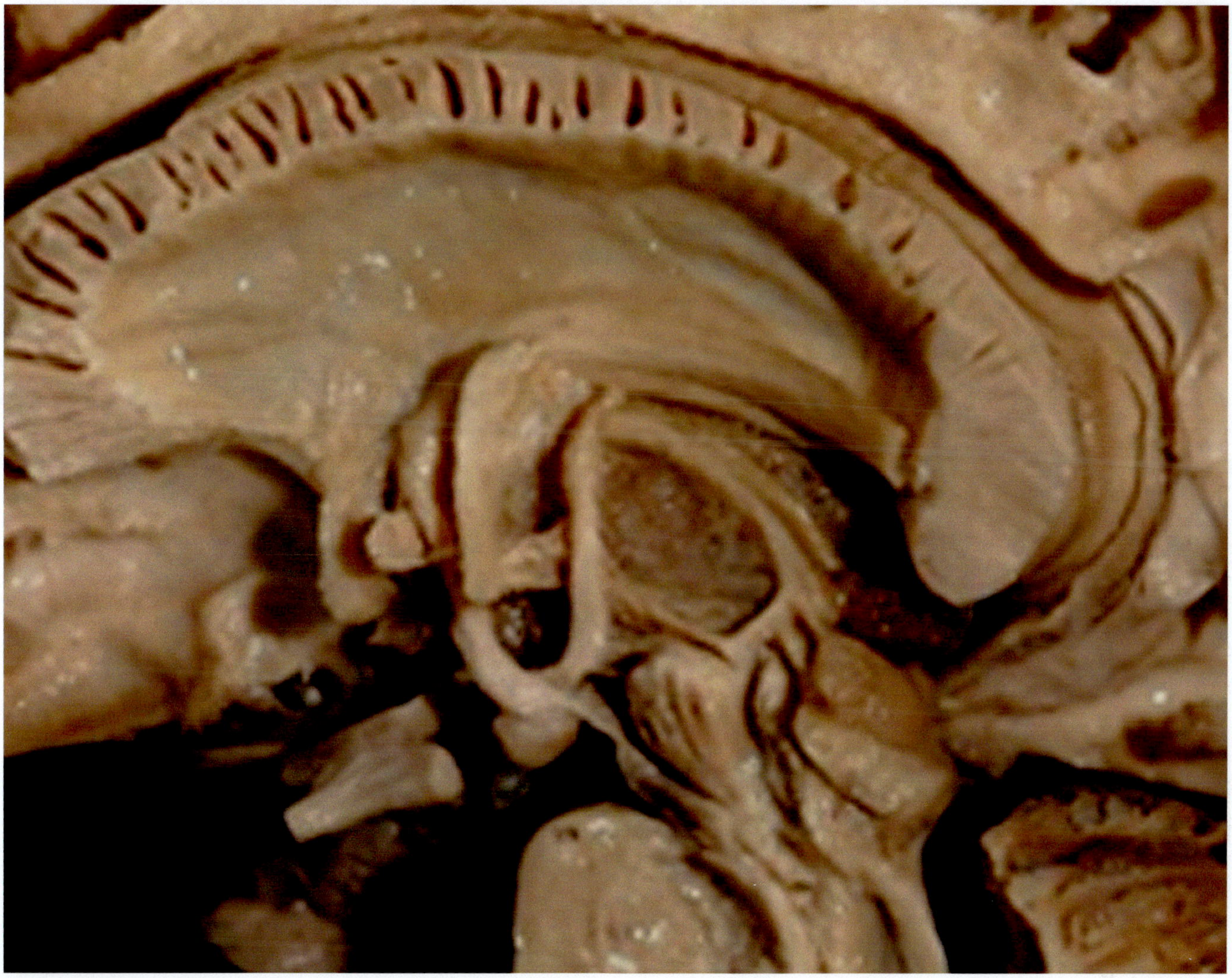

Fig. 5.35 Anatomical preparation of the limbic connections. The inner circuits of the limbic system on the left hemisphere are partially superimposed on the contralateral side. The fibers of the preoptic area, septal nuclei, hypothalamus, and amygdala pass medial to the mammillothalamic fasciculus. The habenula connects with anterior thalamic nuclei and the bed nucleus of the stria terminalis via the stria medullaris and with the interpeduncular nucleus in the ventral tegmental region via the retroflexus fasciculus. This fasciculus is also called the habenula interpeduncular tract. *f* fasciculus, *n* nucleus, *t* tract, *tr* tract (©Kadri 2023. All Rights Reserved)

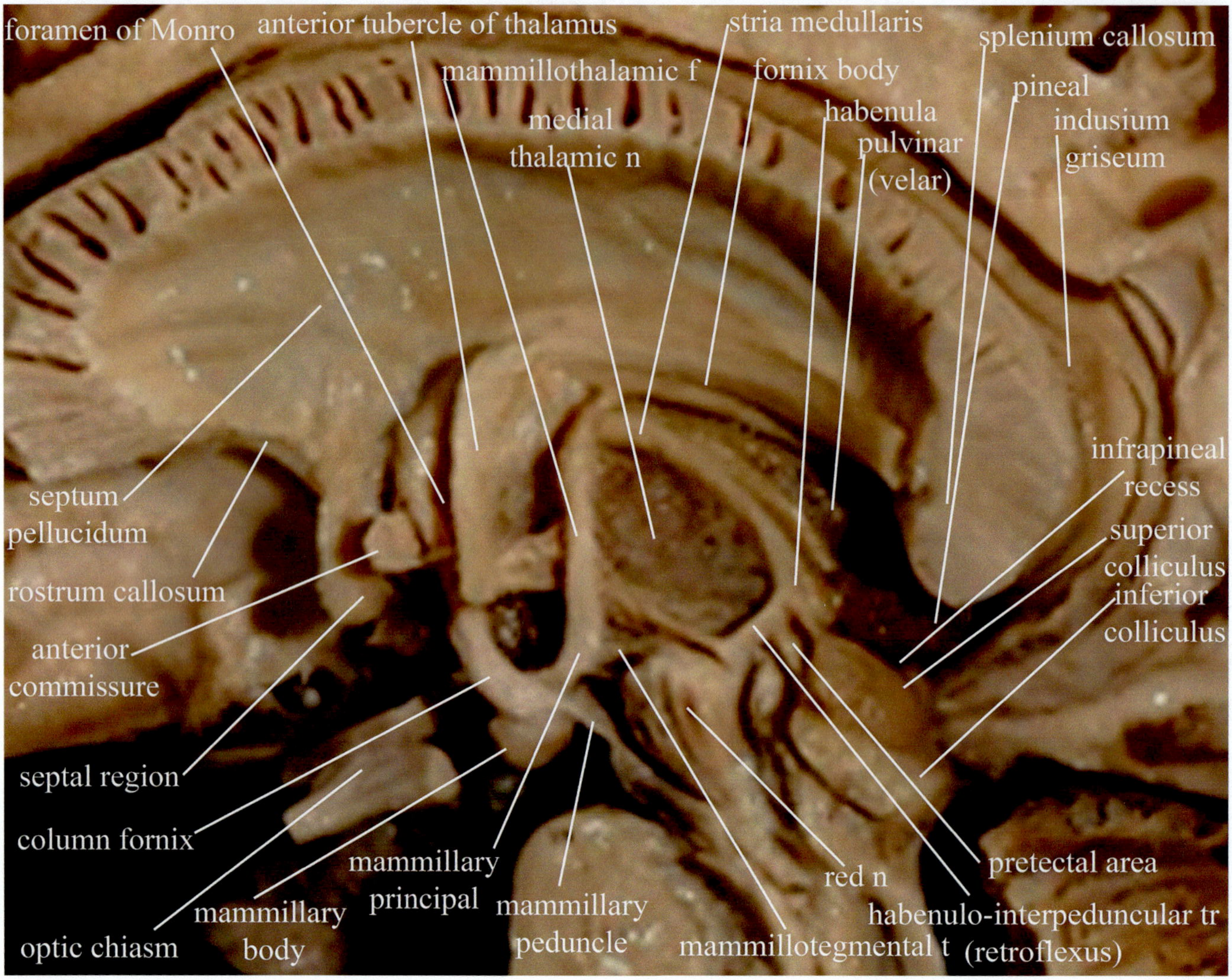

Fig. 5.35 (continued)

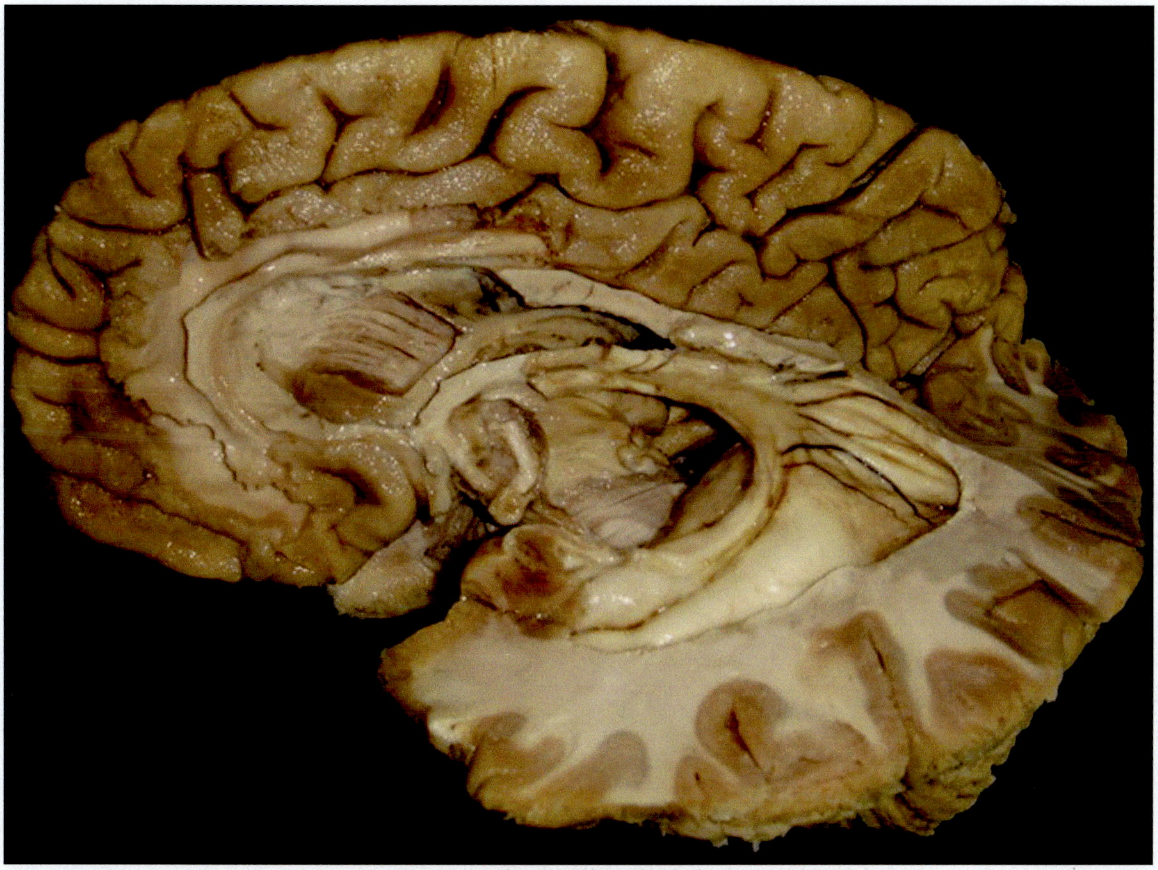

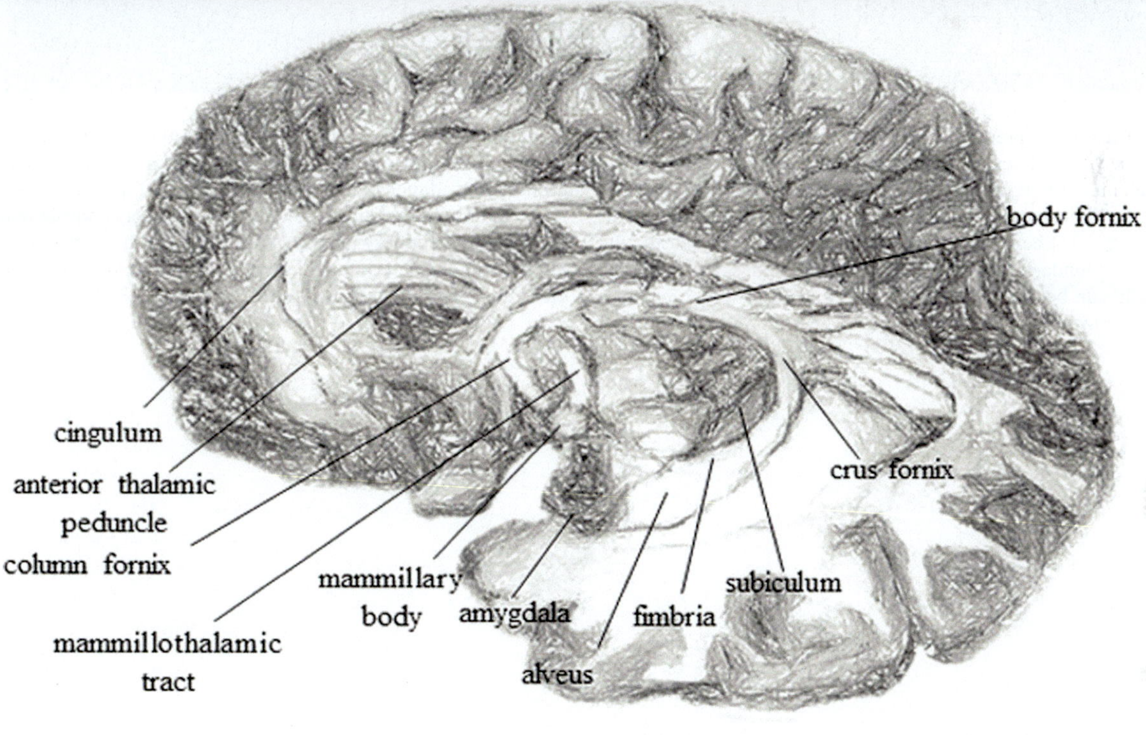

Fig. 5.36 Anatomical preparation and illustration of the limbic connections. The main efferent of the hippocampal formation is the fornix. The fibers of the fornix may be direct or cross to the contralateral side. They divide in relation to the anterior commissure into precommissural and retrocommissural fibers. Precommissural fibers are mainly connections to the septal region. Retrocommissural fibers are connections to the hypothalamus and ventral tegmentum. The main hypothalamic nucleus connection is to the mammillary body, which projects to the anterior nuclei of the thalamus. The anterior nuclei of the thalamus connect with the prefrontal cortex. The cingulum connects the prefrontal area with the entorhinal area of the parahippocampal gyrus and the amygdala. (©Kadri 2023. All Rights Reserved)

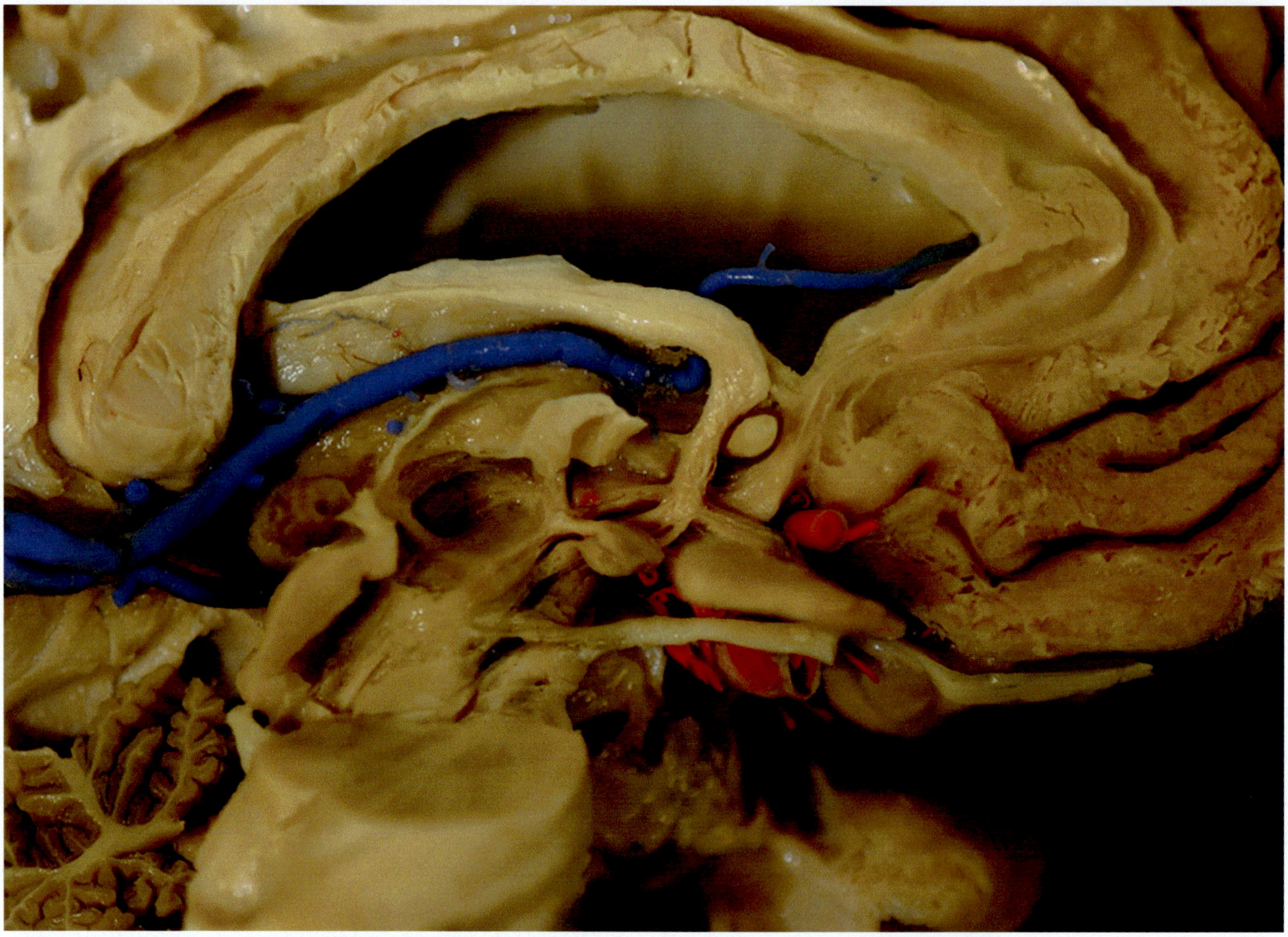

Fig. 5.37 Anatomical preparation of the ventral connections of the amygdala. The ventral fibers of the amygdala connect the septal region, the hypothalamus, the dorsal thalamus, the epithalamus, and the tegmentum of the brainstem. On the surface of the basal aspect of the telencephalon, these fibers are seen as the diagonal band at the posterolateral boundaries of the anterior perforated substance. The diagonal band can be further divided into horizontal and vertical segments. The horizontal segment contains all the ventral fibers of the amygdala, while the vertical segment contains the septal fibers. The fibers directed toward the thalamus, epithalamus, and tegmentum turn around the peduncle with other connections of the basal nuclei, medial to the mammillothalamic fasciculus. These fibers contribute to forming the internal medullary lamina of the thalamus that separates the medial nuclei from the anterior and lateral nuclei of the dorsal thalamus. *f* fasciculus, *g* gyrus, *n* nucleus, *st* stria (©Kadri 2023. All Rights Reserved)

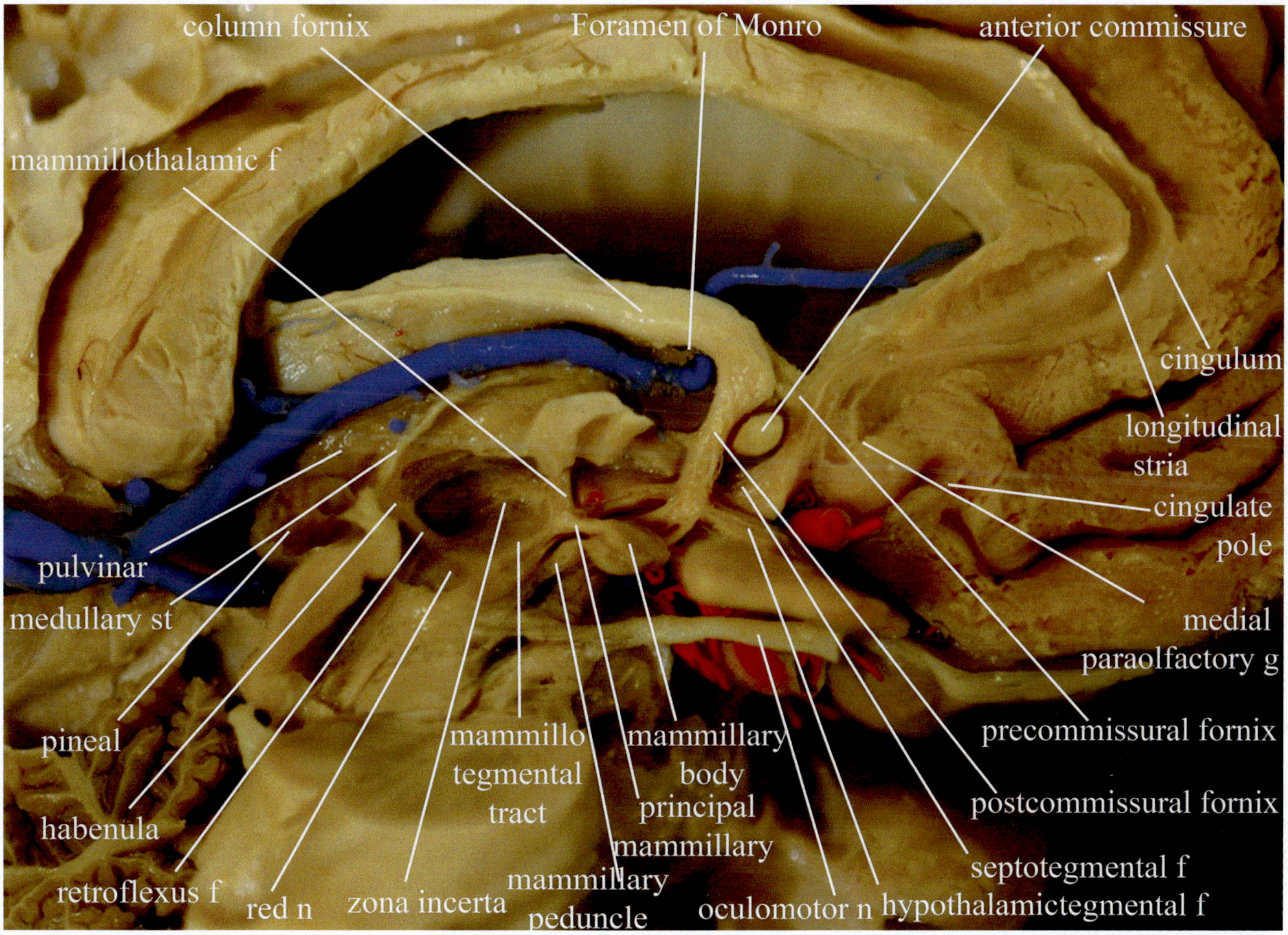

Fig. 5.37 (continued)

The crura continue around the posterior surface of the pulvinar to the medial wall of the atrium and join the inferior surface of the corpus callosum. The dorsal hippocampal commissure can be seen in this region. Anteriorly, at the junction of the atrium and the body of the ventricle, the fornix approaches its contralateral part and forms the body. The body of the fornix is at the medial wall of the lateral ventricle and on the superior surface of the thalamus. It constitutes the inferior limit of the septum pellucidum and the superior lip of the choroidal fissure. The body contains the hippocampal decussation and the ventral hippocampal commissure [14, 16].

Anteriorly, at the level of the anterior tubercles of the thalamus, the body separates from its counterpart and is named the column. The column detaches from the thalamic surface, enlarging the choroidal fissure, and constitutes the superior and anterior limits of the foramen of Monro. Its fibers are anterior (precommissural) and posterior (postcommissural) to the anterior commissure at the anterosuperior part of the third ventricle. The precommissural fibers reach the septal region and were named by Achille-Louis Foville (1799–1878) as the olfactory fasciculus of Ammon's horn [13]. The precommissural fibers, which cross the midline, are designated as the hippocampal decussation [14]. The postcommissural fibers continue inferiorly, are integrated into the lateral wall of the third ventricle, and reach the mammillary body on the floor of the third ventricle.

The Mammillaris Princeps: The Mammillothalamic and Mammillotegmental Tracts

The fasciculus mammillaris princeps, also known as the principal mammillary tract, is a compact bundle that constitutes the main efferents of the mammillary body (Figs. 5.20, 5.24, 5.25, 5.26, 5.27, 5.35, 5.36, and 5.37). It ascends dorsally in the lateral wall of the third ventricle and, after a short distance, splits into two components: the mammillothalamic and the mammillotegmental tracts. The mammillothalamic tract is the larger of the two and courses toward the anterior nuclei of the thalamus. The mammillotegmental tract curves caudally into the tegmentum of the midbrain to reach the dorsal tegmental nucleus of Gudden and the nucleus reticularis of Bechterew in the tegmentum of the pons [5, 6, 14].

The main afferents of the mammillary body are from the subiculum of the hippocampal formation via the fornix and from the paramedian midbrain zone via the mammillary peduncle. The mammillary peduncle originates mainly from the dorsal tegmental nucleus of Gudden and ascends along the ventral surface of the midbrain toward the mammillary body and medial septal nucleus [4–6]. The mammillotegmental tract and the mammillary peduncle are part of the limbic system—midbrain circuit.

Ventral Fibers of the Amygdala

The ventral fibers of the amygdala are both association and projection fibers [4–6, 15, 16]. They are one of the components of the ansa peduncularis [5, 6, 15], a fiber system embedded in the anterior perforated substance that turns around the mesencephalic peduncle (Figs. 5.12, 5.20, 5.22, 5.24, 5.25, 5.26, 5.27, and 5.38) [16]. They run above the optic tract [17] and parallel and posteroinferior to the anterior commissure [18]. These fibers connect the amygdala to the septal area, hypothalamus, thalamus, and tegmentum of the brainstem. The amygdala-septal fibers are association fibers. The ventral projection fibers of the amygdala project to the hypothalamus (amygdalo-hypothalamic), the medial nucleus of the thalamus (amygdalo-thalamic), and tegmental areas of the mesencephalon [15, 16]. These fibers are embedded in the diagonal band of Broca. This diagonal band (diagonal stria or Broca band) is a fiber bundle permeated by neurons that is continuous with the paraterminal gyrus, which corresponds to the cisternal surface of the septal region. The diagonal band can be promptly identified in the macroscopic view as a whitish band that limits the irregular quadrangular space of the anterior perforated substance in its posterolateral, posteromedial, and anteromedial borders. Two segments of the diagonal band can be identified: the horizontal and the vertical. The horizontal segment is formed by ventral fibers connecting the basal telencephalic nuclei (striatum, pallidum, and amygdala) with the septal region, hypothalamus, thalamus, subthalamus, and tegmentum of the brainstem. The fibers of the horizontal portion that extend toward the septal area and hypothalamus are mainly from the amygdala, and they constitute the vertical portion of the diagonal band. The horizontal portion is bordered medially by the optic tract, while the vertical portion is bordered medially by the preoptic area. The preoptic area is discussed with the chapter on the hypothalamus. The diagonal band of Broca is part of the limbic continuum at the midline of the hypothalamus and brainstem [15, 16].

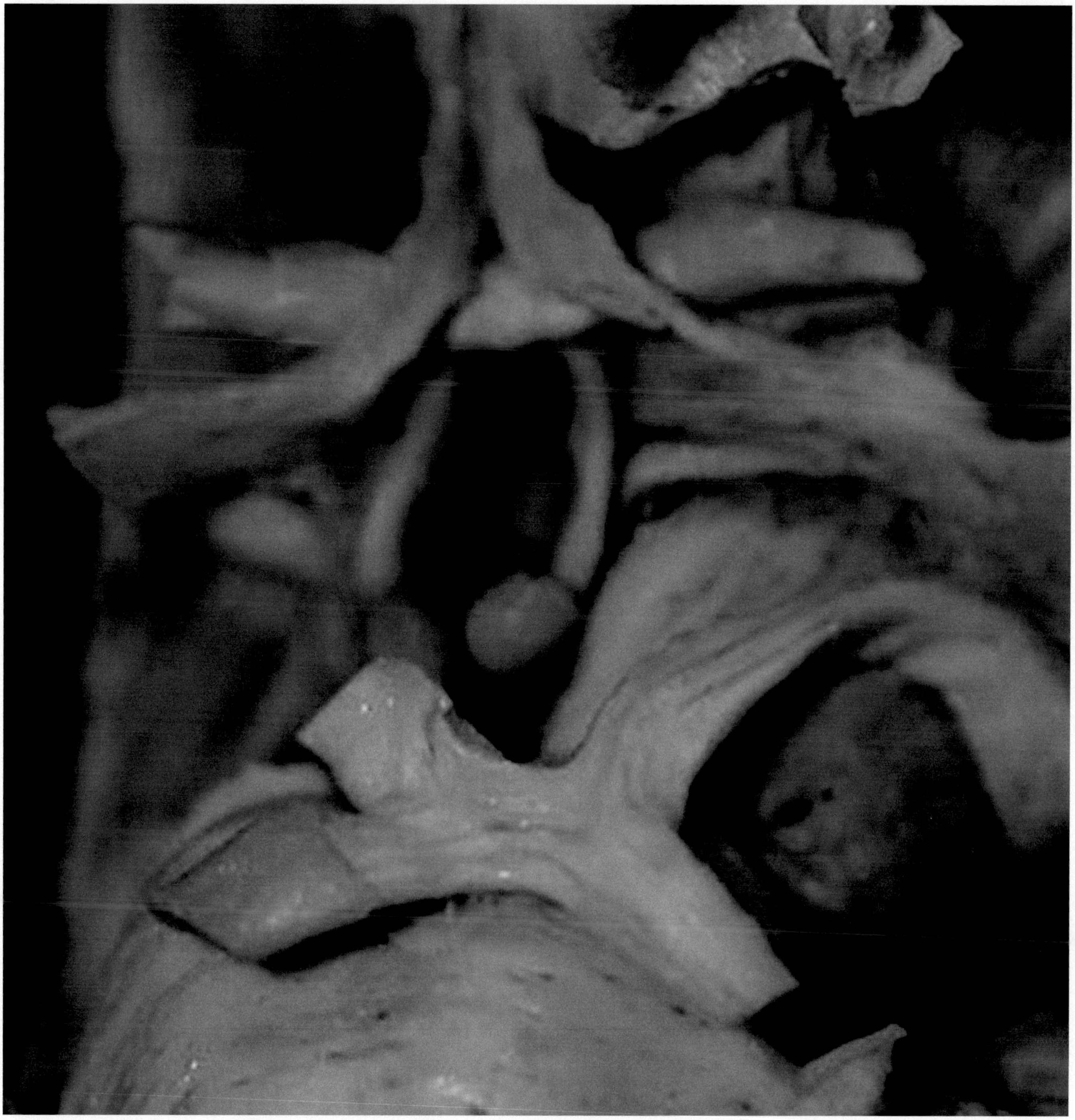

Fig. 5.38 Anatomical preparation and illustration of the ventral fibers of the amygdala. These are association fibers connecting with the septal region and projection fibers connecting with the hypothalamus, thalamus, and tegmentum of the brainstem. (©Kadri 2023. All Rights Reserved)

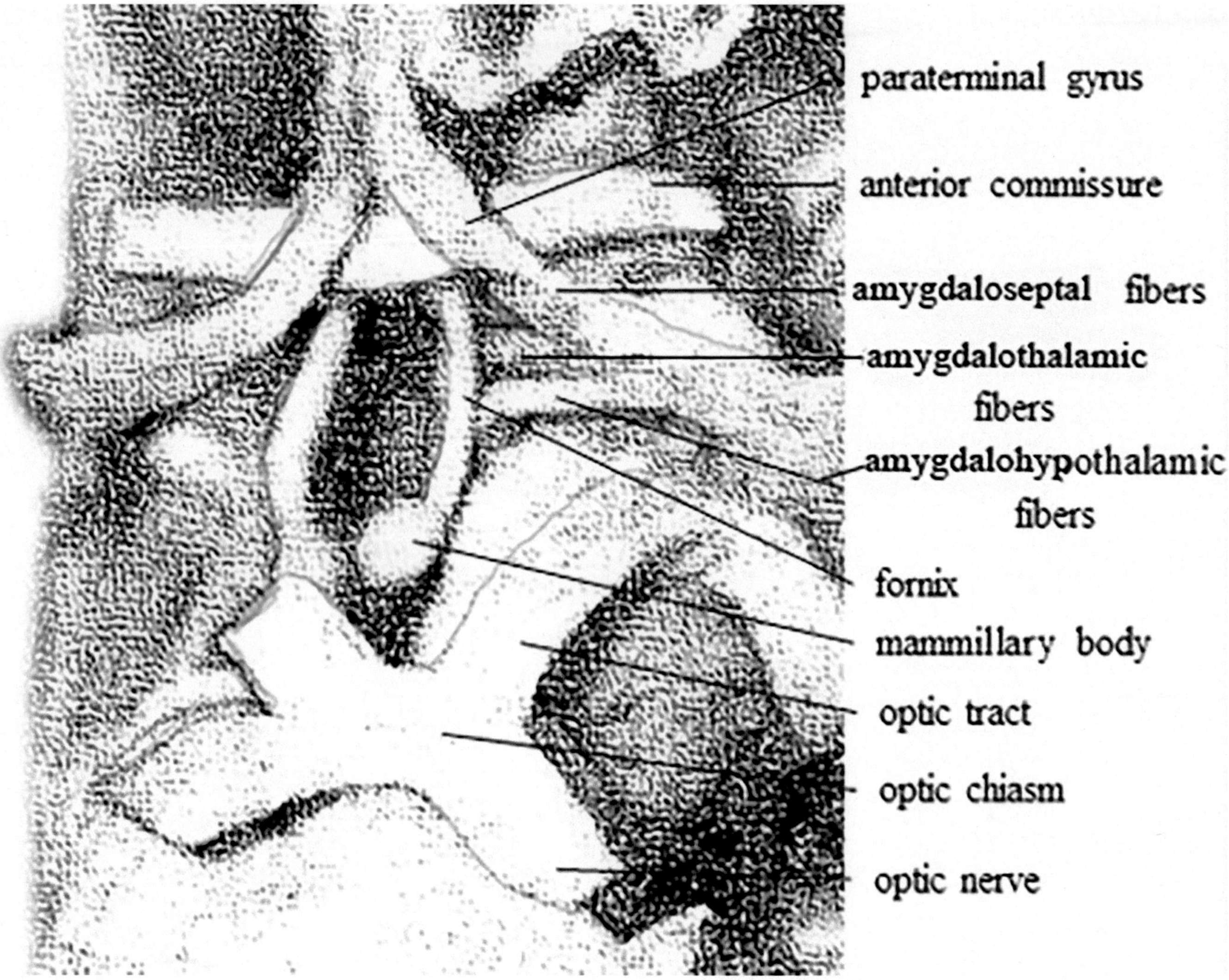

Fig. 5.38 (continued)

Dorsal Fibers of the Amygdala

The stria terminalis, among other names, is also called the taenia of Tarin (Pierre Tarin, a French anatomist, 1725–1761) and is the main dorsal efferent of the amygdala (Figs. 5.20, 5.24, 5.25, 5.26, 5.27, 5.39, and 5.40). It connects the caudomedial aspect of the amygdala, at the roof of the temporal horn of the lateral ventricle. From there, it turns around the caudate and the thalamus within the striothalamic sulcus to reach the anterior commissure, where it splits into precommissural, commissural, and postcommissural fibers. The precommissural fibers project toward the medial preoptic area and the anterior hypothalamic areas. The commissural fibers travel along with the anterior commissural fibers, connecting both amygdalas. The postcommissural fibers project to the bed nucleus of the stria terminalis and the anterior hypothalamic nuclei. The bed nucleus of the stria terminalis is a cluster of neurons that accompanies the stria terminalis in most of its extension, and its presence may explain the appearance of the *lamina affixa*. The lamina affixa is the given name of the appearance of the thickened ependyma overlying the striothalamic sulcus. In the body of the ventricle, the thalamostriate veins course alongside the stria terminalis within the striothalamic sulcus to reach the interventricular foramen of Monro [13].

The Extracapsular Thalamic Peduncle

The extracapsular thalamic peduncle is the thalamic connection with telencephalic structures that do not transit the internal capsule (Figs. 5.20, 5.27, and 5.40 and see figures in Chap. 7). They connect the medial nuclei of the thalamus and epithalamus (habenula) with the septal area, amygdala, pallidum, and striatum (the basal nuclei) [6, 19]. These fibers course under, rather than between, the putamen and caudate nucleus [15]. The extracapsular thalamic peduncle is part, together with the prosencephalic tract, of the midline limbic circuit. Fibers of the preoptic region, septal region, and hypothalamus join the fibers looping around the peduncle of the amygdala, pallidum, and striatum to connect with the medial thalamus, composing the extracapsular thalamic peduncle.

The Prosencephalic Tract

The prosencephalic tract is also known as the medial prosencephalic bundle, medial forebrain fasciculus, and medial telencephalic bundle (Figs. 5.20, 5.27, 5.40, 5.41, and 5.42) [1, 4, 5, 16]. It connects the limbic structures of the telencephalon and diencephalon with the caudal central nervous system, mainly the mesencephalic and pontine segments of the brainstem [5, 6]. It courses at the depth and in the vicinity of the hypothalamic sulcus within the third ventricle and can be identified when the ependyma is removed. Fibers to the tegmentum are mainly medial to the mammillothalamic tract. In fact, the mammillotegmental tract, which has a common trunk with the mammillothalamic tract, forms the principal mammillary tract (*fasciculus mammillaris princes*) and is part of the prosencephalic tract. The fibers that connect the septal region and anterior hypothalamic areas are also termed the dorsal longitudinal fasciculus (of Schutz). These fibers connect with the dorsal tegmental nucleus at the ventral periaqueductal gray matter, and the connection extends to the lower tegmentum of the brainstem. These fibers are distributed to the peripeduncular nucleus, the substantia nigra, the ventral tegmental area, the cuneiform nucleus, the periaqueductal gray matter, the raphe nuclei, the parabrachial nucleus, the locus coeruleus, the rhombencephalic reticular formation, the dorsal nucleus of the vagus, and the nucleus of the solitary tract [5].

The Circuit of Papez

In his article, "A proposed mechanism of emotion" from 1937, James W. Papez (1883–1958) proposed that a circular array of neural components and their connections constitute a mechanism that enhances emotional awareness (Figs. 5.36 and 5.43) [20]. In Papez's conjecture, emotional consciousness has an anatomical parallel to other senses, such as the somesthetic senses or vision. This circuit enwraps the hippocampus, the postcommissural fornix, the mammillary body, the mammillothalamic tract, the anterior thalamic nucleus, the thalamocingulate projections (via the anterior thalamic peduncle), the cingulate gyrus, and the cingulum.

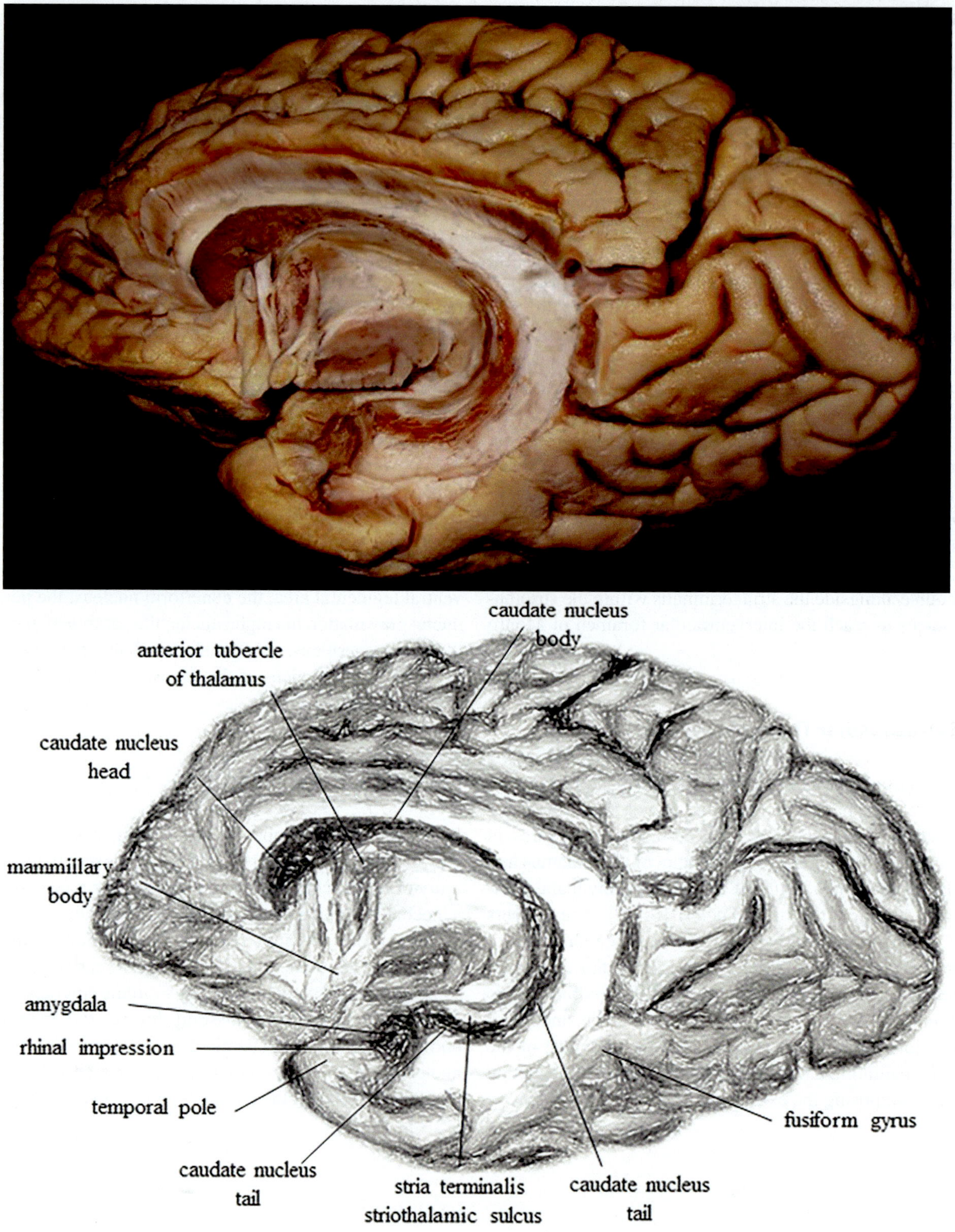

Fig. 5.39 Anatomical preparation and illustration of the dorsal amygdala connections in the left hemisphere, medial view. The corpus callosum and parahippocampal gyrus have been removed to expose the lateral wall of the ventricular cavity. The caudate nucleus is exposed around the thalamus. The dorsal connections of the amygdala form the stria terminalis, which runs with the striothalamic sulcus. The striothalamic sulcus is broader at its temporal horn. At this area in the anterior roof of the temporal horn, the tail of the caudate becomes more distant from the border of the thalamus. The thalamus is represented by the lateral geniculate body and the emerging fibers of the inferior and posterior thalamic peduncles, emanating mainly from the geniculate bodies and the pulvinar. The tail of the caudate seems mostly to be discontinuous macro- and mesoscopically, in an island of clusters of gray substance directed toward the ventricular surface of the amygdala. Sometimes, the tail of the caudate may seem to fuse in continuity with the amygdala. (©Kadri 2023. All Rights Reserved)

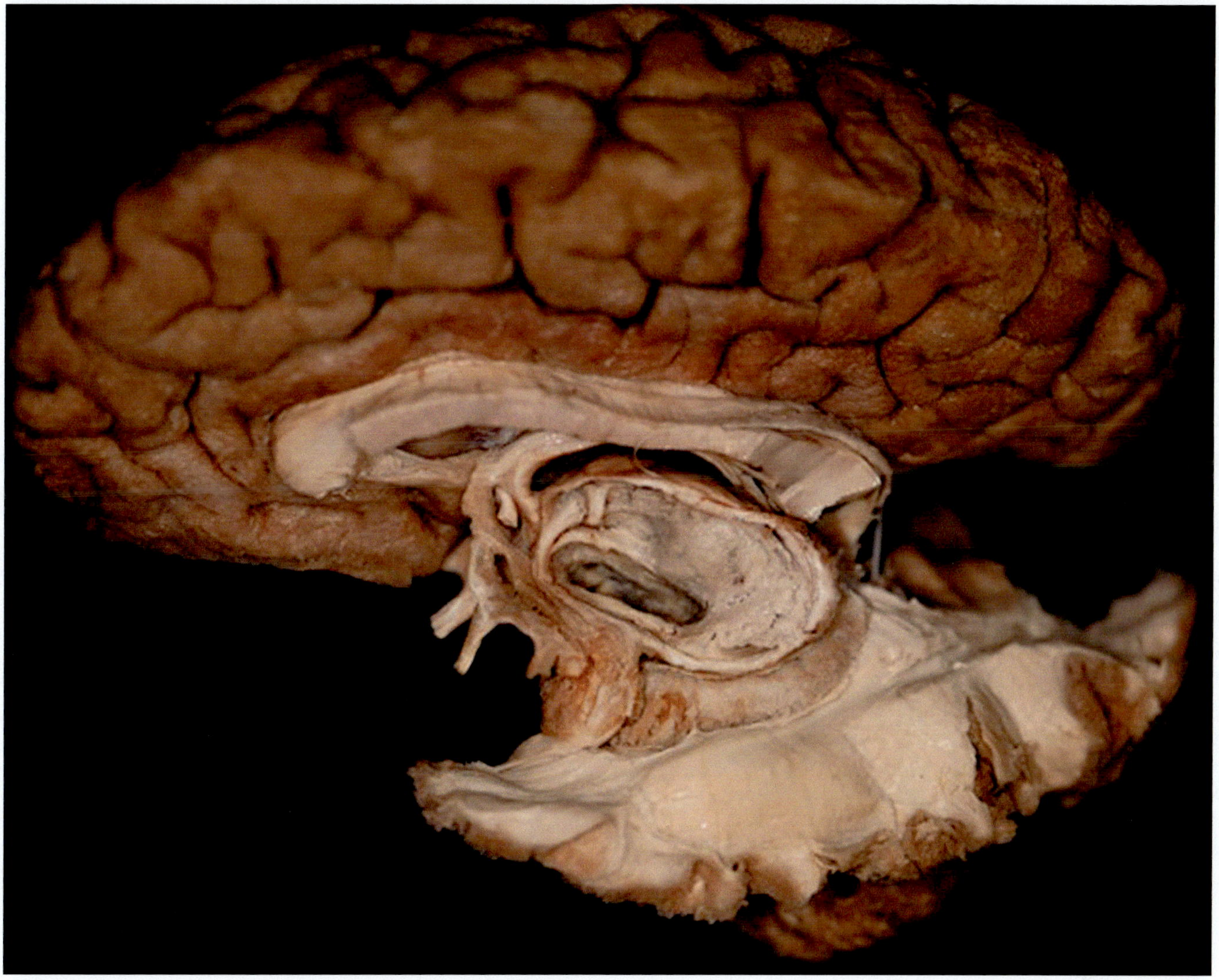

Fig. 5.40 Anatomical preparation and illustration of the telencephalic and diencephalic limbic connections. The internal capsule and cerebral peduncle have been removed, and the capsular surface of the thalamus is exposed. Several fibers connecting the basal nuclei (preoptic area, septal nuclei, amygdala, striatum, and pallidum) to the medial dorsal thalamus, epithalamus, subthalamus, and the tegmentum of the mesencephalon loop around the ventral surface of the crus of the cerebral peduncle (removed) and are named the *ansa peduncularis*. (©Kadri 2023. All Rights Reserved)

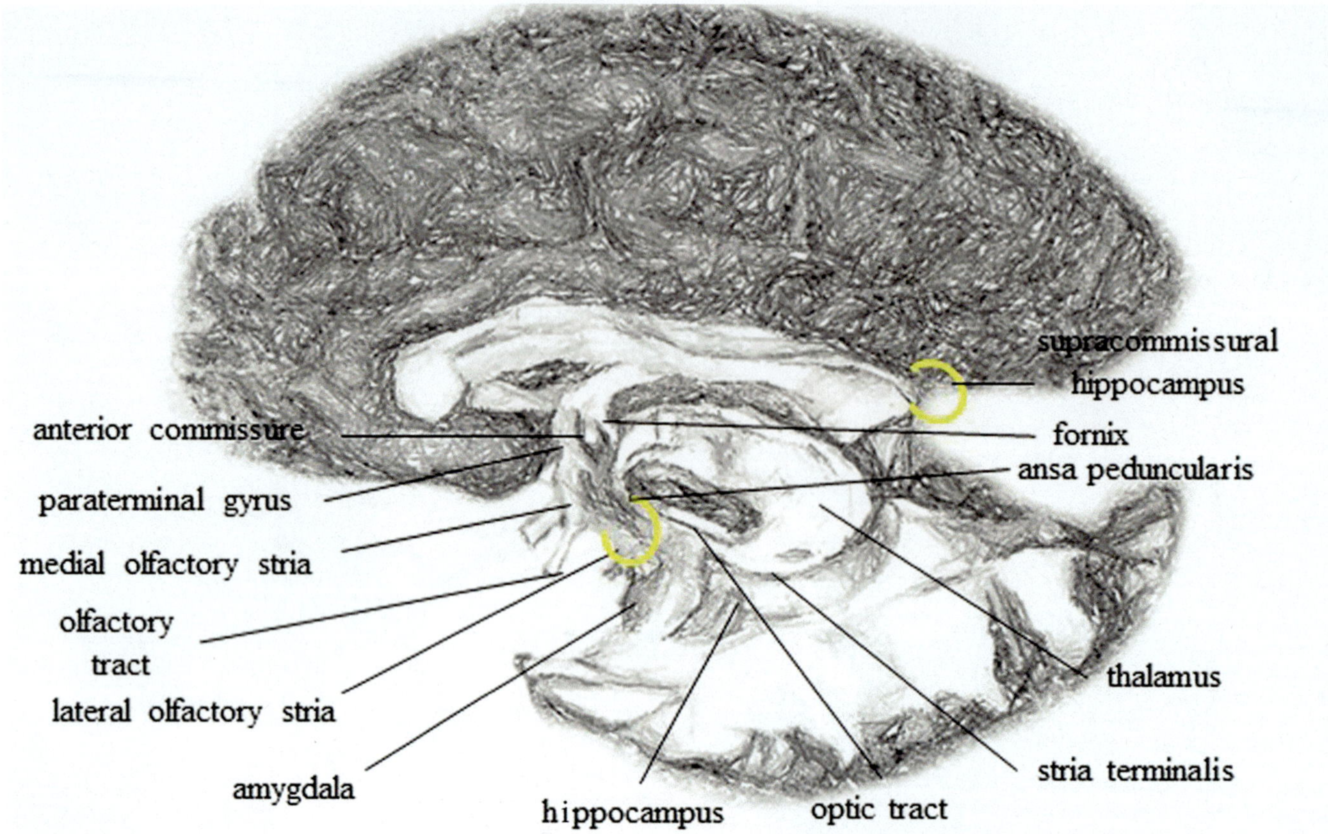

Fig. 5.40 (continued)

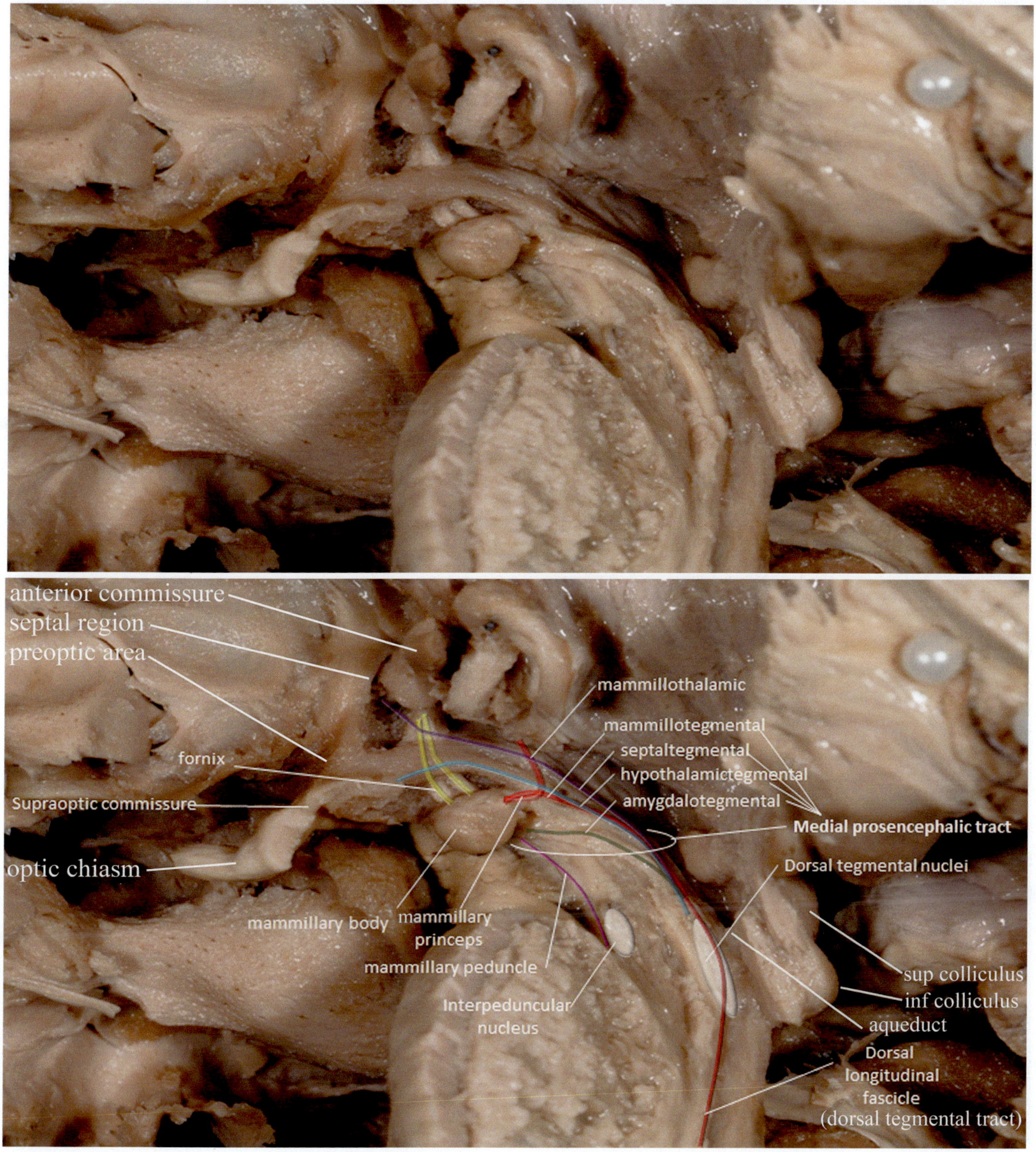

Fig. 5.41 Anatomical preparation of the telencephalic and diencephalic limbic connections to the tegmentum of the brainstem, medial view. The medial prosencephalic tract is the main limbic projection of the prosencephalon. The limbic basal nuclei of the telencephalon (the amygdala and the septal nuclei) and the limbic diencephalic nuclei (anterior nuclei of the thalamus), bed nucleus of the stria terminalis, and hypothalamic nuclei (mammillary body and supraoptic and suprachiasmatic nuclei) connect with the mesencephalic tegmentum via the medial prosencephalic tract. *inf* inferior, *sup* superior (©Kadri 2023. All Rights Reserved)

Fig. 5.42 Anatomical preparation of the telencephalic and diencephalic limbic connections to the tegmentum of the brainstem, medial view. The prosencephalic tract connects the mesencephalic tegmentum. Fibers from the hypothalamus and septal region connecting to the pontine tegmentum are named the dorsal longitudinal fasciculus (of Schutz). This fasciculus is dorsal and parallel to the medial longitudinal fasciculus. *ant* anterior, *f* fasciculus, *g* gyrus, *inf* inferior, *n* nerve, *nn* nucleus, *ped* peduncle, *post* posterior, *s* sulcus, *sup* superior, *t* tract

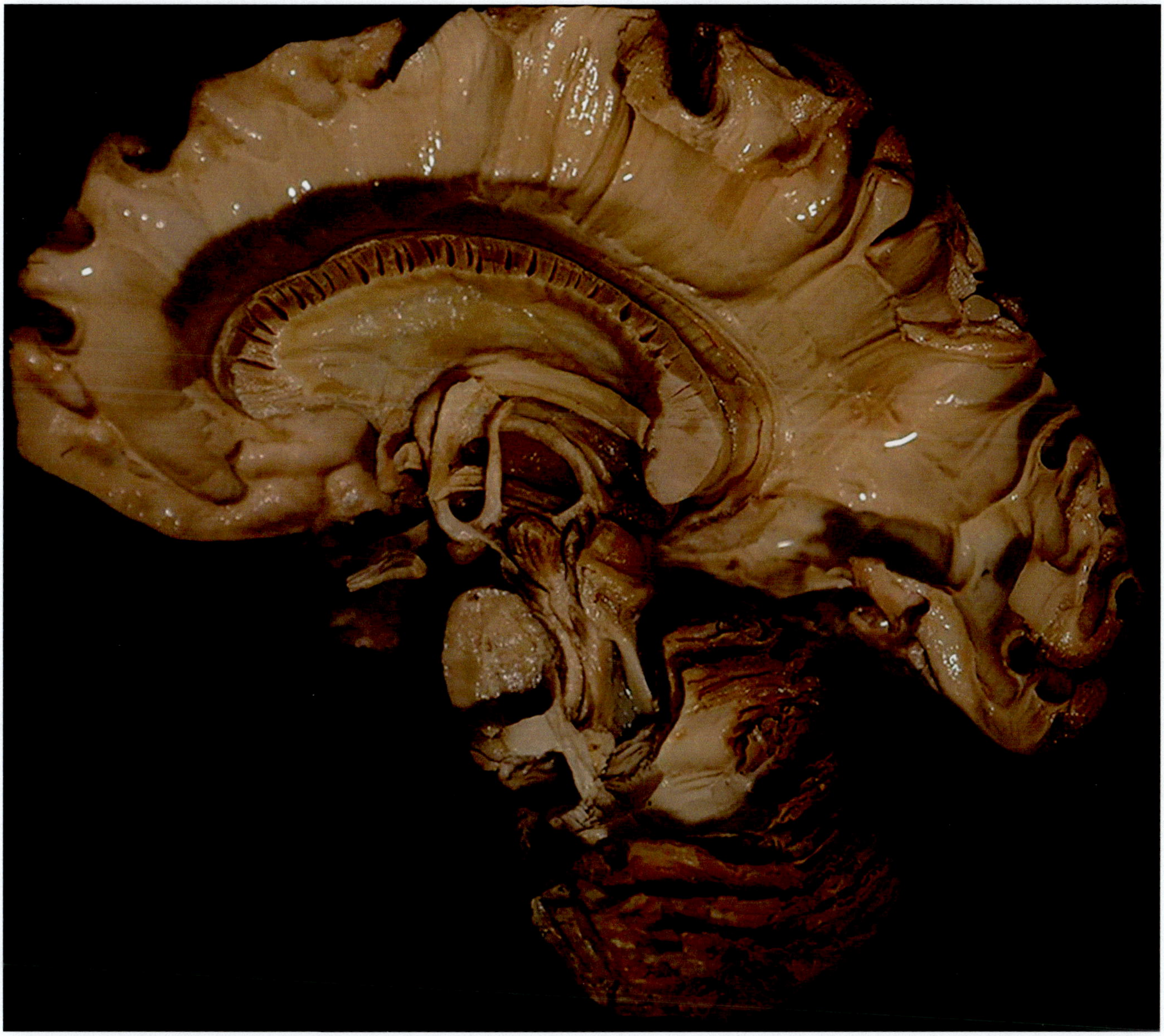

Fig. 5.43 Anatomical preparation of the limbic connections. The fornix is a complex association (septal region), commissural (ventral and dorsal hippocampal commissures), and projection (hypothalamus and mesencephalic tegmentum) fiber system connecting the hippocampal formation. The habenula is at the epithalamus and is connected with the septal region and anterior nuclei of the thalamus and the bed nucleus of the stria terminalis via the stria medullaris and with the interpeduncular nucleus via the retroflexus fasciculus. The mammillary body connects with the hippocampal formation via the postcommissural fibers of the fornix, with the amygdala via the amygdalo-hypothalamic fibers, with the anterior thalamic nuclei via the mammillothalamic tract, with the dorsal tegmentum via the amygdalo-tegmental tract, and with the ventral mesencephalic tegmentum via the mammillary peduncle. *g* gyrus

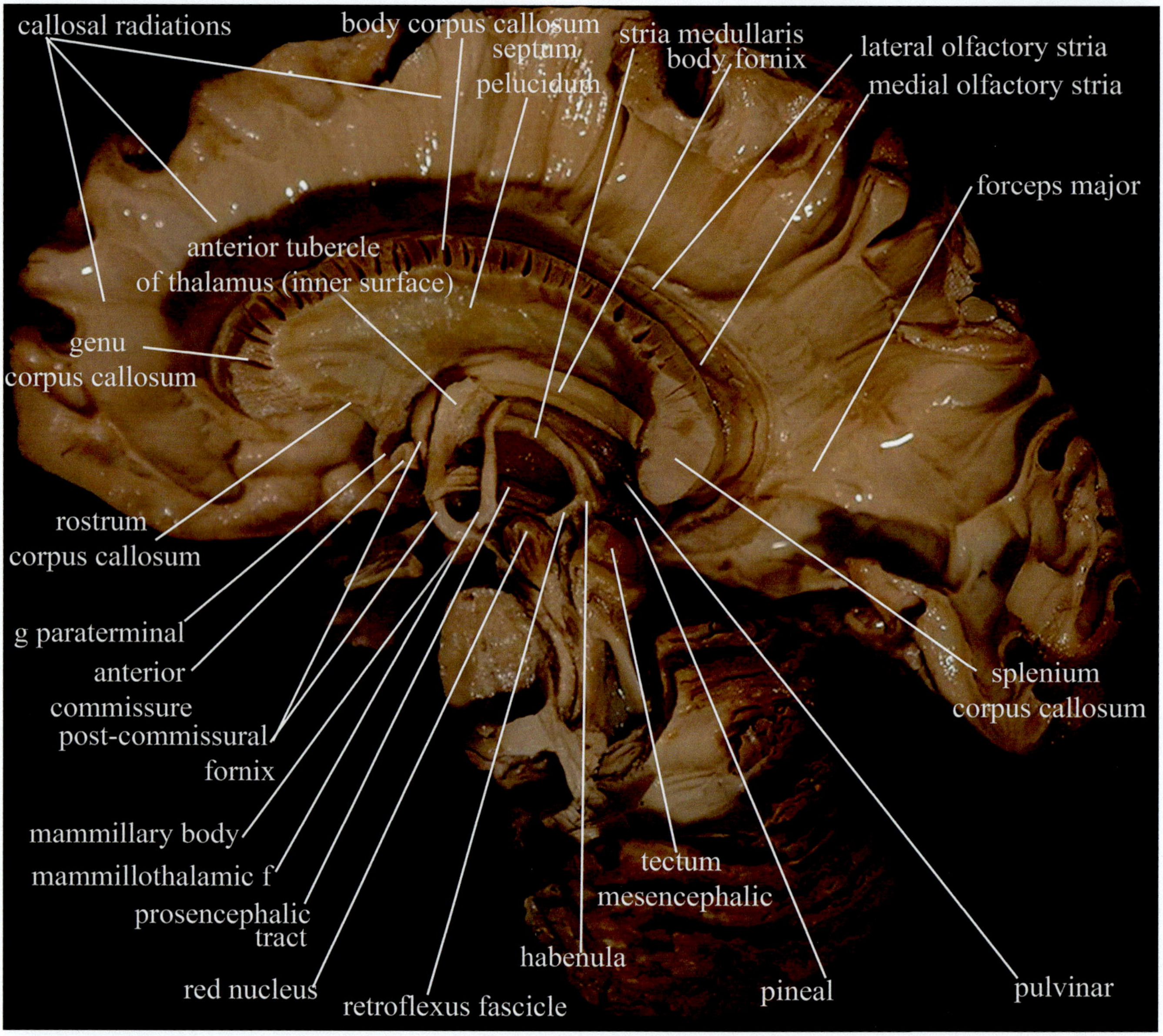

Fig. 5.43 (continued)

References

1. Carpenter MB. Core text of neuroanatomy. Subsequent ed. Baltimore, MD: Williams &Wilkins; 1991.
2. Duvernoy HM. The human brain: surface, three-dimensional sectional anatomy with MRI, and blood supply. 2nd ed. Vienna: Springer; 1999.
3. Duvernoy HM. The human hippocampus: functional anatomy, vascularization and serial sections with MRI. 3rd ed. Berlin: Springer; 2005.
4. Heimer L. The human brain and spinal cord: functional neuroanatomy and dissection guide. 2nd ed. New York: Springer; 1994.
5. Nieuwenhuys R, Voogd J, van Huijzen C. The human central nervous system: a synopsis and atlas. 4th ed. Berlin: Springer; 2006.
6. Standring S. Cerebral hemisphere. In: Gray's anatomy: the anatomical basis of clinical practice. 40th ed. Philadelphia, PA: Elsevier; 2008. p. 335–59.
7. Yaşargil MG, Türe U, Yaşargil DCH. Surgical anatomy of supratentorial midline lesions. Neurosurg Focus. 2005;15(18):E1. https://doi.org/10.3171/foc.2005.18.6.14.
8. Ono M, Kubik S, Abernathey CD. Atlas of the cerebral sulci. New York: Thieme; 1990. p. 81–2.
9. Brodmann K. In: Garey LJ, editor. Brodmann's localisation in the cerebral cortex. 3rd ed. New York: Springer; 2006. https://doi.org/10.1007/b138298.
10. Rhoton AL Jr. The cerebrum. Neurosurgery. 2002;51(4 Suppl):S1–52. https://doi.org/10.1097/00006123-200210001-00002.
11. Turner W. The convolutions of the brain: a study in comparative anatomy. J Anat Physiol. 1890;25(Pt 1):105–53.
12. Yaşargil MG. Microneurosurgery, Vol. IVA. New York: Thieme; 1994.
13. Testut L, Latarjet A. Tratado de anatomia humana, vol. 2. 9th ed. Barcelona: Salvat Editores; 1966.
14. Naidich TP, Krayenbühl N, Kollias S, Bou-Haidar P, Bluestone AY, Carpenter DM. White matter. In: Naidich TP, Castillo M, Cha S, Smirniotopoulos JG, editors. Imaging of the brain. Philadelphia, PA: Saunders; 2013. p. 205–44.
15. Klingler J, Gloor P. The connections of the amygdala and of the anterior temporal cortex in the human brain. J Comp Neurol. 1960;115:333–69. https://doi.org/10.1002/cne.901150305.
16. Gloor P. The temporal lobe and limbic system. New York: Oxford University Press.
17. Peltier J, Travers N, Destrieux C, Velut S. Optic radiations: a microsurgical anatomical study. J Neurosurg. 2006;105:294–300. https://doi.org/10.3171/jns.2006.105.2.294.
18. Peuskens D, van Loon J, Van Calenbergh F, van den Bergh R, Goffin J, Plets C. Anatomy of the anterior temporal lobe and the frontotemporal region demonstrated by fiber dissection. Neurosurgery. 2004;55:1174–84.
19. Türe U, Yaşargil MG, Friedman AH, Al-Mefty O, Yaşargil DG. Fiber dissection technique: lateral aspect of the brain. Neurosurgery. 2000;47:417–27. https://doi.org/10.1097/00006123-200008000-00028.
20. Papez JW. A proposed mechanism of emotion. 1937. J Neuropsychiatry Clin Neurosci. 1995;7:103–12. https://doi.org/10.1176/jnp.7.1.103.

The Basal Nuclei and the Claustrum

6

From an embryological point of view, the basal nucleus is derived from the telencephalic subpallium and is composed of the caudate nucleus, putamen, globus pallidus, and amygdala [1]. The claustrum is derived from the pallium and is therefore not part of the basal ganglia, and the amygdala is considered with the septal region within the limbic system. The basal nuclei concept may be expanded and includes centers that extend from the basal telencephalon into the ventral thalamus and tegmentum of the mesencephalon. It is the main core of the extrapyramidal motor circuit. In a functional rather than morphological concept, it encompasses the striatum complex, the globus pallidus, the subthalamic nucleus, and the substantia nigra [1–3].

The Striatum Complex

During ontogenesis, an increasing number of fibers pass to, through, and from the striatum (Figs. 6.1, 6.2, 6.3, 6.4, 6.5, 6.6, 6.7, 6.8, 6.9, 6.10, 6.11, 6.12, 6.13, and 6.14). These fibers form the internal capsule and divide the striatum into the caudate nucleus, medially, and the putamen, laterally. This separation is not complete, however, and through its entire length, bridges of striatal cells are interposed between the bundles of the internal capsule [4]. These bridges constitute the caudolenticular bridges and are prominent in the anterior limb of the internal capsule [5]. They are less visible as the internal capsule compacts from the genu toward the posterior limb. The anterior limb of the internal capsule separates the putamen from the caudate nucleus, where the posterior limb separates the lentiform nucleus from the thalamus (see below). The caudate-putamen nucleus is the largest subcortical nucleus of the brain. Structurally, the caudate and the putamen are the same [1, 4, 6].

The striatum can be divided into ventral and dorsal parts. The dorsal part is composed of the caudate nucleus, putamen, and globus pallidus. The ventral portion is mainly composed of the nucleus accumbens septi and the olfactory tubercle. The substantia innominata is also included in the ventral striatum [2, 3]. The macroscopic appearance of the substantia innominata is distinguished from the white and gray matter substance on the cisternal surface of the anterior perforated substance. This is how the structure received its name. The amygdala has strong connections with the ventral striatum and parts of the dorsal striatum. Together, these areas are referred to as the limbic striatum. The nucleus accumbens is a ventral expansion of the head of the caudate nucleus that extends into the septal area. The olfactory tubercle receives direct projections of the olfactory bulb. It superficially covers the nucleus accumbens and the head of the caudate and forms a continuum with these structures [1, 2].

The caudate is an elongated mass that protrudes into the lateral ventricle as it wraps around the thalamus. It is separated from the thalamus by the striothalamic sulcus, which contains the stria terminalis and a diffuse arrangement of neurons known as the bed nucleus of the stria terminalis. The caudate is subdivided into head, body, and tail [4]. The head is the largest portion and protrudes into the lateral wall of the frontal horn. Its posterior limit is the foramen of Monro. The head of the caudate forms the anterolateral limit of the foramen of Monro. The body is at the lateral wall of the body of the lateral ventricle. The tail is the longest portion and wraps around the pulvinar to compose the anterior wall of the atrium and the roof of the temporal horn. It extends to the amygdaloid complex and is bordered externally by the tapetum of the corpus callosum [1, 4, 5].

The putamen is situated medial to the insula, extreme capsule, claustrum, and external capsule. Together with the globus pallidus, it constitutes the lentiform nucleus. The lentiform nucleus gained its name because it resembles a lens, with the curvature facing the outer limit and the apex medially. The lentiform nucleus is constituted by the outer putamen and the inner globus pallidus, which are separated by the external medullary lamina. The internal medullary lamina of the globus pallidus separates it into an external and an internal nucleus.

P. Kadri, *The Cartographic Atlas of the Brain*, https://doi.org/10.1007/978-3-031-38062-4_6

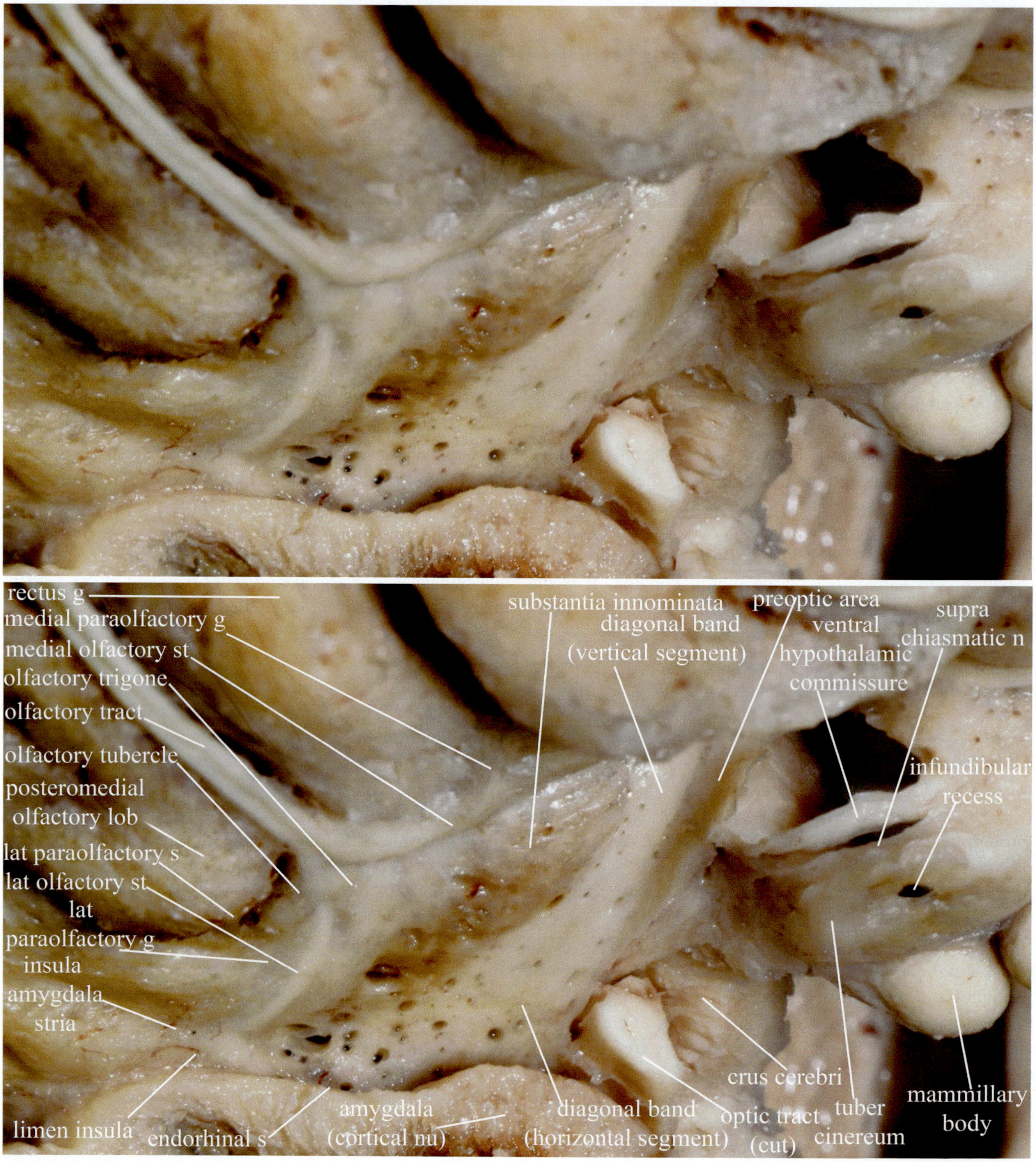

Fig. 6.1 Anatomical preparation of the anterior perforated substance. The optic apparatus has been removed, exposing the lateral and anterior hypothalamus. *g* gyrus, *lat* lateral, *lob* lobule, *n* nerve, *nu* nucleus, *s* sulcus, *st* stria (©Kadri 2023. All Rights Reserved)

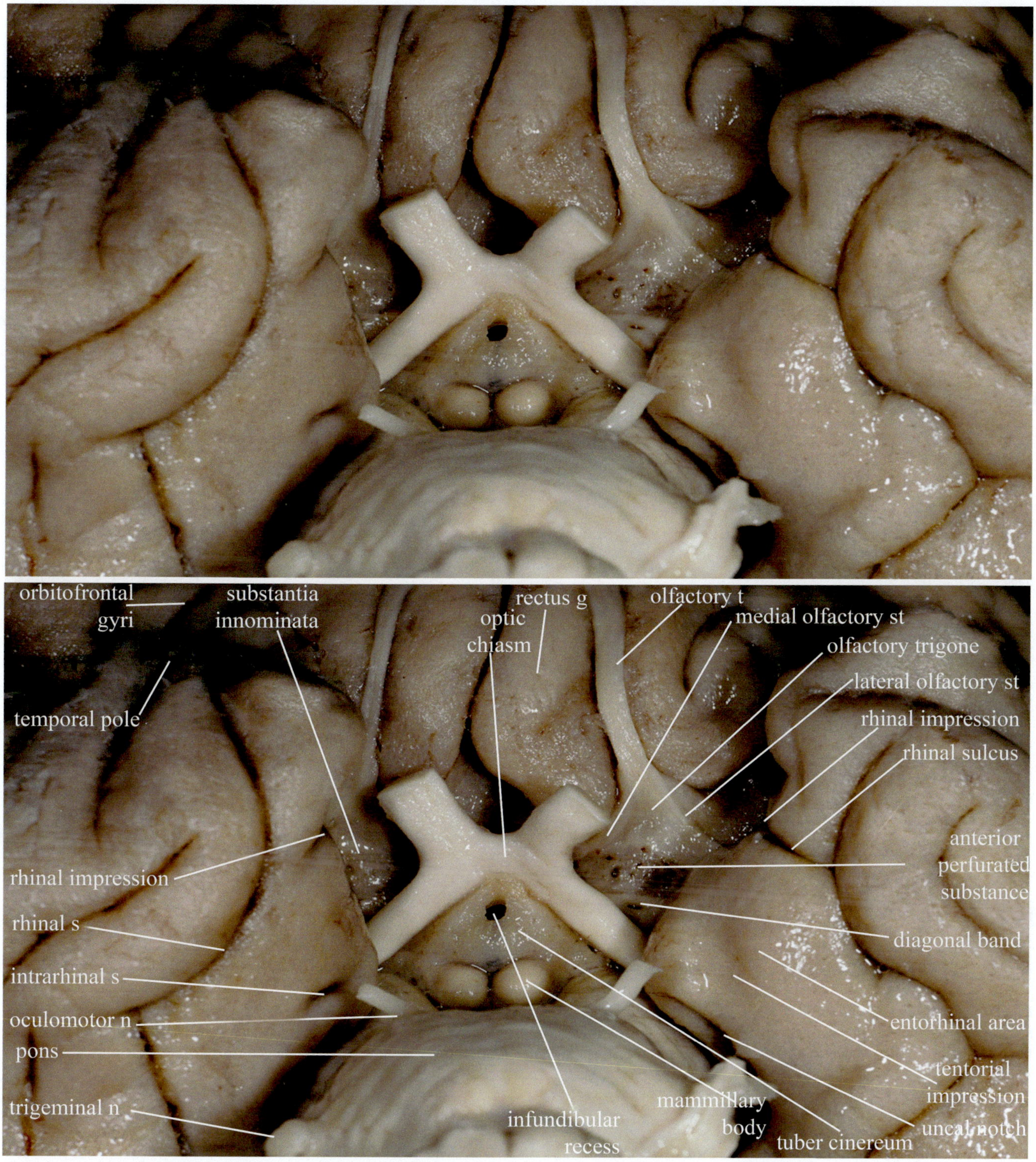

Fig. 6.2 Anatomical preparation of the anterior perforated substance. This substance is bordered by a collar of white matter fibers. Anteriorly, these are the medial and lateral olfactory striae; posteriorly, they are the horizontal and vertical segments of the diagonal band. The ventral striatal and pallidal complex burst at the center of the anterior perforated substance, and because this area does not resemble the typical appearance or consistency of either gray or white matter, it was named the substantia innominata. *g* gyrus, *n* nerve, *s* sulcus, *st* stria, *t* tract (©Kadri 2023. All Rights Reserved)

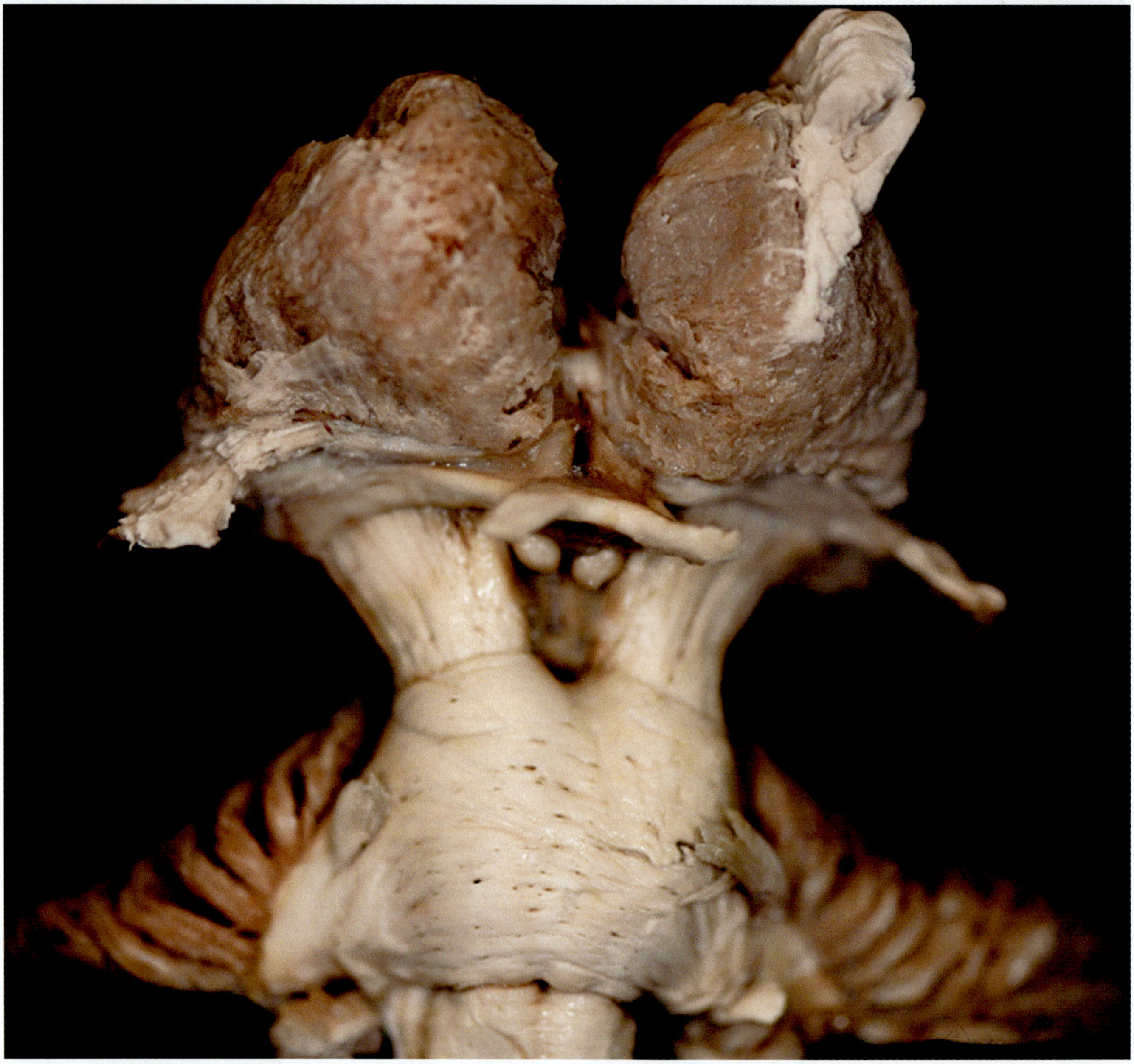

Fig. 6.3 Anatomical preparation of the basal telencephalic nuclei. The striatum is composed of the putamen and caudate nucleus that have been separated by the projection fiber system during ontogenesis. The portion ventral to the anterior commissure is termed the ventral striatum, and its main components are the accumbens nucleus and the olfactory tubercle. *n* nucleus, *s* sulcus (©Kadri 2023. All Rights Reserved)

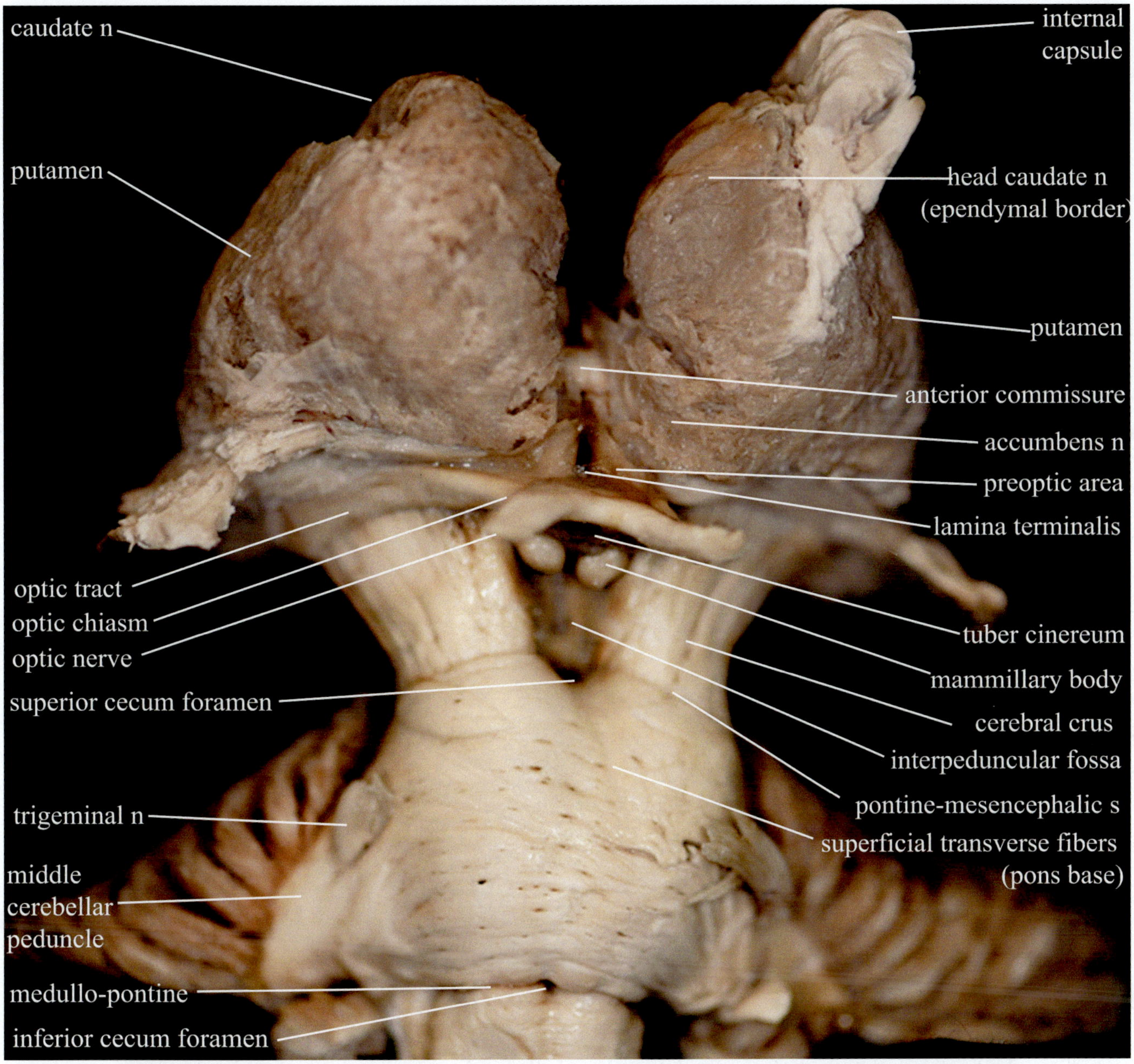

Fig. 6.3 (continued)

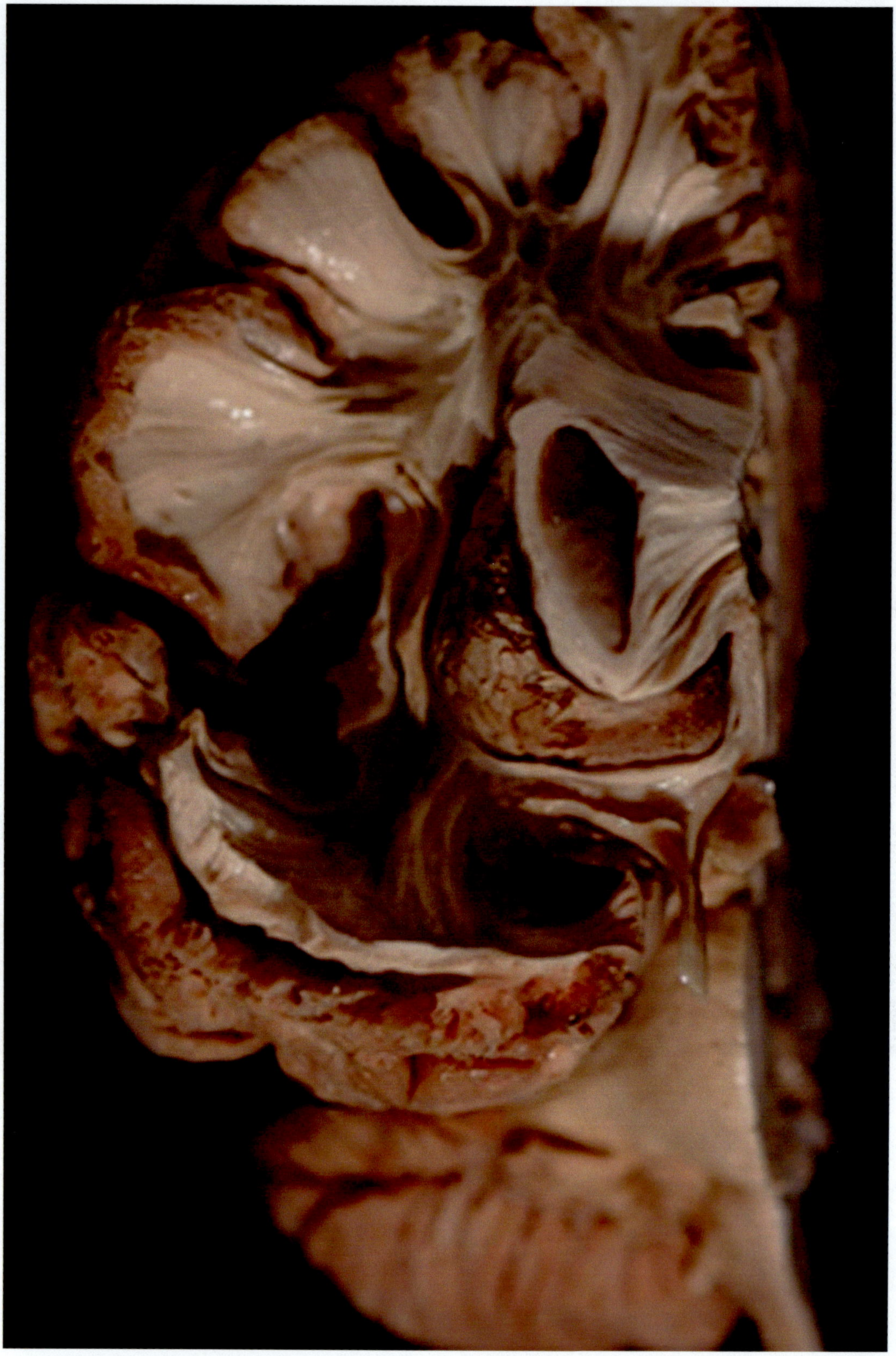

Fig. 6.4 Anatomical preparation of the ventral striatum. The projection fibers incompletely separate the striatum into the caudate nucleus medially and the putamen laterally. Observed in planar cuts (axial, coronal, and sagittal), the gray matter of the striatum is traversed by white strips corresponding to the projection fibers that observers have compared to the cutaneous white striae; hence, it was named striatum. Ventrally, the striatum is not separated and is known as the ventral striatum. The nucleus accumbens and the olfactory tubercle constitute the main neuronal group of the ventral striatum. *lat* lateral, *n* nucleus (©Kadri 2023. All Rights Reserved)

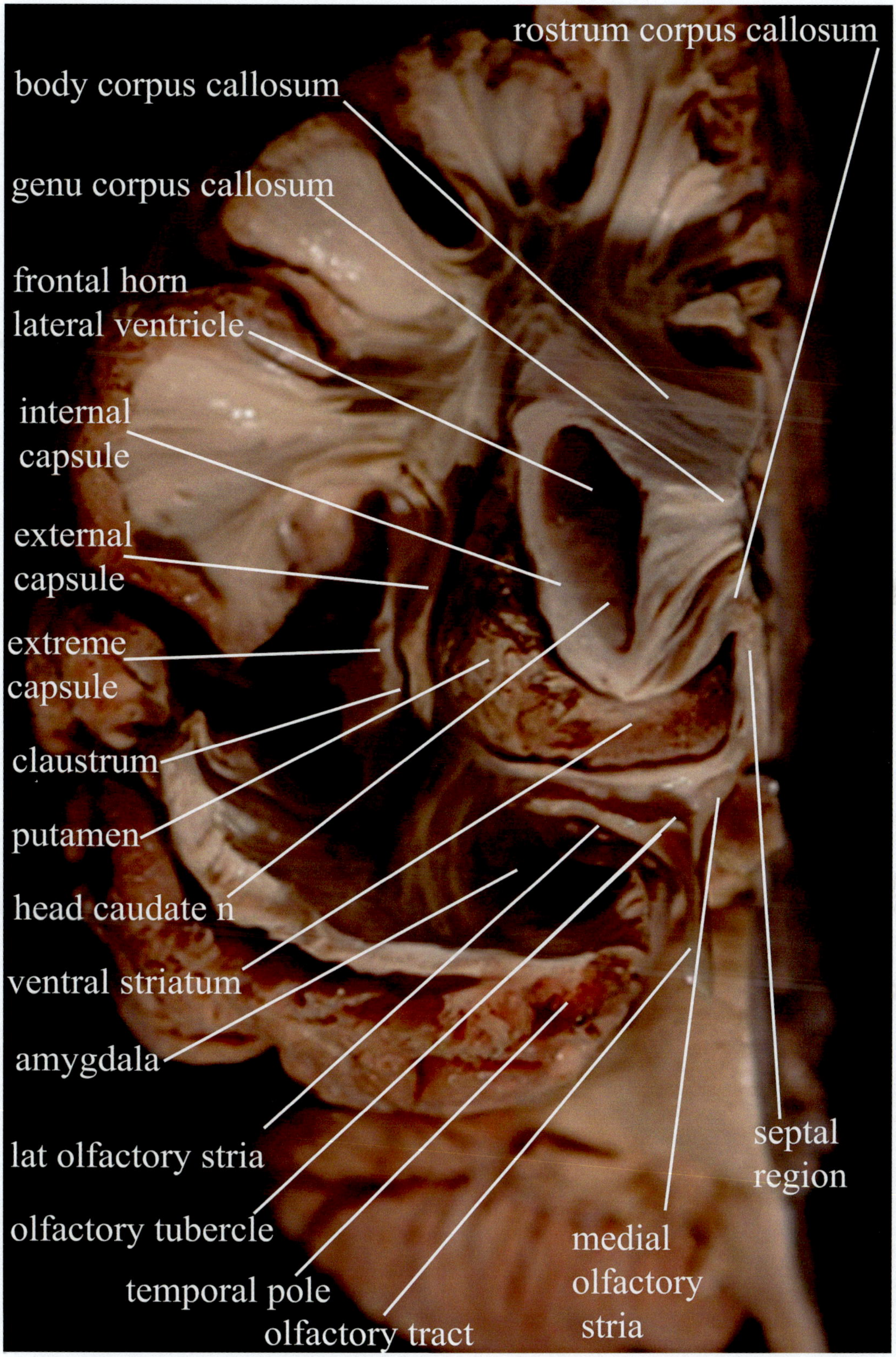

Fig. 6.4 (continued)

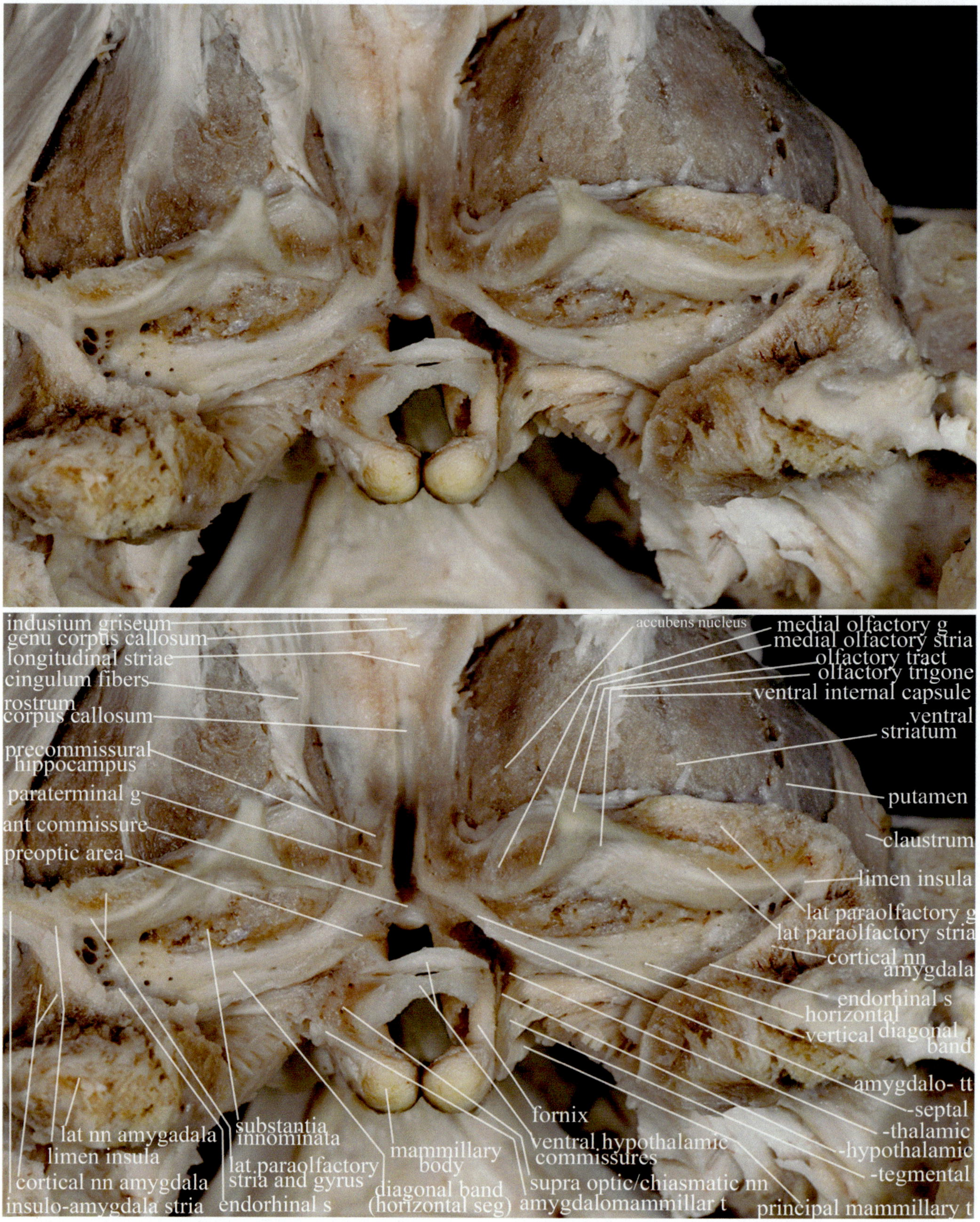

Fig. 6.5 Anatomical preparation of the basal nuclei surrounding the anterior perforated substance. The ventral striatum is mainly composed of the nucleus accumbens and the olfactory tubercule. *g* gyrus, *lat* lateral, *nn* nucleus, *s* sulcus, *seg* segment, *t* tract, *tt* tract (©Kadri 2023. All Rights Reserved)

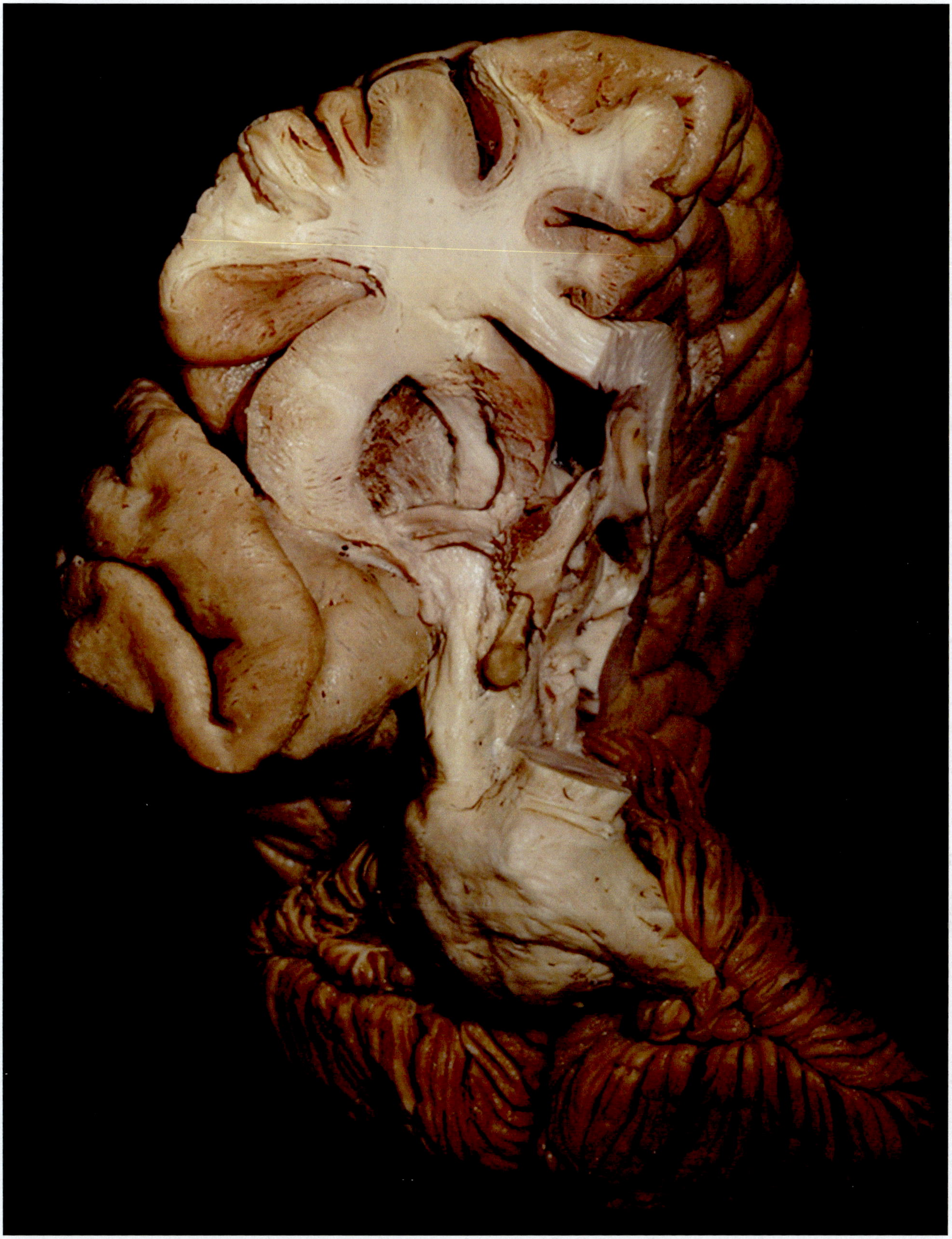

Fig. 6.8 Anatomical preparation of the dorsal striatum and dorsal pallidum. The basal nuclei ventral to the anterior commissure have been removed. The separation of the caudate nucleus and putamen by the internal capsule is evident. The internal globus pallidus is also called the entopeduncular nucleus because of its relationship with the internal capsule and cerebral peduncle. *ant* anterior, *f* fasciculus, *g* gyrus, *n* nerve, *s* sulcus (©Kadri 2023. All Rights Reserved)

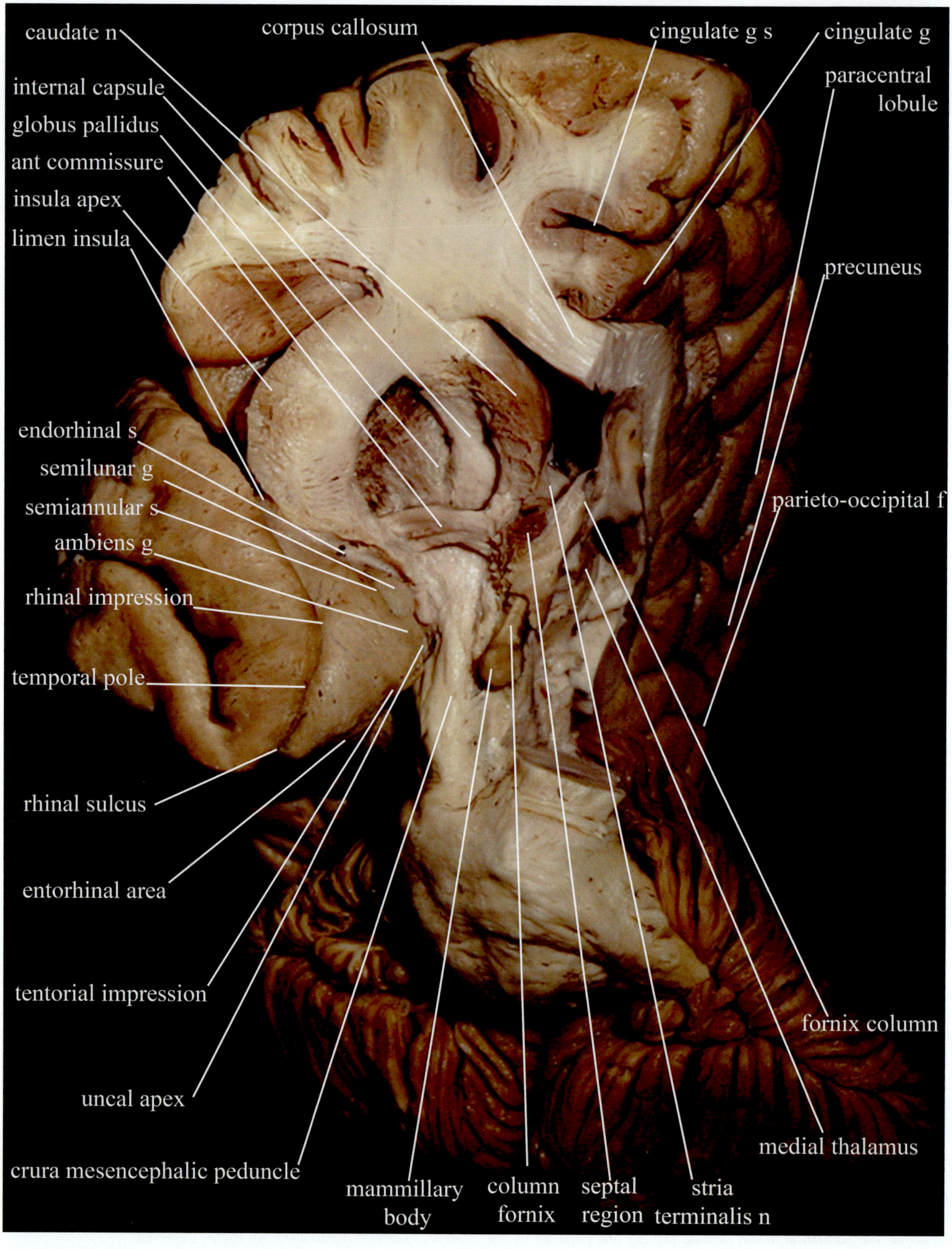

Fig. 6.8 (continued)

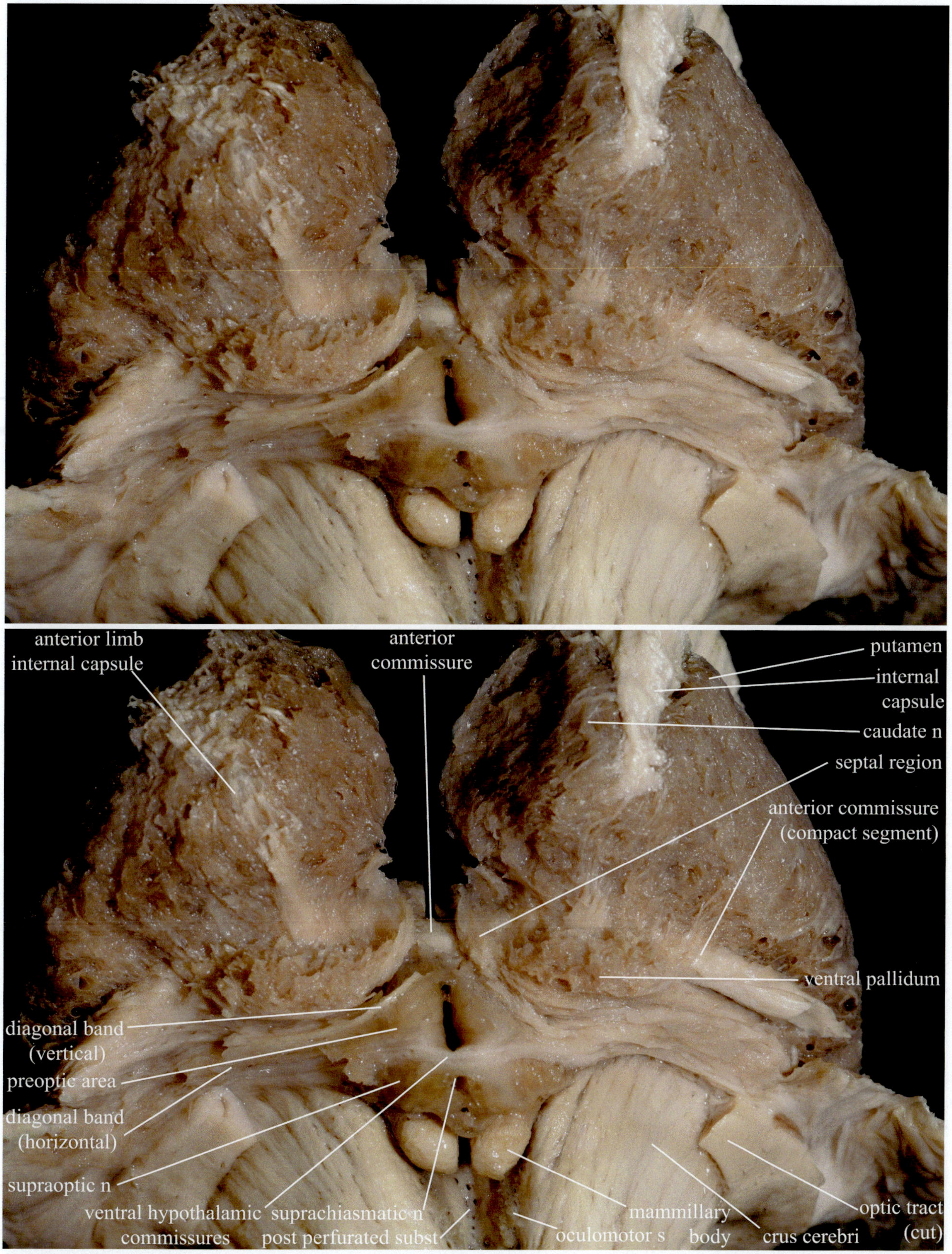

Fig. 6.9 Anatomical preparation of the basal telencephalic nuclei. The ventral striatum has been removed to expose the anterior commissure. This commissure is the limit between the ventral and dorsal regions of the basal nuclei. *n* nerve, *post* posterior, *subst* substance (©Kadri 2023. All Rights Reserved)

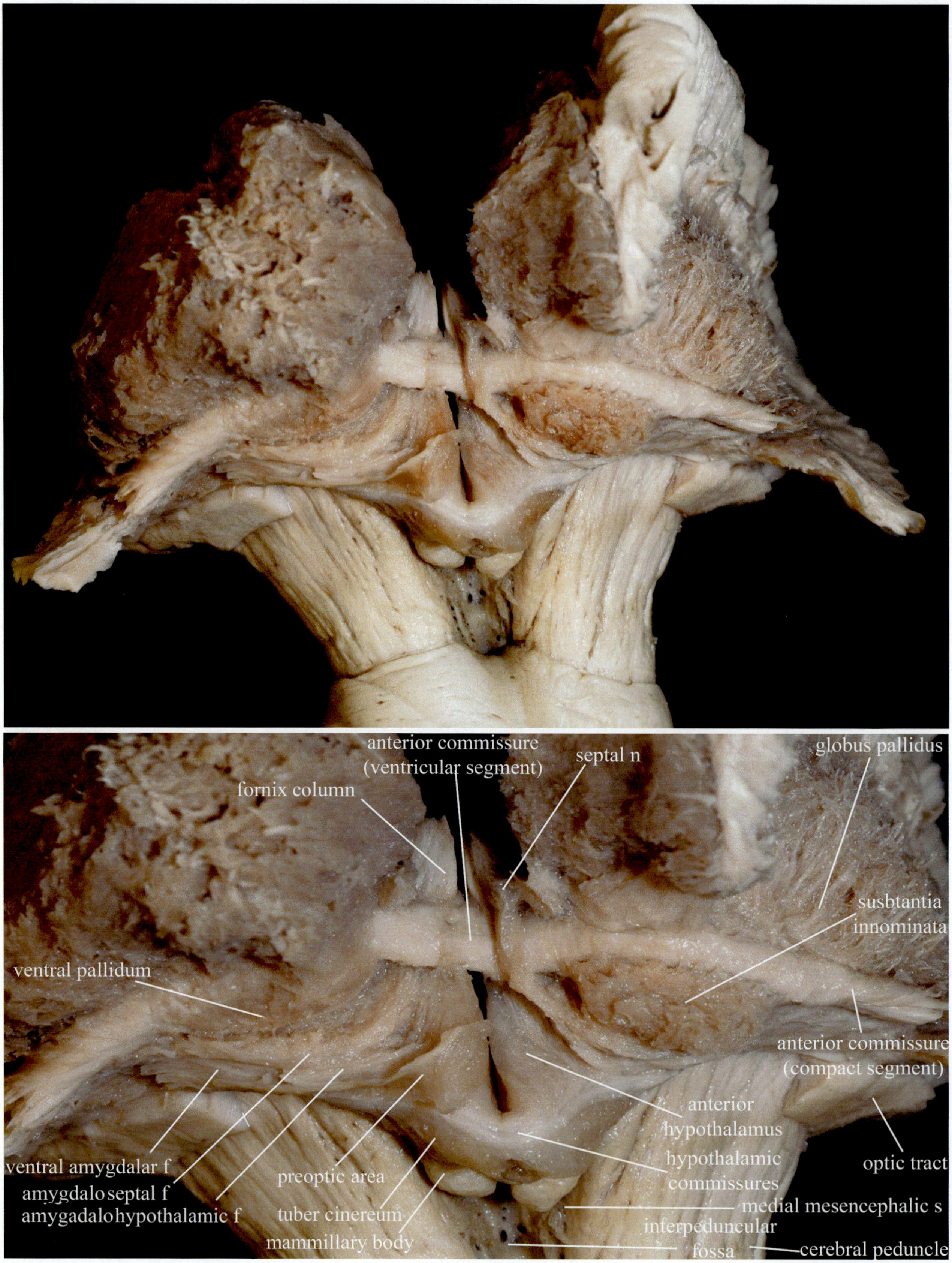

Fig. 6.10 Anatomical preparation of the ventral basal nuclei. The ventral basal nuclei protrude into the anterior perforated substance as a grayish substance that is limited by the diagonal band and the anterior commissure. *f* fasciculus, *n* nerve, *s* sulcus (©Kadri 2023. All Rights Reserved)

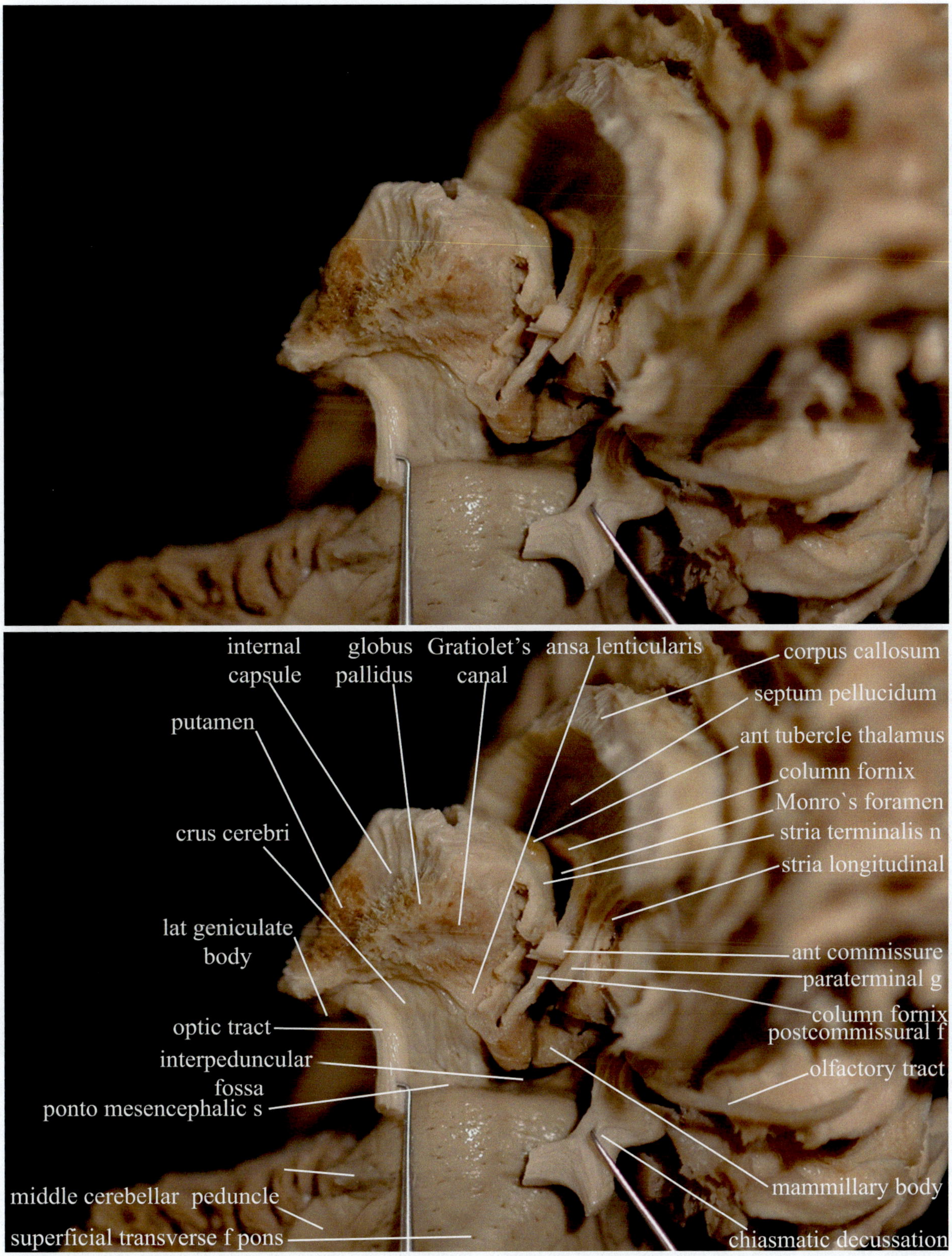

Fig. 6.11 Anatomical preparation of the dorsal pallidum and putamen. The ventral pallidum and striatum have been removed. The anterior commissure is severed at the medial portion of its pars compacta. Gratiolet's canal is exposed. The optic tract was transected at the transition of the internal capsule and crus cerebri. The fibers of the optic nerve, chiasm, and tract were dissected. The nasal fibers cross to the contralateral fibers. The ansa lenticularis loops around the ventral crus cerebri and connects the laterally positioned putamen and the external globus pallidus with the medially located thalamus, subthalamus, and tegmentum of the mesencephalon (mainly with the red nucleus). *ant* anterior, *f* fasciculus, *g* gyrus, *lat* lateral, *n* nerve, *s* sulcus (©Kadri 2023. All Rights Reserved)

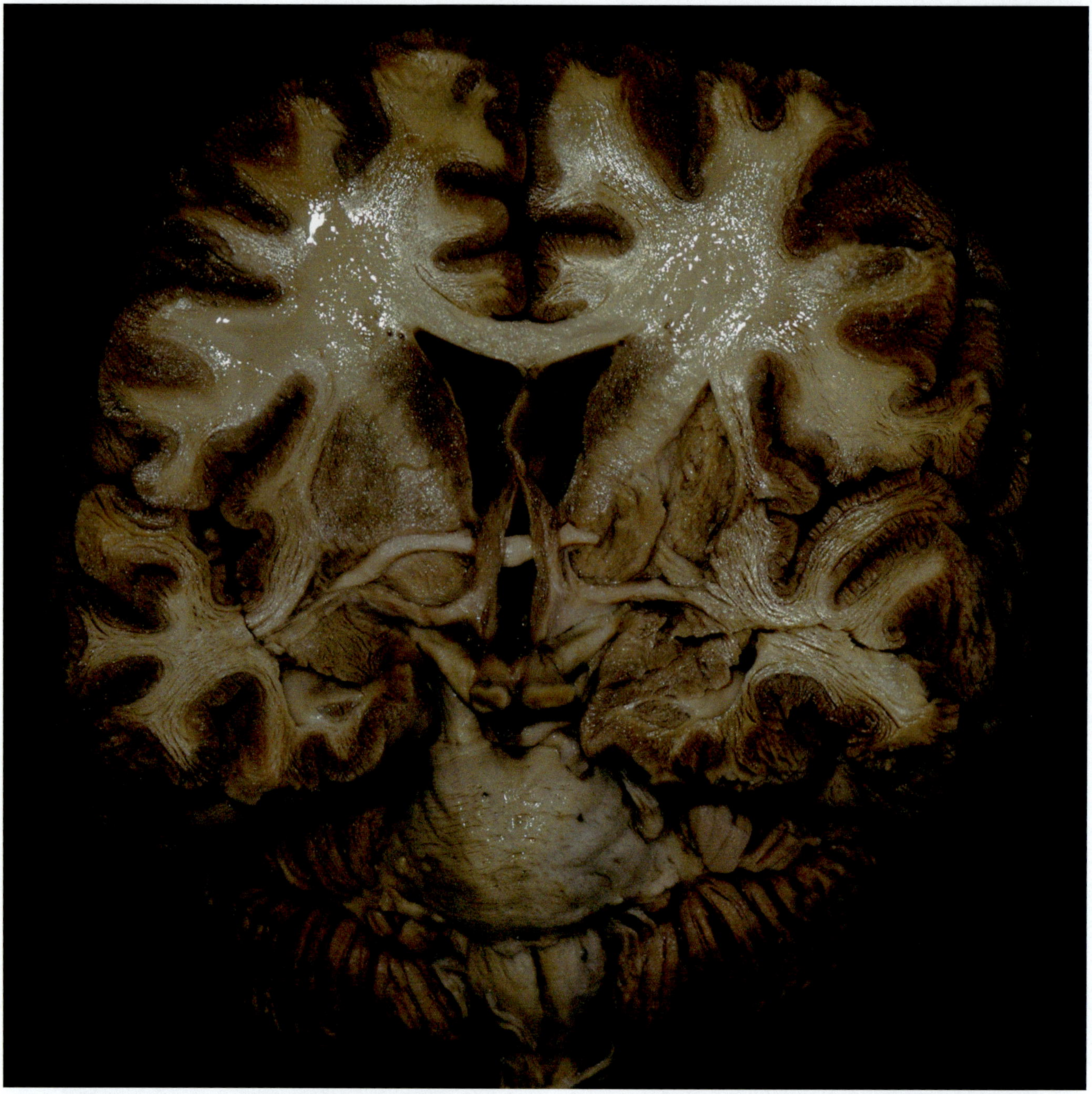

Fig. 6.12 Anatomical preparation of the amygdala, globus pallidus, and lentiform nucleus. The ventral fibers of the amygdala are the most superficial fibers of the basal nuclei fiber system. They loop around the ventral surface of the projection fibers that compose the crus cerebri. The peduncular loop is also known as the ansa peduncularis and is composed of the basal telencephalic nuclei connections with the diencephalon and brainstem. The basal telencephalic nuclei are composed of the putamen, the caudate nucleus, the globus pallidus, the amygdala, and the septal region. The septal region is an important relay of the limbic system situated at the limits of the precommissural hippocampal formation, the hypothalamus, and the ventral dorsal thalamus that harbors the nuclei of the stria terminalis. On the surface, the bed nucleus of the stria terminalis is covered by fibers of the stria terminalis and a hyalin thickness of the ependymal layer called the lamina affixa. The amygdala is a complex of nuclei that can be grouped by using different methods. The method used here followed the context of the current piece and classified the nuclei according to morphology, connection, and relationships with other structures. In the connective context, the ventral fibers of the amygdala converge to form the horizontal segment of the diagonal band. The most superficial of these fibers separate and connect the amygdala with the septal region. The fibers connecting the amygdala with the hypothalamus are the middle layer of fibers, connecting to the mammillary and chiasmatic areas. Therefore, they are not part of the ansa peduncularis as they do not loop around the peduncle. A deeper layer of fibers of the amygdala loops around the cerebral peduncle to reach the limbic dorsal thalamus, the epithalamus, and the tegmentum mesencephalic. *f* fasciculus, *lat* lateral, *m* medial, *n* nerve, *seg* segment

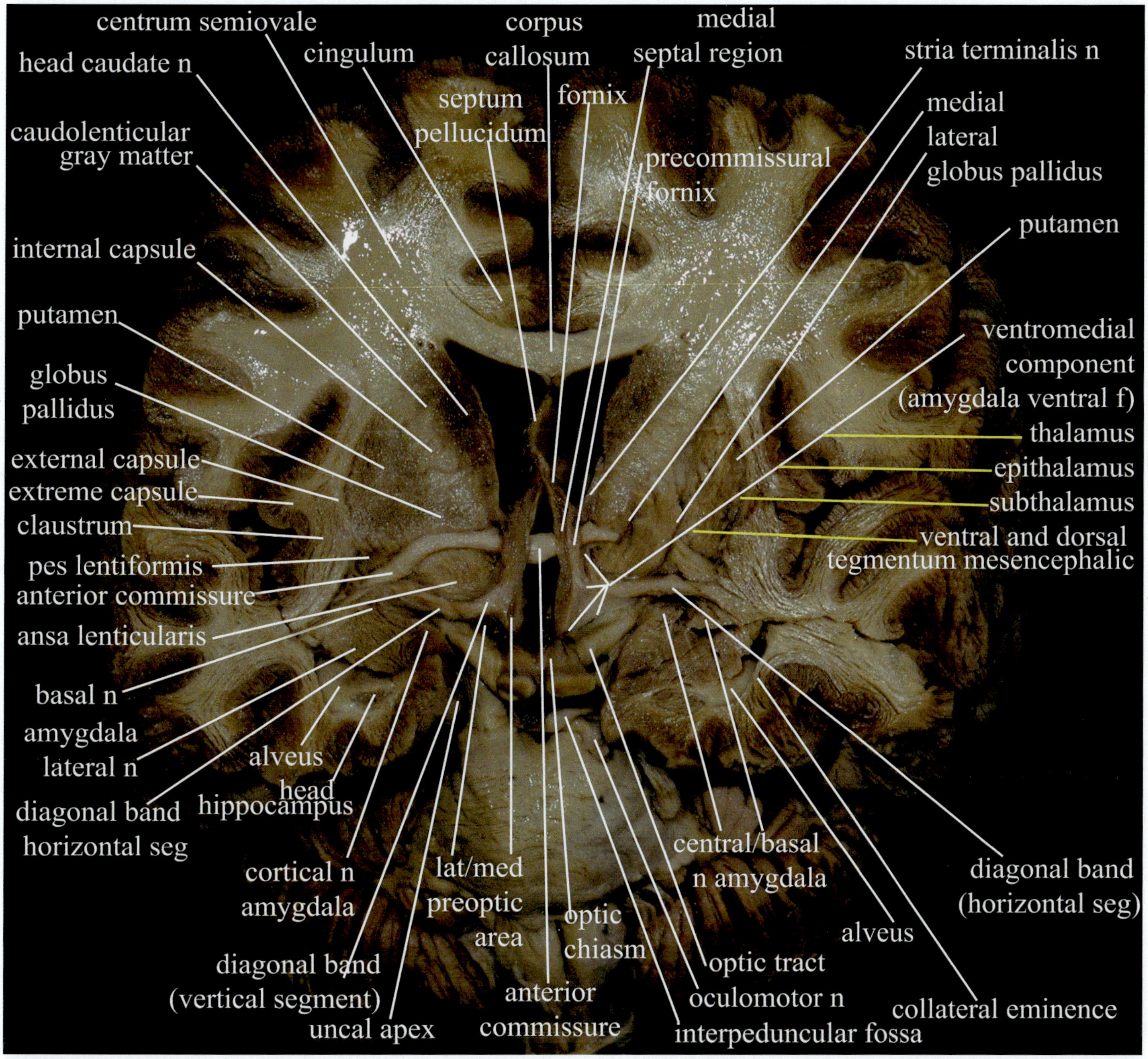

Fig. 6.12 (continued)

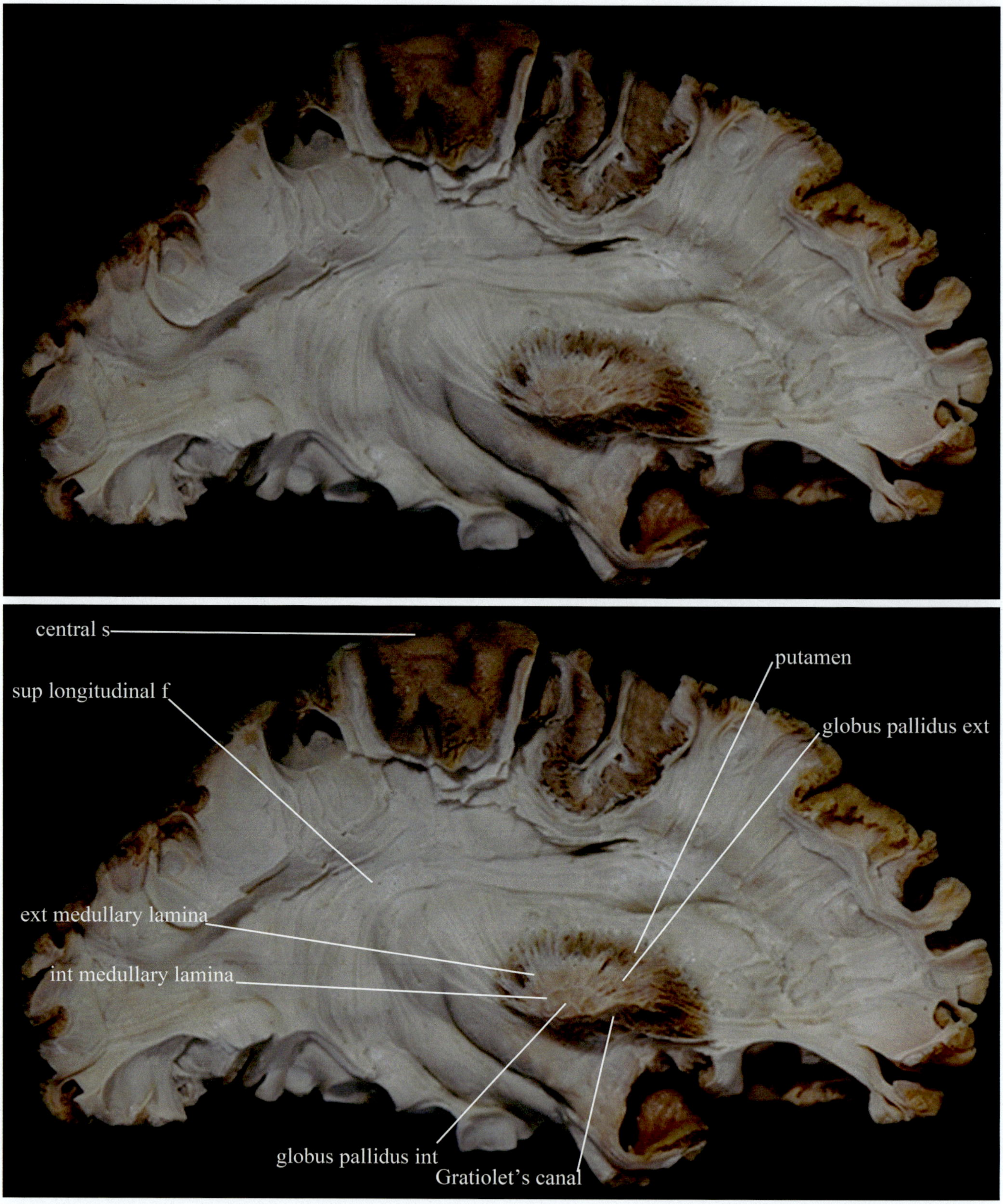

Fig. 6.13 Anatomical preparation of the capsular surface of the lentiform nucleus. The lentiform nucleus is composed of the putamen and the globus pallidus. *ext* external, *f* fasciculus, *int* internal, *s* sulcus, *sup* superior (©Kadri 2023. All Rights Reserved)

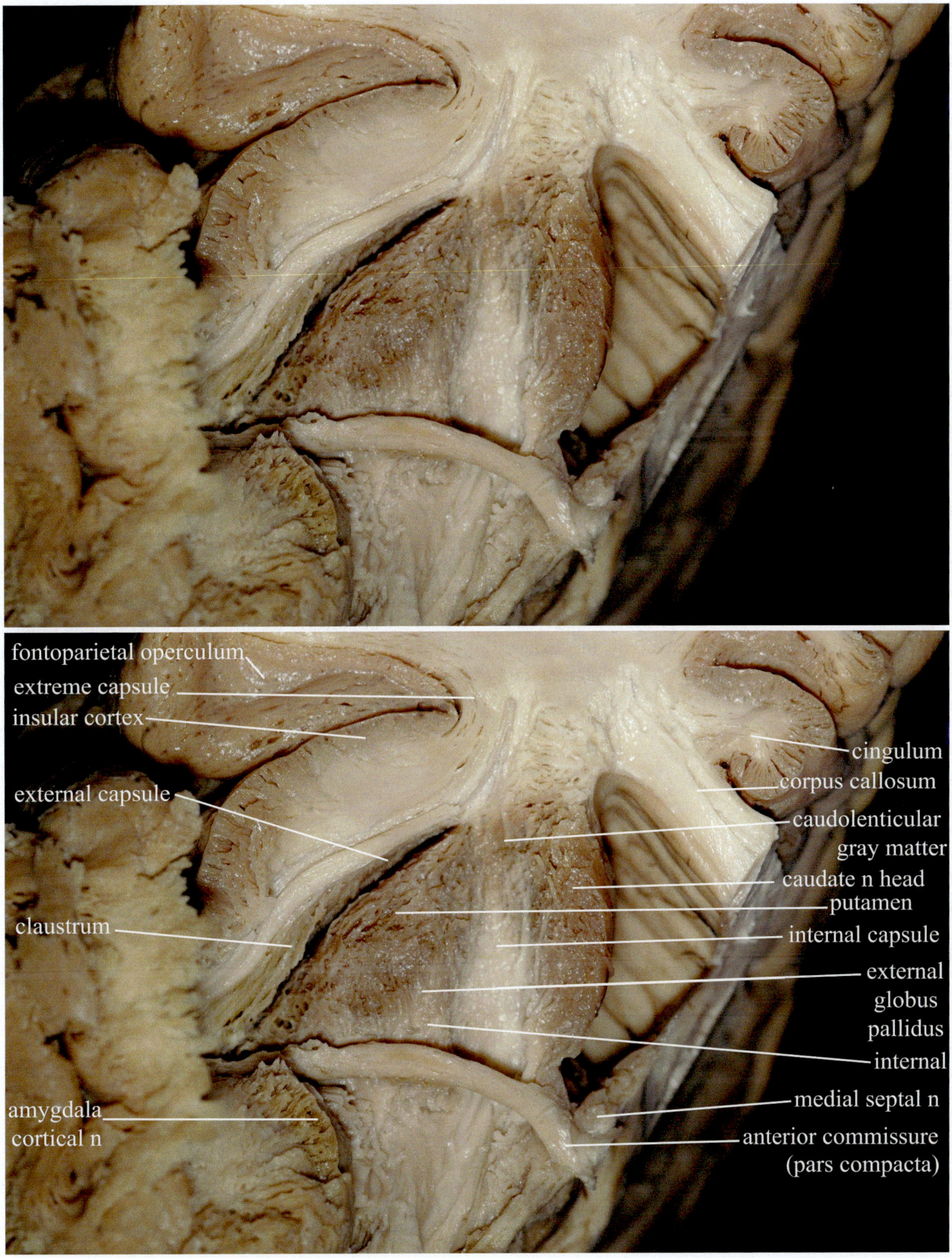

Fig. 6.14 Anatomical preparation of the striatal gray matter bridges traversing the internal capsule. The caudolenticular (or striatal) gray matter connections are broader and more numerous ventrally, resulted in the thinner and sparsely fibrous aspect of the anterior limb of the internal capsule. *n* nucleus (©Kadri 2023. All Rights Reserved)

The main afferents of the striatum are from the cortex, the intramedullary nuclei of the thalamus, the substantia nigra, and the dorsal raphe nuclei. The cortical connections form the external capsule and the subcallosal stratum (or Muratoff bundle) [1, 2]. The connections of the lentiform nucleus with the intermedullary nucleus of the thalamus, substantia nigra, and dorsal raphe nuclei do not pass through the internal capsule. Instead, they turn around the crus of the mesencephalon and constitute the ansa lenticularis. Together with the ventral fibers of the amygdala, which also turn around the peduncle (see below), the ansa lenticularis constitutes the ansa peduncularis. The fibers from the amygdala and lentiform nucleus that are projected toward the thalamus do not pass through the internal capsule and constitute part of the extracapsular thalamic peduncle [7].

The Globus Pallidus

The globus pallidus is a nucleus on the mediobasal aspect of the putamen (Figs. 6.7, 6.8, 6.9, 6.10, 6.11, 6.12, 6.13, 6.14, 6.15, 6.16, 6.17, 6.18, 6.19, 6.20, 6.21, 6.22, 6.23, and 6.24). Compared with the putamen, it has more consistency and a paler appearance on macroscopic view; hence, it received its name. It is separated from the putamen by the external medullary lamina, of the globus pallidus. The internal medullary lamina separates the globus pallidus into internal and external parts. The main afferent connection of the globus pallidus is the striatum. The main efferent connection of the external globus pallidus projects to the subthalamic nucleus of Luys. The fibers of the external globus pallidus converge and constitute the internal medullary lamina of the globus pallidus, in coronal and axial cuts. The internal medullary lamina constitutes the ansa lenticularis, part of the ansa peduncularis. The internal segment projects toward the thalamus, habenula, and tegmental mesencephalic and pontine nuclei. These fibers traverse the internal capsule toward the tegmental field ventral to the zona incerta (the H1 field of Forel). They are named the lenticular fasciculus, which merges with the ansa lenticularis around the zona incerta (or H field of Forel) and ascends as the thalamic fasciculus. The thalamic fasciculus is also known as the H2 field of Forel and projects toward the anterior ventral lateral nucleus [2, 8].

Other structures related to the basal ganglia are the substantia innominata, the basal nucleus of Meynert, and the magnocellular basal forebrain. The substantia innominata is a flat mass located ventral to the globus pallidus and separated from it by the anterior commissure. It is also termed the ventral pallidum [2]. Anteriorly, it covers the olfactory tubercle and is bordered medially by the hypothalamus. The basal nucleus of Meynert and the magnocellular basal forebrain nuclei are not clearly identified on macroscopic brain dissection, even with the aid of the surgical microscope. The ventral portion of the globus pallidus, which is intimately related to the substantia innominata, is also referred to as the ventral pallidum. The dorsal pallidum is related to the striatum.

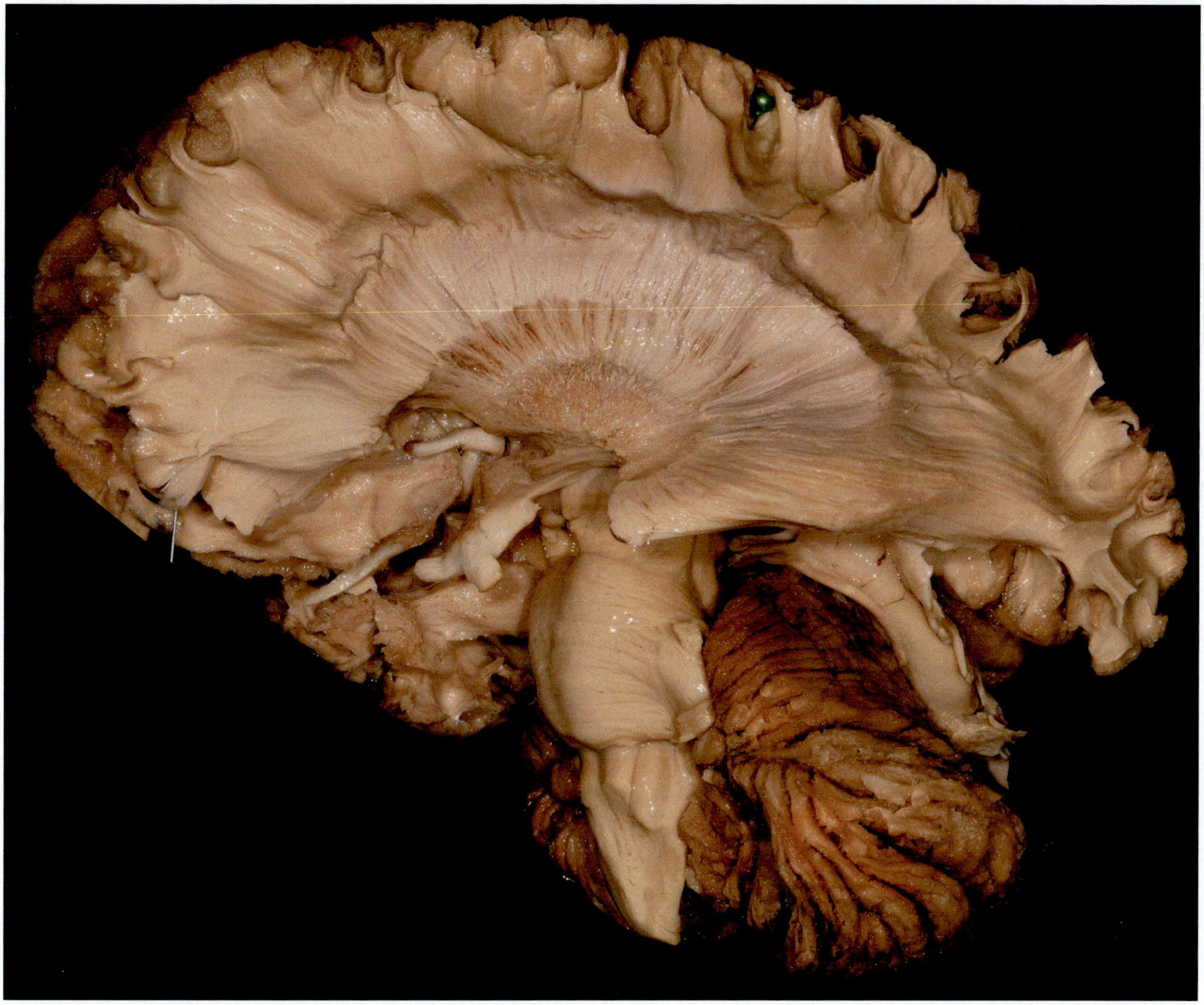

Fig. 6.15 Anatomical preparation of the left external globus pallidus. The association and commissural fibers of the left hemisphere have been removed. The external boundary of the globus pallidus is the external medullary lamina, which has a "hairy" aspect because of the fibers' convergence of the pallidum-striatal connections. *g* gyrus, *n* nucleus, *ped* peduncle, *s* sulcus, *tr* tract (©Kadri 2023. All Rights Reserved)

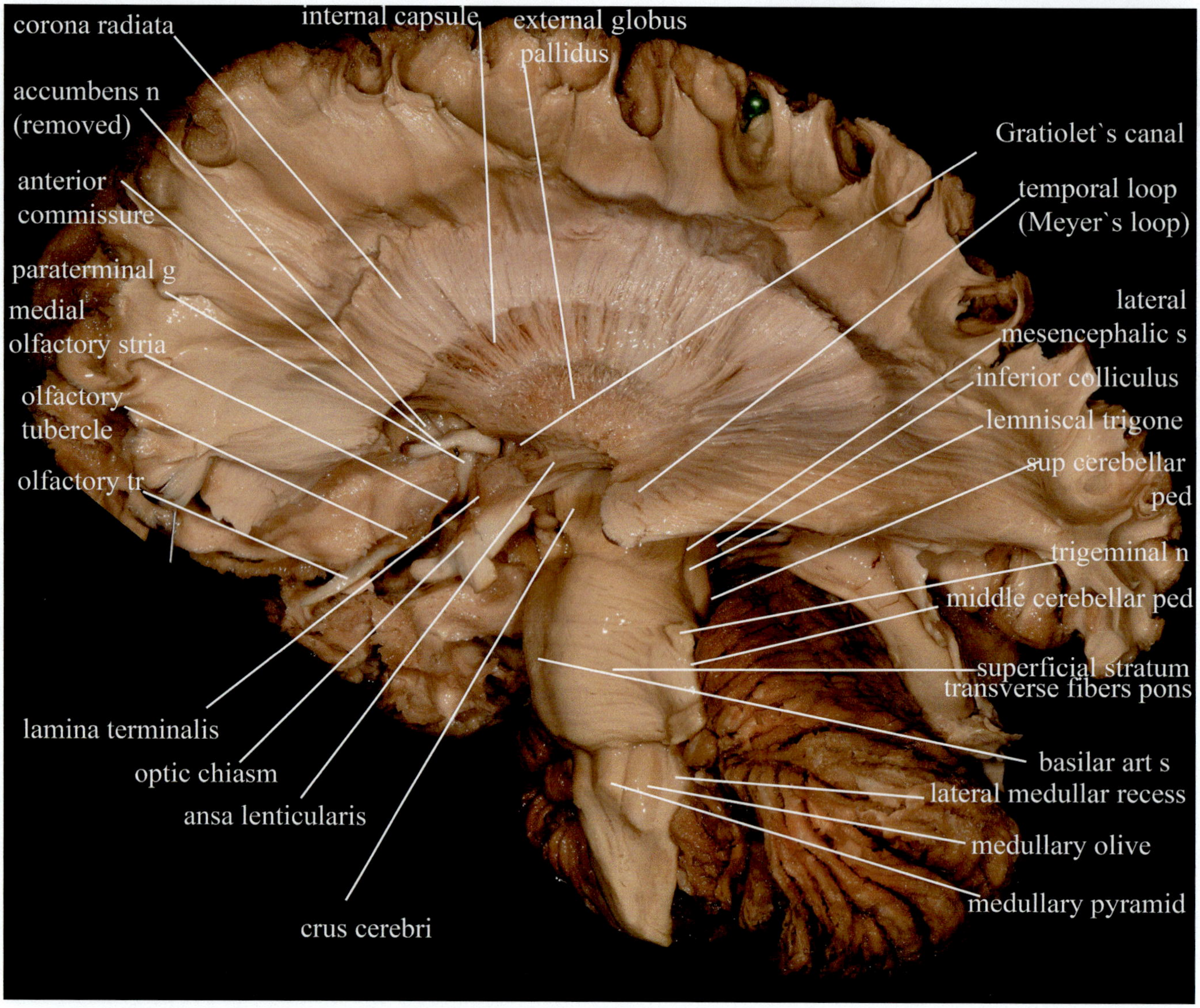

Fig. 6.15 (continued)

Fig. 6.16 Anatomical preparation of the external globus pallidus and Gratiolet's canal. Displacing the lateral extension of the anterior commissure from the canal exposes the amygdala, ventral striatum, and pallidum continuum.

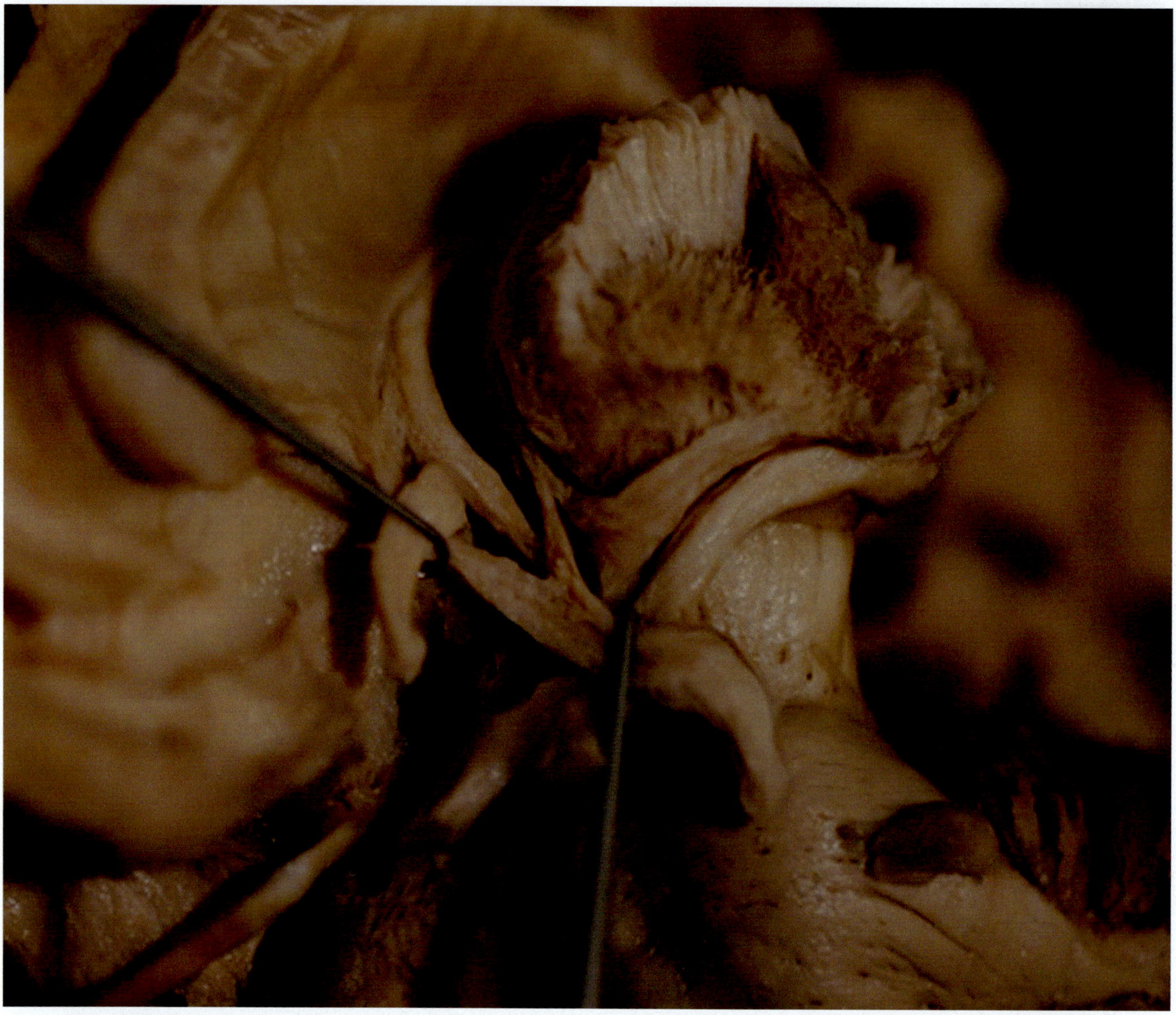

Fig. 6.17 Anatomical preparation of the ansa peduncularis. Ventral amygdala connections reach the septal region, hypothalamus (anterior and mammillary regions), thalamus (anterior and medial), epithalamus (habenula nuclei), dorsal tegmental mesencephalic region, and ventral mesencephalic region. The fibers connecting to the thalamus, epithalamus, and tegmentum mesencephalic loop around the ventricle and compose, with the ansa lenticularis, the so-called ansa peduncularis. *ant* anterior, *b* body, *f* fasciculus, *lat* lateral, *n* nerve, *ped* peduncle, *s* sulcus, *t* tract (©Kadri 2023. All Rights Reserved)

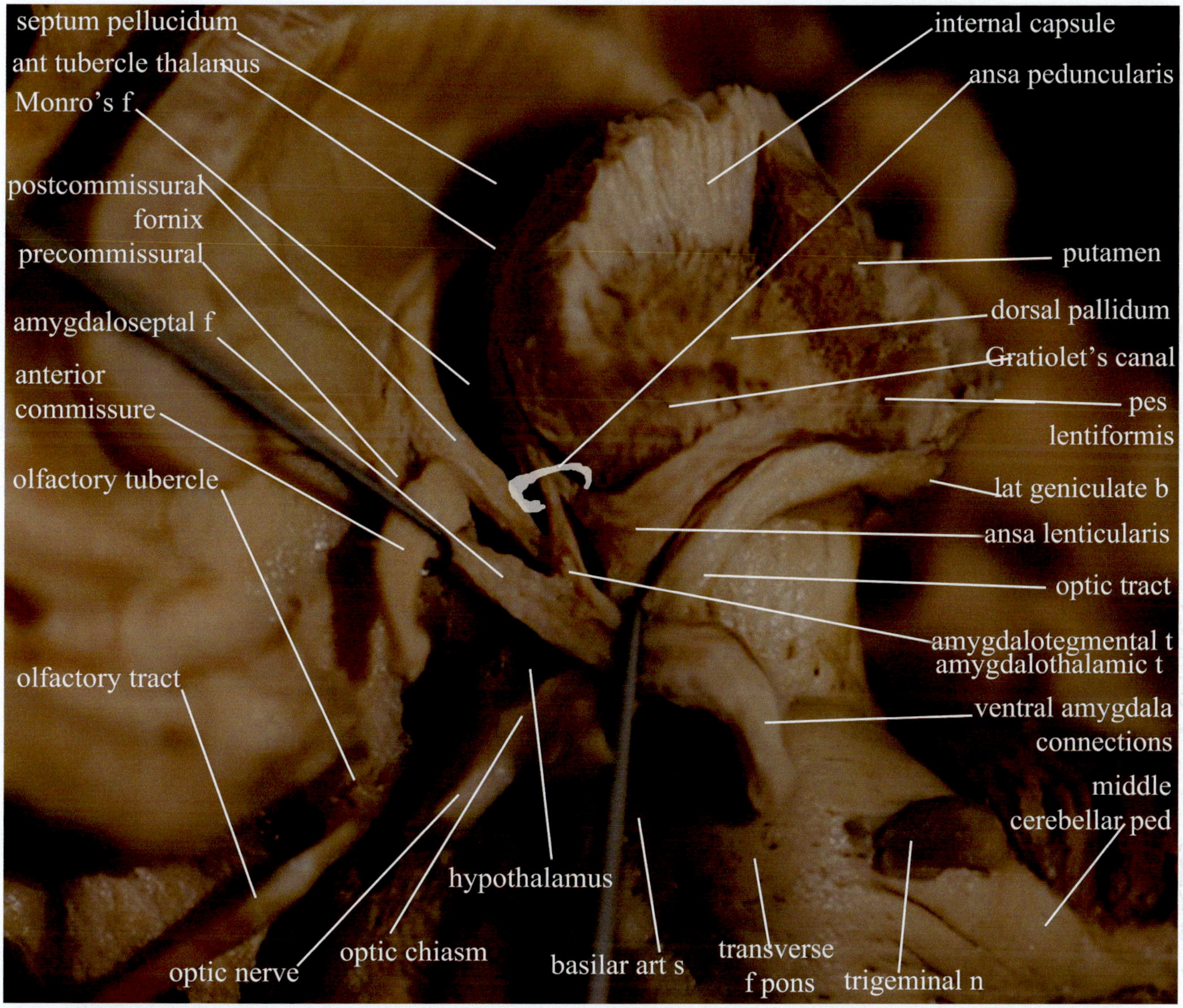

Fig. 6.17 (continued)

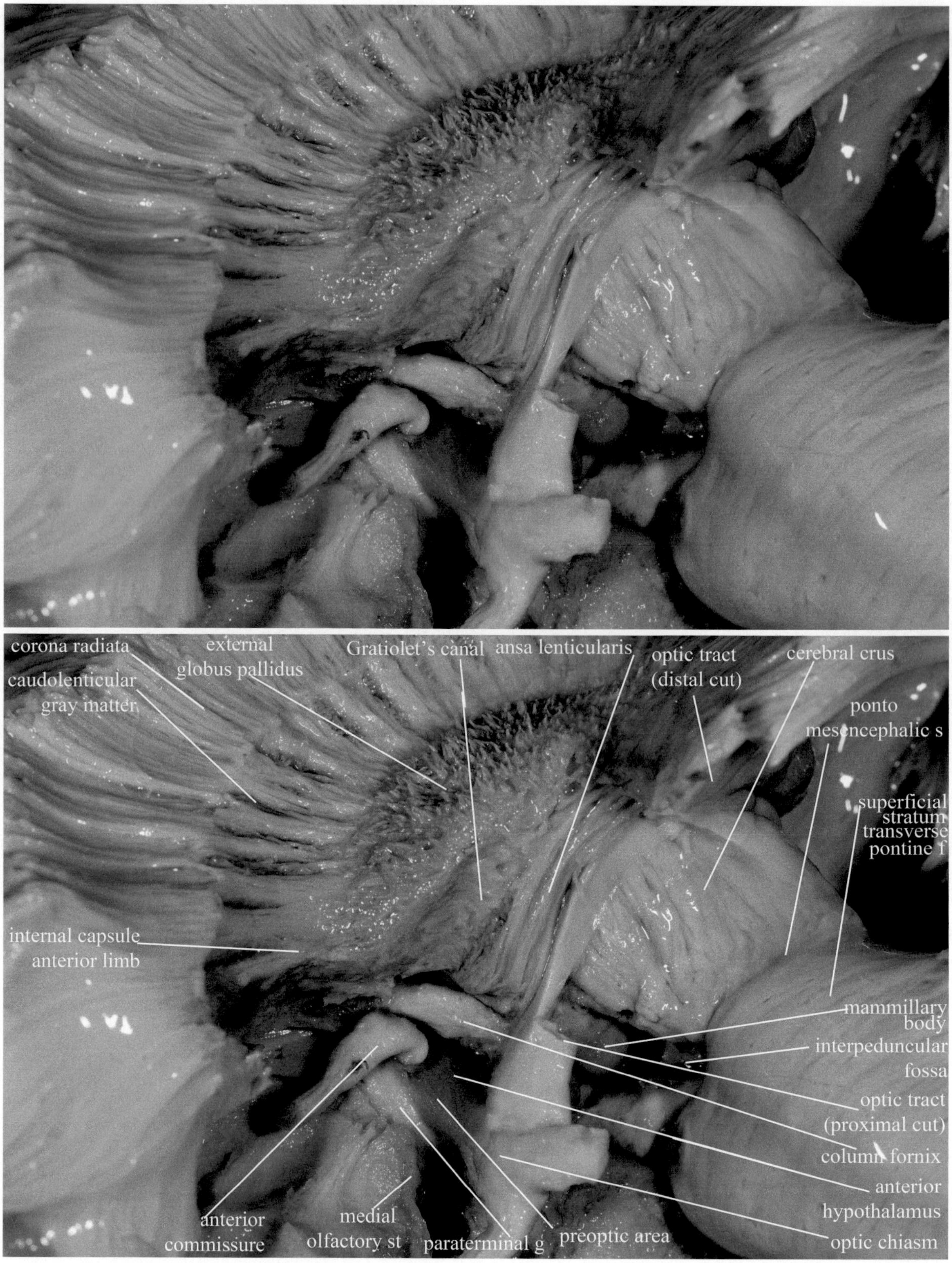

Fig. 6.18 Anatomical preparation of the ansa lenticularis. Removing the amygdala, its connections, the anterior commissure, and the optic tract exposes the connections of the dorsal lentiform that loop around the ventral internal capsule and peduncle, called the ansa lenticularis. *f* fasciculus, *g* gyrus, *s* sulcus, *st* stria (©Kadri 2023. All Rights Reserved)

Fig. 6.19 Anatomical preparation of the dorsal striatum. The posterior thalamic peduncle and temporopontine fibers compose the projection fibers that assume the characteristic loop at the roof and lateral wall of the temporal horn of the lateral ventricle. *n* nucleus (©Kadri 2023. All Rights Reserved)

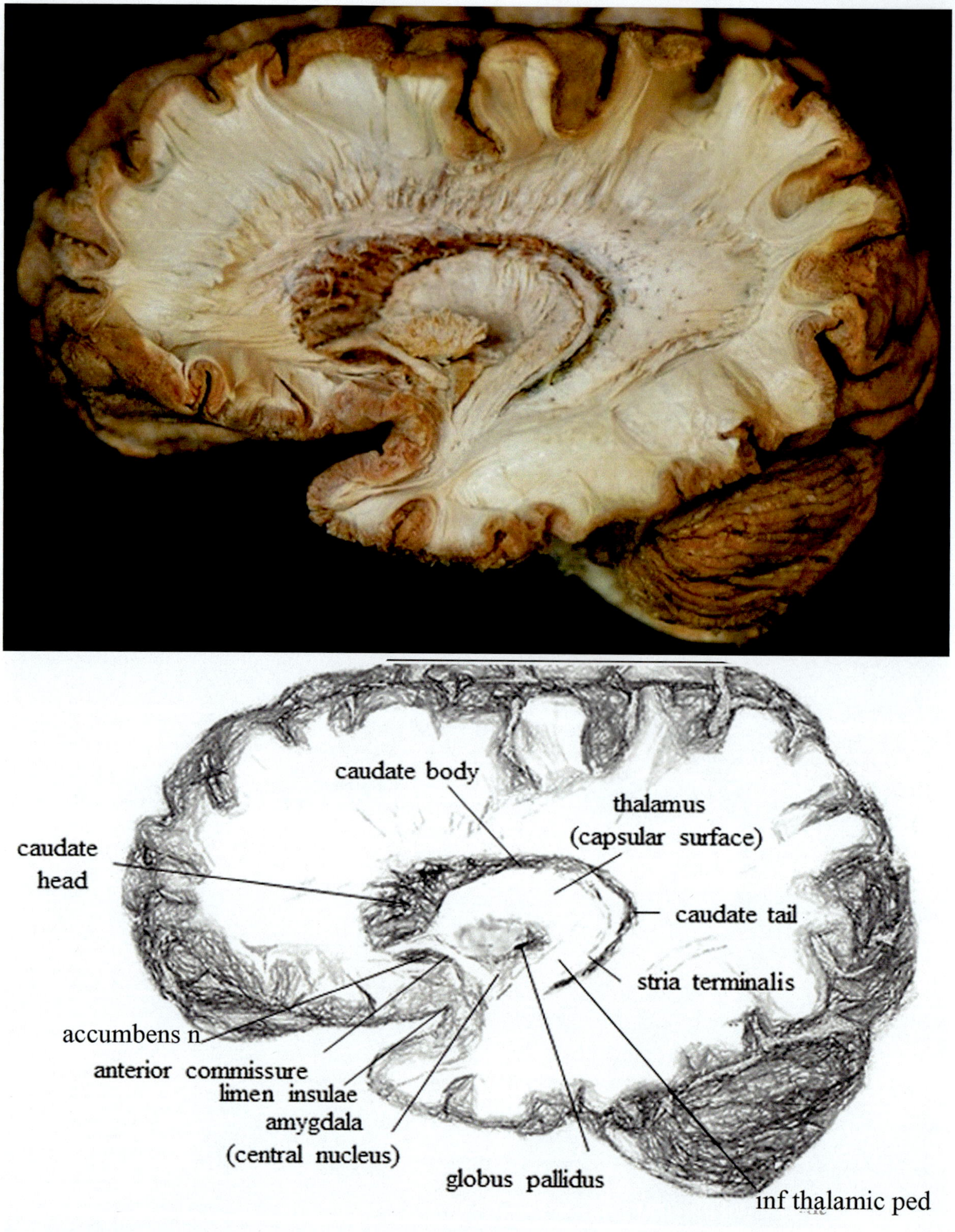

Fig. 6.20 Anatomical preparation and illustration of the dorsal pallidum (external globus pallidus) and nucleus accumbens. The association, commissural, and projection fibers of the left hemisphere have been removed. *inf* inferior, *n* nucleus (©Kadri 2023. All Rights Reserved)

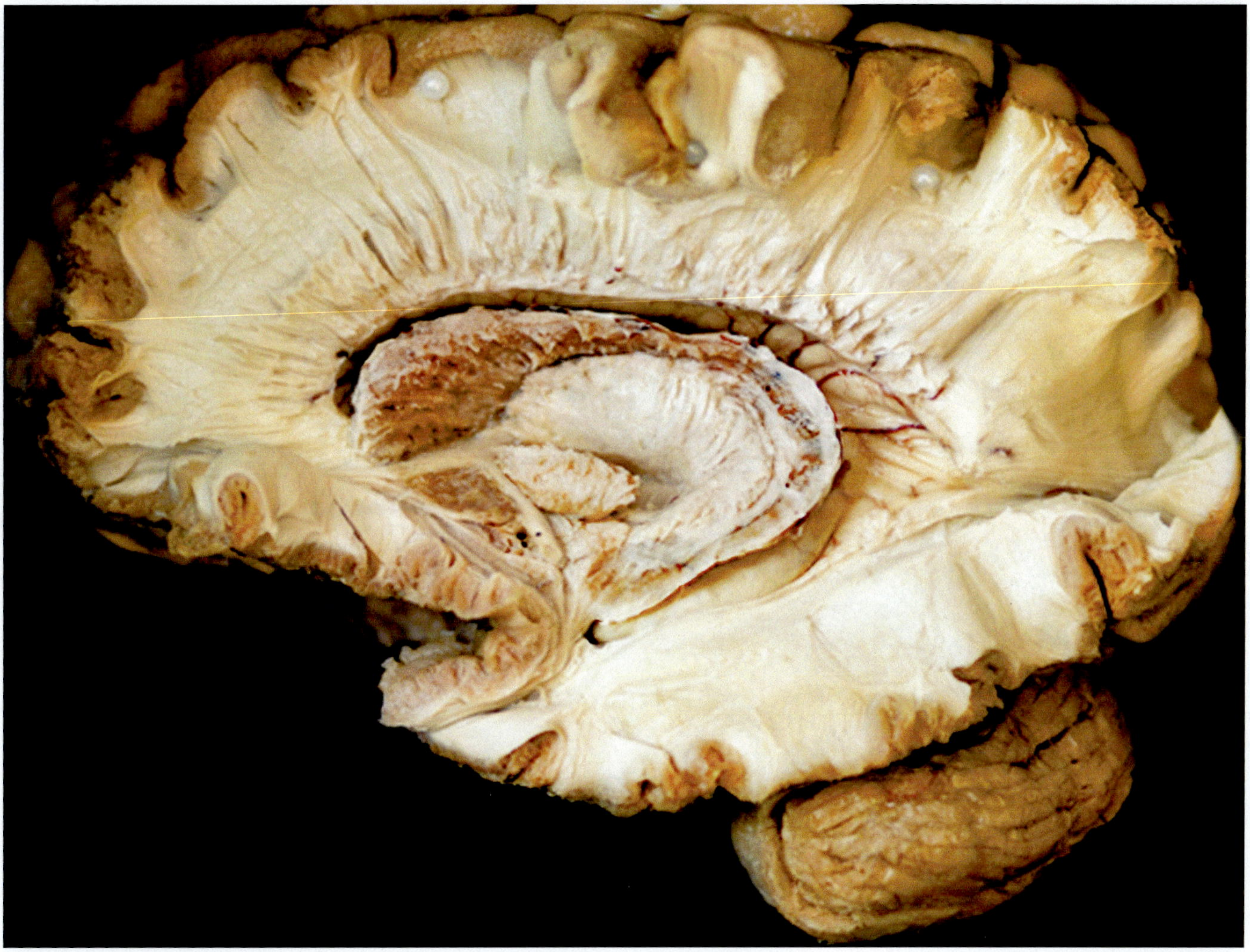

Fig. 6.21 Anatomical preparation of the basal nuclei of the telencephalon. The putamen and internal capsule have been removed, and the capsular surface of the thalamus is exposed. The caudate nucleus wraps around the thalamus, separated by the striothalamic sulcus. The exposed lateral ventricles depict the frontal and uncal recesses. *s* sulcus (©Kadri 2023. All Rights Reserved)

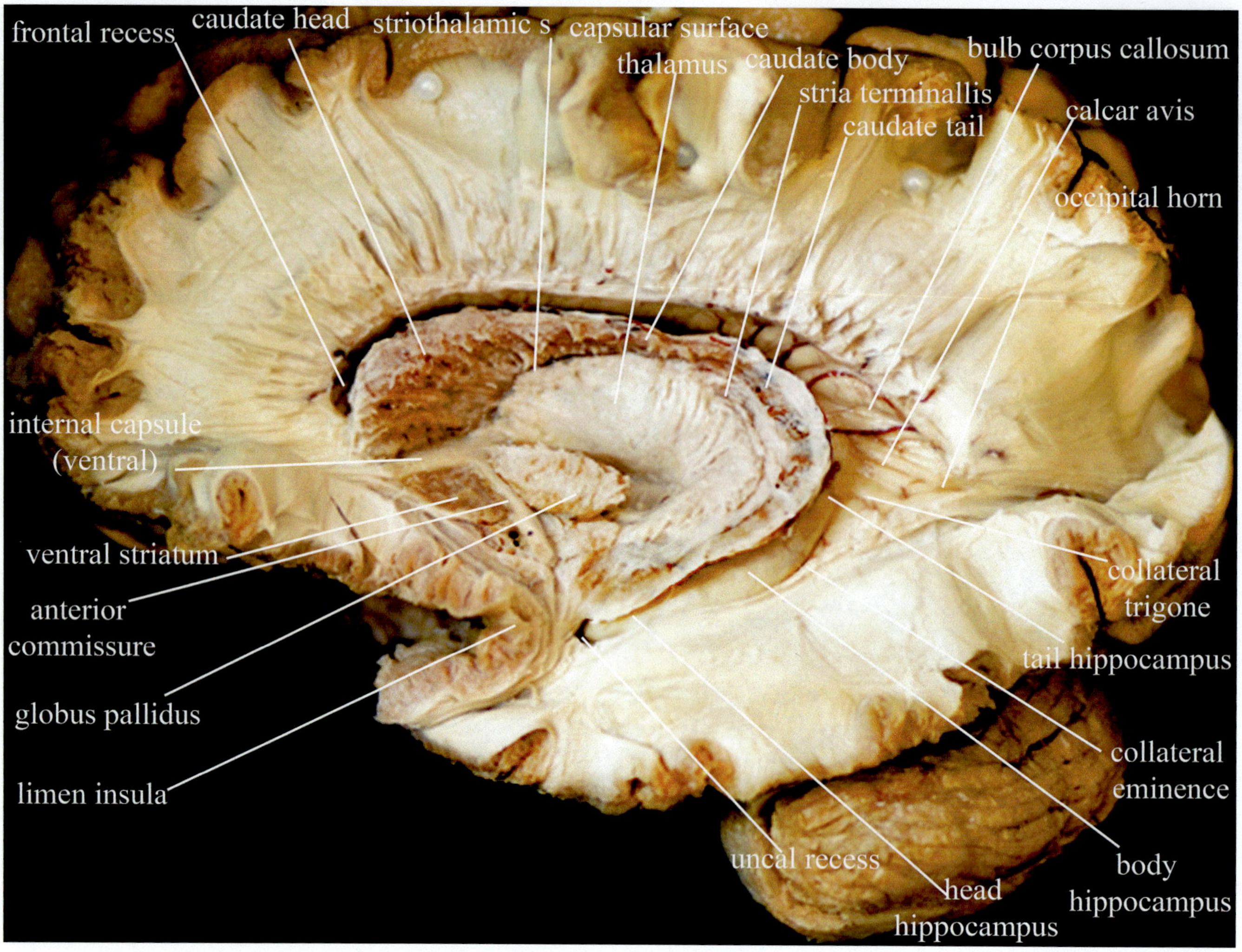

Fig. 6.21 (continued)

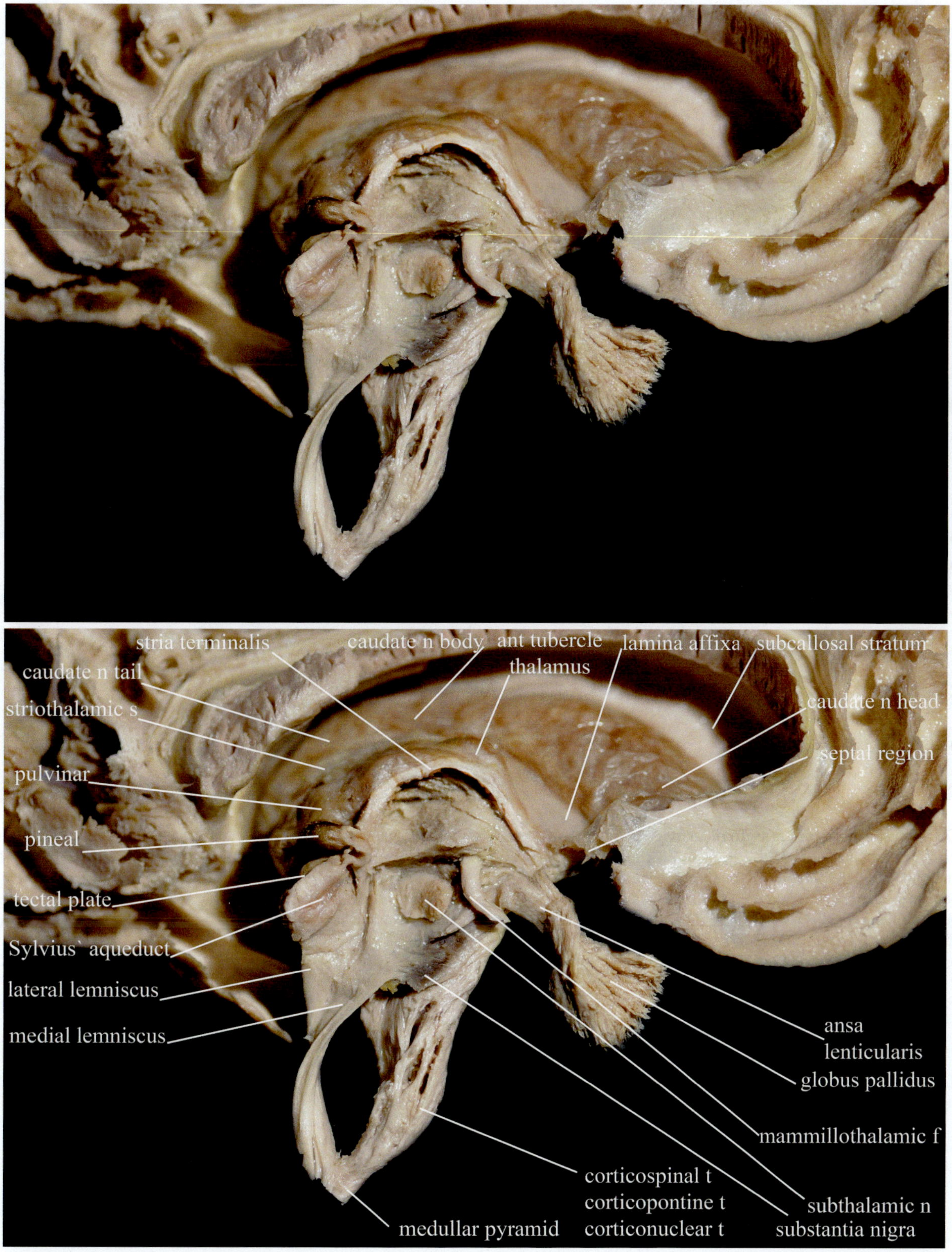

Fig. 6.22 Anatomical preparation of the ansa lenticularis. The external globus pallidus has been dislodged with the connections of the lentiform nucleus that loop around the peduncle. The connections of this nucleus loop around the peduncle and are named the ansa lenticularis. On the diencephalon, the column of the fornix and mammillary body were removed, but the mammillothalamic tract, zona incerta, and subthalamic nucleus were left in place. On the mesencephalon, the red nucleus was removed, but the substantia nigra and the ascending lemniscal fibers were preserved. On the pons, the cerebellar peduncles were removed, while the ascending and descending fibers were preserved. *ant* anterior, *f* fasciculus, *n* nucleus, *s* sulcus, *t* tract (©Kadri 2023. All Rights Reserved)

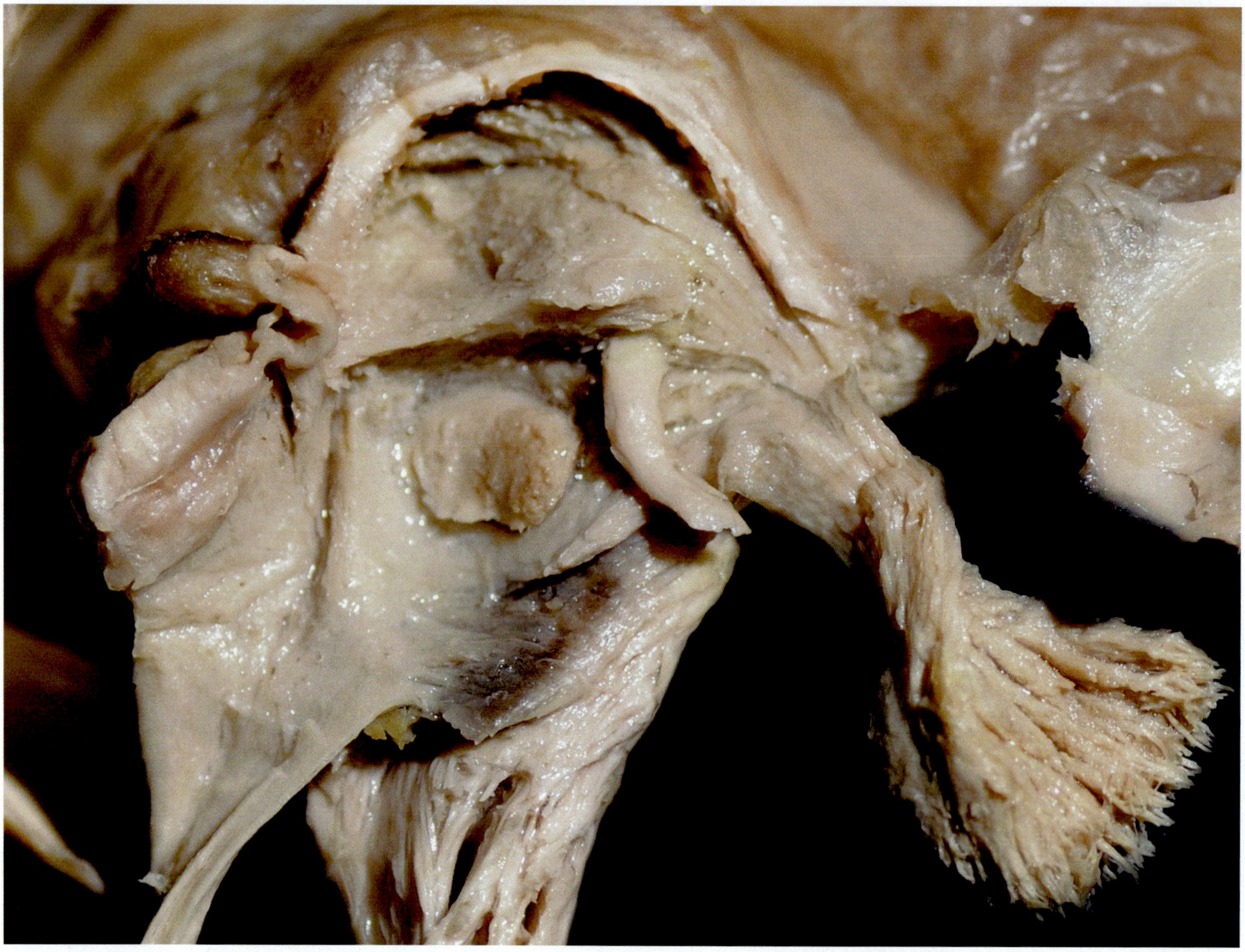

Fig. 6.23 Close-up view of the anatomical preparation of the ansa lenticularis. This structure is part of a broader bundle of fibers that also contains the amygdalo-thalamic and amygdalo-tegmental fibers. Together, these fibers are called the ansa peduncularis. Dorsal striatal and dorsal pallidal connections with the diencephalon and tegmentum mesencephalic are through the ansa lenticularis. The internal globus pallidus is intimately related to the projection fibers at the junction of the internal capsule and crus cerebri; thus, it is also termed the entopeduncular portion. Connections of the internal globus pallidus that traverse the internal capsule with the subthalamic nucleus are called the fasciculus, and connections with the zona incerta are named the fasciculus lenticularis. *ant* anterior, *f* fasciculus, *inf* inferior, *n* nucleus, *sup* superior (©Kadri 2023. All Rights Reserved)

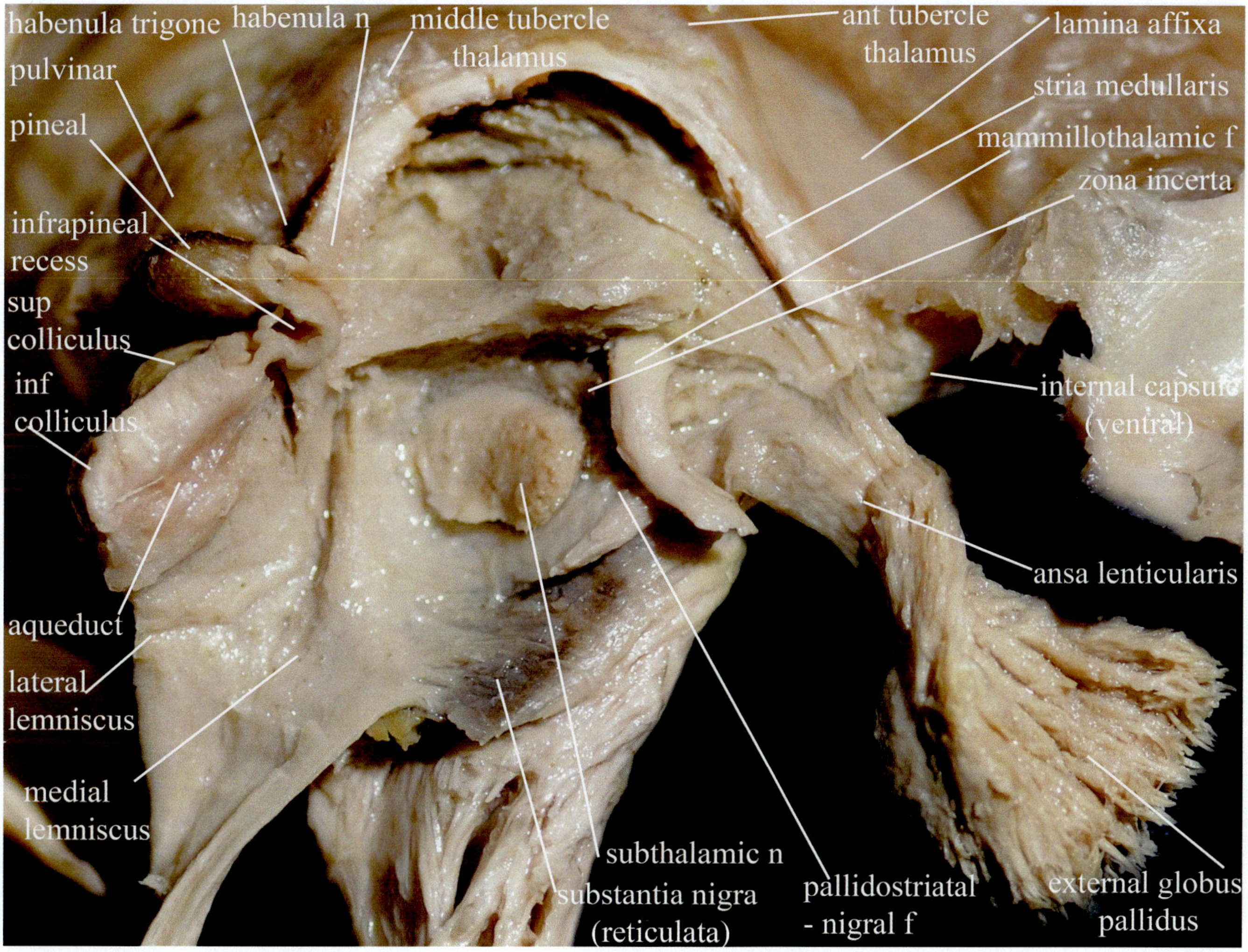

Fig. 6.23 (continued)

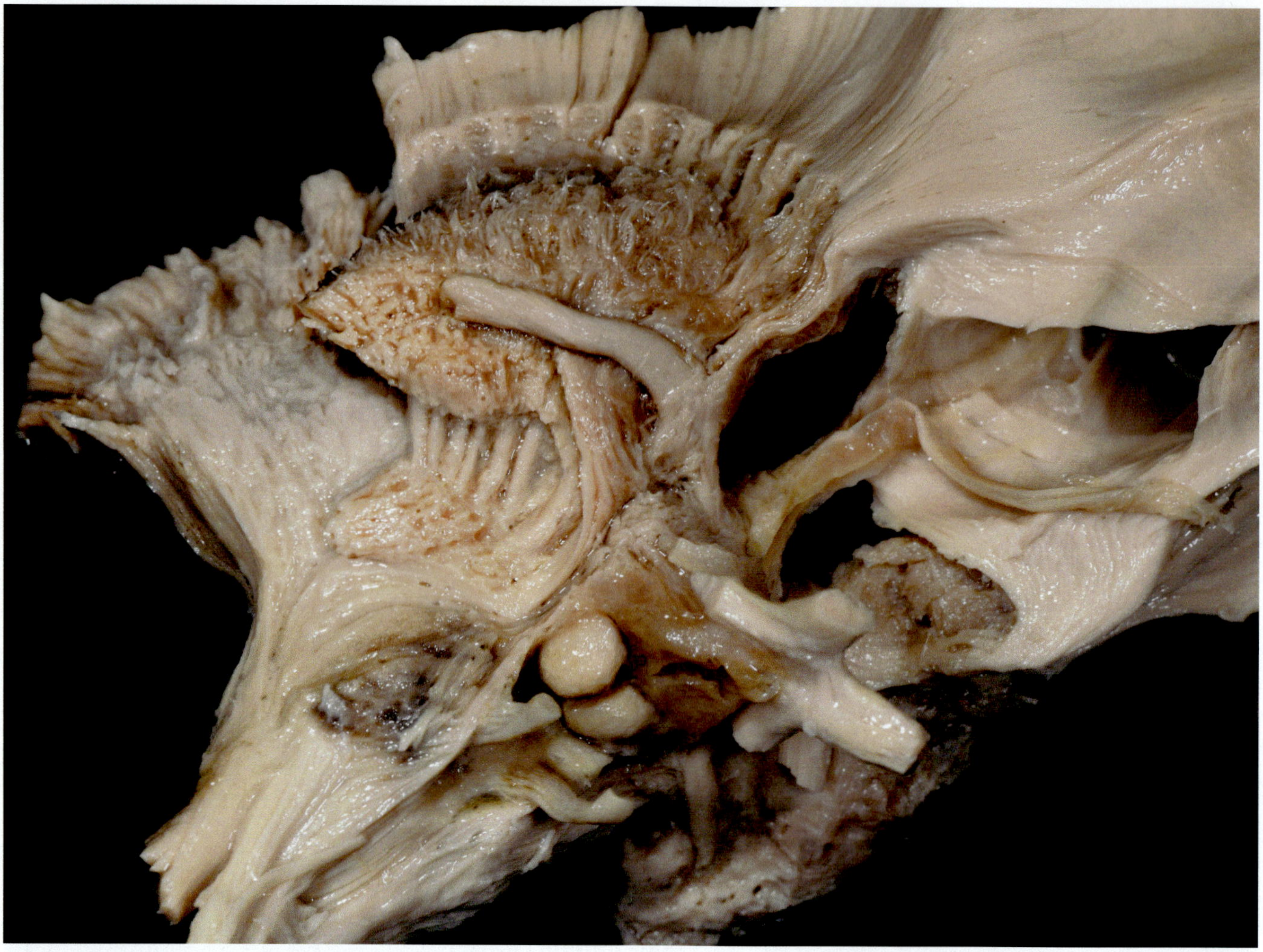

Fig. 6.24 Anatomical preparation of the striatal and pallidal connections with the subthalamus via the ansa lenticularis and fasciculus lenticularis. The ansa lenticularis connects the striatum and pallidum with the substantia nigra, the subthalamus (subthalamic nucleus and zona incerta), and the thalamus. The fasciculus lenticularis traverses the projection fibers at the transition of the internal capsule to the crus cerebri to connect the globus pallidus with the subthalamus and dorsal thalamus. Fibers connecting with the subthalamic nucleus are called fasciculus subthalamic. *f* fasciculus, *g* gyrus, *int* internal, *s* sulcus, *seg* segment, *str* stria (©Kadri 2023. All Rights Reserved)

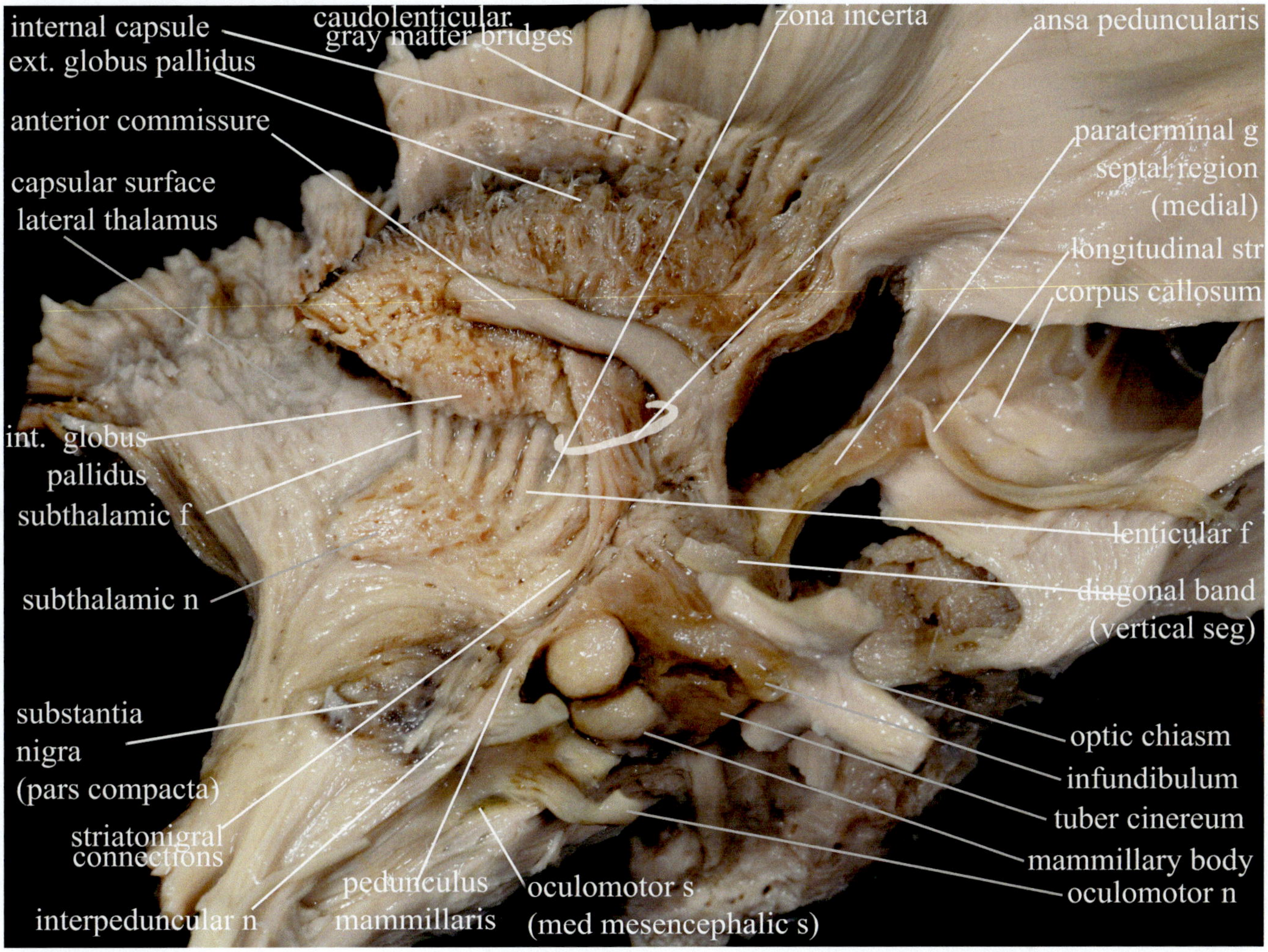

Fig. 6.24 (continued)

The Subthalamic Nucleus

The subthalamic nucleus of Luys is in the caudal part of the diencephalon and is part of the subthalamus (Figs. 6.22, 6.23, 6.24, and 6.25). It is dorsolateral to the zona incerta and dorsal to the transition of the internal capsule to the crus of the mesencephalon. It receives afferents from the cortex and globus pallidus externus. Its main efferents project toward the internal, external, and ventral pallidus and the substantia nigra [1, 2].

The Substantia Nigra and Tegmental Area

The substantia nigra separates the crus cerebri from the tegmentum of the mesencephalon (Figs. 6.22, 6.23, 6.24, 6.25, and 6.26). It extends into the diencephalon, almost reaching the globus pallidus and the substantia innominata. It is divided into a dorsal *pars compacta* and a ventral *pars reticulata*. The pars compacta is the rich, darkly pigmented region that gives the characteristic appearance to the substantia nigra. Its main afferents come from the striatum, subthalamic nucleus, and pedunculopontine tegmental nucleus. The main efferent is toward the striatum, forming the striatonigral and nigrostriatal reciprocating system. Other efferents include connections to the ventral anterior and mediodorsal nuclei (nigrothalamic), to the superior colliculus (nigrotectal), and to the pedunculopontine nucleus (nigrotegmental) [1, 2].

The tegmental area integrated into the circuit of the basal ganglia contains the ventral tegmental area, the nuclei parabrachialis pigmentosus, and the nucleus pedunculopontine tegmentum.

The ventral tegmental area is in the ventromedial tegmentum of the mesencephalon and is continuous with the hypothalamus. This area contains the interpeduncular nucleus. The pedunculopontine tegmental nucleus is the most caudal component of the basal nuclei circuit.

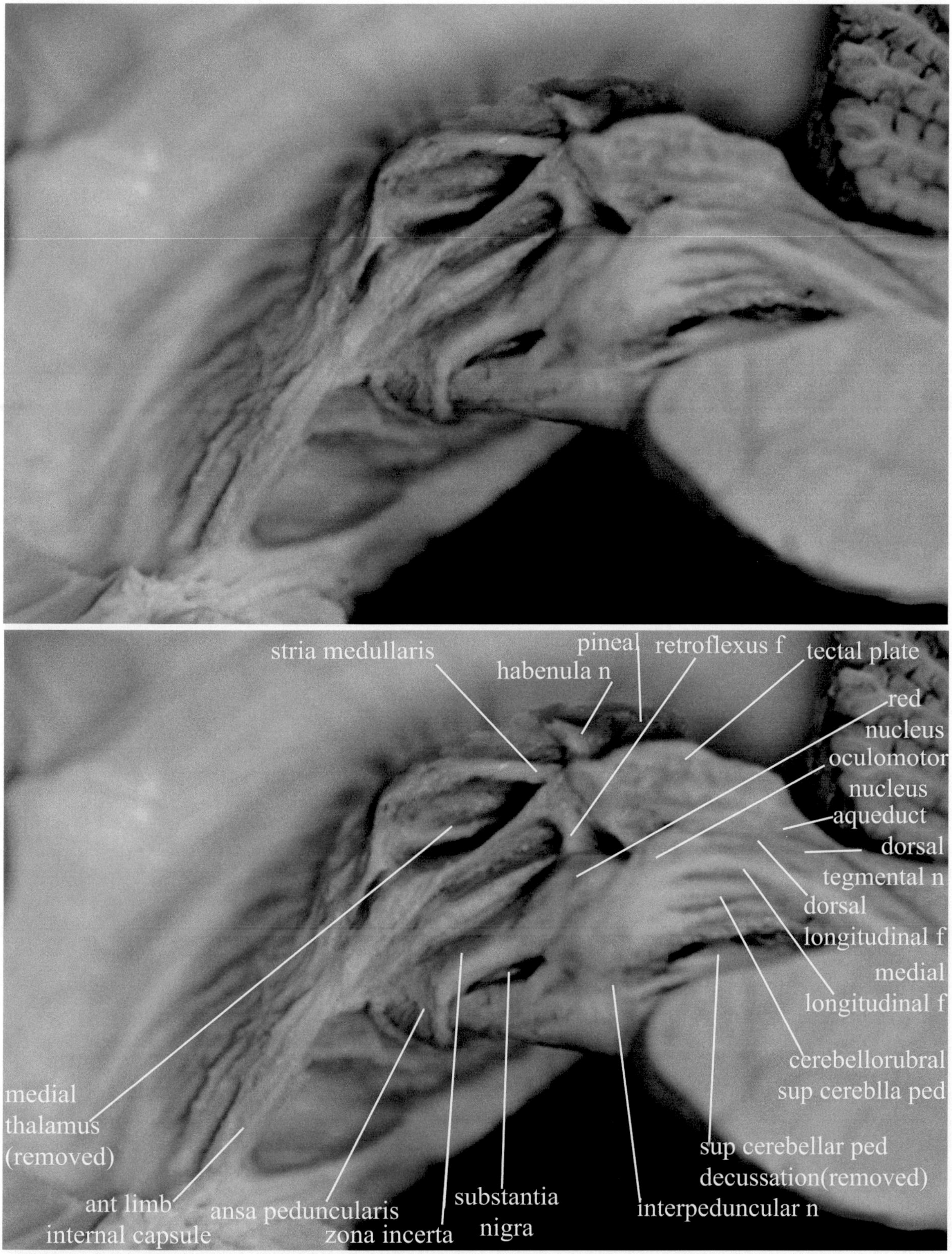

Fig. 6.25 Anatomical preparation of the connections of the basal telencephalic nuclei and the thalamus and tegmental mesencephalic. The intrinsic transition between the diencephalon and mesencephalon is the fiber system passing rostral to the red nucleus. The intrinsic transition between the mesencephalon and the pons is the superior cerebellar peduncle decussation. *ant* anterior, *f* fasciculus, *n* nerve, *ped* peduncle, *sup* superior (©Kadri 2023. All Rights Reserved)

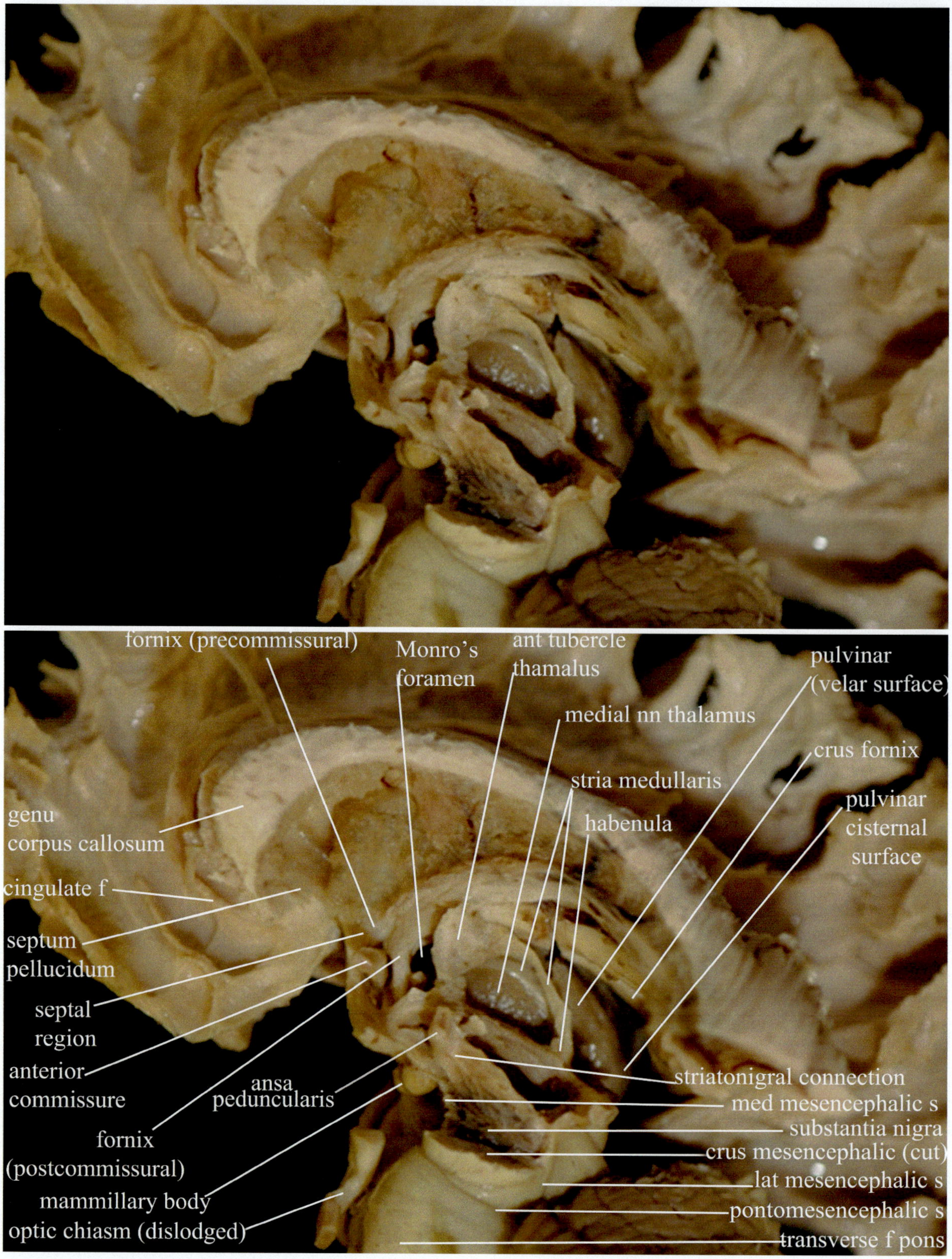

Fig. 6.26 Anatomical preparation of the mesencephalic connections of the substantia nigra. The substantia nigra is dorsal to the crus cerebri and is a limit between the base and tegmentum of the mesencephalic peduncle. The substantia nigra extends from the medial to the lateral mesencephalic peduncle, which is the superficial limit of the ascending (lemniscal triangle) and descending (crus cerebri) projection fibers. The medial mesencephalic sulcus is also named the oculomotor nerve sulcus and is the cisternal appearance of the oculomotor nerve rootlets. The posterior perforated substance is situated between the paired medial mesencephalic sulci. *ant* anterior, *f* fasciculus, *lat* lateral, *nn* nucleus, *s* sulcus. (©Kadri 2023. All Rights Reserved)

The Claustrum

As described above, the claustrum is derived from the pallium and is therefore not part of the basal ganglia, which is derived from the subpallium (Figs. 6.4, 6.12, 6.14, 6.27, and 6.28). The claustrum is a thin layer of gray matter interposed between the putamen and the extreme capsule. It is embedded within the external capsule and is divided into a compact dorsal or insular part and a fragmented ventral or temporal part [9]. The ventral part is spaced between the fibers of the anterior commissure and the uncinate fasciculus and extends laterally, covering the amygdala. The claustrum has reciprocal connections with several cortical areas, resembling the pulvinar of the thalamus. One hypothesis is that the claustrum is involved with consciousness [10].

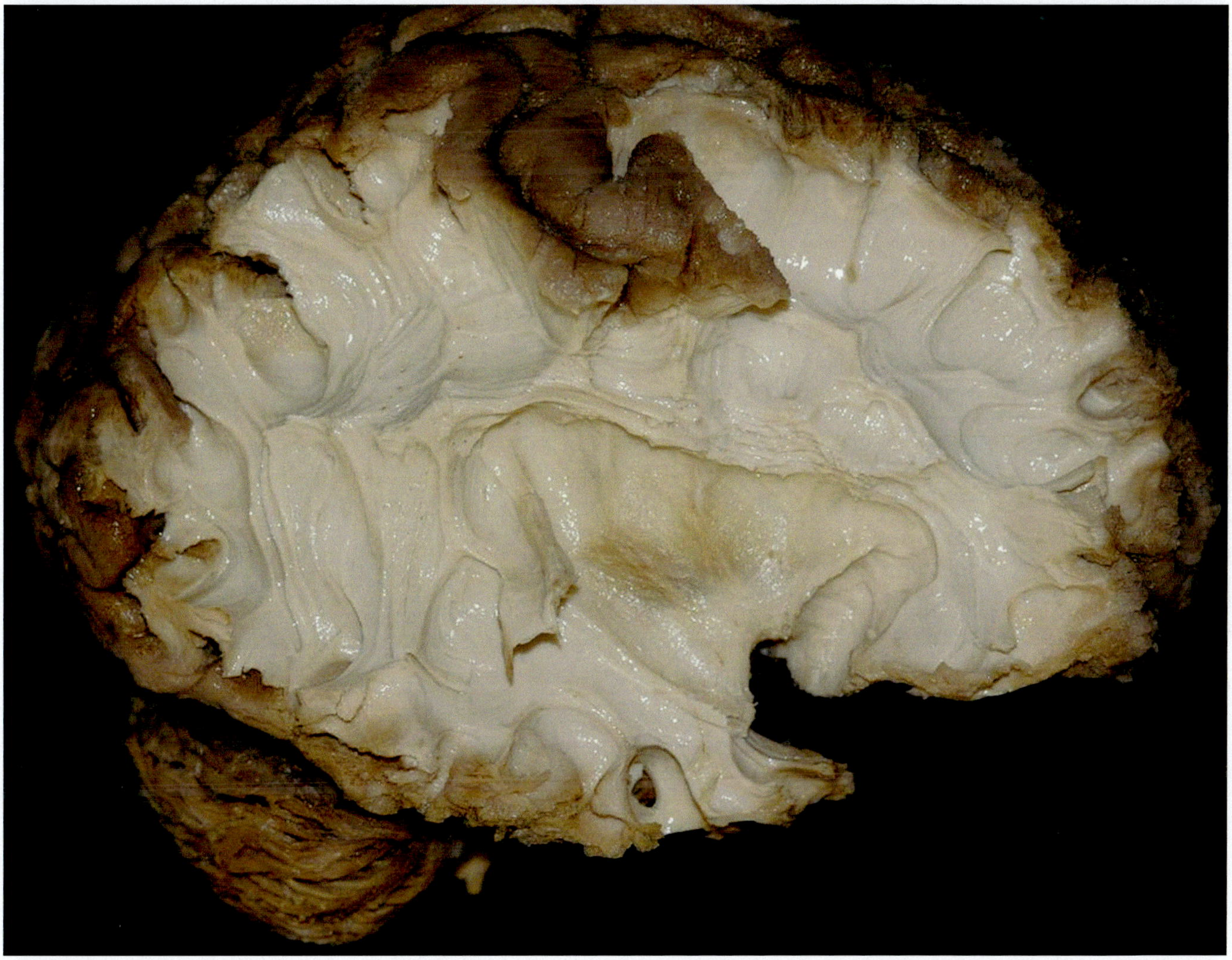

Fig. 6.27 Anatomical preparation of the left hemisphere exposing the association fibers. The claustrum is embedded in the fibers of the external capsule; therefore, the internal capsule is divided into ventral and dorsal parts. *f* fasciculus, *g* gyrus, *s* sulcus (©Kadri 2023. All Rights Reserved)

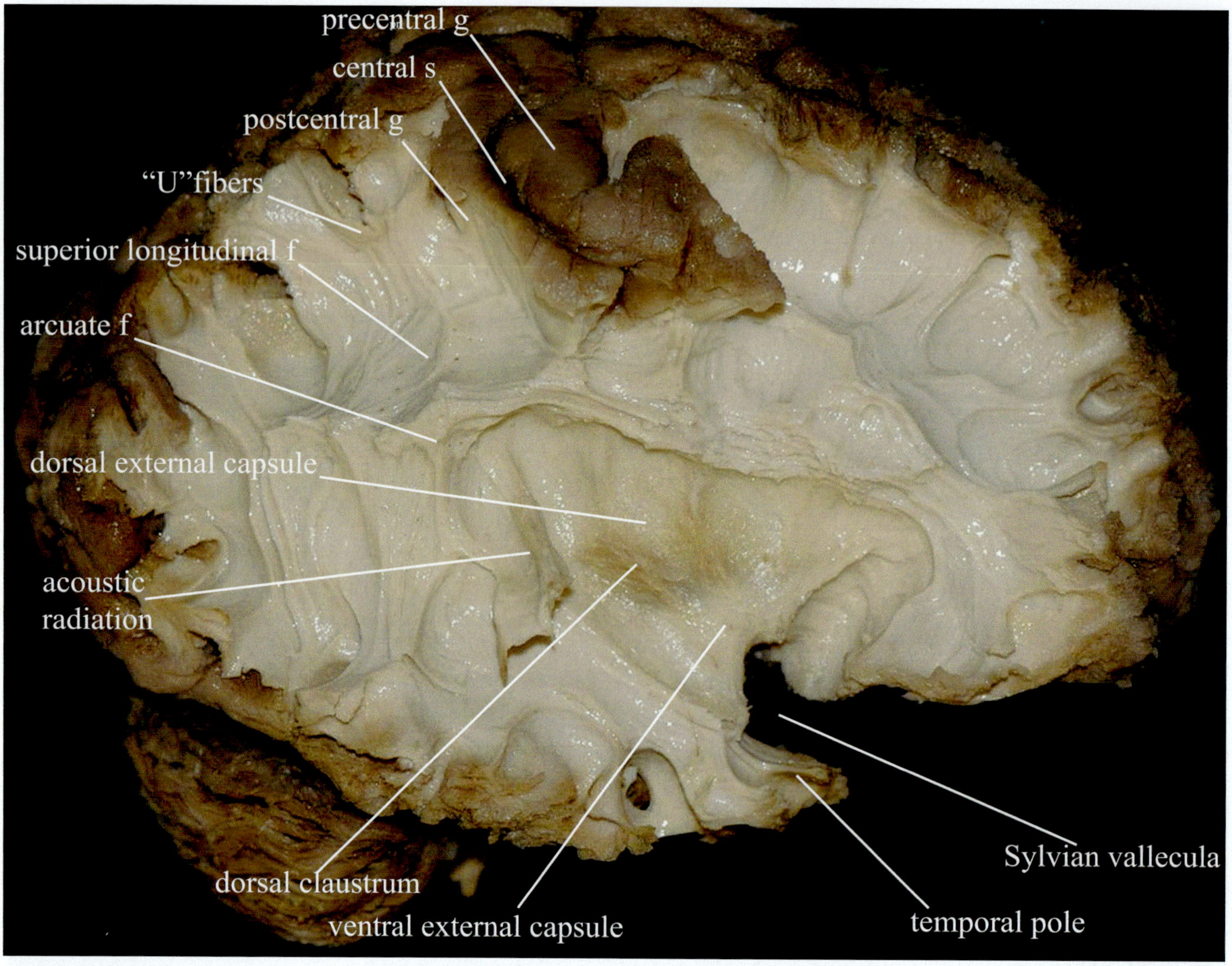

Fig. 6.27 (continued)

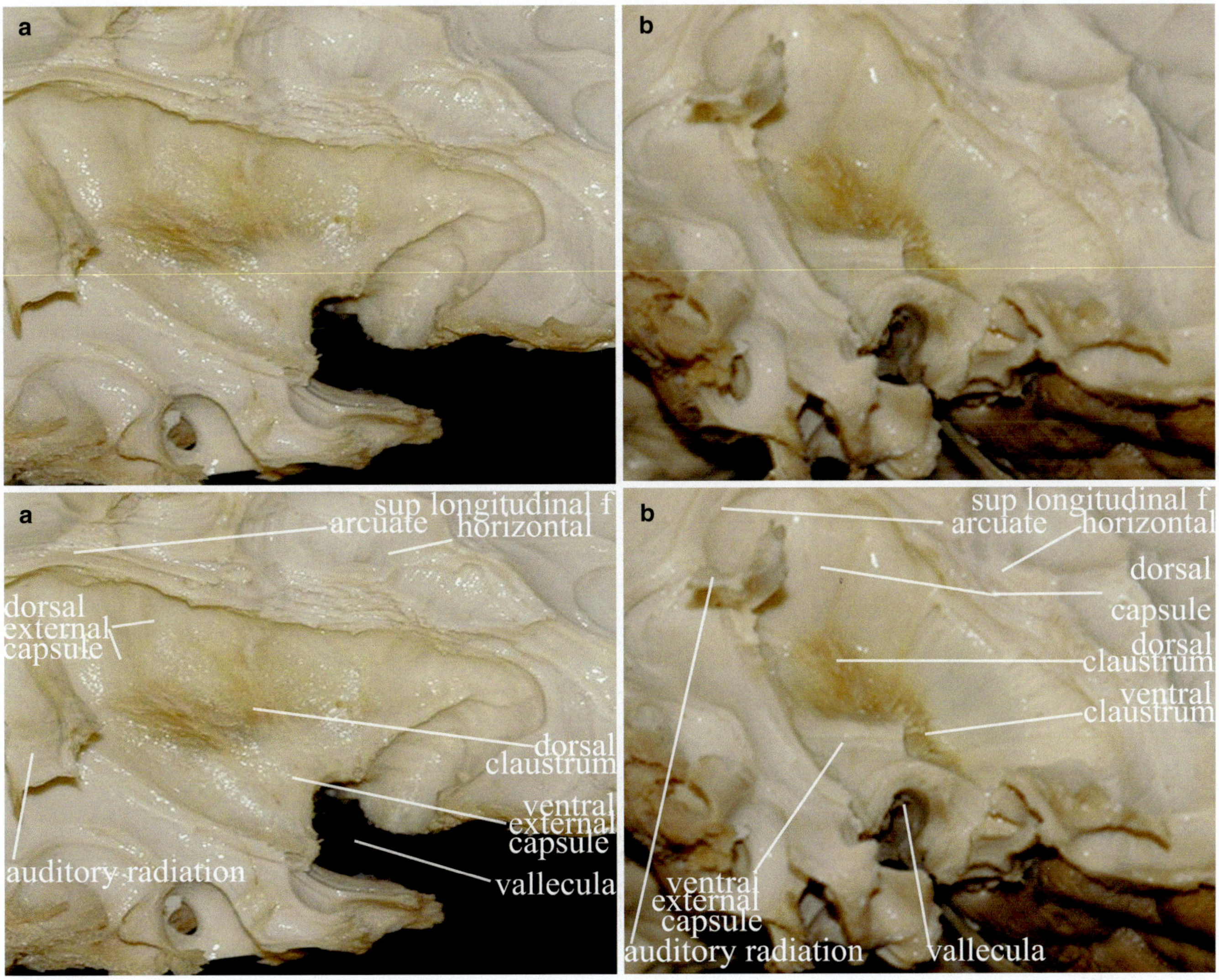

Fig. 6.28 Anatomical preparation of the claustrum. Although a neuronal group is not on the cortical surface, the claustrum is not part of the basal nuclei of the telencephalon. The detailed ontogenesis of the claustrum has a different embryological origin related more to the insula. (**a**) The dorsal claustrum is embedded in the fibers of the external capsule. (**b**) The ventral claustrum is embedded between the layers of the uncinate and occipitofrontal fasciculus. The external capsule is divided into ventral and dorsal parts; consequently, the claustrum is separated into the dorsal and ventral claustrum. The dorsal capsule is composed of the association fibers of the cortex with the claustrum and the striatum. The ventral external capsule is composed of the long association fibers of the uncinate fasciculus and occipitofrontal fasciculus as they converge at the limen insula to connect the frontal lobe to the temporal and occipital lobes. The claustrum is visible when the extreme capsule is removed. The superficial layer of the dorsal external capsule is composed of cortico-claustral connections, and as its fibers are removed, the dorsal claustrum is removed as well. Once the dorsal claustrum has been completely removed, a thin remnant layer of fibers is exposed. This layer connects the cortex to the striatum; thus, they are called the cortico-striatal fibers. *f* fasciculus, *sup* superior (©Kadri 2023. All Rights Reserved)

References

1. Nieuwenhuys R, Voogd J, van Huijzen C. The human central nervous system: a synopsis and atlas. 4th ed. Berlin: Springer; 2008.
2. Heimer L, Switzer RD, Van Hoesen GW. Ventral striatum and ventral pallidum: components of the motor system? Trends Neurosci. 1982;5:83–7. https://doi.org/10.1016/0166-2236(82)90037-6.
3. Heimer L. The human brain and spinal cord: functional neuroanatomy and dissection guide. 2nd ed. New York: Springer; 1994.
4. Testut L, Latarjet A. Tratado de anatomia humana, vol. 2. 9th ed. Barcelona: Salvat Editores; 1966.
5. Türe U, Yaşargil MG, Friedman AH, Al-Mefty O, Yaşargil DG. Fiber dissection technique: lateral aspect of the brain. Neurosurgery. 2000;47:417–27. https://doi.org/10.1097/00006123-200008000-00028.
6. Standring S. Cerebral hemisphere. In: Gray's anatomy: the anatomical basis of clinical practice. 40th ed. Philadelphia, PA: Elsevier; 2008. p. 335–59.
7. Serra C, Türe U, Krayenbühl N, Şengül G, Yaşargil DCH, Yaşargil MG. Topographic classification of the thalamus surfaces related to microneurosurgery: a white matter fiber microdissection study. World Neurosurg. 2017;97:438–52. https://doi.org/10.1016/j.wneu.2016.09.101.
8. Carpenter MB. Core text of neuroanatomy. Subsequent ed. Baltimore, MD: Williams &Wilkins; 1991.
9. Fernández-Miranda JC, Rhoton AL Jr, Kakizawa Y, Choi C, Alvarez-Linera J. The claustrum and its projection system in the human brain: a microsurgical and tractographic anatomical study. J Neurosurg. 2008;108:764–74. https://doi.org/10.3171/JNS/2008/108/4/0764.
10. Crick FC, Koch C. What is the function of the claustrum? Philos Trans R Soc Lond Ser B Biol Sci. 2005;360:1271–9. https://doi.org/10.1098/rstb.2005.1661.

7 The Diencephalon

The diencephalon and the telencephalon constitute the prosencephalon or cerebrum. The diencephalon occupies the center of the prosencephalon and is divided into four zones: the hypothalamus, the ventral thalamus, the dorsal thalamus, and the epithalamus. (In this classification, the metathalamus is part of the dorsal thalamus.) All the diencephalic zones relate to the third ventricle. These nuclei are interposed between the telencephalon and the mesencephalon (Figs. 7.1, 7.2, and 7.3) [1].

P. Kadri, *The Cartographic Atlas of the Brain*, https://doi.org/10.1007/978-3-031-38062-4_7

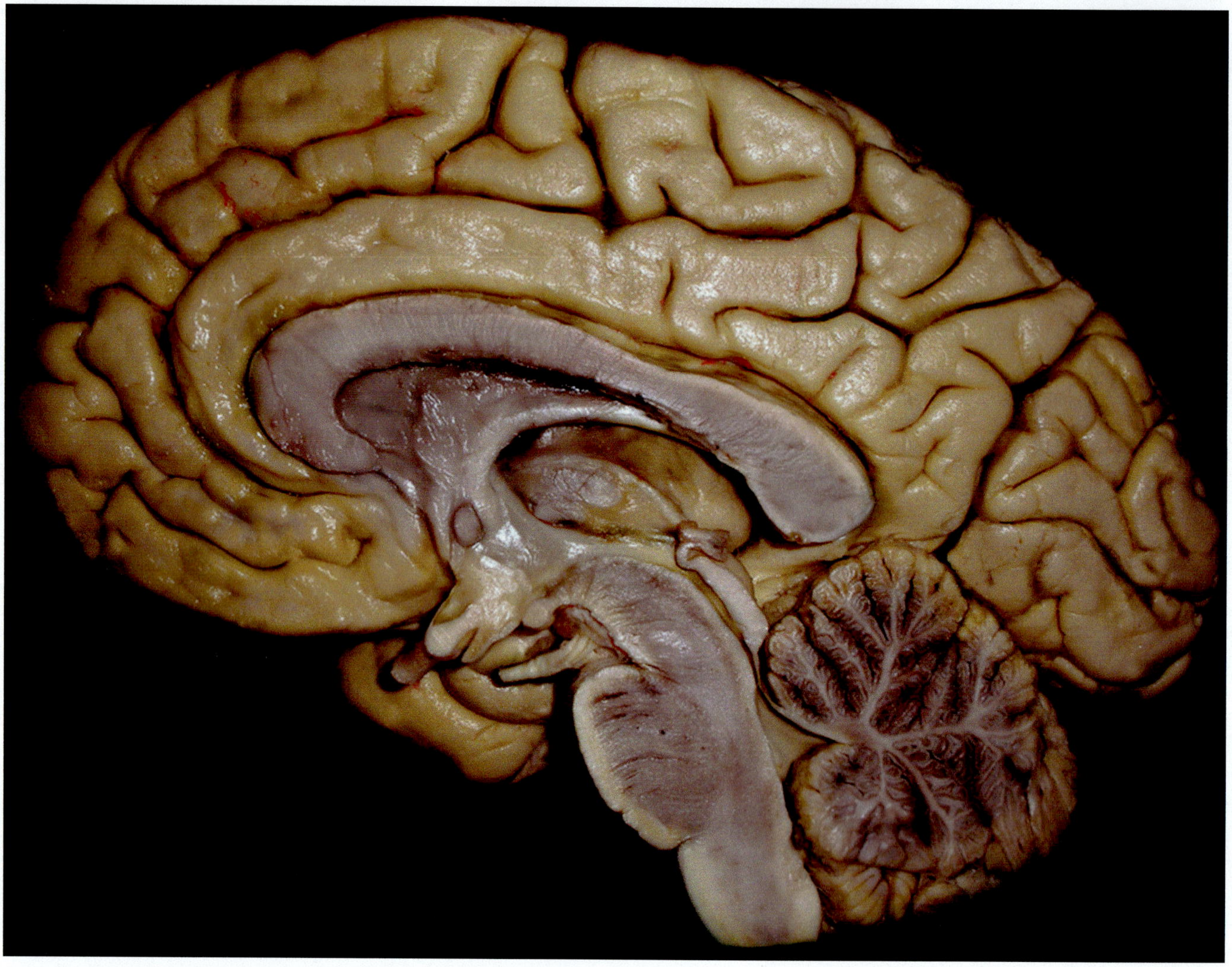

Fig. 7.1 Anatomical preparation of the midline view of the encephalon. The prosencephalon is divided into the diencephalon and telencephalon. The diencephalon is further divided into the hypothalamus, ventral thalamus or subthalamus, dorsal thalamus, metathalamus, and epithalamus. *ant* anterior, *f* fasciculus, *g* gyrus, *inf* inferior, *n* nerve, *nn* nucleus, *ped* peduncle, *post* posterior, *s* sulcus, *sup* superior, *t* tract (©Kadri 2023. All Rights Reserved)

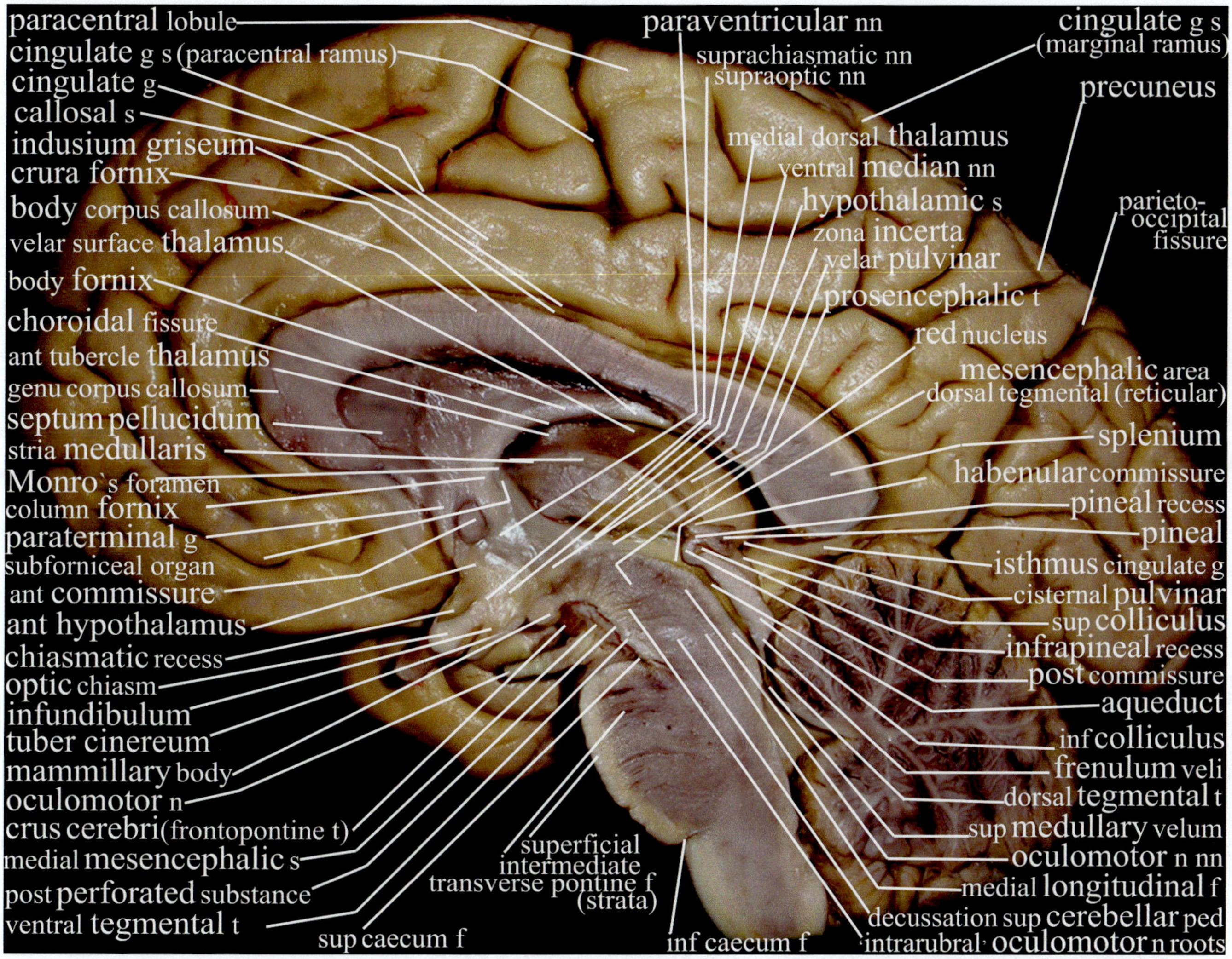

Fig. 7.1 (continued)

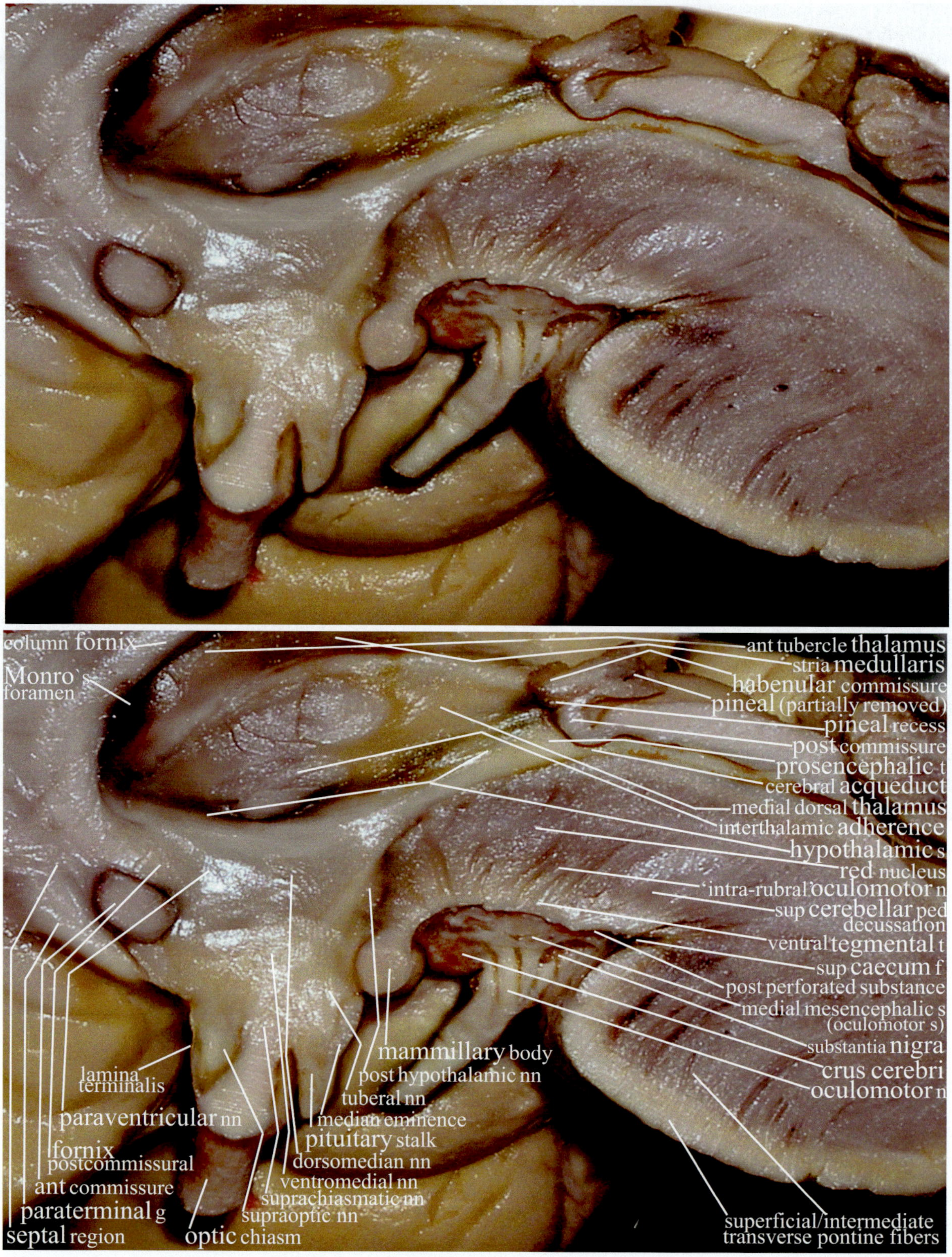

Fig. 7.2 Anatomical preparation of the diencephalon. The hypothalamus, ventral thalamus, dorsal thalamus, and epithalamus relate to the third ventricle. The hypothalamic sulcus is the center of the cleavage plane that separates the diencephalon and the dorsal mesencephalon and roughly extends from the foramen of Monro to the cerebral aqueduct. *ant* anterior, *f* fasciculus, *g* gyrus, *n* nerve, *nn* nucleus, *ped* peduncle, *post* posterior, *s* sulcus, *sup* superior, *t* tract (©Kadri 2023. All Rights Reserved)

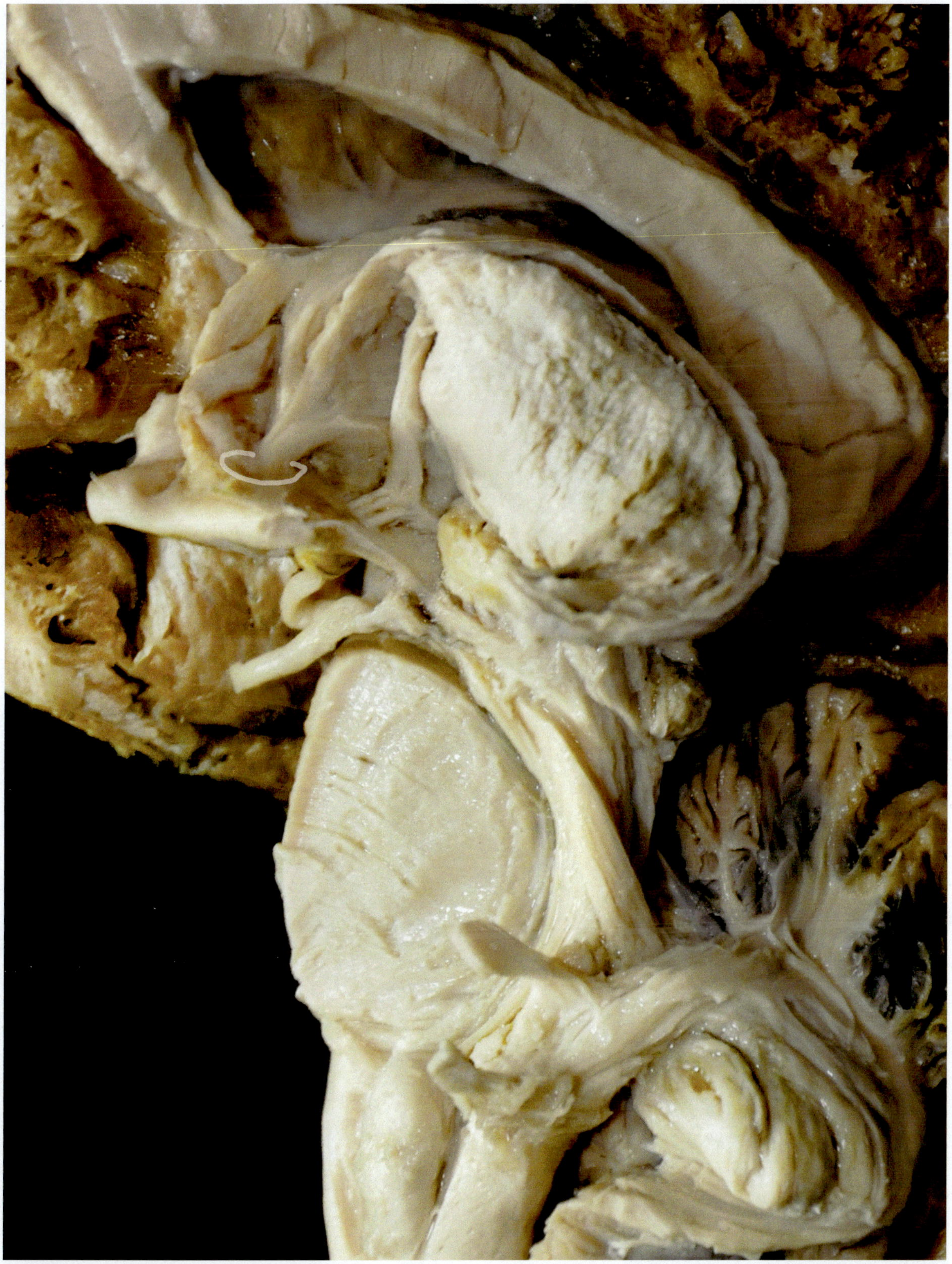

Fig. 7.3 Anatomical preparation of the diencephalon and the tegmentum mesencephalic after the removal of the projection fibers. The mammillothalamic tract is used as a landmark for the internal medullary lamina and is exposed after the anterior ventral nuclei of the thalamus are removed. *ant* anterior, *ext* external, *f* fasciculus or foramen, *g* gyrus, *inf* inferior, *med* medial, *n* nerve, *nn* nucleus, *ped* peduncle, *s* sulcus, *sup* superior, *t* tract (©Kadri 2023. All Rights Reserved)

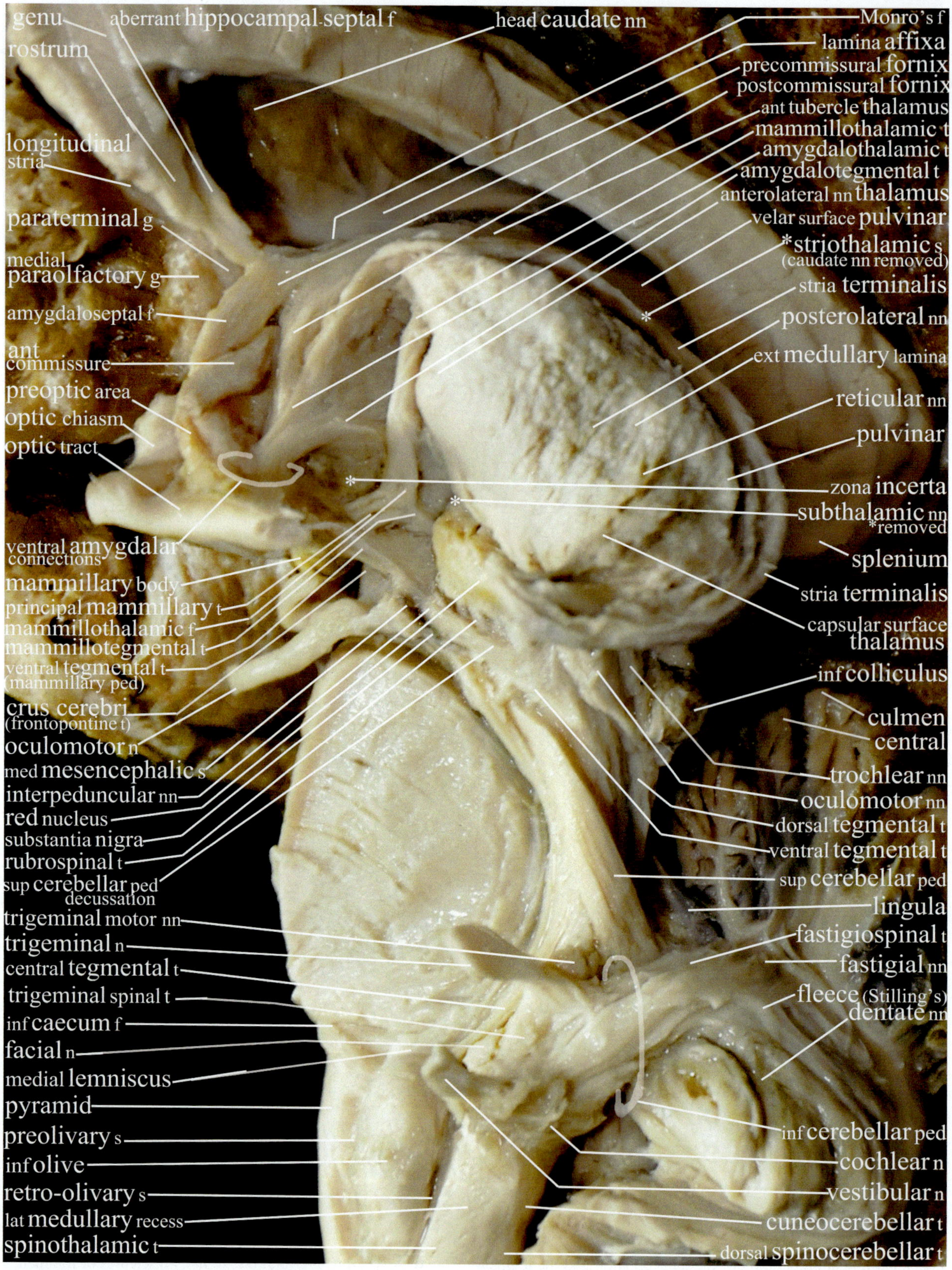

Fig. 7.3 (continued)

The Epithalamus

The epithalamus is at the posterior limit of the third ventricle. It contains the pineal gland and the habenular complex (Figs. 7.1, 7.2, 7.3, 7.4, and 7.5).

The pineal gland, also called the epiphysis, is a midline piriform-shaped structure located cephalic to the tectum of the mesencephalon (Fig. 7.6). It is connected to the posterior third ventricle by its attachments to the habenular commissure superiorly and the posterior commissure inferiorly (Fig. 7.7). The pineal recess is a tiny extension of the ventricular cavity between the two commissures (Fig. 7.8). Above the habenular commissure, another posterior extension of the third ventricle is referred to as the suprapineal recess.

The posterior commissure is the posterior limit between the diencephalon and the mesencephalon and the posterior uppermost limit of the cerebral aqueduct (Figs. 7.9 and 7.10). The aqueduct interconnects the third and fourth ventricles.

The habenula consists of nuclear masses situated at the habenular trigone (Figs. 7.11 and 7.12). The habenular trigone comprises two small triangular eminences at each side of the pineal gland as seen from a dorsal view of the third ventricle (Figs. 7.13 and 7.14). Its main afferent is the stria medullaris thalamus, and its main efferent is the retroflexus fasciculus (Figs. 7.15 and 7.16) [1]. The stria medullaris connects the bed nucleus of the stria terminalis, the septal region, and the hypothalamus with the habenular nuclei (Figs. 7.16 and 7.17). The stria medullaris is the site where the tela choroidea implants in the roof of the third ventricle. Its insertion into the thalamus is at the stria medullaris through a delicate pial-ependymal projection known as the taenia. The taenia thalamus is the insertion of the tela choroidea into the dorsomedial surface of the thalamus. The retroflexus fasciculus is also called the habenulo-interpeduncular tract and is projected toward the interpeduncular nucleus at the base of the interpeduncular fossa (Figs. 7.4, 7.5, 7.15, 7.16, and 7.17). This circuit is part of the limbic system.

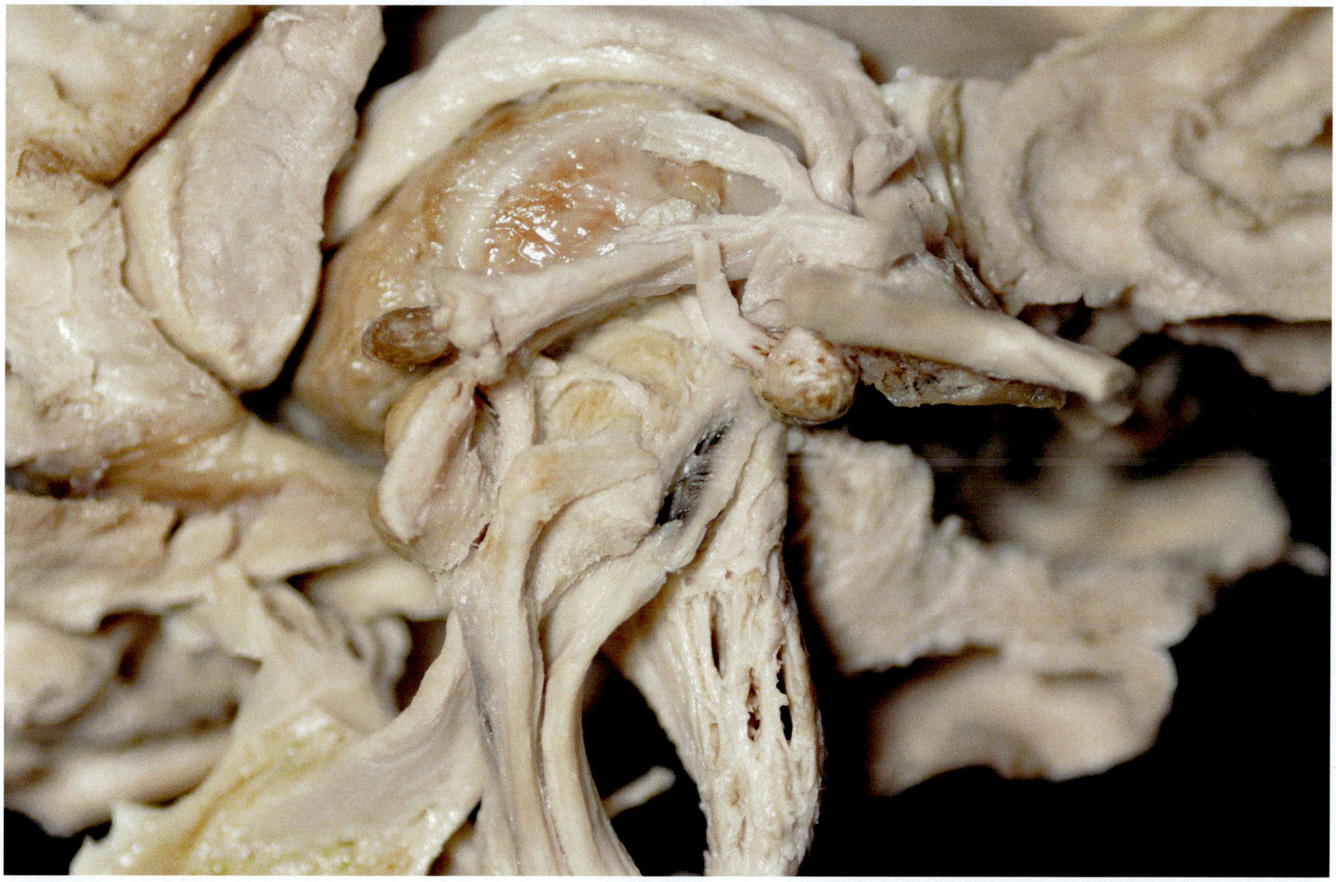

Fig. 7.4 Anatomical preparation of the white matter connections at the base of the third ventricle. The most superficial medial fiber bundle is the prosencephalic tract, extending from the midline to the hypothalamic sulcus. It is at the floor of the third ventricle. These fibers connect the telencephalic (amygdala, hippocampal formation, and septal region), thalamic (bed nuclei stria terminalis, paraventricular nuclei, and medial nuclei), hypothalamic, and epithalamic limbic diencephalic areas to the dorsal reticular area of the ventricular surface of the tegmentum of the mesencephalon. *ant* anterior, *f* fasciculus or foramen, *g* gyrus, *inf* inferior, *lat* lateral, *n* nerve, *nn* nucleus, *ped* peduncle, *post* posterior, *sup* superior, *t* tract (©Kadri 2023. All Rights Reserved)

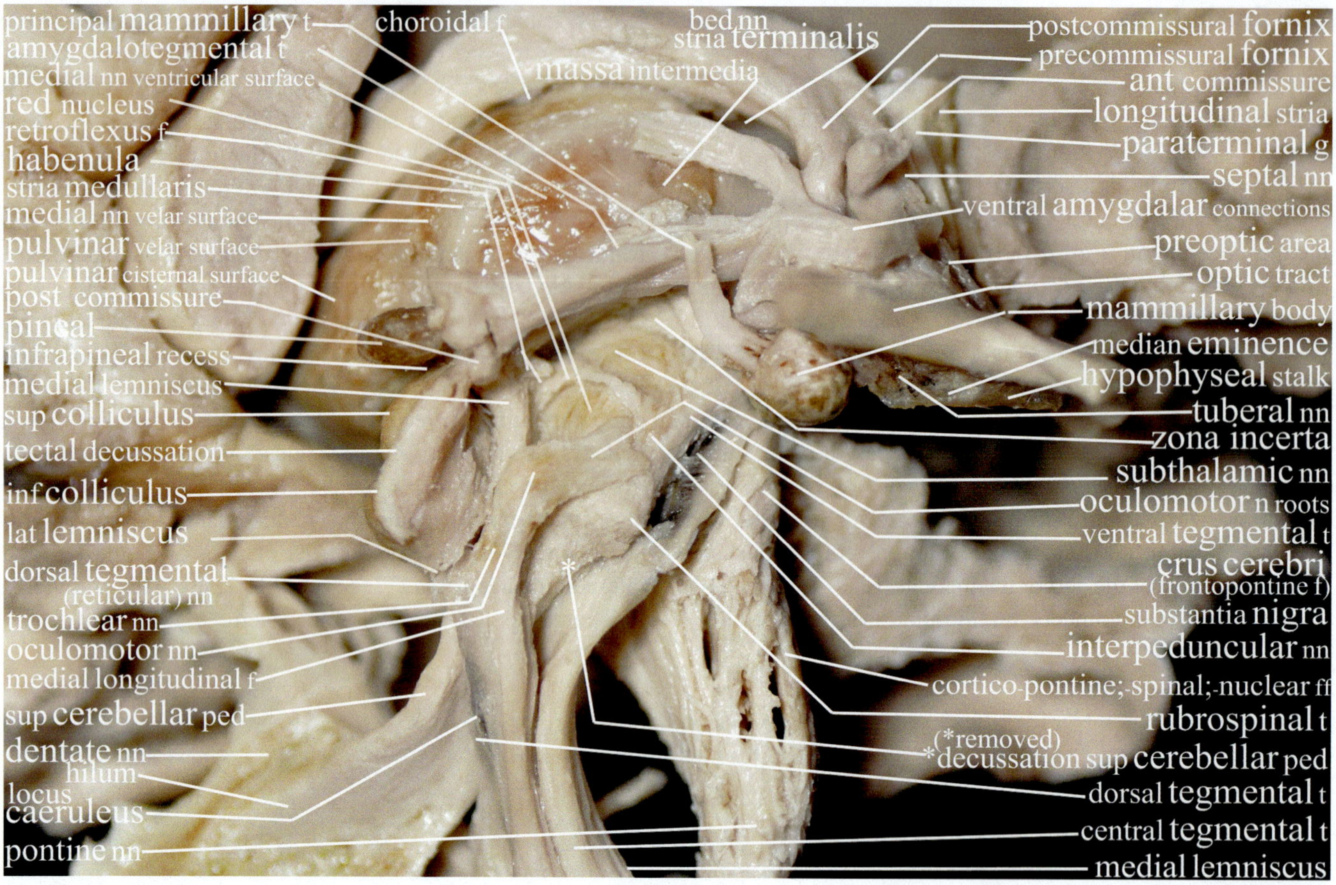

Fig. 7.4 (continued)

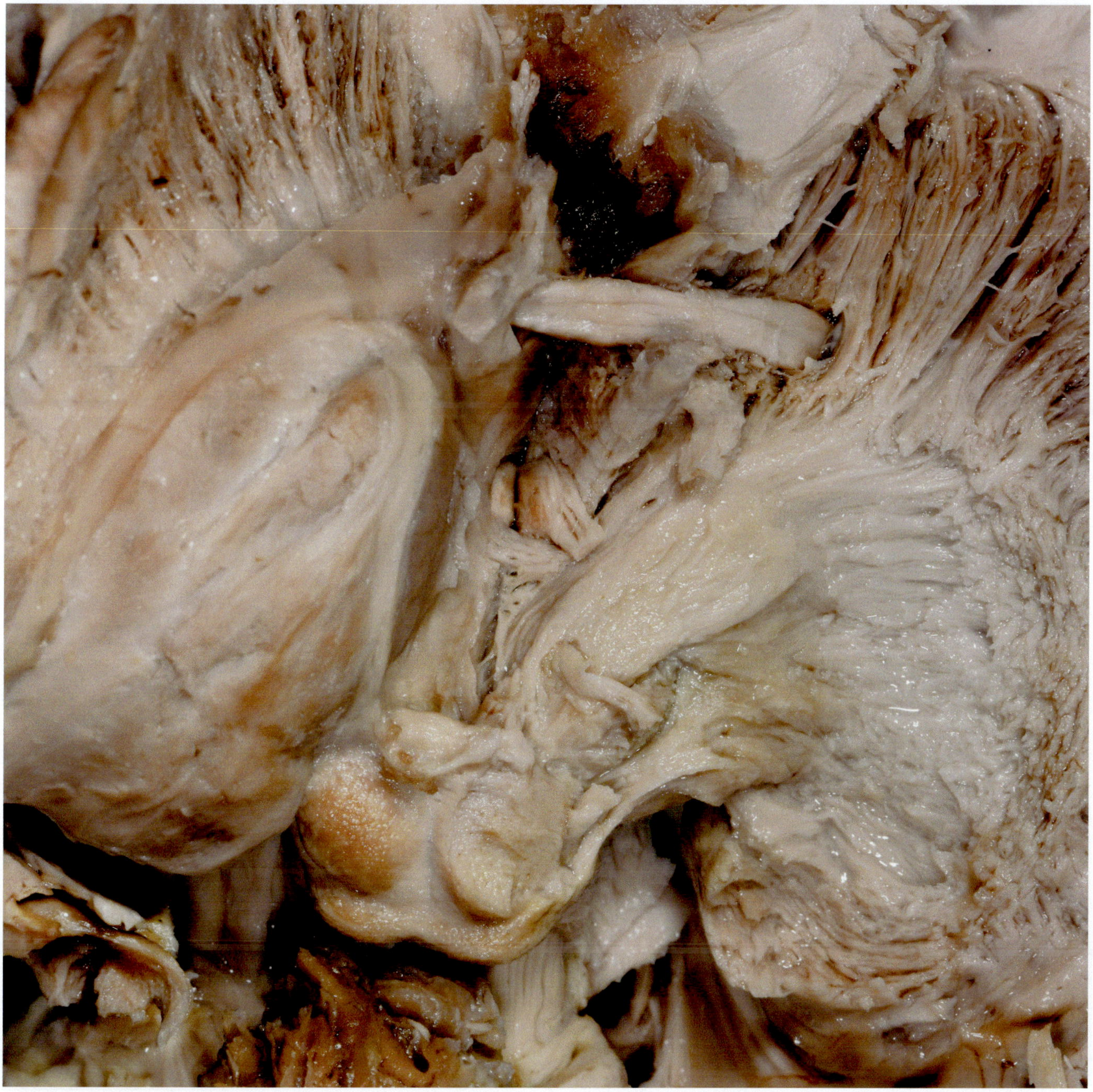

Fig. 7.5 Anatomical preparation of the tegmentum of the mesencephalon. The dentato-rubro-thalamic connections are centered at the red nucleus. The habenular–interpeduncular connections through the retroflexus tract are a reliable reference that points the limbic connections of the prosencephalic tract and the extrapeduncular thalamic peduncle fibers to the epithalamus. *ant* anterior, *ext* external, *g* gyrus, *inf* inferior, *nn* nucleus, *post* posterior, *s* sulcus, *sup* superior, *t* tract ()

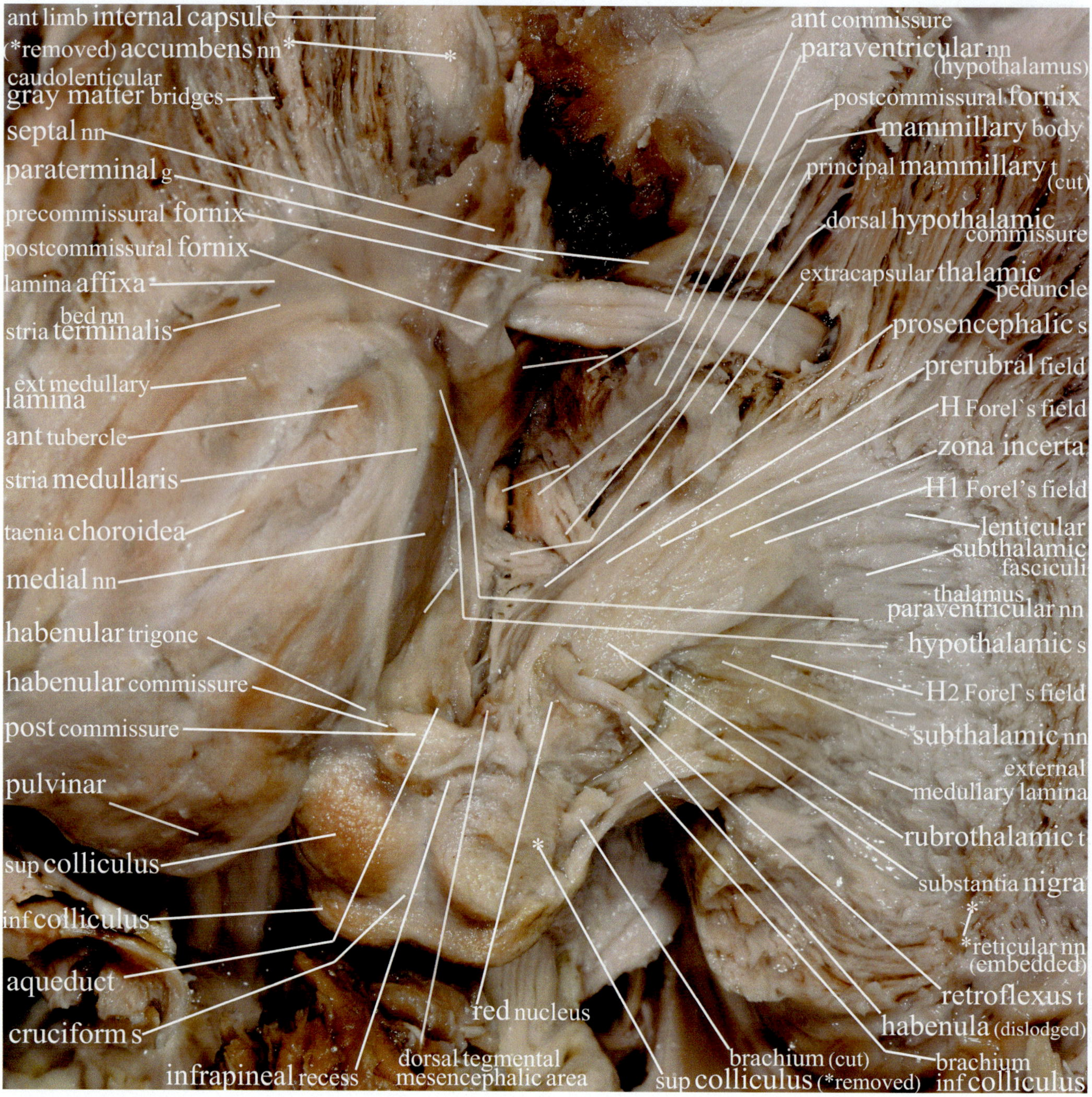

Fig. 7.5 (continued)

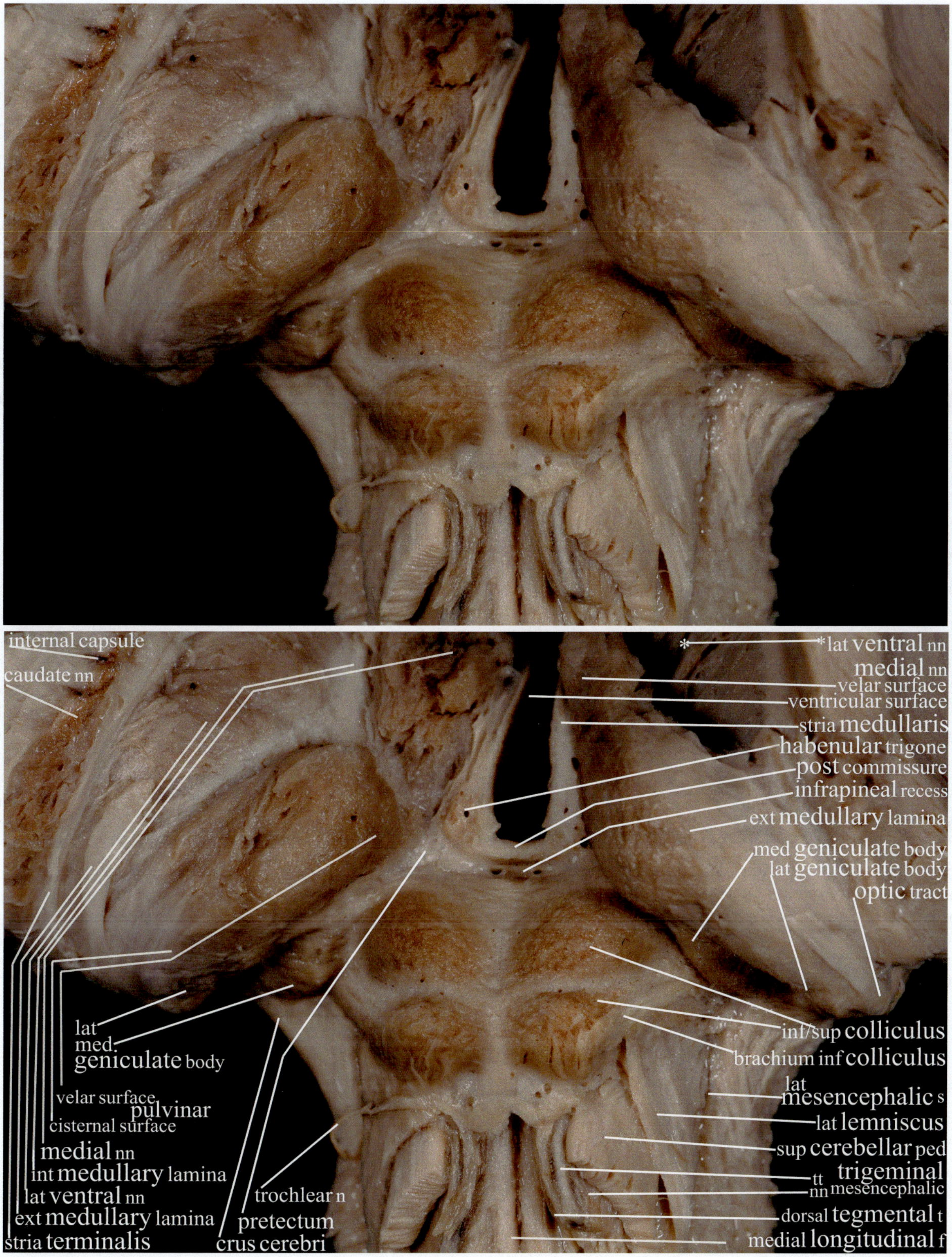

Fig. 7.6 Anatomical preparation of the posterior third ventricle and the ventricular, velar, and cisternal surfaces of the thalamus. The cisternal surface of the tectal plate is also known as the collicular surface of the tectum and the quadrigeminal plate because of the disposition of the paired superior and inferior colliculi. *ext* external, *f* fasciculus, *inf* inferior, *int* internal, *lat* lateral, *med* medial, *n* nerve, *nn* nucleus, *ped* peduncle, *post* posterior, *s* sulcus, *sup* superior, *t* tract, *tt* tract (©Kadri 2023. All Rights Reserved)

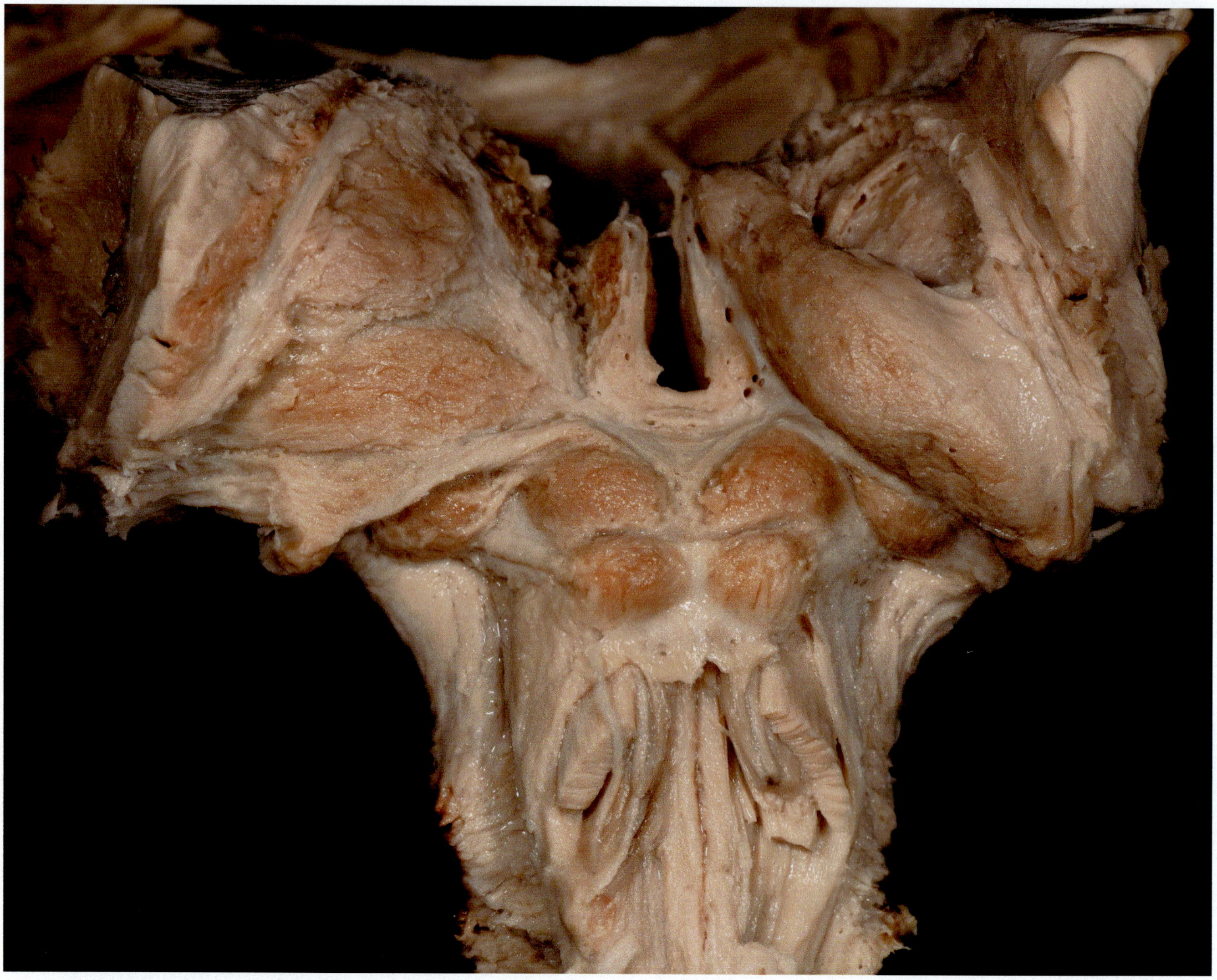

Fig. 7.7 Anatomical preparation of the dorsal thalamus and epithalamus and the mesencephalic tectum and tegmentum of the pons. The pulvinar has been removed from the left hemisphere, and the ventral nuclei of the thalamus are exposed. The ventral nuclei can be subdivided into anterior, lateral, and posterior. The posterior ventral nuclei can be further divided into medial and lateral. The lateral posterior ventral nucleus is connected via the medial lemniscus, and the medial posterior ventral nucleus is connected via the spinothalamic and trigeminothalamic tracts. *f* fasciculus, *inf* inferior, *int* internal, *lat* lateral, *med* medial, *n* nerve, *nn* nucleus, *ped* peduncle, *post* posterior, *s* sulcus, *sup* superior, *t* tract, *tt* tract

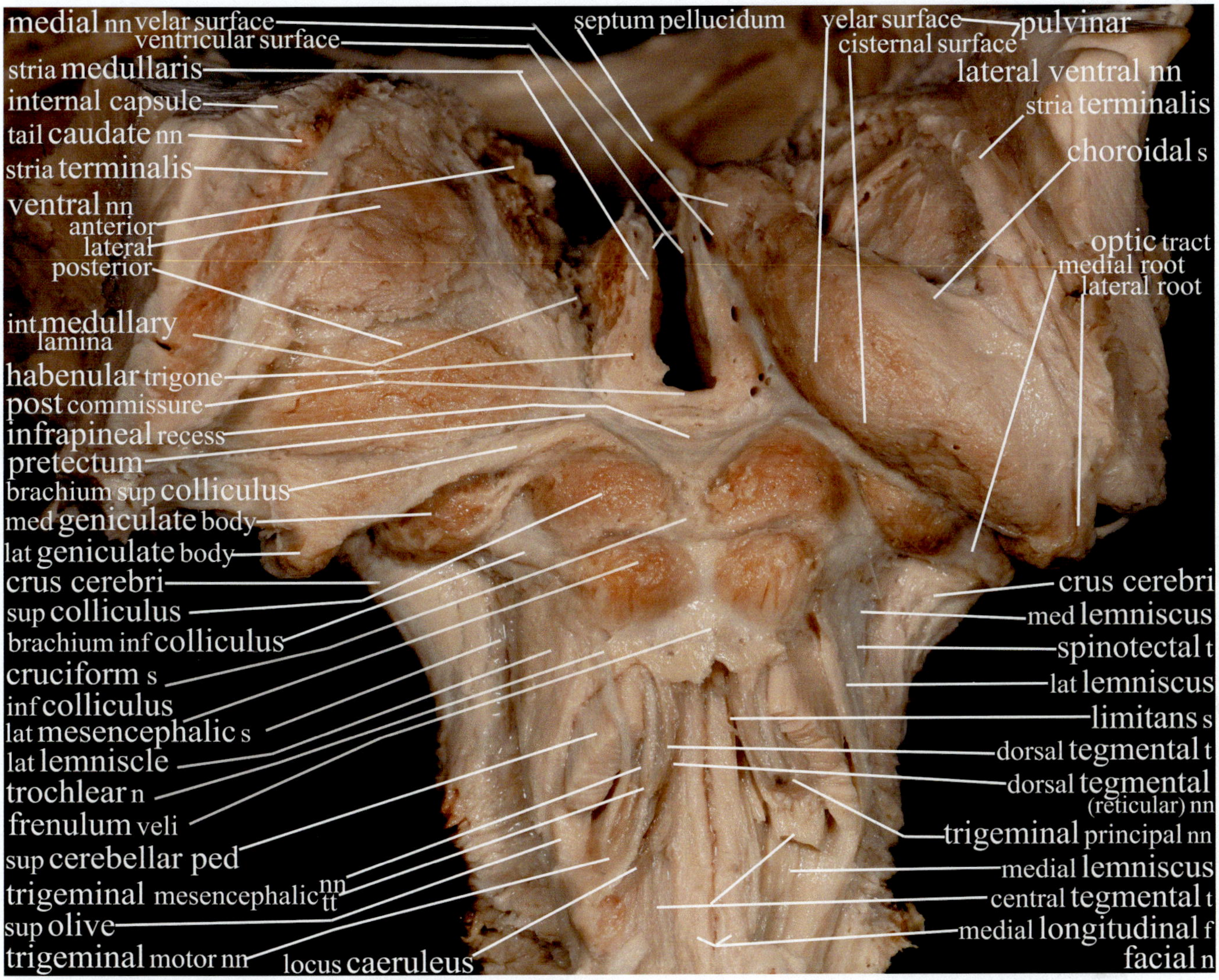

Fig. 7.7 (continued)

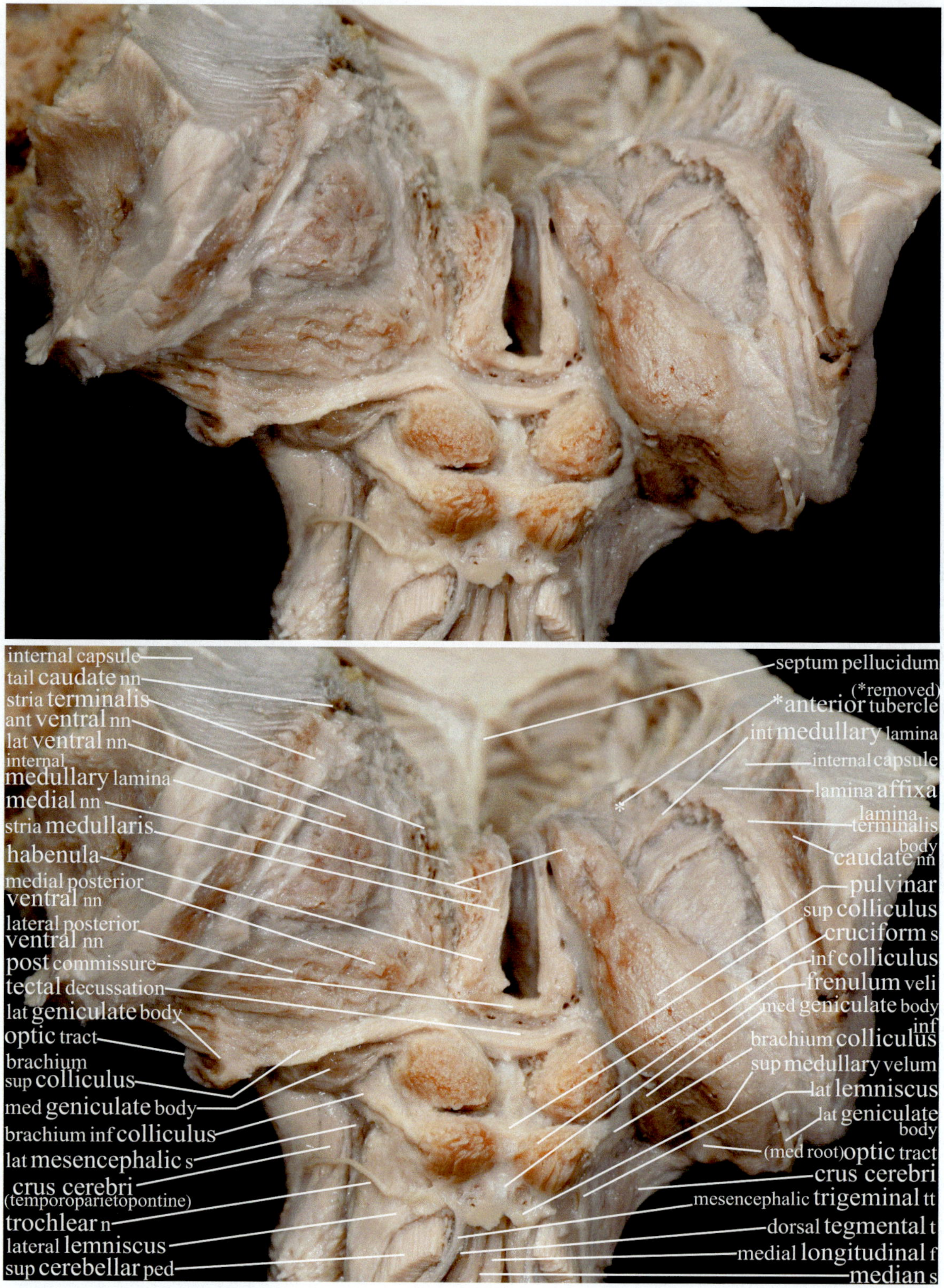

Fig. 7.8 Anatomical preparation of the cisternal surface of the epithalamus and the quadrigeminal plate. The posterior commissure is a complex arrangement of different fasciculi that cross the midline at the cephalic portion of the tectal plate. It contains several connections: diencephalic–diencephalic (habenular commissure, lateral geniculate body, and posterior thalamus), diencephalic–tectal (crossed fibers of the brachium of the superior colliculus), and tectal–tectal (periaqueductal nuclei related to the pupillary light reflex). The posterior commissure is also known as the epithalamic commissure. Its paramedian extensions are limited ventrally by the habenular commissure and nuclei and dorsally by the pretectum. The pretectum is formed in part by the tectal connections with the metathalamus and a reticular nucleus embedded in this net, known as the pretectal nucleus. *ant* anterior, *f* fasciculus, *inf* inferior, *int* internal, *lat* lateral, *med* medial, *n* nerve, *nn* nucleus, *ped* peduncle, *post* posterior, *s* sulcus, *sup* superior, *t* tract, *tt* tract (©Kadri 2023. All Rights Reserved)

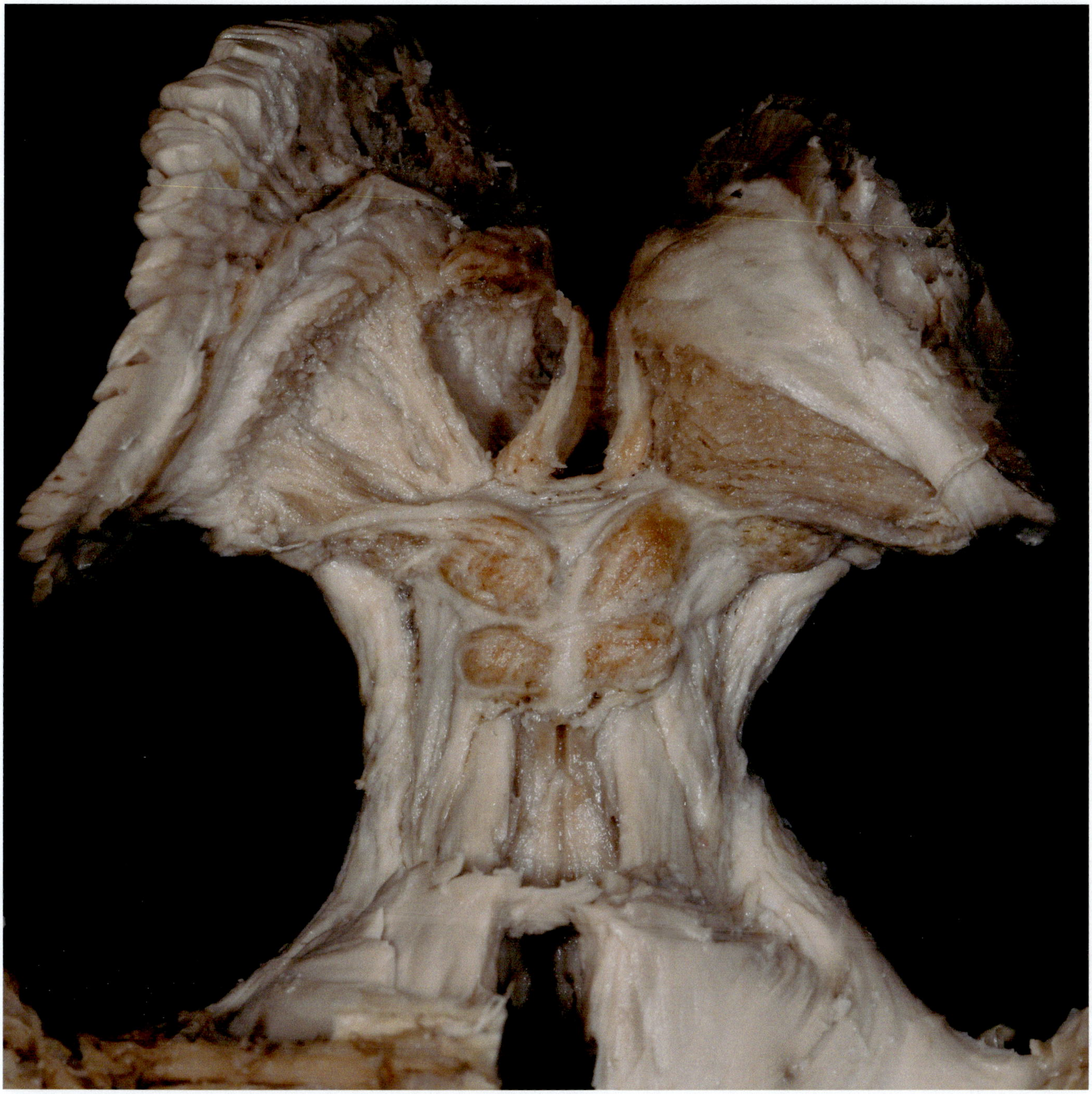

Fig. 7.9 Anatomical preparation of the ventral nuclei of the dorsal thalamus. The posterior tubercle of the thalamus, more widely known as the pulvinar, has been removed. The pulvinar is the dorsal representation of the posterior nuclei of the thalamus. Removing the posterior nuclei of the thalamus discloses the radiating fibers of the internal medullary lamina of the thalamus. This lamina separates the anterior and medial nuclei of the thalamus from the lateral and posterior nuclei. The arrangement of the internal medullary lamina is extremely complex, as it comprises the relationship and proportion of the reticular and intralaminar nuclei. *ant* anterior, *ext* external, *inf* inferior, *int* internal, *lat* lateral, *med* medial, *mid* middle, *n* nerve, *nn* nucleus, *ped* peduncle, *post* posterior, *s* sulcus, *sup* superior, *t* tract

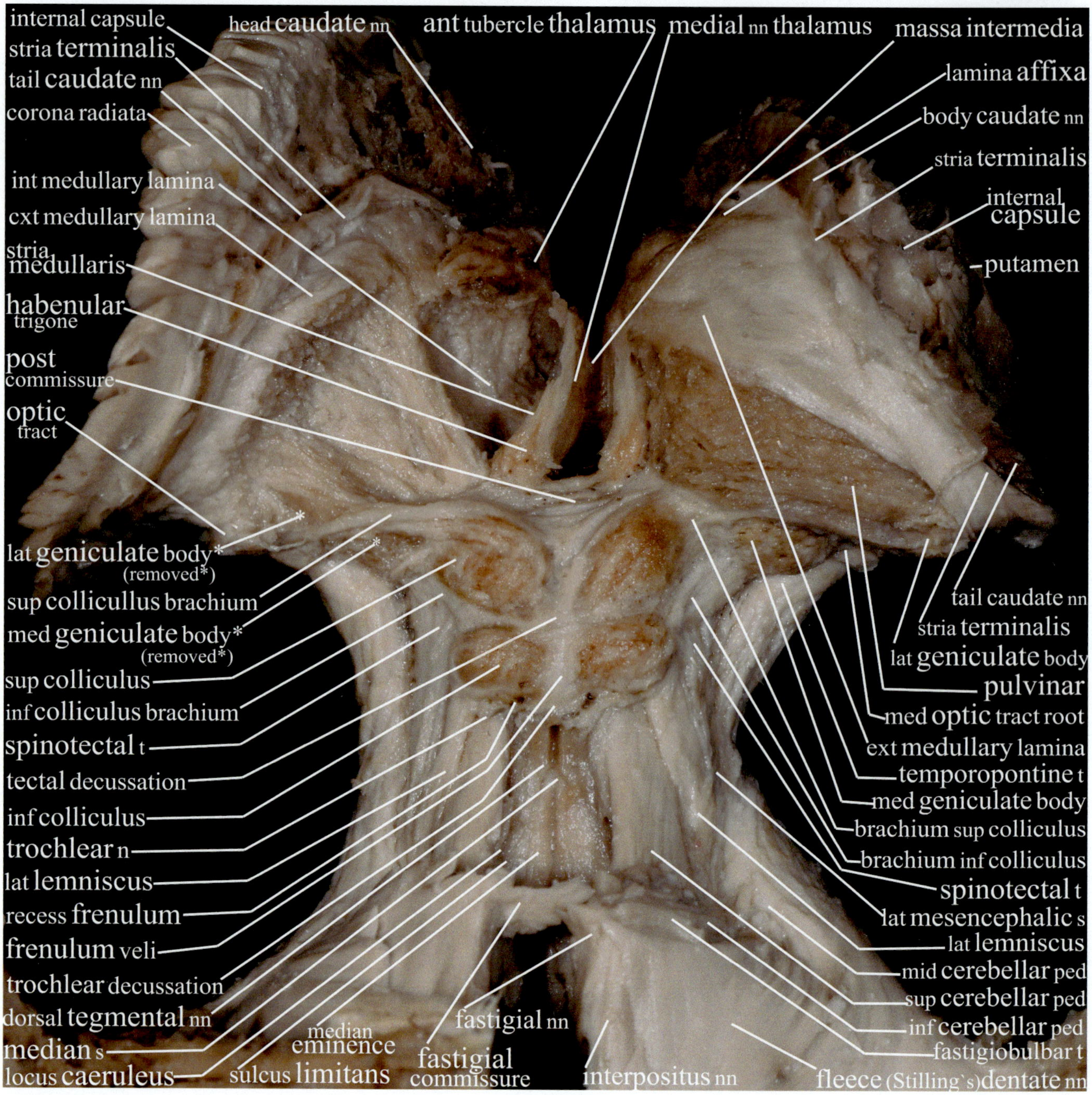

Fig. 7.9 (continued)

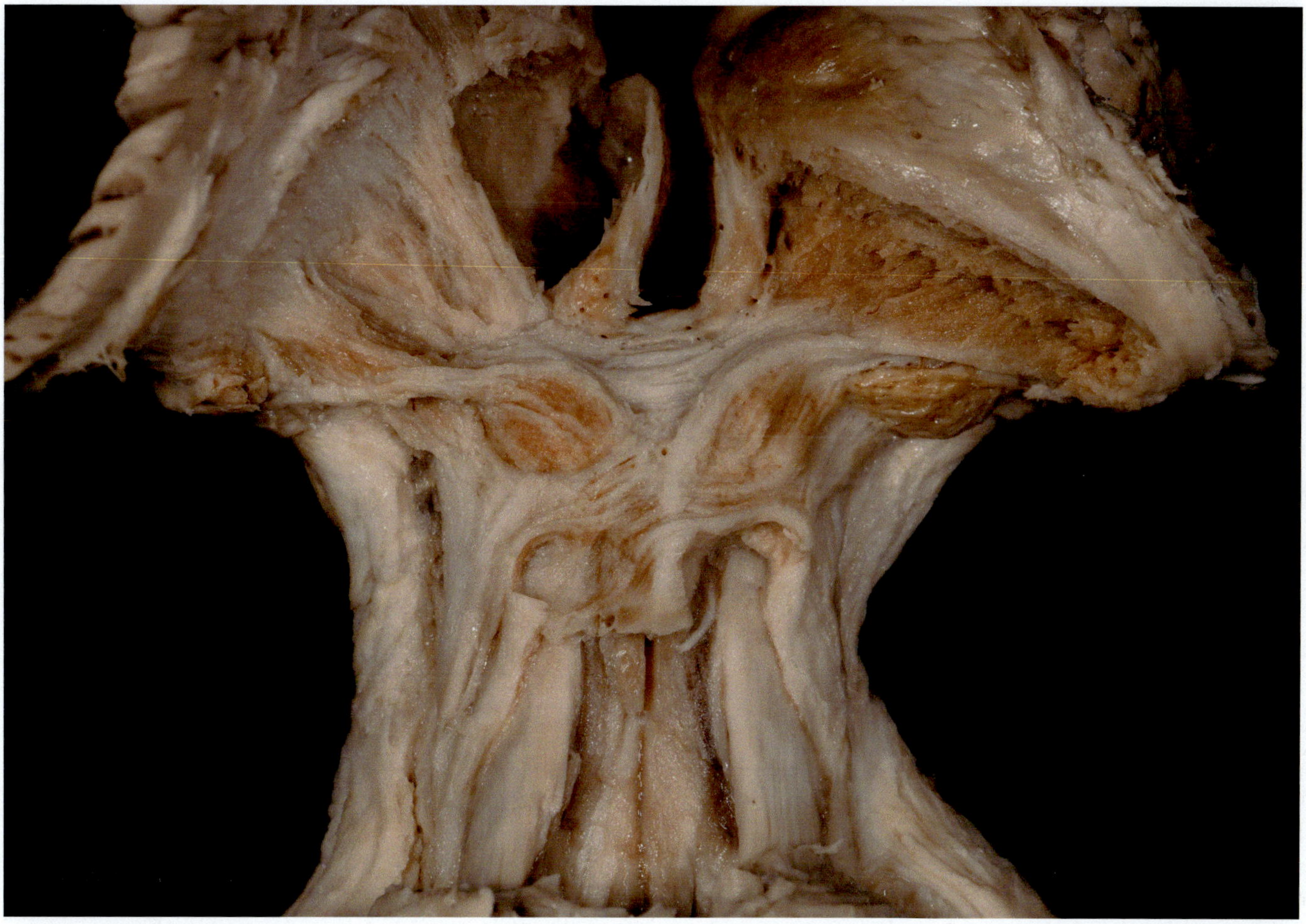

Fig. 7.10 Anatomical preparation of the internal medullary lamina and epithalamus and dissection of the mesencephalic tectum and pontine tegmentum. The lateral lemniscus is the layer of fibers that emerge on the lateral surface of the pons dorsal to the lateral mesencephalic sulcus. These fibers traverse the surface of the superior cerebellar peduncle to connect to the quadrigeminal plate. The lateral lemniscus contains the connections of the inferior colliculus and auditory and other medullotectal and spinotectal tracts. The colliculi are stratified into a structure of superimposed gray and white matter. The main cephalic connection of the tectal colliculi is with the metathalamus (lateral and medial geniculate bodies) via the brachium colliculi. Some of these connections are contralateral and contribute to the tectal decussation and posterior commissure. *ant* anterior, *ext* external, *f* fasciculus, *inf* inferior, *lat* lateral, *med* medial, *n* nerve, *nn* nucleus, *ped* peduncle, *post* posterior, *s* sulcus, *sup* superior, *t* tract (©Kadri 2023. All Rights Reserved)

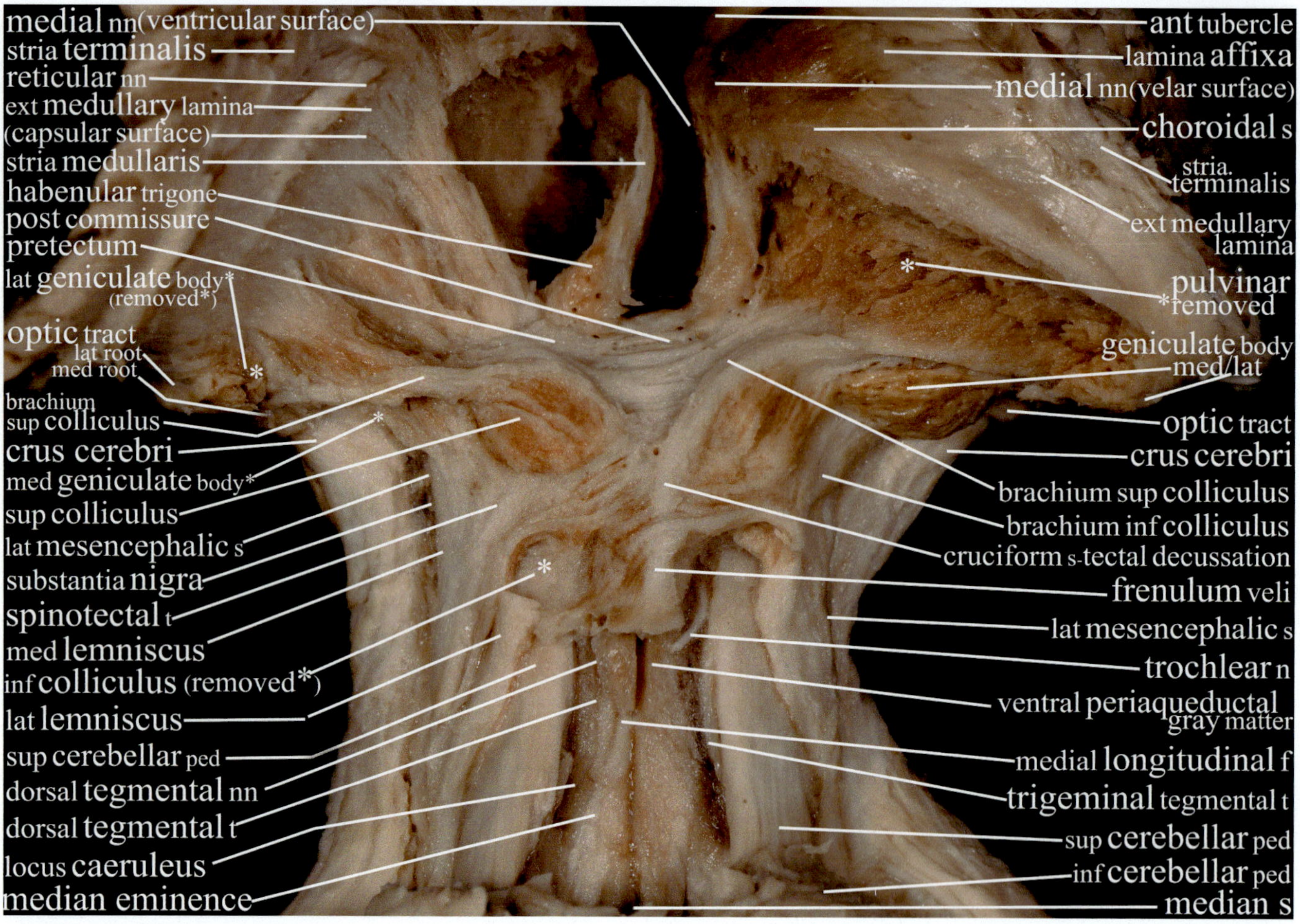

Fig. 7.10 (continued)

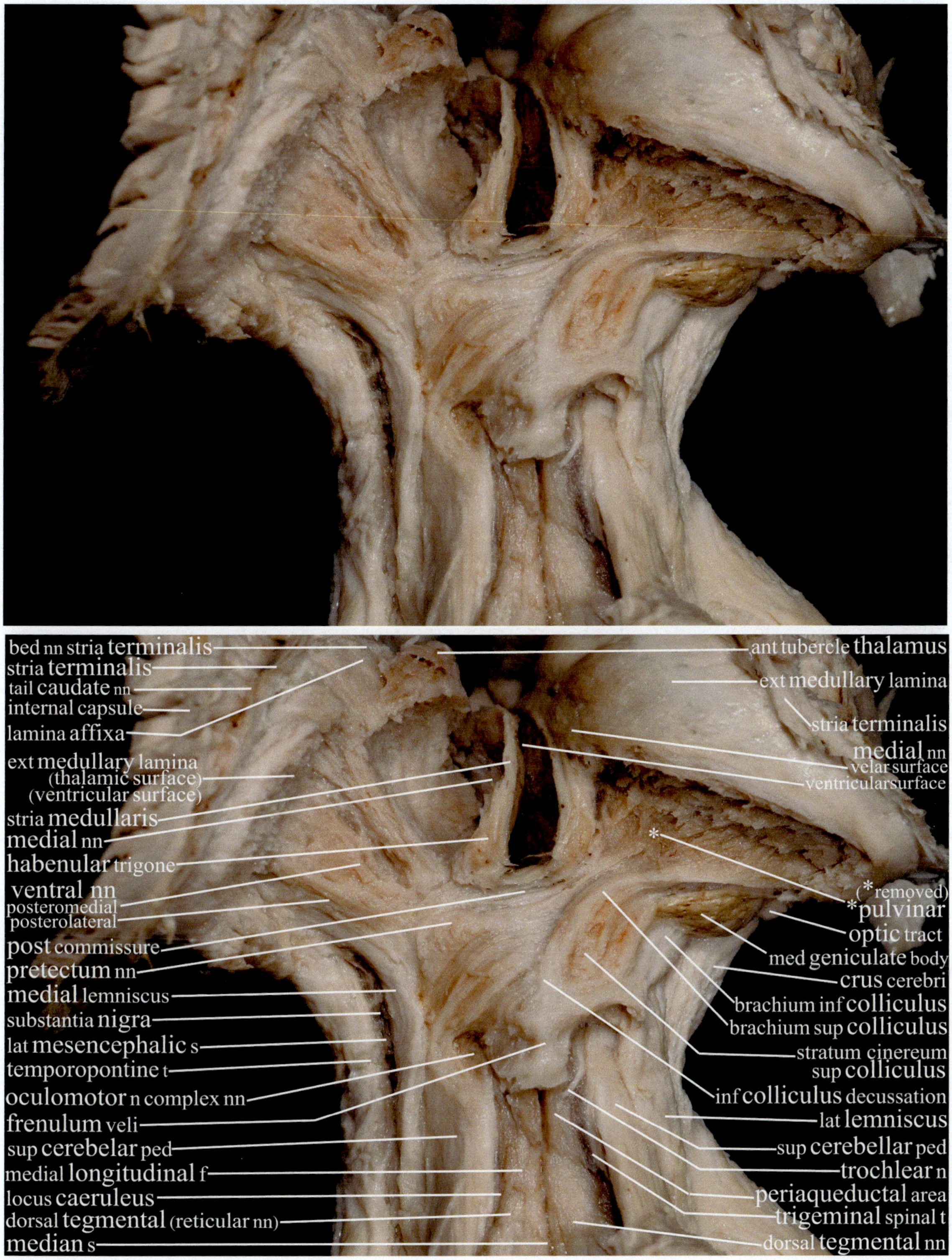

Fig. 7.11 Anatomical preparation of the internal medullary lamina of the thalamus, epithalamus and the pretectal area of the mesencephalon. The inner fibers of the lateral lemniscus connect to the pretectum. The posterior commissure is the cephalic limit of the tectal plate. The lateral mesencephalic sulcus and the substantia nigra surfacing on it are landmarks to identify the medial lemniscus, which is placed dorsal to them. The medial lemniscus can be identified at the pial surface at the ventral border of the lemniscal trigone. *ant* anterior, *ext* external, *f* fasciculus, *inf* inferior, *lat* lateral, *med* medial, *n* nerve, *nn* nucleus, *ped* peduncle, *post* posterior, *s* sulcus, *sup* superior, *t* tract (©Kadri 2023. All Rights Reserved)

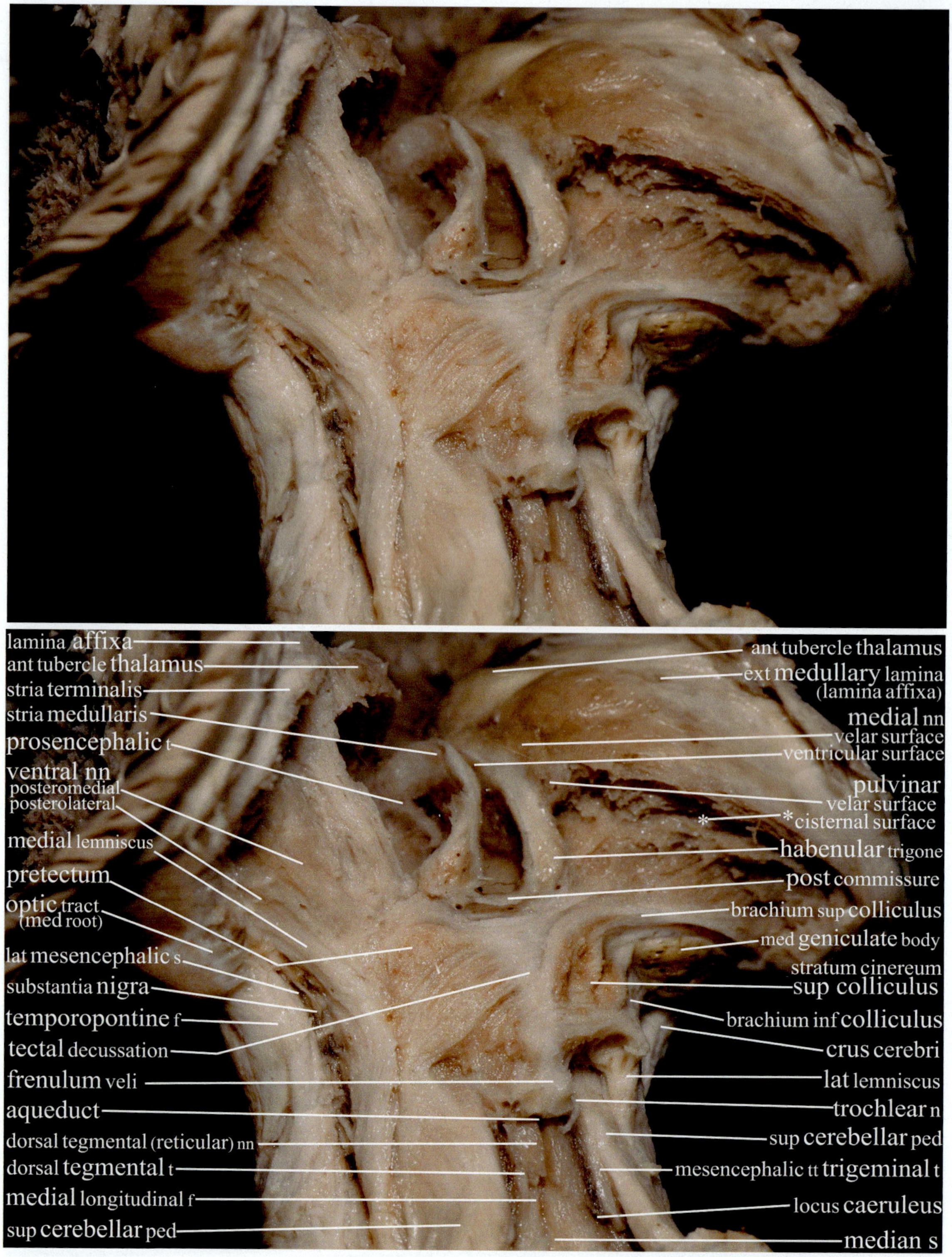

Fig. 7.12 Close-up view of the epithalamus and pretectum on an anatomical preparation of the internal medullary lamina, epithalamus, and pretectal area of the mesencephalon. The longitudinal arm of the cruciform sulcus is crossed by fibers of the superior and inferior colliculi and fibers of the medullary and spinal connections of the tectal plate. *ant* anterior, *ext* external, *f* fasciculus, *inf* inferior, *lat* lateral, *med* medial, *n* nerve, *nn* nucleus, *ped* peduncle, *post* posterior, *s* sulcus, *sup* superior, *t* tract, *tt* tract (©Kadri 2023. All Rights Reserved)

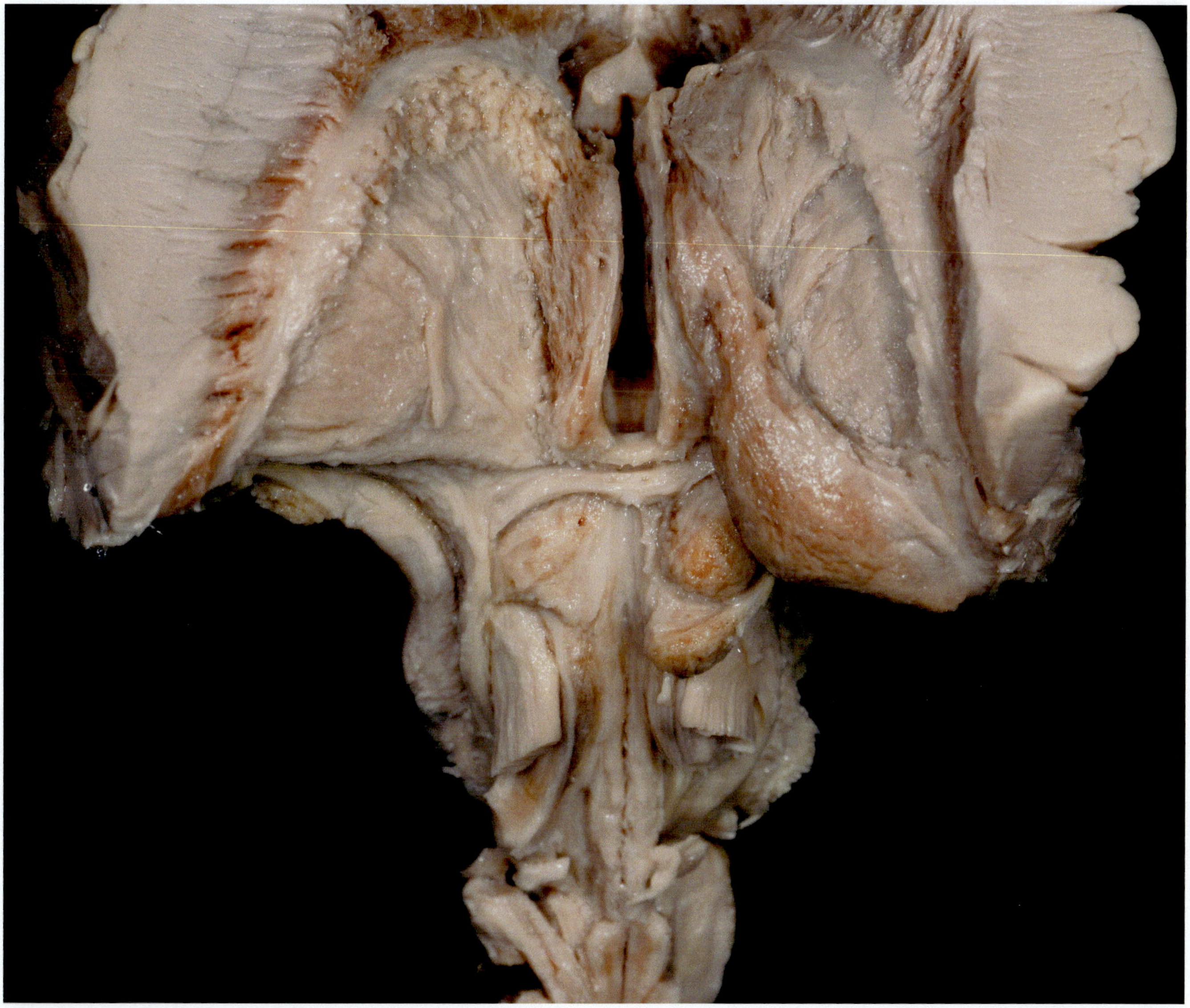

Fig. 7.13 Anatomical preparation of the ventral and medial nuclei of the dorsal thalamus, the epithalamus, pretectum, and quadrigeminal plate of the tectum mesencephalic and the tegmentum of the brainstem at the floor of the fourth ventricle. *ant* anterior, *ext* external, fasciculus, *inf* inferior, *int* internal, *lat* lateral, *n* nerve, *nn* nucleus, *ped* peduncle, *post* posterior, *s* sulcus, *sup* superior, *t* tract (©Kadri 2023. All Rights Reserved)

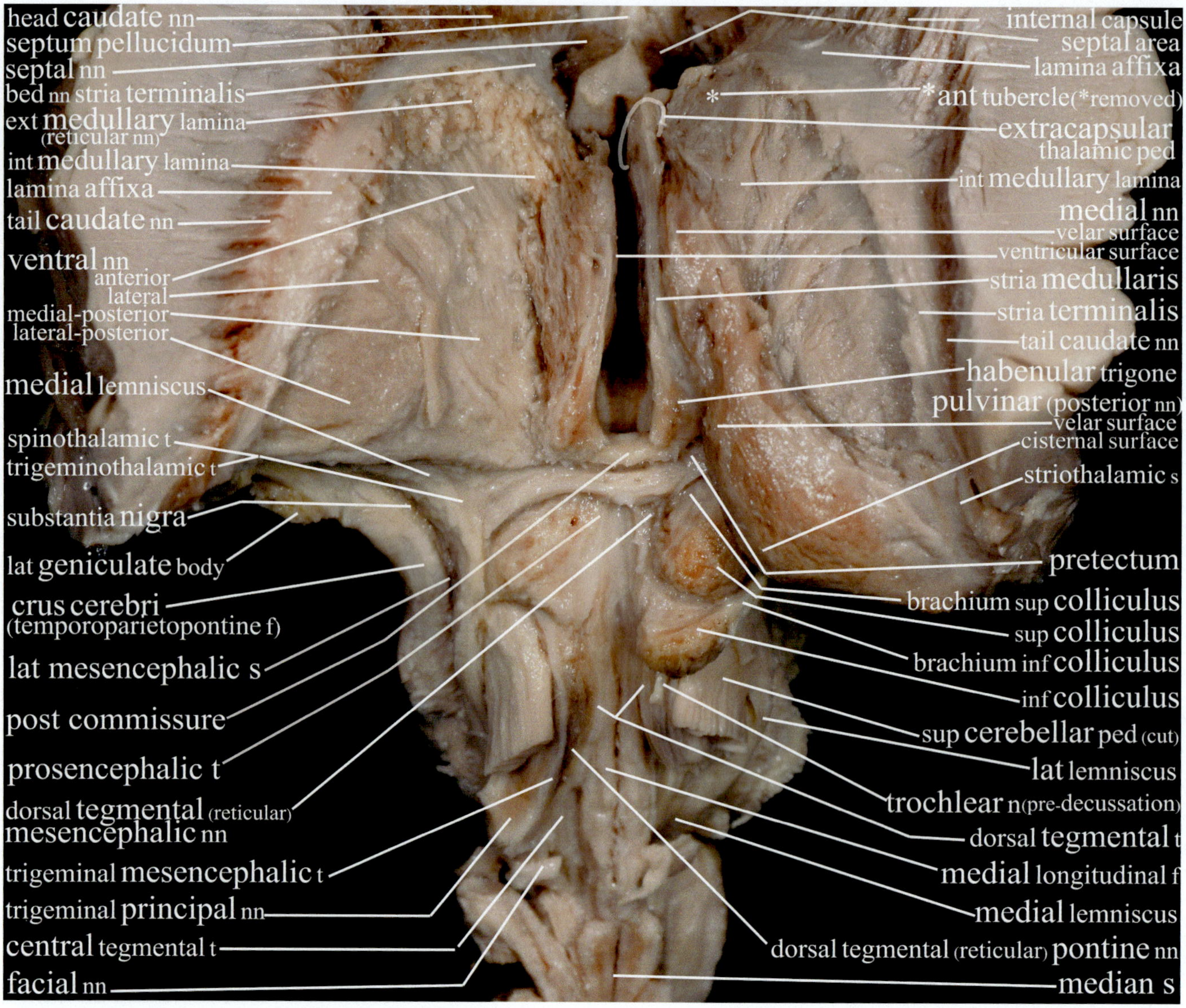

Fig. 7.13 (continued)

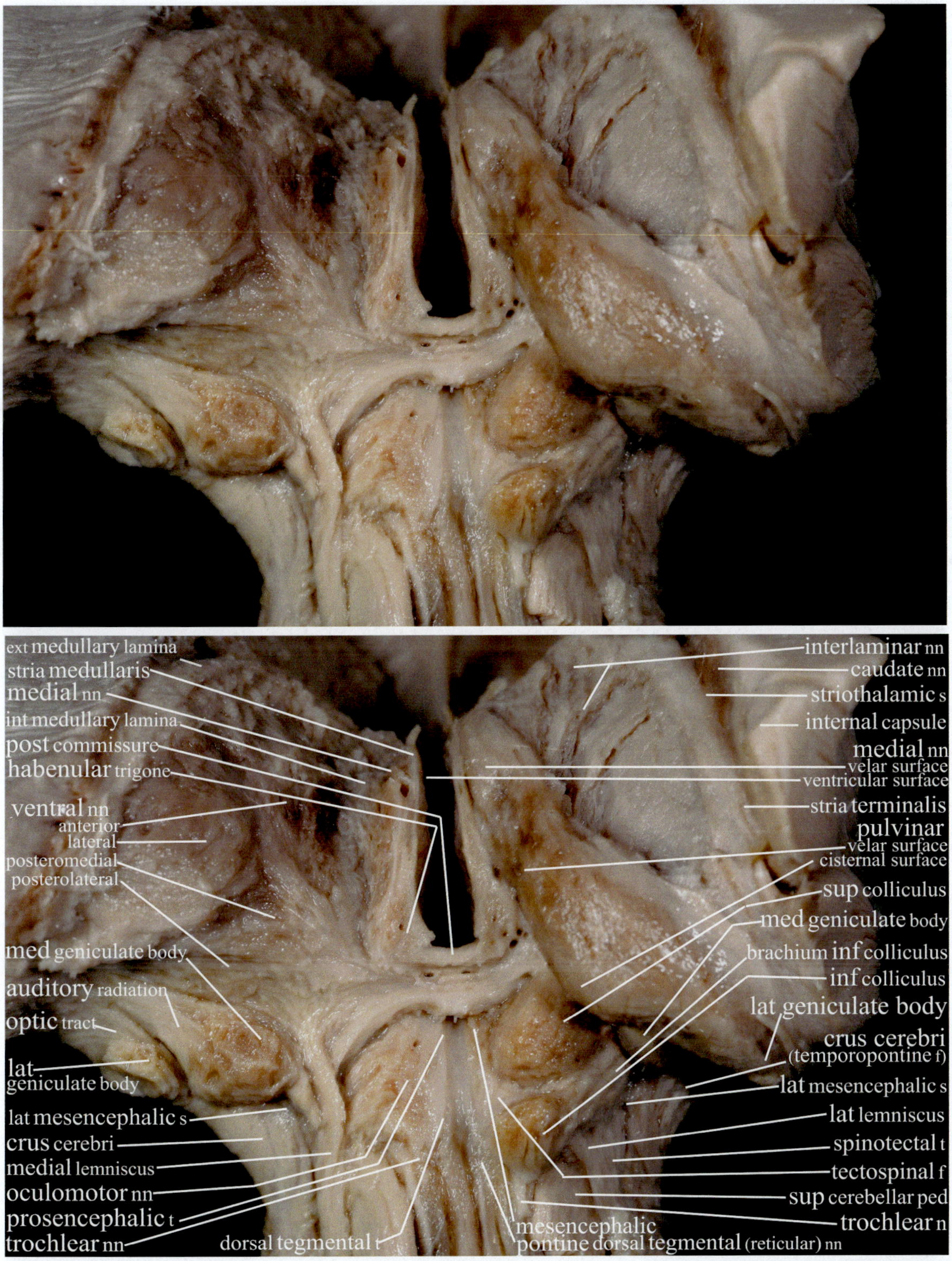

Fig. 7.14 Close-up view of the epithalamus and ventral nuclei of the dorsal thalamus. Removing the posterior nuclei of the thalamus (the pulvinar) unveils the ventral nuclei. *ext* external, *f* fasciculus, *inf* inferior, *int* internal, *lat* lateral, *med* medial, *nn* nucleus, *ped* peduncle, *post* posterior, *s* sulcus, *sup* superior, *t* tract (©Kadri 2023. All Rights Reserved)

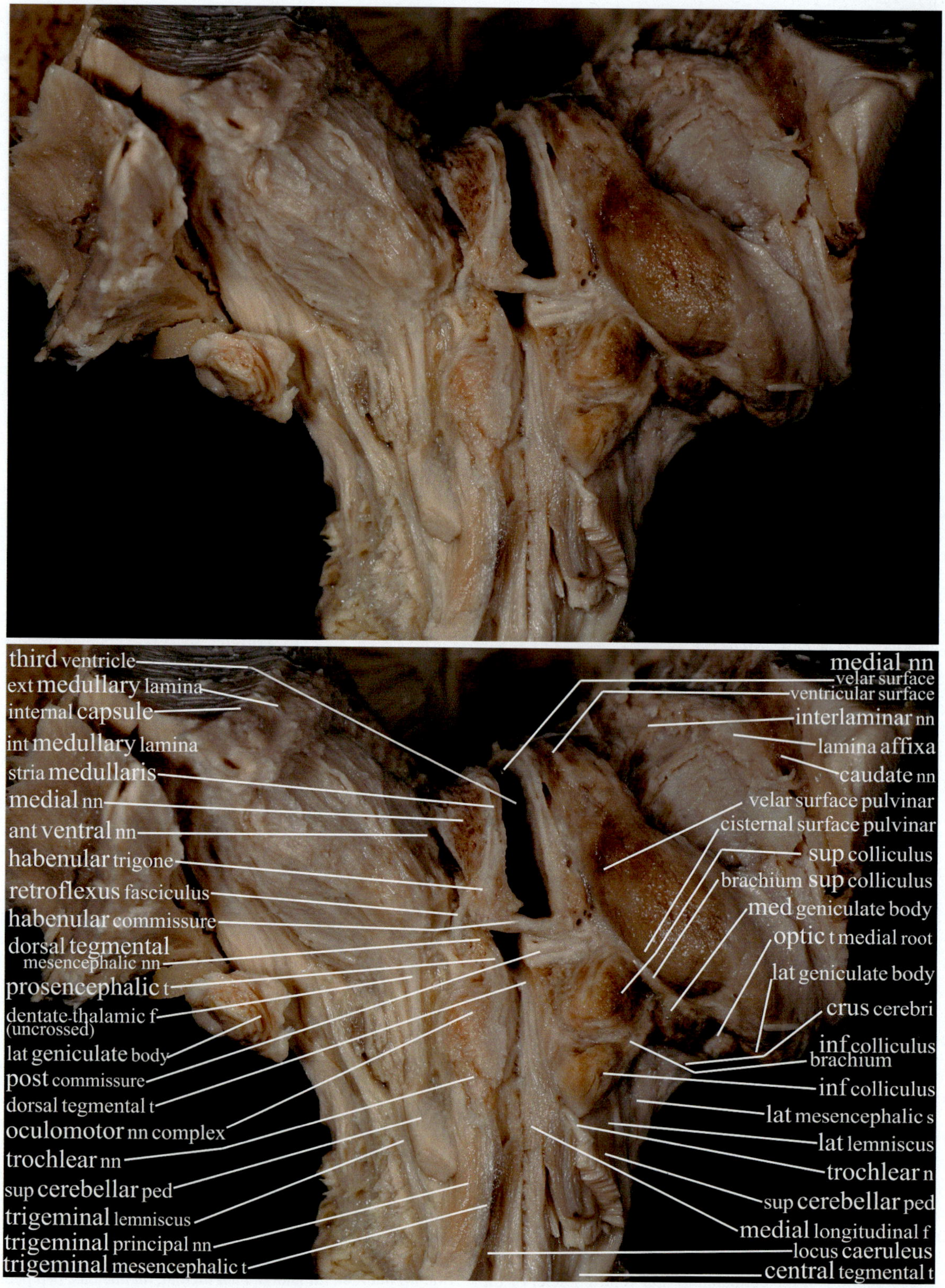

Fig. 7.15 Anatomical preparation of the dorsal thalamus, epithalamus, and dorsal tegmentum of the mesencephalon. The posterior commissure is the dorsal limit of the diencephalon and is composed of diencephalic and mesencephalic fibers that cross the midline. *ant* anterior, *ext* external, *f* fasciculus, *inf* inferior, *int* internal, *lat* lateral, *med* medial, *n* nerve, *nn* nucleus, *ped* peduncle, *post* posterior, *s* sulcus, *sup* superior, *t* tract (©Kadri 2023. All Rights Reserved)

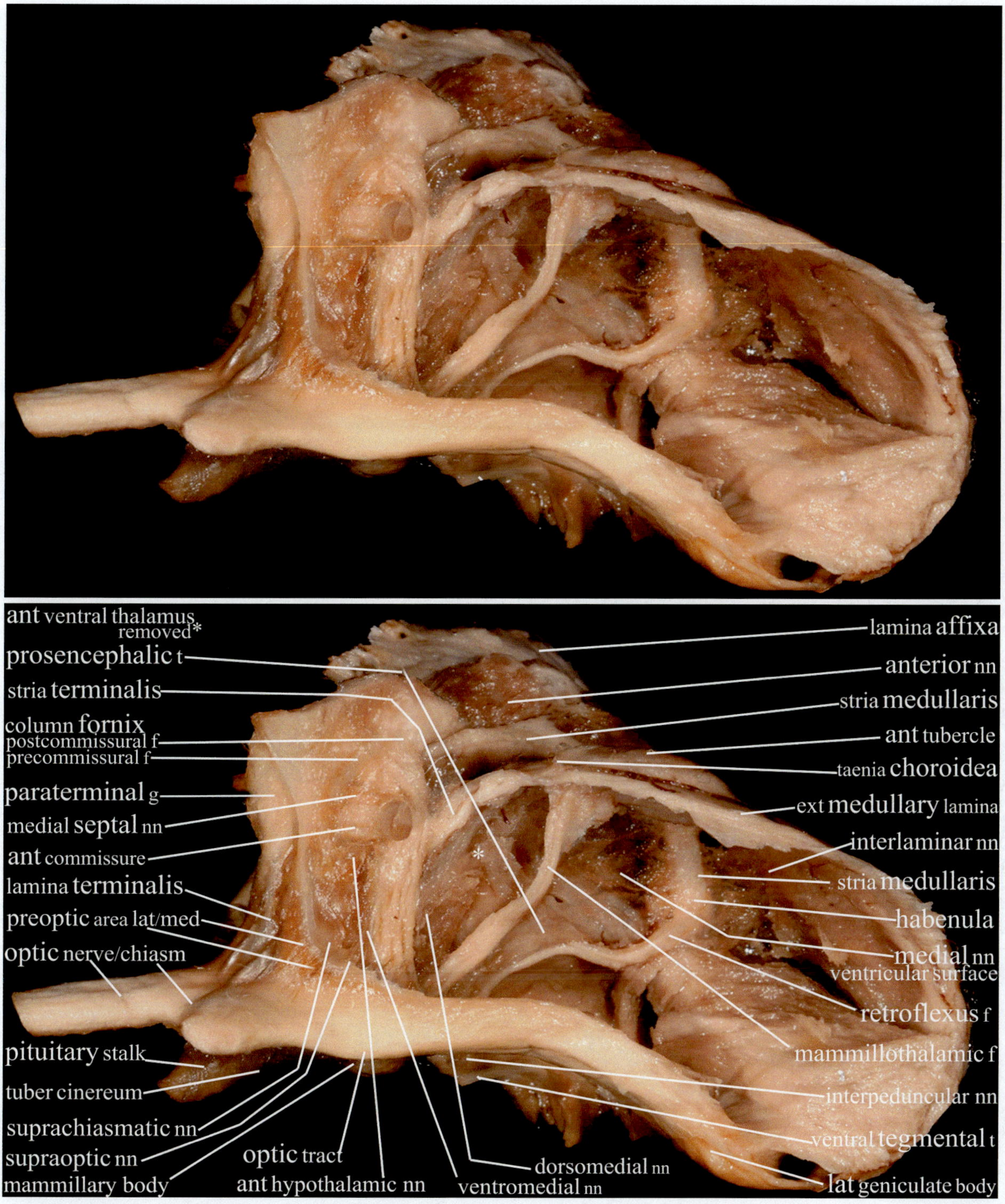

Fig. 7.16 Anatomical preparation of the lateral hypothalamus and the diencephalic limbic connections. The preoptic area is at the cisternal surface, ventral to the insertion of the lamina terminalis. The precommissural fibers of the column of the fornix connect with the lateral septal region, which also connects with the amygdala dorsally via the stria terminalis. The medial septal region connects with the amygdala ventrally via the amygdalo-septal fibers embedded within the diagonal band of the anterior perforated substance. The medial septal region also connects with the median olfactory region via the medial olfactory stria and with the supra- and precommissural hippocampus via the longitudinal striae. This convergence of white tracts manifests as a characteristic whitish surface of the medial septal nuclei at the cisternal surface that is known as the paraterminal gyrus. The mammillothalamic and retroflexus tracts are the main limbic diencephalic connections. These tracts are a reliable landmark to separate limbic connections located medial to them, directed to the tegmentum. Therefore, they are part of the prosencephalic tract, from the limbic thalamic and epithalamic connections, located lateral to the mammillothalamic and retroflexus tracts. These limbic thalamic and epithalamic connections are part of the extracapsular thalamic peduncle. *ant* anterior, *ext* external, *f* fasciculus, *g* gyrus, *lat* lateral, *med* medial, *nn* nucleus, *t* tract (©Kadri 2023. All Rights Reserved)

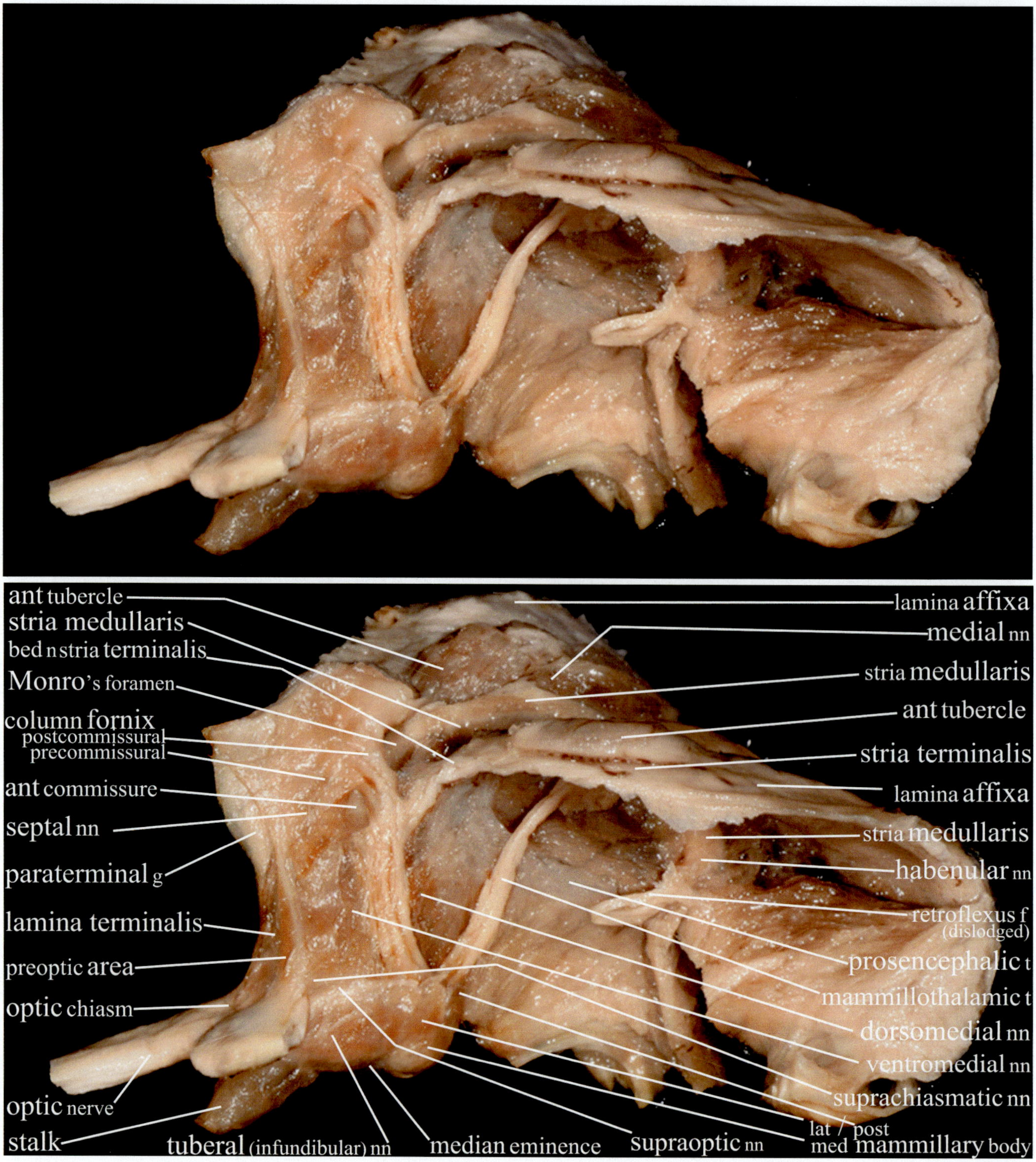

Fig. 7.17 Anatomical preparation of the sequential dissection shown in Fig. 7.16. The optic tract has been removed, and the retroflexus tract is deflected. *ant* anterior, *f* fasciculus, *g* gyrus, *lat* lateral, *med* medial, *n* nerve, *nn* nucleus, *post* posterior, *t* tract (©Kadri 2023. All Rights Reserved)

The Dorsal Thalamus

The dorsal thalamus is the main mass of the thalamus (Fig. 7.18). It is an ovoid nuclear mass extending from the interventricular foramen to the aqueduct and is separated from the ventral thalamus and internal capsule by the external medullary lamina (Fig. 7.19). The external medullary lamina is the layer of myelinated fibers from the thalamic peduncles, also embedded with the reticular nuclei, which covers the dorsal thalamus (Figs. 7.20 and 7.21). At the ventricular surface, the external medullary lamina is also known as the stratum zonale. The dorsal thalamus is the site of the reciprocal connection of the thalamus with the cortex and basal nuclei [1]. The different corticothalamic and thalamocortical connections are distributed through the anterior, superior, posterior, and inferior thalamic peduncles. The basal ganglia interconnections are mainly through the extracapsular thalamic peduncle (Fig. 7.22), which are fibers that originate in the medial aspect of the thalamus and join the ansa peduncularis to reach the amygdala, the lentiform nucleus, and the anterior temporal lobe. The anterior thalamic peduncle travels within the anterior limb of the internal capsule and connects to the frontal cortex. The superior thalamic peduncle connects with the central lobe (precentral and postcentral gyri). The posterior thalamic peduncle connects the middle and posterior temporal, inferior parietal, and occipital areas. The posterior thalamic peduncle harbors its peculiar loop on the roof and lateral wall of the temporal horn. This loop is also known as Meyer's loop. The inferior thalamic peduncle connects the anterior temporal region.

Parcellation of the Dorsal Thalamus

The thalamic subdivision has several classifications. For simplicity and according to gross anatomical preparations, we simplified these classifications.

The external medullary lamina penetrates the thalamus and constitutes the internal medullary lamina, which runs anteroposterior and divides the thalamus incompletely into medial and lateral nuclei groups. The posterior portion, undivided by the lamina, constitutes the posterior nuclei group. Ventrally, the internal medullary lamina splits into the Y-shaped medial and lateral laminas. Between these lies the anterior nuclei group. Within the internal medullary lamina, clusters of cells constitute the interlaminar group (Figs. 7.3, 7.4, 7.5, 7.6, 7.7, 7.8, 7.9, 7.13, 7.18, 7.19, and 7.23).

The anterior group is the first classification and is located within the anterior tubercle of the thalamus, on the posterior border of the foramen of Monro. It constitutes the limbic thalamus with the dorsomedial nucleus of the medial group. Its main afferent is the mammillothalamic tract, and its main efferent is the anterior thalamic peduncle (Figs. 7.1, 7.2, 7.3, 7.5, 7.9, 7.11, 7.12, 7.16, 7.17, 7.18, 7.20, and 7.22).

The posterior group comprises the pulvinar and the medial and lateral geniculate bodies (Figs. 7.1, 7.4, 7.5, 7.6, 7.7, 7.8, 7.9, 7.10, 7.11, 7.12, 7.13, 7.14, 7.15, 7.18, and 7.21). Reciprocal pulvinar–cortical connections are distributed via the posterior and inferior thalamic peduncles. The geniculate bodies are also referred to as the metathalamus. The medial geniculate body is the auditory relay of the thalamus. It receives inputs through the brachium of the inferior colliculi, from its nuclei, or directly from the lateral lemniscus, which is the ascending projection of the cochlear nucleus. The medial lemniscus projects through the auditory radiation toward the anterior transverse temporal gyrus (Heschl's gyrus). The lateral geniculate body receives retinal inputs through the optic tract. It projects through the geniculocalcarine tract toward the posterior calcarine fissure. The geniculocalcarine tract is also known as the optic radiation and is part of the posterior thalamic peduncle.

The lateral group is considered to be the most important and is subdivided into ventral and dorsal subgroups. The ventral subgroup is the sensory–motor relay. The ventral anterior nucleus receives afference from the globus pallidus and projects toward the motor cortex. The ventral lateral receives cerebellar inputs and projects toward motor areas. The ventral posterolateral is the sensory relay and receives inputs from the medial and spinal lemnisci. The spinal lemniscus is formed by the union of the lateral and anterior spinothalamic tracts. The ventral posterolateral nucleus projects to the postcentral gyrus. The ventral posteromedial nucleus receives inputs from the trigeminal lemniscus and projects toward the postcentral gyrus and gustatory areas (Figs. 7.1, 7.3, 7.5, 7.9, 7.10, 7.11, 7.12, 7.13, 7.14, 7.15, 7.18, and 7.20).

The medial and interlaminar groups comprise the fourth classification. The interlaminar group is distributed within the internal medullary lamina and maintains connections with the reticular formation of the brainstem. It is related to cortical activation. The dorsomedial nucleus comprises the main nuclei of the medial group. It is part of the limbic thalamus. Its main afferents are the amygdala and hypothalamus, and its main efferent is the frontal lobe (Figs. 7.1, 7.2, 7.3, 7.4, 7.5, 7.7, 7.8, 7.13, 7.14, 7.15, 7.16, 7.19, 7.20, 7.21, 7.22, and 7.23).

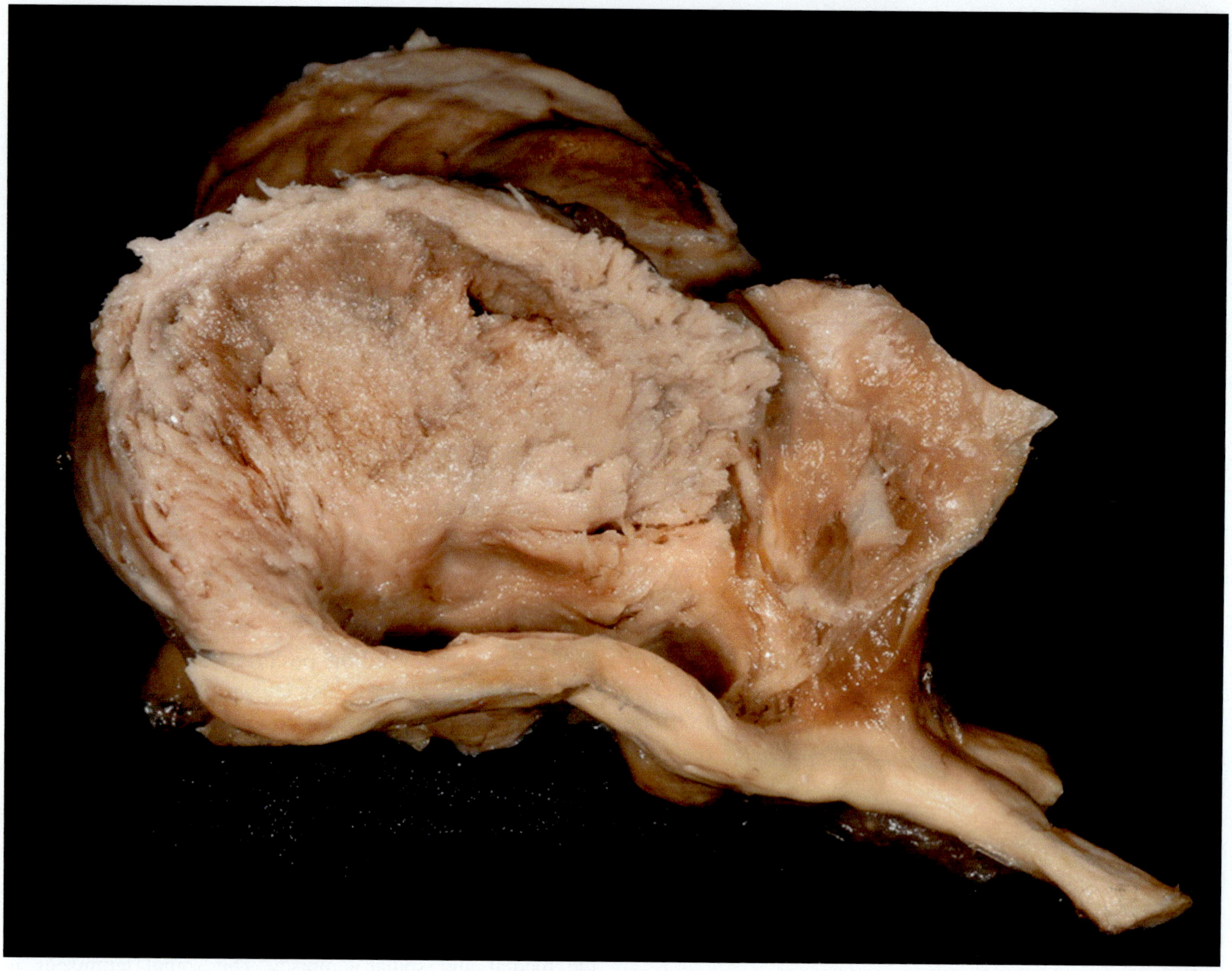

Fig. 7.18 Anatomical preparation of the capsular surface of the thalamus. Removing the projection fibers exposes the capsular surface of the dorsal thalamus, with the reticular nuclei embedded in the external medullary lamina and the lateral hypothalamic and lateral subthalamic regions. *ant* anterior, *lat* lateral, *n* nerve, *nn* nucleus (©Kadri 2023. All Rights Reserved)

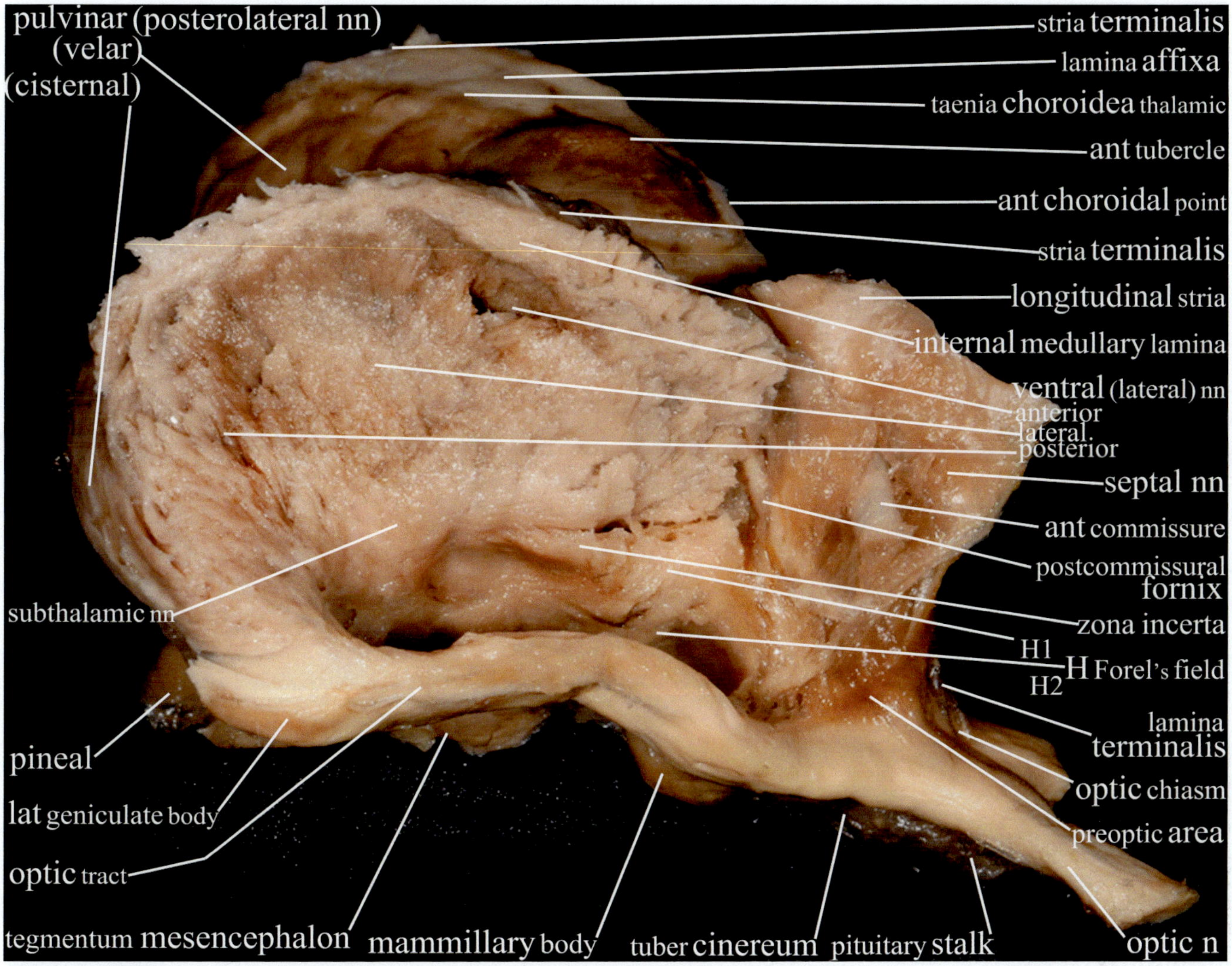

Fig. 7.18 (continued)

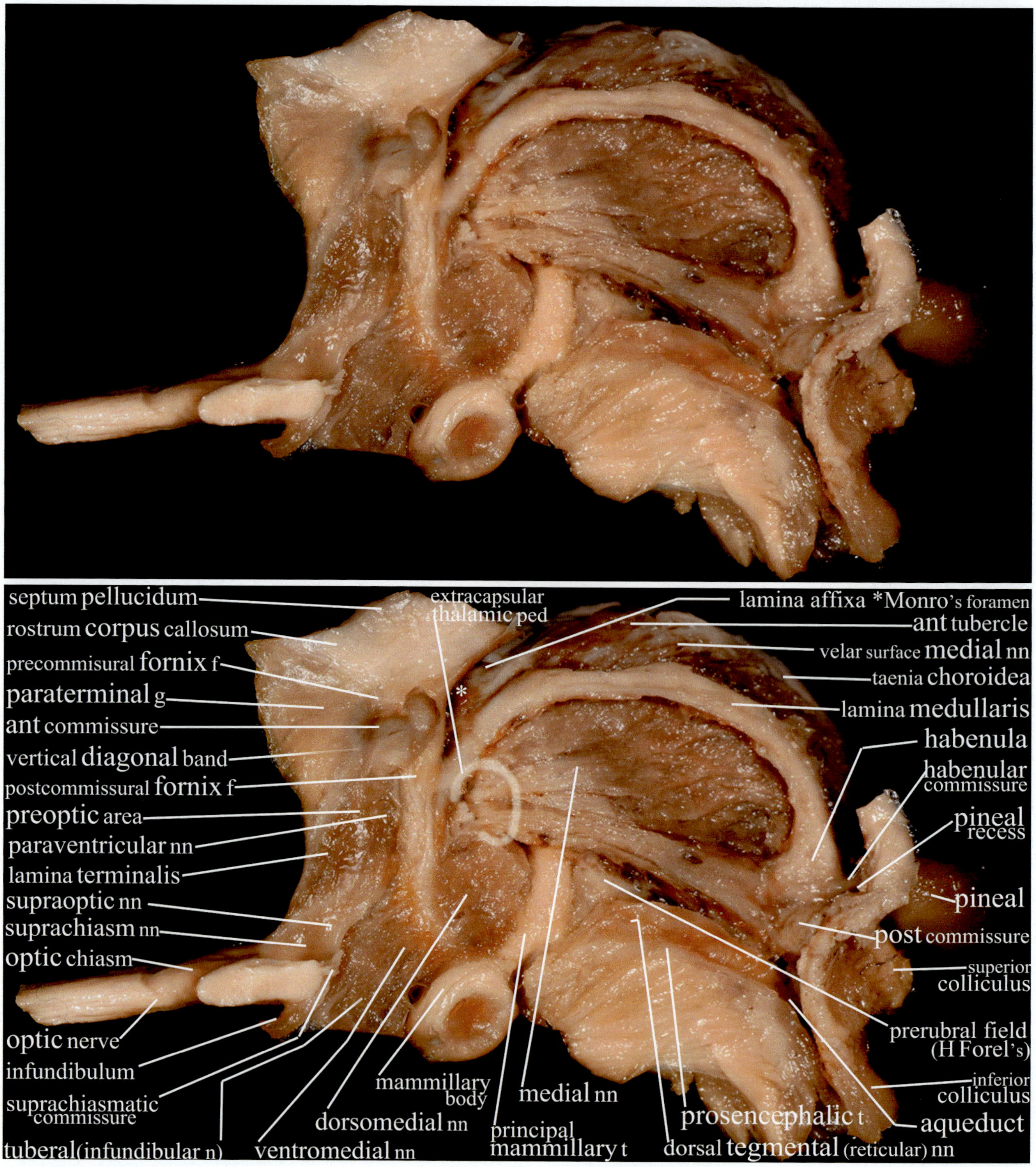

Fig. 7.19 Anatomical preparation of the limbic connections at the diencephalic mesencephalic regions. The medial thalamic surface (third ventricular and velar surfaces) harbors the medial nuclei of the thalamus, the medial limit of which is the internal medullary lamina, and is the main limbic surface of the thalamus. *ant* anterior, *f* fasciculus, *g* gyrus, *n* nerve, *nn* nucleus, *ped* peduncle, *post* posterior, *t* tract (©Kadri 2023. All Rights Reserved)

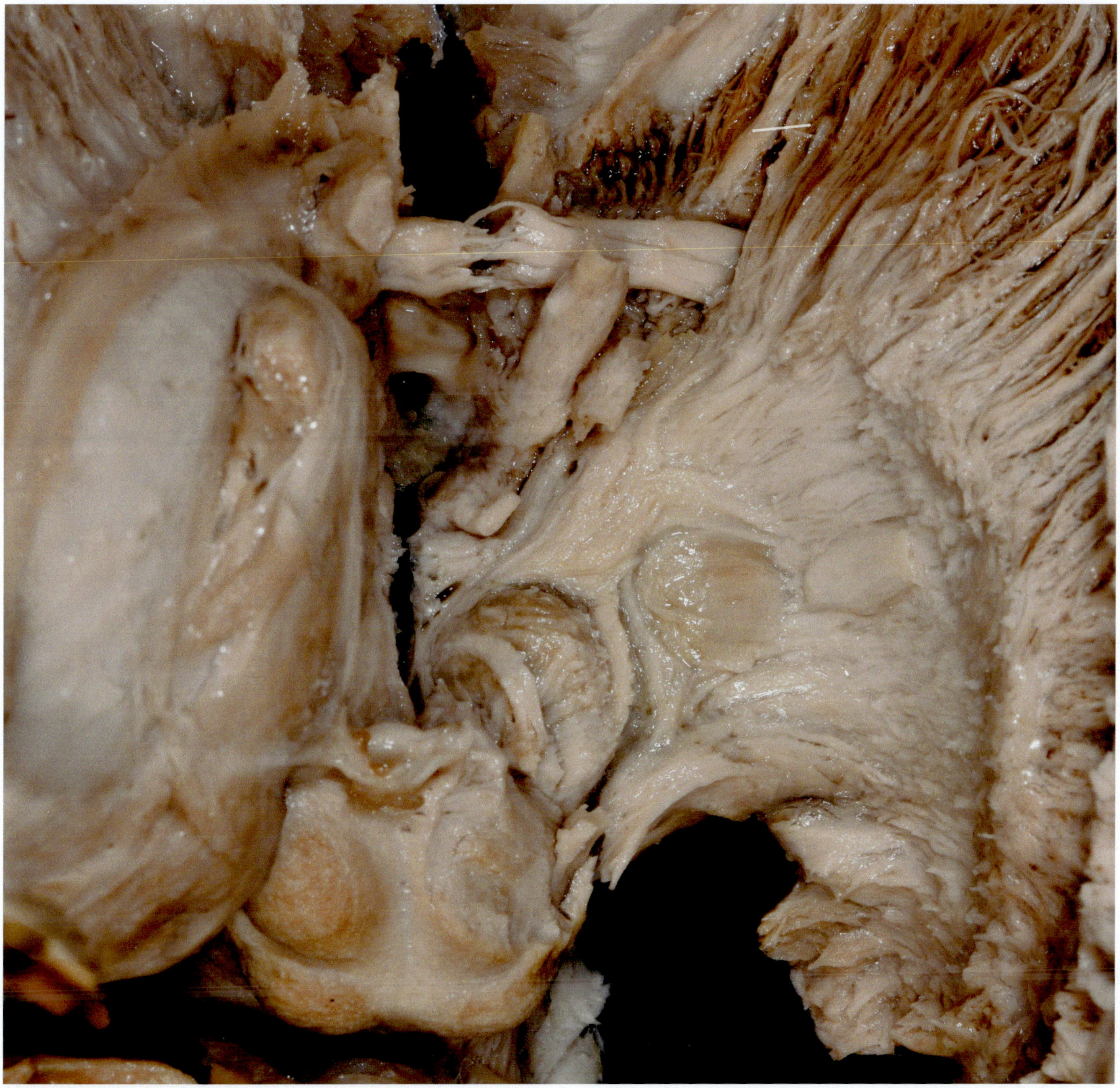

Fig. 7.20 Anatomical preparation of the diencephalic–mesencephalic transition. The transition is centered in the subthalamic area, on the diencephalic side, and in the red nucleus and its partially removed white matter envelope on the tegmental mesencephalic tract. *ant* anterior, *ext* external, *g* gyrus, *nn* nucleus, *t* tract, *tt* tract

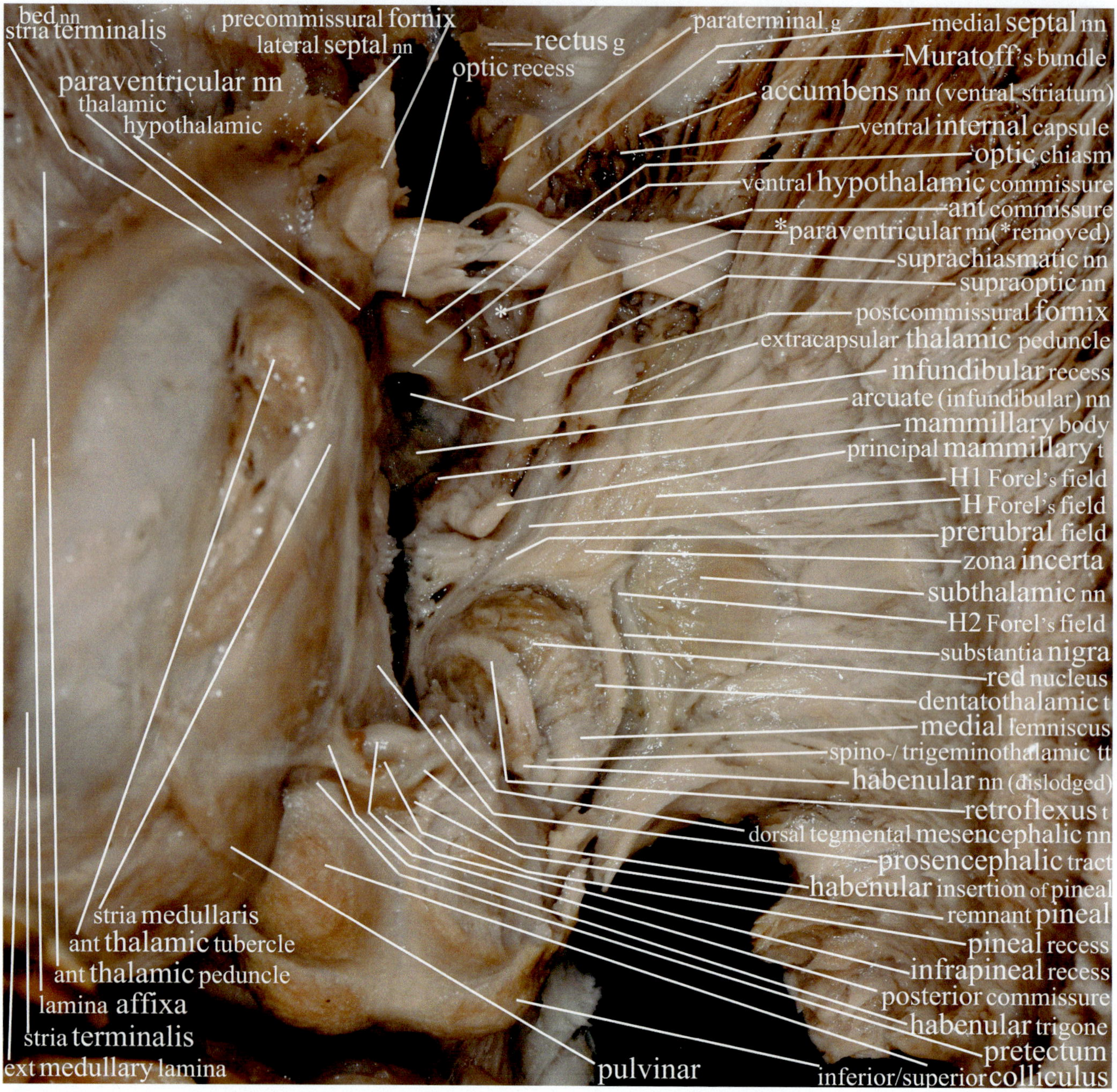

Fig. 7.20 (continued)

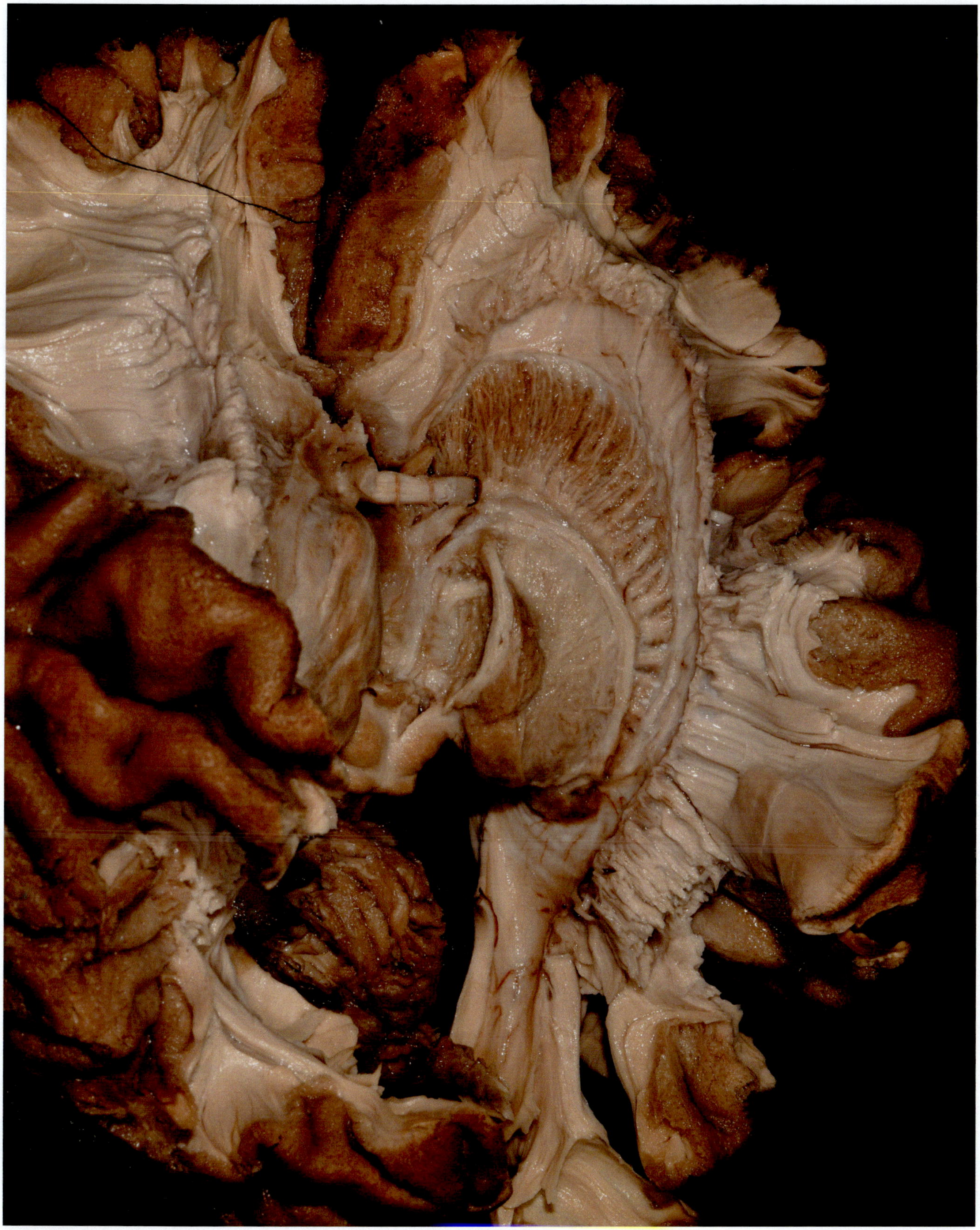

Fig. 7.21 Anatomical preparation of the inner aspect of the internal capsule after the removal of the caudate nucleus. The lamina affixa is on the ventricular surface of the dorsal thalamus, where the fibers of the stria terminalis spread like a fan to its bed nuclei, to the septal nuclei, to the hypothalamus, and to the epithalamus. Also shown is the internal medullary lamina of the thalamus. *ant* anterior, *f* fasciculus, *g* gyrus, *inf* inferior, *lat* lateral, *nn* nucleus, *ped* peduncle, *s* sulcus, *sup* superior, *t* tract

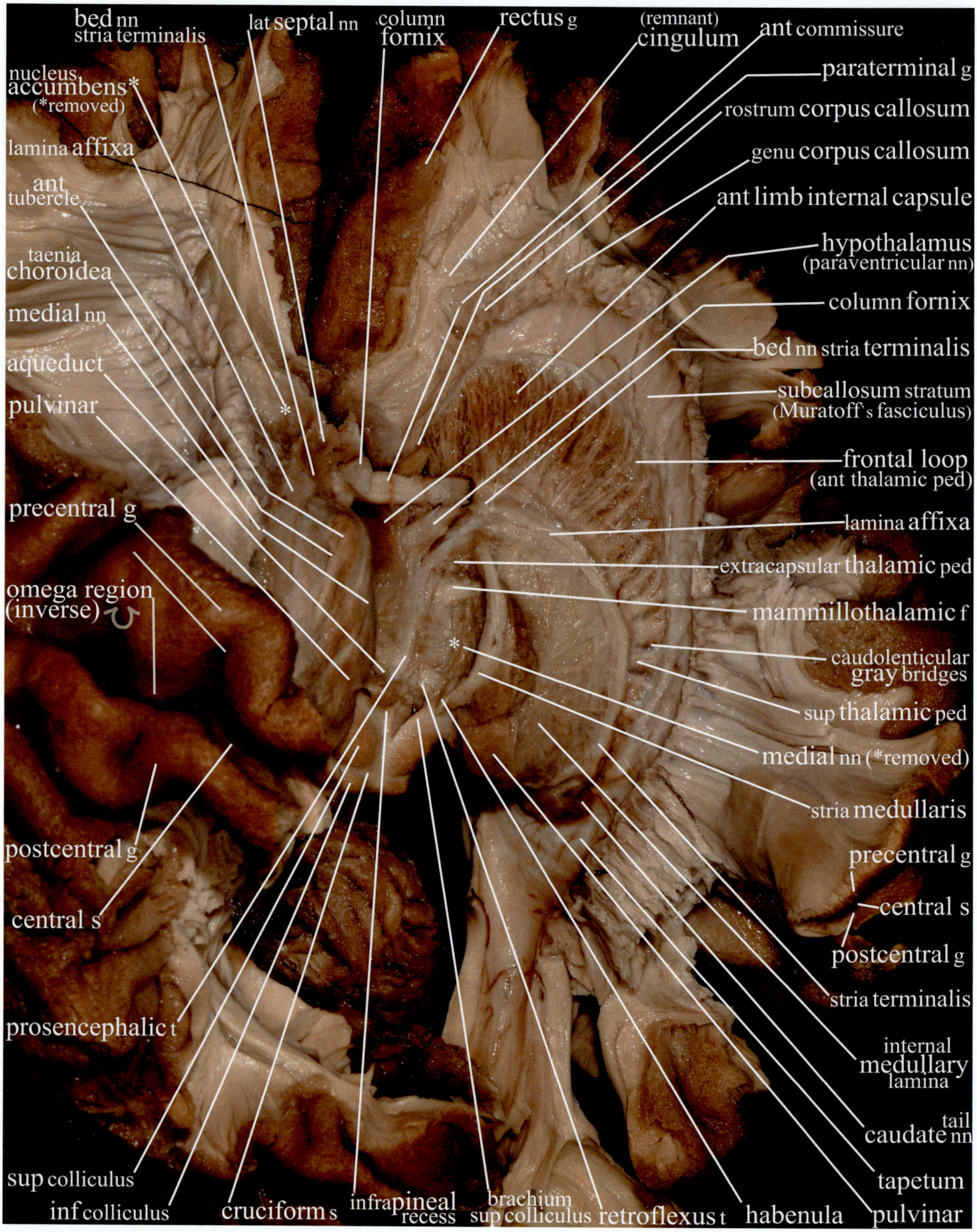

Fig. 7.21 (continued)

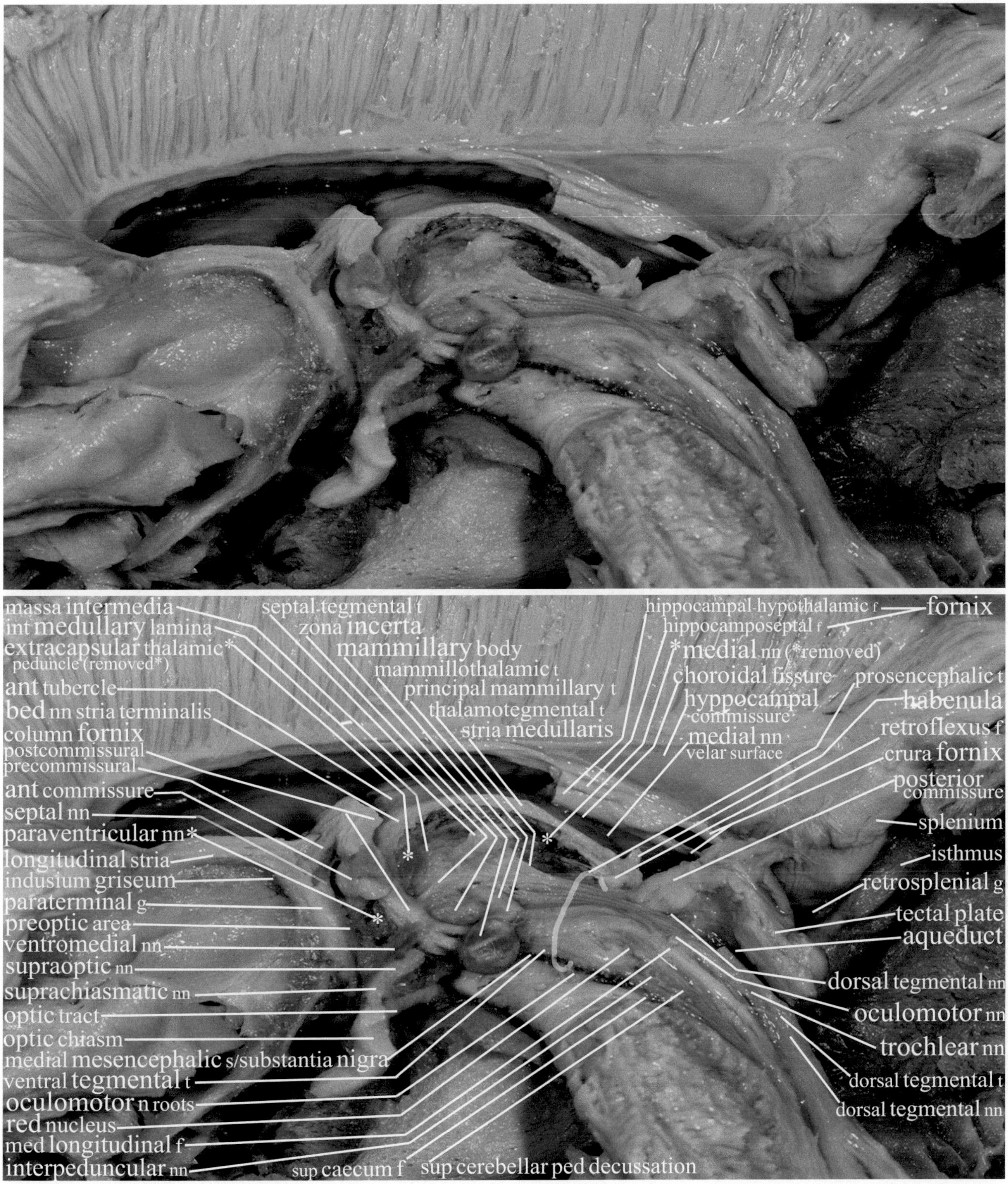

Fig. 7.22 Anatomical preparation of the limbic connections displayed on the lateral wall and floor of the third ventricle. The intermediate mass of the thalamus is not always present. When it is, as shown here, it is a reliable landmark to identify the border between fibers connecting diencephalic (medial thalamic nuclei and habenular nuclei) and mesencephalic structures (dorsal tegmental nuclei and tectal nuclei). *ant* anterior, *f* fasciculus, *g* gyrus, *int* internal, *med* medial, *n* nerve, *nn* nucleus, *ped* peduncle, *s* sulcus, *sup* superior, *t* tract (©Kadri 2023. All Rights Reserved)

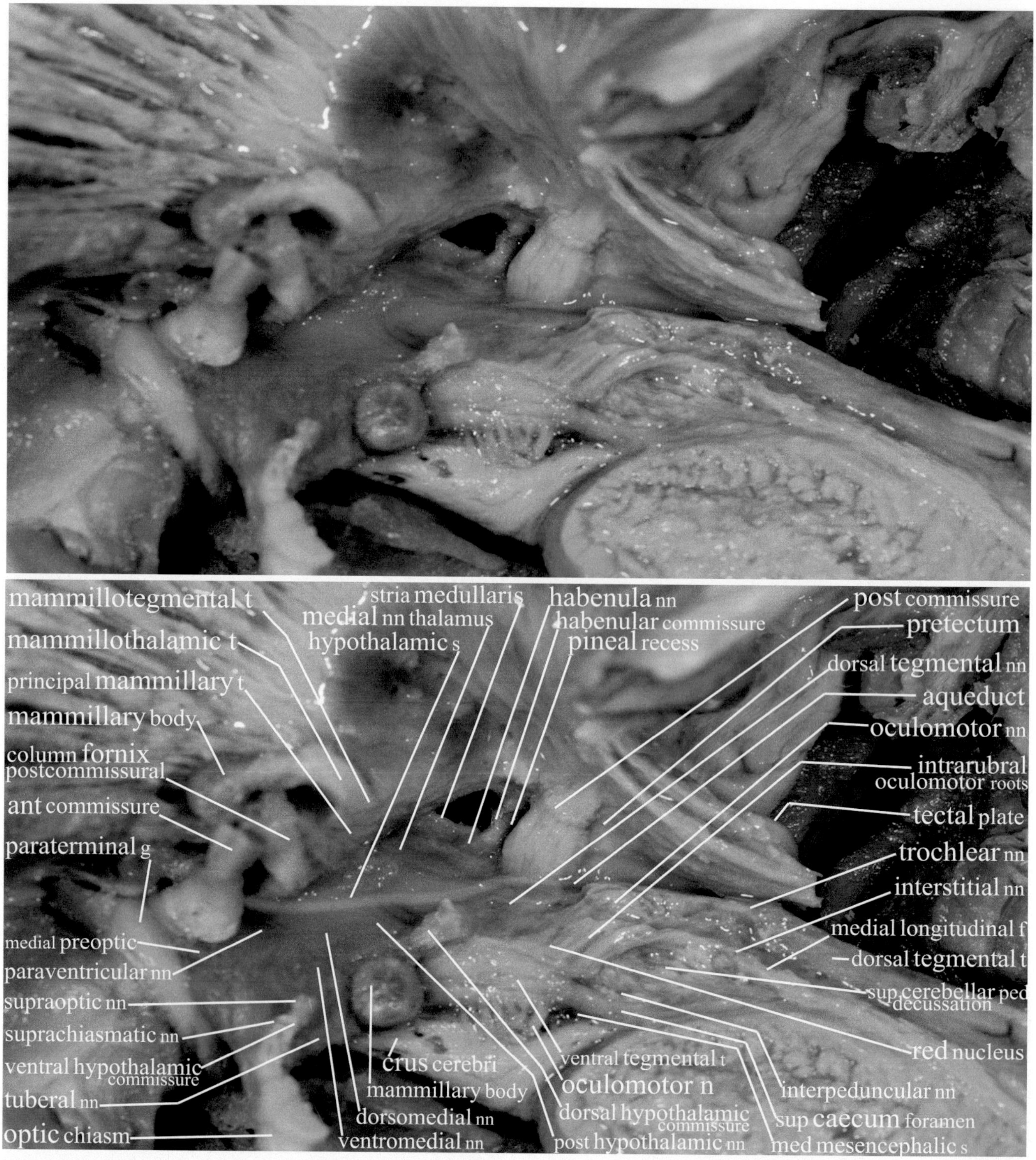

Fig. 7.23 Anatomical preparation of the gray matter limbic continuum along the diencephalic–mesencephalic transition at the floor of the third ventricle. At a precise point, this is the boundary of the posterior hypothalamic nuclei and the dorsal tegmental reticular mesencephalic gray matter that extends dorsal to the mammillary bodies and to the boundaries of the aqueduct. *ant* anterior, *f* fasciculus, *g* gyrus, *med* medial, *n* nerve, *nn* nucleus, *ped* peduncle, *post* posterior, *s* sulcus, *sup* superior, *t* tract (©Kadri 2023. All Rights Reserved)

The Ventral Thalamus (Subthalamus)

The ventral thalamus is mostly known as the subthalamus in reference to the subthalamic nucleus, which is its principal structure. It is a small area situated at the transition of the mesencephalon and diencephalon (Figs. 7.3, 7.4, 7.5, 7.18, 7.20, 7.22, and 7.24). Therefore, its superior limit is the thalamus, its inferior limit is the diencephalon, its lateral limit is the internal capsule, and its anteromedial limit is the hypothalamus. The transition between the mesencephalon and the subthalamus is not distinct. The extensions of the red nucleus, substantia nigra, and reticular substance of the mesencephalon are denoted as the zona incerta (uncertain zone). The main structure of the subthalamus is the subthalamic nucleus (of Luys), which has important connections with the globus pallidus and is an important center of the extrapyramidal motor system. Other nuclear centers contained in the subthalamus are the pregeniculate nucleus and the nucleus of the zona incerta. The reticular nucleus of the dorsal thalamus is also considered with the subthalamus.

The medial zone of the subthalamus contains a dense network of fibers. Among them are connections of the globus pallidus toward the subthalamus and thalamus and cerebellar connections toward the thalamus and subthalamus. These fibers constitute the tegmental fields and are named after the German term for tegmentum, *Haube*, as the H field (H field of Forel). The zona incerta divides the H field into ventral and dorsal segments. The ventral segment is also denoted as the H1 field and the dorsal segment as the H2 field (Figs. 7.4, 7.5, 7.12, 7.19, 7.20, 7.21, and 7.24).

The reticular nucleus is a thin nuclear mass interposed between the dorsal thalamus and the internal capsule. It is separated from the dorsal capsule by the external medullary lamina. It derives its name from the multiple interruptions of its continuity by the dorsal thalamic connections.

The pregeniculate nucleus is a visual nucleus that receives inputs mainly from the contralateral retina. It also contains reciprocal connections with the pretectum, superior colliculi, and brainstem [1].

The zona incerta can be considered the rostral continuation of the reticular formation of the brainstem. The reticular formation is a discontinuous cluster of different types of neurons aligned in the central tegmentum of the brainstem. These neuronal centers are interrupted by ascending and descending tracts that pass through or connect to different areas of the brain, the brainstem, or the spinal cord. The fiber system passing through the zona incerta, as considered above, constitutes the tegmental field. The nuclear component of the zona incerta receives inputs from several limbic and motor areas from cortical, basal ganglia, and diencephalic centers. It also receives inputs from several areas of the brainstem and from cerebellar nuclei. It projects toward the thalamus, hypothalamus, and several areas of the brainstem (Figs. 7.3, 7.4, 7.5, 7.18, 7.20, 7.21, 7.22, and 7.24).

The subthalamic nucleus, or corpus Luysi, is dorsal to the zona incerta at the transition of the internal capsule to the mesencephalic peduncle. Its main afferent is the globus pallidus. Its main efference is toward the globus pallidus and substantia nigra.

The Fiber System of the Subthalamus

The arrangement of fibers of the subthalamus is named the tegmental field (H, H1, and H2 fields of Forel). The H field is also known as the prerubral field and is separated by the zona incerta into dorsal and ventral segments. The ventral segment is referred to as the H1 field, while the dorsal segment is referred to as H2. The globus pallidus contributes to all fields and is divided into internal and external parts. The connections of the internal globus pallidus pass through and around the internal capsule. The connections that pass through the internal capsule are the subthalamic fasciculus and the lenticular fasciculus. The connections that turn around the peduncle form the ansa lenticularis, which turns ventrally around the mesencephalic peduncle. The ansa lenticularis together with the ventral fibers of the amygdala that follow the same trajectory constitutes the ansa peduncularis. The subthalamic fasciculus is made up of the straightforward ventral connections of the globus pallidus with the subthalamic nucleus through the internal capsule. The lenticular fasciculus is the dorsal connection of the globus pallidus to the thalamus through the internal capsule.

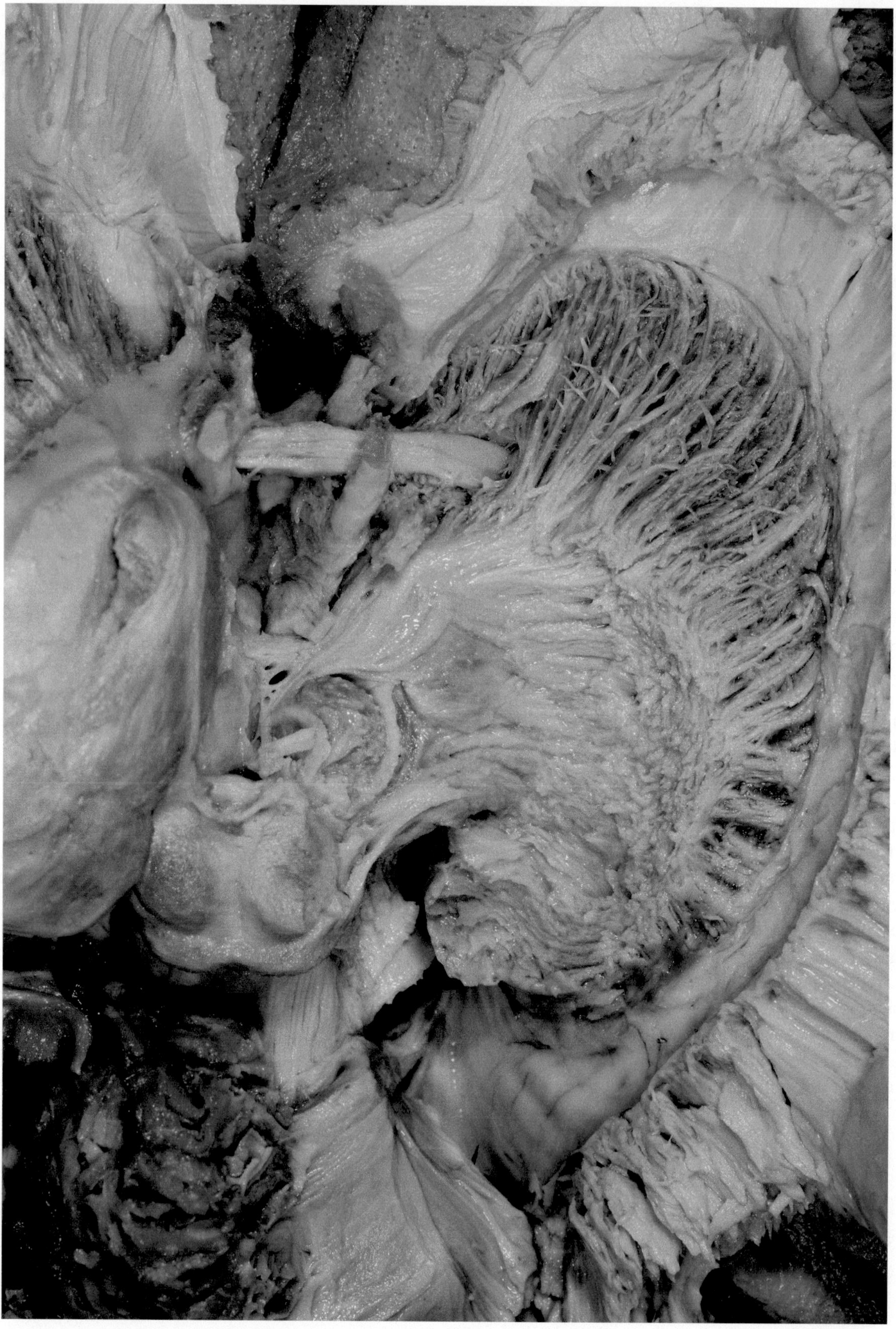

Fig. 7.24 Anatomical preparation of the subthalamic connections. The main subthalamic gray matter clusters are the subthalamic nucleus and the reticular nuclei of the zona incerta. *ant* anterior, *ext* external, *f* fasciculus, *g* gyrus, *inf* inferior, *lat* lateral, *med* medial, *nn* nucleus, *ped* peduncle, *post* posterior, *s* sulcus, *sup* superior, *t* tract (©Kadri 2023. All Rights Reserved)

Fig. 7.24 (continued)

The Hypothalamus

The hypothalamus is a tiny area on the wall of the third ventricle that is critical for regulating endocrine functions, controlling autonomic reactions, and generating basic behavior responses (Figs. 7.1, 7.2, 7.16, 7.17, 7.18, 7.19, 7.22, 7.23, 7.24, and 7.25). On the ventricular surface, it is separated from the thalamus by the hypothalamic sulcus (Fig. 7.26). This sulcus extends from the foramen of Monro to the cerebral aqueduct. The optic chiasm, the tuber cinereum, the infundibulum, and the mammillary body are visible parts of the hypothalamus on the inferior surface of the brain (Fig. 7.27). The hypothalamus is mainly composed of nuclear arrangements of gray matter, but some fiber tracts connect the hypothalamus to different areas of the encephalon. The fornix connects the mammillary body with the hippocampal formation. It demarcates the hypothalamus into a medial and a lateral area. The medial area is predominantly composed of gray matter and is the location of the main hypothalamic nuclei. The lateral area is mainly composed of fiber tracts that interconnect the hypothalamus or pass through it in a sagittal orientation. The hypothalamus can also be divided into coronal planes in preoptic, supraoptic, tuberal, and mammillary parts (Figs. 7.28, 7.29, 7.30, 7.31, and 7.32).

On the most anterior part of the third ventricle, in the vicinity of the lamina terminalis, a small region is denoted as the preoptic area. This area is derived from the telencephalic vesicle and therefore is not part of the diencephalon (Figs. 7.2, 7.3, 7.16, 7.17, 7.18, 7.19, 7.20, 7.21, 7.22, and 7.33). However, because of its anatomical relations and functions, it is grouped with the hypothalamic nuclei. The preoptic nuclei are the periventricular and the medial. The supraoptic nuclei are the suprachiasmatic, supraoptic, and paraventricular. The tuberal nuclei are the arcuate, dorsomedial, and ventromedial. The mammillary nuclei are the medial, lateral, and posterior.

The tuber cinereum is an eminence visible on the inferior surface of the brain, anterior to the mammillary body and continuous with the infundibulum. The infundibulum gives rise to the median eminence and the neurohypophysis. The median eminence forms the neurohematological interface between the hypothalamus and the adenohypophysis.

Hypothalamic Connections

The principal connections of the hypothalamus are the limbic system and the hypophysis. The hypothalamus also has connections with the prefrontal cortex, autonomic centers of the brainstem and the spinal cord, and direct connections from the retina and olfactory cortex.

The limbic connections have reciprocal relations with the hippocampal formation, thalamus, septal area, and amygdala. They receive information from the hippocampal formation via the fornix and project toward the anterior nuclei of the thalamus via the mammillothalamic tract. The mammillary body also projects toward the reticular formation via the mammillotegmental tract. The common origin of the mammillothalamic and mammillotegmental tracts is denoted as the principal mammillary tract. The hypothalamic inputs from the amygdala are ventral, from the amygdalo-hypothalamic tract, and dorsal, from the stria terminalis. The interconnections with the septal region are via the prosencephalic tract (Fig. 7.33). The prosencephalic tract, also known as the medial telencephalic fasciculus or medial forebrain bundle, is one of the components of this fiber system that interconnects the septal region and the reticular formation of the tegmentum of the brainstem. It passes through the lateral hypothalamus and may be considered the central connection of the limbic system. It extends through the mesencephalic and pontine tegmentum connecting several nuclei. Efferent connections toward the neurohypophysis from the supraoptic and paraventricular nuclei form the hypothalamic-hypophyseal tract. These axons transport oxytocin and vasopressin to the neurohypophysis. Efferents from the arcuate nucleus toward the median eminence form the tuberoinfundibular tract [2].

The prefrontal inputs are received through direct connections or indirectly through connection with the medial thalamus. The paraventricular nucleus is considered the major controller of the autonomic system. It has direct projections to the dorsal nucleus of the vagus, nucleus ambiguous, thoracic sympathetic, and sacral parasympathetic preganglionic columns [1].

Visceral inputs to the hypothalamus are received via connections with the solitary tract nucleus. Visceral outputs are direct or indirect and project toward the sympathetic and parasympathetic nuclei. The autonomic centers of the lower medulla may connect directly or indirectly with the basal forebrain and the hypothalamus via the dorsal tegmental longitudinal tract. The dorsal tegmental longitudinal tract is also known as the periventricular fiber system or fasciculus of Schutz. The indirect pathway is the relay at the central reticular gray matter and at the dorsal tegmental nuclei of Gudden, which is also connected with the mammillary body via the mammillotegmental tract. This tract extends caudally to the tegmental nucleus of Bekhterev within the reticular formation of the pons.

In addition to the endocrinological influence, the hypothalamus also plays a key role in circadian rhythm (the wake-and-sleep cycle), the stress response (paraventricular nucleus), thermoregulation (preoptic nuclei), food intake (ventral nuclei), water intake (lateral hypothalamus), diuresis (supraoptic and paraventricular nuclei), sexual behavior, and the defensive response [1].

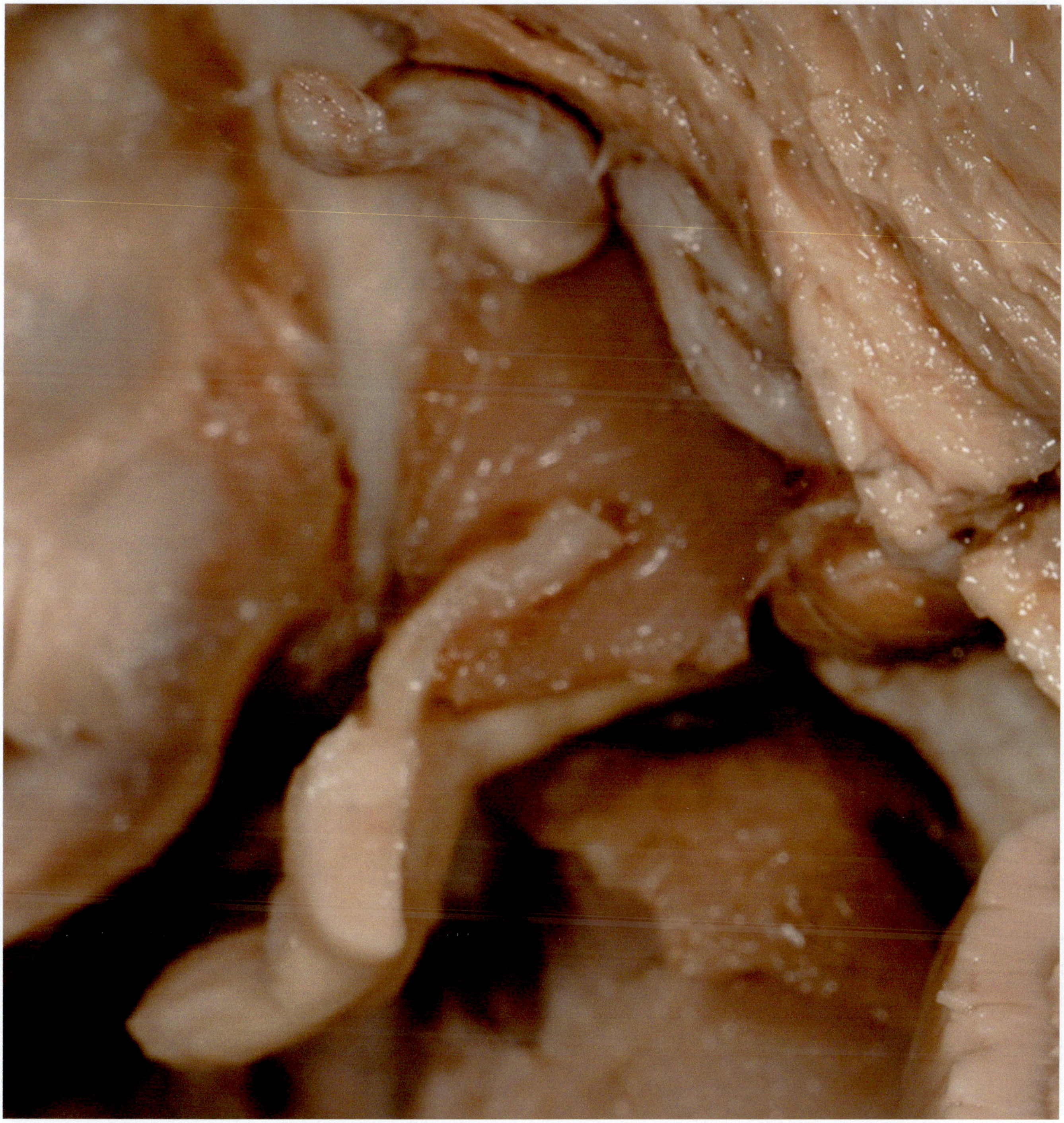

Fig. 7.25 Anatomical preparation of the ventral hypothalamic commissure at the anterior hypothalamic region surrounding the optic apparatus. *g* gyrus, *nn* nucleus, *s* sulcus (©Kadri 2023. All Rights Reserved)

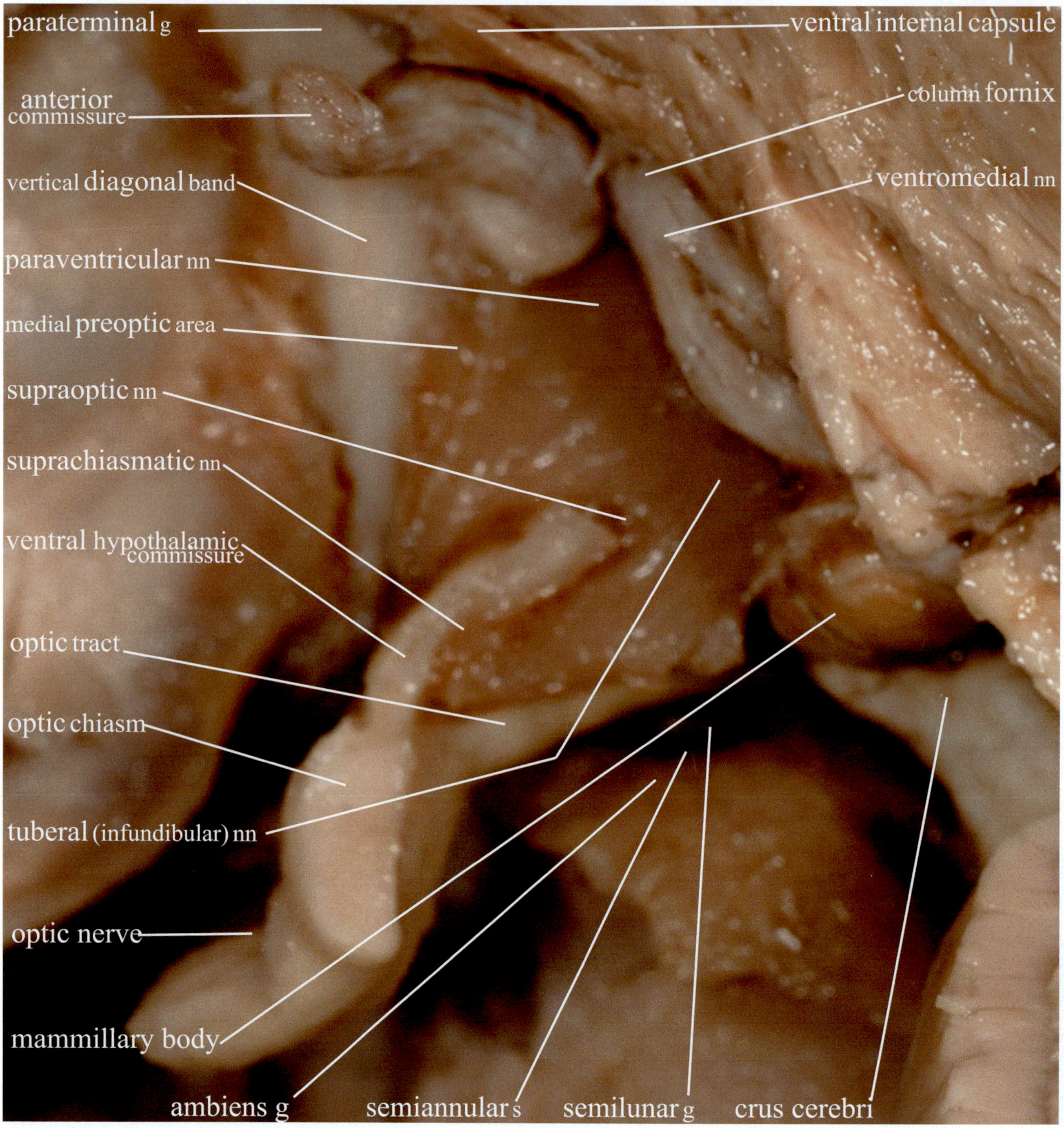

Fig. 7.25 (continued)

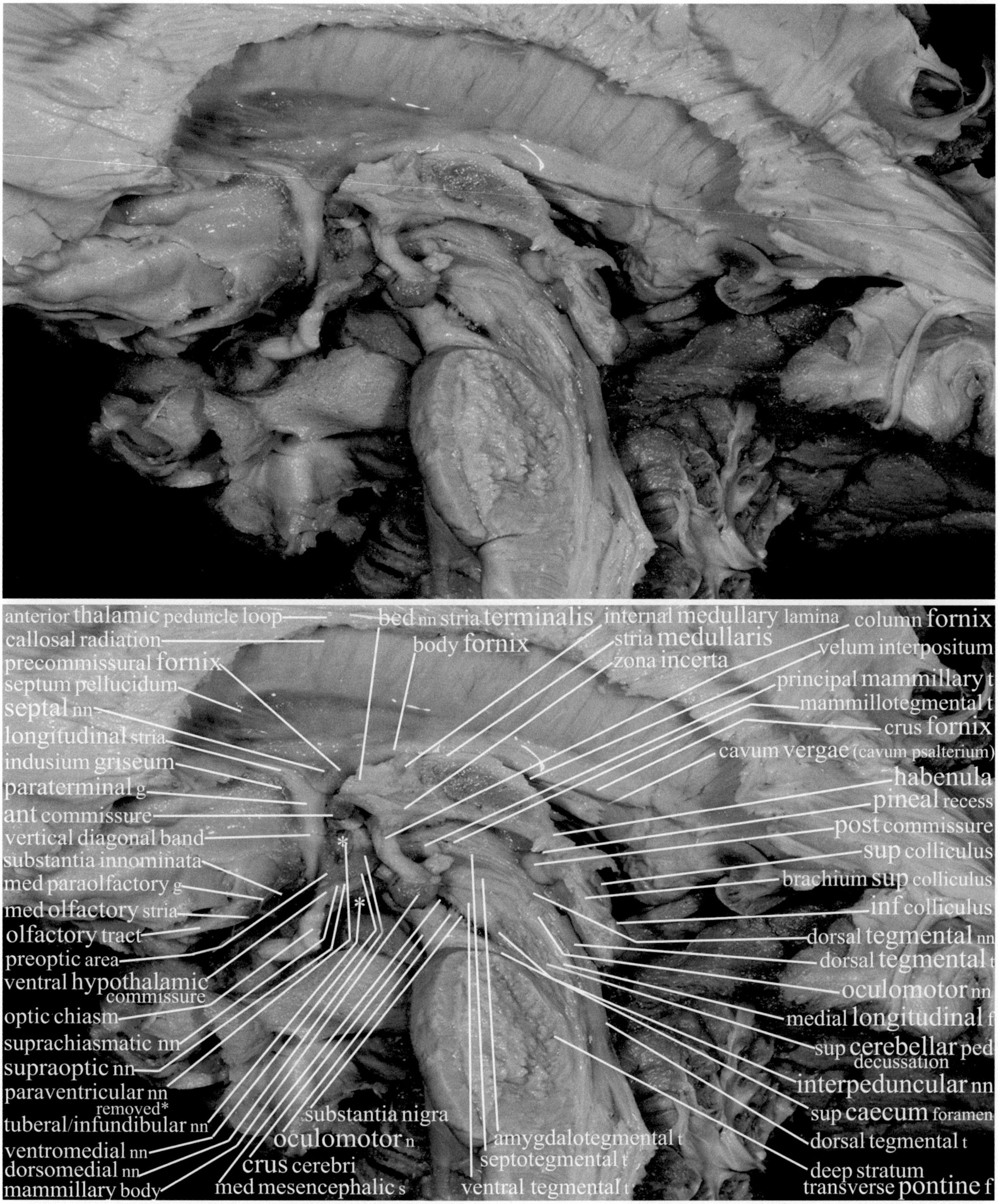

Fig. 7.26 Anatomical preparation of the internal medullary lamina of the thalamus and the hypothalamic connections to the epithalamus and dorsal mesencephalic tegmentum. *ant* anterior, *f* fasciculus, *g* gyrus, *inf* inferior, *med* medial, *n* nerve, *nn* nucleus, *ped* peduncle, *post* posterior, *s* sulcus, *sup* superior, *t* tract (©Kadri 2023. All Rights Reserved)

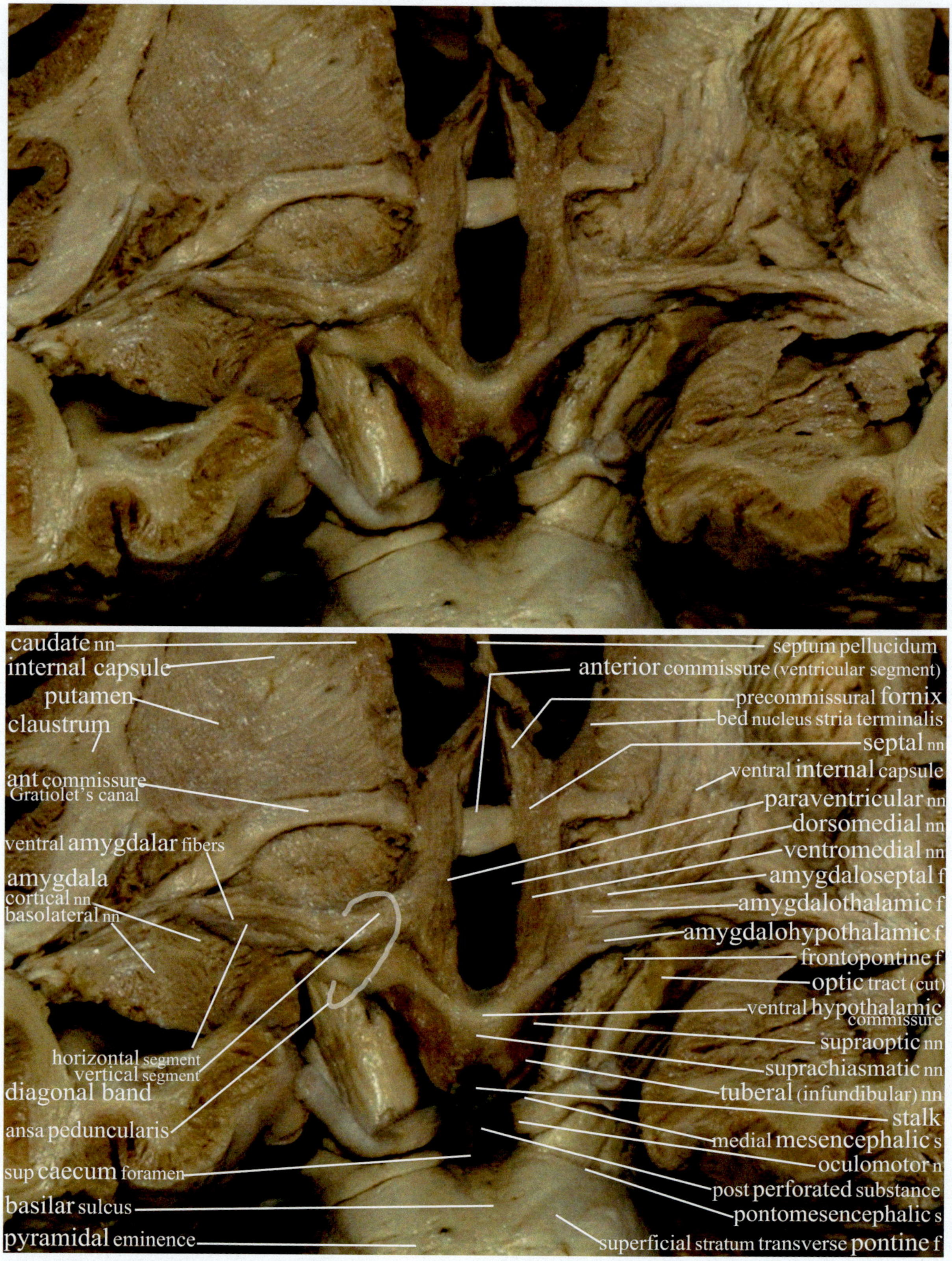

Fig. 7.27 Anatomical preparation of the hypothalamus after the removal of the optic apparatus. The ventral view displays the paired paramedian clusters of gray matter on the roof of the interpeduncular fossa, the tuber cinereum. *ant* anterior, *f* fasciculus, *n* nerve, *nn* nucleus, *p* posterior, *s* sulcus, *sup* superior (©Kadri 2023. All Rights Reserved)

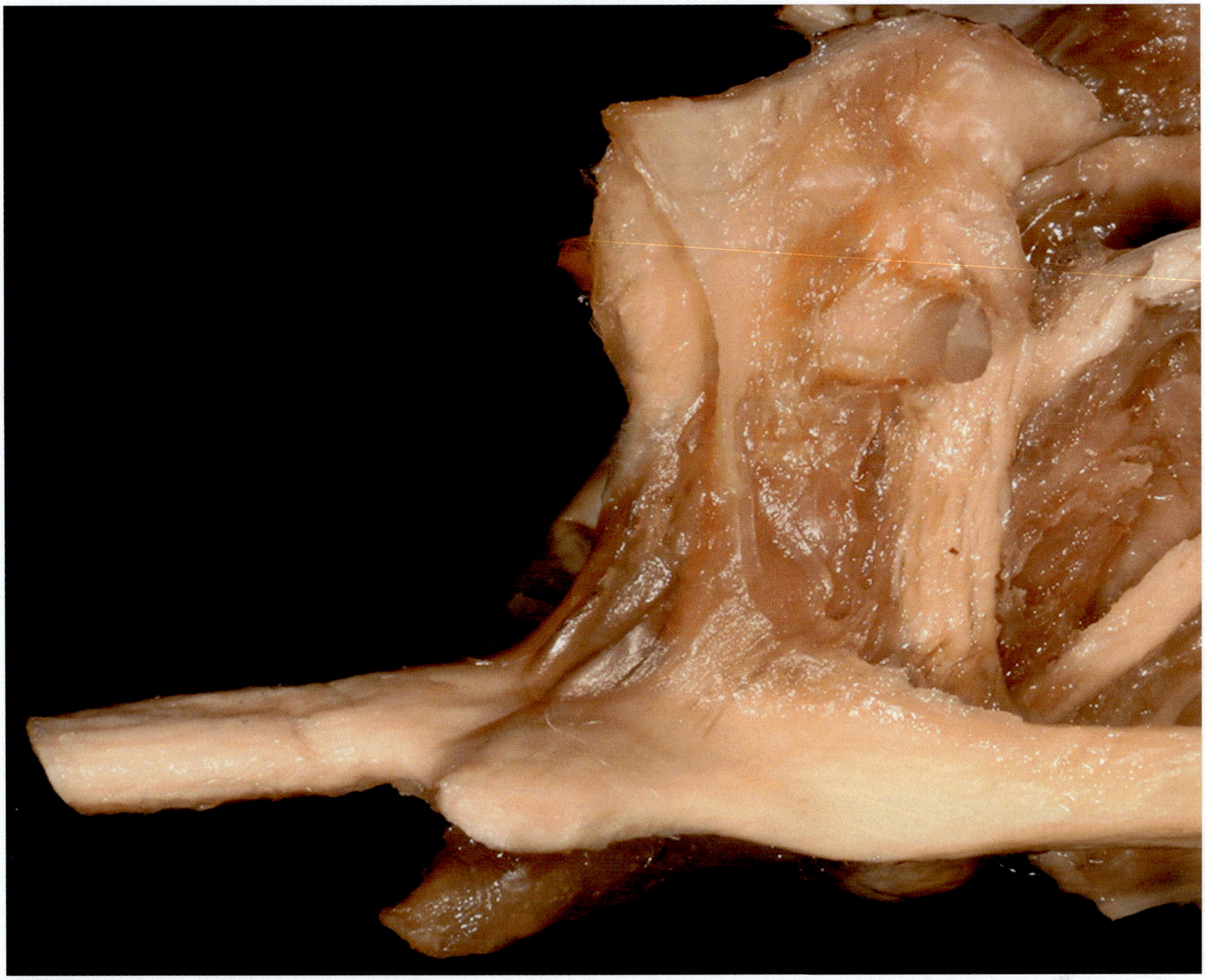

Fig. 7.28 The first of the five serial images: anatomical preparation of the hypothalamus from a lateral view with the optic apparatus. *ant* anterior, *f* fasciculus, *g* gyrus, *nn* nucleus, *t* tract (©Kadri 2023. All Rights Reserved)

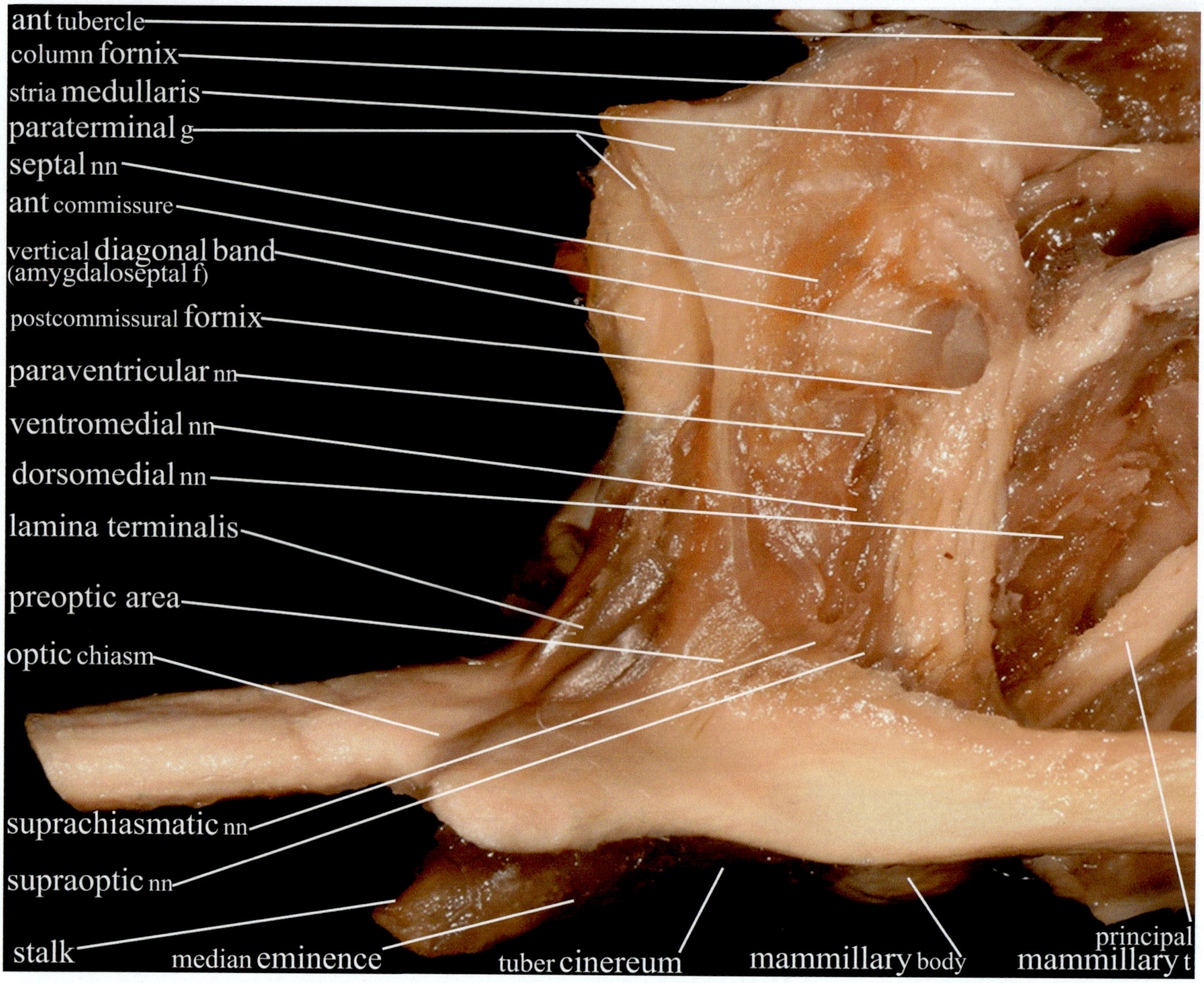

Fig. 7.28 (continued)

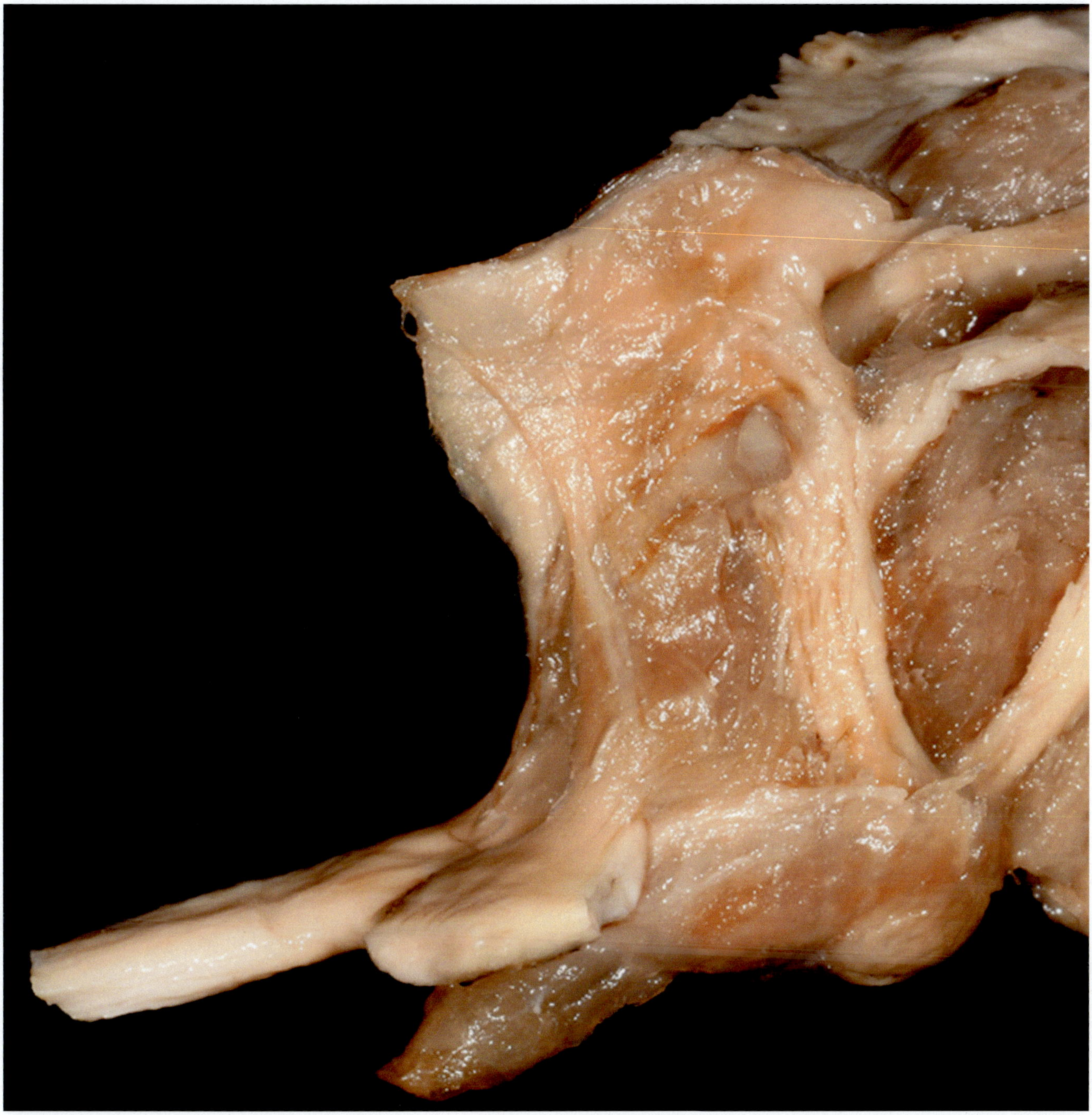

Fig. 7.29 The second of the five serial images: anatomical preparation of the hypothalamus from a lateral view after the removal of the optic tract. The lateral surfaces of the hypothalamic nuclei are exposed. *ant* anterior, *f* fasciculus, *f** foramen, *g* gyrus, *nn* nucleus ()

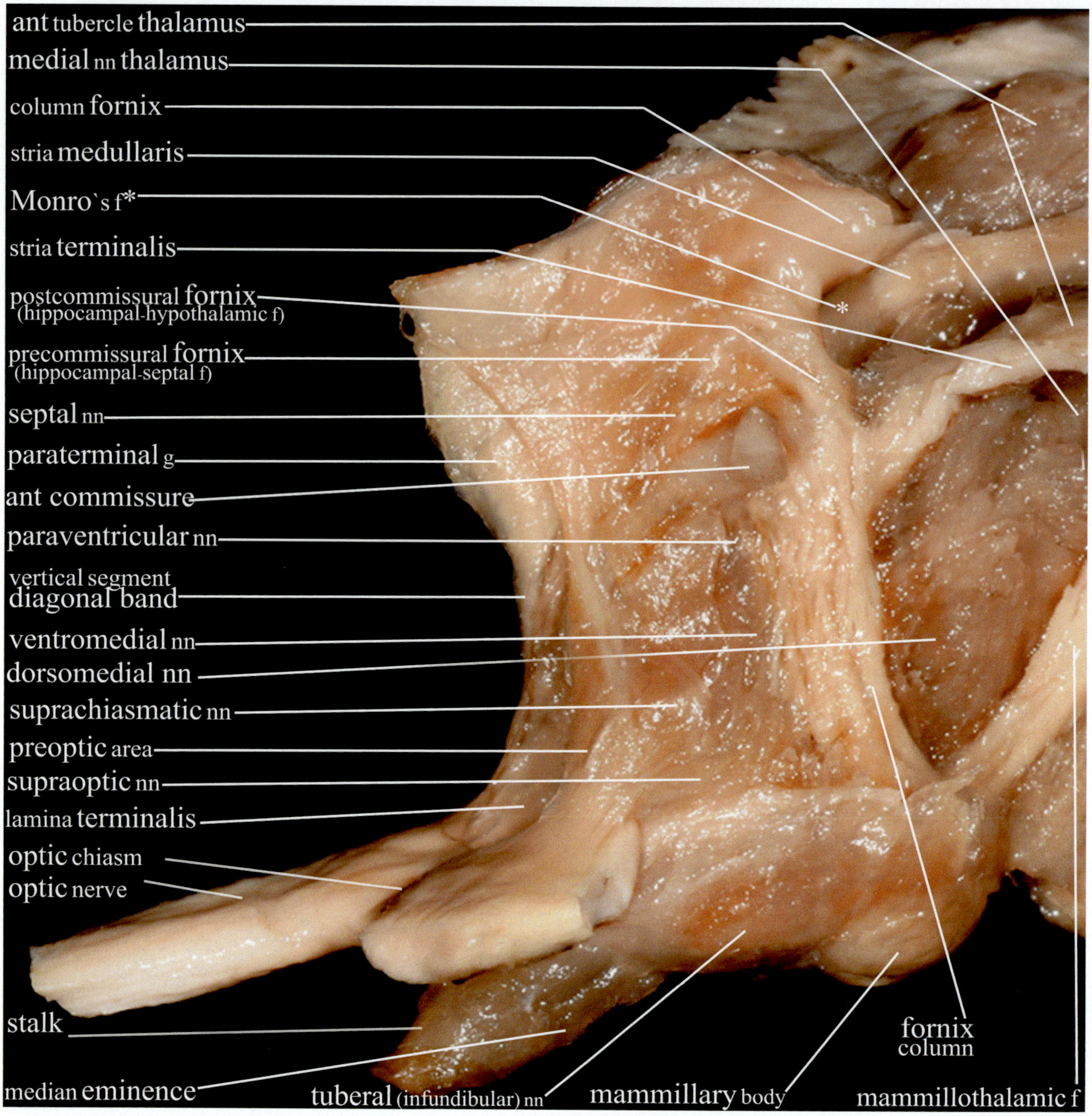

Fig. 7.29 (continued)

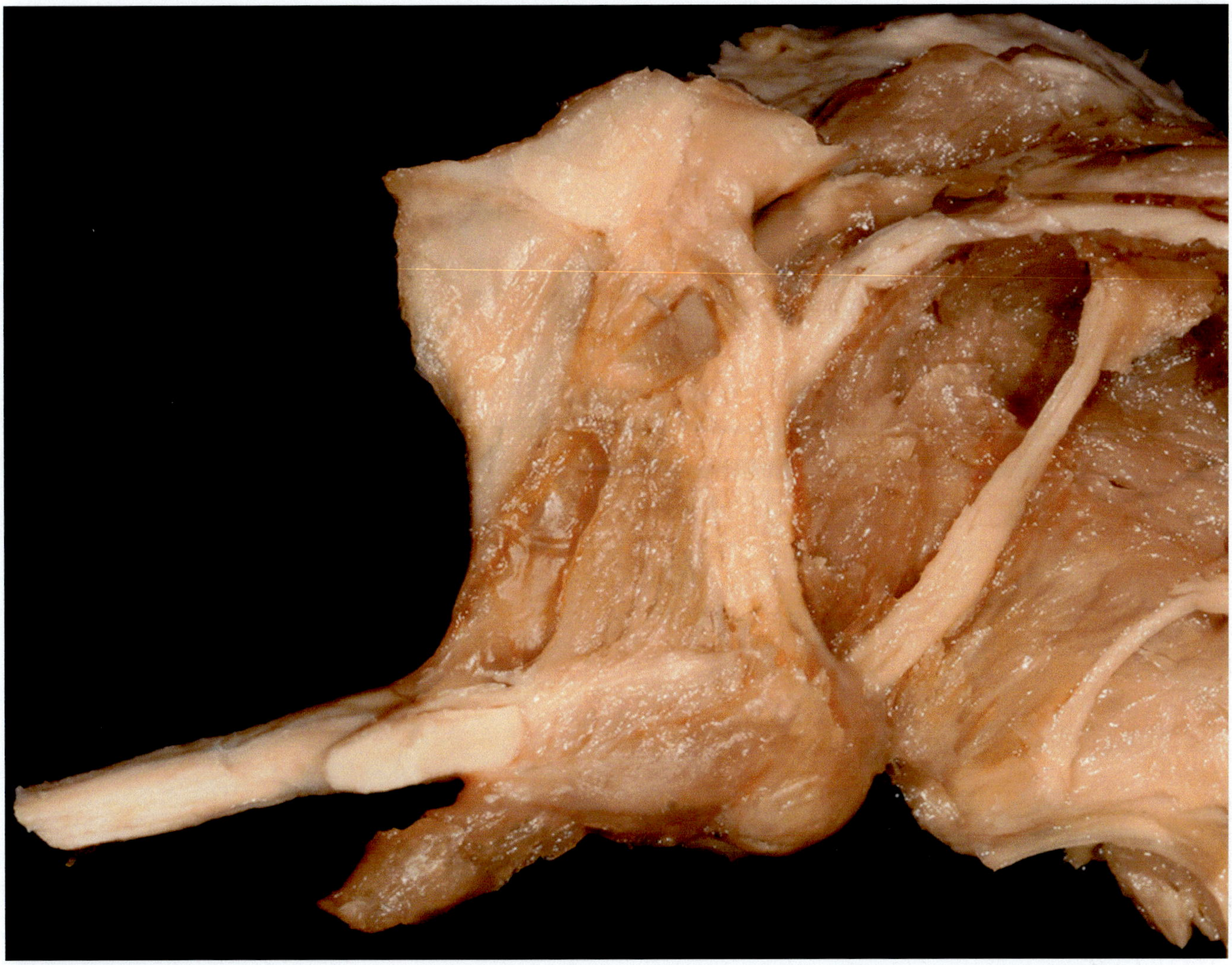

Fig. 7.30 The third of the five serial images: anatomical preparation of the hypothalamus from a lateral view after the removal of the ipsilateral preoptic area. The contralateral anterior hypothalamus is exposed. *ant* anterior, *f* fasciculus, *f** foramen, *g* gyrus, *lat* lateral, *med* medial, *nn* nucleus, *post* posterior, *t* tract (©Kadri 2023. All Rights Reserved)

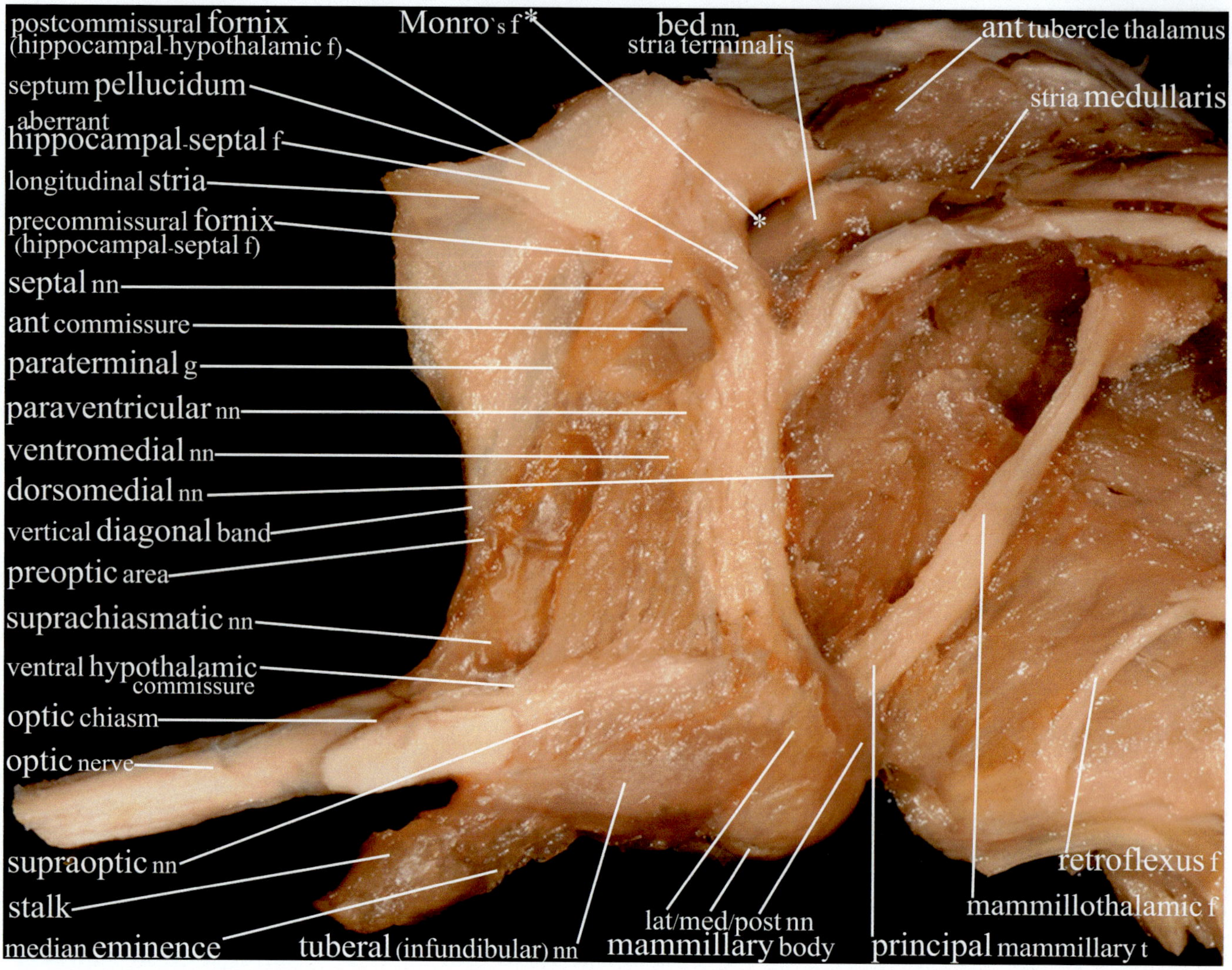

Fig. 7.30 (continued)

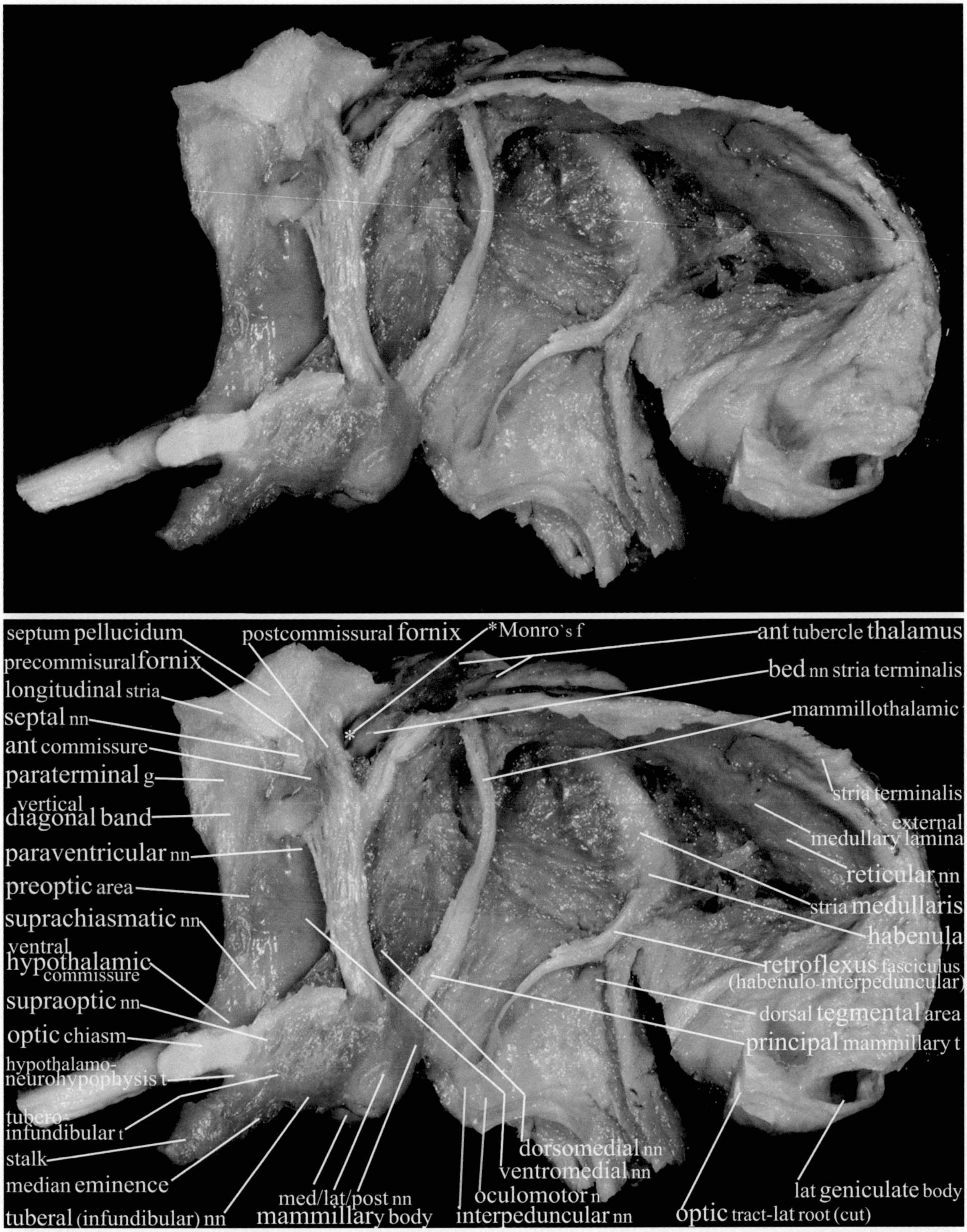

Fig. 7.31 The fourth of the five serial images: anatomical preparation of the hypothalamus from a lateral view after the removal of the anterior hypothalamus and the ventral and dorsal medial nuclei of the ipsilateral hypothalamus. *ant* anterior, *f** foramen, *g* gyrus, *lat* lateral, *med* medial, *n* nerve, *nn* nucleus, *post* posterior, *t* tract (©Kadri 2023. All Rights Reserved)

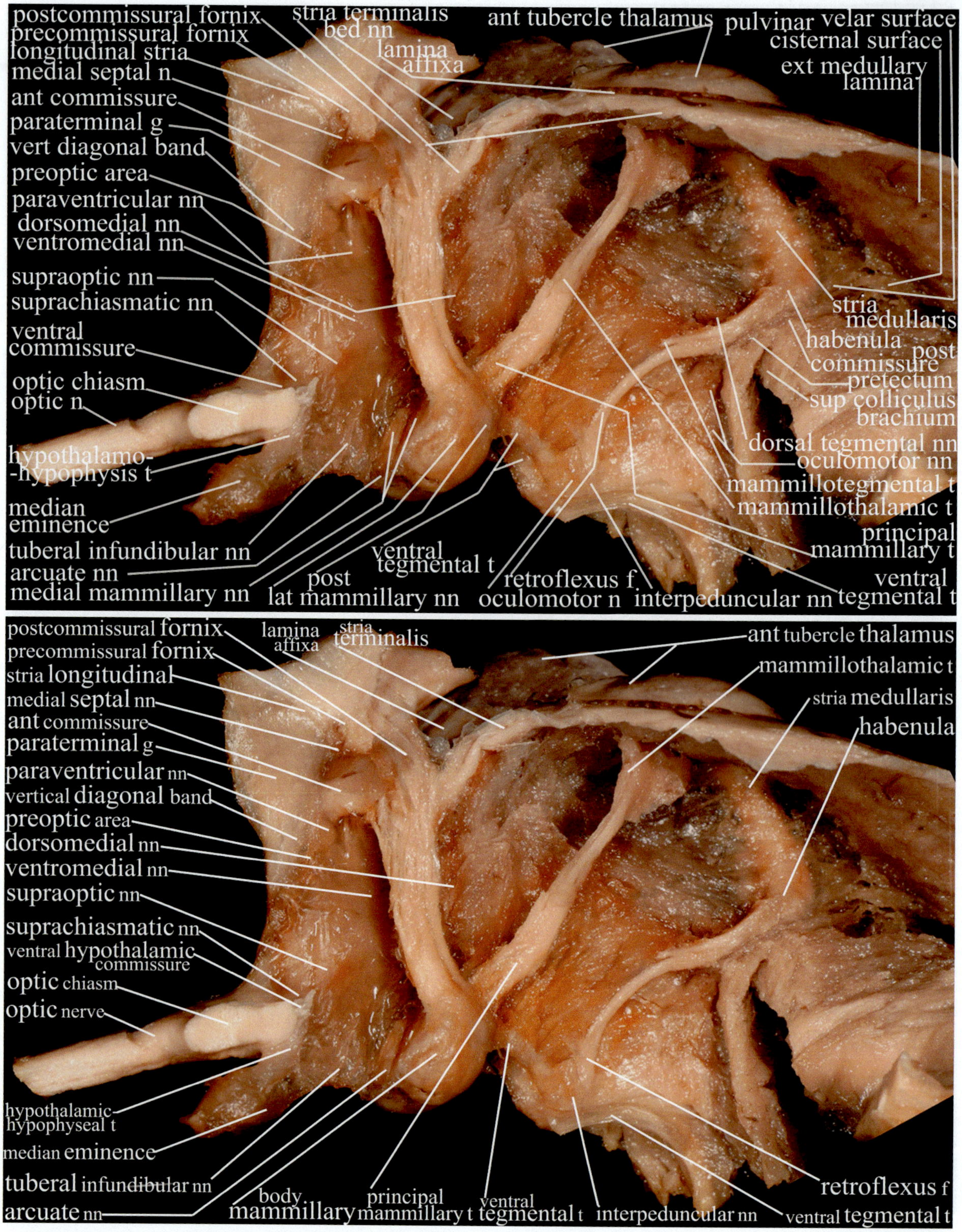

Fig. 7.32 The fifth of the five serial images: anatomical preparation of the hypothalamus from a lateral view after the removal of the tuberal region of the ipsilateral hypothalamus. *ant* anterior, *ext* external, *f* fasciculus, *g* gyrus, *lat* lateral, *n* nerve, *nn* nucleus, *post* posterior, *sup* superior, *t* tract, *vert* vertical (©Kadri 2023. All Rights Reserved)

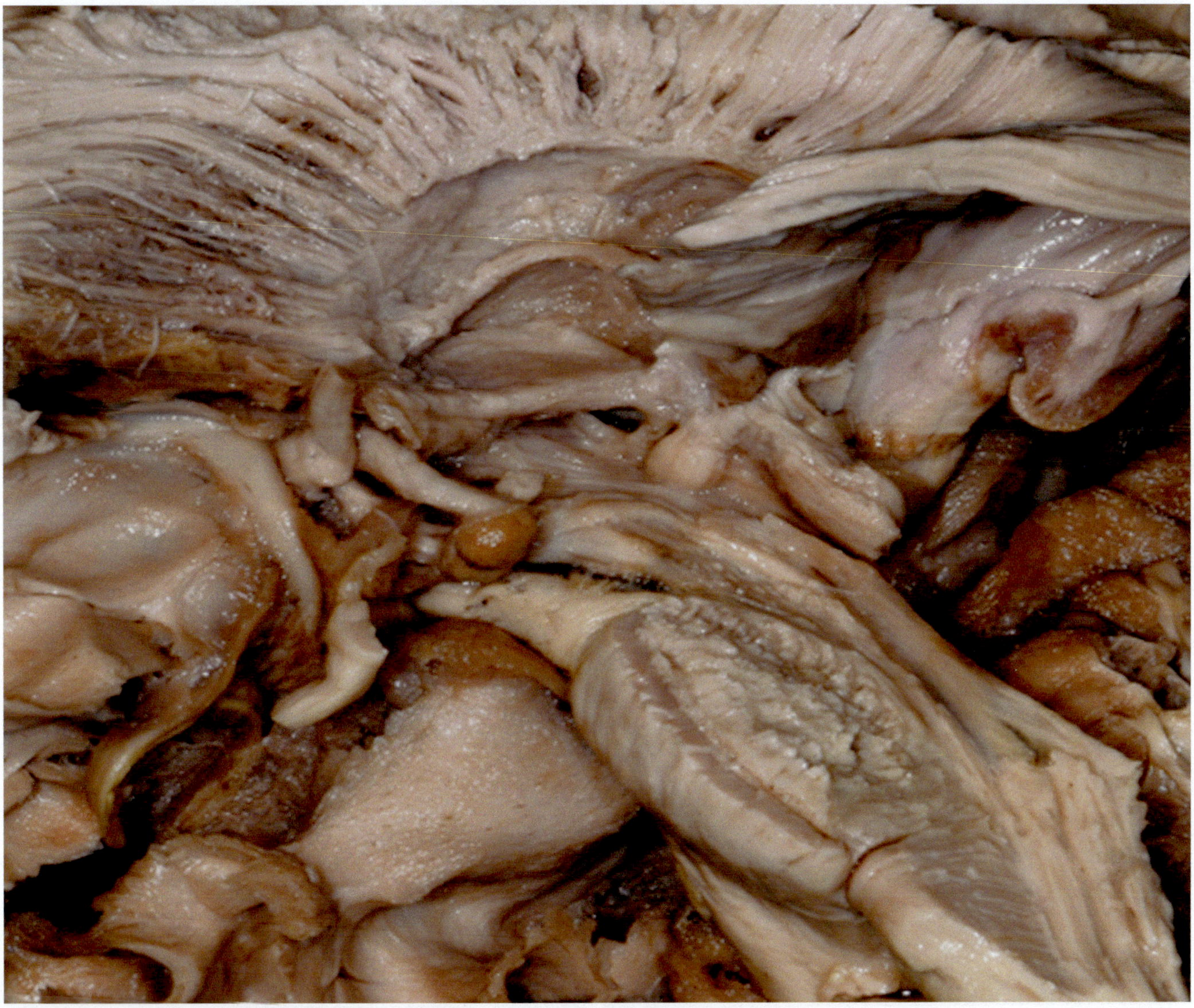

Fig. 7.33 Anatomical preparation of the midline limbic connection of the septal region, hypothalamus, medial thalamus, epithalamus, and the tegmentum of the brainstem. *ant* anterior, *ext* external, *f* fasciculus, *g* gyrus, *inf* inferior, *int* internal, *med* medial, *n* nervve, *nn* nucleus, *ped* peduncle, *post* posterior, *s* sulcus, *sup* superior, *t* tract (©Kadri 2023. All Rights Reserved)

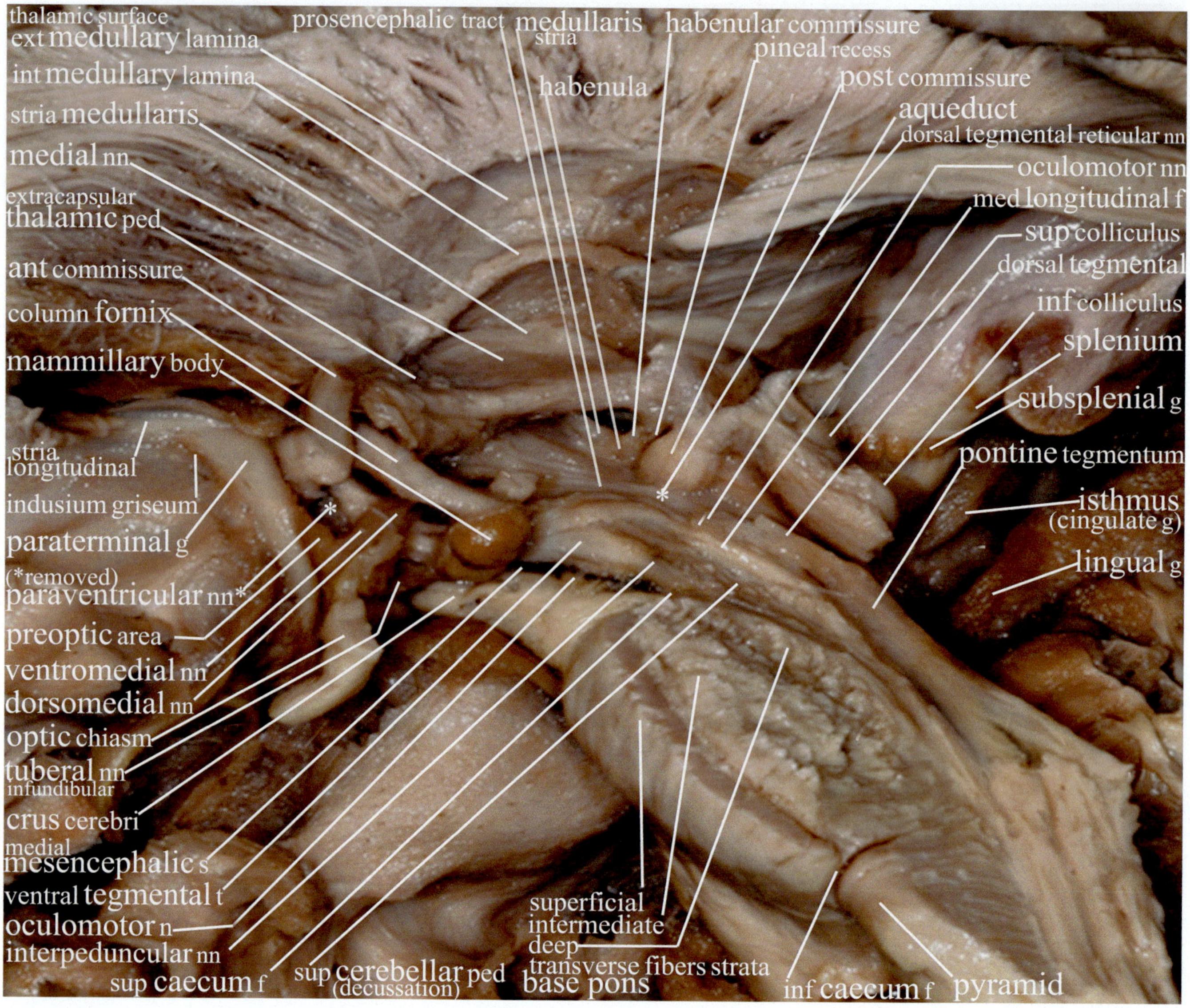

Fig. 7.33 (continued)

References

1. Nieuwenhuys R, Voogd J, van Huijzen C. The human central nervous system: a synopsis and atlas. 4th ed. Berlin: Springer; 2008.
2. Lang J. Surgical anatomy of the hypothalamus. Acta Neurochir. 1985;75(1–4):5–22. https://doi.org/10.1007/BF01406320.

The Brainstem and the Cerebellum

8

Introduction

The brainstem is the central axis of the encephalon (Fig. 8.1) and is interposed between the spinal cord and the diencephalon (Fig. 8.2). It is divided according to its morphology and ontogenesis into the mesencephalon, pons (with the cerebellum), and medulla oblongata (Figs. 8.3 and 8.4). The optic tract is the most distinguishable surface limit between the diencephalon and the mesencephalon (Fig. 8.5). The most distinguishable surface limit between the mesencephalon and the pons is the pontomesencephalic sulcus. The pontomedullary sulcus is the most distinguishable surface limit between the pons and the medulla. The limit between the medulla and the spinal cord is the plane between the inferior limit of the decussation of the corticospinal tract and the origin of the first spinal nerve root [1, 2].

The brainstem can be divided into a base and a tegmentum (Fig. 8.6). At the mesencephalon, the limit between these two structures is the reticulate portion of the substantia nigra (Fig. 8.7). At the pons, this limit is between the trapezoid body and the medial lemniscus. At the medulla, the limit between the base and the tegmentum is the sensory decussation of the cuneate and gracile nuclei (Fig. 8.8).

Along the midline, on the depth of the median sulcus, is a continuum of the crossing fibers of the anterior white commissure of the spinal cord. These fibers are subsequently substituted by the massive decussation of the pyramids and the gracile and cuneate nuclei decussation fibers. Following cranially are the decussation of the vestibular and acoustic radiations and the trigeminal decussation. The oculomotor system decussation is part of the dorsal tegmental decussation of the mesencephalon (Fig. 8.9), which is composed of fibers of the dorsal medial tegmentum of the mesencephalon, pretectum fibers, and the superficial decussation of the fibers of the prosencephalic tract. The dorsal tegmental decussation of the mesencephalon is dorsal to the parvocellular part of the red nucleus. The ventral tegmental decussation comes from the rubrospinal tract, connecting the magnocellular red nucleus with the contralateral spinal cord. Caudal to it, the superior cerebellar decussation is composed of crossed dentatorubral and dentatothalamic connections.

P. Kadri, *The Cartographic Atlas of the Brain*, https://doi.org/10.1007/978-3-031-38062-4_8

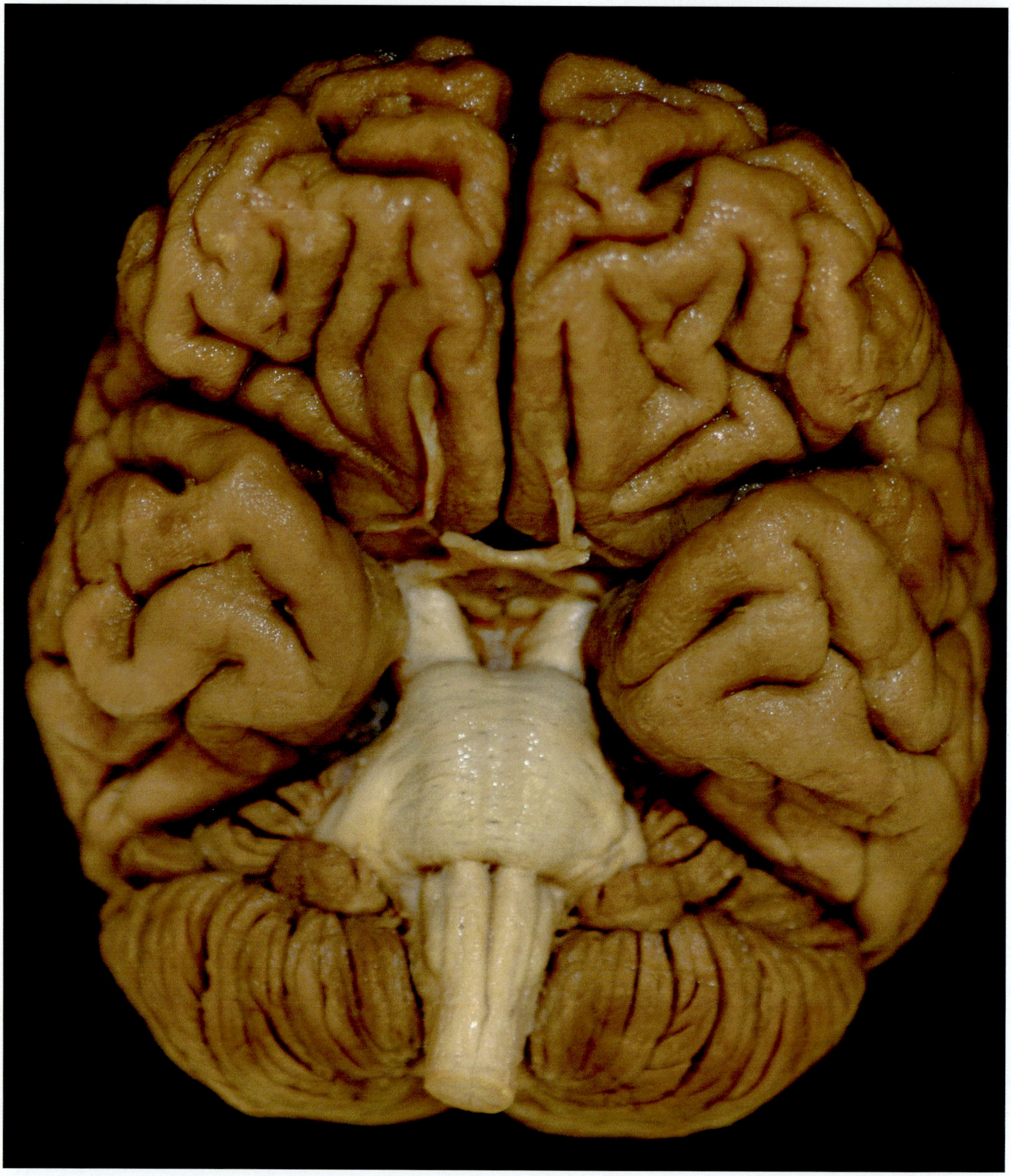

Fig. 8.1 Anatomical preparation of the encephalon. The encephalon was named by Aristotle (c. 384–322 BC) and is the rostral topographic division of the central nervous system in vertebrates. *ant* anterior, *f* fasciculus, *g* gyrus, *inf* inferior, *lat* lateral, *med* medial, *n* nerve, *ped* peduncle, *post* posterior, *s* sulcus, *sup* superior (©Kadri 2023. All Rights Reserved)

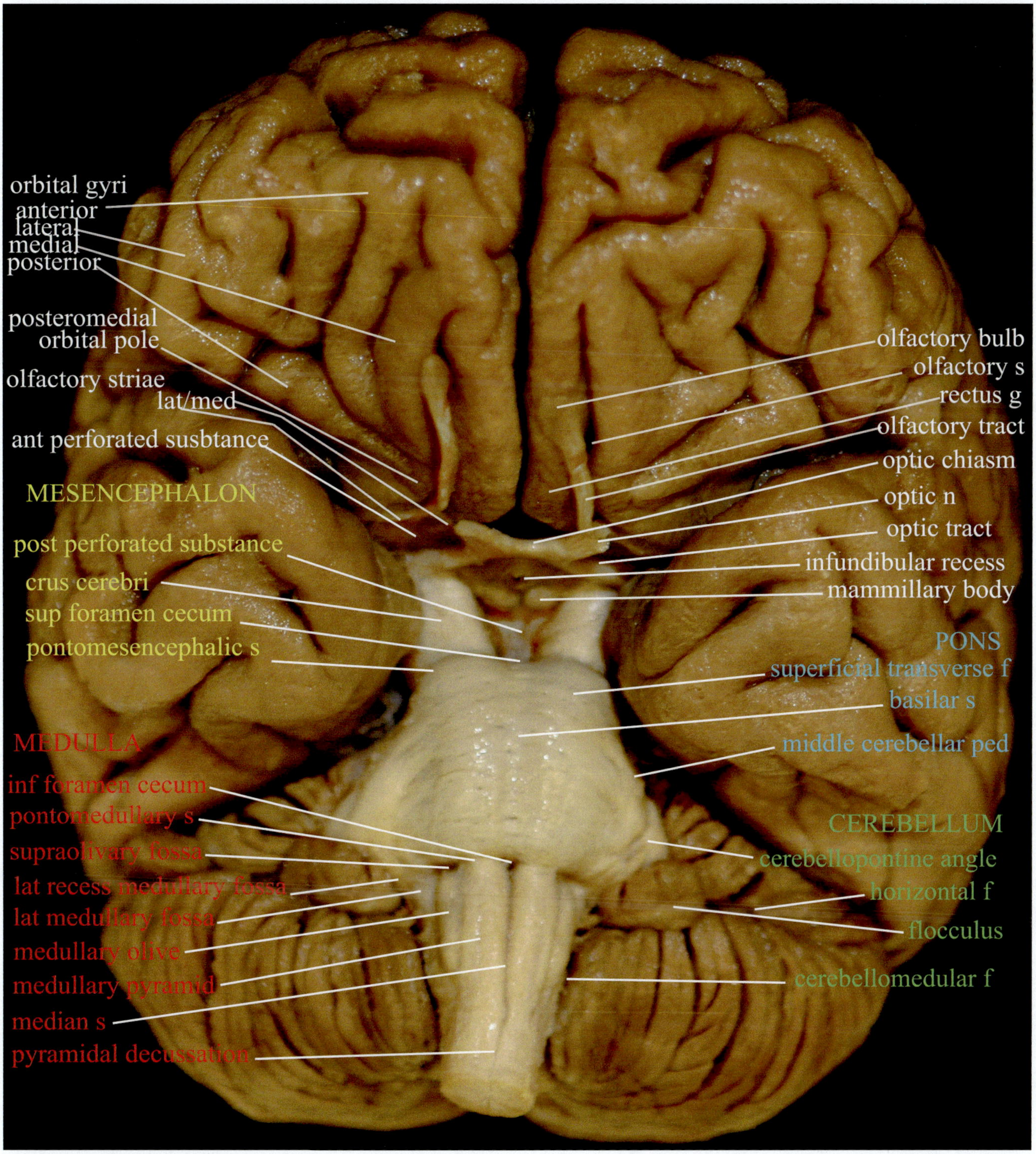

Fig. 8.1 (continued)

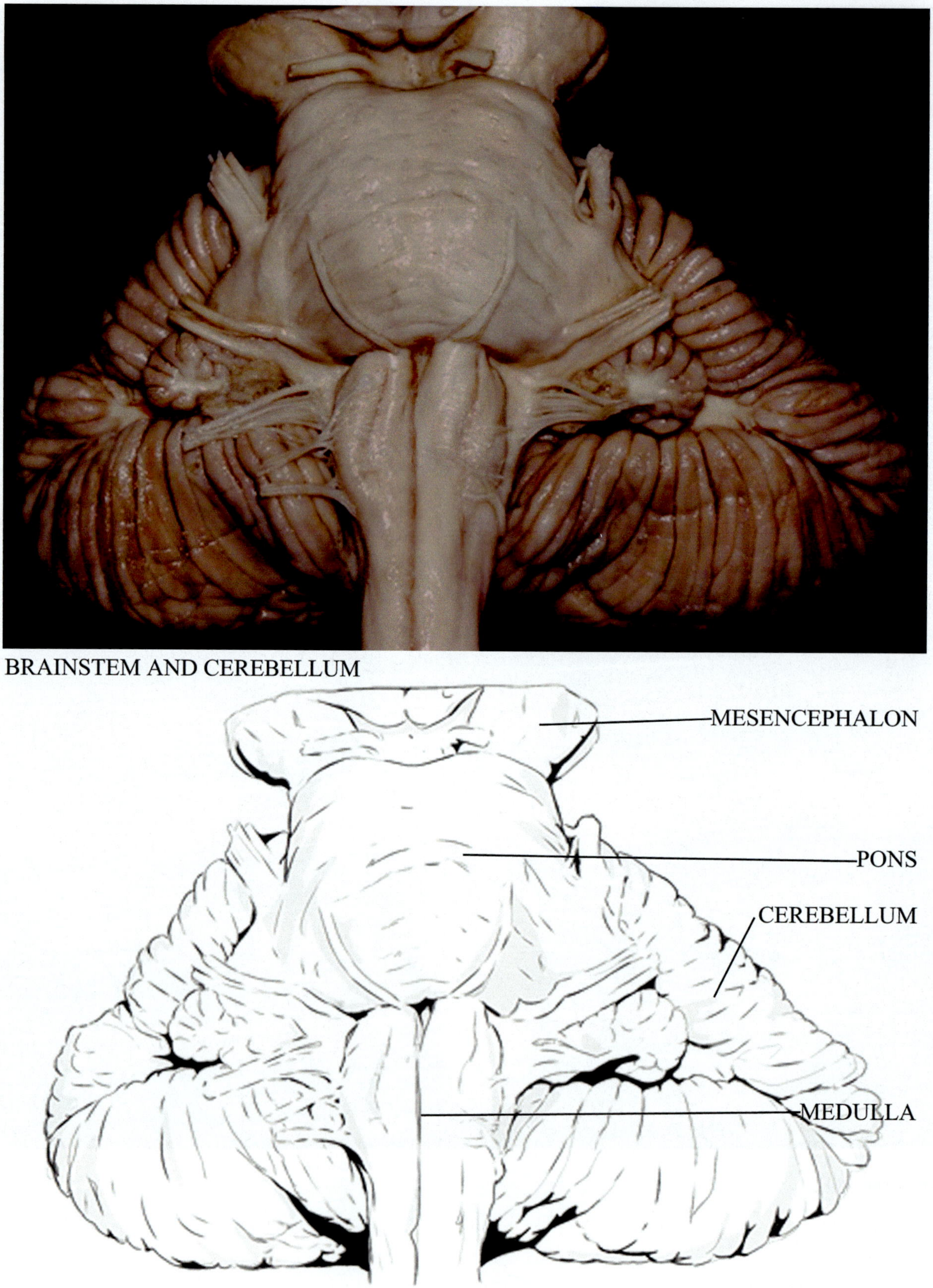

Fig. 8.2 Anatomical preparation and illustration by Renata Kadri of the brainstem, anterior view. The brainstem is morphologically and ontogenetically divided into the mesencephalon, rhombencephalon, and myelencephalon. The rhombencephalon is subdivided into the pons and cerebellum. (©Kadri 2023. All Rights Reserved)

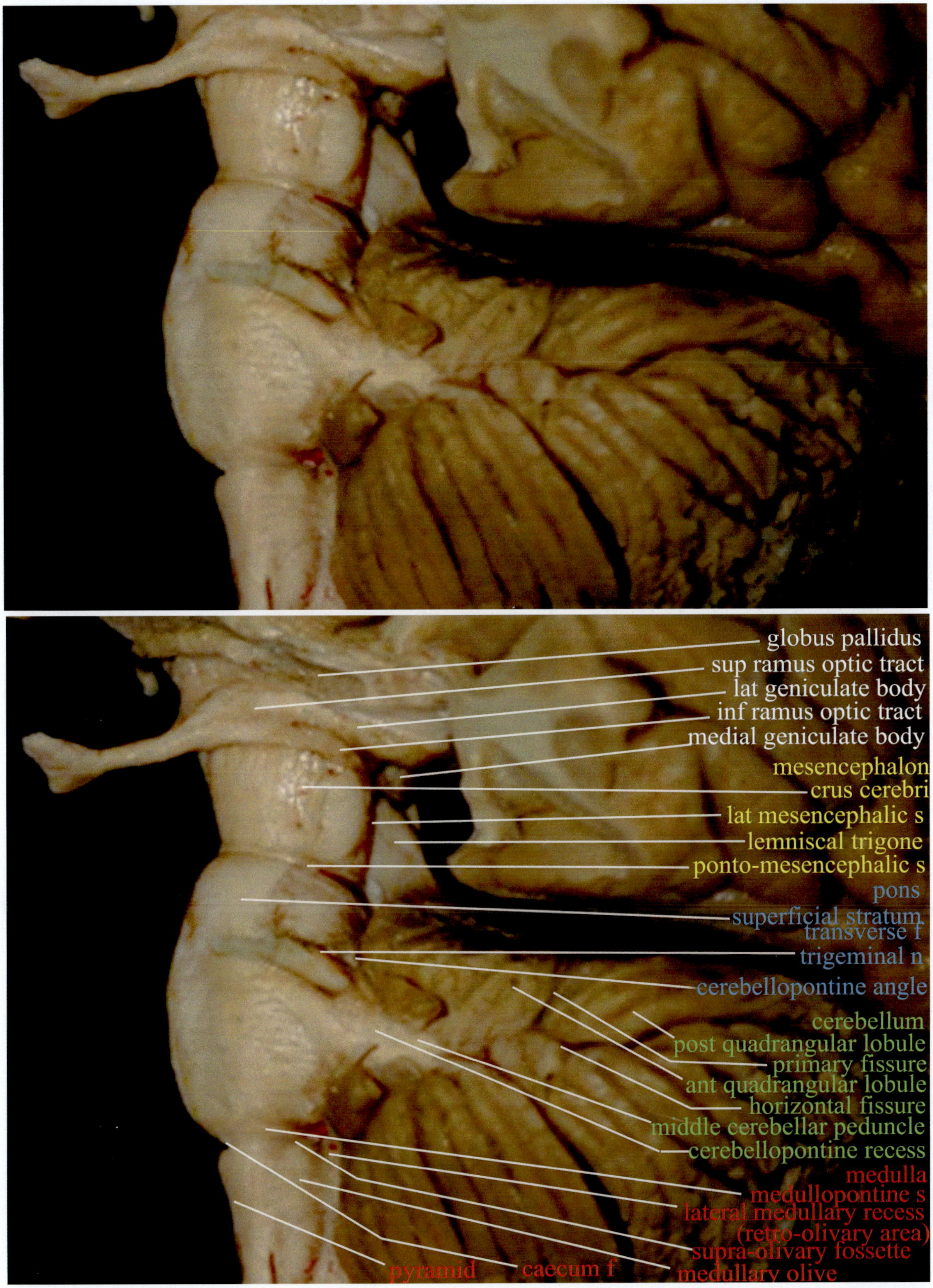

Fig. 8.3 Lateral left view of the anatomical preparation of the brainstem. *ant* anterior, *f* fasciculus, *inf* inferior, *lat* lateral, *n* nerve, *post* posterior, *s* sulcus, *sup* superior (©Kadri 2023. All Rights Reserved)

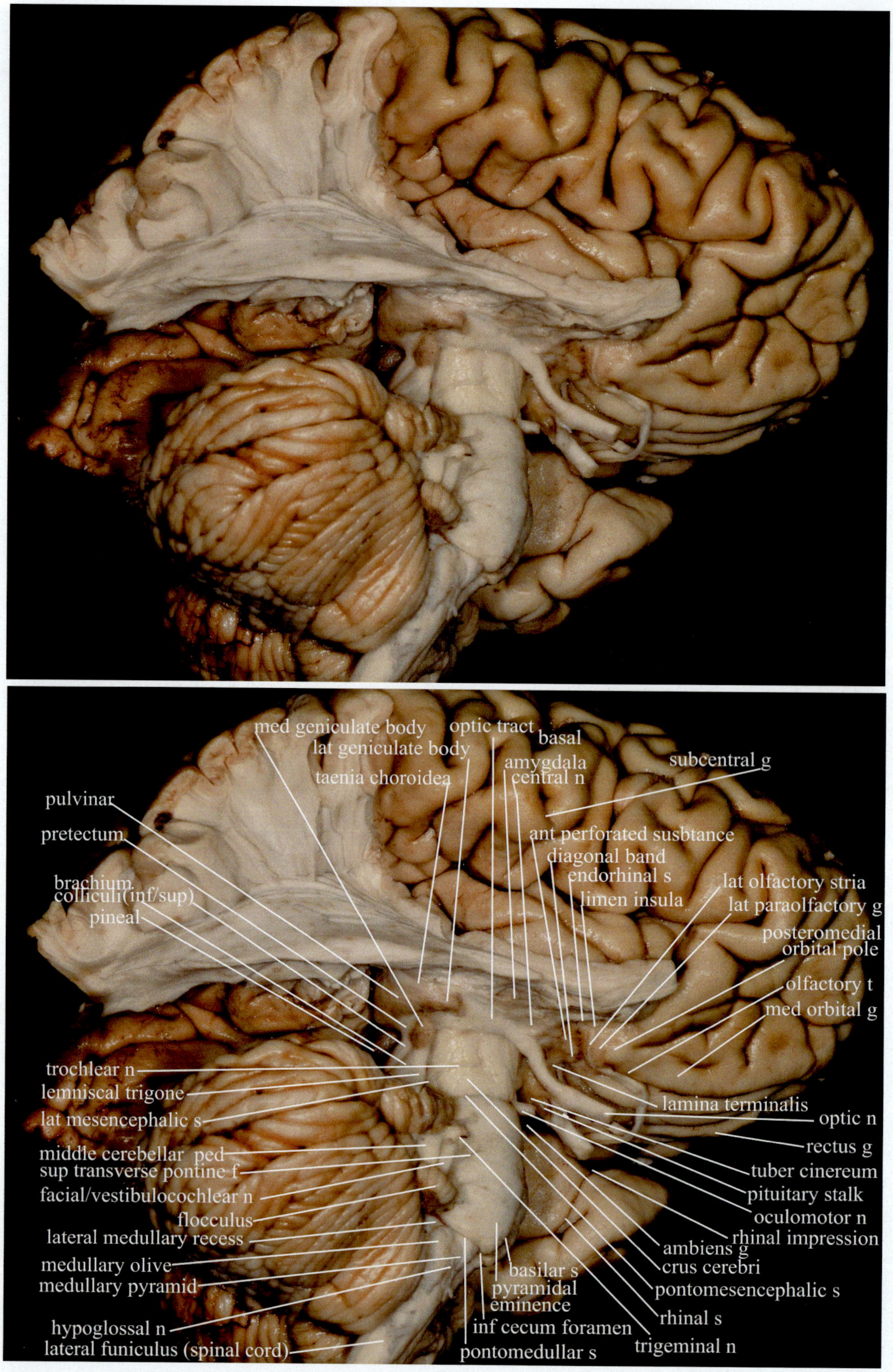

Fig. 8.4 Lateral right view of the anatomical preparation of the brainstem. The neocortical and limbic components of the temporal lobe have been removed. The brainstem, thalamus, inferior ventricular horn, basal nuclei, posterior insula, subopercular region, orbitofrontal area, and related structures are exposed. *ant* anterior, *f* fasciculus, *g* gyrus, *inf* inferior, *lat* lateral, *med* medial, *n* nerve, *ped* peduncle, *s* sulcus, *sup* superior, *t* tract (©Kadri 2023. All Rights Reserved)

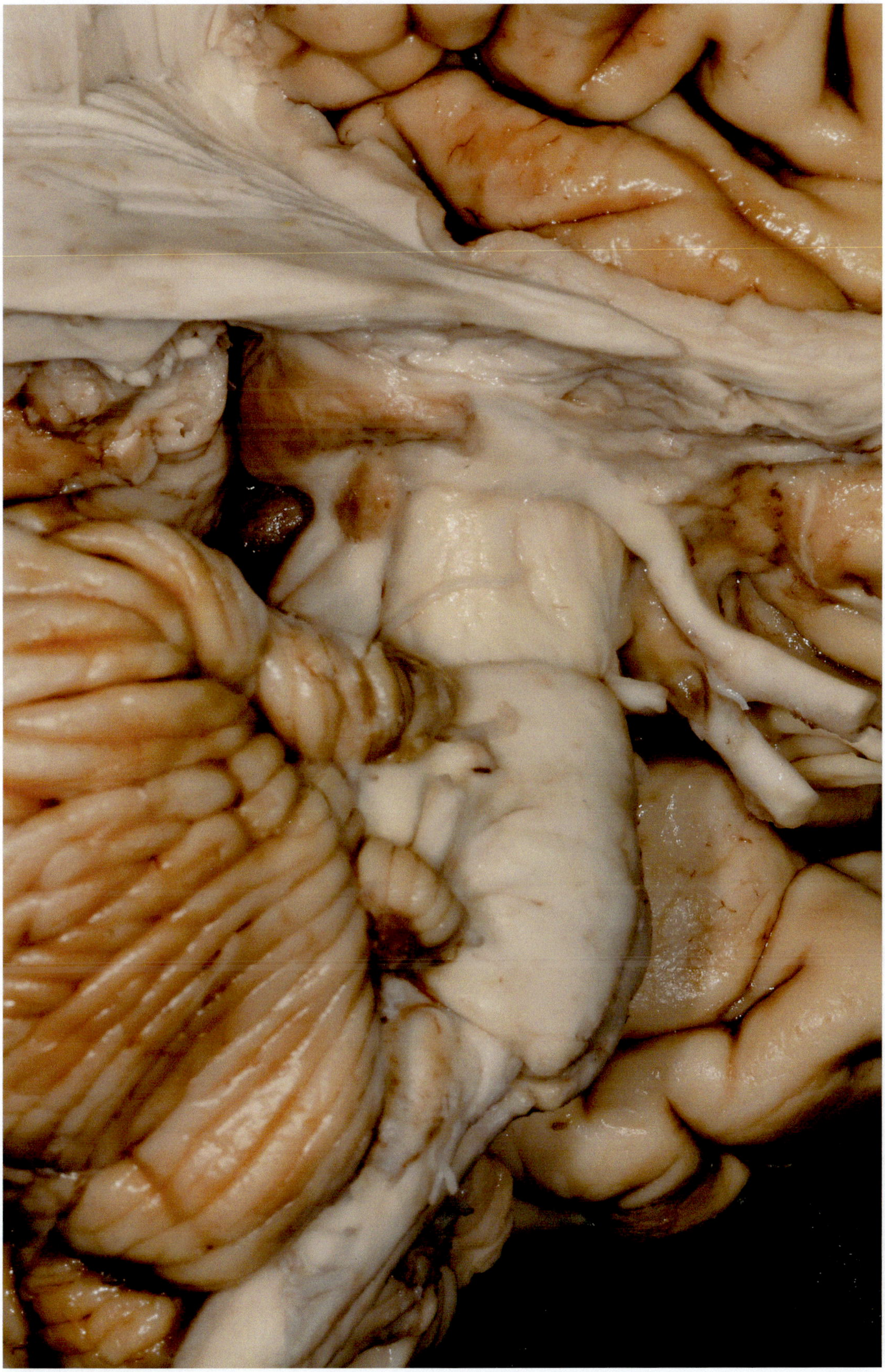

Fig. 8.5 Lateral right close-up view of the anatomical preparation of the brainstem. *f* fasciculus, *g* gyrus, *inf* inferior, *lat* lateral, *med* medial, *n* nerve, *nn* nucleus, *s* sulcus, *sup* superior (©Kadri 2023. All Rights Reserved)

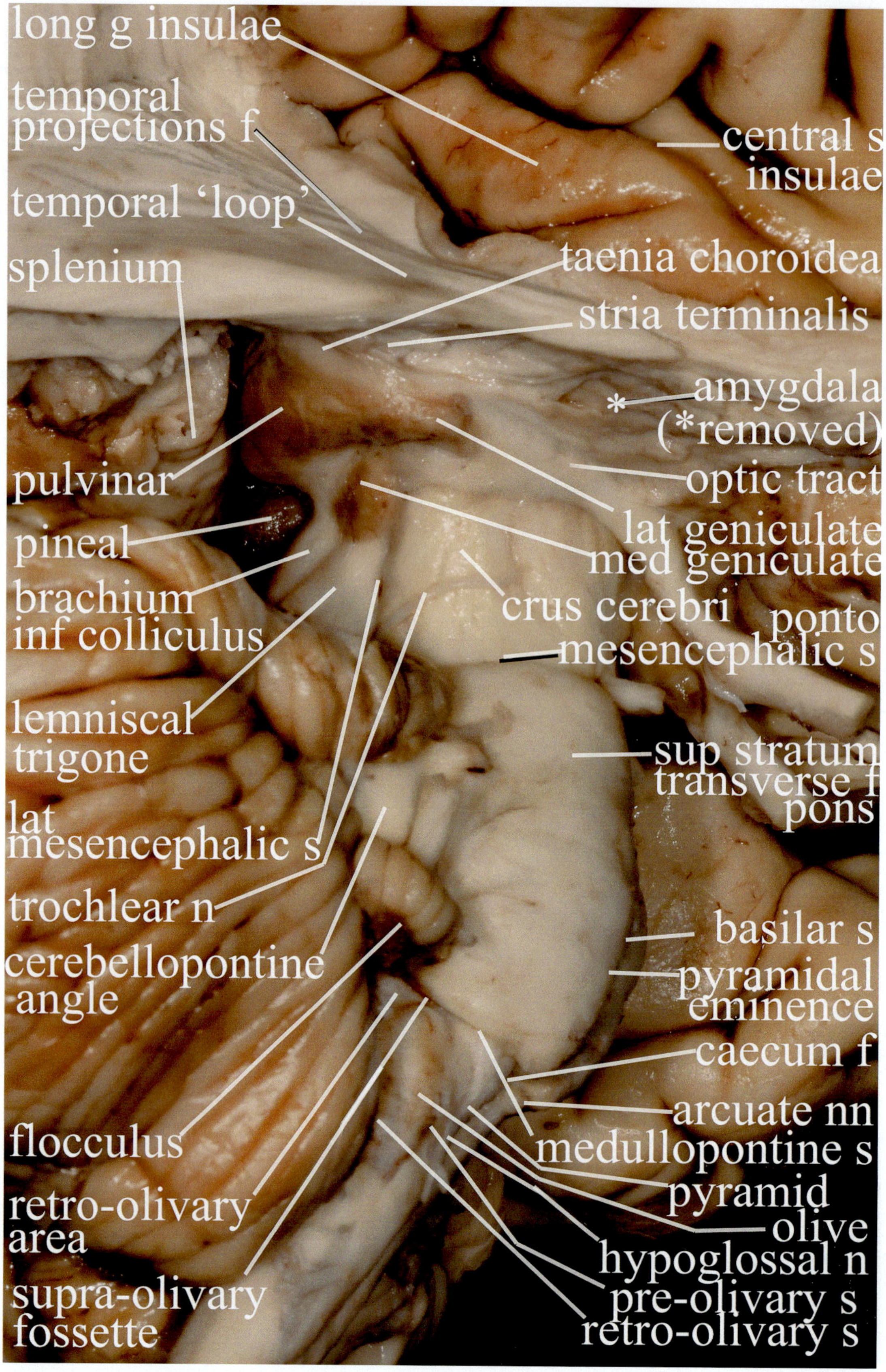

Fig. 8.5 (continued)

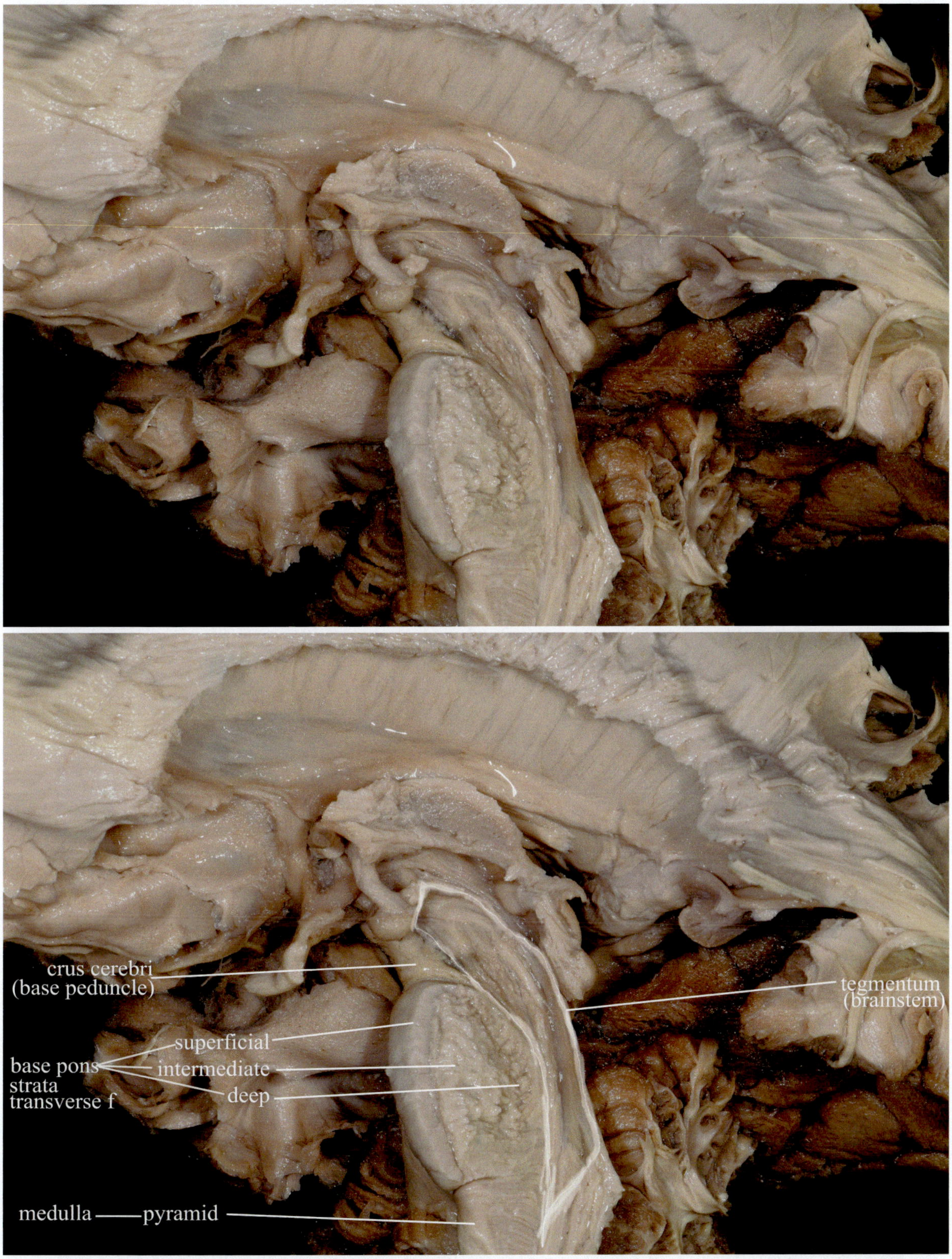

Fig. 8.6 Anatomical preparation of the base and tegmentum of the brainstem. The projection fibers, the only component of the base at the mesencephalon and medulla, constitute the crus cerebri and pyramid. The crus cerebri is mainly composed of the frontopontine, corticospinal, temporoparietopontine, corticonuclear, and corticotectal tracts, arrayed in this sequence from ventral to dorsal. The pyramid is composed of the corticospinal tract. The corticonuclear tracts are the inner center fibers and connect with the ipsilateral and contralateral sides. The frontopontine tract is composed of the most ventromedial fibers, and the parietotemporopontine tract harbors the most dorsolateral fibers at the proximal crus cerebri. Some fibers from the dorsolateral proximal crus traverse ventromedially and are called the transverse peduncular tract. *f* fasciculus (©Kadri 2023. All Rights Reserved)

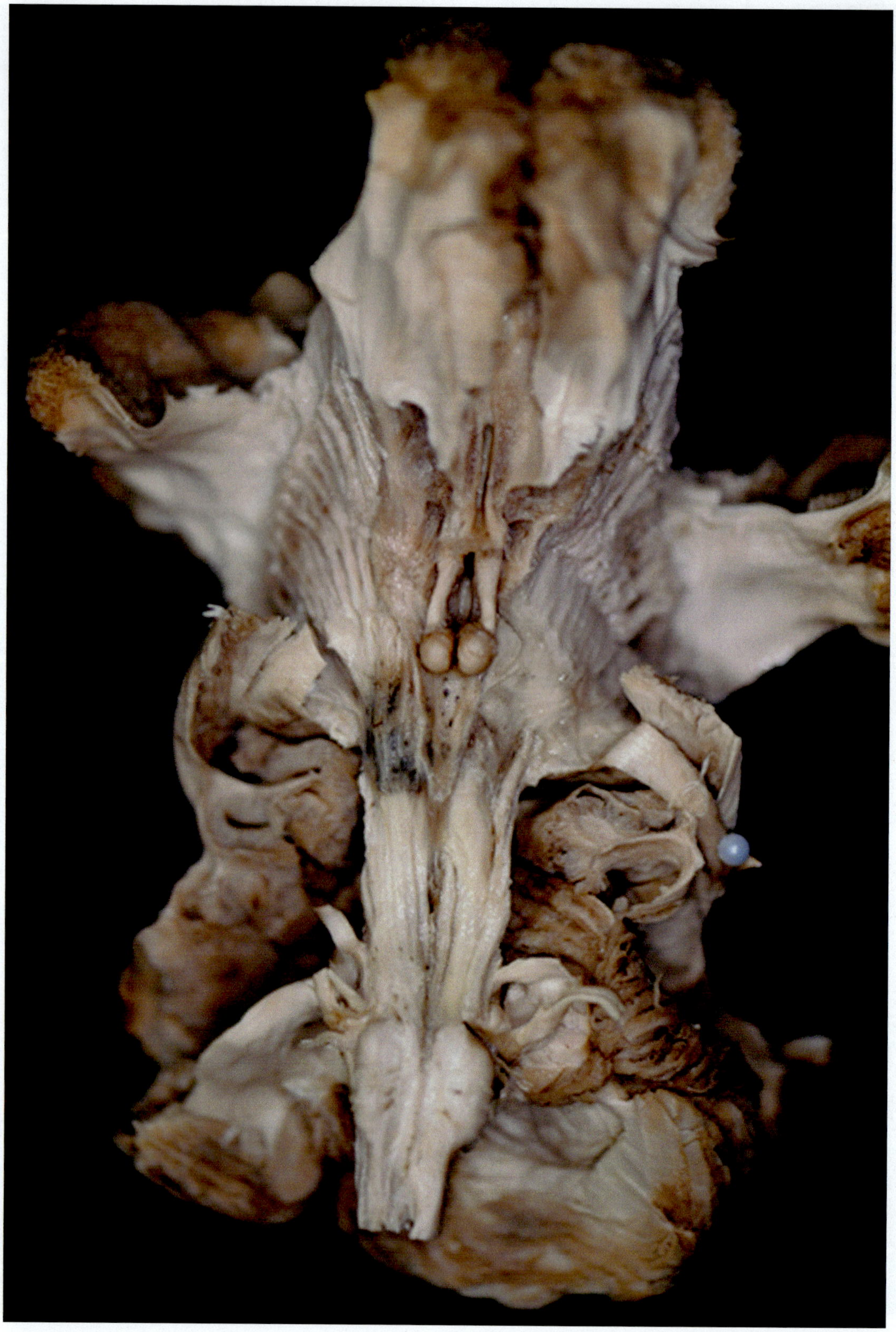

Fig. 8.7 Anatomical preparation of the extrapyramidal fibers of the brainstem and cerebellum. The substantia nigra and its connections with the tegmentum of the pons and medulla cover the ventral tegmental mesencephalic decussation. The fibers of the superior cerebellar peduncle that cross to connect with the red nucleus and thalamus, passing through or around the red nucleus, and constituting part of the pars reticulata, or parvocellular part, of this nucleus are an important component of the ventral decussation. Connections of the red nucleus and the inferior olive of the medulla are an important component of the central tegmental tract. This tract envelops the inferior olivary nuclei in a dense network of fibers named the amiculum. *ant* anterior, *f* fasciculus, *g* gyrus, *inf* inferior, *lat* lateral, *n* nerve, *ped* peduncle, *post* posterior, *sup* superior, *t* tract (©Kadri 2023. All Rights Reserved)

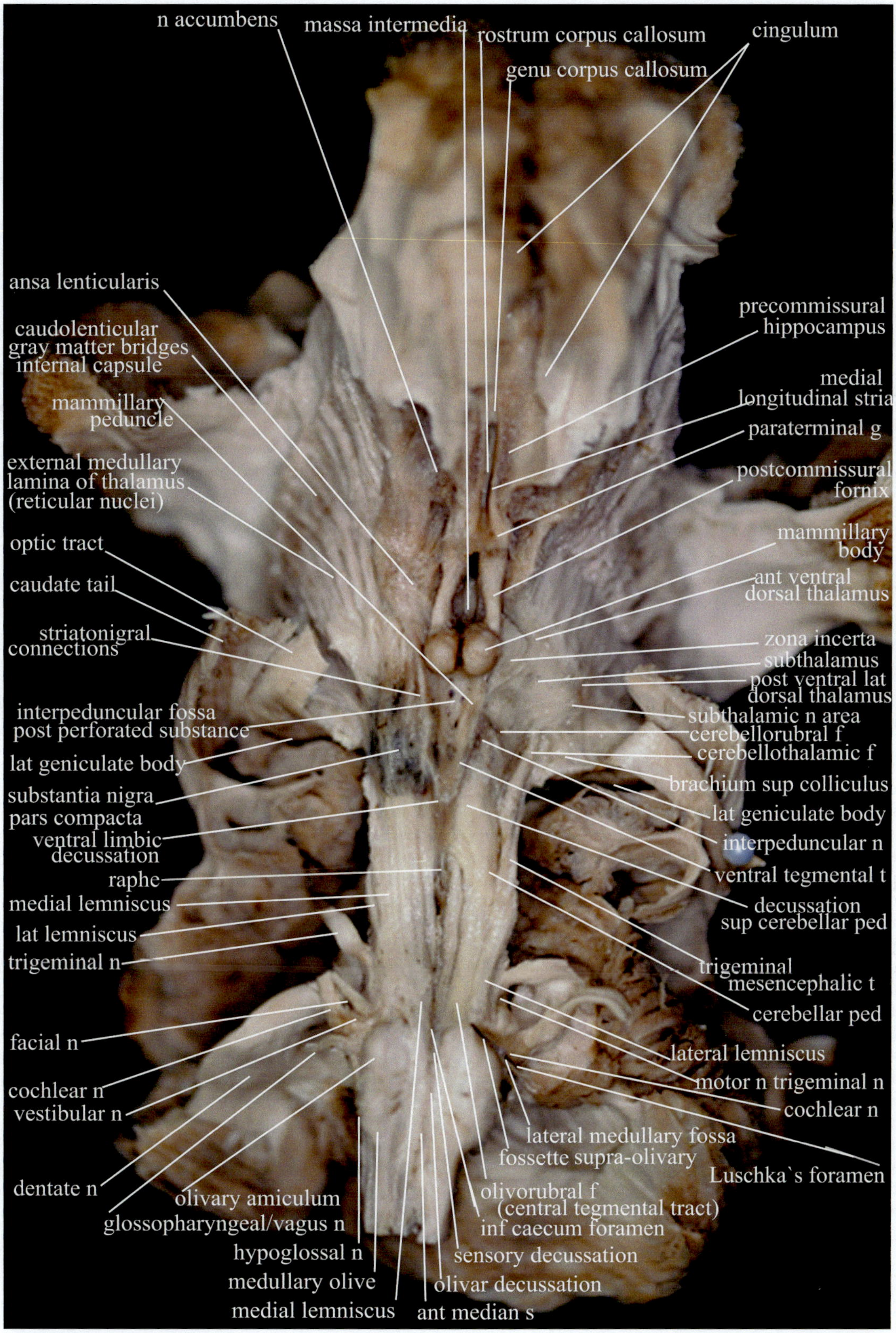

Fig. 8.7 (continued)

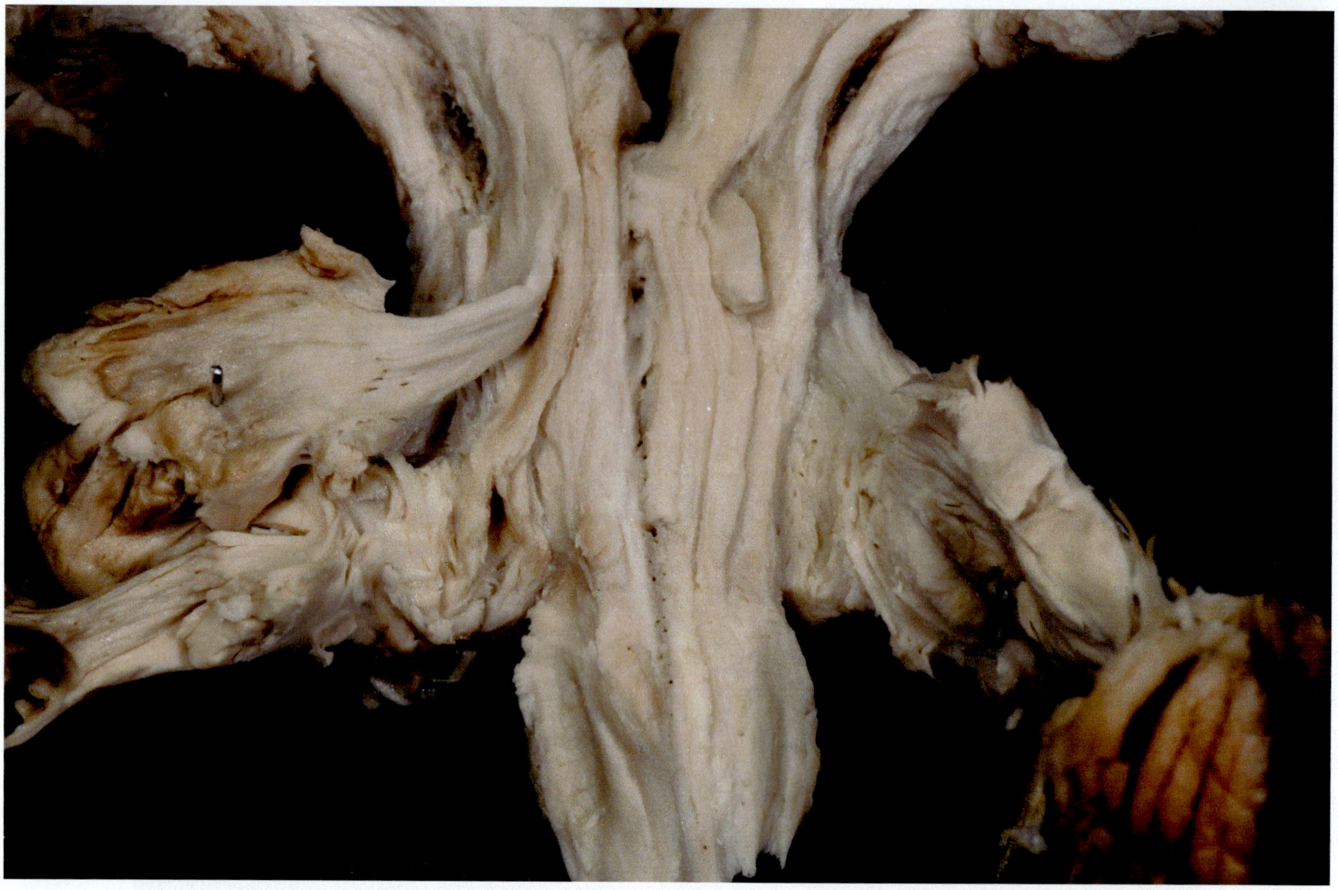

Fig. 8.8 Anatomical preparation of the tegmentum of the brainstem. Four groups of longitudinal oriented fibers are distinguishable. The medial longitudinal fasciculus is the main connection of the special motor and sensory nuclei. The motor hypoglossal, dorsal vagus, ambiguus, facial, abducens, trigeminal motor, trochlear, and oculomotor complex nuclei, along with the special sensory cochlear and vestibular nuclei, are connected by the medial longitudinal fasciculus. Alongside the medial longitudinal fasciculus are crossing connections between them. These crossing fibers at the midline were named the *raphe* by the French anatomist Felix Vicq d'Azyr (1748–1794). The medial longitudinal fasciculus is a landmark to subdivide the brainstem tegmentum into dorsal and ventral regions. These dorsal limbic–autonomic tegmental nuclei are at the dorsal mesencephalon and dorsal pons. The dorsal tegmental mesencephalic nuclei are covered by the prosencephalic tract on the floor of the third ventricle and ventral and lateral aqueductal areas. The prosencephalic tract is formed by connections of the telencephalon and diencephalon with the tegmentum of the brainstem. The dorsal mesencephalic nuclei connect to the septal region, amygdala, bed nucleus of the stria terminalis, anterior thalamus, preoptic area, anterior hypothalamus, and mammillary body. One of these connections is the mammillotegmental tract, which is the dorsal mesencephalic connection of the mammillary nuclei, mainly the lateral mammillary, and has a common trunk with the mammillothalamic fasciculus, named the principal mammillary tract. The ventral connection of the mammillary body connects with the ventral tegmental mesencephalic nuclei and is named the mammillary peduncle. The ventral tegmental mesencephalic nuclei are cranial and ventral to the mesencephalic decussations. One of the ventral tegmental nuclei is the interpeduncular nucleus. This nucleus connects with the amygdala, septal region, hypothalamus (mainly ipsilateral and contralateral mammillary body, but also the anterior nuclei), hippocampal formation (via the fornix) and the epithalamus (via the retroflexus tract). The connections of the telencephalon and hypothalamus are grouped and named the ventral tegmental fasciculus. One of the components of the ventral tegmental tract is the connection of the mammillary bodies with the ventral tegmentum, named the mammillary peduncle. The retroflexus tract is also named the habenulo-interpeduncular tract. The dorsal tegmental mesencephalic nucleus is a reticular nucleus embedded within the fibers of the prosencephalic tract. It is also limited dorsally and externally by the pretectum. *f* fasciculus, *inf* inferior, *lat* lateral, *med* medial, *mid* middle, *n* nerve, *nn* nucleus, *ped* peduncle, *s* sulcus, *sup* superior, *t* tract ()

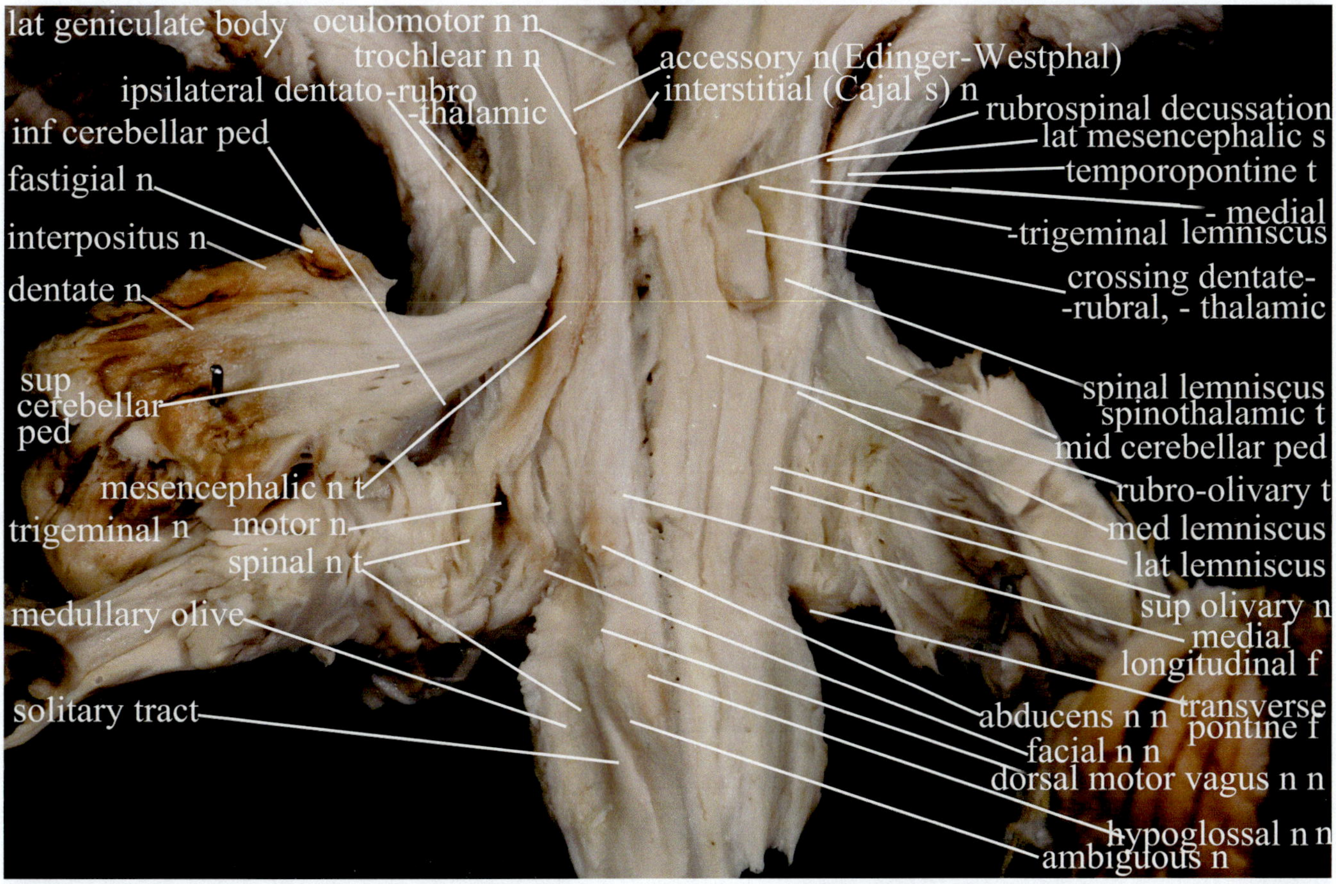

Fig. 8.8 (continued)

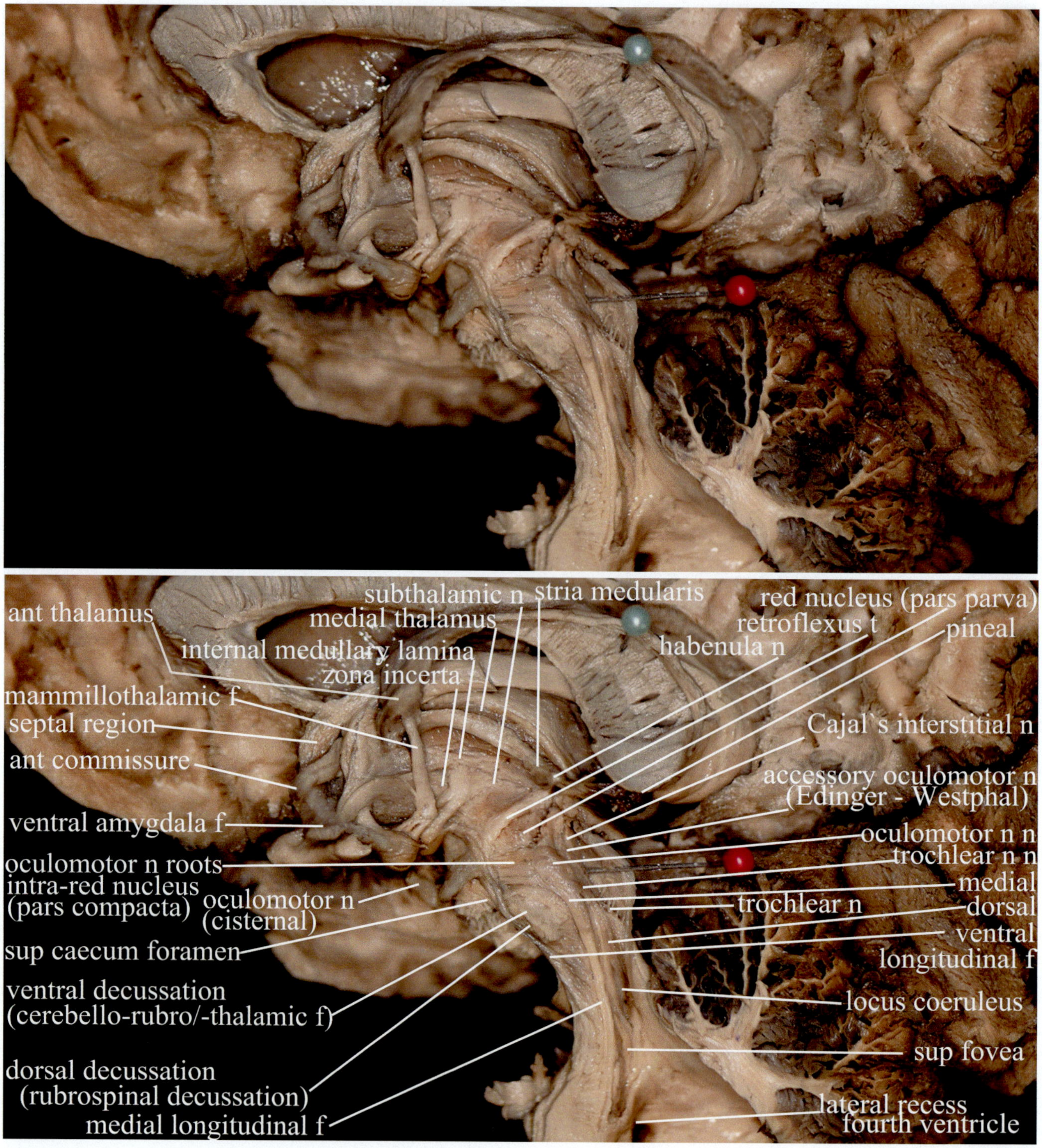

Fig. 8.9 Anatomical preparation of the longitudinally oriented fiber bundles of the tegmentum of the brainstem. In this view, the fibers are classified as the medial, dorsal, and ventral longitudinal fasciculi. Composed of different fiber bundles, the medial longitudinal fasciculus is the connection of the motor neurons of the cranial nerves. The ventral and dorsal longitudinal fasciculi are the ventral and dorsal tegmental limbic connections. *ant* anterior, *f* fasciculus, *n* nerve, *nn* nucleus, *sup* superior, *t* tract (©Kadri 2023. All Rights Reserved)

The Mesencephalon

The mesencephalon is the only one of the three primitive vesicles that does not subdivide during ontogenesis [1]. It is composed of two equal halves known as the cerebral peduncles (Fig. 8.10). Each peduncle can be subdivided into three portions: the crus cerebri, the tegmentum, and the tectum. Between the crura is the interpeduncular fossa (Fig. 8.11). The crus cerebri and the tegmentum are separated by the substantia nigra (Fig. 8.12); together they compose the base of the peduncle. The pes peduncular is also known as the crus cerebri and contains the substantia nigra, which extends from the lateral to the medial mesencephalic sulci. The medial mesencephalic sulcus is also known as the oculomotor nerve sulcus (Figs. 8.13 and 8.14). The caudal limit of the mesencephalon is the decussation of the trochlear nerve within the superior medullary vellum dorsally and the pontomesencephalic sulcus ventrally [2]. The decussation of the superior cerebellar peduncle is the limit between the mesencephalic and the pontine tegmentum (Fig. 8.15). The superior limit of the mesencephalon is a line traced between the mammillary body and the posterior commissure at the midline. This line accompanies the inferior border of the optic tract laterally (Fig. 8.16). The periaqueductal gray matter is partially in the tectum and partially at the tegmentum, surrounding the aqueduct. It is continuous through the floor of the third ventricle with the gray matter of the hypothalamus and inferiorly with the gray matter of the fourth ventricle. It is connected by the dorsal longitudinal fasciculus of Schutz, which connects the hypothalamus with the visceral centers of the brain stem (Figs. 8.17, 8.18, and 8.19) [1, 2]. Throughout this chapter, the dorsal longitudinal fasciculus is called the dorsal tegmental tract.

The mesencephalic tectum is also known as the quadrigeminal plate and consists of the paired superior and inferior colliculi (Figs. 8.20 and 8.21). These are separated by the longitudinal and transverse segments of the cruciform sulcus [2]. On the superior part, between the two superior colliculi, the cruciform sulcus harbors a triangular depression known as the infrapineal recess. The inferior colliculus is part of the

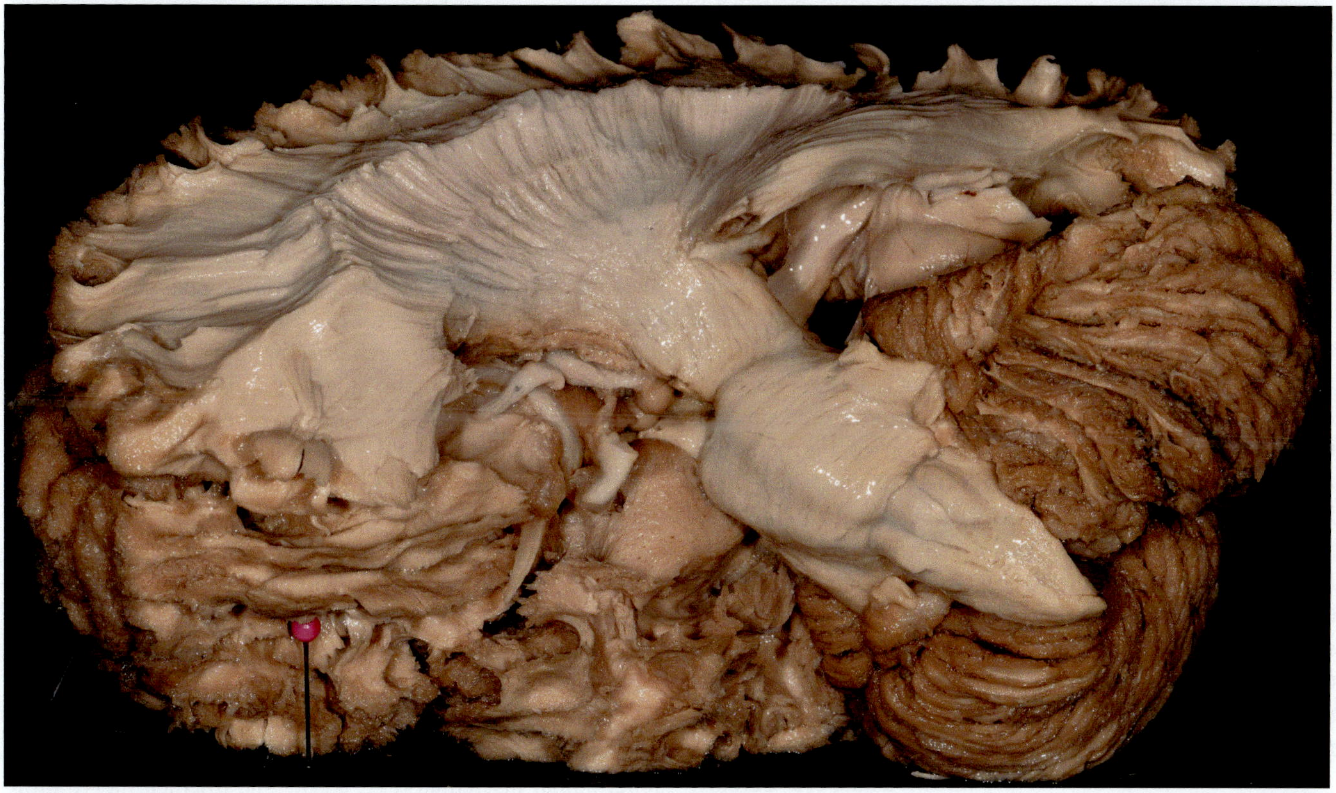

Fig. 8.10 Anatomical preparation of the crus cerebri of the left peduncle of the mesencephalon. Removing the lateral basal nuclei and optic chiasm exposes the continuum of the telencephalic projection fibers through the corona radiata, internal capsule, and crus cerebri of the mesencephalic peduncle. At the crus cerebri, the parietotemporopontine fibers occupy the lateral portion and are limited with the lemniscus fibers by the lateral mesencephalic sulcus. Some of the external or dorsal fibers on the proximal portion of the crus cerebri may have an aberrant dorsal-to-ventral course, as they connect with the pontine nuclei at the base of the pons. These are designated as the transverse peduncular tract or taenia pons. *f* fasciculus, *g* gyrus, *lat* lateral, *med* medial, *n* nerve, *ped* peduncle, *seg* segment, *s* sulcus (©Kadri 2023. All Rights Reserved)

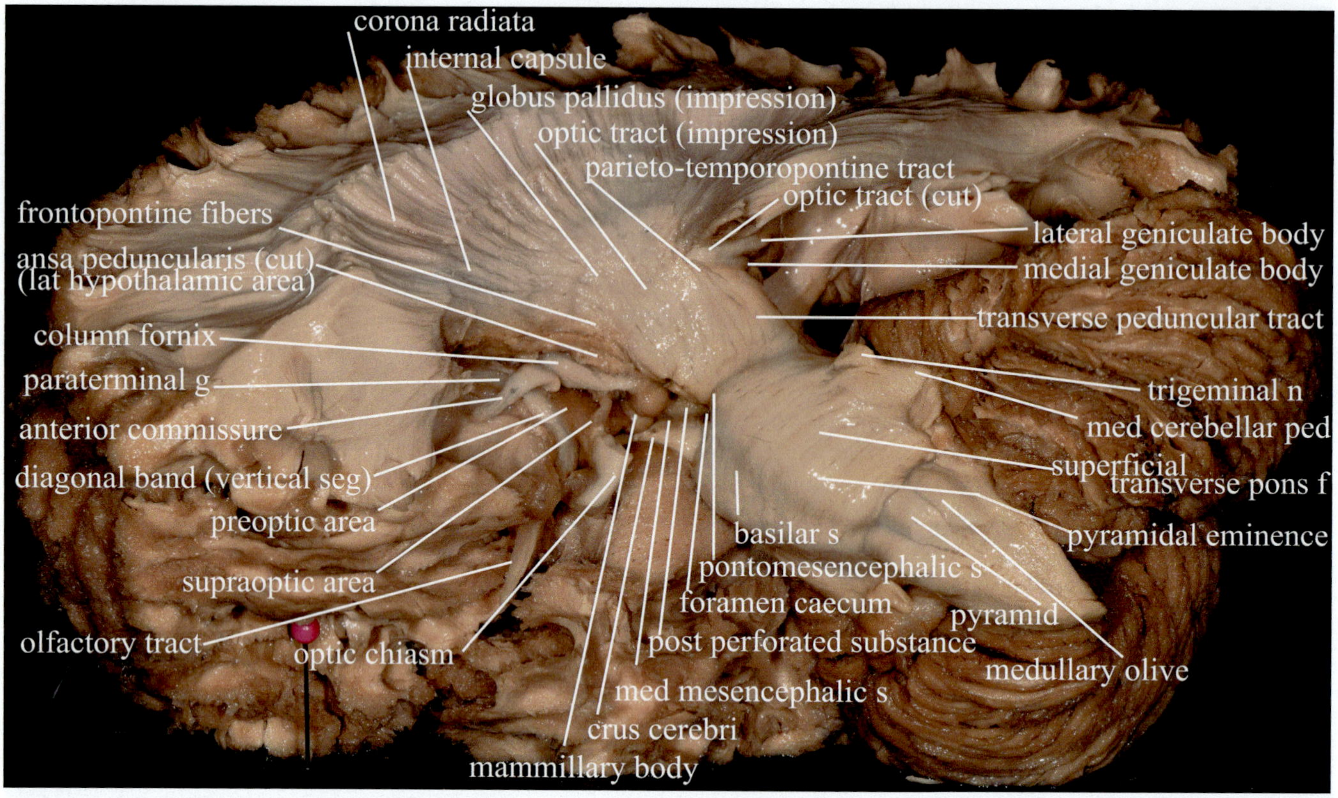

Fig. 8.10 (continued)

acoustic system and connects via the lateral lemniscus with the medullar and pontine auditory centers and with the medial geniculate body via the brachium of the inferior colliculus. The superior colliculus is a multilayered structure that receives afferents from the optic tract; visual cortex; and somatosensory, acoustic, cerebellar, and nonvisual cortical areas [1]. The superior colliculus connects to the pretectum and visual thalamic centers via the brachium of the superior colliculus and to the reticular formation of the pons and medulla and to the spinal cord via the tectomedullary (also known as the tectobulbar) and tectospinal tracts. The superior colliculus is one of the major centers to coordinate eye and head movements [1]. The tectospinal tract is a prominent efferent of the superior colliculi, and it decussates in the dorsal tegmental decussation. The pretectum extends between the superior colliculi and the thalamus. It contains important nuclei for the pupillary light reflex and optokinetic ocular movements (Figs. 8.22, 8.23, and 8.24).

The base of the peduncle, also known as the crus cerebri, is the main projection of the cortex to the brain stem and the spinal cord. The frontopontine, corticospinal, and parieto-temporopontine fibers are arrayed from medial to lateral. Some fibers may turn around the peduncle from dorsal to ventral directions. These aberrant fibers are designated as the bundle of Feres, the taenia pons, and the transverse peduncular tract [2]. The substantia nigra is divided into a dorsal dark pigmented pars compacta and a ventral scarcely pigmented pars reticulata. The lateral mesencephalic sulcus divides the lateral surface of the midbrain. Anterior to it is the corticospinal tract at the base of the peduncle, with the descending fibers. Posterior to it, the ascending fibers of the lateral tegmental area form a triangular space bordered by the lateral mesencephalic sulcus, the brachium of the inferior colliculi, and the superior cerebellar peduncle. This space is known as Reil's triangle, the acoustic triangle, or the lemniscal triangle (or trigone), and its superficial layer is the lateral lemniscus [2].

The tegmentum contains the motor nuclei of the oculomotor and trochlear nerves; the crossed fibers of the superior cerebellar peduncle; the red nucleus; and the ascending somatosensory fibers of the medial, trigeminal, and spinal lemnisci (Fig. 8.25). The nuclei of the oculomotor and trochlear nerves are located near the midline, ventral to the periaqueductal gray matter. They are enveloped by the medial longitudinal fasciculus (Fig. 8.26). The trochlear nerve projects dorsally to cross in the superior medullary velum and emerge inferior to the inferior colliculi, lateral to the frenulum veli. The oculomotor nerve traverses the mesencephalic tegmentum to emerge in the interpeduncular fossa. The accessory oculomotor nucleus of Edinger–Westphal is located within the pretectum, dorsal and rostral to the main nucleus, and carries the parasympathetic fibers for the pupillary reflex [1, 2].

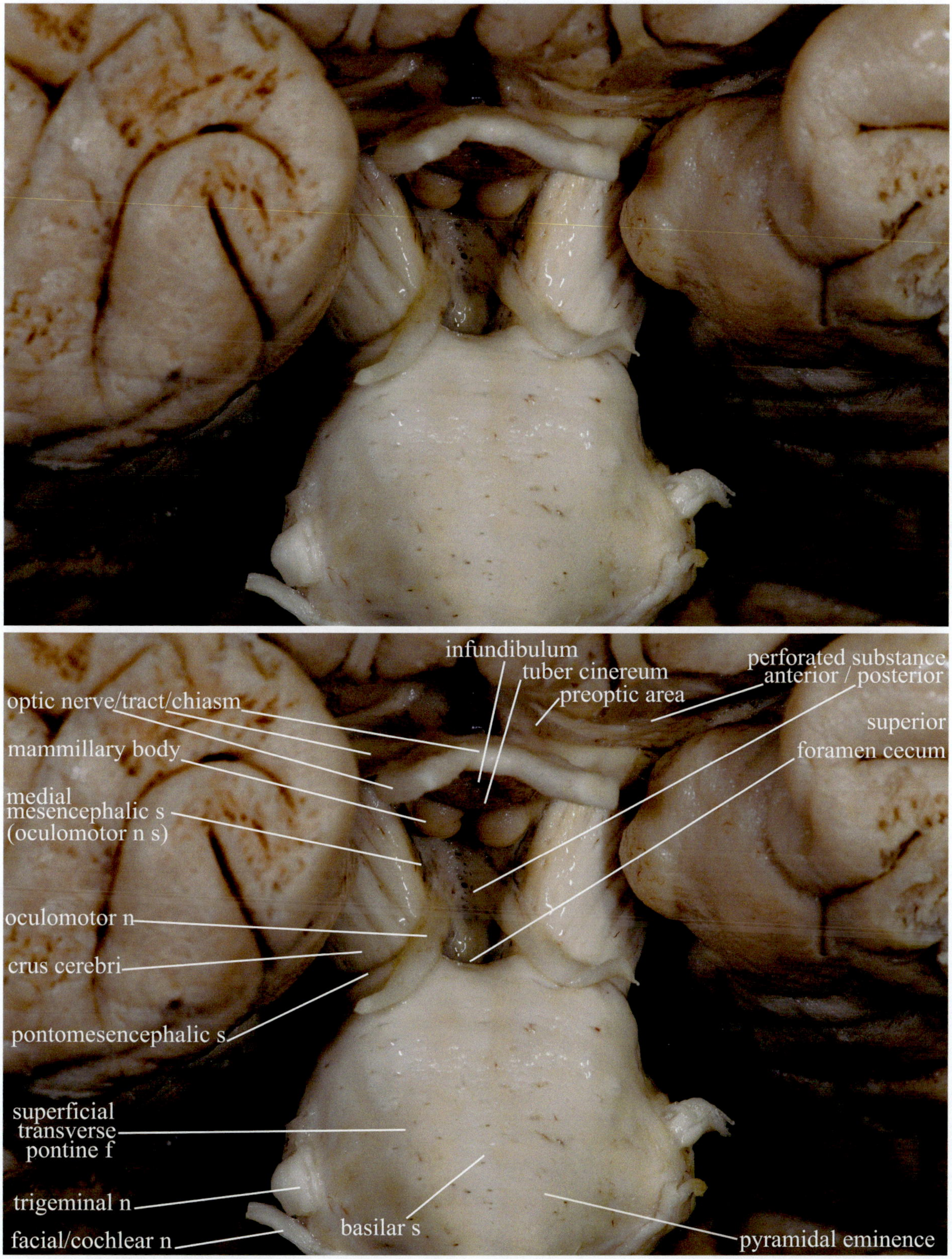

Fig. 8.11 Anatomical preparation of the ventral peduncle of the mesencephalon and the interpeduncular fossa. The posterior perforated substance is the floor of the interpeduncular fossa, which is limited laterally by the medial mesencephalic sulcus. At the limit of the interpeduncular fossa with the pons, the pontomesencephalic sulcus intercepts the medial mesencephalic sulcus at the foramen cecum. The medial mesencephalic sulcus is also named the oculomotor nerve sulcus. *f* fasciculus, *n* nerve, *s* sulcus (©Kadri 2023. All Rights Reserved)

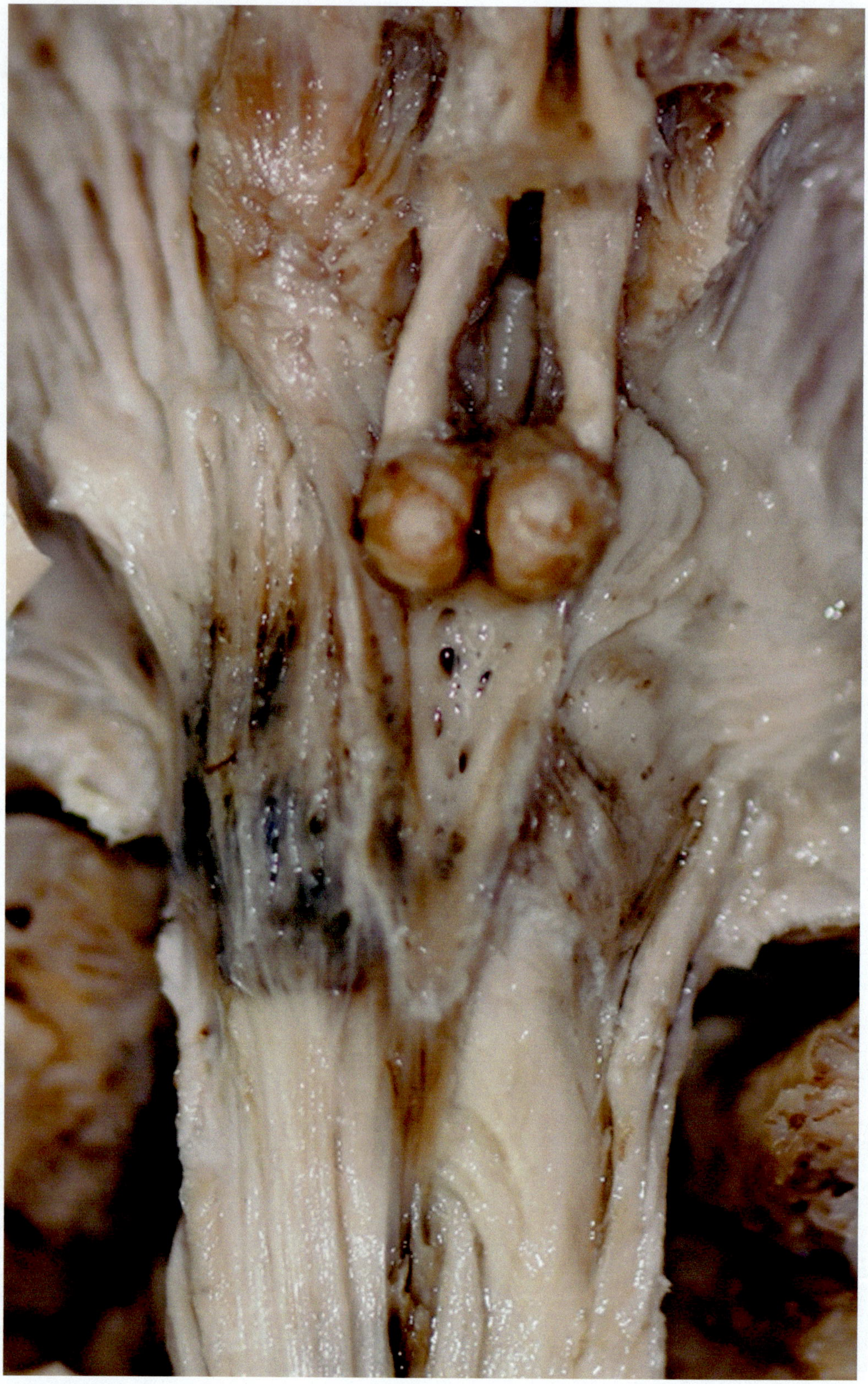

Fig. 8.12 Anatomical preparation of the pars compacta of the substantia nigra and the ventral tegmentum of the mesencephalon. Removing the substantia nigra exposes the interpeduncular nucleus at the level where the oculomotor nerve emerges into the cisternal space at the medial mesencephalic sulcus. The medial border of the medial mesencephalic sulcus is created by a limbic connection between the amygdala, septal region, hippocampus (via the fornix), mammillary body, and the interpeduncular nucleus and through and with the connections with the interpeduncular nucleus and the dorsal mesencephalic and dorsal pontine nuclei. *f* fasciculus, *lat* lateral, *n* nerve, *sup* superior, *tr* tract (©Kadri 2023. All Rights Reserved)

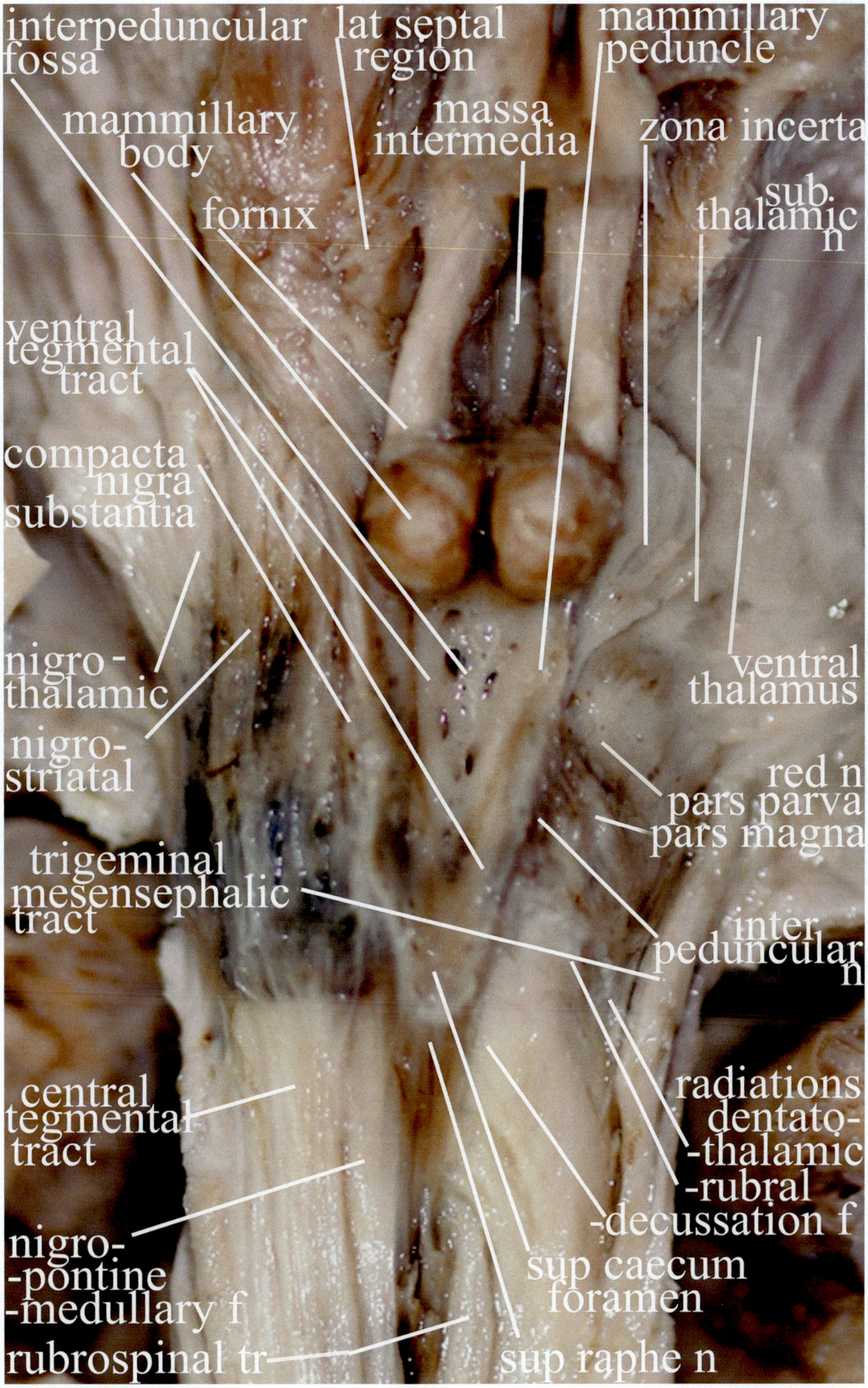

Fig. 8.12 (continued)

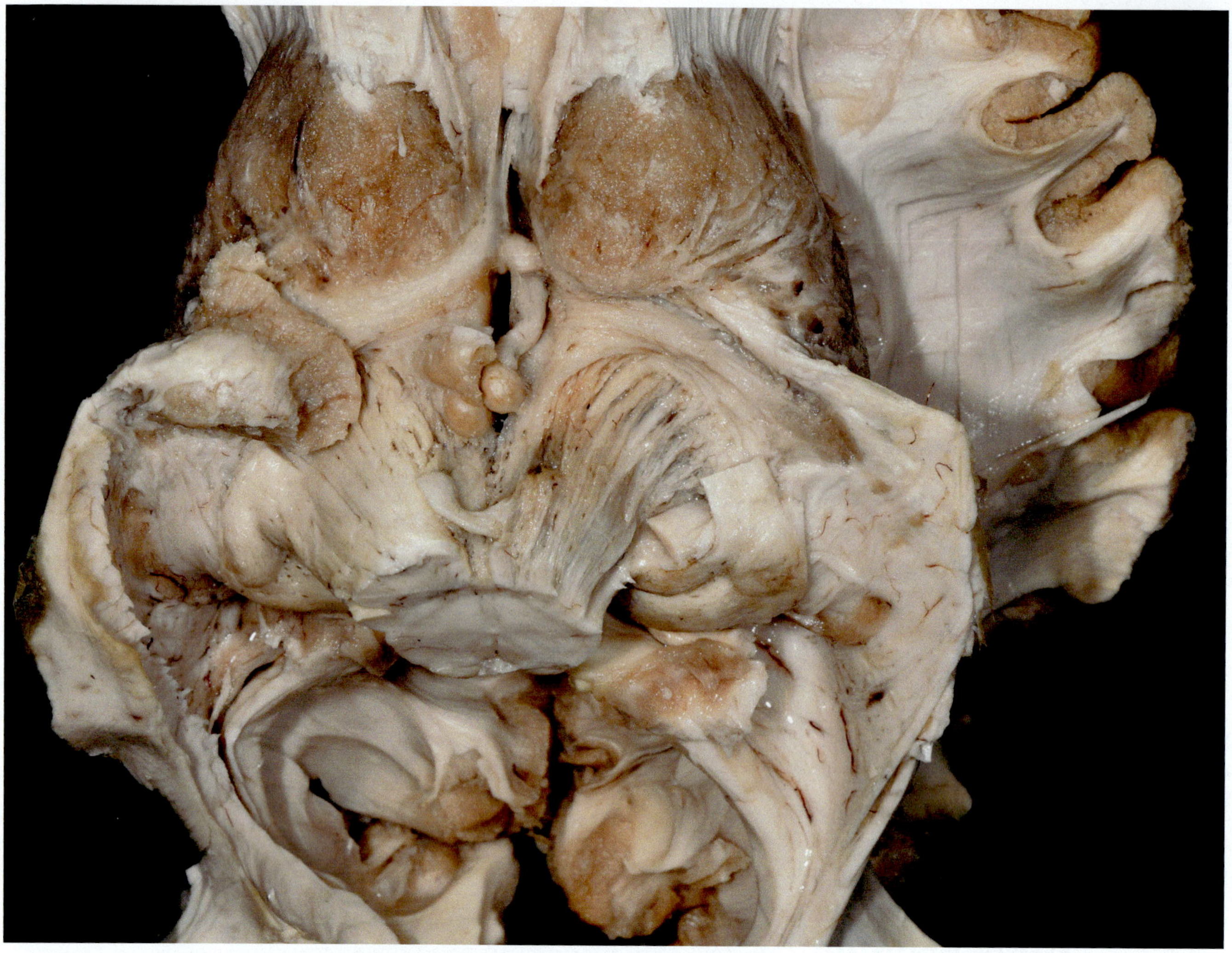

Fig. 8.13 Anatomical preparation of the ventral tegmentum of the mesencephalon. Removing the corticopontine, corticonuclear, and corticospinal tracts exposes the pars reticulata of the substantia nigra. *ant* anterior, *f* fasciculus, *g* gyrus, *lat* lateral, *med* medial, *n* nerve, *post* posterior, *seg* segment, *sup* superior (©Kadri 2023. All Rights Reserved)

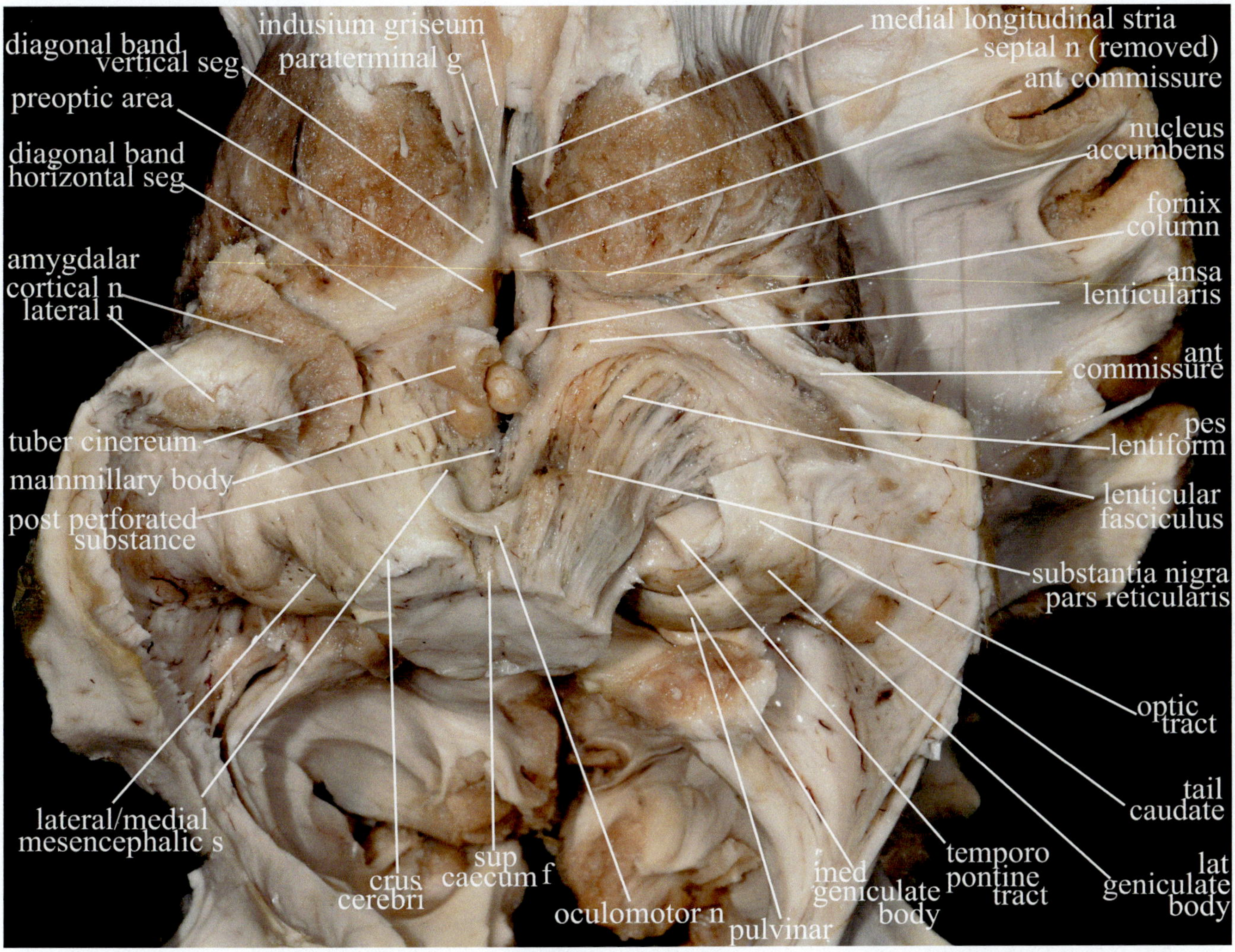

Fig. 8.13 (continued)

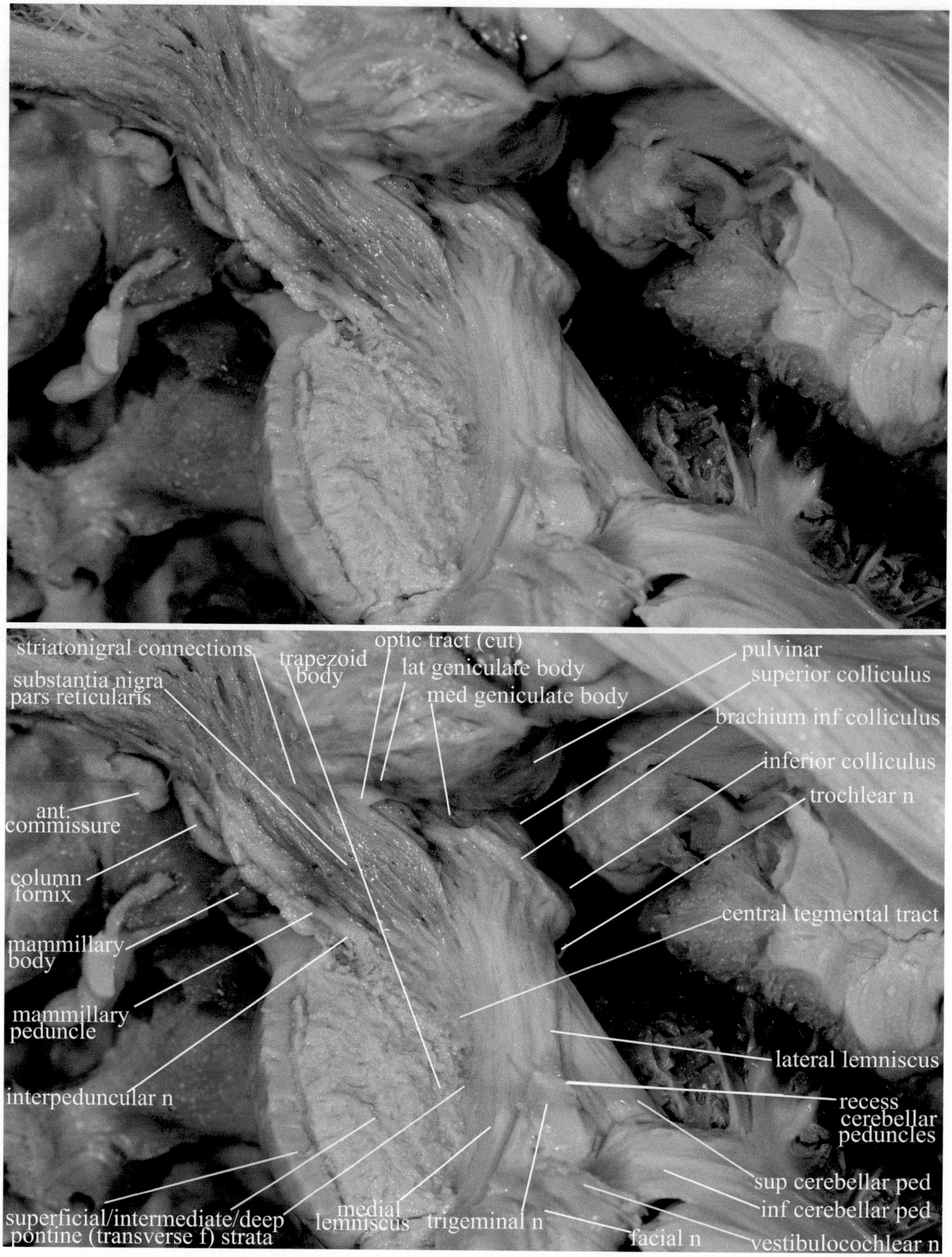

Fig. 8.14 Anatomical preparation of the ventral tegmentum of the mesencephalon and pons. The substantia nigra is the limit of the crus and the tegmentum. The ventral substantia nigra is rich in connections with the striatum and pallidum and, based on this fiber appearance, is called the pars reticulata. Fibers connecting and passing through the pars reticulata to the caudal central nervous system are part of the central tegmental tract. The fibers from the posterior spinal funiculus connect, in the dorsal lower medulla, with the cuneatus and gracile nuclei. The connections of the cuneatus and gracile nuclei cross at the medullar sensory decussation and ascend as the medial lemniscus in the ventral tegmentum of the pons. As the medial lemniscus ascends to connect to the lateral dorsal thalamus, it passes through the dorsal tegmentum of the mesencephalon. *ant* anterior, *f* fasciculus, *inf* inferior, *lat* lateral, *med* medial, *n* nerve, *ped* peduncle, *sup* superior (©Kadri 2023. All Rights Reserved)

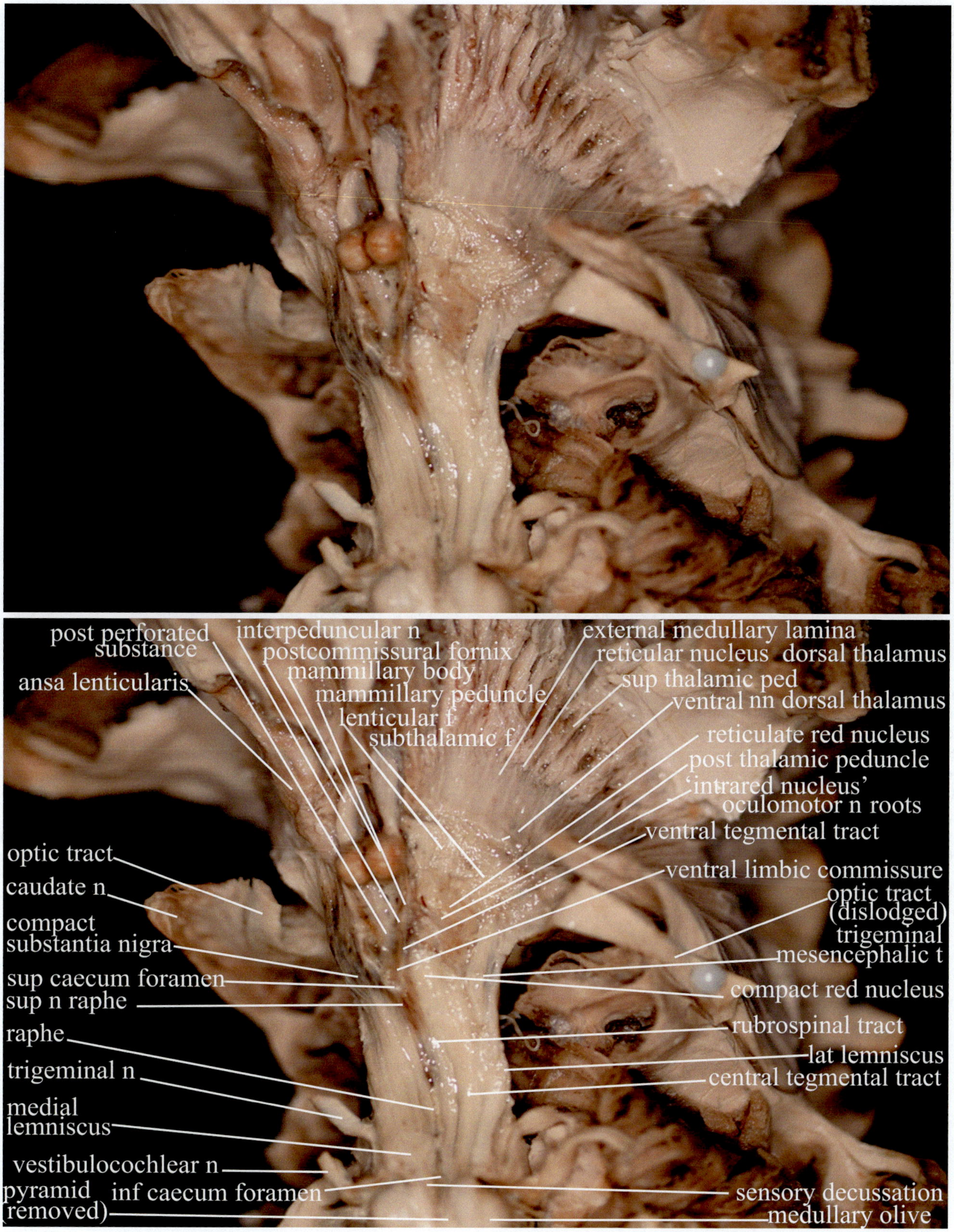

Fig. 8.15 Anatomical preparation of the ventral tegmentum of the mesencephalon, ventral subthalamus (covering the ventral aspect of the dorsal thalamus), and the capsular surface of the external medullary lamina of the lateral dorsal thalamus. *f* fasciculus, *inf* inferior, *lat* lateral, *n* nerve, *nn* nucleus, *ped* peduncle, *post* posterior, *sup* superior, *t* tract

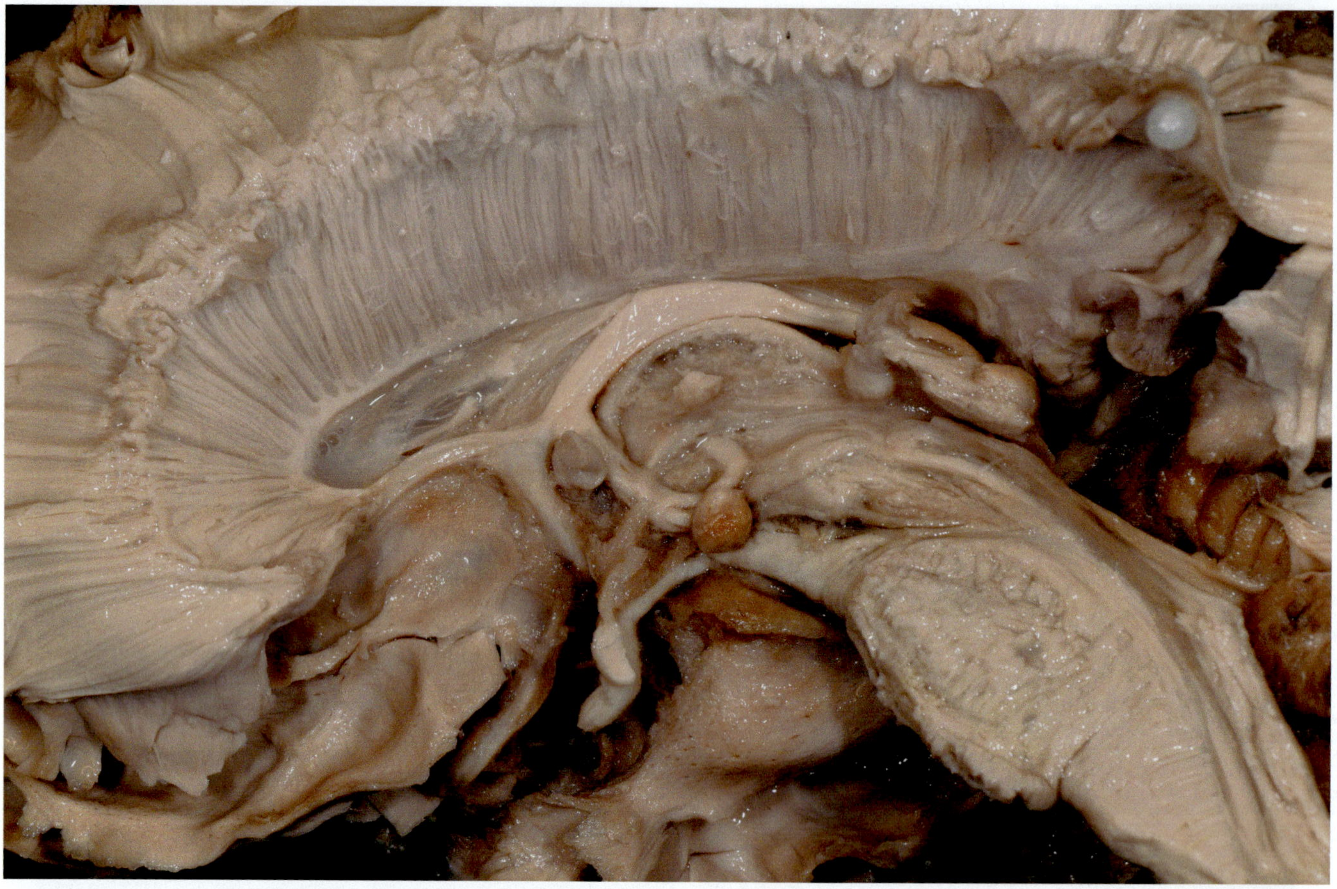

Fig. 8.16 Anatomical preparation of the tegmentum of the mesencephalon. Along the dorsal and ventral limits of the tegmentum, the limbic autonomic connections form the ventral and dorsal tegmental tracts. Removing the medial thalamic nuclei exposes the internal medullary lamina of the thalamus. Fibers connecting the medial and lateral nuclei of the thalamus that do not pass through the internal capsule compose the extrathalamic peduncle. These fibers connect the septal region and preoptic area, the amygdala (part of the ansa peduncularis), and the striatum and pallidum through the ansa lenticularis. Fibers that connect the thalamus with the pallidum through the internal capsule are part of the lenticular fasciculus (not visible in this preparation). Limbic connections with the ventral and dorsal tegmental mesencephalic areas are through the ventral and dorsal tegmental tract. The tegmental tracts are also called the dorsal and ventral longitudinal fasciculus and connect the dorsal and ventral tegmental nuclei. One of the ventral tegmental nuclei is the interpeduncular nucleus. The ventral tegmental tract connects the interpeduncular nucleus with the hypothalamus, mainly its mammillary nuclei (hence also termed the mammillary peduncle), the septal area, the amygdala, and the hippocampal formation (via extensions of the column of the fornix). *ant* anterior, *f* fasciculus, *g* gyrus, *med* medial, *n* nerve, *post* posterior, *s* sulcus, *seg* segment, *sup* superior, *t* tract (©Kadri 2023. All Rights Reserved)

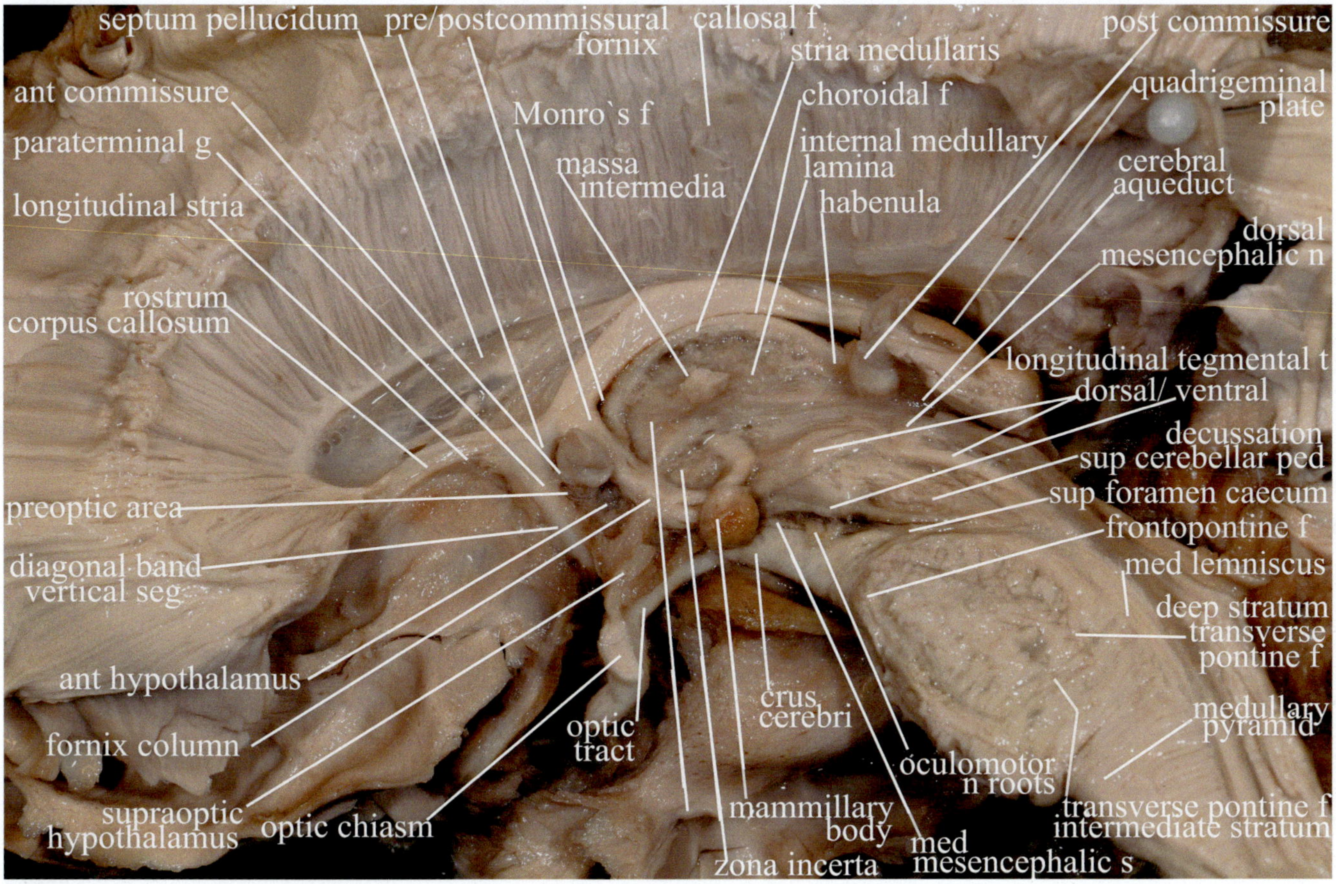

Fig. 8.16 (continued)

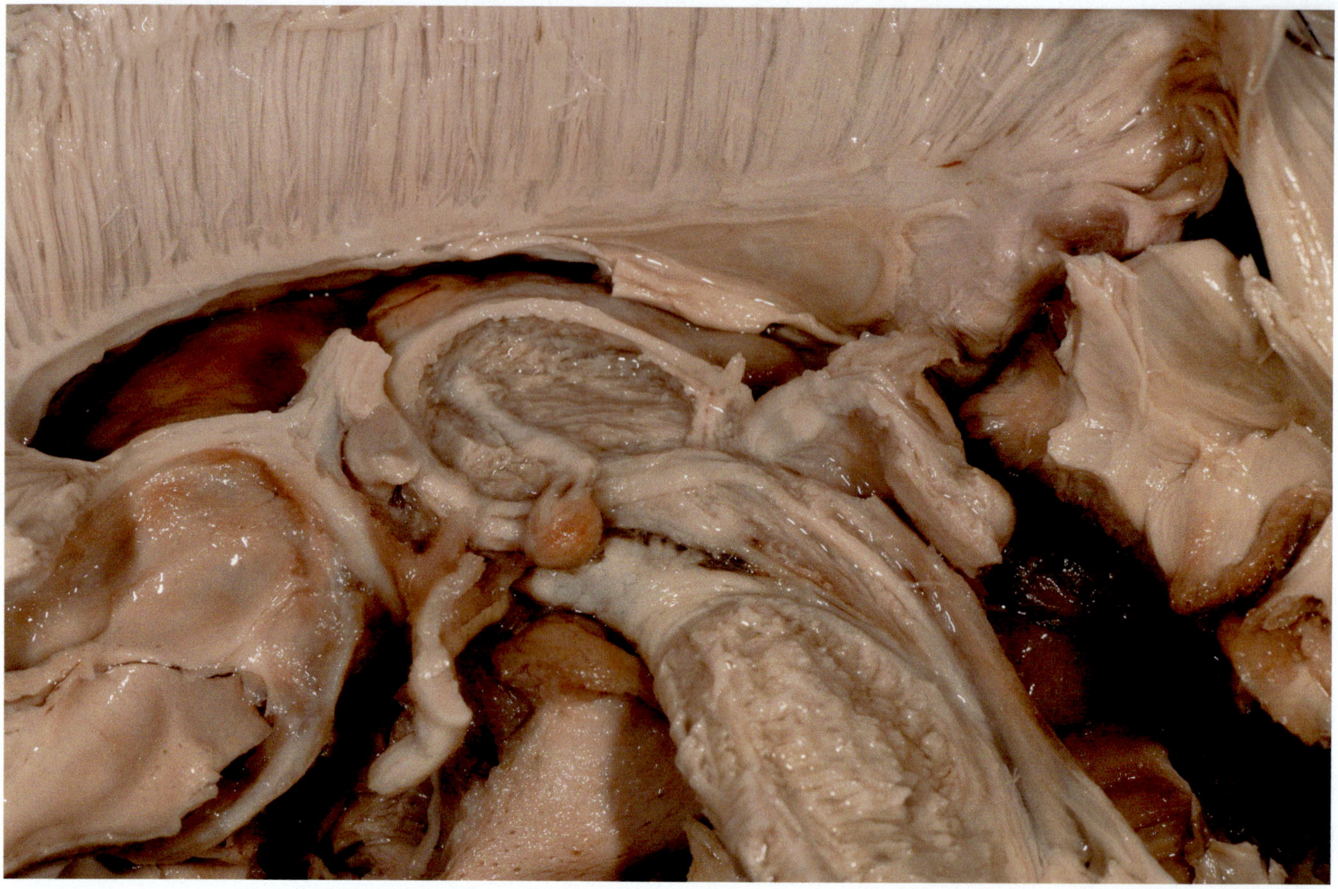

Fig. 8.17 Anatomical preparation of the extracapsular thalamic peduncle, internal medullary lamina of the thalamus, subthalamus with the zona incerta, and the subthalamic area. The ventral and dorsal tegmental tracts are the limbic connections with the tegmental ventral and dorsal limbic nuclei. The limbic nuclei of the tegmentum are embedded in the longitudinal (ventral and dorsal tegmental tracts) and transverse fibers of the limbic system in the tegmentum of the brain stem. This fibrous nuclear arrangement is also known as the reticular substance. The interpeduncular nucleus is one of the ventral tegmental nuclei of the limbic complex and connects with the medial habenular nuclei of the epithalamus via the retroflexus fasciculus. The retroflexus fasciculus is an important landmark separating the thalamic connections and the tegmental connections. Connections lateral to the retroflexus fasciculus are mainly thalamic and subthalamic. Connections medial to the retroflexus fasciculus are mainly limbic. These limbic connections extend from the hypothalamic sulcus, in the floor of the third ventricle, to the medial mesencephalic sulcus, in the external limit of the posterior perforated substance, into the floor of the interpeduncular fossa. In the floor of the third ventricle, these fibers are mainly from the septal region, anterior hypothalamus, amygdala, stria terminalis, and anterior thalamus. These telencephalic diencephalic limbic connections to the tegmentum of the brain stem are named the prosencephalic, medial telencephalic, or medial forebrain band. Part of these connections extend to the pontine tegmentum limbic nuclei embedded in the substantia reticulata. These fibers extend from the hypothalamus to the tegmentum of the pons, passing through or connecting with the dorsal mesencephalic nucleus, and are named the dorsal tegmental tract, dorsal longitudinal fasciculus, or posterior longitudinal fasciculus. *ant* anterior, *g* gyrus, *med* medial, *n* nerve, *ped* peduncle, *s* sulcus, *st* stria, *sup* superior, *t* tract (©Kadri 2023. All Rights Reserved)

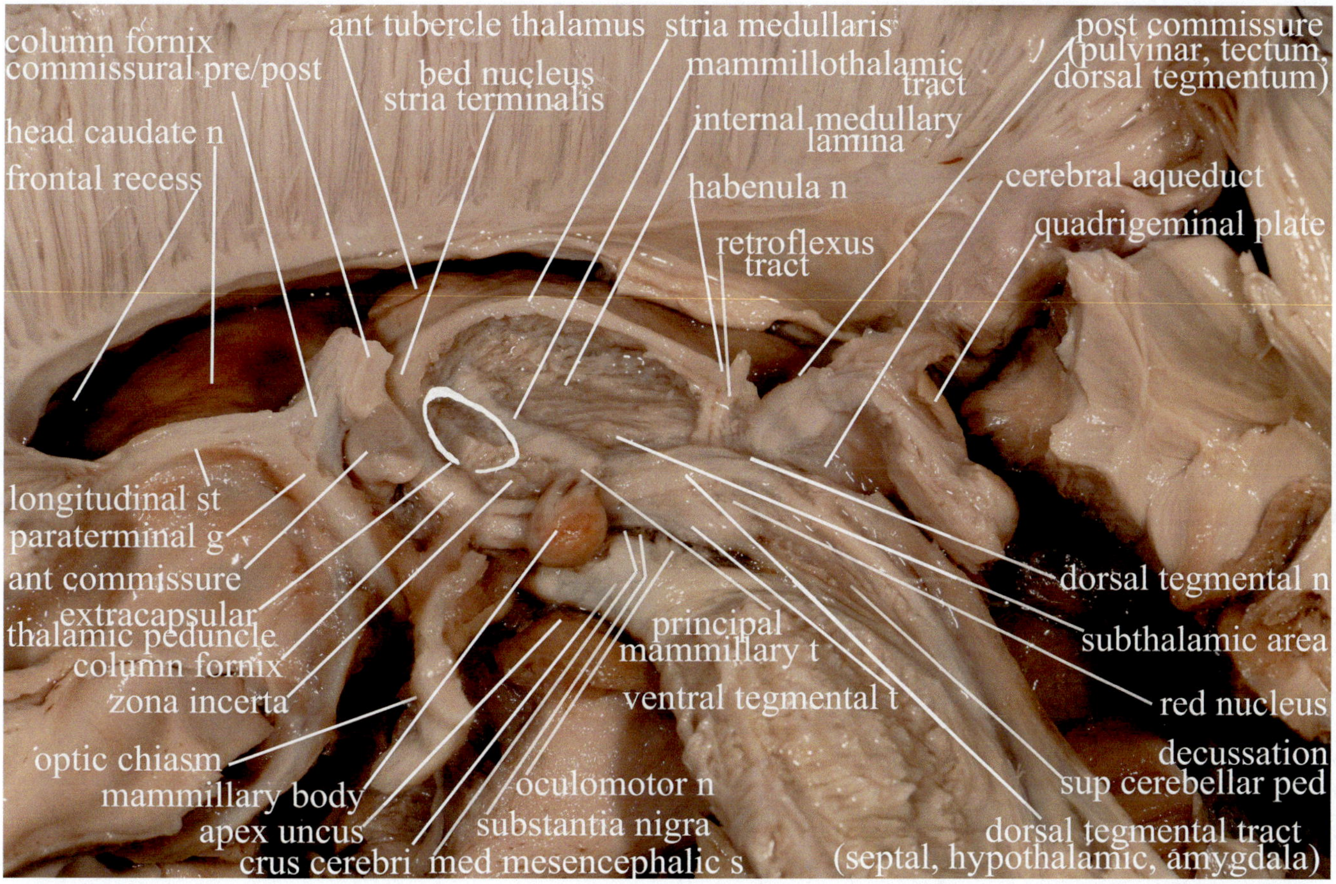

Fig. 8.17 (continued)

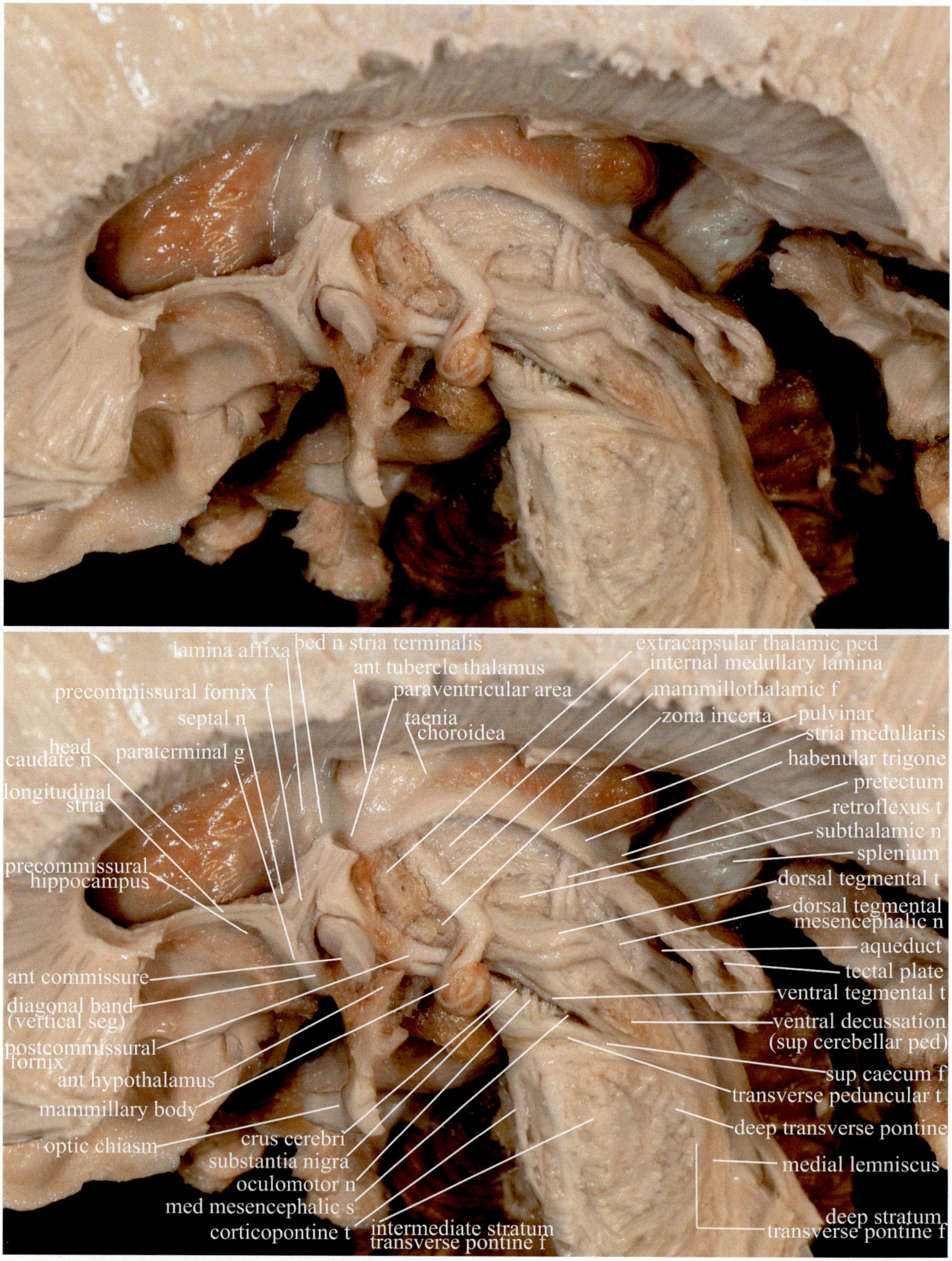

Fig. 8.18 Anatomical preparation of the transitional diencephalic–tegmental area. *ant* anterior, *f* fasciculus, *g* gyrus, *med* medial, *n* nerve, *ped* peduncle, *s* sulcus, *seg* segment, *sup* superior, *t* tract (©Kadri 2023. All Rights Reserved)

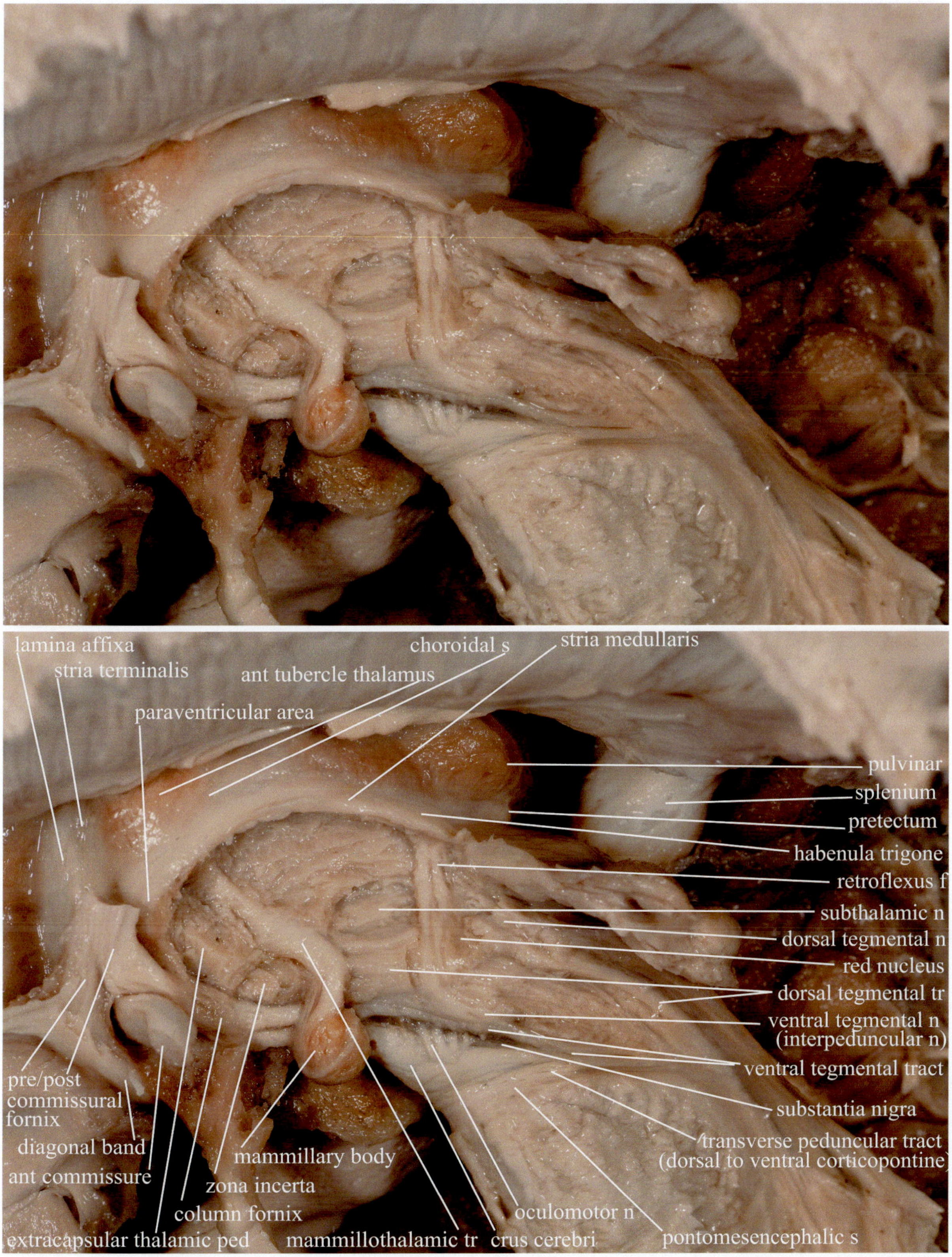

Fig. 8.19 Anatomical preparation of the tegmentum of the mesencephalon. The retroflexus fasciculus is a connection between the epithalamus and the ventral tegmentum of the mesencephalon and is also named the habenulo-interpeduncular nucleus tract. Fibers superficial to the retroflexus fasciculus are mainly limbic and autonomic fibers. Among these fibers, there are connections of the telencephalic basal nuclei (the septal nuclei, the amygdala, and the limbic portions of the ventral striatum [nucleus accumbens and olfactory tubercle]) and the diencephalon with the autonomic centers of the tegmentum of the brain stem. *ant* anterior, *f* fasciculus, *n* nerve, *ped* peduncle, *s* sulcus, *tr* tract (©Kadri 2023. All Rights Reserved)

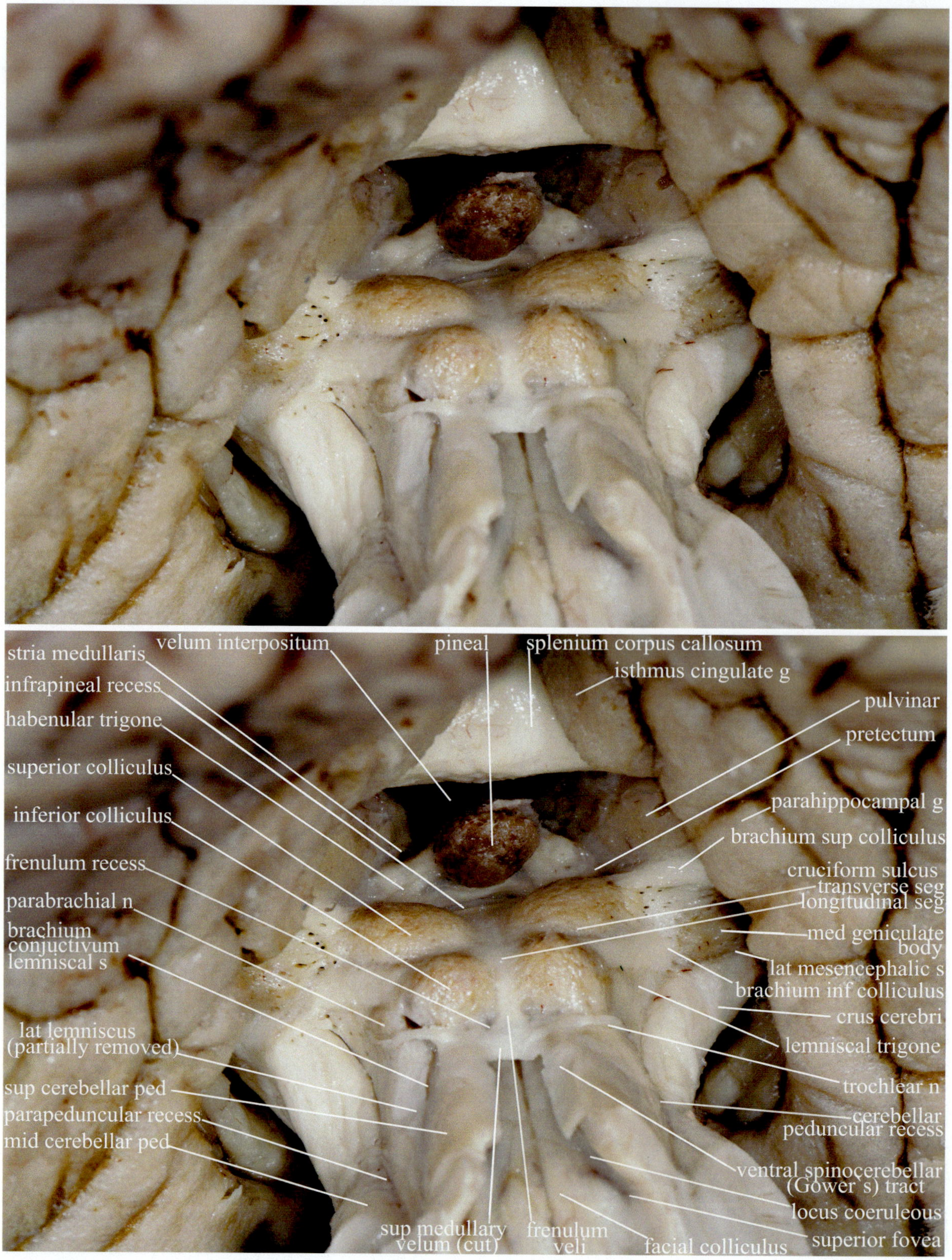

Fig. 8.20 Anatomical preparation of the tectum of the mesencephalon and posterior epithalamus. The tectum of the mesencephalon is composed of the dorsal periaqueductal gray matter and the tectal plate. The tectal plate is composed of the superior and inferior colliculi and their connection. The superior and inferior colliculi are separated by the transverse and longitudinal segments of the cruciform sulcus. The longitudinal segment fuses with the superior medullary velum of the fourth ventricle through the frenulum velum. A depression on each side of the frenulum is known as the frenulum recess. The superior colliculus connects with the medial geniculate body through the brachium of the superior colliculus. The inferior colliculus connects with the lateral geniculate body through the brachium of the inferior colliculi. *g* gyrus, *inf* inferior, *lat* lateral, *med* medial, *mid* middle, *n* nerve, *ped* peduncle, *s* sulcus, *seg* segment, *sup* superior (©Kadri 2023. All Rights Reserved)

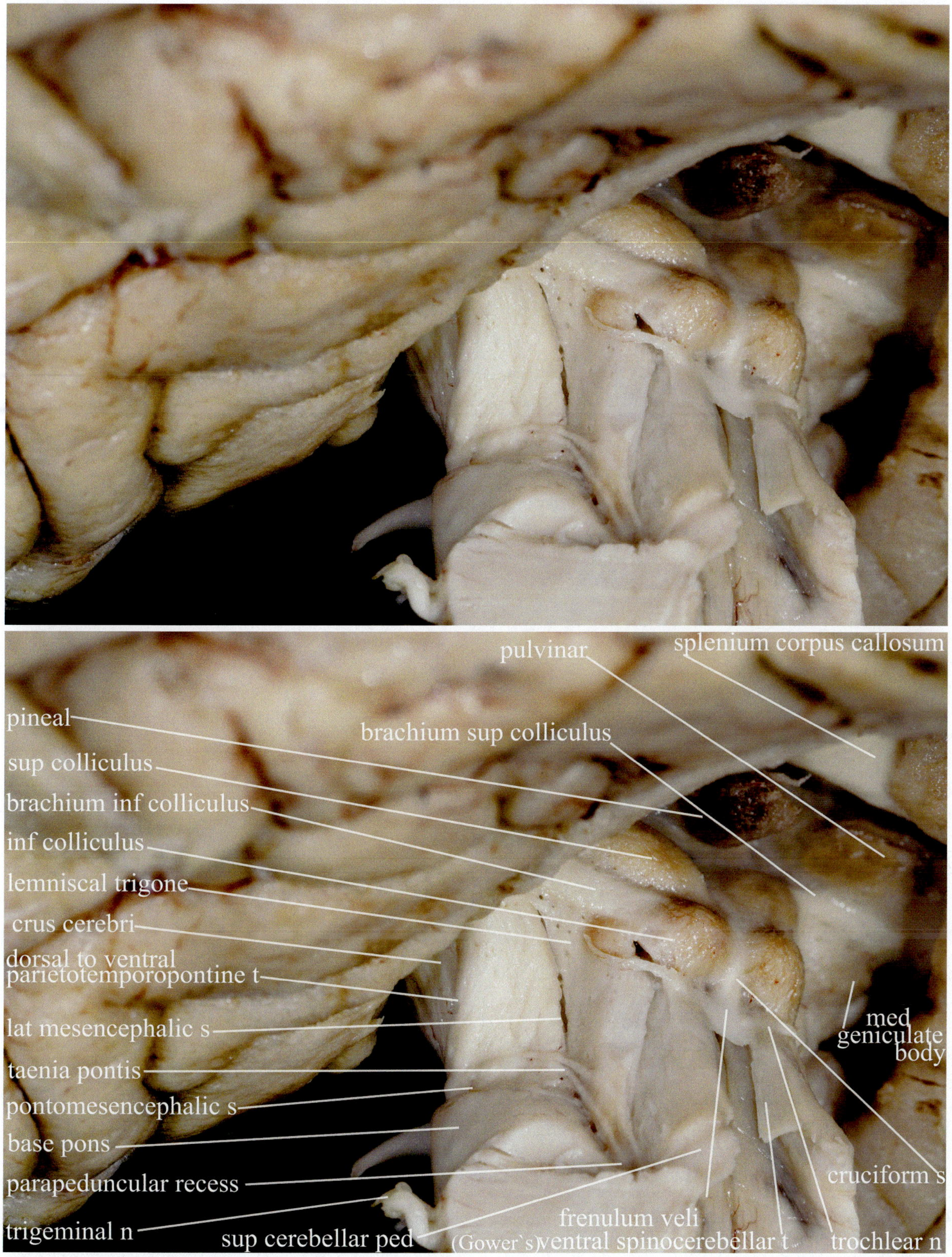

Fig. 8.21 Anatomical preparation of the mesencephalic tectum, lateral view. The fasciculus of Gowers that appears in this dissection is the fibers of the crossed ventral spinocerebellar tract. *inf* inferior, *lat* lateral, *med* medial, *n* nerve, *ped* peduncle, *s* sulcus, *sup* superior, *t* tract (©Kadri 2023. All Rights Reserved)

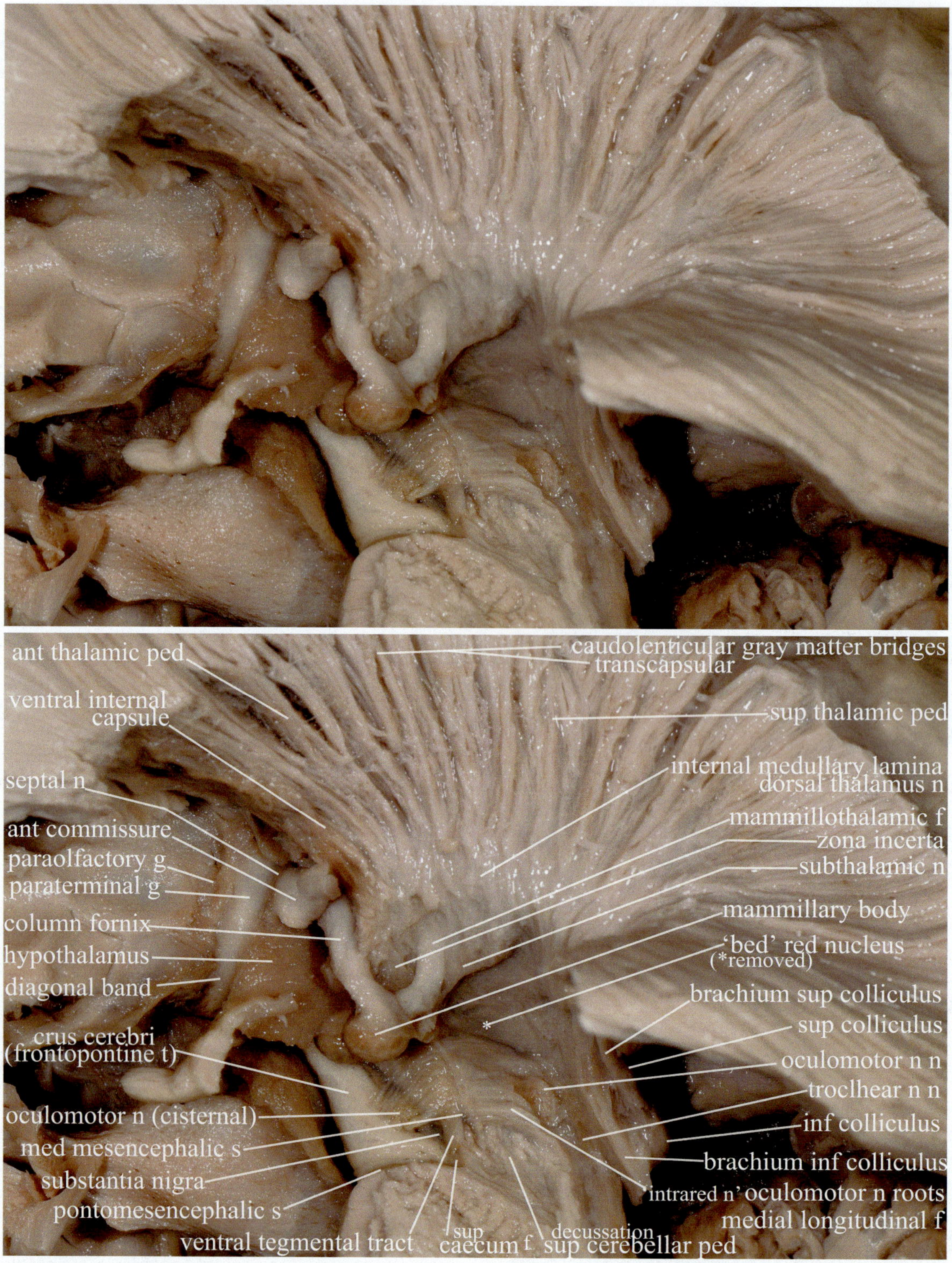

Fig. 8.22 Anatomical preparation of the dorsal mesencephalic tegmentum. The ventromedial substantia nigra presents at the surface on the lateral border of the posterior perforated substance at the floor of the interpeduncular fossa. It delineates the medial mesencephalic sulcus. The intrinsic roots of the oculomotor nerve pass through the ventromedial substantia nigra, and therefore, the medial mesencephalic sulcus is also termed the oculomotor nerve sulcus. As the intrinsic fibers of the oculomotor nerve connect with the oculomotor nuclei at the dorsal tegmentum of the mesencephalon, they pass through the red nuclei. The oculomotor nuclei are at the dorsal mesencephalic tegmentum, at the level of the superior colliculus. The trochlear nuclei, also at the dorsal mesencephalic tegmentum, are at the level of the inferior colliculus. The nuclei of the oculomotor and trochlear nerves are embedded in and by the fibers that form the medial longitudinal fasciculus. *ant* anterior, *f* fasciculus, *g* gyrus, *inf* inferior, *med* medial, *n* nerve, *nn* nucleus, *ped* peduncle, *s* sulcus, *sup* superior, *t* tract

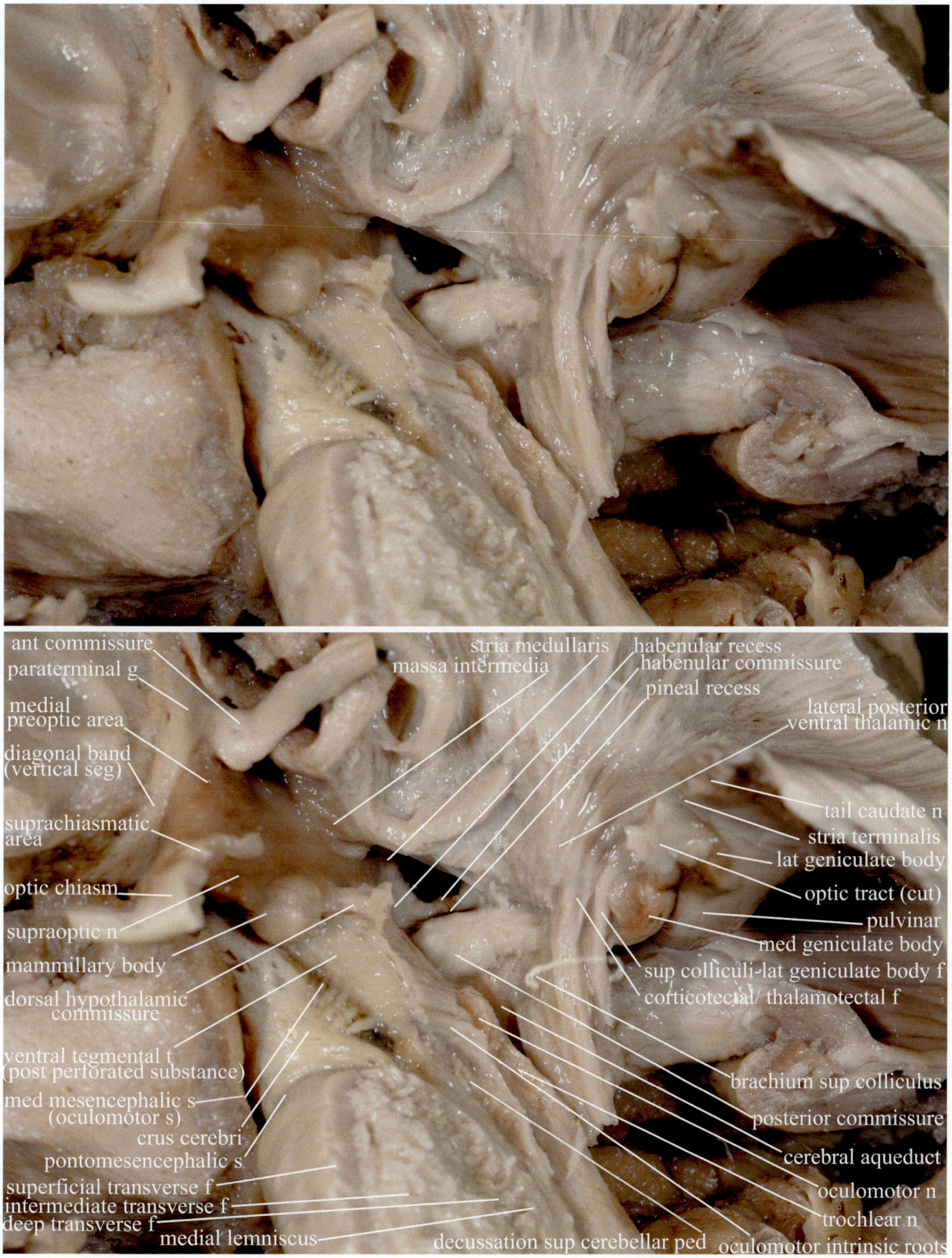

Fig. 8.23 Anatomical preparation of the posterior commissure and tectal plate. The mesencephalic tegmentum is dominated by the red nucleus and the decussation of the superior cerebellar peduncle. The main connections of the red nucleus with the cerebellum are ipsilateral and crossed via the superior cerebellar peduncle. The superior cerebellar peduncle is also connected via the cerebellothalamic connections. *ant* anterior, *f* fasciculus, *g* gyrus, *lat* lateral, *med* medial, *n* nerve, *ped* peduncle, *post* posterior, *s* sulcus, *seg* segment, *sup* superior, *t* tract (©Kadri 2023. All Rights Reserved)

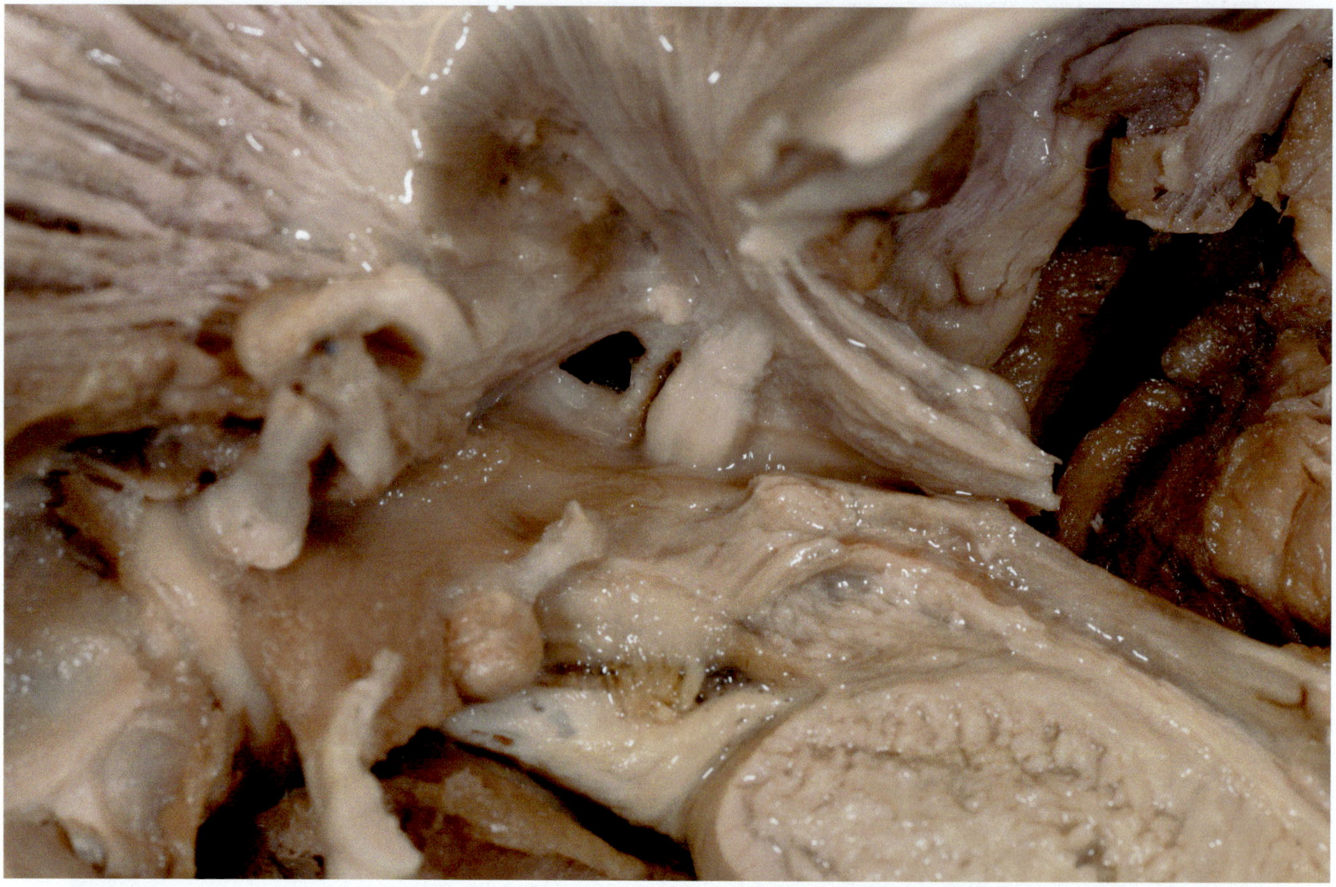

Fig. 8.24 Anatomical preparation of the medial diencephalon and medial mesencephalon along the surfaces of the third ventricle and interpeduncular fossa. The floor of the third ventricle is divided into anterior (ventral and hypothalamic related) and posterior (dorsal or tegmental mesencephalic) surfaces. The posterior floor of the third ventricle is divided by a shallow groove, the continuation of the ventral medial sulcus of the medulla. At the midline, the autonomic fibers of the medial prosencephalic bundle connecting the telencephalic and diencephalic centers with the tegmentum of the brain stem are divided by the red nucleus and the ventral decussations of the superior cerebellar and red nucleus into ventral and dorsal bundles, known as the dorsal and ventral tegmental tracts. The dorsal tegmental tract is composed of connections with the amygdala, ventral striatum, septal region, and hypothalamus. The ventral tegmental tract is composed of connections with the amygdala, septal region, bed nucleus of the stria terminalis, hippocampus (via the fornix), and mainly the mammillary body of the hypothalamus that extends to the ventral tegmentum of the mesencephalon. At the surface, the ventral tegmental tract occupies the floor of the interpeduncular fossa from the midline to the medial mesencephalic sulcus. This sulcus is the lateral surface limit of the tegmentum of the mesencephalon and the origin of many of the oculomotor nerve fibers, therefore, it is also known as the oculomotor nerve sulcus. The floor and lateral walls of the third ventricle are limited by the hypothalamic sulcus, which separates the dorsal thalamus from the hypothalamus and the subthalamus (or ventral thalamus). The main structures of the subthalamus are the subthalamic nucleus, and both are embedded in a dense network of connections of the basal telencephalic nuclei with the subthalamus and dorsal thalamus and epithalamus. *ant* anterior, *f* fasciculus, *g* gyrus, *inf* inferior, *lat* lateral, *med* medial, *n* nerve, *ped* peduncle, *post* posterior, *s* sulcus, *sup* superior, *t* tract

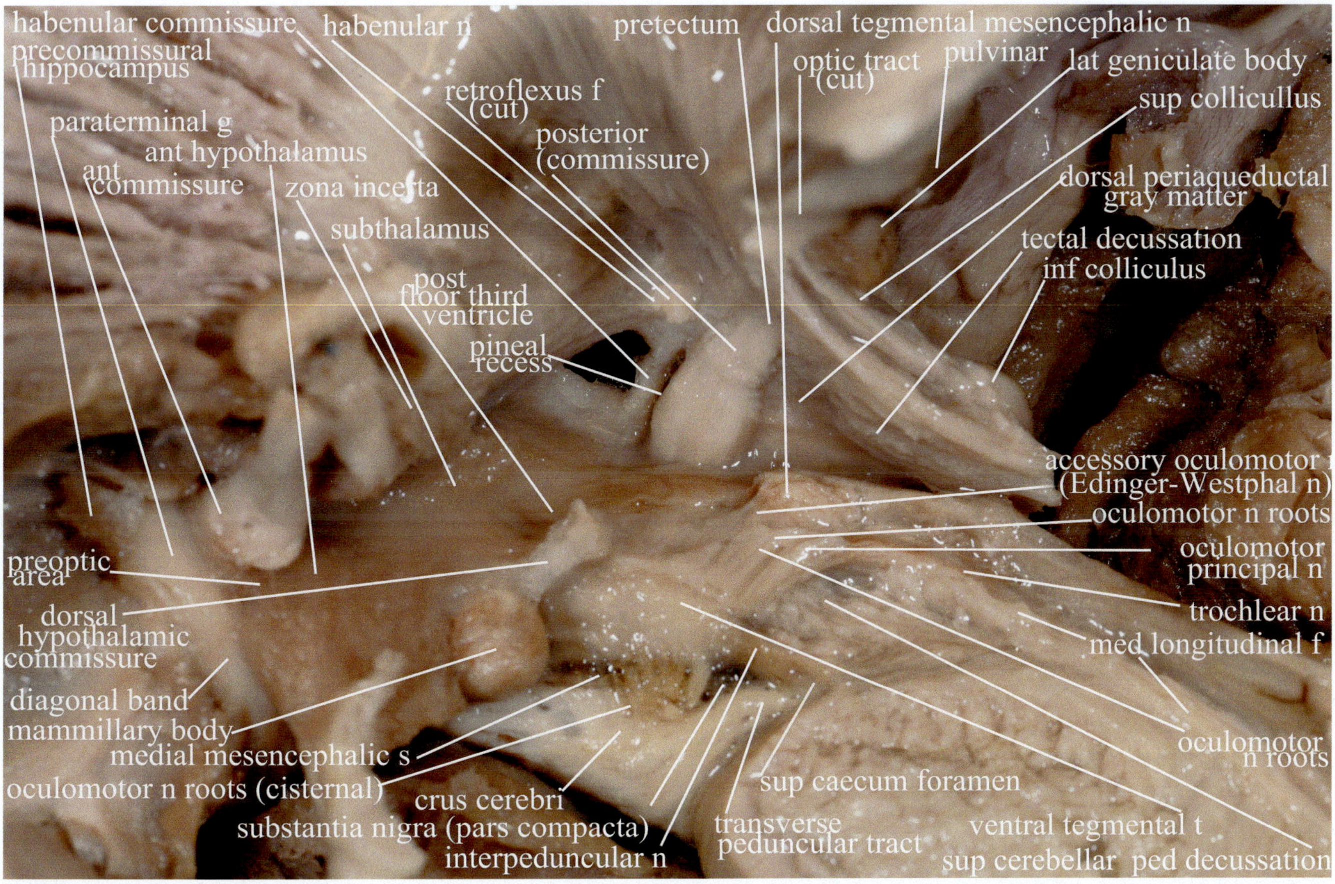

Fig. 8.24 (continued)

The red nucleus is enveloped and traversed by the fibers of the superior cerebellar peduncle (Figs. 8.27 and 8.28). The red nucleus receives its afferents from the cerebellum, striatum, and cerebral cortex. It projects its efferents to the thalamus, the olive, and the spinal cord [1, 2]. The cerebellar afferents are via the superior cerebellar peduncle. The superior cerebellar peduncle decussates at the transition of the pons and mesencephalon in the ventral tegmentum. Its decussation is also known as Wernicke's decussation (Fig. 8.29). The striatal afferents are via the ansa lenticularis and the lenticular fasciculus (Fig. 8.30) [1]. The thalamic efferents constitute the tegmental radiation (also known as the rubrothalamic tract or peduncle) and project toward the lateral ventral thalamus (Fig. 8.31). The rubrospinal tract decussates in the ventral decussation of Forel. The central tegmental tract originates in the thalamus and subthalamus, but a great contingent of its fibers is from the red nucleus. These rubro-olivar projections are ipsilateral to the inferior olive [1].

The oculomotor nerve traverses the magnocellular part of the red nucleus. The habenulo-interpeduncular tract traverses on the medial border of the parvocellular part of the red nucleus. The main efferents of the red nucleus are the central tegmental tract and rubrobulbar and rubrospinal tracts. The rubrobulbar and rubrospinal tracts decussate in the ventral tegmental decussation. The central tegmental tract projects toward the ipsilateral inferior olive in the medulla.

A smaller tract connects the tegmentum of the brain stem. The medial tegmental tract extends from the nuclei of Darkschewitsch, at the superior periaqueductal gray matter, to the inferior olive. It is described as ventral to the medial longitudinal fasciculus. The dorsal longitudinal fasciculus is dorsolateral to the medial longitudinal fasciculus [1].

An important center for coordinated movements of the eye and head is situated between the red nucleus and the central periaqueductal gray matter. It is known as the interstitial nucleus of Cajal and harbors bilateral connections to the nuclei of the oculomotor and trochlear nerves, the reticular formation, and the vestibular nuclei. The interstitiospinal tract is part of the medial longitudinal fasciculus system.

The habenulo-interpeduncular tract is also known as the retroflexus fasciculus of Meynert and connects the habenular nuclei of the diencephalon to the interpeduncular nuclei of the mesencephalon [2].

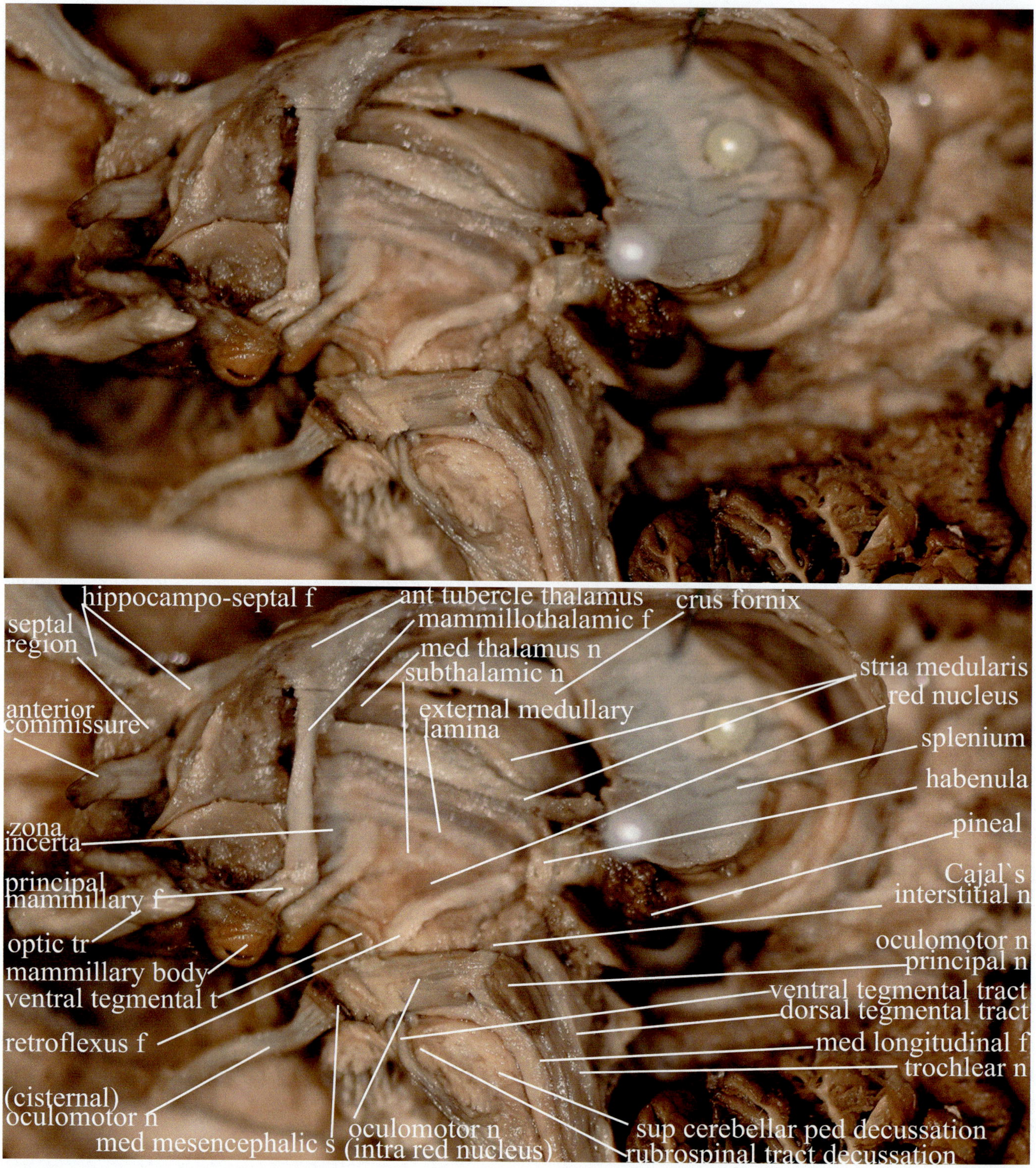

Fig. 8.25 Anatomical preparation of the principal nuclei of the oculomotor and trochlear nerves after the removal of the ipsilateral red nucleus. The intrinsic fibers of the oculomotor nerve pass through the medial reticulata (also named the parva) portion (pars) of the red nucleus. These fibers emerge in the lateral limit of the posterior perforated substance on the floor of the interpeduncular fossa, which is the medial mesencephalic sulcus—thus, it was named oculomotor nerve sulcus. The oculomotor nerve fibers are dorsolateral to the limbic connections of the ventral tegmentum of the mesencephalon. The ventral limbic tegmental connections are the ventral tegmental tract and the retroflexus fasciculus, with the ventral tegmental nuclei. The interpeduncular nucleus is one of these nuclei. The ventral tegmental tract contains septal, amygdala, hippocampal (via the fornix), and diverse hypothalamic connections. However, one of these recognized connections with the mammillary body, the mammillary peduncle, is also used as a synonym. *ant* anterior, *f* fasciculus, *med* medial, *n* nerve, *nn* nucleus, *ped* peduncle, *s* sulcus, *sup* superior, *tr* tract (©Kadri 2023. All Rights Reserved)

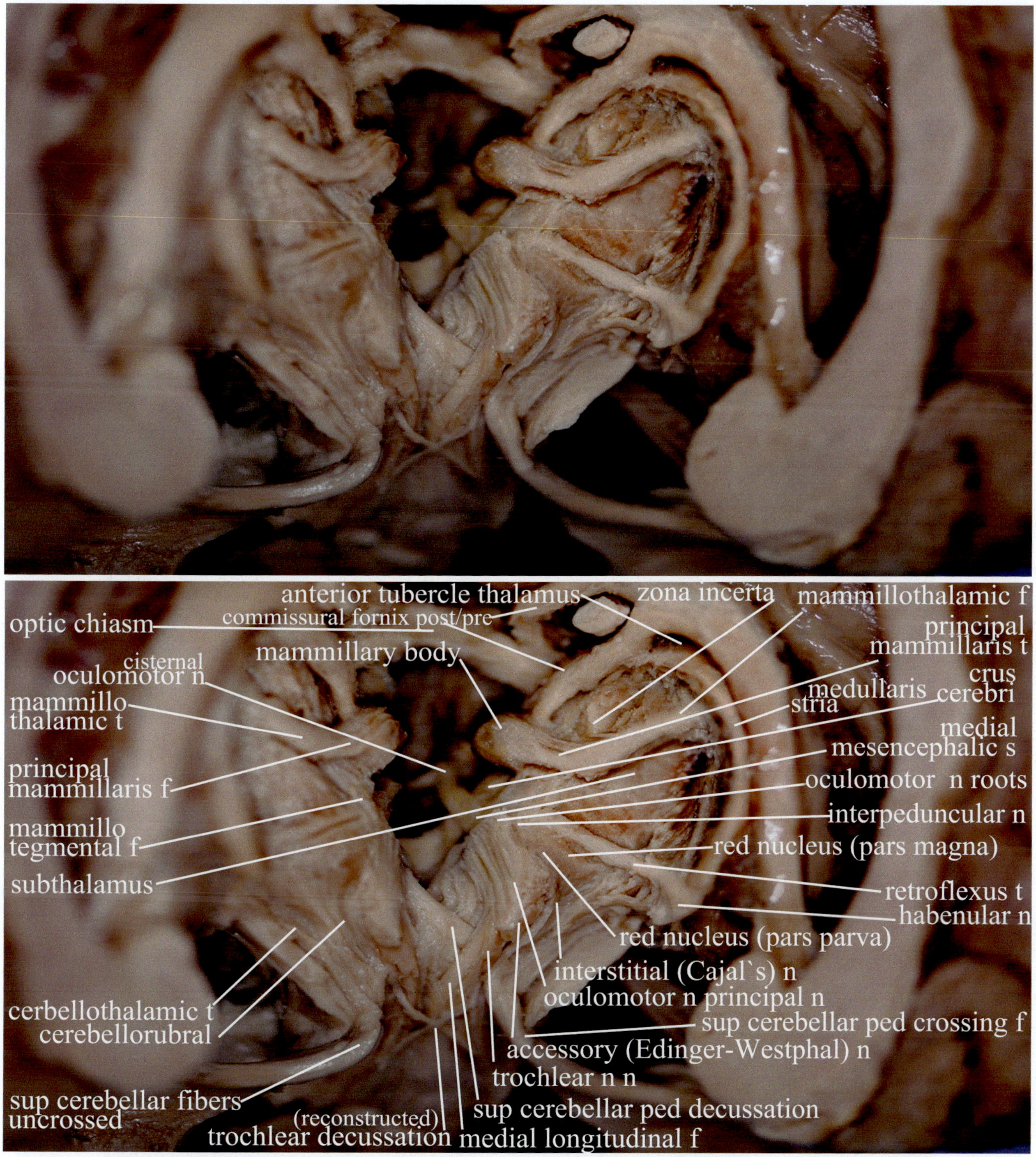

Fig. 8.26 Anatomical preparation of the commissural fibers of the tegmentum of the brain stem. The superior cerebellopeduncle connects the cerebellum (mainly the dentate nucleus) with the mesencephalon (red nucleus, oculomotor nerve nuclei, pretectum, and superior colliculi) and thalamus (subthalamus and ventral posterior nuclei of the lateral thalamus and epithalamus). Most of the fibers of the superior cerebellar peduncle that englobe or pass through the red nucleus connect with the red nucleus. These fibers extend connections to the ventral lateral thalamus and pass through Forel's fields of the rostral tegmentum of the mesencephalon and subthalamus transition. There, they connect with the zona incerta nucleus but mainly with the subthalamic nucleus in its extension to the posterior ventral nuclei of the lateral dorsal thalamus. *f* fasciculus, *n* nerve, *nn* nucleus, *ped* peduncle, *s* sulcus, *sup* superior, *t* tract (©Kadri 2023. All Rights Reserved)

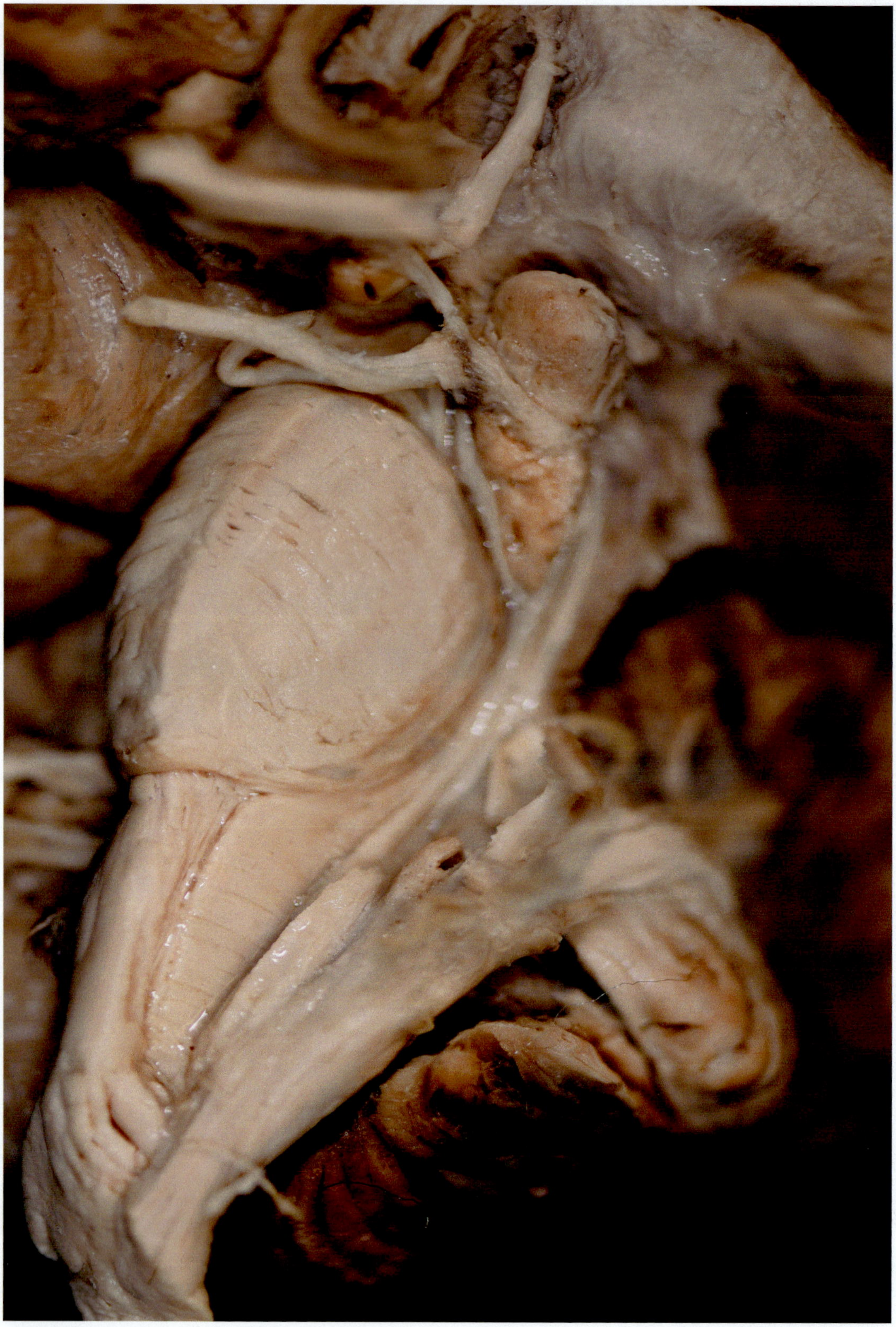

Fig. 8.27 Anatomical preparation of the base and tegmentum of the brain stem. The compact and reticulate portions of the red nucleus are exposed through the removal of the ipsilateral fibers of the superior cerebellar peduncle. The superior cerebellar peduncle is composed of crossed and uncrossed dentate-rubral and dentate-thalamic fibers. The compact portion is traversed by roots of the principal nucleus of the oculomotor nerve. *ant* anterior, *ext* external, *f* fasciculus, *g* gyrus, *inf* inferior, *med* medial, *n* nerve, *nn* nucleus, *ped* peduncle, *post* posterior, *s* sulcus, *sup* superior, *t* tract (©Kadri 2023. All Rights Reserved)

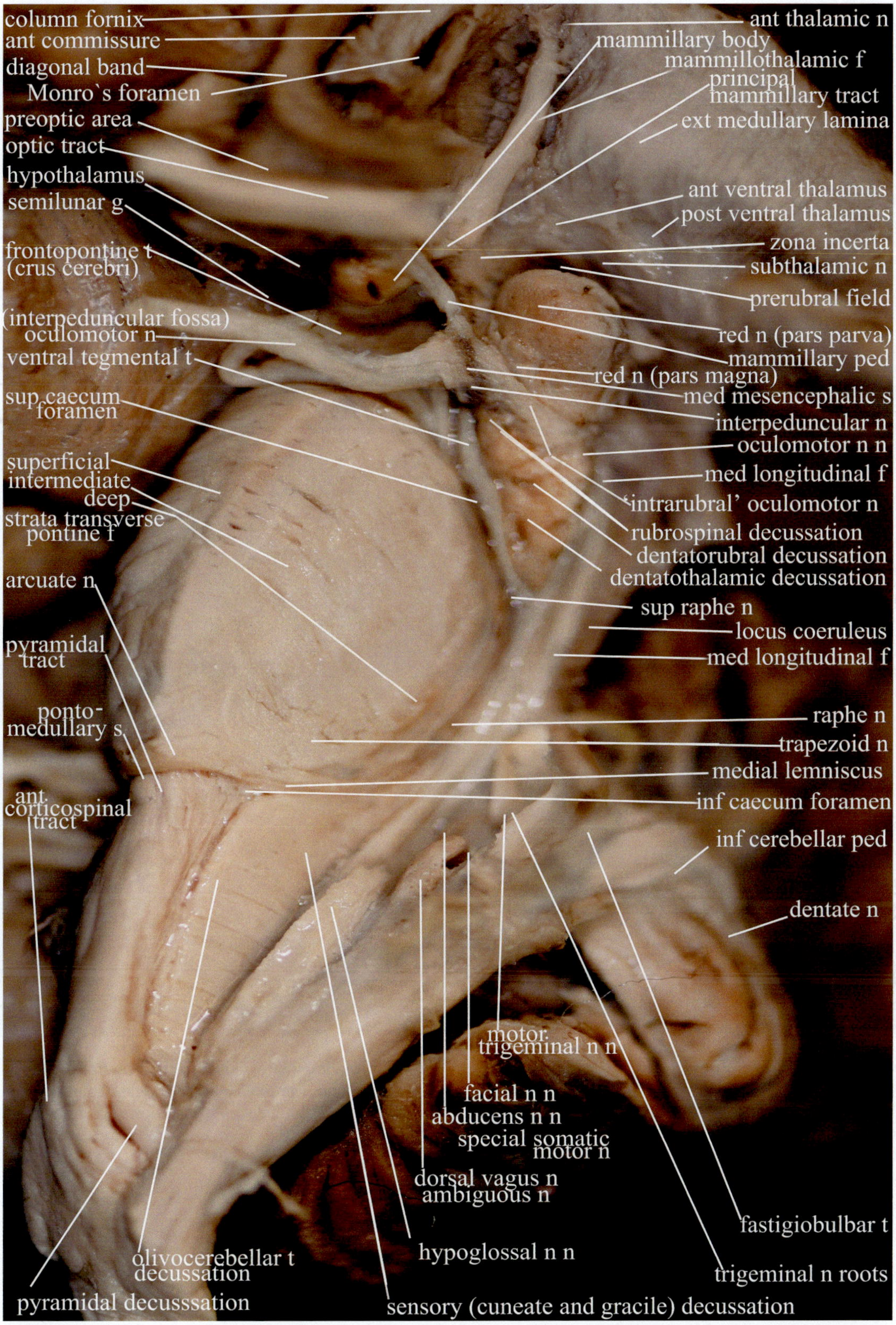

Fig. 8.27 (continued)

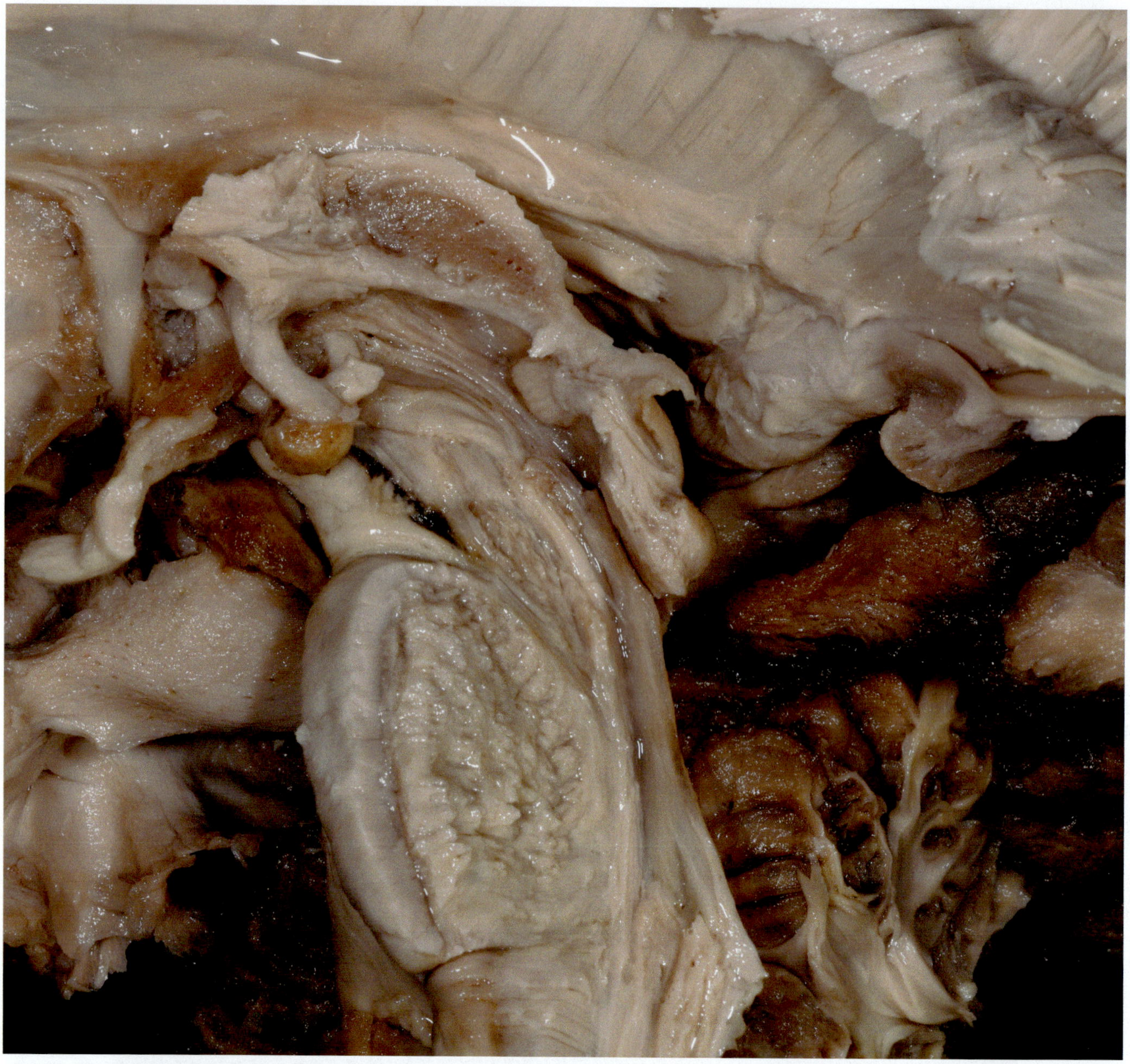

Fig. 8.28 Anatomical preparation of the diencephalic-mesencephalic and the mesencephalic-pontine intrinsic midline limits. The transition of the fibers of the extracapsular thalamic peduncle to the fibers of the prosencephalic tract are the most superficial medial border of the fiber system between the thalamus and the mesencephalon. The most proximal fibers of the superior cerebellar decussation are the intrinsic mesencephalic–pontine border. *ant* anterior, *f* fasciculus, *med* medial, *n* nerve, *nn* nucleus, *ped* peduncle, *post* posterior, *s* sulcus, *t* tract (©Kadri 2023. All Rights Reserved)

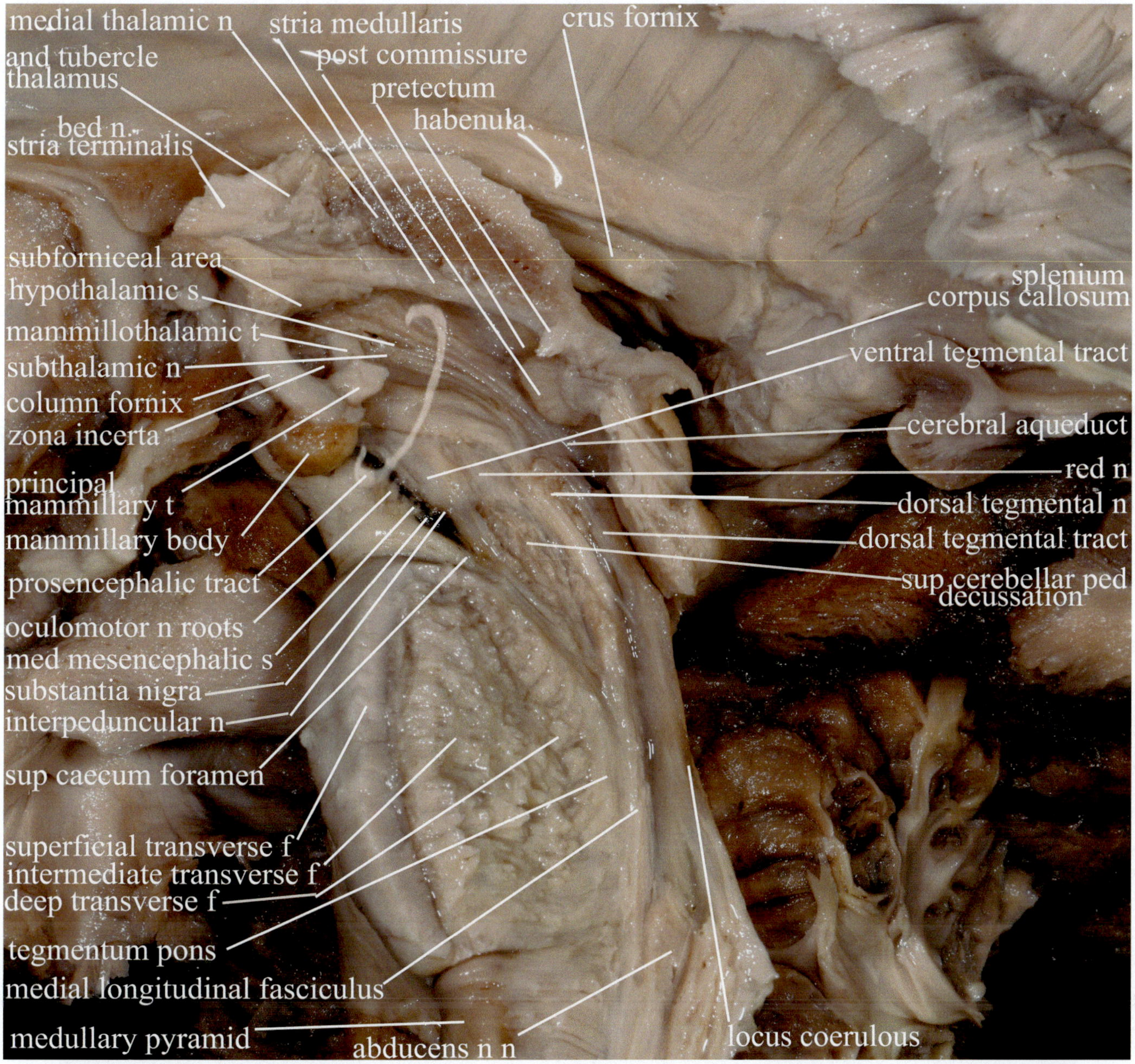

Fig. 8.28 (continued)

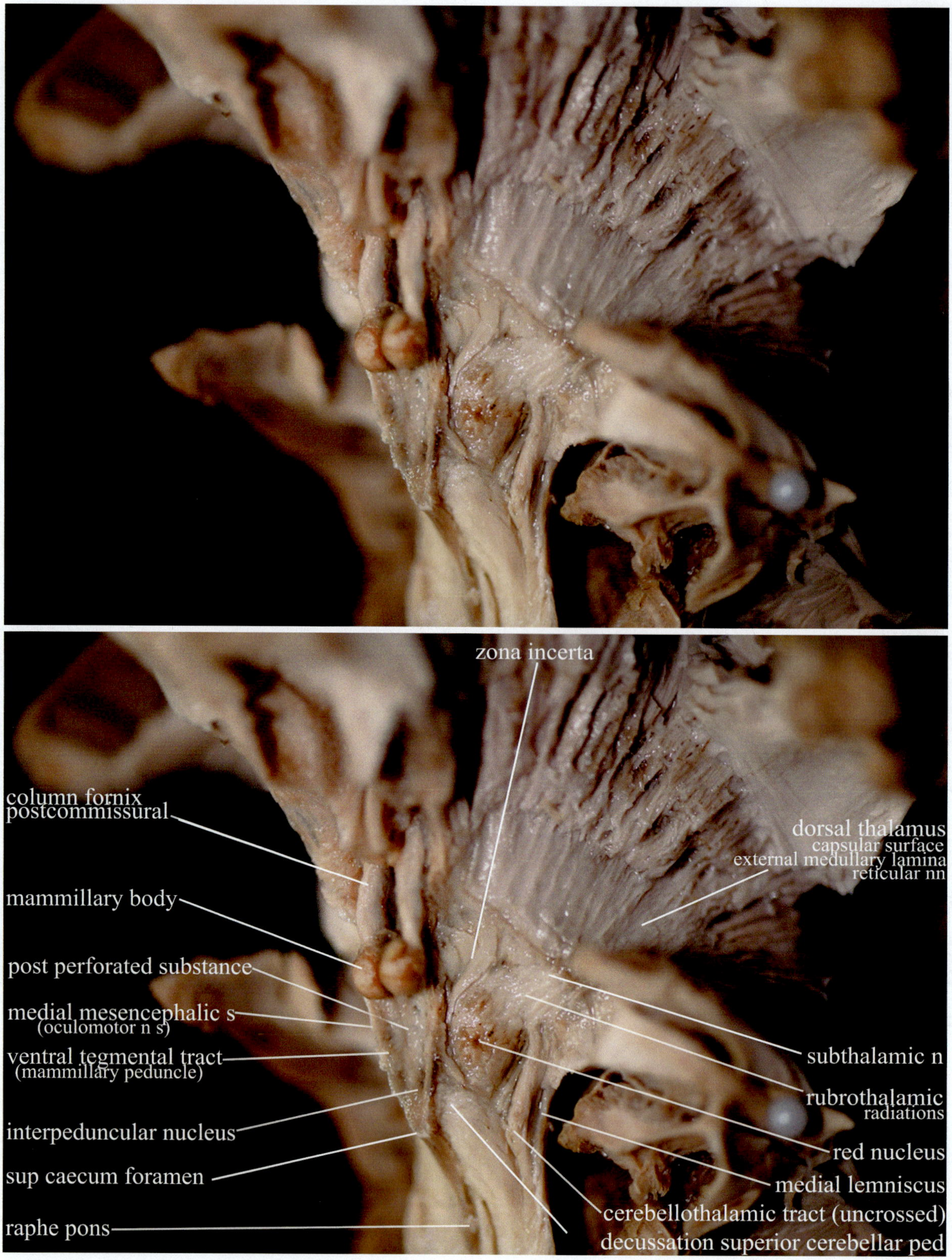

Fig. 8.29 Anatomical preparation of the mesencephalic tegmentum. The crus cerebri and the substantia nigra have been removed, and the ventral surface of the tegmentum is exposed. The red nucleus is the largest compact neuronal mass in the mesencephalon, and its connections dominate the tegmentum of the mesencephalon. The main connections with the caudal brain stem are via the central tegmental tract, while the main connections with the cerebellum are via the superior cerebellar peduncle, the brachium conjunctivum. These connections are ipsilateral and crossed. The superior cerebellar peduncle also contains ipsilateral and crossed cerebellothalamic connections. *n* nerve, *nn* nucleus, *ped* peduncle, *post* posterior, *s* sulcus, *sup* superior (©Kadri 2023. All Rights Reserved)

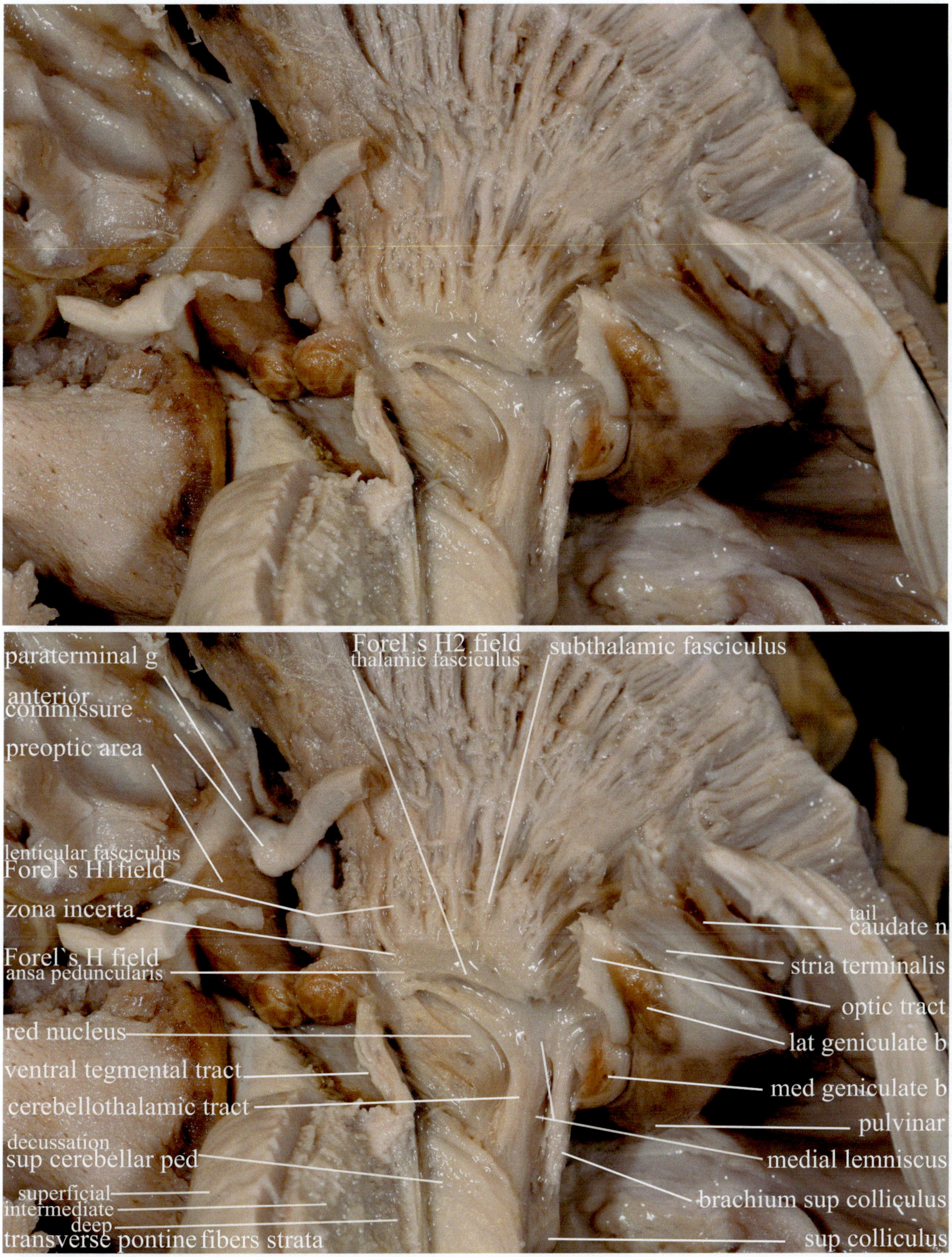

Fig. 8.30 Anatomical preparation of the transition of the subthalamus and the mesencephalic tegmentum. The connections of the cerebellum and thalamus, relaying or passing through the red nucleus, dominate the ventral mesencephalic tegmentum. *b* body, *g* gyrus, *lat* lateral, *med* medial, *n* nerve, *ped* peduncle, *sup* superior (©Kadri 2023. All Rights Reserved)

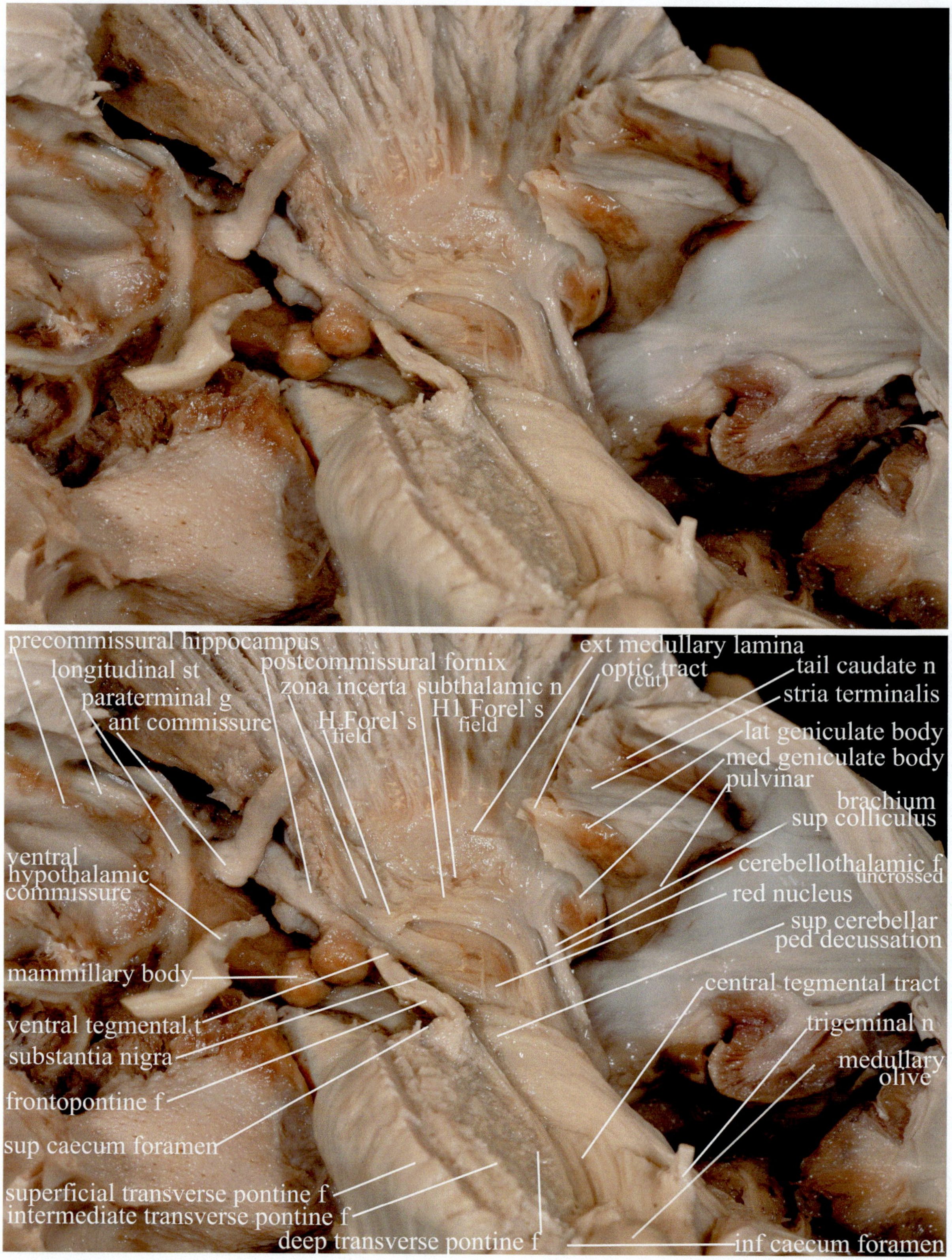

Fig. 8.31 Anatomical preparation of the subthalamus and ventral tegmental mesencephalic region. The removal of the pars reticulata and compacta of the substantia nigra and its connections reveals the crossing fibers of the superior cerebellar peduncle connecting with the contralateral red nucleus and thalamus, as well as the ipsilateral cerebellar thalamic connections. The cerebellar, medullary (inferior olive), and spinal thalamic connections are part of the anterolateral ascending system. The zona incerta is continuous with the external medullary lamina, which is embedded in the reticular nucleus of the thalamus. *ant* anterior, *ext* external, *f* fasciculus, *g* gyrus, *inf* inferior, *lat* lateral, *med* medial, *n* nerve, *ped* peduncle, *st* stria, *sup* superior, *t* tract (©Kadri 2023. All Rights Reserved)

The Pons

The pons is also named the protuberance, the annular protuberance, the pons Varolii, and the metencephalon. It is situated between the mesencephalon and the medulla and derives its name from the arrangement of its superficial fibers. Its superficial transverse fibers bridge over the longitudinal fibers that transverse the pons from the medulla and mesencephalon and resemble a bridge (Fig. 8.32).

The pons is separated inferiorly from the medulla by the pontomedullary sulcus, also called the inferior pontine sulcus. This sulcus presents, from midline to lateral, the inferior caecum foramen, the supra-olivary fossae, and the lateral medullary recess. Superiorly, the pons is separated from the mesencephalon by the mesencephalic pontine sulcus. This sulcus is also known as the superior pontine sulcus. The pons has an irregular conical shape, the cephalic larger than the caudal portion, with anterior and posterior surfaces (Fig. 8.32).

The anterior surface is concave and known as the clival or basilar surface. This surface presents a median sulcus, the basilar sulcus. This sulcus is commonly related to the basilar artery, although not produced by its impression. Lateral to the basilar sulcus, on both the sides, the pyramidal eminence corresponds to the descending fibers of the pyramidal tract. Lateral to the pyramidal eminence, the apparent origin of the trigeminal nerve is the arbitrary limit between the pons and the middle cerebellar peduncle. The trigeminal nerve consists of a sensory and a motor root. The sensory root is larger and more medial. The motor root is smaller and located lateral and slightly superior. The superficies of the anterior surface is formed by the fibers of the middle cerebellar peduncle that cross the midline. These fibers are known as transverse pontine fibers (Fig. 8.33). The superficial transverse pontine fibers can be divided into superior, middle, and inferior groups. The superior group passes over the trigeminal emergence and runs toward the posterior portion of the middle cerebellar peduncle. The inferior group passes under the trigeminal emergence and runs toward the anterior and inferior surface of the middle cerebellar peduncle. The middle group turns inferior and posterior in the direction of the facial nerve as the fibers reach the inferior surface of the trigeminal emergence (Fig. 8.34). The intermediate transverse fibers intercalate with the longitudinal projection fibers of the telencephalon. The deep transverse fibers are mainly the decussation of the acoustic radiation that passes through the trapezoid body (Figs. 8.35 and 8.36) [1, 2].

The posterior surface is the ventricular surface and is known as the superior, pontine, or protuberance triangle of the floor of the fourth ventricle. This surface is covered by the cerebellum. The limit between the pons and the medulla on the ventricular surface is a line passing through the upper limit of the foramen of Luschka at the lateral recess of the fourth ventricle. (The fourth ventricle is presented separately in this atlas.)

The internal organization of the pons is quite complex. It is divided into ventral, intermediate, and dorsal zones. The ventral zone is also known as the pes or base of the pons and contains the superficial and intermediate transverse fibers and the pontine nuclei. The intermediate zone is also known as the tegmentum and contains the pontine reticular substance and the upper portion of the fourth ventricle structures. The dorsal zone corresponds to the cerebellum. At the midline, the superior medullary velum constitutes the superior roof of the fourth ventricle. The superior velum contains the lingula, which is the first of the vermin lobules without hemispheric correspondence. The decussation of the fourth nerve within the superior medullary velum marks the posterior limit between the pons and the mesencephalon posteriorly. (The cerebellum is discussed separately in this atlas.)

The base of the pons is more compact and continuous with the cerebral peduncle superiorly and with the medullar pyramids inferiorly (Fig. 8.37). It is composed of longitudinal fibers, transverse fibers, and pontine nuclei (Fig. 8.38). The longitudinal fibers are the corticospinal, corticopontine, and corticonuclear tracts. Corticospinal fibers traverse the base of the pons; corticopontine fibers connect with the ipsilateral pontine nuclei, a relay to cerebellar connections. The corticonuclear tract connects with the ipsilateral and contralateral nuclei of the cranial nerves. The transverse fibers are the fibers of the middle cerebellar peduncle and vestibular nuclei. The pontine nuclei are dispersed around the transverse fibers. They are the motor nuclei of the pons that connect via ipsilateral corticopontine fibers and the contralateral neocerebellum via the transverse intermediate fibers (Fig. 8.39). The corticobulbar fibers are dorsal to the corticospinal tract and connect in similar proportions to the ipsilateral and contralateral nuclei of the cranial nerve (Figs. 8.40 and 8.41) [1, 2].

The tegmentum of the pons is more complex (Figs. 8.42 and 8.43). It is occupied by the reticular formation of the pons (also known as the substantia of Deiters) [1, 2]. The reticular formation is composed of a net of columns of gray matter interposed by longitudinal fibers. These fibers are crossed by arcuate and radiated fibers. The longitudinal fibers are the longitudinal fasciculus of the reticular formation, the medial longitudinal fasciculus, the posterior longitudinal fasciculus, and the central tegmental tract (Fig. 8.44). The posterior longitudinal fasciculus is also named the dorsal tegmental tract. The arcuate fibers form the trapezoid body. The radiated fibers penetrate the pons and correspond mainly to the fibers of the cranial nerves. The radiated fibers of the pons are the intrinsic fibers of the trigeminal, abducens, facial, vestibular, and, in some ways, the cochlear

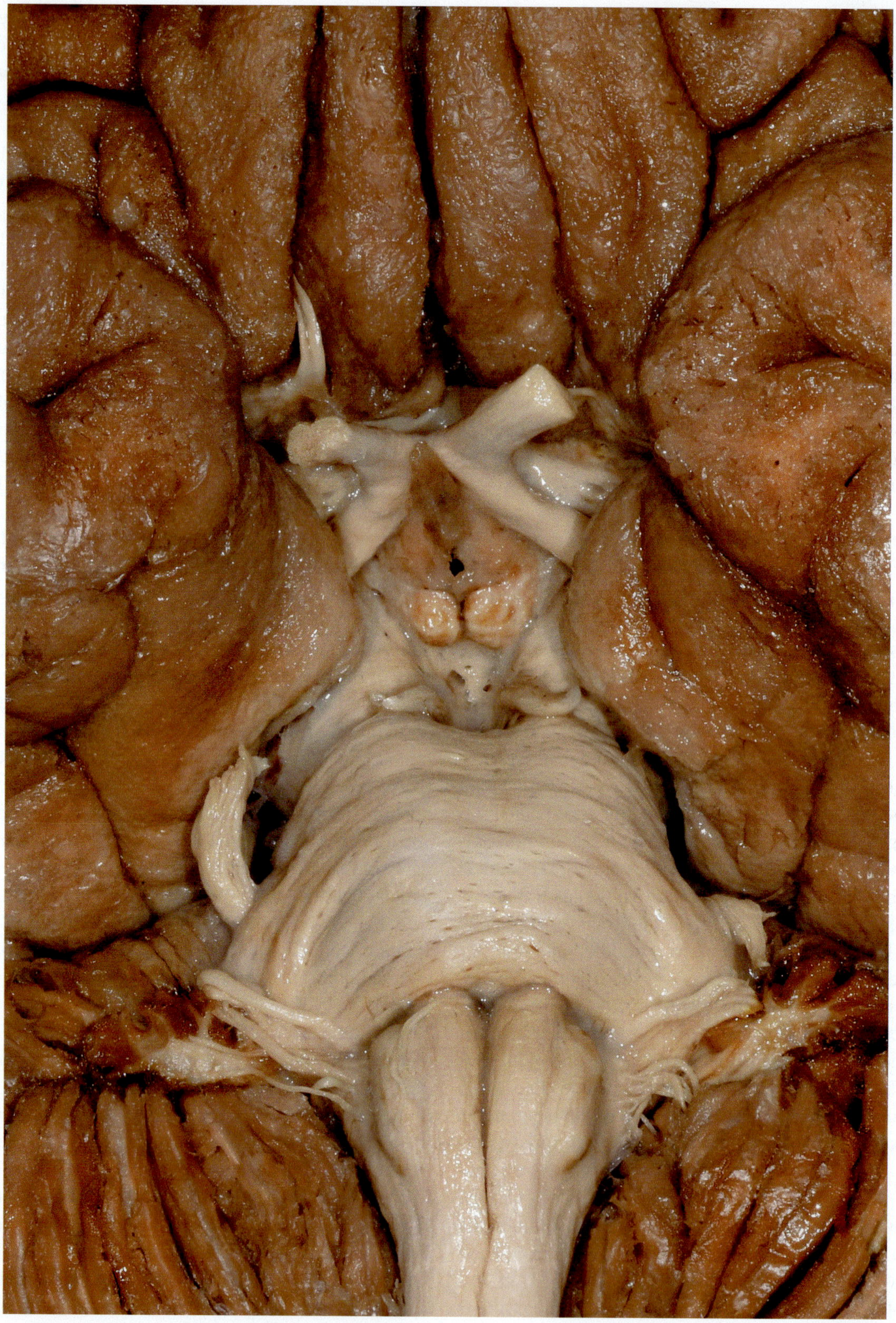

Fig. 8.32 Anatomical preparation of the ventral brainstem. The ventral pons has a conical shape. On the midline, the shallow basilar sulcus extends from the inferior to the superior caecum foramina and separates the paired paramedian pyramidal eminences. The superficial transverse pontine fibers can be classified into superior, middle, and inferior groups. *ant* anterior, *f* fasciculus, *g* gyrus, *gg* gyri, *inf* inferior, *lat* lateral, *med* medial, *n* nerve, *ped* peduncle, *post* posterior, *s* sulcus, *seg* segment, *sup* superior, *t* tract (©Kadri 2023. All Rights Reserved)

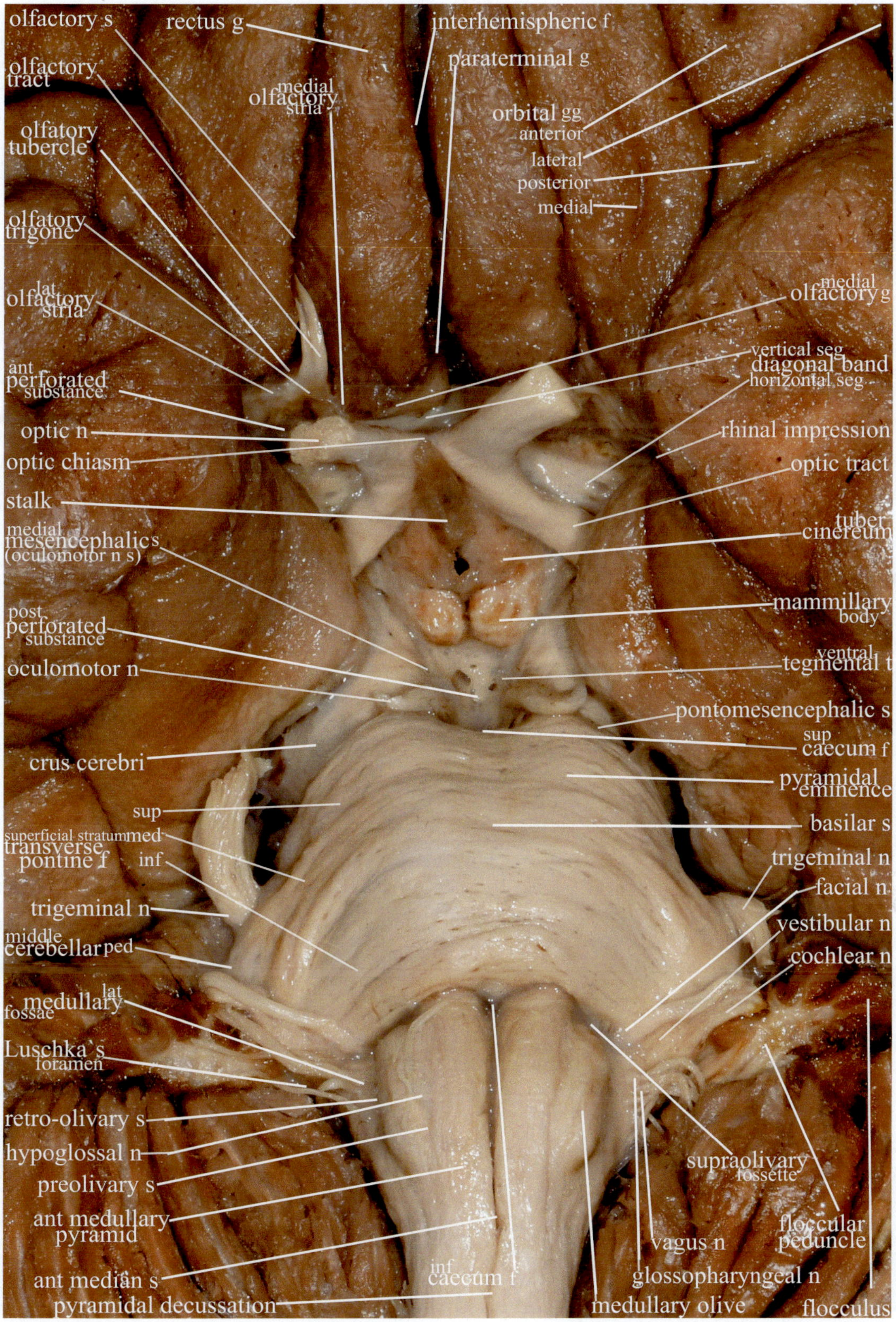

Fig. 8.32 (continued)

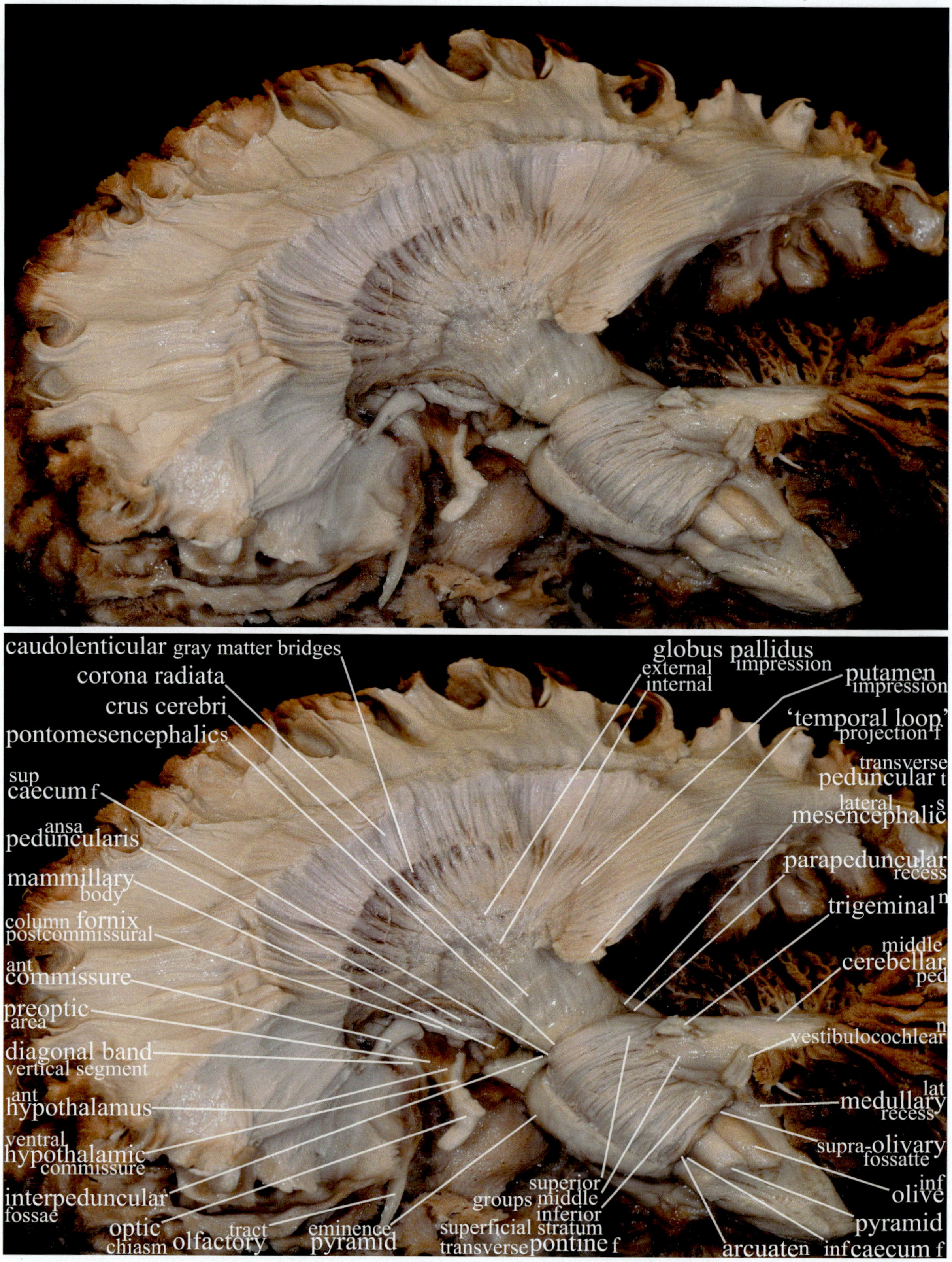

Fig. 8.33 Anatomical preparation of the superficial transverse fibers of the pons. The superficial transverse fibers cover the longitudinal telencephalic projection fibers and are formed by the pontine–cerebellar connections that cross the midline into the middle cerebellar peduncle. *ant* anterior, *f* fasciculus, *inf* inferior, *lat* lateral, *ped* peduncle, *s* sulcus, *sup* superior, *t* tract (©Kadri 2023. All Rights Reserved)

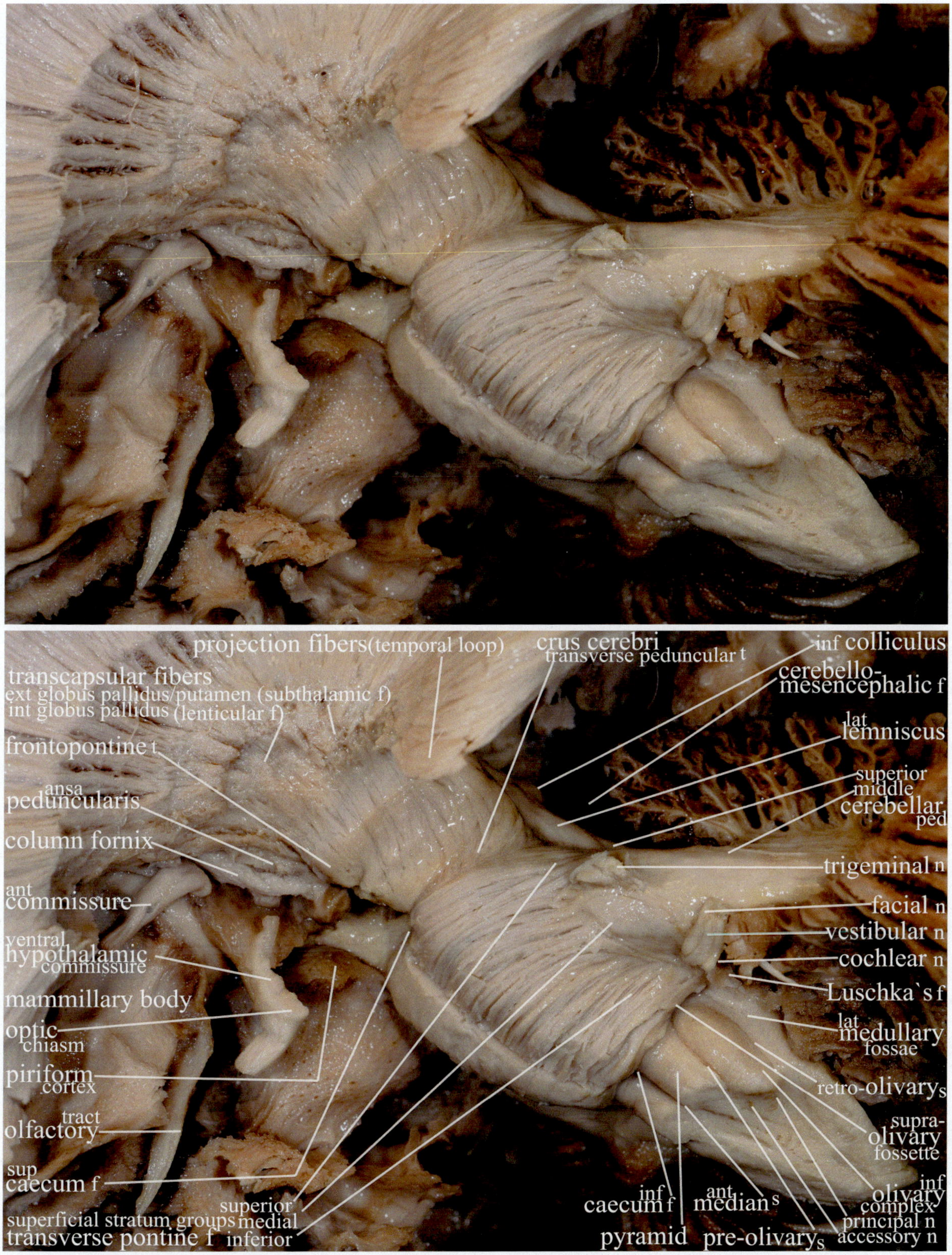

Fig. 8.34 Anatomical preparation of the superficial transverse fibers of the pons, close-up view. *ant* anterior, *ext* external, *f* fasciculus or foramen, *inf* inferior, *int* interior, *lat* lateral, *n* nerve, *ped* peduncle, *s* sulcus, *sup* superior, *t* tract (©Kadri 2023. All Rights Reserved)

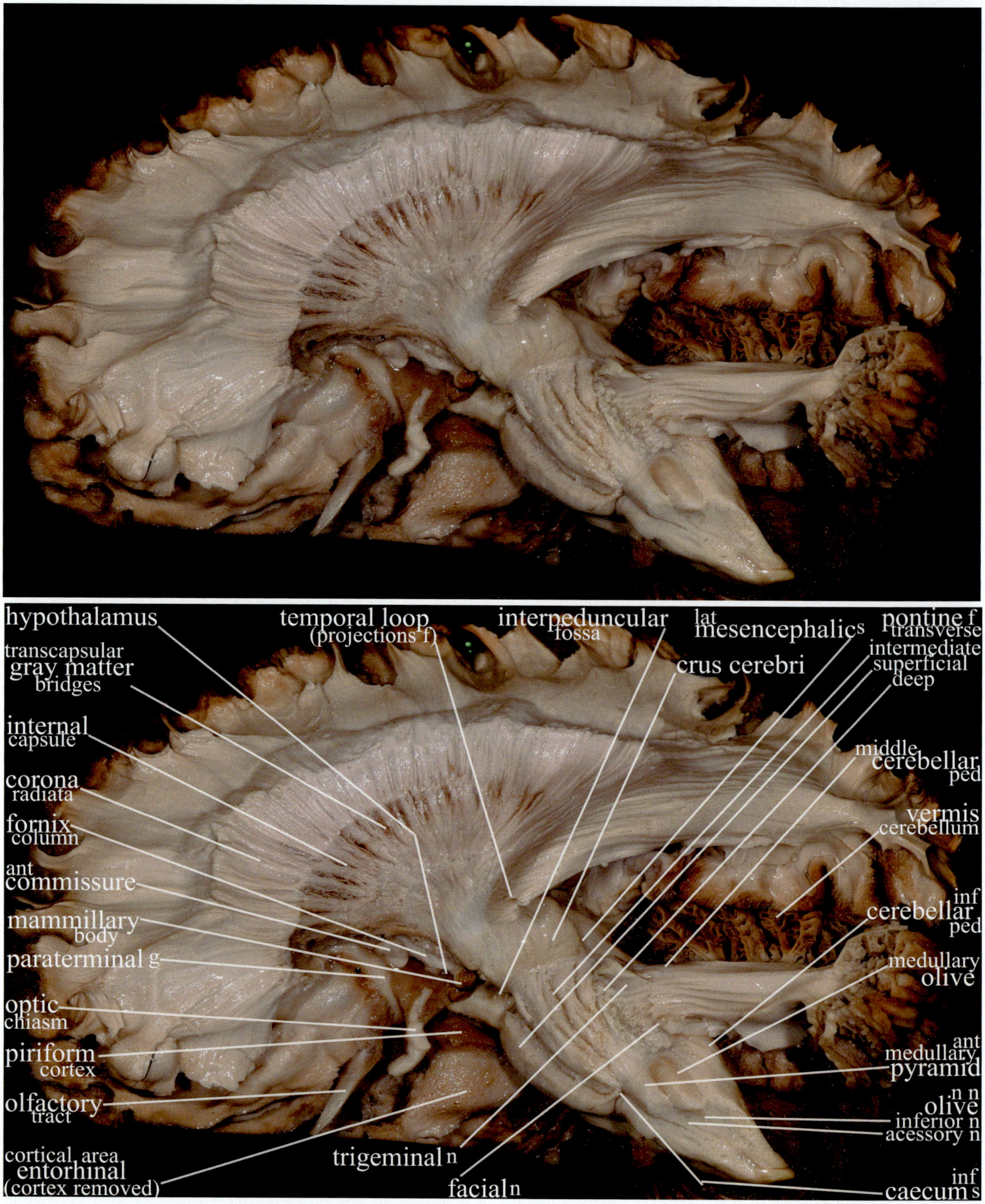

Fig. 8.35 Anatomical preparation of the base of the brainstem. The intermediate transverse fibers of the pons are intercalated with the longitudinal projection fibers of the telencephalon in their connections from the crus cerebri to the pyramid of the medulla. *ant* anterior, *f* fasciculus, *g* gyrus, *inf* inferior, *lat* lateral, *n* nerve, *nn* nucleus, *ped* peduncle, *s* sulcus (©Kadri 2023. All Rights Reserved)

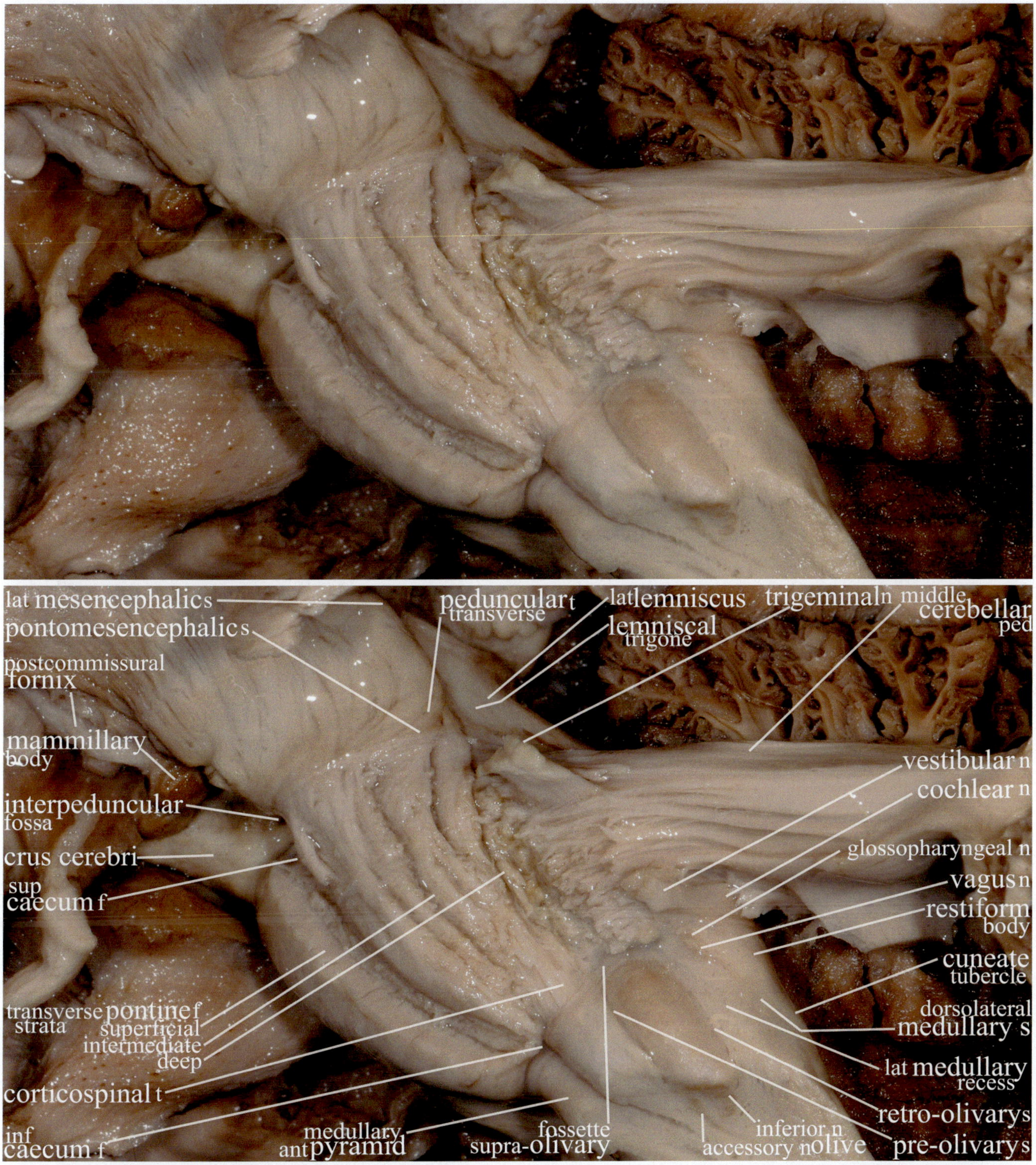

Fig. 8.36 Anatomical preparation of the transverse fibers of the pons. The limit of the transverse pontine fibers and the middle cerebellar peduncle is arbitrarily named as the apparent origin of the cisternal trigeminal nerve. The transverse fibers of the pons, the longitudinal projection fibers of the telencephalon (corticospinal, corticopontine, and corticobulbar tracts), and the pontine nuclei are the main components of the base of the pons. According to their relationship with the corticospinal tract, the transverse pontine fibers can be classified into superficial, intermediate, and deep fibers. *ant* anterior, *f* fasciculus, *inf* inferior, *lat* lateral, *n* nerve, *ped* peduncle, *s* sulcus, *sup* superior, *t* tract (©Kadri 2023. All Rights Reserved)

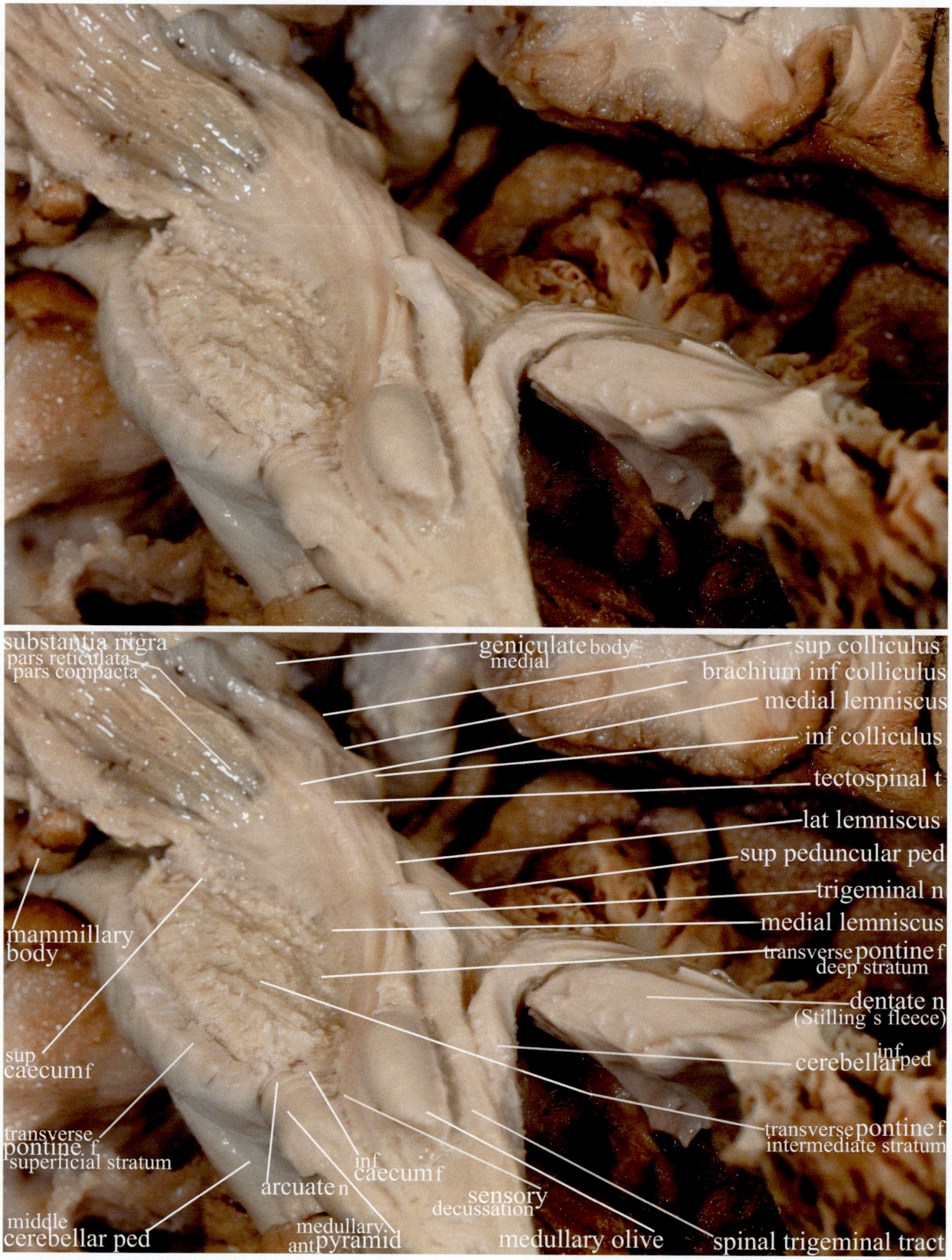

Fig. 8.37 Anatomical preparation of the tegmentum of the brainstem. The ventral lamina of the tegmentum is the substantia nigra and its caudal connections in the mesencephalon and the lemniscal fibers of the medial lemniscus at the pons. *ant* anterior, *f* fasciculus, *inf* inferior, *lat* lateral, *n* nerve, *ped* peduncle, *sup* superior, *t* tract (©Kadri 2023. All Rights Reserved)

Fig. 8.38 Anatomical preparation of the tegmentum of the brainstem. The lateral lemniscus radiates from the superior olivary nucleus to the inferior colliculus. The parapeduncular cerebellar recess is the reference for the motor nuclei of the trigeminal nerve and the sensory ascending trigeminal mesencephalic nuclei and tract. The ascending trigeminal tract decussates and ascends on the contralateral side as the trigeminal lemniscus. The trigeminal lemniscus is dorsal to the medial lemniscus and, therefore, does not abut the surface of the lemniscal triangle. *ant* anterior, *f* fasciculus, *inf* inferior, *lat* lateral, *n* nerve, *ped* peduncle, *sup* superior, *t* tract (©Kadri 2023. All Rights Reserved)

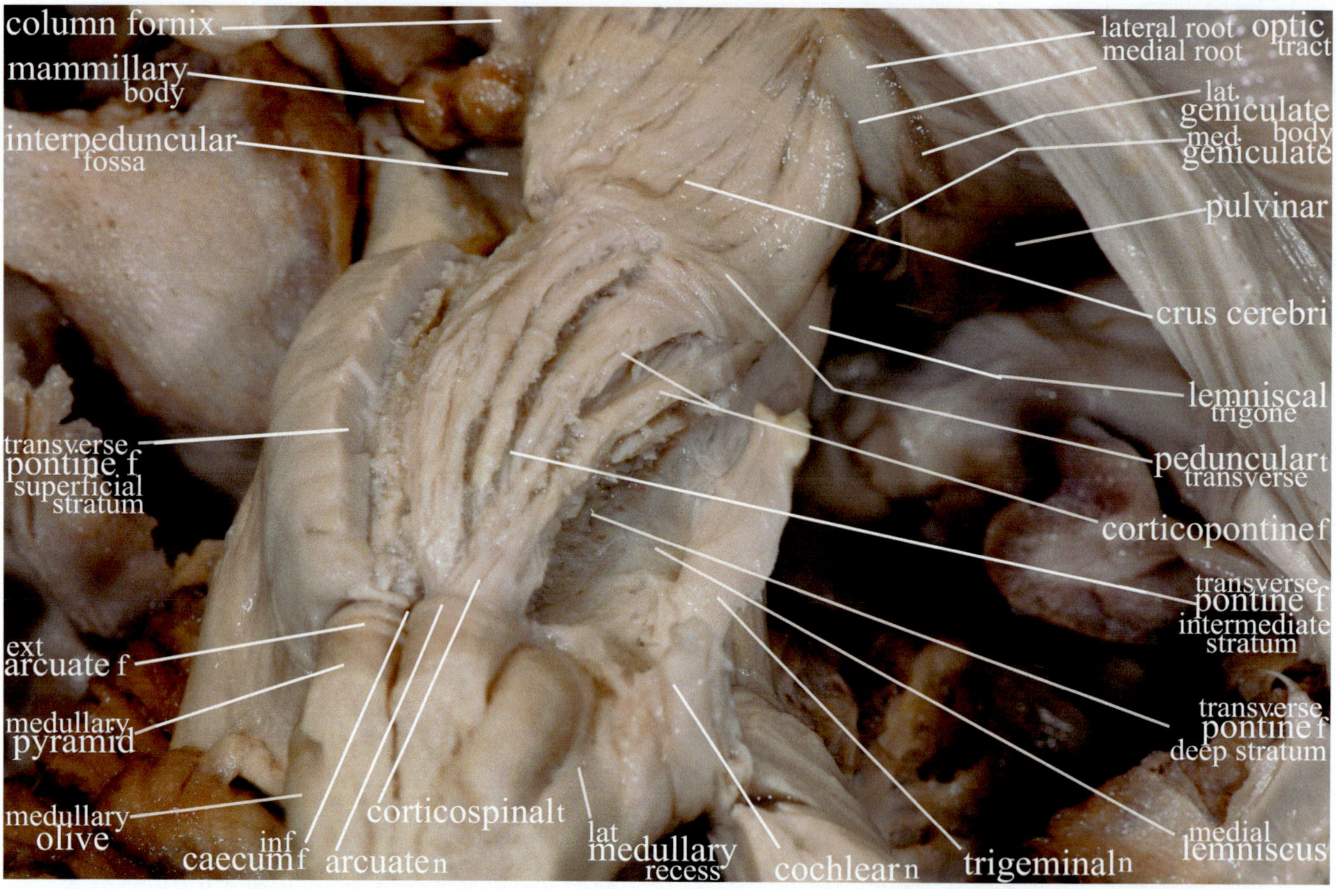

Fig. 8.39 Anatomical preparation of the base and tegmentum transition of the brainstem. The corticopontine tracts are directed to the ipsilateral and pontine nuclei dispersed and embedded between the arcuate (or transverse) pontine fibers. The pontine nuclei connect ipsilaterally and contralaterally with the cerebellum, primarily the neocerebellum within the cerebellar hemispheres. *ext* external, *f* fasciculus, *inf* inferior, *lat* lateral, *med* medial, *n* nerve, *ped* peduncle, *t* tract (©Kadri 2023. All Rights Reserved)

nerves. The six intrinsic fibers of the cranial nerve originate from the abducens nucleus at the lower pontine triangle of the fourth ventricle, and its fibers traverse the tegmentum of the pons, between the medial longitudinal fasciculus and the central tegmental tract, and the base of the pons, between the longitudinal and transverse pontine fibers, to emerge at the pontomedullary sulcus. At the pontine portion of the superior triangle of the fourth ventricle, the reference point of the abducens nerve nuclei is the facial colliculus, a protuberance on the median eminence. The intrinsic facial fibers from the facial nuclei, situated slightly ventrolateral to the abducens, turn around the abducens nuclei, connect fibers of the superior gustatory nucleus, and traverse ventrally between the medial lemniscus and the spinal tract of the trigeminal nerve to emerge at the pontomedullary sulcus between the deep transverse fibers of the pons and the ascending fibers of the anterolateral fasciculus. The apparent cisternal origin of the facial nerve is at the lateral medullary recess, more medially at the supra-olivary fossette. The intermediate nerve (Wrisberg's nerve) follows the same path of the motor facial nerve but emerges lateral to it in the pontomedullary sulcus. The vestibular nerve fibers are lateral to the intermediate nerve. These fibers enter the pontomedullary sulcus, many times between the fibers of the inferior group of the superficial pontine fibers, to traverse dorsally. They are bounded medially by the trigeminal tract of the trigeminal nerve and laterally by the fibers of the inferior cerebellar peduncle. Although not a group of radiated fibers of the pons, the fibers of the cochlear nerve do not enter the pontomedullary sulcus, but instead turn dorsally around the restiform body to connect with the ventral and dorsal cochlear nucleus. The dorsal cochlear nucleus reference at the lateral recess of the fourth ventricular floor is the acoustic tubercle. Dorsal to the cochlear nerve fibers, the peduncle of the floccule connects the floccule with the nodule and the superior vestibular area.

The gray matter of the reticular formation is grouped into the nuclei of the cranial nerves, the substantia propria of the pons, and the central gray matter. The cranial nerve nuclei are placed externally and comprise an internal and external column. The internal column corresponds to the facial nerve and motor nucleus of the trigeminal nerve. The external column is the main sensitive trigeminal nerve nucleus. The gray

Fig. 8.40 Anatomical preparation of the pontine nuclei. The pontine nuclei are the corticocerebellar circuit obligatory relays of the ipsilateral and contralateral corticopontine nuclei. *f* fasciculus, *inf* inferior, *med* medial, *n* nerve, *ped* peduncle, *s* sulcus, *sup* superior, *t* tract (©Kadri 2023. All Rights Reserved)

Fig. 8.40 (continued)

matter proper of the pons is the superior olive, the trapezoid body nuclei, and the nuclei of the lateral lemniscus. The central gray matter is also disposed as an internal and external column at the floor of the fourth ventricle. The internal column contains the abducens nerve nucleus. The external column contains the connection of the vestibular nerve.

The transverse fibers from the middle cerebellar peduncle are arranged in compact bundles separated by the nuclei of the gray matter (Fig. 8.45). They can be further divided into superficial, intermediate, and deep strata. The superficial stratum is anterior to the pyramidal tract and is present in the entire extension of the pons. The intermediate stratum is also known as the *stratum complexus* and is located mainly in the superior two-thirds of the pons. It divides the pyramidal tract into several small fasciculi and terminates when the corticopontine fibers reach the corresponding pontine nuclei. The deep stratum is located in the deep third, between the pyramidal tract and the medial lemniscus. The transverse fibers may also cross the midline at the raphe. They may change their stratum or direction and continue as arcuate fibers [2].

The transverse fibers from the vestibular and cochlear nerves form the trapezoid body and the lateral lemniscus (Figs. 8.39, 8.42, and 8.44). The trapezoid body is formed by the transverse fibers of the ventral cochlear nuclei. It also contains the trapezoid nucleus. The fibers of the dorsal cochlear nucleus that form the acoustic tubercle at the lateral recess of the fourth ventricle are disposed as the acoustic stria at the transition point from the pons and the medulla (Fig. 8.44). The fibers that cross the midline ascend toward the superior olive and from there to the lateral lemniscus toward the inferior colliculi. Clusters of neurons separate the medial and lateral lemnisci. These cell masses are known as the nuclei of the lateral lemniscus [1].

The longitudinal fibers are the motor pathway, the sensory pathway, and the longitudinal association fasciculi. The motor pathway is composed of the pyramidal, corticobulbar, and corticopontine fibers (Figs. 8.41, 8.42, and 8.45). The

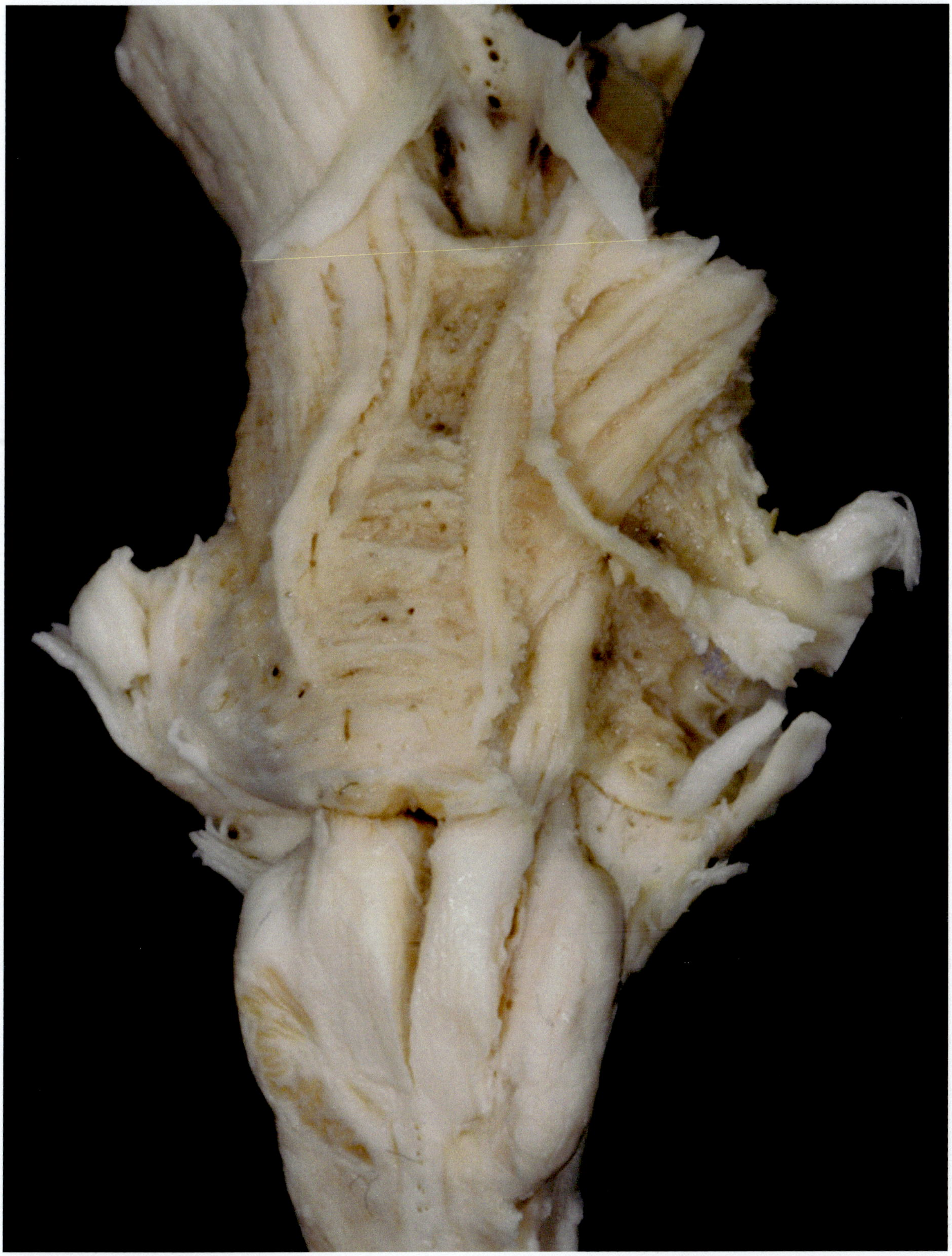

Fig. 8.41 Anatomical preparation of the base of the pons. The intermediate transverse fibers of the pons are exposed. The base of the pons is stratified into superficial, intermediate, and deep strata. The intermediate transverse pontine fibers intersect the corticospinal tract in the intermediate stratum. The superficial stratum has been removed but corresponds to the fibers superficial to the corticospinal fibers. It is most developed at the superior pons, superior to the intrinsic nerve roots of the trigeminal nerve. The deep or profound stratum is composed of the deep transverse fibers that connect the pontine nuclei with the cerebellum. *f* fasciculus, *inf* inferior, *lat* lateral, *med* medial, *n* nerve, *post* posterior, *s* sulcus, *t* tract (©Kadri 2023. All Rights Reserved)

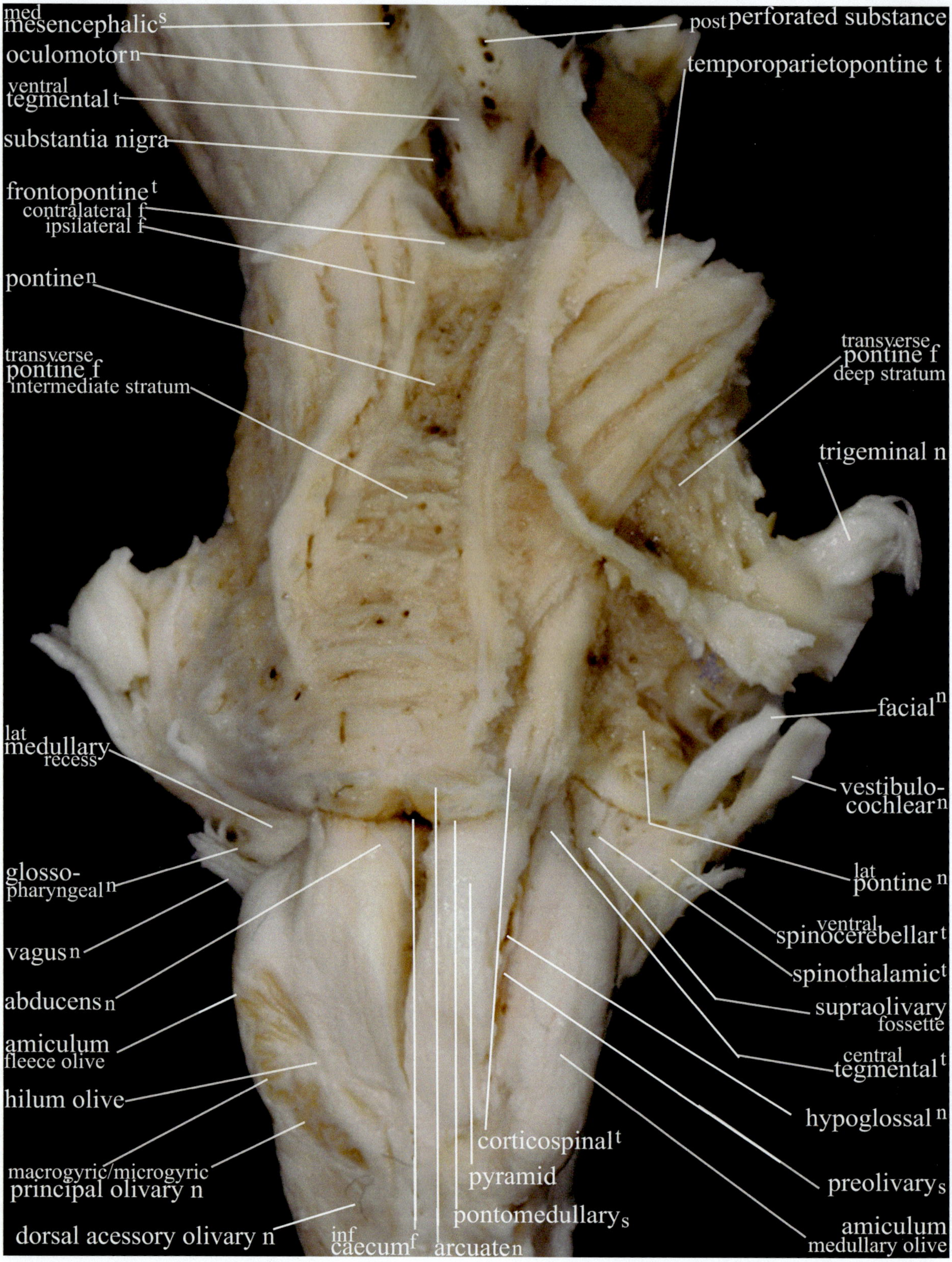

Fig. 8.41 (continued)

corticopontine fibers are also known as the geniculate tract because these fibers pass through the genu of the internal capsule [2]. The extrapyramidal motor connections between the medullary olive and the ipsilateral and contralateral mesencephalic red nucleus and substantia nigra form the central tegmental tract. This tract is composed ventrodorsally of a sequence of connections of the medullary olive, which densely envelop the principal and accessory inferior olivary nuclei and are named the amiculum (Figs. 8.41, 8.42, 8.43, and 8.44). The connections of the substantia nigra are the most ventral to the decussation of the superior cerebellar peduncle (Figs. 8.37, 8.40, and 8.41). The fibers that leave the mesencephalic tectum (tectopontine, tectobulbar, and tectospinal) are part of the visual–oculomotor and auditory–vestibular reflexes. These fibers traverse between the fibers of the lateral and medial lemnisci. The decussation of the superior cerebellar peduncle (Figs. 8.38, 8.43, and 8.44), with its mainly dentate contralateral red nucleus and ventral thalamus connections, dominates the ventral tegmental mesencephalic pontine transition. The superior cerebellar decussation is followed by the decussation of the superior transverse pontine fibers (Figs. 8.33, 8.34, 8.35, and 8.36). The sensory pathway is formed by the decussation fibers of the cuneiform and gracile fasciculi. The crossing fibers form a triangular field posterior to the pyramid known as the piriform decussation (Fig. 8.37) [2]. From this decussation, the ascending fibers flatten and constitute the medial lemniscus (Figs. 8.37, 8.38, 8.42, and 8.45). The medial lemniscus separates the base of the pons from the tegmentum. It is separated from the lateral lemniscus by the superior olive and the nuclei of the lateral lemniscus. The medial lemniscus is mainly composed of the fibers of the cuneiform and gracile that convey conscious proprioception and epicritic tactile perception. It also receives the fibers of the anterolateral fasciculus. The anterolateral fasciculus contains the spinothalamic tract that transmits the inputs of pain and temperature (Fig. 8.45). The medial lemniscus also receives the fibers that traverse from the nuclei of the reticular formation to carry the sensory information of the trigeminal, intermediary of Wrisberg, glossopharyngeal, and vagus nerves.

The longitudinal association fibers are composed of fibers of varying lengths that contribute to different integration of the systems. The visual, oculomotor, vestibular, auditory, and limbic motor nuclei integrate via the medial longitudinal fasciculus (Figs. 8.38, 8.43, and 8.44). The dorsal and ventral tegmental mesencephalic tracts are the main limbic–autonomic longitudinal connections within the tegmentum. The posterior (or dorsal) longitudinal (or tegmental) fasciculus (or tract) is the main limbic connection between the dorsal tegmental pontine nucleus and the hypothalamus (Figs. 8.43, and 8.44). The dorsal tegmental tract connects the septal region and the hypothalamus with the reticular nuclei of the dorsal tegmentum of the brainstem. It is part of the midline telencephalic and diencephalic limbic projections that are collectively named the prosencephalic tract. The prosencephalic tract is also named the *fasciculus medialis telencephalic* and medial forebrain bundle [3]. Caudal to the dorsal tegmental nuclei of the mesencephalon, the limbic connections of the dorsal tegmental tract are reinforced by the crossed fibers of the dorsal tegmental decussation (fibers connecting the tectum of the mesencephalon with the medulla and the spinal cord, the tectobulbar and tectospinal tracts). The other association is the anterior (or ventral) longitudinal (or tegmental) fasciculus (or tract). The ventral tegmental tract is the main limbic connection between the amygdala, the septal region, the hypothalamic area (mainly the ipsi- and contralateral mammillary bodies), and the hippocampal formation (via the fornix), with the ventral tegmental nuclei of the mesencephalon. The interpeduncular nucleus is one of the ventral tegmental nuclei (Figs. 8.40, 8.41, 8.42, 8.44, and 8.45). The main extrapyramidal connection is the central tegmental tract (Figs. 8.41, 8.42, 8.44, 8.45, and 8.46). This tract extends from the red nucleus toward the inferior olive. It runs in the central part of the tegmentum and is dorsal to the medial lemniscus.

As in the medulla, the pons also harbors gray matter homologous to the medulla and gray matter proper of the pons. Seven columns or nuclei are identified in relation to the gray matter of the medulla. These are the nuclei of the facial, abducens, trochlear, and oculomotor nerves; the motor nuclei of the trigeminal nerve; and the locus coeruleus.

The facial nerve has a motor and a sensitive nucleus (Figs. 8.33, 8.41, 8.42, and 8.45). The motor nucleus is between the medulla and the pons. It is located medial to the trigeminal root and extends from the superior olive to the top of the nucleus ambiguus. The sensory fibers of the facial nerve originate at the geniculate ganglion and enter the medulla to pass through the spinal tract of the trigeminal nerve and reach the solitary tract and nucleus.

The abducens nerve has two nuclei (Figs. 8.43 and 8.44). The principal or dorsal nucleus protrudes inside the ventricle and is involved with the fibers of the facial nerve. This protrusion, lateral to the median sulcus and in the base of the pontine surface of the fourth ventricle, is known as the facial eminence, facial colliculi, or eminence teres. The accessory nucleus is interposed between the principal and the facial nerve nuclei.

The trigeminal nerve is a mixed nerve and has two roots (Figs. 8.32, 8.37, 8.39, 8.41 and 8.42). The sensory root is the most voluminous. The principal motor root is located within the reticular formation of the pons and lateral to the superior fovea of the fourth ventricle (see below). The sensory fibers originate from the Gasserian ganglion, and as they reach the pons, they bifurcate into ascending and descending fibers. The ascending fibers are shorter and are situated medial to the restiform body. The descending fibers

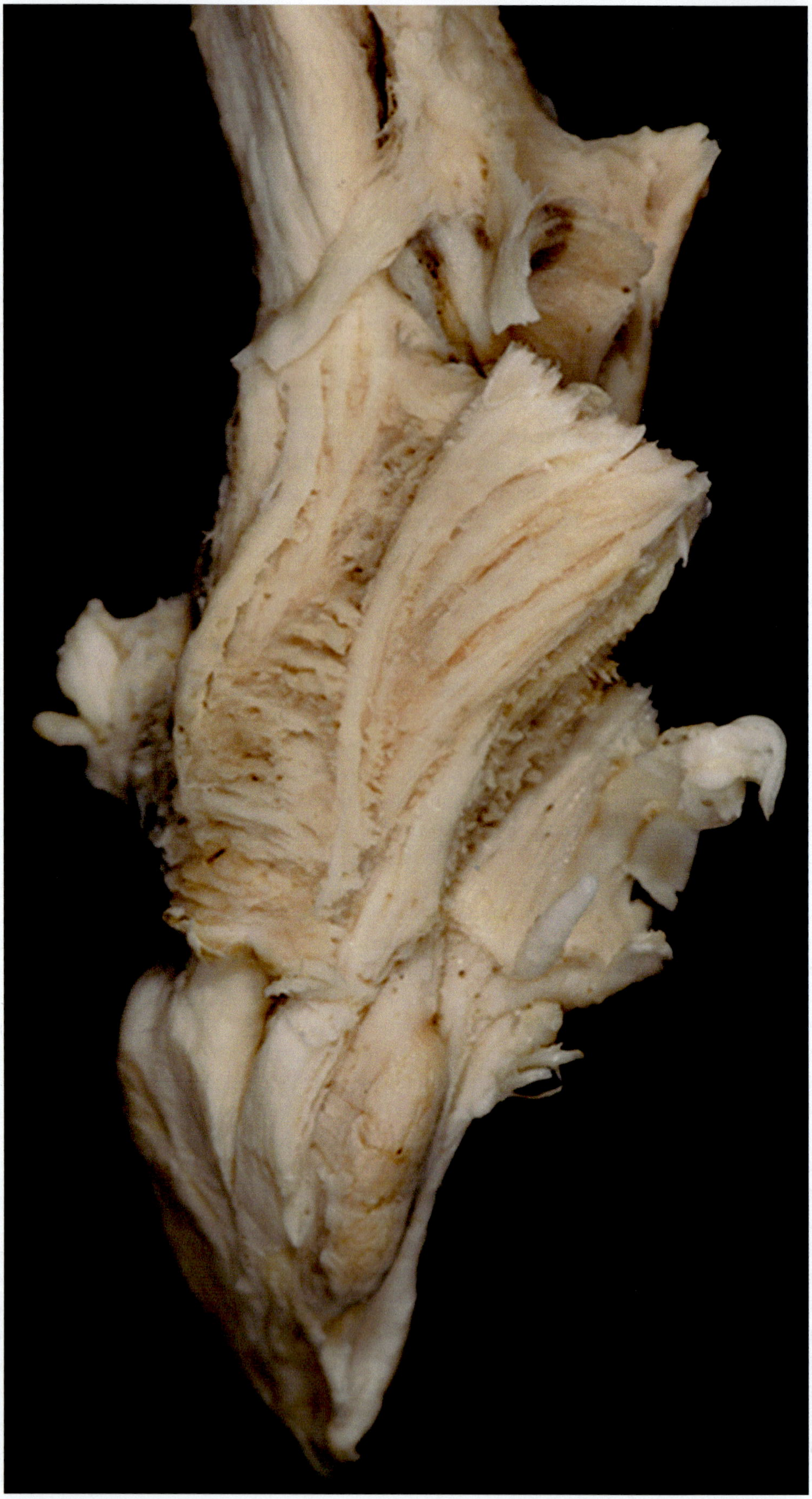

Fig. 8.42 Anatomical preparation of the base of the pons. The projection fibers of the base of the pons can be categorized into corticospinal, corticopontine, and corticonuclear fibers. The corticospinal tract passes through the intermediate stratum. The corticopontine tract (frontopontine and temporoparietopontine tracts) connects with ipsilateral basal pontine nuclei. The corticonuclear, also called the corticobulbar, tract is equally connected ipsilaterally and contralaterally. *ant* anterior, *f* fasciculus, *inf* inferior, *lat* lateral, *n* nerve, *ped* peduncle, *s* sulcus, *t* tract (©Kadri 2023. All Rights Reserved)

Fig. 8.42 (continued)

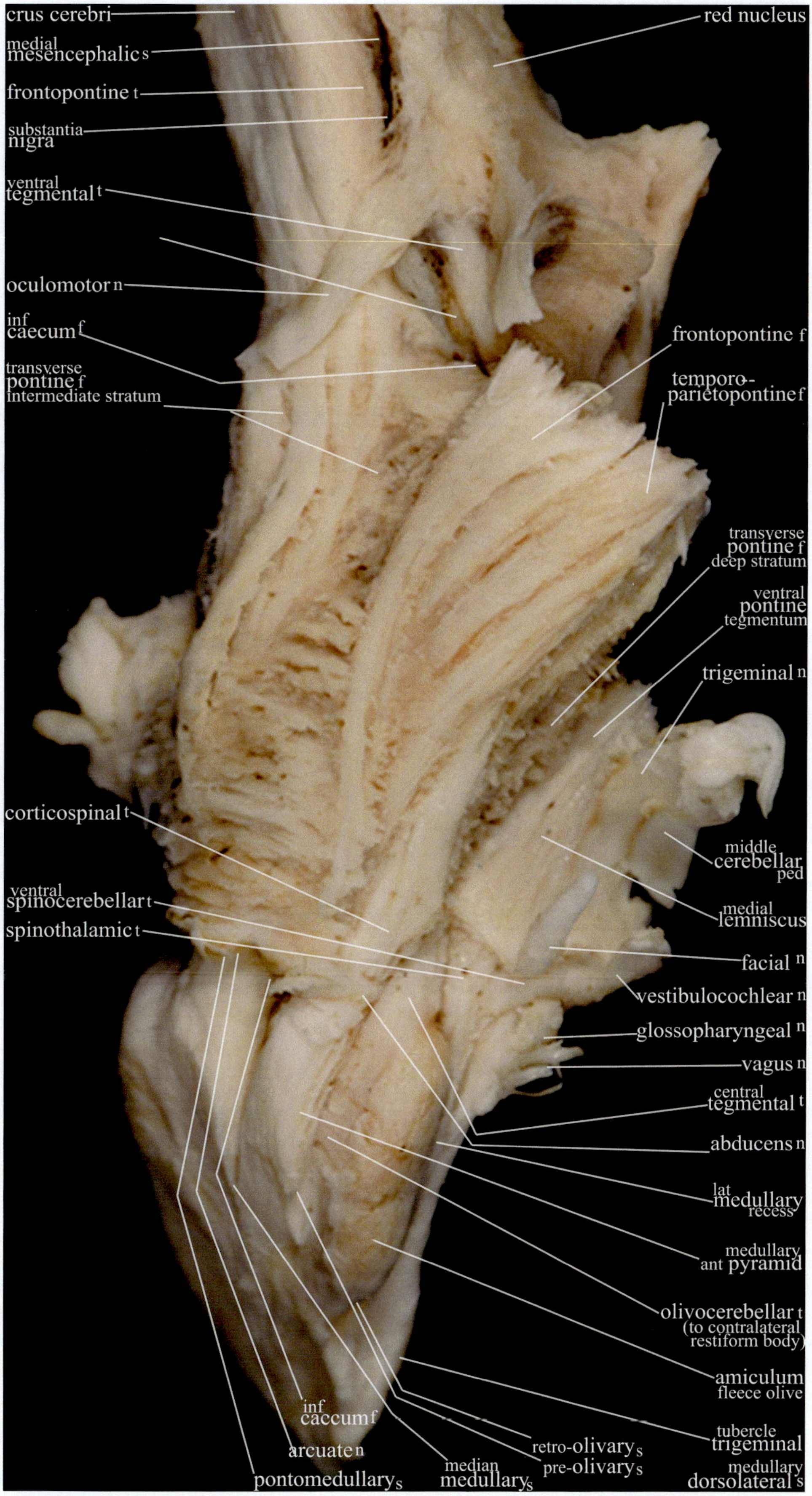

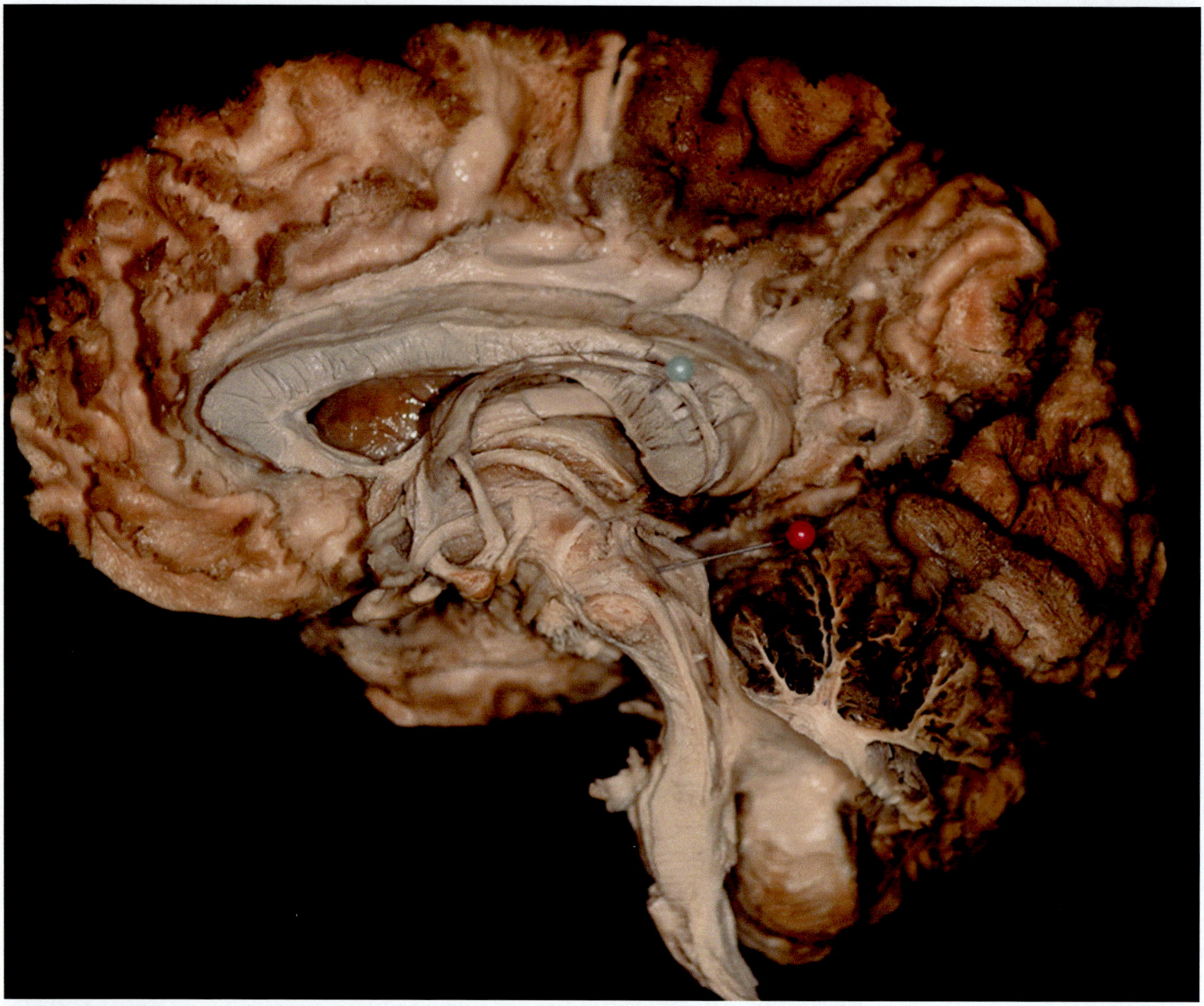

Fig. 8.43 Anatomical preparation of the tegmentum of the brainstem. *ant* anterior, *f* fasciculus, *g* gyrus, *inf* inferior, *n* nerve, *nn* nucleus, *ped* peduncle, *post* posterior, *sup* superior, *t* tract

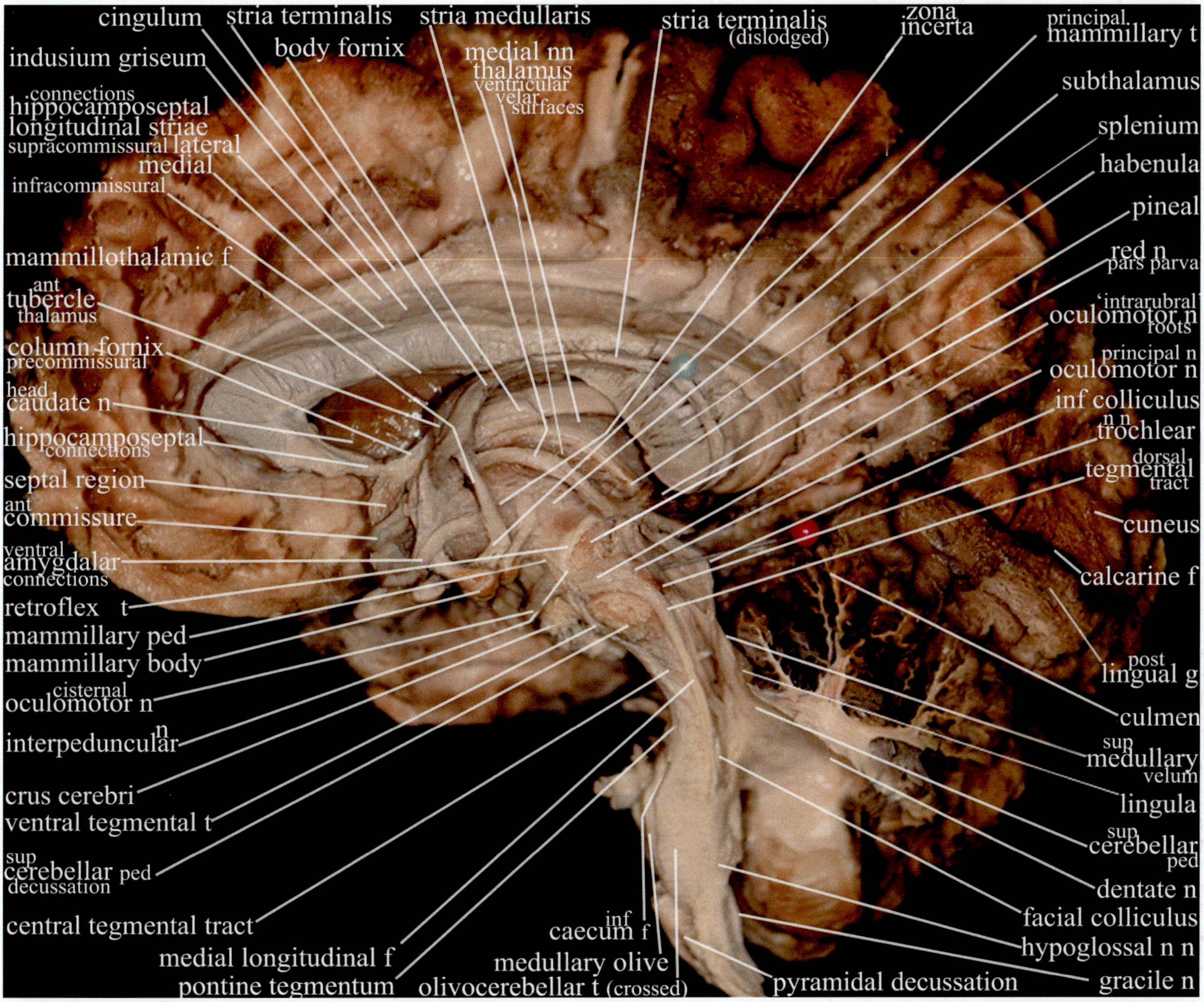

Fig. 8.43 (continued)

are longer, are more prominent, and extend through the medulla and upper part of the cervical cord to reach the spinal nucleus of the trigeminal nerve along its side. The locus coeruleus is a small bluish strip at the superior border of the fourth ventricle related to the trigeminal nerve (Figs. 8.44 and 8.46).

The gray matter proper of the pons is formed by the pontine nuclei, the superior olive, and the nuclei of the reticular formation (Fig. 8.44). The pontine nuclei are dispersed through the base of the pons, permeating the longitudinal, transverse, and arcuate fibers. These nuclei interconnect the cortex and the cerebellum. The superior olive is posterior to the trapezoid body and slightly anterior and medial to the facial nucleus. The superior olive is one of the stations of the acoustic pathways. The gray matter of the reticular substance is dispersed through its extension. Two midline nuclei masses, however, are recognized: the reticulate nucleus and the central superior nucleus. The reticulate nucleus is inferior and continuous with the inferior central nuclei of the medulla. The superior central nucleus extends from the trapezoid body to the superior cerebellar decussation.

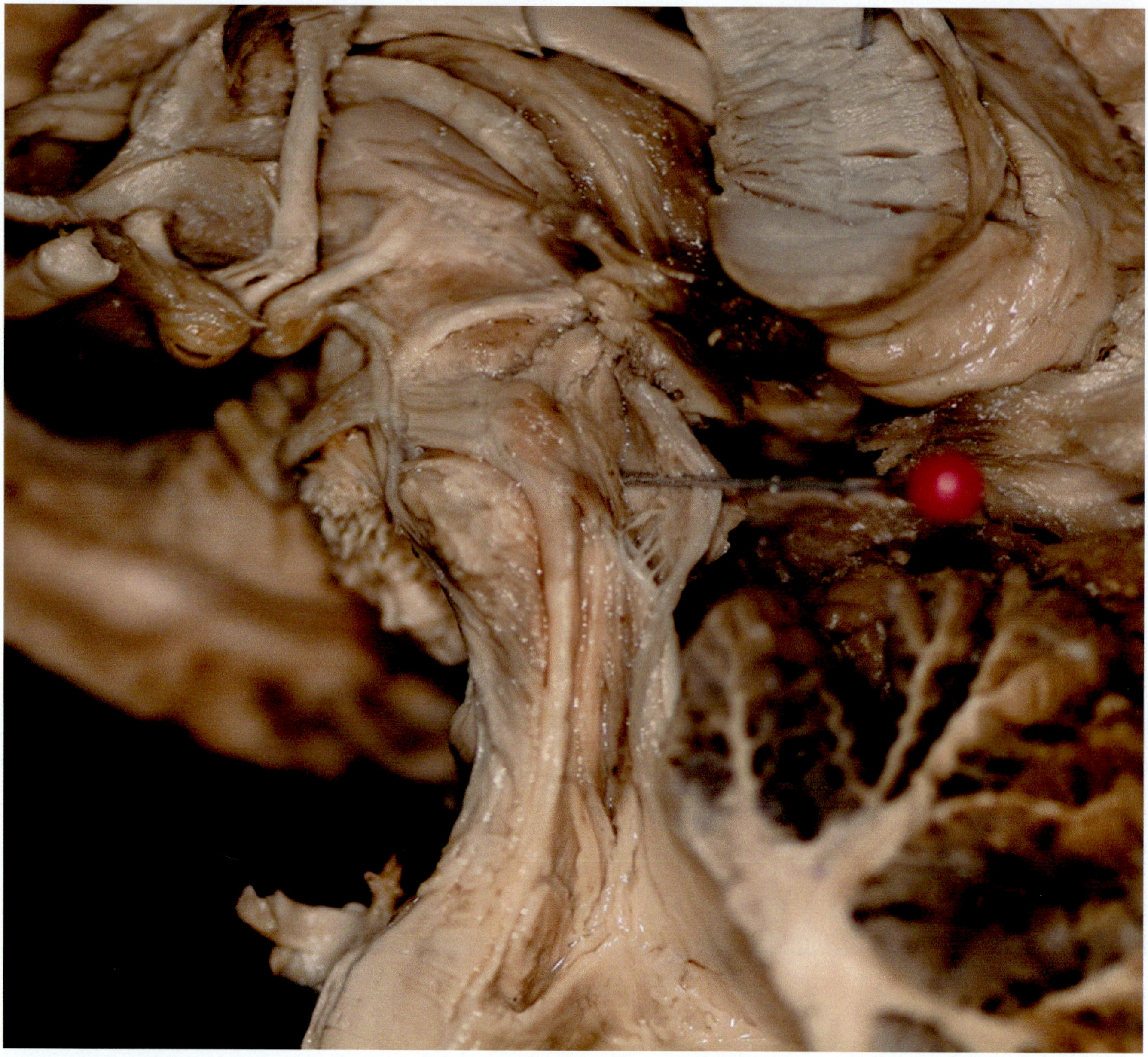

Fig. 8.44 Anatomical preparation of the tegmentum of the brainstem. The medial longitudinal fasciculus is the main interconnection of the nuclei of the cranial nerves. *f* fasciculus, *inf* inferior, *n* nerve, *nn* nucleus, *post* posterior, *ped* peduncle, *sup* superior, *t* tract (©Kadri 2023. All Rights Reserved)

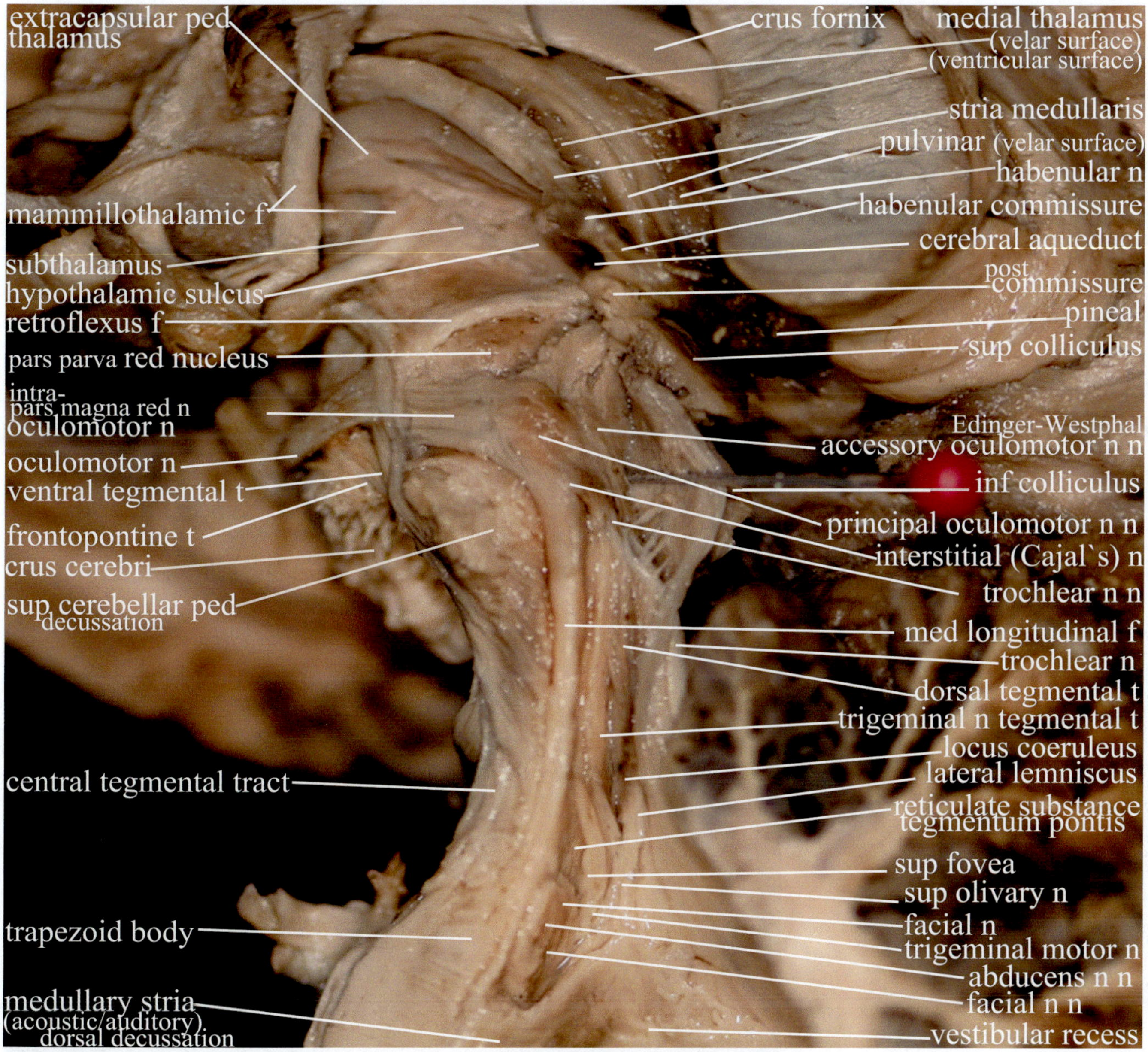

Fig. 8.44 (continued)

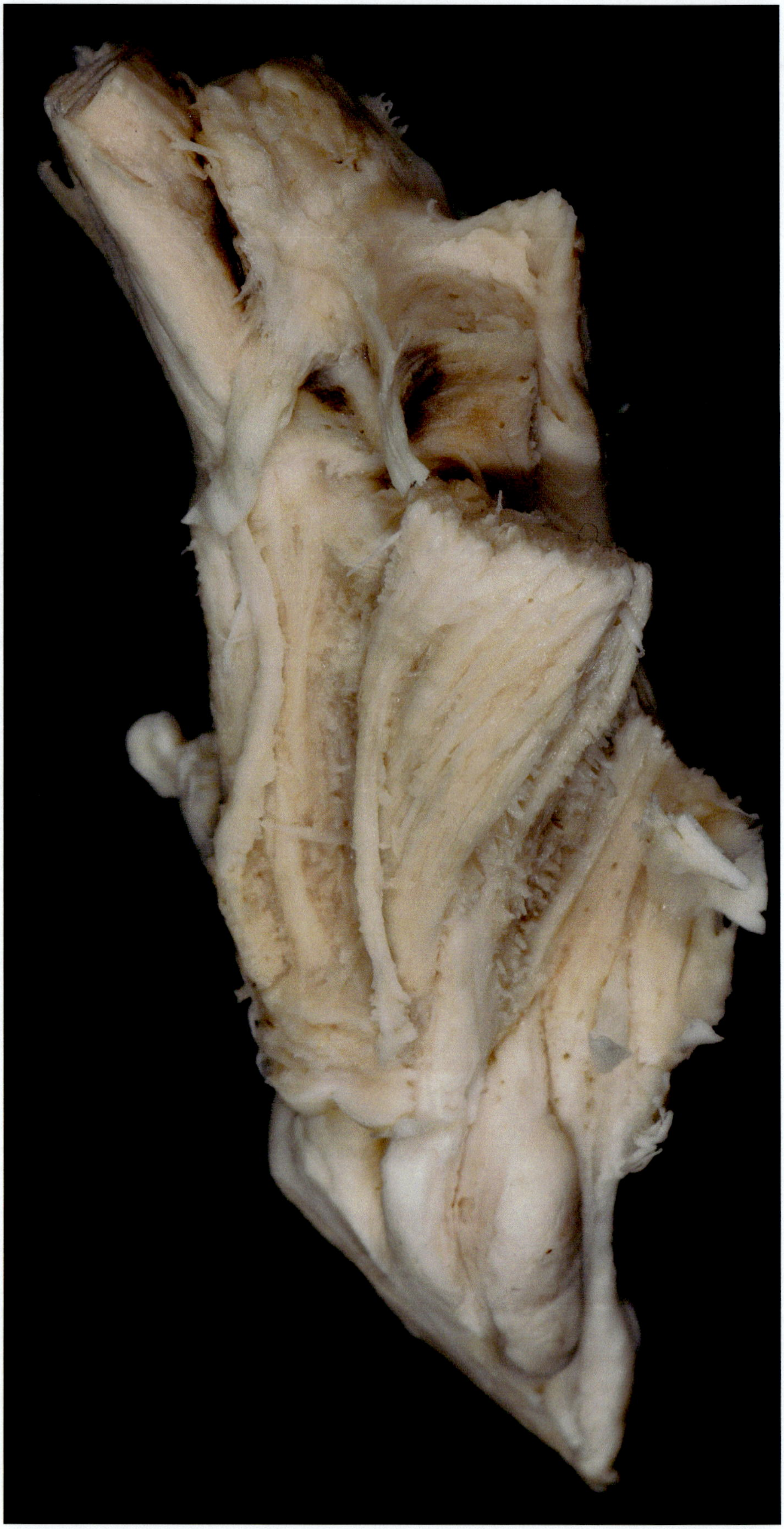

Fig. 8.45 Anatomical preparation of the base and tegmentum of the pons. The base of the pons is sorted into the superficial, intermediate, and deep strata. The intermediate stratum is also known as the *stratum complexus*. The superficial stratum is formed by the superficial transverse fibers of the pons. *ant* anterior, *f* fasciculus, *inf* inferior, *lat* lateral, *n* nerve, *s* sulcus, *sup* superior, *t* tract (©Kadri 2023. All Rights Reserved)

Fig. 8.45 (continued)

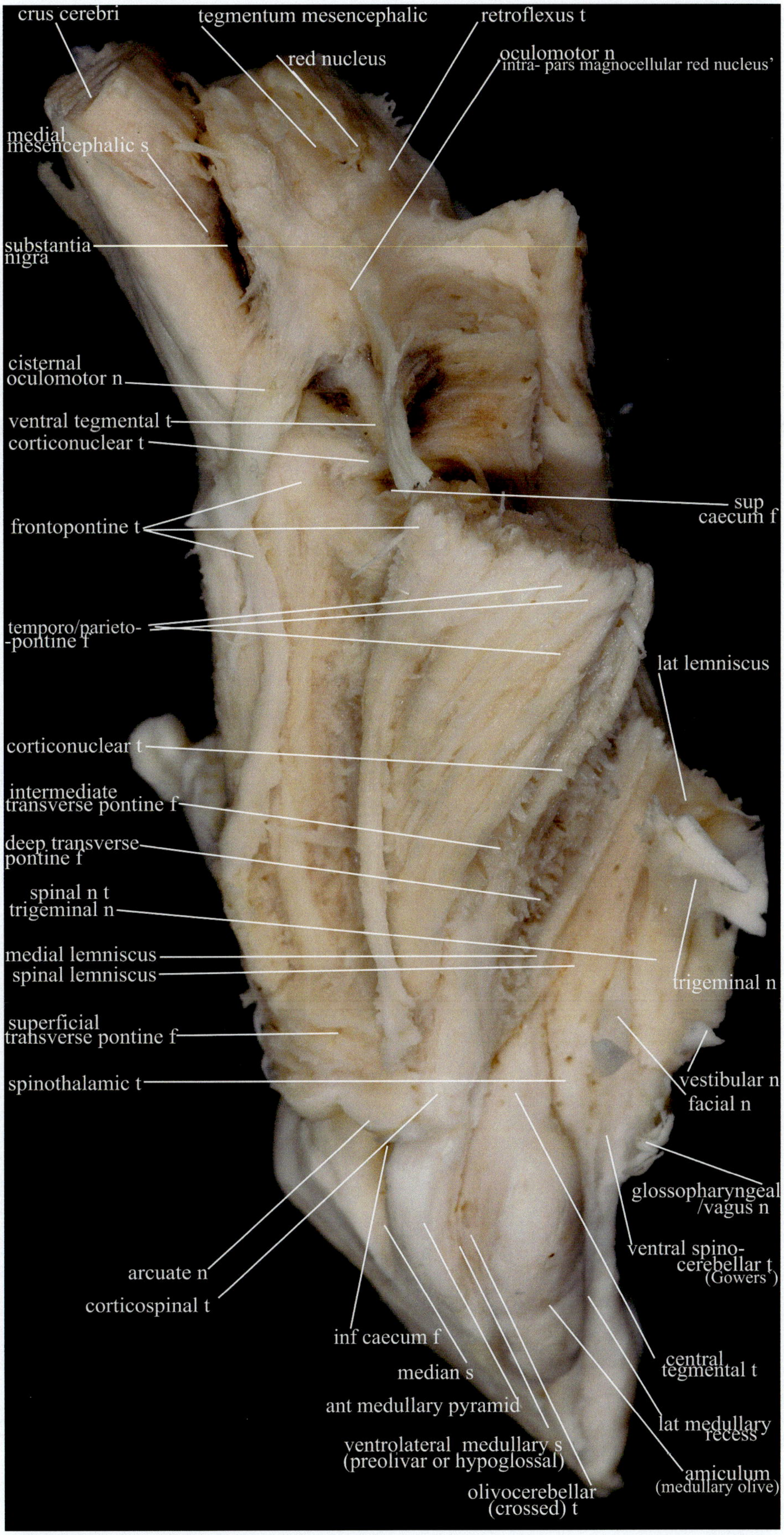

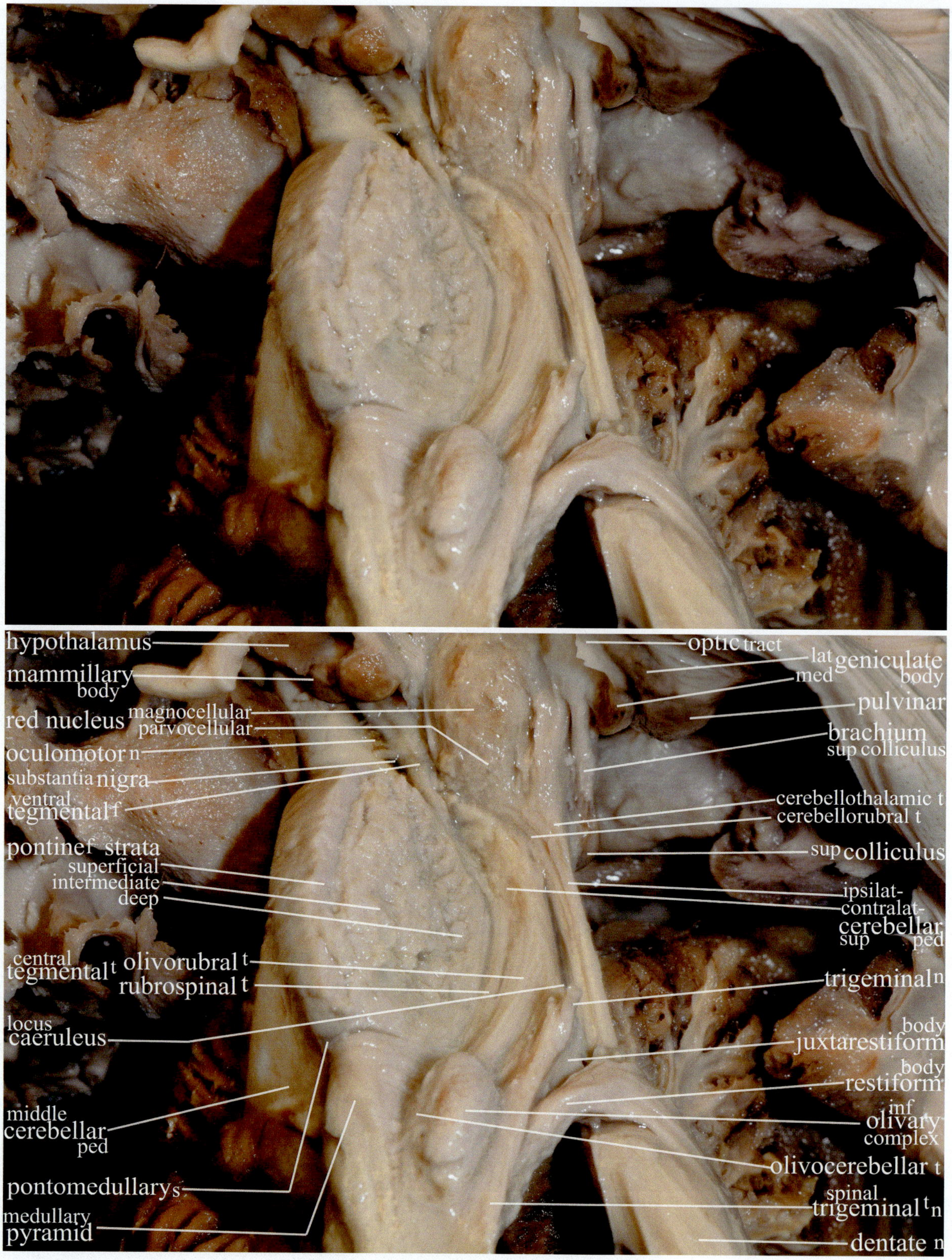

Fig. 8.46 Anatomical preparation of the tegmentum of the brainstem. The central tegmental tract is mainly an extrapyramidal connection of the red nucleus and principal inferior olive nucleus. *f* fasciculus, *inf* inferior, *lat* lateral, *med* medial, *n* nerve, *ped* peduncle, *s* sulcus, *sup* superior, *t* tract (©Kadri 2023. All Rights Reserved)

Medulla Oblongata

The medulla oblongata is also known as the spinal bulb or myelencephalon, and it is in the shape of a cone. The inferior limit is the decussation of the corticospinal tract or the emergence of the first root of the spinal nerve. The medulla is divided into lower (closed) and upper (open) portions according to the presence of the central canal or the fourth ventricle. The central canal of the spinal cord elongates cranially to the inferior half of the medulla and opens to constitute the inferior limit of the fourth ventricle. The internal arrangement of the lower medulla is quite similar to the spinal cord; however, several nuclei that differ from the spinal cord appear at this level. The motor nuclei are the hypoglossal, the dorsal nucleus of the vagus, and the ambiguus. The sensory nuclei are the gracile and cuneiform, the vestibular and cochlear nuclei, the spinal nucleus of the trigeminal nerve, and the solitary tract nuclei.

External Surface of the Medulla

The medulla has anterior, lateral, and posterior surfaces.

Anterior Surface

On the anterior of the external surface of the medulla, the continuation of the median sulcus of the spinal cord is called the median sulcus of the medulla (Fig. 8.47). This sulcus extends from the decussation of the corticospinal tract to the inferior cecum foramen, a tiny triangular fossette at the junction of the median and pontomedullary sulci. The median sulcus of the medulla is interrupted inferiorly by the decussation of the pyramid. The median sulcus is as deep as the analogous portion of the spine, and its depth is interrupted by a layer of white matter. The anterior pyramids, or simply pyramids, are located on each side of the medial sulcus. Although it seems continuous with the anterior funiculus of the spinal cord, the pyramid is totally distinct. Cranially, the pyramid disappears in the pontomedullary sulcus covered by the superficial and intermediate stratum of the transverse fibers of the pons. The cisternal emergence of the abducens nerve is from the pontomedullary sulcus, on top of the pyramid. The lateral limit of the pyramid is the ventrolateral sulcus, which is, at the spinal cord, the apparent origin of the ventral motor roots of the spinal nerves. At the medulla, the ventrolateral sulcus is where the cisternal hypoglossal nerve roots emerge. Therefore, it is also known as the hypoglossal sulcus at this point. The ventrolateral sulcus limits the pyramid and the olive, and because of this, it is also known as the pre-olivary sulcus (Fig. 8.48) [1].

Lateral Surface

The lateral surface is limited by the ventral and dorsal lateral sulci. It contains the medullary olive ventrally and the extension of the lateral funiculus of the medulla dorsally. The olives are paired, whitish, elongated, oval-shaped eminences that house the inferior olivary complex, which is composed of the principal, dorsal, and accessory nuclei. The medullary olive is also known as the inferior olivary complex and is limited by the pre-olivary and retro-olivary sulci. Its whitish appearance is due to the amiculum, a dense layer of fiber inputs from the central tegmental tract that cover the surface of the inferior olivary nuclei (Figs. 8.49 and 8.50). The inferior pole of the olive is partially covered by the external ventral arcuate fibers, which are connections from the restiform body (part of the inferior cerebellar peduncle). These are the fibers that cross the surface of the olive and the lateral recess of the medulla to join the dorsal spinocerebellar tract and the cuneate tract to form the restiform body (Figs. 8.51, 8.52, and 8.53). The dorsal external arcuate fibers are the fibers of the cuneate tract that run aberrantly on the surface. The olive is separated from the pons by the supra-olivary fossa, a small recess where the cisternal facial nerve and the intermediate nerve of Wrisberg emerge from the pontomedullary sulcus. The dorsolateral surface of the medulla is also known as the lateral recess and is limited by the retro-olivary sulcus and the dorsolateral sulcus (Fig. 8.51). On the rostral portion of the lateral recess, lateral to the supra-olivary fossa and therefore lateral to the cisternal emergence of the facial nerve, is an entrance called the lateral fossette. It follows the dorsolateral sulcus cranially and is the origin of the cisternal segment of the vestibular and cochlear nerves. The dorsolateral sulcus of the medulla is continuous with its analogous spinal structure and is the origin of the glossopharyngeal and vagus nerve roots.

Posterior Surface

The medulla is divided into upper and lower parts by the transition of the central canal of the medulla into the fourth ventricle. The inferior region is similar to the spinal cord and is divided in half by the median posterior sulcus (Fig. 8.54). This sulcus is very superficial, and its depth is formed by the posterior gray commissure of the spinal cord. The most rostral portion of the gray commissure of the spinal cord is the obex. The posterior median sulcus is the medial limit of the posterior funiculus, which contains the gracile and cuneiform tracts separated by the posterior intermediate sulcus. The gracile is medial and is also named the fasciculus of Goll. The cuneiform is lateral and is also called the fasciculus of Burdach.

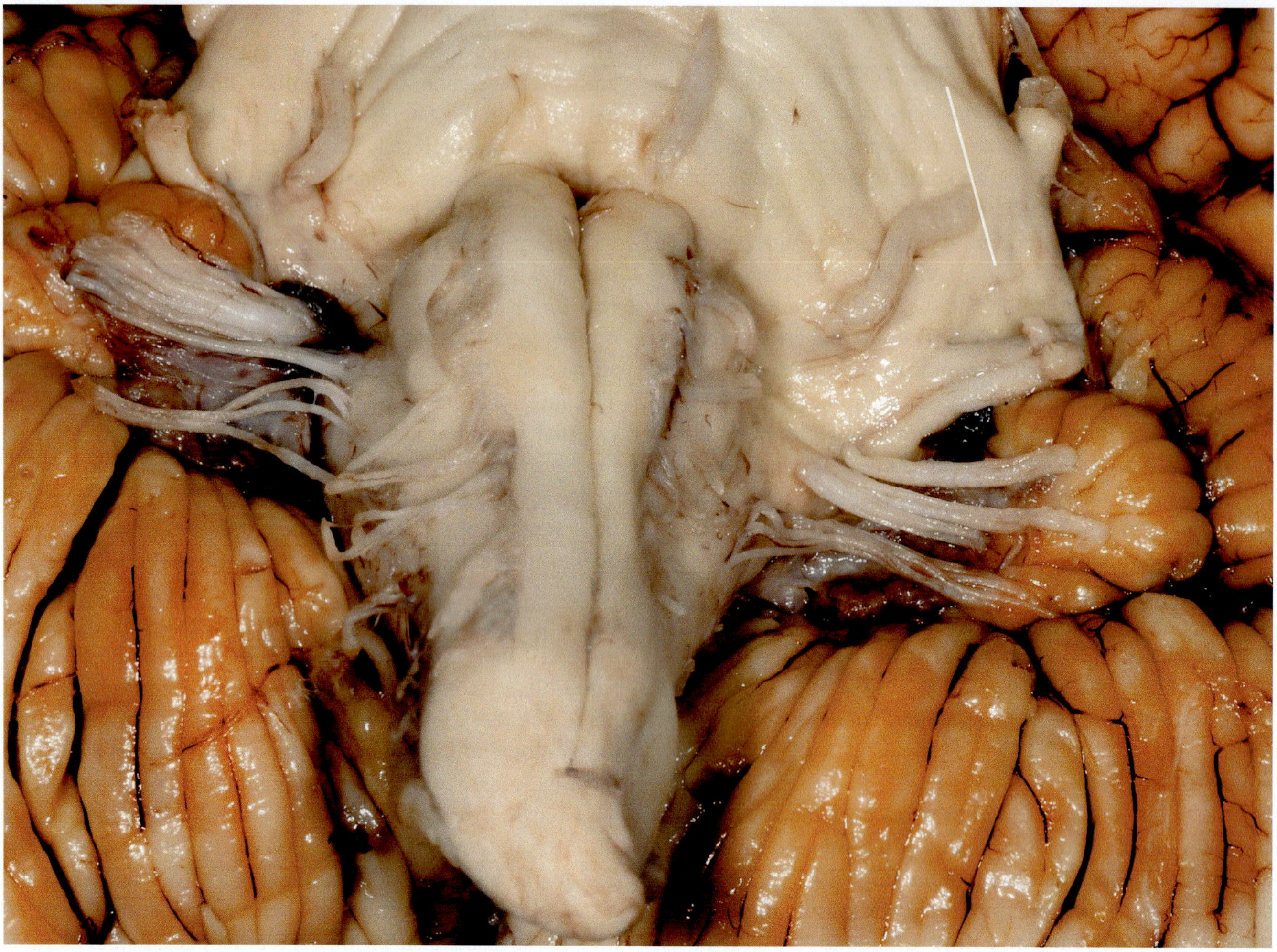

Fig. 8.47 Anatomical preparation of the ventral medulla oblongata. The protusion of the choroid plexus through the foramen of Luschka resembles a flower basket, as it is covered on its ventral (cisternal) surface by the lateral extension of the inferior medullary velum, with rootlets of the glossopharyngeal and vagus nerves resting more ventral. This structure was first described by Bochdalek in 1833 [4]. *ant* anterior, *f* fasciculus, *inf* inferior, *lat* lateral, *n* nerve, *ped* peduncle, *s* sulcus, *t* tract

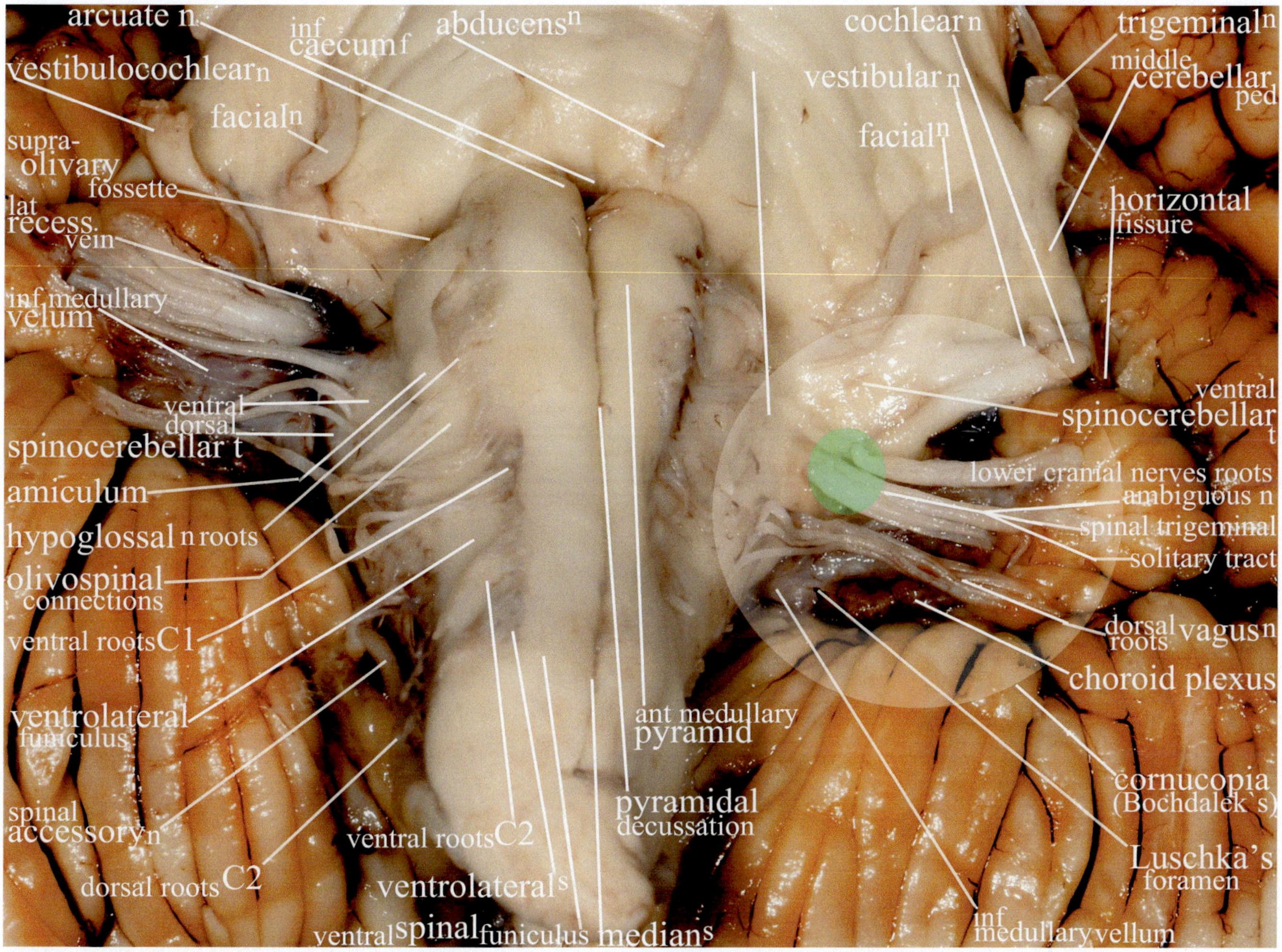

Fig. 8.47 (continued)

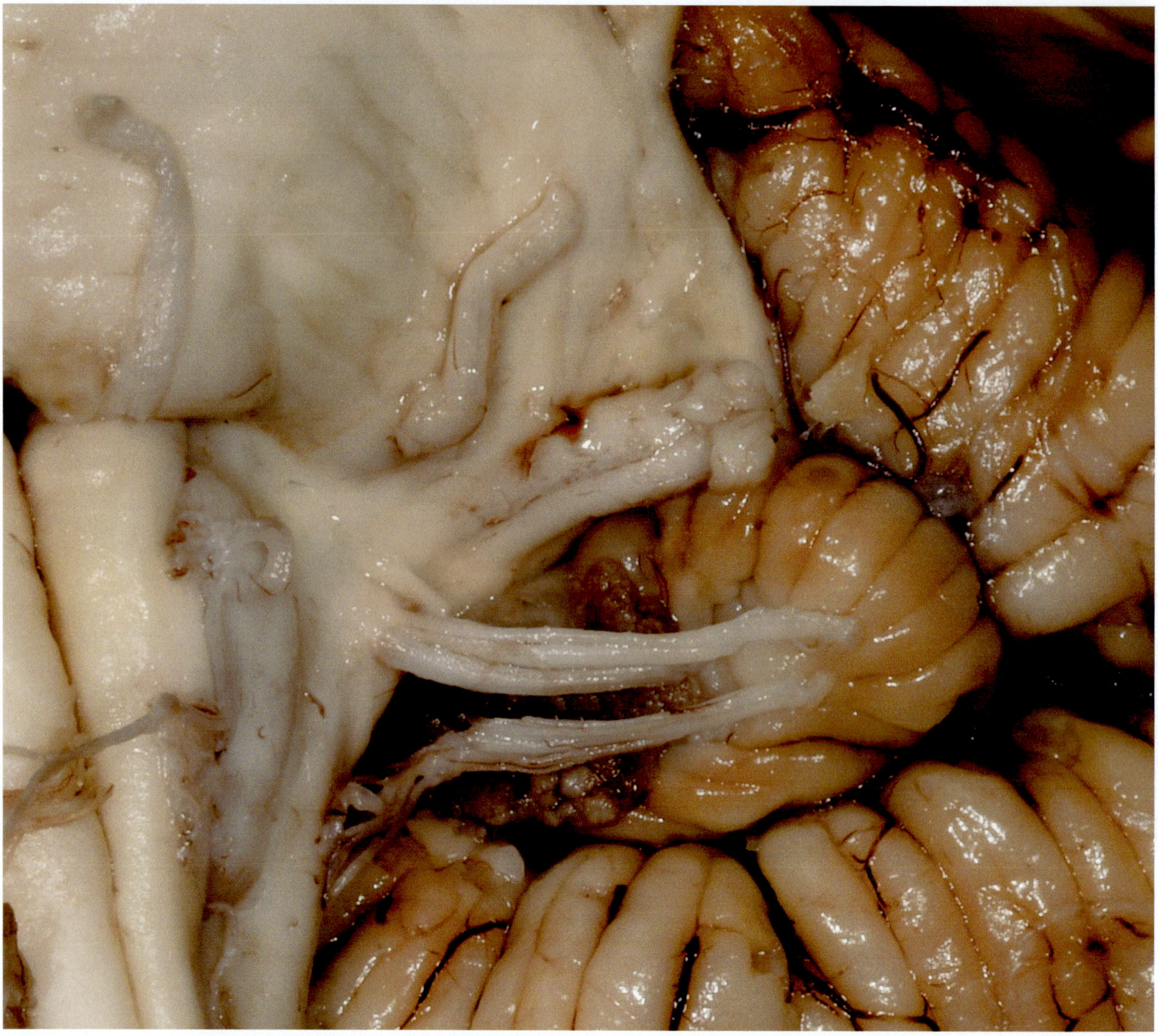

Fig. 8.48 Anatomical preparation of the left lateral medullary recess. This is a close-up view of the "flower basket" of the lateral recess of the fourth ventricle. *ant* anterior, *f* fasciculus, *inf* inferior, *lat* lateral, *n* nerve, *nn* nucleus, *ped* peduncle, *s* sulcus, *t* tract (©Kadri 2023. All Rights Reserved)

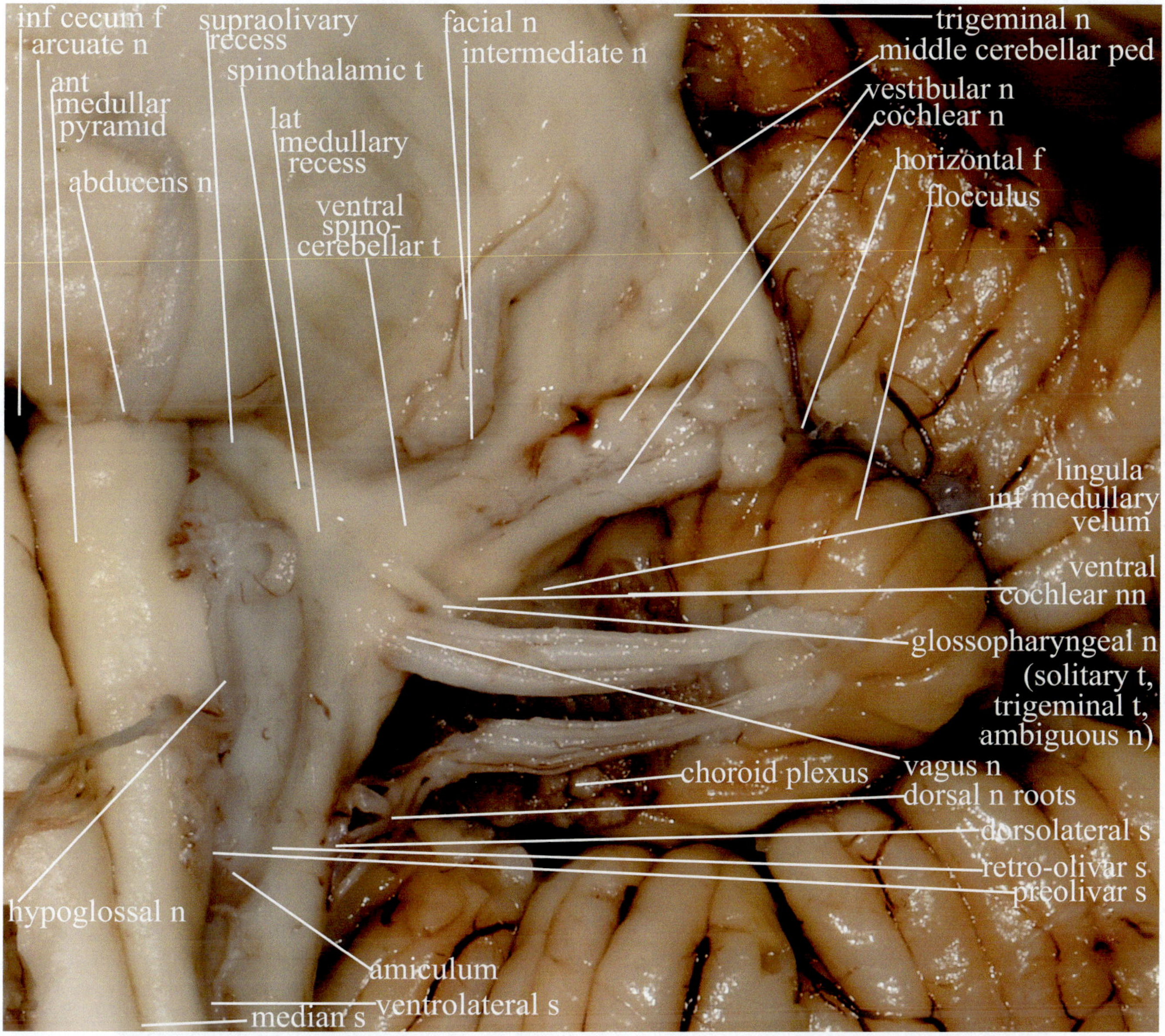

Fig. 8.48 (continued)

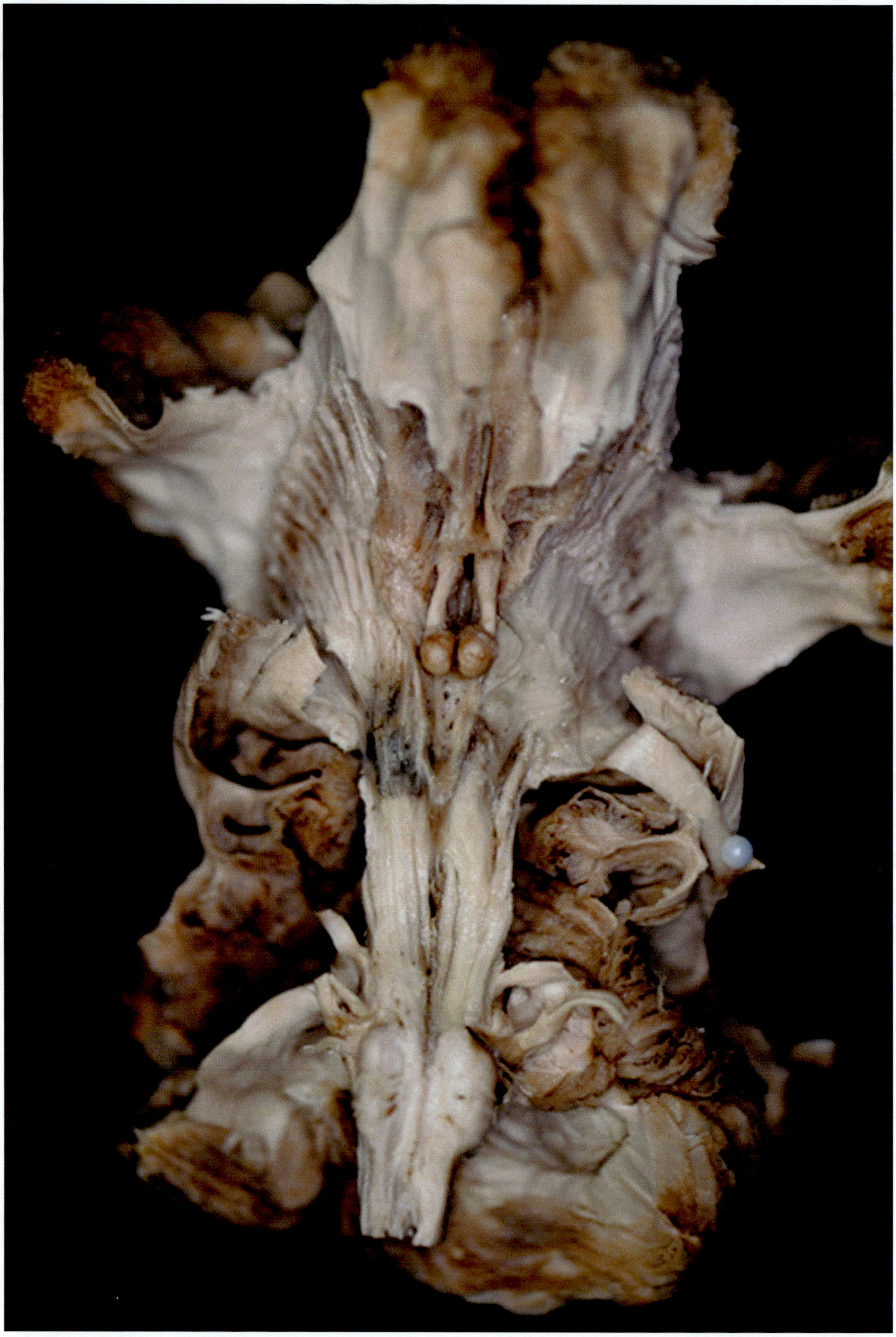

Fig. 8.49 Anatomical preparation of the tegmentum of the brainstem. The removal of the corticospinal tract at the base of the pons and medullary pyramids exposes the medial surface of the inferior olive and intra-axial roots of the hypoglossal nerve and the medial lemniscus (right side). Removing the medial lemniscus exposes the central tegmental tract and the spinothalamic tract (left side). *f* fasciculus, *inf* inferior, *n* nerve, *ped* peduncle, *post* posterior, *sup* superior, *t* tract (©Kadri 2023. All Rights Reserved)

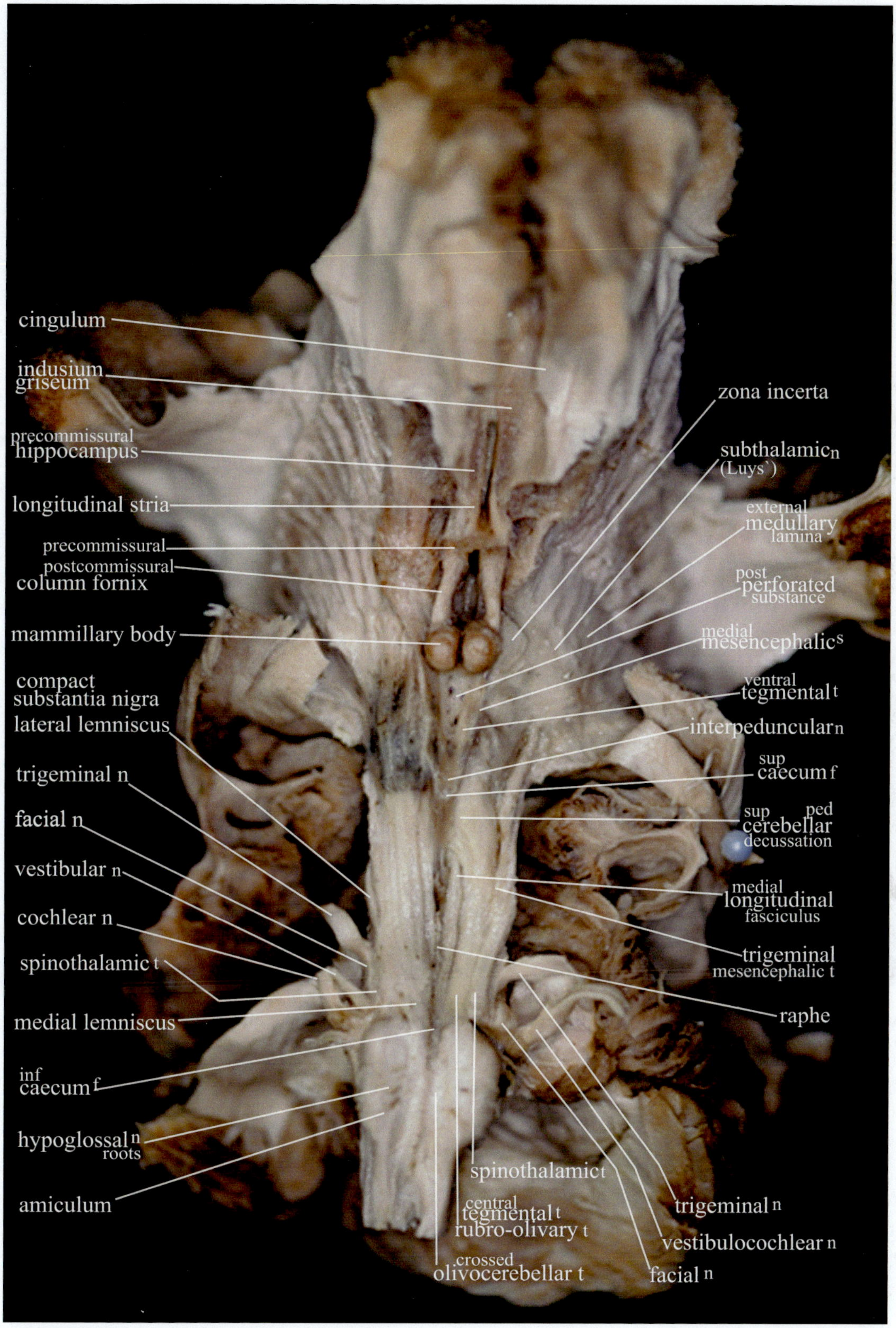

Fig. 8.49 (continued)

Fig. 8.50 Anatomical preparation of the tegmentum of the medulla (close-up anterior view). *f* fasciculus, *inf* inferior, *n* nerve, *s* sulcus, *t* tract (©Kadri 2023. All Rights Reserved)

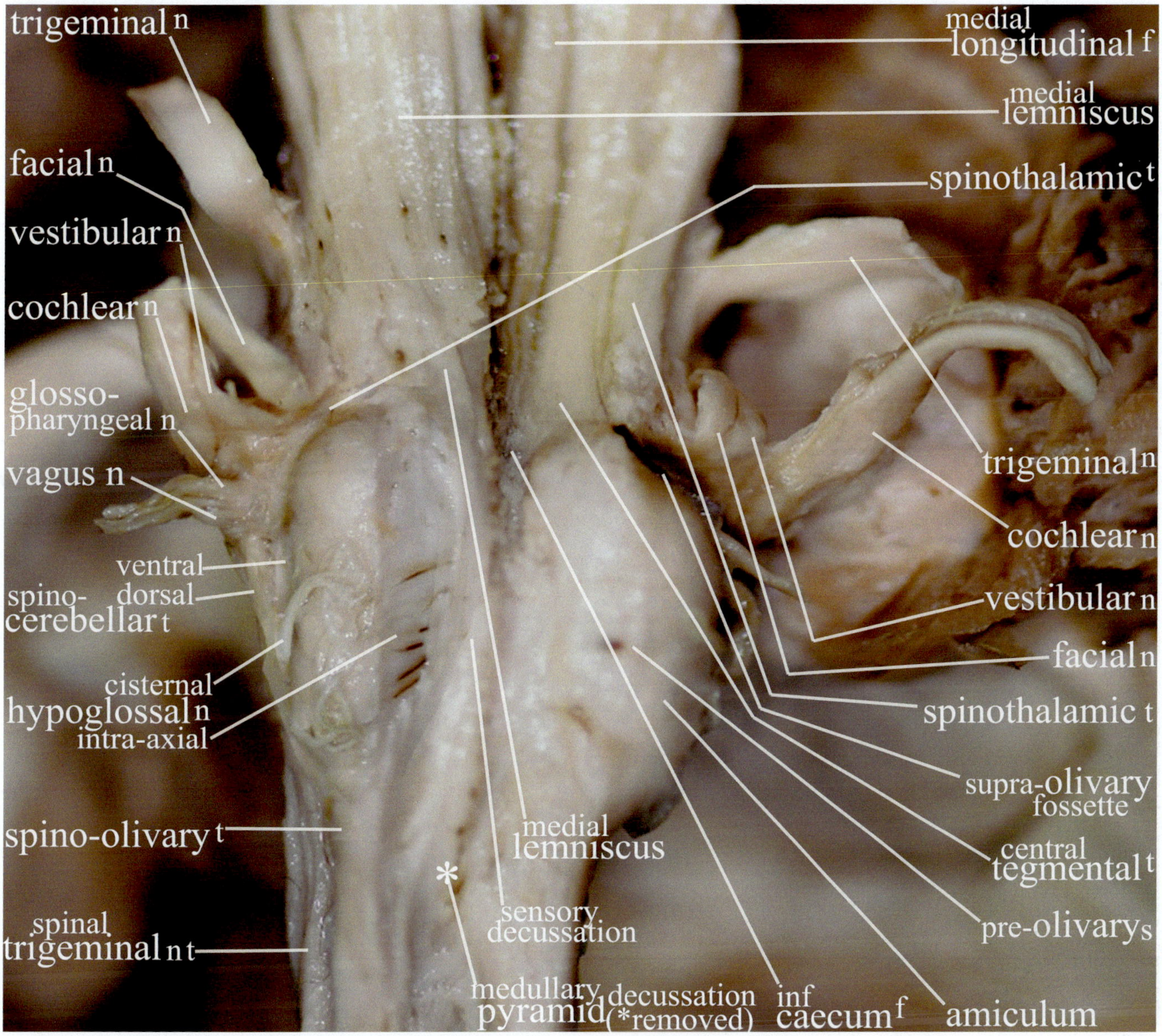

Fig. 8.50 (continued)

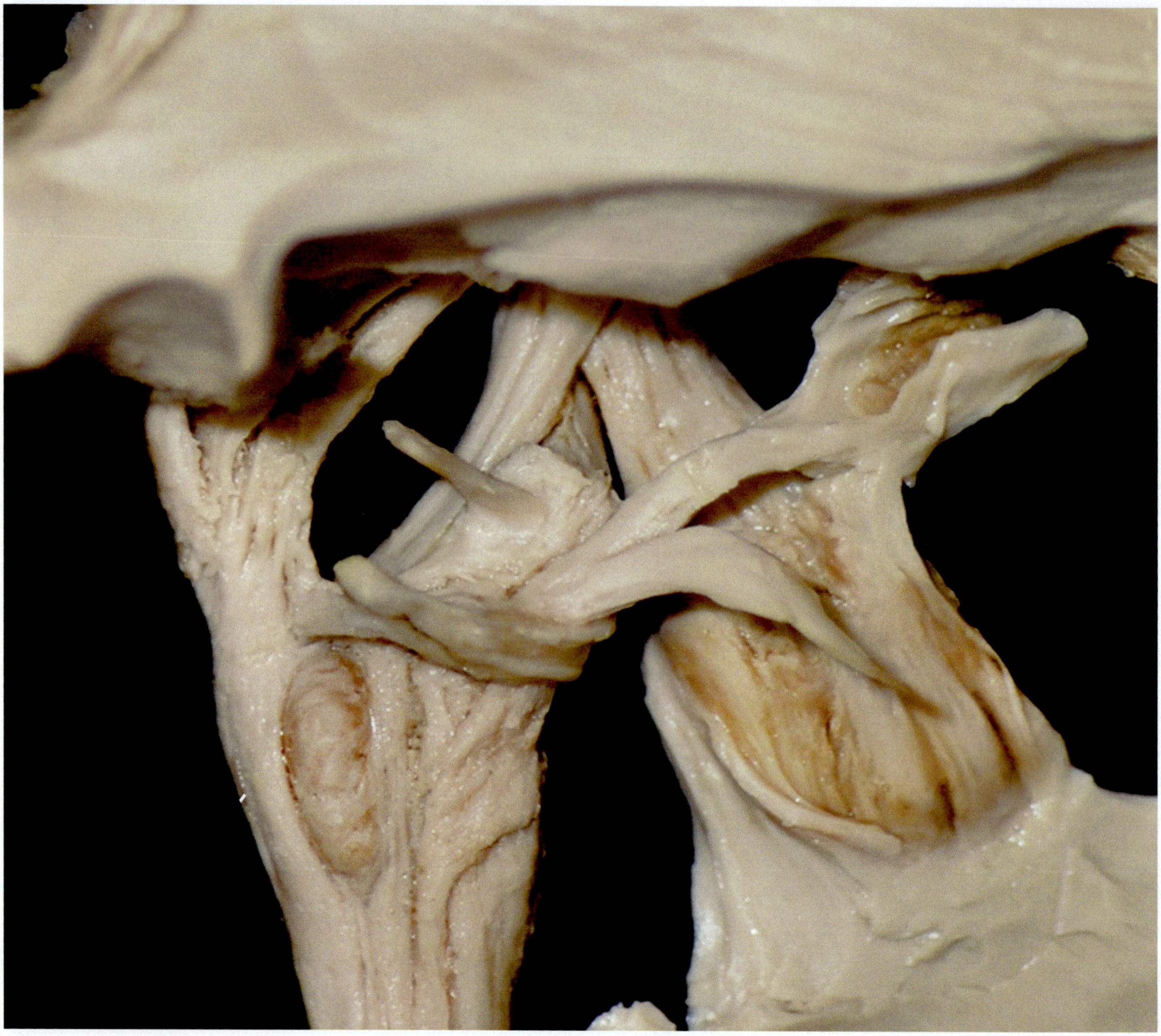

Fig. 8.51 Anatomical preparation of the anterolateral medulla, pons, and cerebellar connections. The spinothalamic tract is dorsolateral to the amiculum of the medullary olive. The spinocerebellar tracts are dorsal to the spinothalamic tract at the inferior medulla and are divided into ventral and dorsal tracts. The ventral spinocerebellar tract ascends parallel to the spinothalamic tract and ventral to the roots of the glossopharyngeal nerve. The dorsal spinothalamic tract turns dorsolateral and joins the contralateral olivocerebellar tract to form the restiform body with the more dorsal fibers of the accessory cuneate nucleus. The crossed olivocerebellar tract, if present on the surface of the amiculum or pyramid and/or traversing the lateral recess of the medulla, is made up of the ventral superficial arcuate fibers [2]. The fibers connecting the accessory cuneate nuclei via the restiform body, if present at the surface of the restiform body, are named the dorsal superficial arcuate fibers [2]. *ant* anterior, *f* fasciculus, *inf* inferior, *lat* lateral, *n* nerve, *nn* nucleus, *ped* peduncle, *sup* superior, *t* tract, *tt* tract (©Kadri 2023. All Rights Reserved)

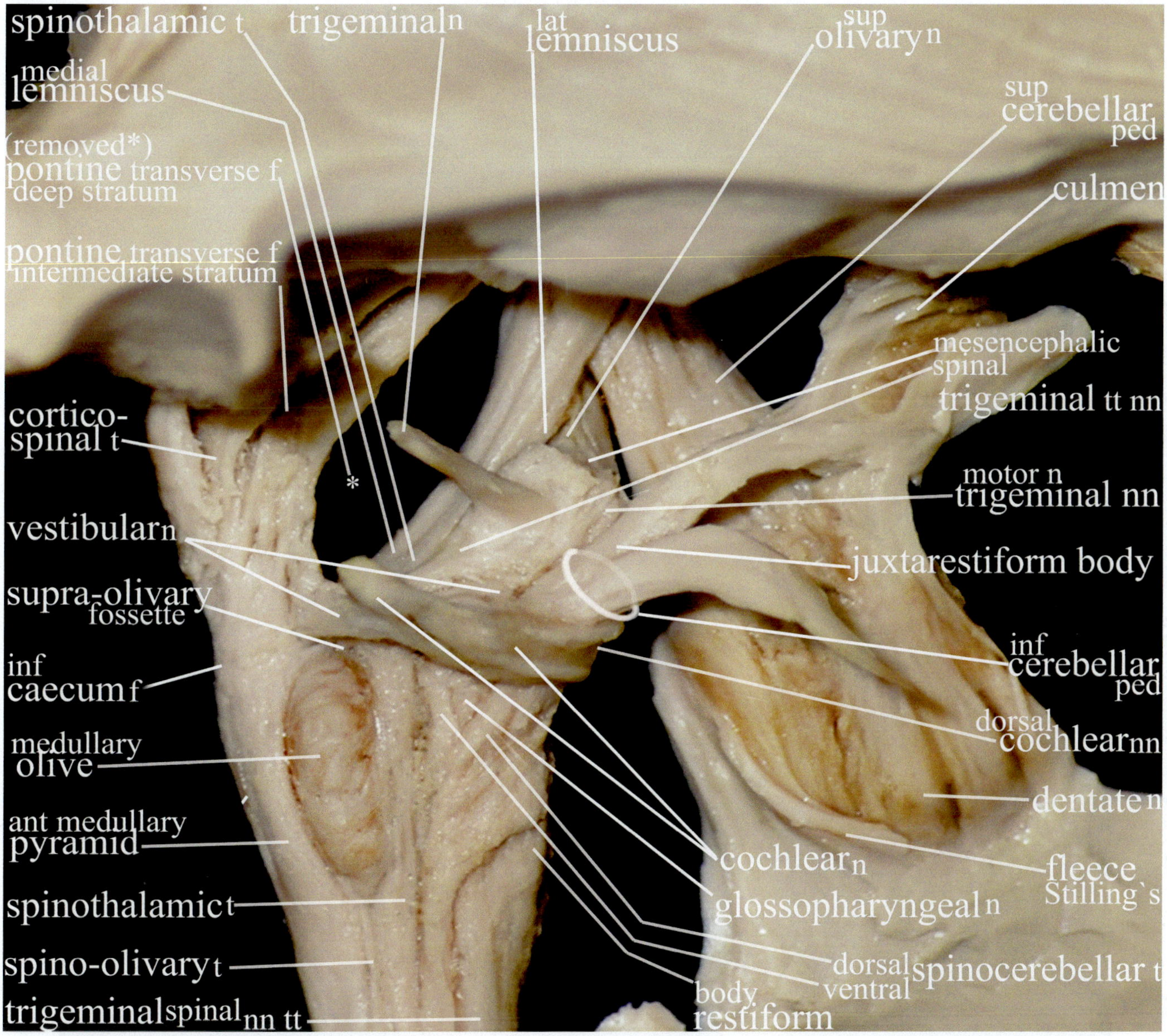

Fig. 8.51 (continued)

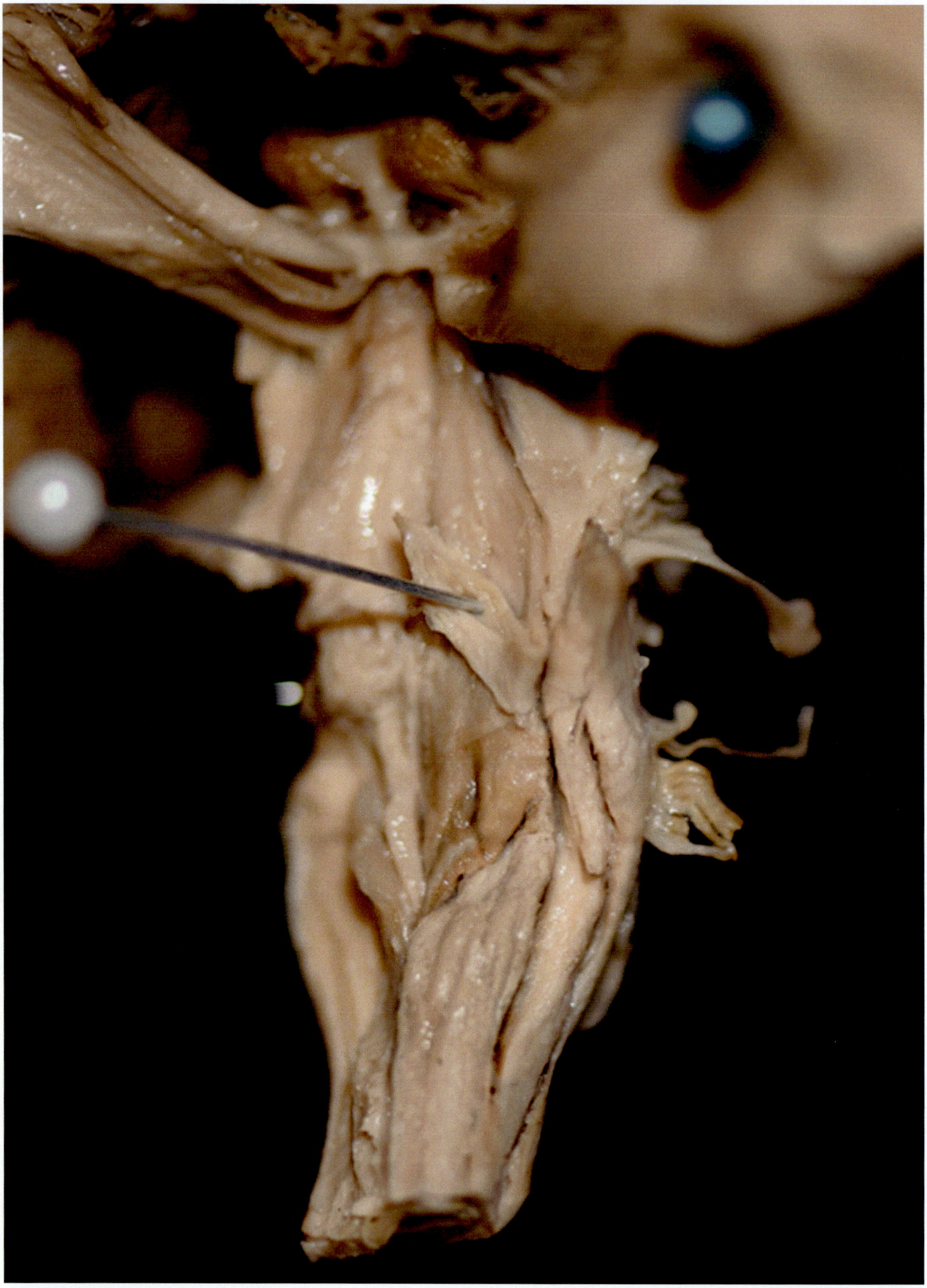

Fig. 8.52 Anatomical preparation of the lateral funiculus of the medulla and the restiform body of the inferior cerebellar peduncle. *f* fasciculus, *inf* inferior, *n* nerve, *nn* nucleus, *ped* peduncle, *s* sulcus, *sup* superior, *t* tract, *tt* tract

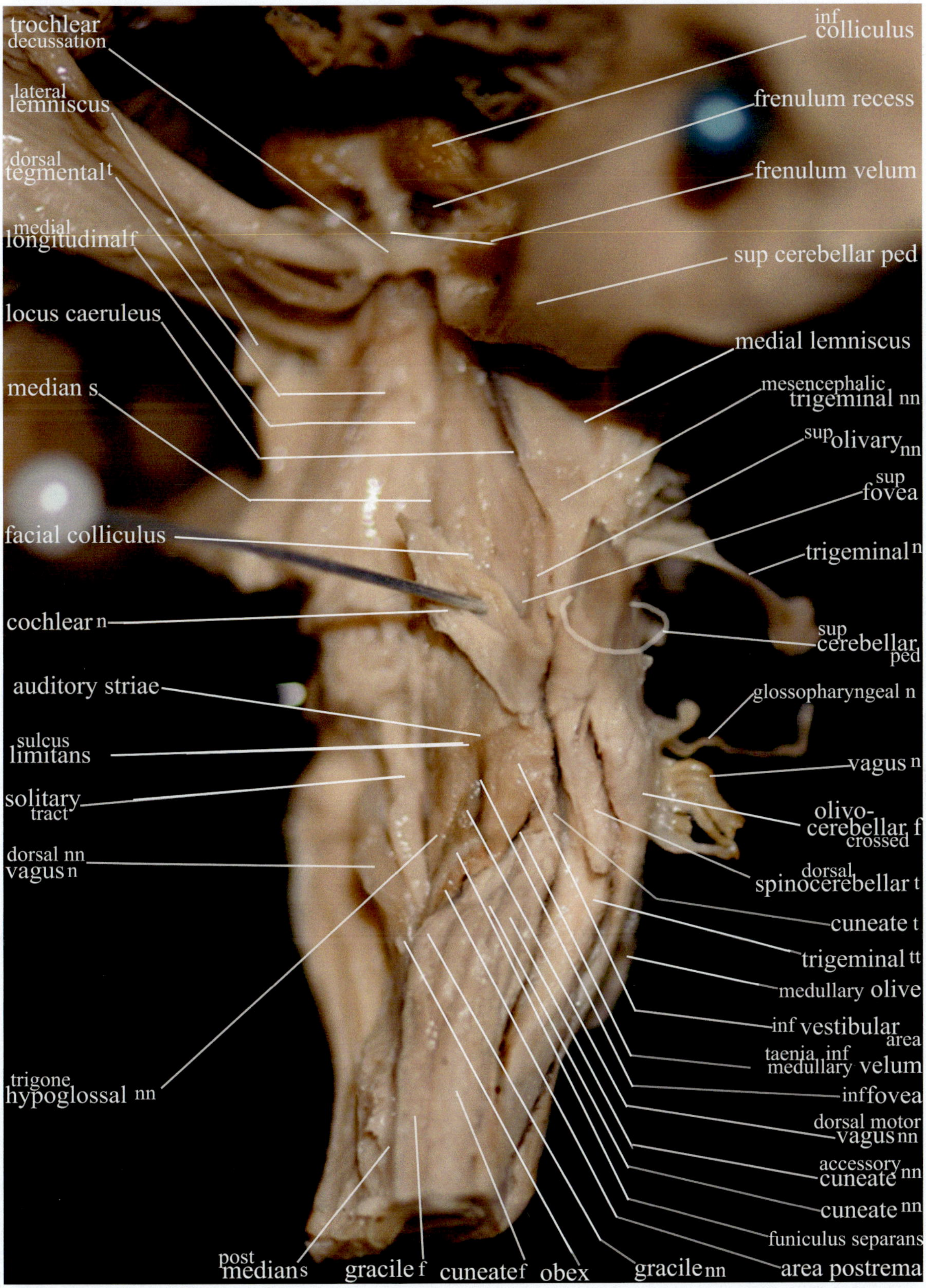

Fig. 8.52 (continued)

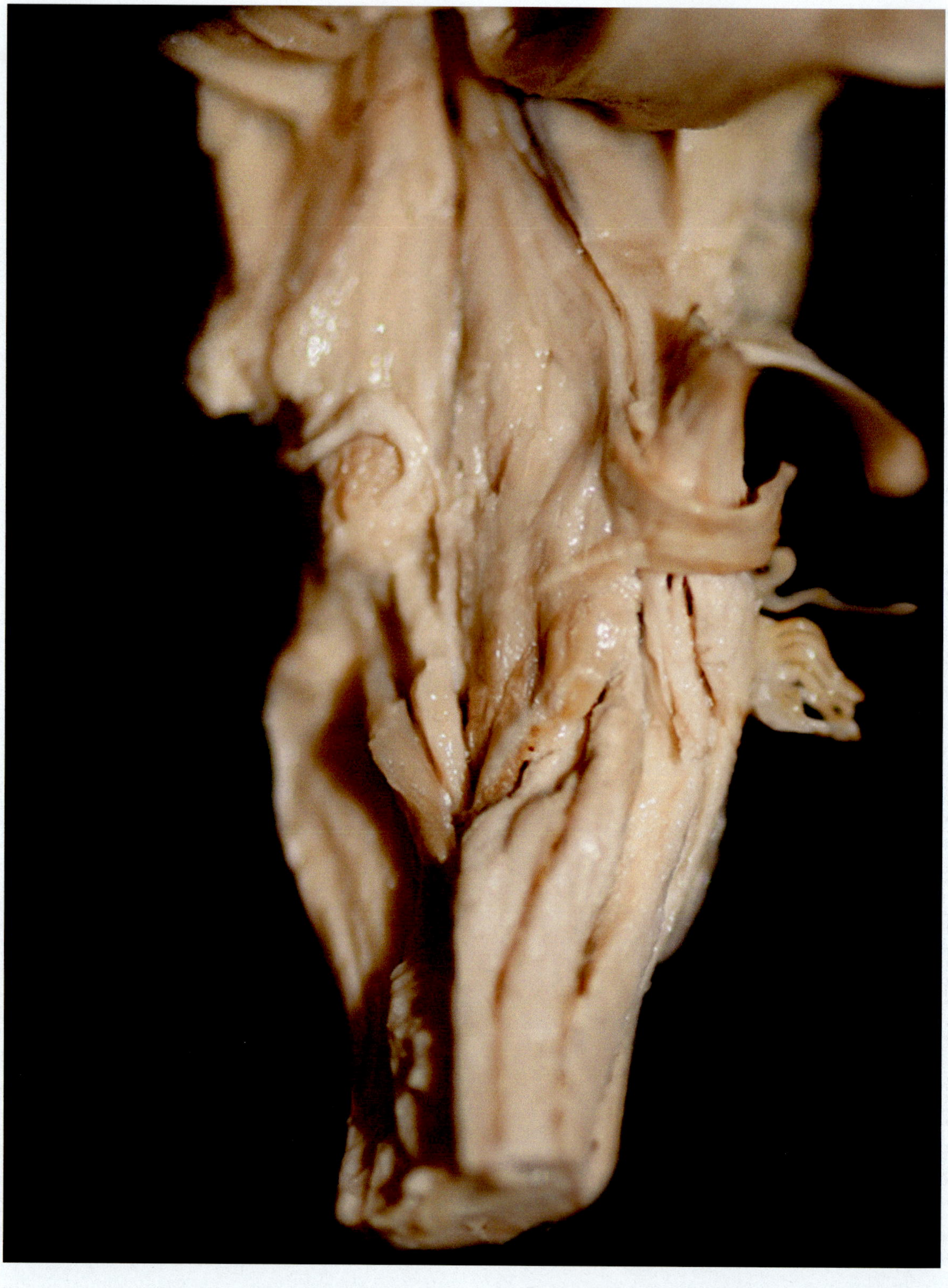

Fig. 8.53 Anatomical preparation of the posterior medulla. The restiform and juxtarestiform bodies that constitute the inferior cerebellar peduncle are identified. The cochlear nerve in its ventral-to-dorsal trajectory turns around the restiform body to reach the dorsal and ventral cochlear nuclei. The dorsal cochlear nucleus is sometimes recognized at the surface as the acoustic tubercle. *f* fasciculus, *inf* inferior, *n* nerve, *nn* nucleus, *s* sulcus, *sup* superior, *t* tract, *tt* tract

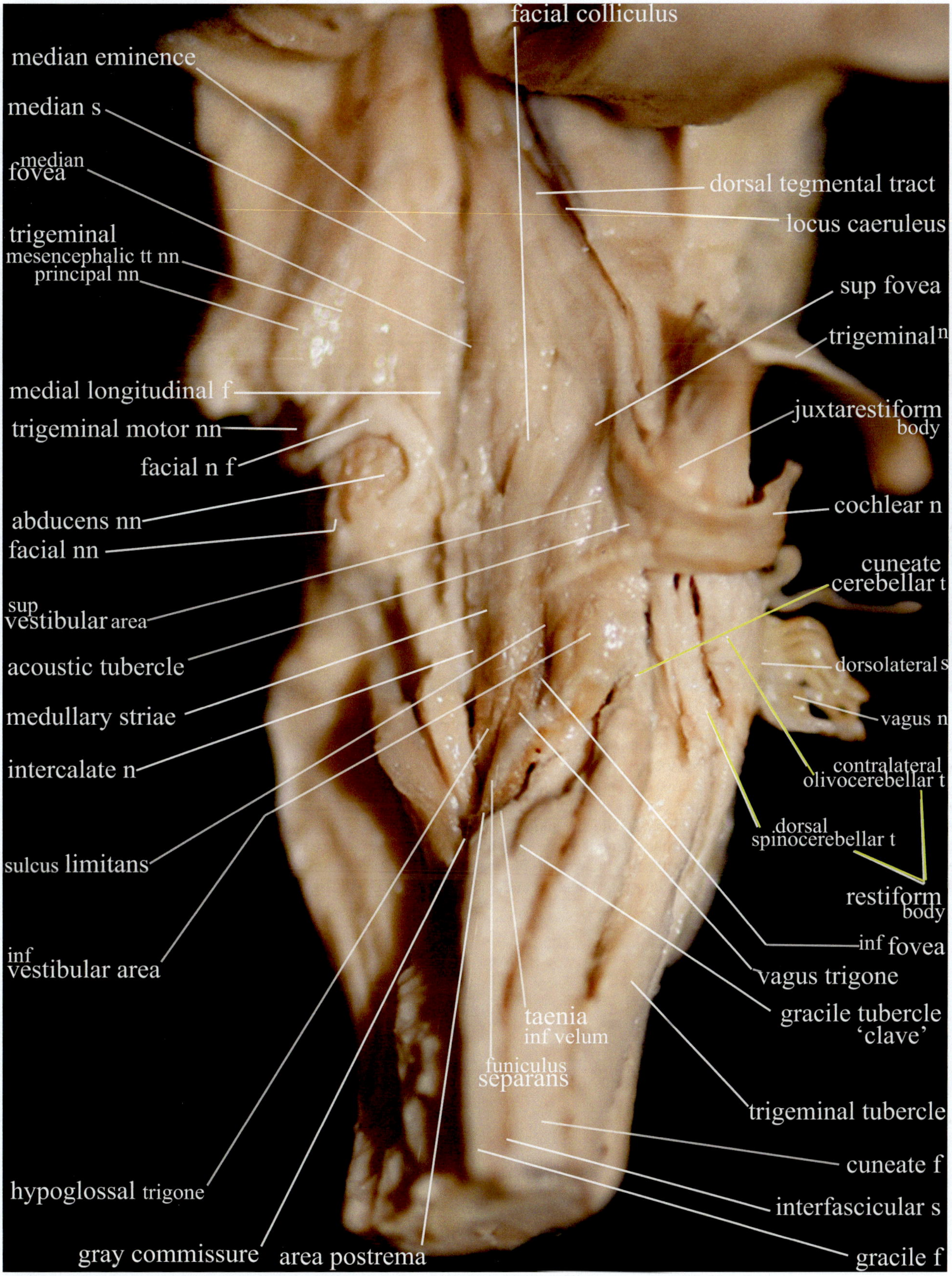

Fig. 8.53 (continued)

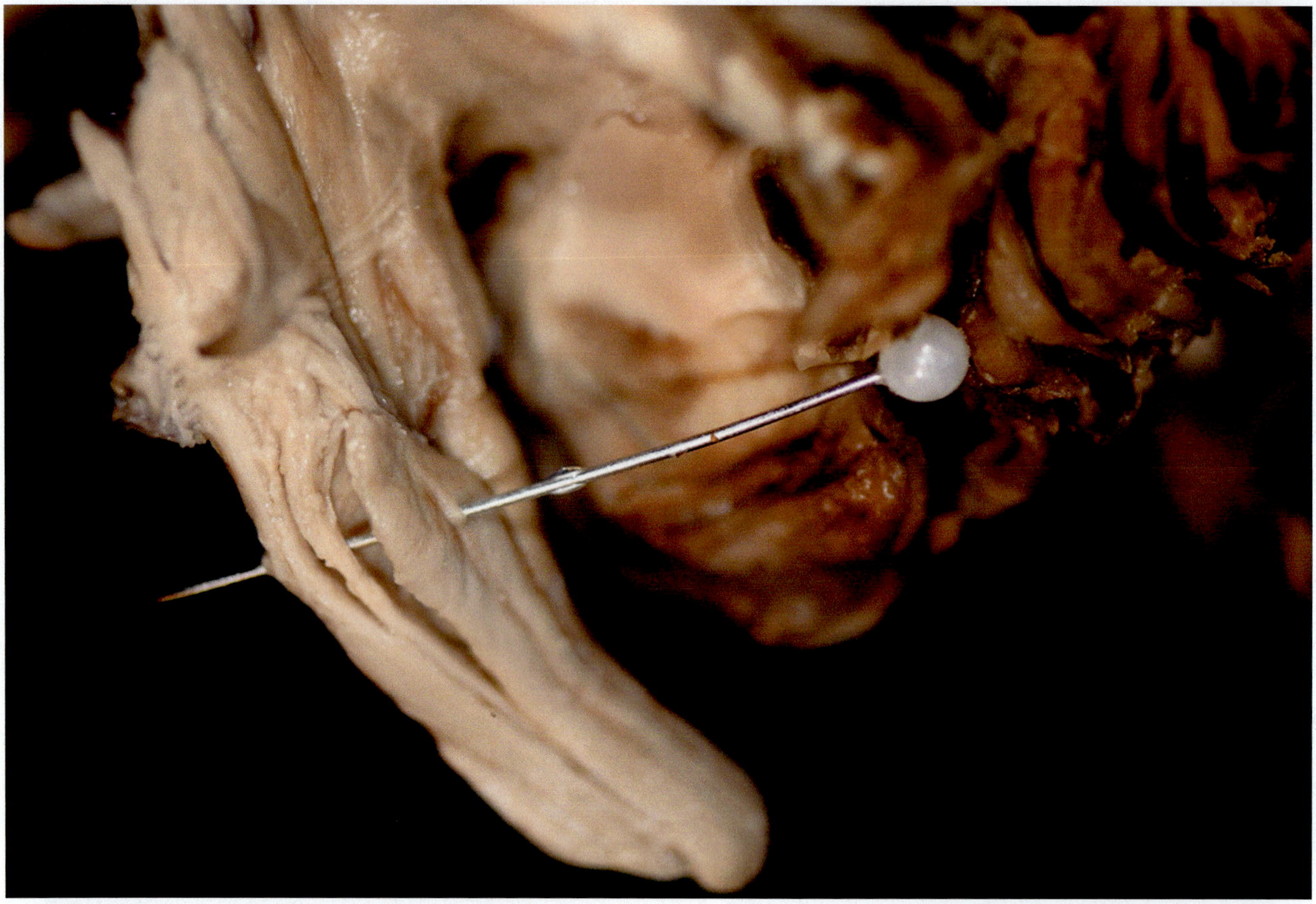

Fig. 8.54 Anatomical preparation of the posterior funiculus of the lower medulla and the restiform body. The restiform body is mainly composed of fibers of the cuneate, dorsal spinocerebellar, and contralateral olivocerebellar tracts. The cuneate tract is also known as the fastigiobulbar tract and connects the accessory cuneate nucleus with the contralateral fastigial nucleus of the cerebellum. The dorsal spinocerebellar tract ascends through the lateral funiculus, ventral to the spinal trigeminal nucleus, and is separated from the fibers of the ventral spinocerebellar tract by the intra-axial nerve roots of the dorsal nucleus of the vagus. It is also separated from the spinocerebellar tract by the roots of the solitary tract, nucleus ambiguus, and sensory trigeminal parts that compose the vagus, glossopharyngeal, and intermedial nerves. The contralateral olivary fibers, dorsal to the spinothalamic and ventral spinocerebellar tracts at the lateral recess of the medulla, join the cuneate tract, located dorsally, and the dorsal spinocerebellar tract, located laterally, to constitute the restiform body. *f* fasciculus, *inf* inferior, *n* nerve, *nn* nucleus, *ped* peduncle, *post* posterior, *s* sulcus, *sup* superior, *t* tract

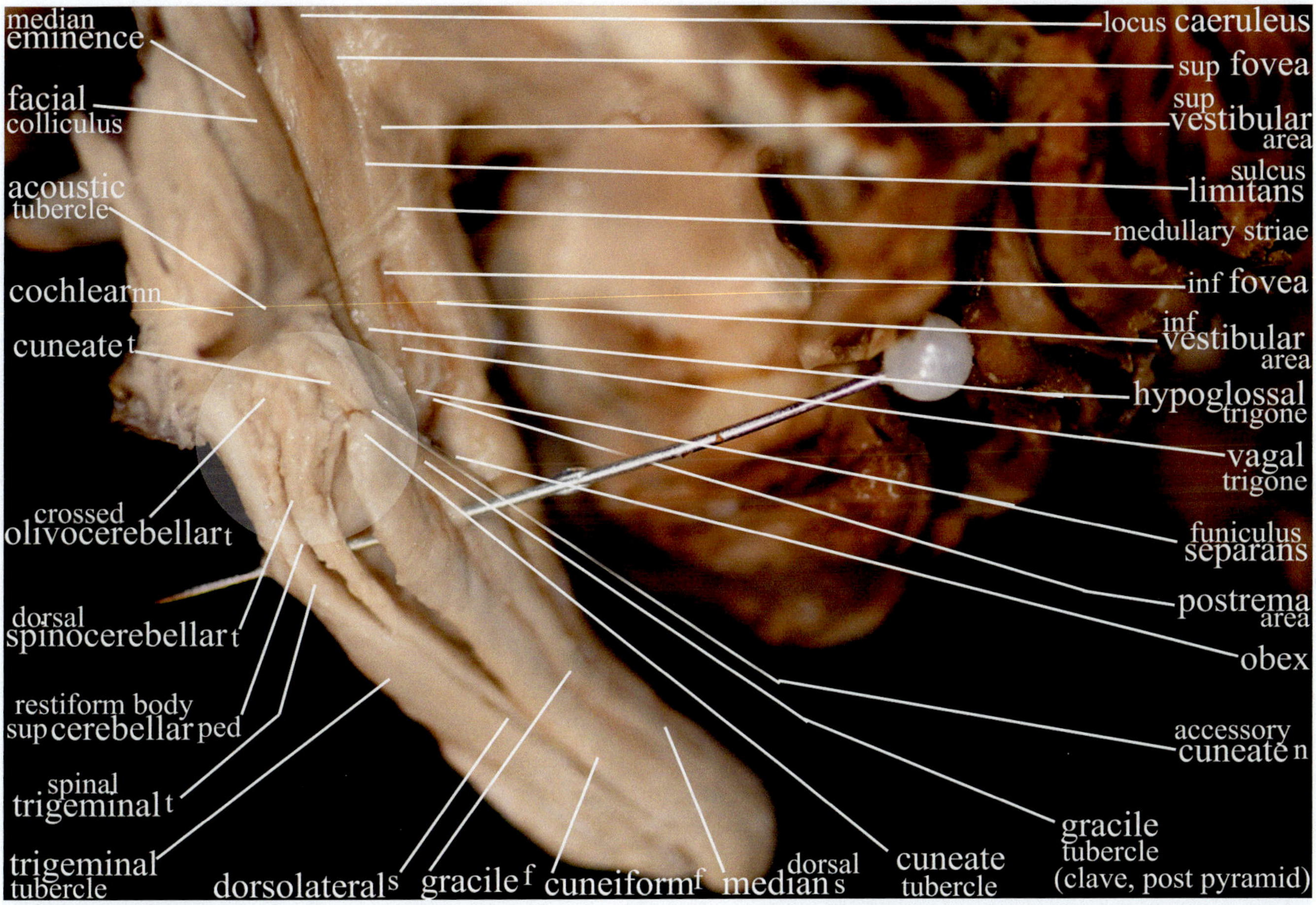

Fig. 8.54 (continued)

The cranial portion is very distinct from the caudal. The opening of the central canal constitutes the inferior limit of the fourth ventricle. The medial posterior sulcus opens, the posterior funiculus separates from the contralateral side and ascends obliquely, and the posterior gray commissure disappears. The tubular form of the central canal flattens and expands to form the inferior half of the floor of the fourth ventricle (Fig. 8.55).

The caudal floor of the fourth ventricle has a triangular shape with the vertex facing inferiorly (Figs. 8.52, 8.53, and 8.55). It is limited laterally by the extension of the gracile and cuneate fasciculi. At the superior part of the medulla, the gracile fasciculus is also named the posterior pyramid, and the cuneate fasciculus is the restiform body. The gracile fasciculus harbors a small oval protuberance at the level of the obex called the gracile tubercle or clave. The superior part of the cuneate fasciculus seems to merge with the inferior cerebellar peduncle at the surface [2]. Hence, the superior cerebellar peduncle is also denoted the restiform body. Ventrolateral to the restiform body, the dorsal lateral sulcus is the emergence of the ninth, tenth, and eleventh cranial nerves. The dorsolateral sulcus separates the posterior from the lateral funiculus.

Internal Organization of the Medulla

The medulla is separated into two exactly symmetric halves. At the caudal medulla, as noted previously, the white and gray matter resembles the arrangement of the cervical spinal cord. At the superior medulla, the internal organization differs substantially.

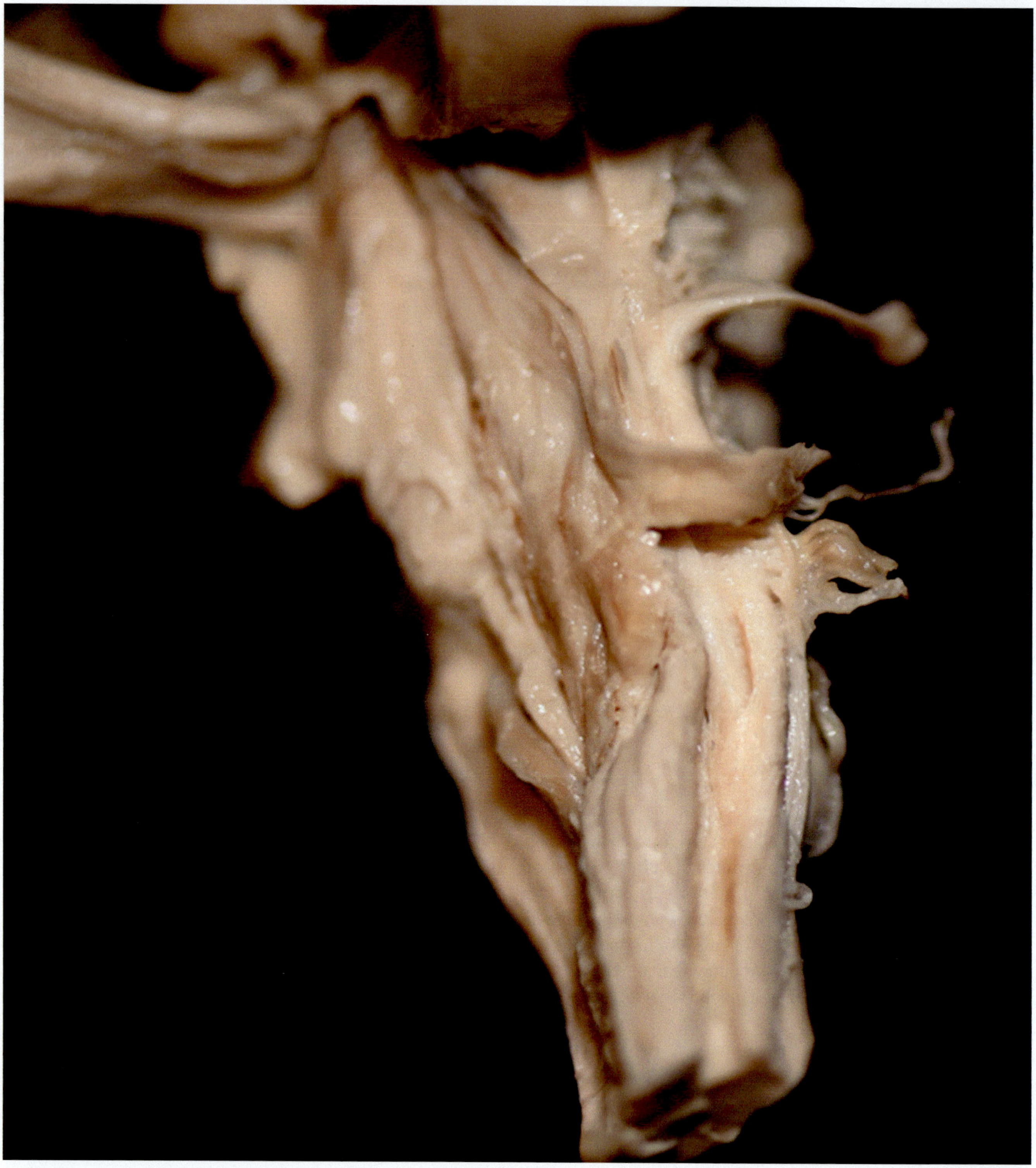

Fig. 8.55 Anatomical preparation of the dorsal brainstem after the removal of the cerebellar peduncles. The dorsolateral sulcus separates the dorsal and lateral funiculus. *f* fasciculus, *inf* inferior, *lat* lateral, *n* nerve, *ped* peduncle, *sup* superior, *t* tract, *tt* tract (©Kadri 2023. All Rights Reserved)

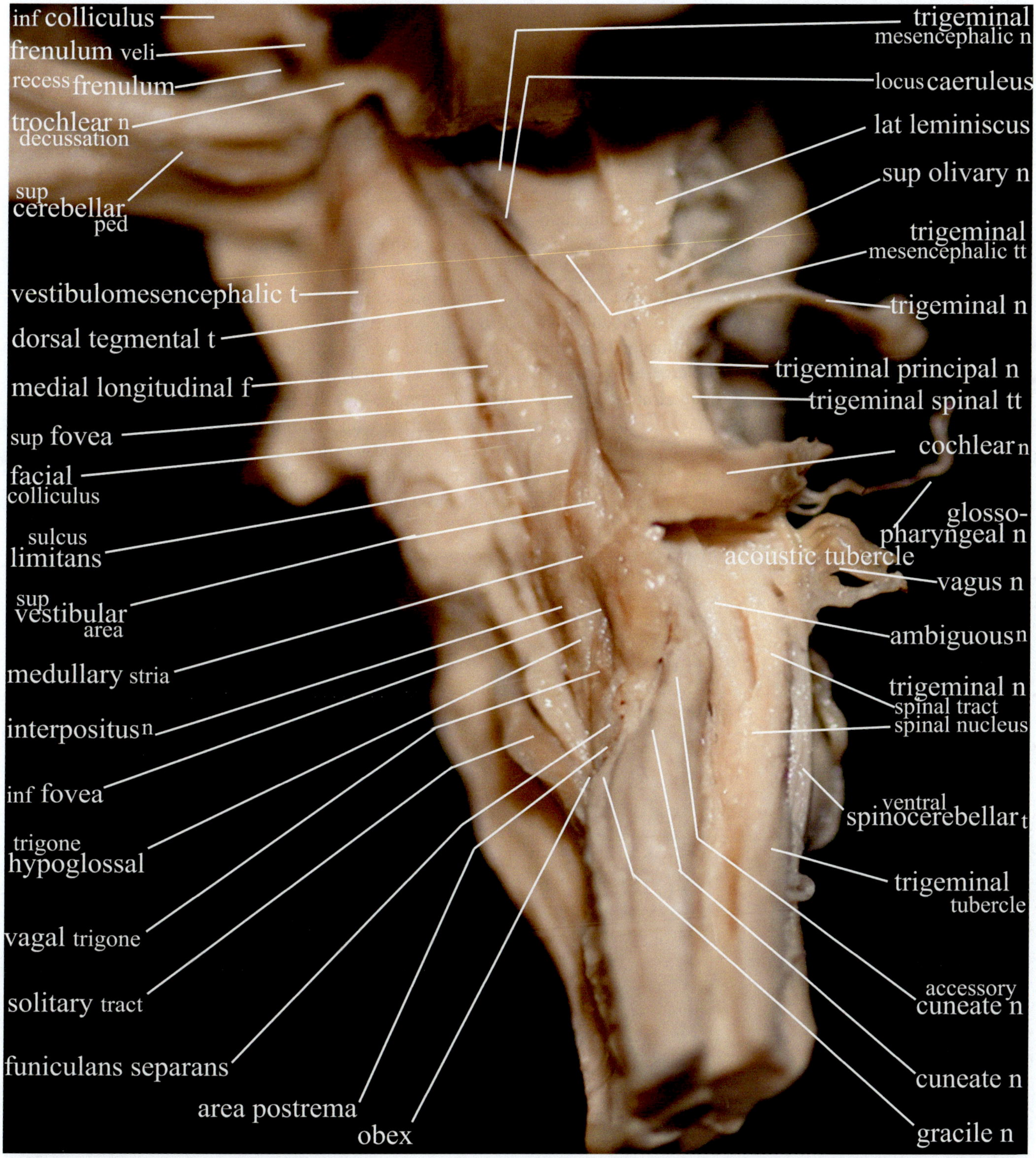

Fig. 8.55 (continued)

Equivalent White Matter Spinal Organization within the Medulla

The medulla contains eight major fiber bundles. These fibers are divided according to their location in the anterolateral and posterior funiculus. The anterolateral funiculus contains the direct and crossed pyramidal tracts, the direct dorsal cerebellar tract, the crossed ventral cerebellar tract (Gowers' fasciculus), the anterolateral ascendent fasciculus, and the medial longitudinal fasciculus. The posterior funiculus contains the gracile and cuneiform fasciculi (Figs. 8.52, 8.53, 8.54, and 8.55) [2].

The direct pyramidal tract is the most anterior and continues within the anterior funiculus of the medulla. Its fibers cross at the level of its termination through the anterior commissure of the spinal cord. The crossed pyramidal tract is most important. It crosses at the pyramidal decussation toward the posterior part of the lateral funiculus (Fig. 8.56) [2].

The gracile and cuneate fasciculi are derived from the sensory components of the spinal nerves, which enter the posterior column through the dorsal nerve roots at the dorsolateral sulcus. The caudal fibers are the most internal. The cuneate fasciculus fibers arrive from the first thoracic and cervical dorsal roots and are separated from the gracile fibers coming from the intermediate sulcus. Together, the cuneiform and gracile fasciculi are named the posterior sensory fasciculus. These fasciculi ascend uninterrupted to synapses in their respective nuclei, named after them, within the posterior medulla. From the gracile and cuneate nuclei, the fibers cross the midline and constitute the sensory decussation (Fig. 8.57) [1, 2].

The spinocerebellar connections can be direct or crossed (Fig. 8.58). At the lower part of the medulla, the direct and crossed fasciculi unite. The ventral spinocerebellar fasciculus is also known as the fasciculus of Gowers and is referred to as the double-crossed fasciculus. Its initial decussation is at the anterior white commissure of the spinal cord, from which it ascends on the surface of the lateral funiculus to connect with the lateral nucleus of Bechterew in the medulla and decussates to enter the cerebellum via the superior cere-

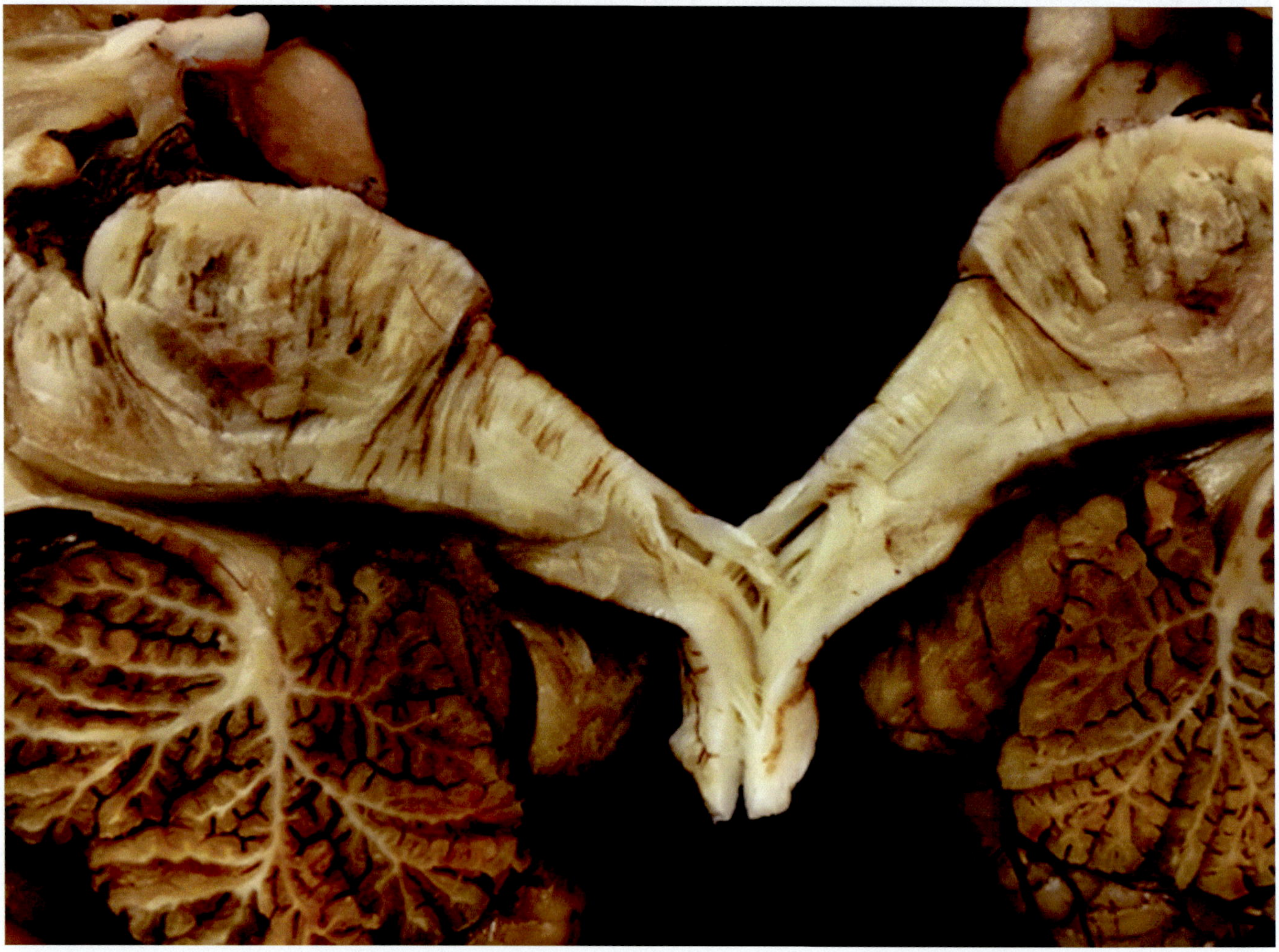

Fig. 8.56 Anatomical preparation of the decussation of the pyramidal tracts. This is a posterior view of the cerebellum and brainstem split along the midline. *f* fasciculus, *inf* inferior, *lat* lateral, *post* posterior, *sup* superior (©Kadri 2023. All Rights Reserved)

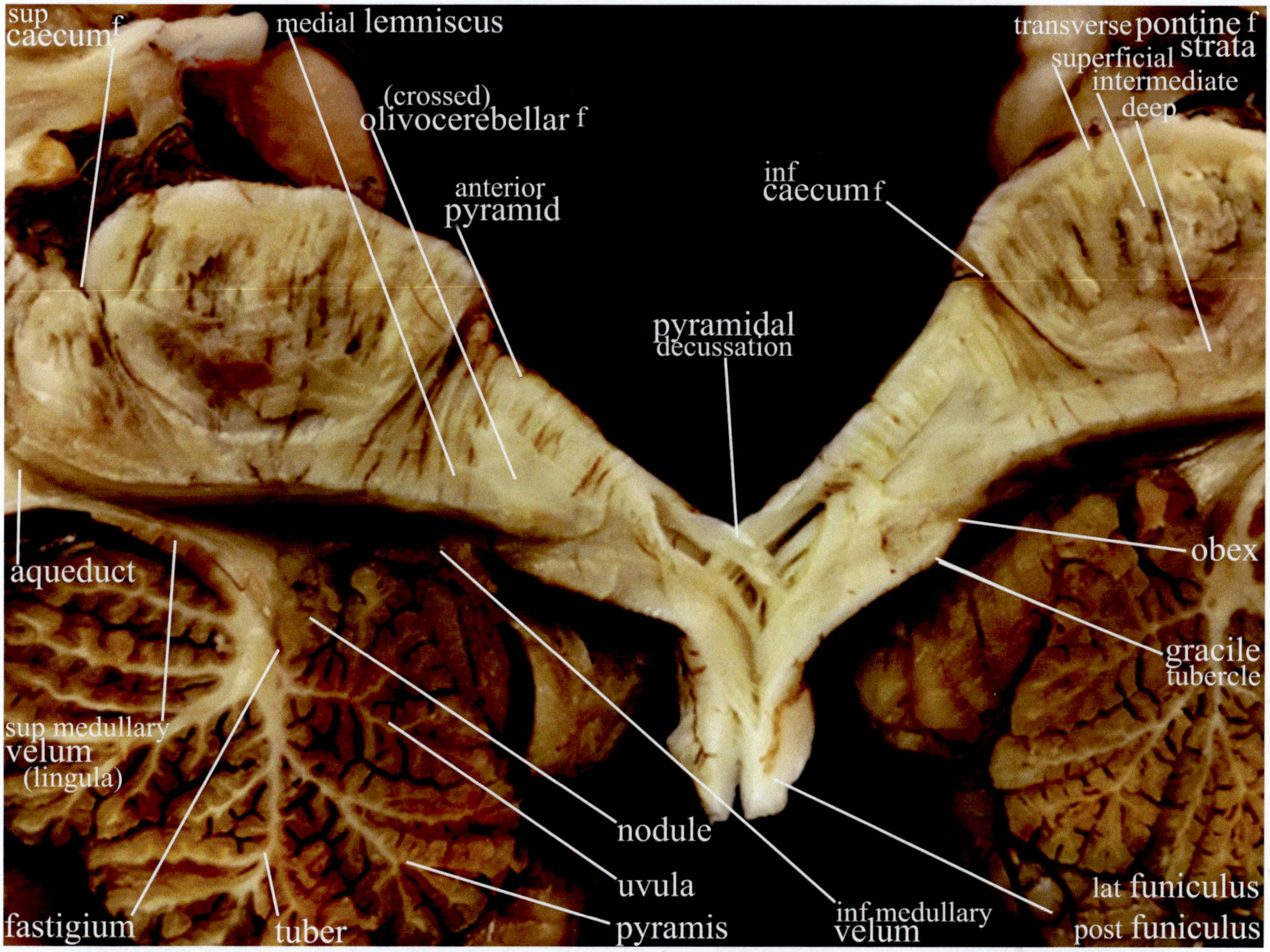

Fig. 8.56 (continued)

bellar peduncle [1]. The direct dorsal spinocerebellar fasciculus ascends toward the vermis cerebellar through the inferior cerebellar peduncle [1, 2]. The inferior cerebellar peduncle turns on top of the superior cerebellar peduncle.

The anterolateral ascendent fasciculus is mainly composed of spinal fibers directed toward the nuclei of the reticular formation within the medulla [1]. This fasciculus is divided into anterior and posterior segments. The anterior segment is smaller and mainly connects to the interolivary reticular formation within the medulla (spinomedullary). The posterior segment is more voluminous, and its fibers project to the olive (spino-olivary fibers), to the pontine (spinopontine) and mesencephalic (spinomesencephalic) reticular formations, to the inferior colliculi (spinotectal), and to the ventral nuclei of the dorsal lateral thalamus (spinothalamic). The fibers of the anterolateral ascending fasciculus run toward the midline, posterior to the sensory decussation, but do not cross it [1]. They constitute a lamina of fibers permeated by several nuclei as they ascend toward the pons and mesencephalon. This netlike arrangement of white and gray matter is denoted as the reticular formation (Latin *retis*, meaning net). At the midline, the reticular formation is denoted the raphe because its appearance is similar to a suture line.

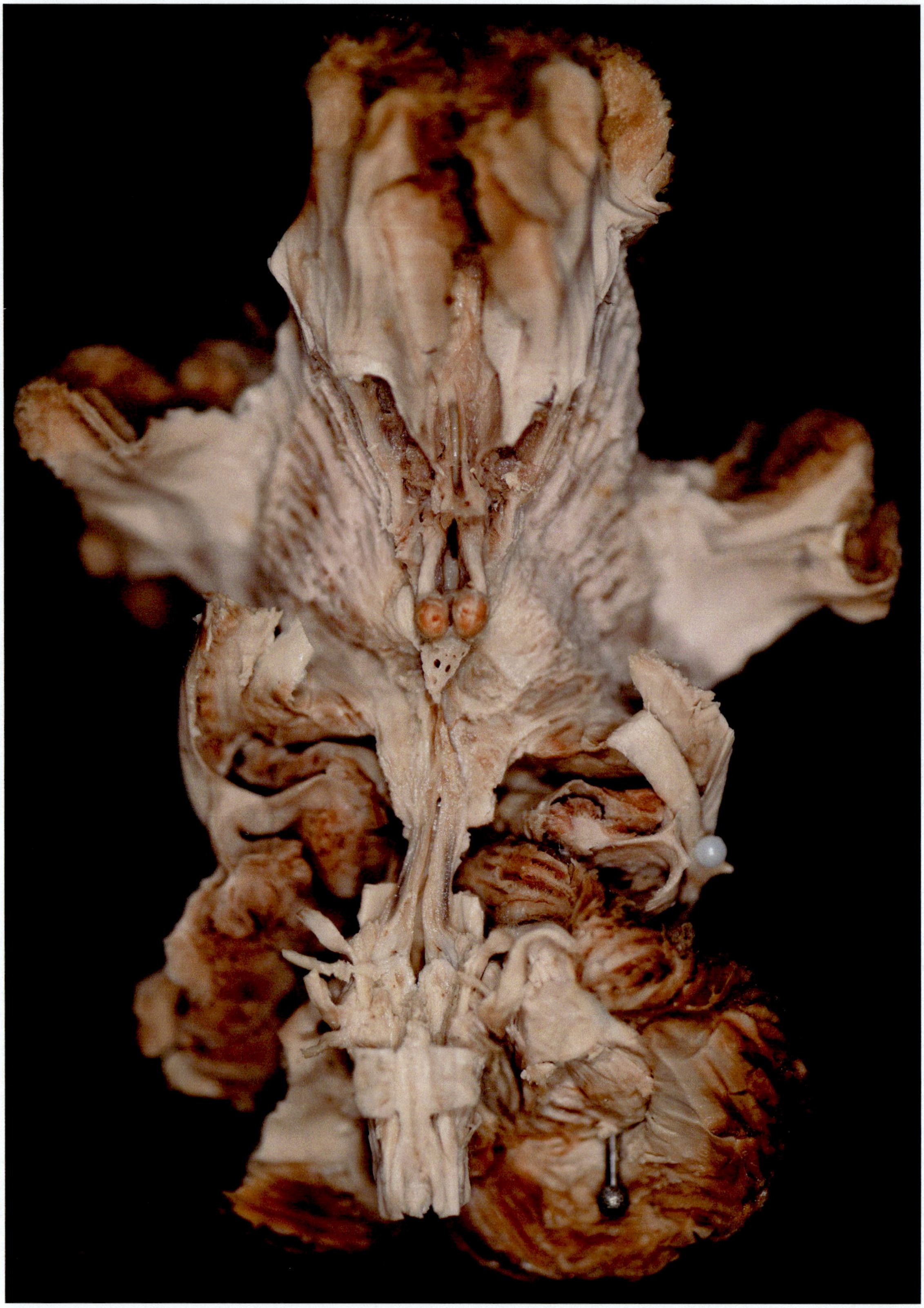

Fig. 8.57 Anatomical preparation of the dorsal tegmentum of the brainstem. The base of the brainstem, the extrapyramidal connections of the cerebellum, the red nucleus, and the medullary olives have been removed. *ant* anterior, *f* fasciculus, *g* gyrus, *inf* inferior, *lat* lateral, *med* medial, *n* nerve, *nn* nucleus, *ped* peduncle, *post* posterior, *sup* superior, *t* tract (©Kadri 2023. All Rights Reserved)

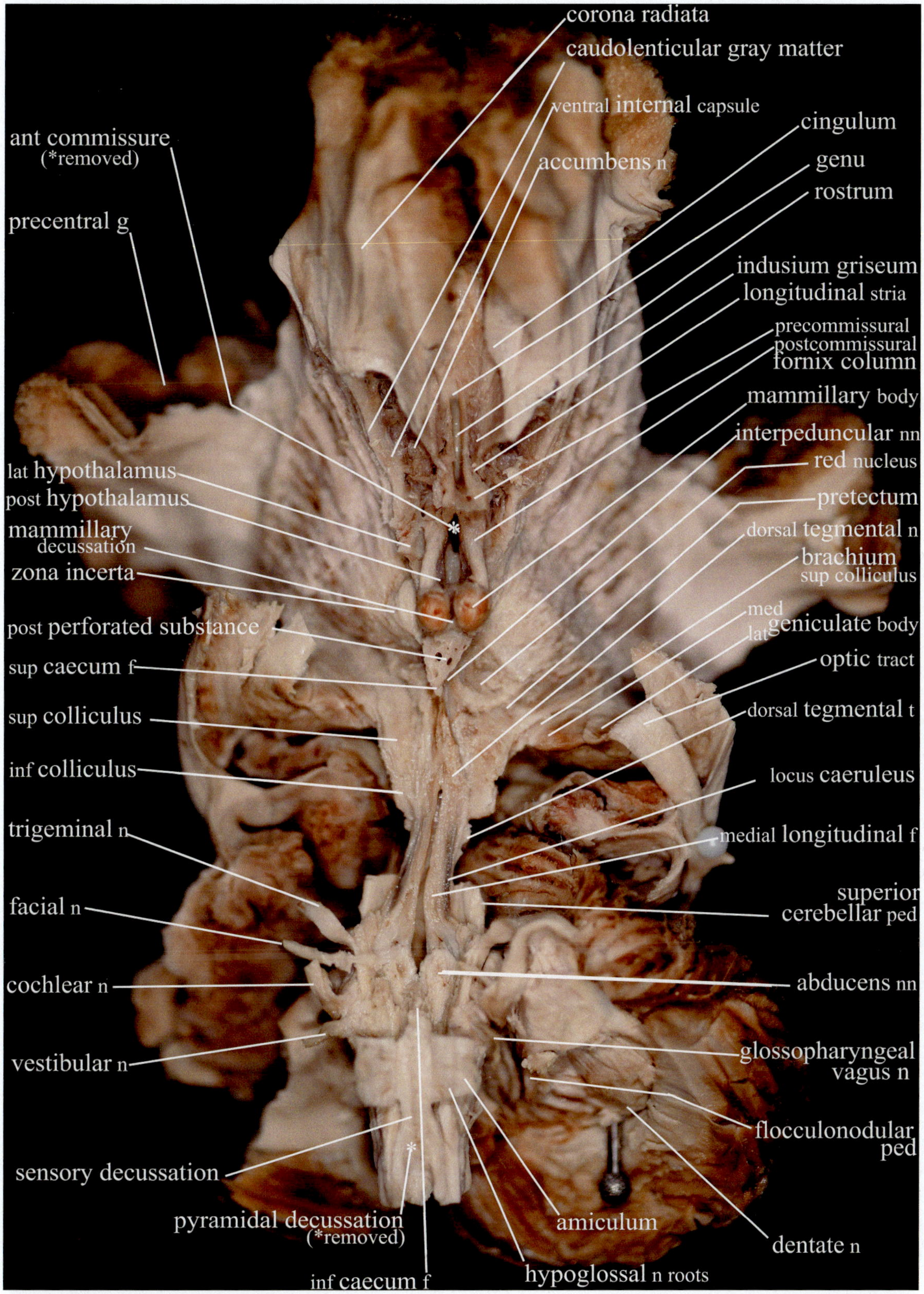

Fig. 8.57 (continued)

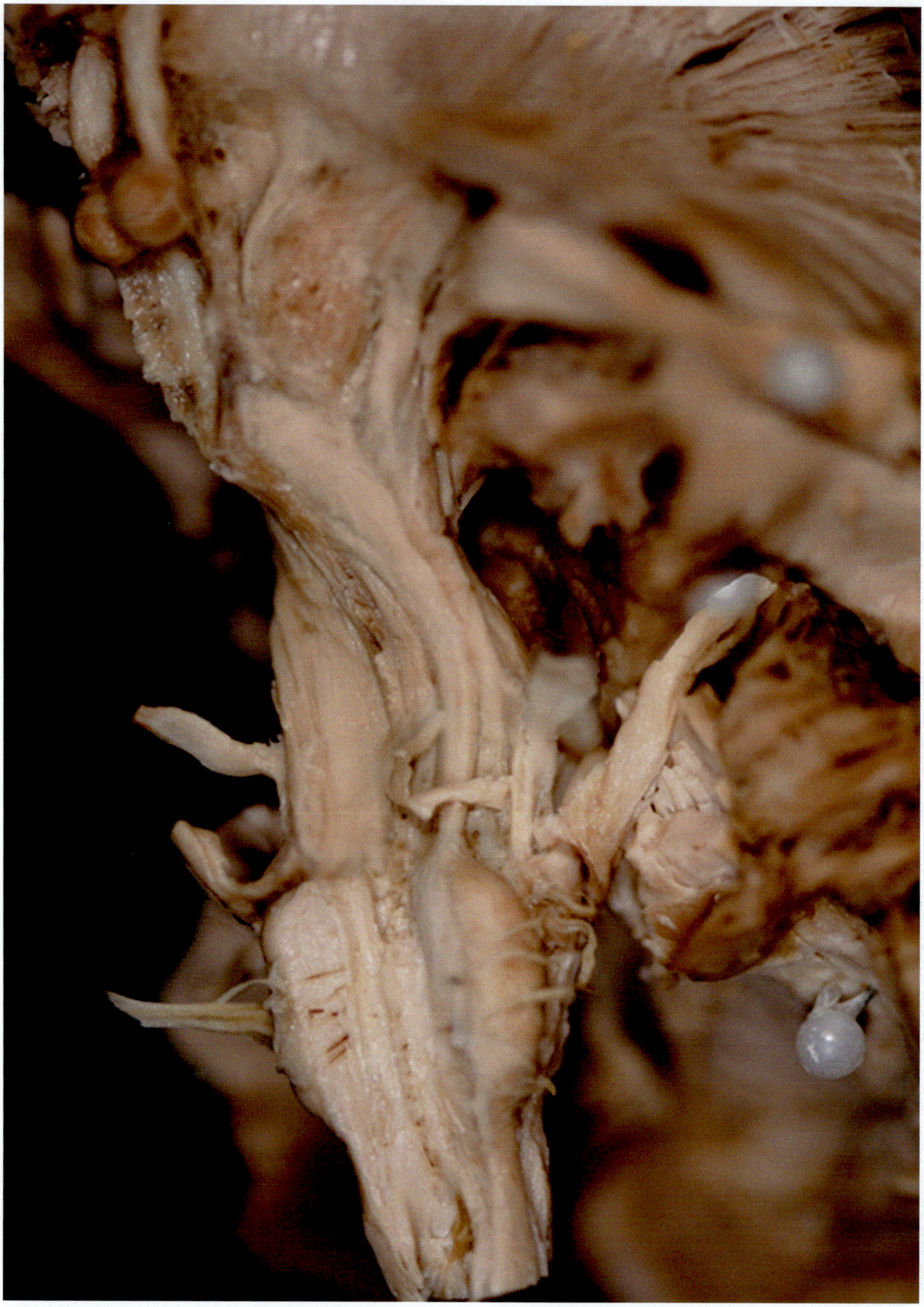

Fig. 8.58 Anatomical preparation of the ventral tegmentum of the brainstem. Ventral to dorsal at the pontomedullary junction, the medial lemniscus is sequentially followed by the central tegmental tract, the spinothalamic tract, and the ventral spinocerebellar tract. *f* fasciculus, *inf* inferior, *n* nerve, *nn* nucleus, *sup* superior, *t* tract, *tt* tract (©Kadri 2023. All Rights Reserved)

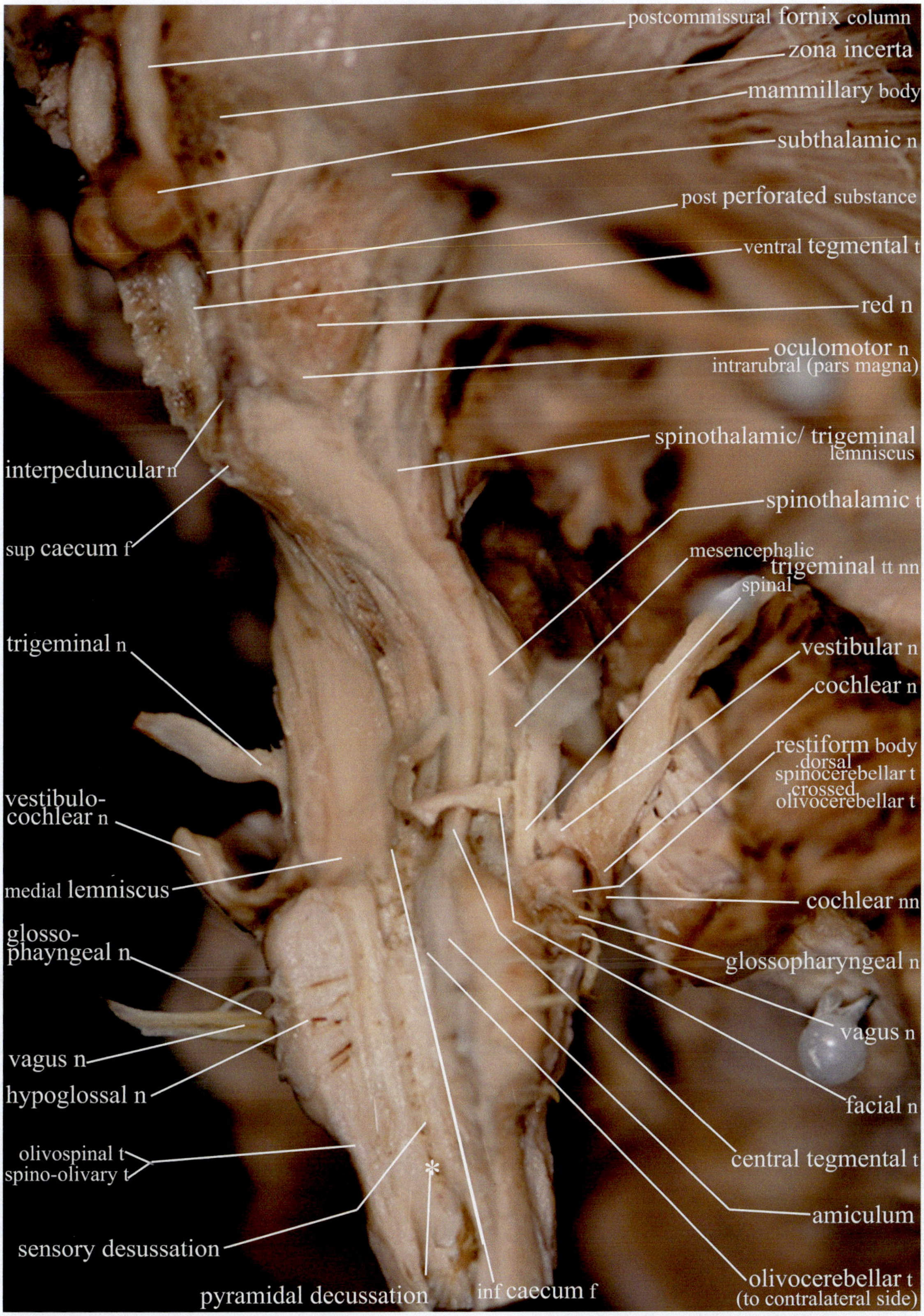

Fig. 8.58 (continued)

Equivalent Gray Matter Spinal Organization within the Medulla

The gray matter of the spinal cord extends toward the medulla but undergoes profound changes because of the pyramidal decussation, the sensory decussation (of Reil), the rise of the fourth ventricle, and the arcuate fibers. The ventral horn is also known as the motor column. The dorsal horn is also known as the sensory column [2, 3].

The motor and sensory decussations divide the central gray matter of the lower medulla into four columns [2]. The pyramidal decussation divides the anterior horn of the spinal cord into a ventral and a dorsal motor column (Fig. 8.59). In an analogous manner, the sensory decussation divides the posterior horn into an internal and an external sensory column. The dorsal motor and the internal sensory columns maintain their relations around the central canal of the spinal cord through the lower medulla (Fig. 8.60).

The opening of the central canal of the spinal cord into the fourth ventricle enlarges the ependymal surface and modifies the relationships of the columns. The dorsal motor column is placed on the surface and assumes a midline position in the floor of the fourth ventricle. As the posterior funiculus separates from the midline, the internal sensory column is placed lateral to the dorsal motor column. The ventral motor column and the external sensory column assume a more ventral position. Because of this topographical realignment, the internal and external sensory columns are also designated as posterior and anterior sensory columns, respectively (Figs. 8.50, 8.58, and 8.59).

The four columns are interrupted by the arcuate fibers. The fibers of sensory decussation of the posterior column are called internal arcuate fibers. The olivocerebellar fibers, connecting with the contralateral cerebellum via the inferior cerebellar peduncle, may run at the surface and cover the

Fig. 8.59 Anatomical preparation of the tegmentum of the medulla after the removal of the pyramid and extrapyramidal connections. The motor nuclei of the trigeminal, abducens, facial, and ambiguus nuclei are exposed. *f* fasciculus, *inf* inferior, *n* nerve, *nn* nucleus, *ped* peduncle, *sup* superior, *t* tract, *tt* tract (©Kadri 2023. All Rights Reserved)

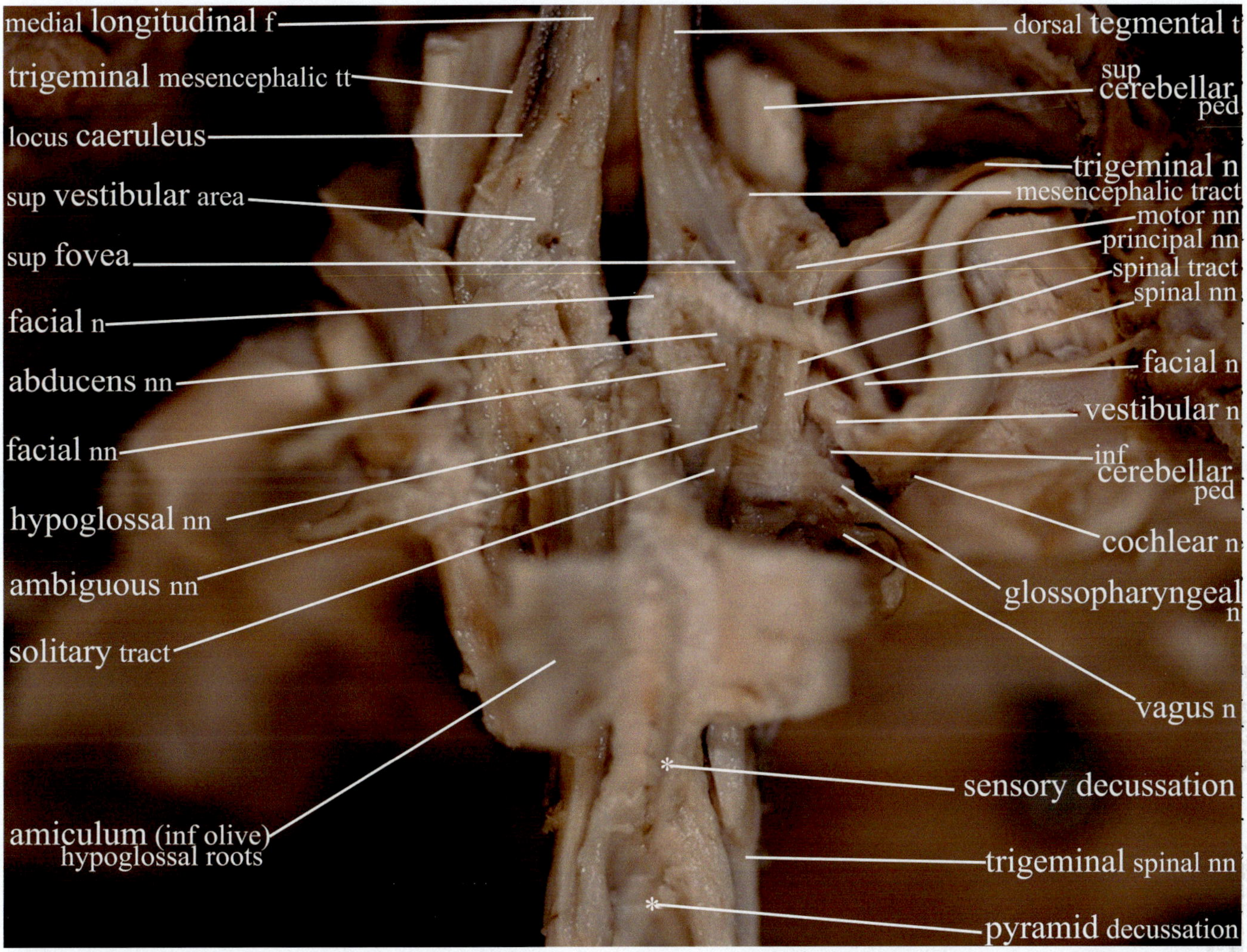

Fig. 8.59 (continued)

ventricle, where the cochlear nerve projects. The vestibular nerve terminates in the external region. At the superior part of the fourth ventricle, the posterior sensory column forms the locus coeruleus, which receives some trigeminal fibers (see below). The anterior sensory column is large and extends from the sensory decussation to the pons [2].

The Medullary Cranial Nerves

The cranial nerves, the nuclei of which are in the medulla, are the hypoglossal, accessory, vagus, glossopharyngeal, acoustic, vestibular, facial, and sensory nuclei of the trigeminal nerve. The hypoglossal nerve is exclusively motor, and its fibers emerge from the pre-olivary sulcus (Figs. 8.47, 8.48, 8.49, 8.50, 8.58, 8.59, and 8.60). Its nucleus is situated in the floor of the fourth ventricle at the internal region. The accessory nerve is exclusively motor, and its origin is from the first to the sixth segments of the spinal cord.

The vagus nerve is mixed, and its fibers emerge from the dorsolateral sulcus of the medulla (Figs. 8.47, 8.48, 8.50, 8.53, 8.58, and 8.60). The sensory fibers are afferents of the than any other sensory system, through the medial vestibulocerebellar fasciculus, the medial longitudinal fasciculus, and the vestibulospinal tract. The inferior, medial, and superior vestibular nuclei project mainly toward the flocculonodular lobule and the uvula through the medial vestibulocerebellar fasciculus of the juxtarestiform body. The flocculonodular lobule is also referred to as the vestibular cerebellum. The juxtarestiform body is the medial part of the inferior cerebellar peduncle formed by the medial and lateral vestibulocerebellar fasciculi. The direct and crossed vestibulospinal and vestibulomesencephalic connections are some of the main components of the medial longitudinal fasciculus. The bilateral medial vestibulospinal tracts mainly project from the medial vestibular nuclei. The ipsilateral lateral vestibulospinal tract is the main projection of the lateral vestibular nuclei and extends to the sacral levels of the spinal cord [2].

The cochlear nerve originates at the organ of Corti and projects toward the ventral and dorsal cochlear nuclei. Each of its fibers emits an ascending and a descending branch. The ascending branch reaches the ventral nucleus, which is

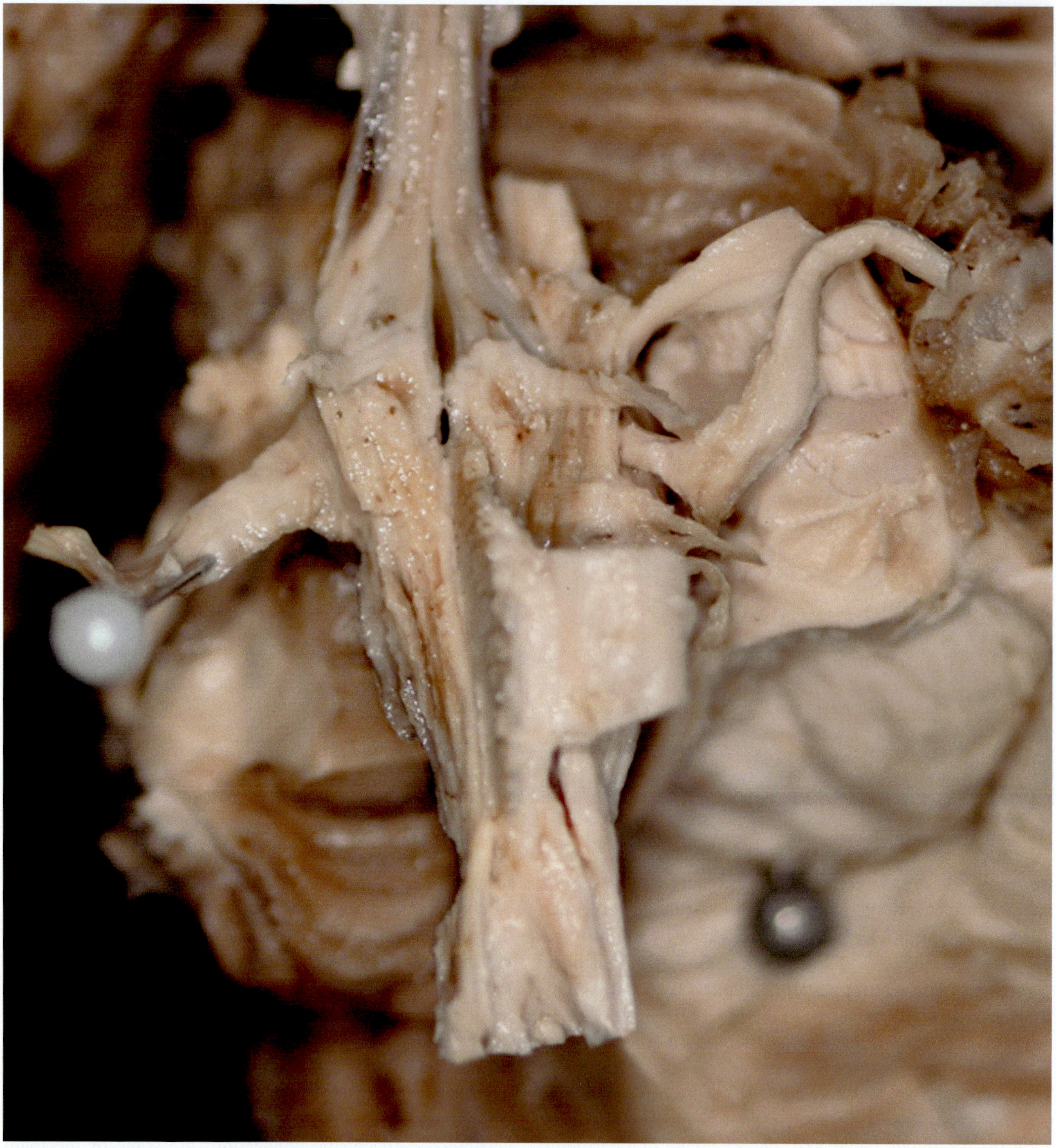

Fig. 8.60 Anterior view of an anatomical preparation of the tegmentum of the medulla showing the cranial nerve nuclei on the tegmentum of the medulla. *f* fasciculus, *inf* inferior, *n* nerve, *nn* nucleus, *ped* peduncle, *sup* superior, *t* tract

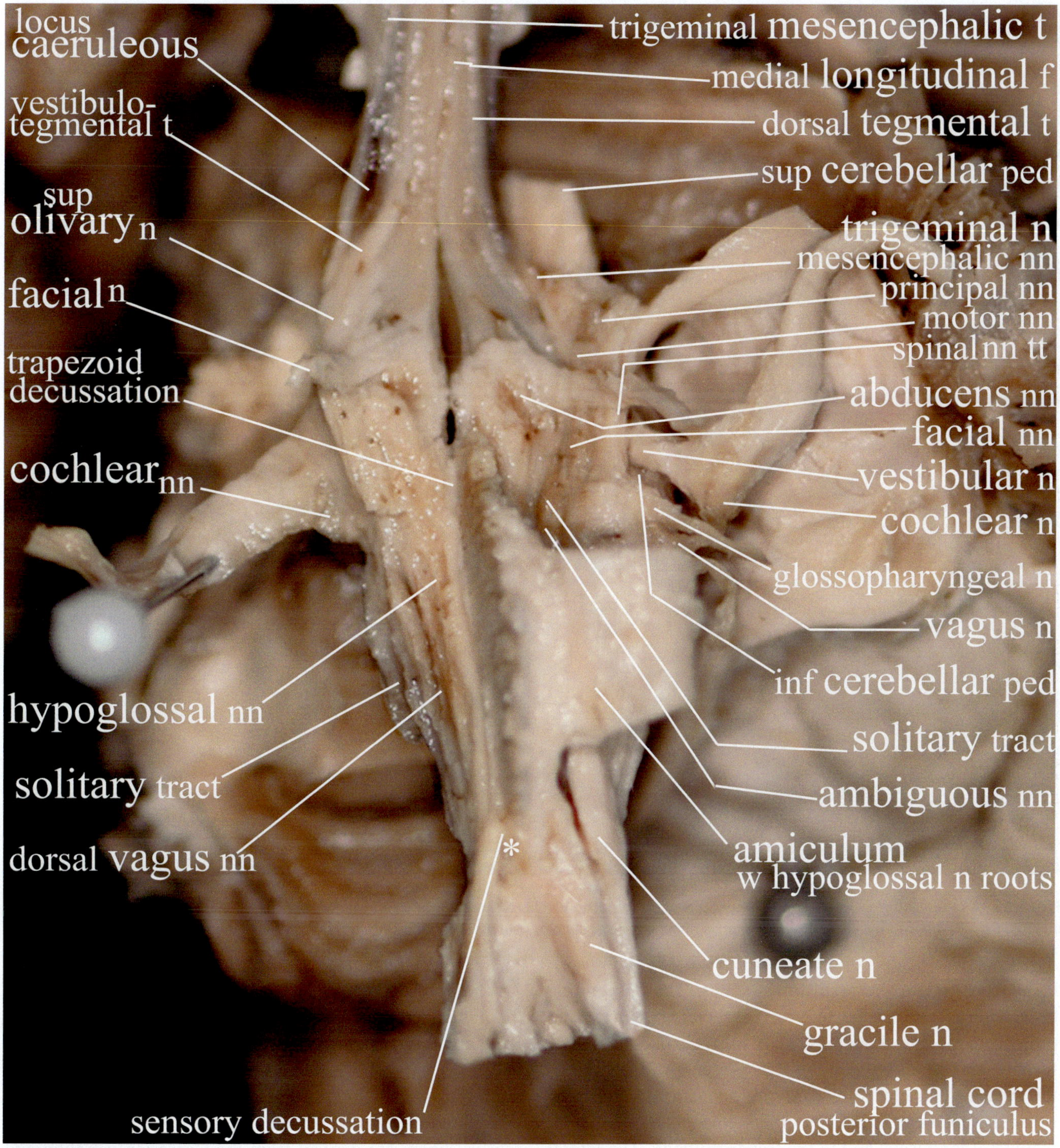

Fig. 8.60 (continued)

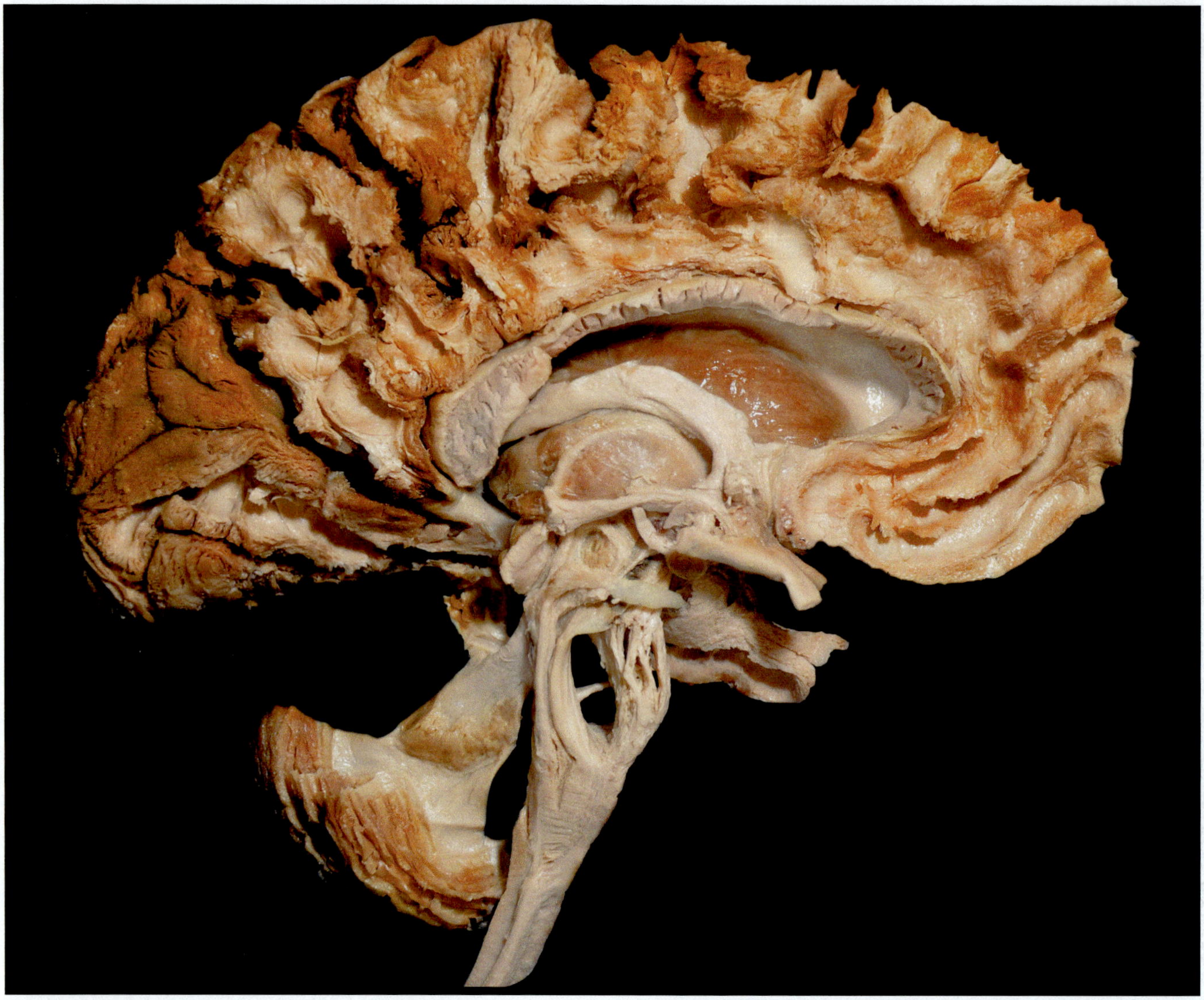

Fig. 8.61 Anatomical preparation of the long tracts of the brainstem. The spinomedullary limit is the pyramidal decussation. *ant* anterior, *f* fasciculus, *g* gyrus, *inf* inferior, *lat* lateral, *n* nerve, *nn* nucleus, *ped* peduncle, *s* sulcus, *sup* superior, *t* tract, *tt* tract (©Kadri 2023. All Rights Reserved)

located ventral to the restiform nucleus. The descending branch turns around the inferior cerebellar peduncle and projects to the dorsal nucleus [1]. In turn, the dorsal nucleus forms a small eminence at the lateral recess of the floor of the fourth ventricle, known as the acoustic tubercle [2]. From the cochlear nuclei, the fibers emanate to the lateral ipsilateral lemniscus and to the trapezoid body. The lateral lemniscus projects toward the inferior colliculi at the mesencephalic tectum. The trapezoid body contains the decussation of auditory fibers and the trapezoid nucleus (Fig. 8.61). The nuclei of the trapezoid body and the superior olivary complex are part of the integration of the reflex involving the acoustic radiation and the nucleus of the sixth nerve. The superior olive is also known as the acoustic olive [2].

The facial nerve nucleus is located at the transition between the medulla and the pons (Figs. 8.47, 8.48, 8.49, 8.50, 8.58, 8.59, and 8.60). The trigeminal nerve has sensory and motor nuclei (Figs. 8.47, 8.50, 8.51, 8.52, 8.58, 8.59, and 8.60), and the motor nucleus is in the pons. The largest portion of the extension of the sensory nuclei is located in the medulla. The sensory fibers are derived from the Gasserian ganglion and reach the pons on its ventrolateral surface. The emerging fibers of the trigeminal nerve are the boundary between the pons and the middle cerebellar peduncle. These fibers cross the transverse fibers of the pons and bifurcate into ascending and descending branches. The descending branches constitute the spinal tract of the trigeminal nerve. This tract extends to the cervical spine below the pyramidal decussation. The ascending branches constitute the mesencephalic tract.

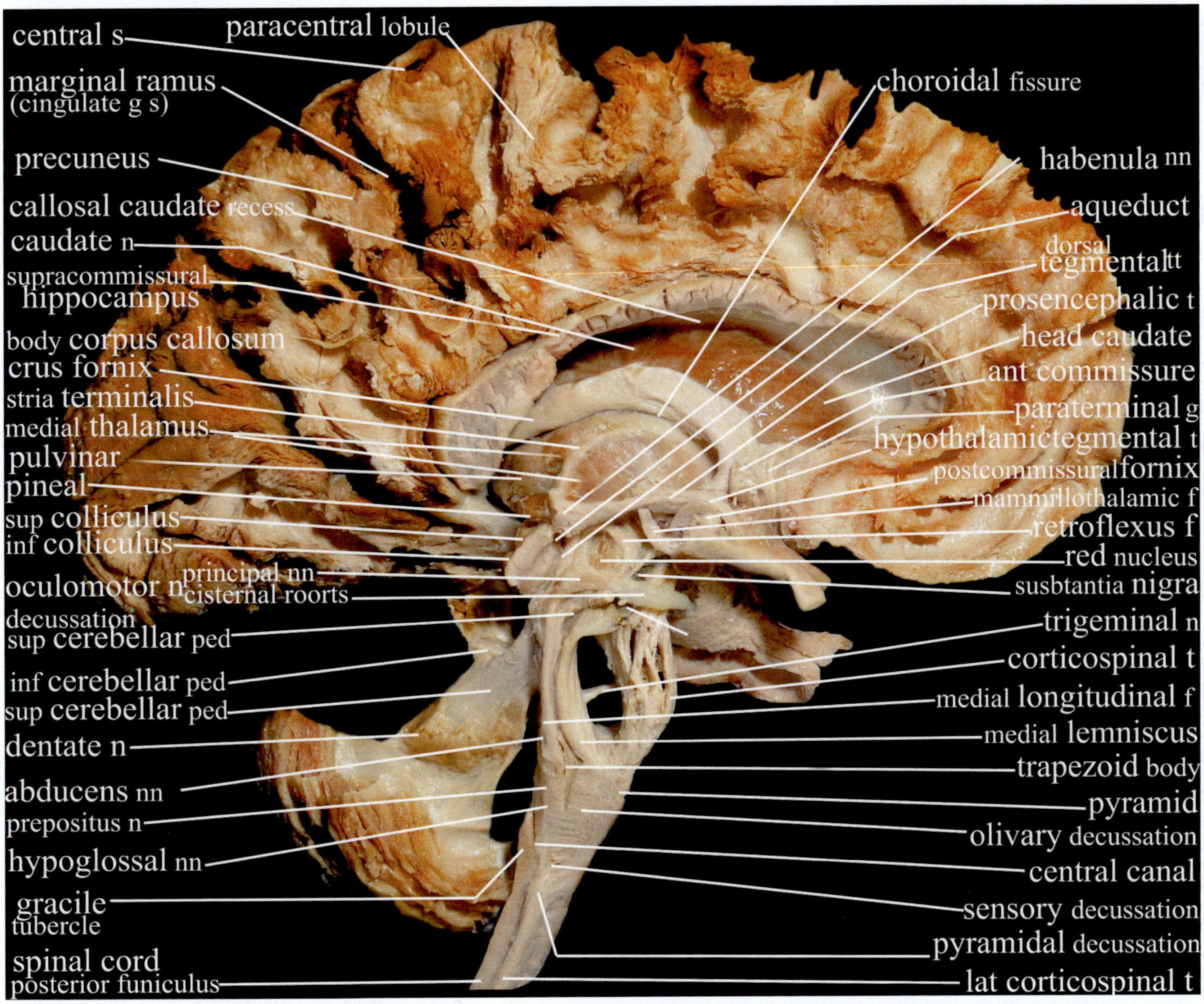

Fig. 8.61 (continued)

Individual Medullary Structures

Some structures of the medulla have no equivalents in the spinal cord. These are the cuneiform and gracile nuclei, the medial lemniscus, the lateral lemniscus, the inferior olive, the accessory nuclei of the olive, the restiform body, the arcuate fibers, and the reticular formation.

The paired gracile nuclei are limited by the posterior median sulcus at the midline. The gracile nucleus is also known as the nucleus of Goll, and its protrusion on the surface is known as the gracile tubercle [3] or the posterior pyramid and the medullary clave. It serves as the relay of the gracile fasciculus within the medulla. The intermediate sulcus limits the gracile and cuneate tubercles. The cuneate nucleus is also known as the nucleus of Burdach and the restiform nucleus. It is the relay of the cuneate fasciculus within the medulla, and its superior part seems to fuse with the inferior cerebellar peduncle; thus, it was named *restiform nucleus* [2]. The second-order neuronal axons of the cuneate and gracile nuclei cross the midline as the internal arcuate fibers, constituting the sensory decussation before ascending toward the ventral lateral thalamus as the medial lemniscus. The three-dimensional arrangement of the decussation fibers is also referred to as the piriform body [2].

The medial lemniscus is the projection of the contralateral cuneiform and gracile nuclei. It is also known as the fasciculus of Rey and the medullothalamic tract [2]. The sensory decussation extends from the level of the pyramidal decussation to the midpoint of the inferior olivary nucleus. The crossing fibers are known as the internal arcuate fibers. After the decussation, the fibers are located posterior to the pyramidal tract.

The inferior olive is also known as the inferior olivary complex [3] and medullary olive. It is located between the pyramid and the anterolateral fasciculus. It is composed of the principal, accessory medial, and accessory dorsal nuclei and is covered by a dense white matter layer known as the amiculum [3]. The main connections of the olive are to the cerebellum, the red nucleus, the thalamus, and the spinal cord. The olivocerebellar pathway is crossed. Its fibers constitute the deep arcuate fibers of the pons and a major portion of the inferior cerebellar peduncle [1]. The rubro-olivary and thalamo-olivary tracts are ipsilateral. These tracts are part of the central tegmental tract. The olivospinal and spino-olivary tracts are connections of the inferior olive with the cervical spinal cord. The triangular space formed by the connections of the ipsilateral red nucleus, the inferior olive, and the contralateral dentate nucleus is known as Guillain–Mollaret triangle, which forms the dentato-rubro-olivary pathway.

The restiform body is a voluminous bundle composed of fibers of the inferior cerebellar peduncle. It is partially covered by the acoustic striae and tubercle and is posterior to the spinal trigeminal tract. The acoustic striae are also known as the medullary striae [3]. The juxtarestiform body is composed of the vestibular nuclei and their connections to the cerebellum.

The netlike reticular formation occupies the central portion of the medulla. This structure extends from the posterior aspect of the pyramid to the nuclei of the floor of the fourth ventricle. Transversely, it extends from the midline to the inferior cerebellar peduncle. The hypoglossal radicular fasciculi divide the reticular formation into internal and external portions [2]. The internal portion is mainly composed of white matter, while the external portion is composed of several nuclei and white matter fasciculi. The fasciculus comprises the arcuate fibers and the longitudinal fibers. These longitudinal fibers form the central tegmental tract, the posterior longitudinal fasciculus, and the solitary tract. The central tegmental tract is mainly composed of connections to the inferior olive with the red nucleus. The central tegmental tract is also known as the Bekhterev tract and is an important connection of the extrapyramidal system [2]. The posterior longitudinal fasciculus extends from the mesencephalon to the reticular formation and is an important connection of the autonomous system between the hypothalamus and the reticular formation. The posterior longitudinal fasciculus [3] is also known as the fasciculus of Schutz [1, 2] or the dorsal tegmental tract. The solitary tract is formed by afferents from the facial, glossopharyngeal, and vagus nerves that terminate in the nuclei of the solitary tract. Aside from gustatory input, the solitary tract is also important for visceral reflexes. The solitary tract projects toward the ventral posterolateral nuclei of the thalamus [2]. Alongside the tract, several different clusters of cells are grouped in nuclei. The nuclei of the solitary tract are divided into superior and inferior portions. The superior portion is the gustative portion. The inferior portion receives visceral input from several nerves: the vagus (the heart and abdominal organs), glossopharyngeal (the middle ear, pharynx, and esophagus), and facial (the external acoustic meatus and tongue). The gray matter of the reticular substance of the medulla can be interpreted as a diffuse nucleus.

The Cerebellum

The cerebellum is a unique, an unpaired, and a symmetrical structure located at the posterior cranial fossa, dorsal to the brain stem (Fig. 8.62). It has a median and two lateral lobules. The median lobule, or median eminence, is known as the vermis. The lateral lobules are also known as hemispheres. The cerebellum has a superior, an inferior, and an anterolateral surface. The superior surface is also known as the tentorial surface, the inferior surface is also known as the suboccipital surface, and the anterolateral surface is also known as the petrosal surface (Figs. 8.63, 8.64, and 8.65). The vermis is limited laterally by the paramedian sulcus on each side. These paramedian sulci are shallow in the superior surface and deep in the inferior surface. In the superior surface, the cortex of the vermis and hemispheres is continuous. In the inferior surface, the cortex is interrupted at the bottom of the paramedian fissure. The vermis is subdivided by transverse sulci, creating the appearance of a silkworm; thus, it was named vermis [2].

Several transverse sulci divide the cerebellum into lobes, lobules, and folios. There are also several subdivisions of the cerebellum based on different points of view. Some of these subdivisions take into account not only the morphology but also the morphogenesis, the ontogenesis, and the phylogenesis and its related function in the proposed division of the cerebellum. Herein is a humble attempt to explain the anatomy of the cerebellum.

The first sulcus to develop is the posterolateral fissure, which divides the cerebellum into the flocculonodular lobe and the body of the cerebellum. The second fissure to develop is the primary fissure, which divides the body into anterior and posterior lobes. The flocculonodular lobe is related to the vestibular system. It is also known as the vestibulocerebellum or oculomotor cerebellum. Aside from the flocculus and nodule, it also contains parts of the ventral uvula and the tonsils. The anterior lobe, with parts of the dorsal uvula and pyramis, is also known as the spinocerebellum. The posterior lobe is mostly connected to the pontine nuclei and therefore is known as the cortical cerebellum or the pontine cerebellum.

The lobules are given different nomenclature in the vermis and hemispheres (Figs. 8.63, 8.64, 8.64, and 8.66). The vermis is divided into the lingula, central lobule, culmen, declive, folium, tuber, pyramis, uvula, and nodule (Fig. 8.66). The lingula is only on the midline and composes the gray lamina of the superior medullary velum, a very tiny lateral extension of the lingula known as the frenulum. The superior medullary vellum is also known as Vieussens' valve [2]. The central lobule is an eminence just behind the lingula with a small lateral extension known as the wing of the central lobule. Together, the culmen and declive are known as the anterior eminence of the vermis or monticulus, and they occupy the largest portion of the superior vermis. The culmen is the highest point, and the declive is the descendent portion. On the hemispheres, the culmen is continuous with the anterior quadrangular lobule and the declive with the posterior quadrangular lobule. The folium is composed of a unique cerebellar folium on the midline and is separated from the tuber by the horizontal fissure. The horizontal fissure is also known as the petrosal fissure on the anterolateral aspect of the cerebellum. On the hemispheres, the folium is related to the superior semilunar lobule and the tuber to the inferior semilunar lobule. The pyramis is the most prominent midline lobule of the inferior vermis. It is also known as the cruciform eminence of Malacarne or pyramis [2]. On the hemispheres, it is related to the biventer lobule. The uvula is the lowest portion of the vermis and is related to the tonsils laterally. They are separated by a sulcus named the uvulotonsillar sulcus. The nodule is related to the floccule and separated from the rest of the cerebellum by the posterolateral fissure.

On each side of the uvula, two whitish strips diverge laterally toward the lateral recess and are known as the inferior medullary velum or valves of Tarin [2]. The inferior medullary velum is covered by the amygdala cerebellar, also known as the tonsils. The inferior medullary velum is continuous with the inferior tela choroidea. It is the caudal roof of the fourth ventricle and extends laterally to the white center of the floccules. The posterior angle between the superior and inferior medullary velum is the most posterior point on the roof of the fourth ventricle and is named the fastigium.

The gray matter of the cerebellum is divided into superficial and central portions. The superficial portion forms the cortex of the cerebellum. The central portion is arranged in paired and symmetrical masses known as the deep cerebellar nuclei. These are divided into dentate, globose, emboliform, and fastigial nuclei (Figs. 8.67 and 8.68). The dentate nuclei are the largest and are also known as the rhomboid body, cerebellar olives, or ciliary body of the cerebellum. Its skewed arrangement in the gray and white matter resembles the inferior olivary complex of the medulla, and its hilum is the main source of the fibers of the superior cerebellar peduncle. Together, the globose and emboliform nuclei are also referred to as the dentate accessory nuclei or interpositus nuclei. The emboliform is external and elongated, while the globose is internal and rounded. The fastigial nuclei are also known as the nuclei of the roof; they are the most medial and considered to be at the vermis. The fastigial fibers are projected toward the contralateral side by the fastigial or cerebellar commissure. The fastigiobulbar tract is also known as the arcuate tract [2] (Figs. 8.68, 8.69, and 8.70).

The cerebellum is connected with the brain stem and the spinal cord through the inferior, superior, and middle cerebellar peduncles (Figs. 8.65, 8.69, 8.71, and 8.72). The inferior cerebellar peduncle connects the medulla to the cerebellum and is part of the inferior limit of the fourth ventricle (Fig. 8.73). This peduncle is composed of the restiform and juxtarestiform bodies. The restiform body is a spinal and medullary afferent. The juxtarestiform body is medial to the restiform, is mainly part of the vestibular network, and contains cortical and fastigial efferents. The middle cerebellar peduncle is the largest connection to the cerebellum and is also known as the brachium pontis. The superior cerebellar peduncle, also known as the brachium conjunctivum, is the main efferent and contains the connections of the dentate, emboliform, and globose nuclei with the ipsilateral and contralateral red nucleus and thalamus (Figs. 8.74, 8.75, and 8.76). The superior cerebellar peduncle may also contain aberrant fibers of the ventral spinal cerebellar tract, also known as Gower's fascicle [2].

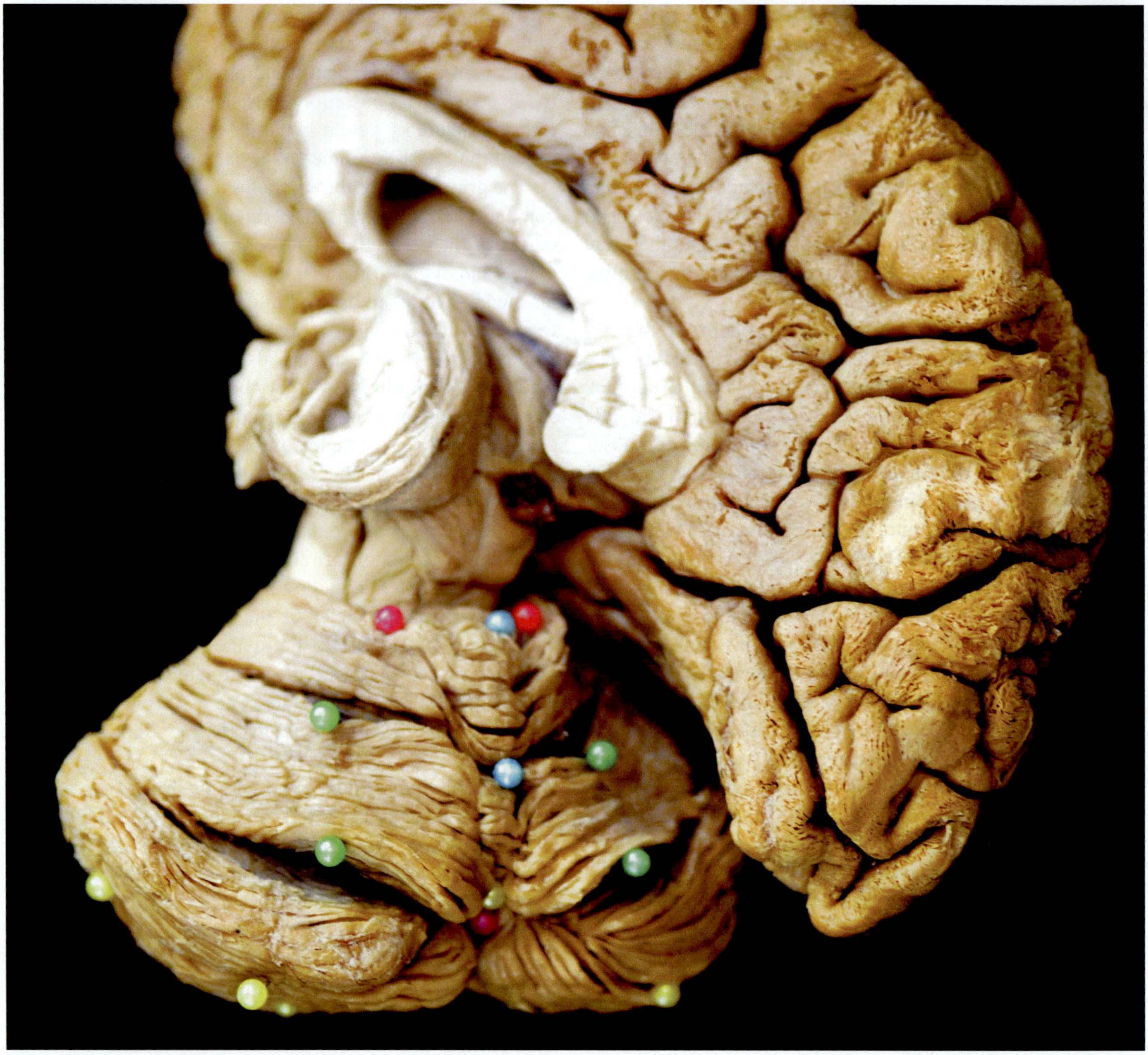

Fig. 8.62 Anatomical preparation of the cerebellum, posterosuperior view. *ant* anterior, *ext* external, *f* fascicle, *g* gyrus, *inf* inferior, *lat* lateral, *post* posterior, *s* sulcus, *sup* superior (©Kadri 2023. All Rights Reserved)

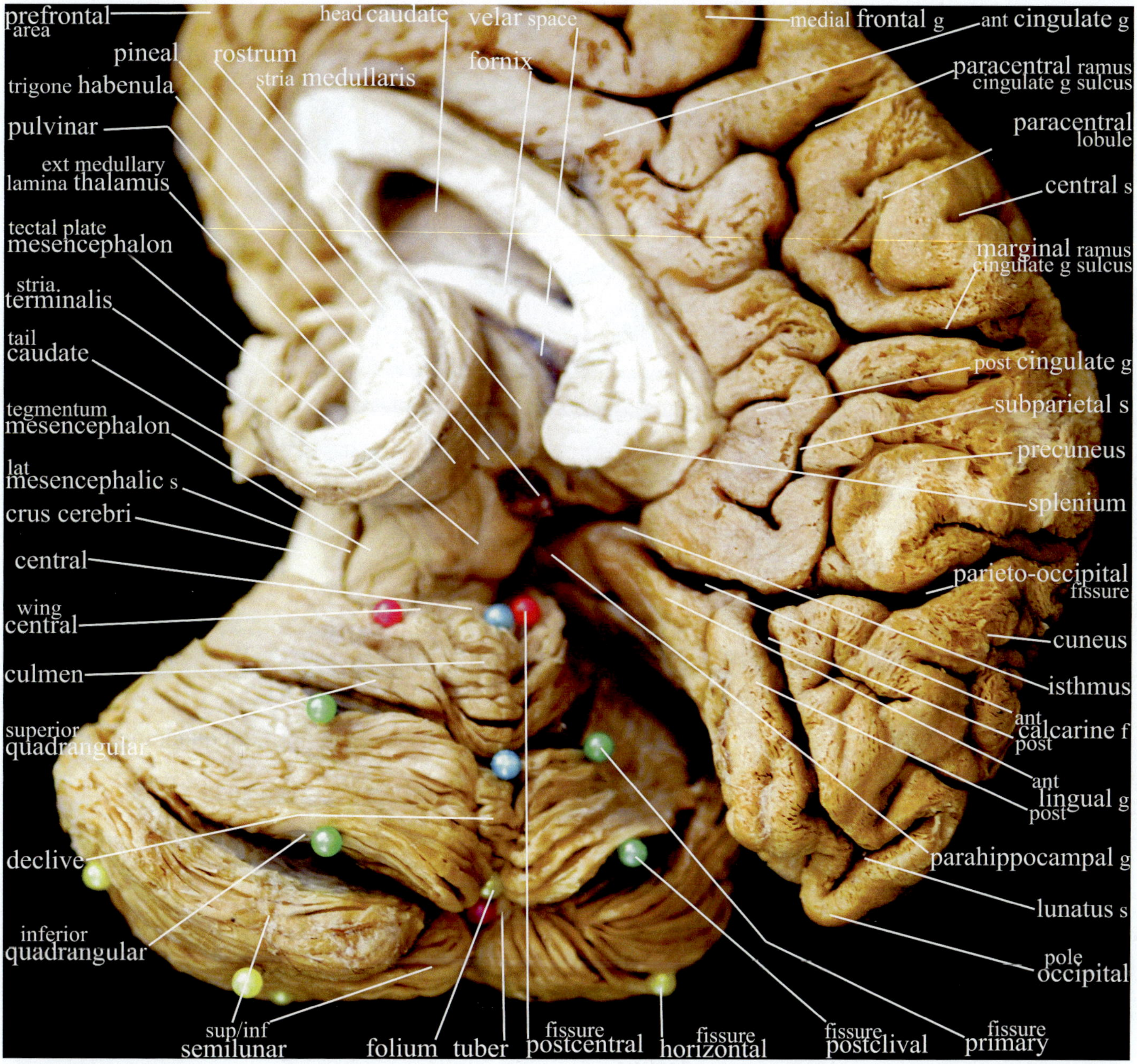

Fig. 8.62 (continued)

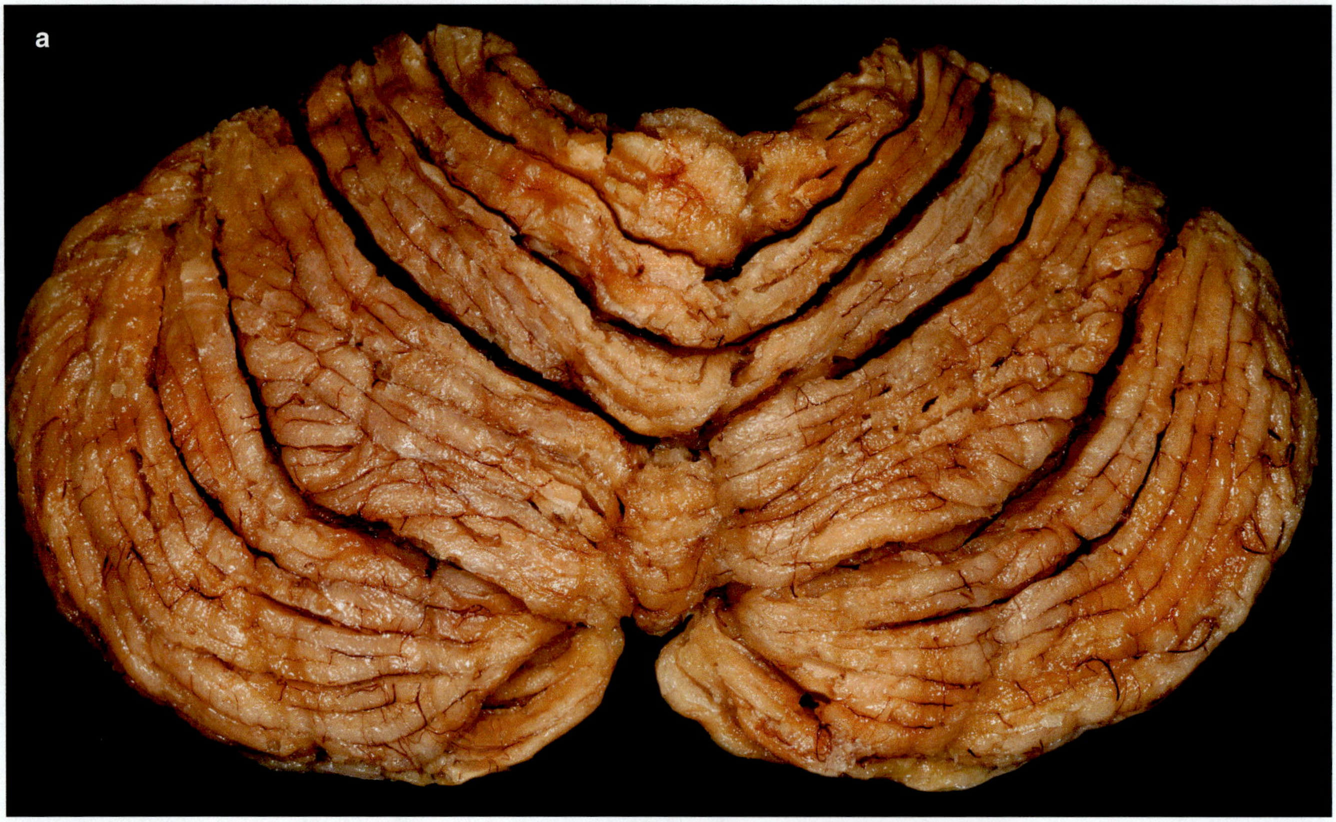

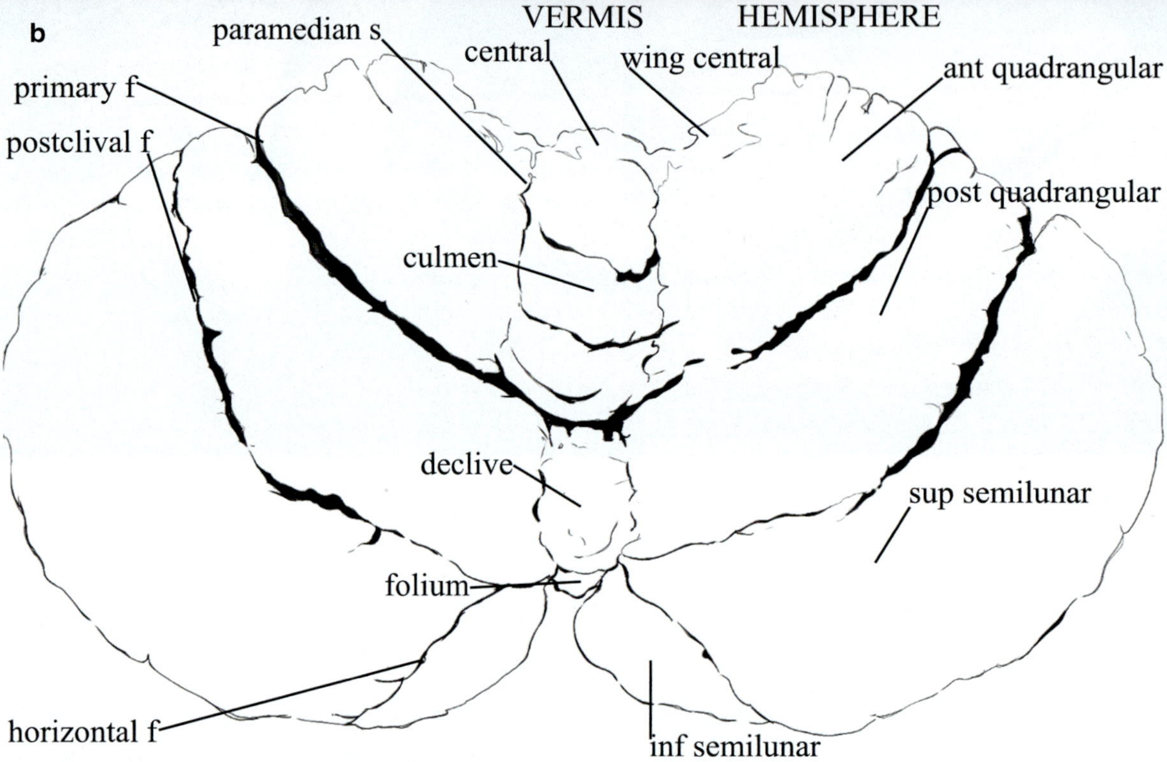

Fig. 8.63 Anatomical preparation and illustration of an isolated cerebellum. (**a**) Superior view. (**b**) Illustration by Renata Kadri. *ant* anterior, *f* fascicle, *inf* inferior, *post* posterior, *s* sulcus, *sup* superior (©Kadri 2023. All Rights Reserved)

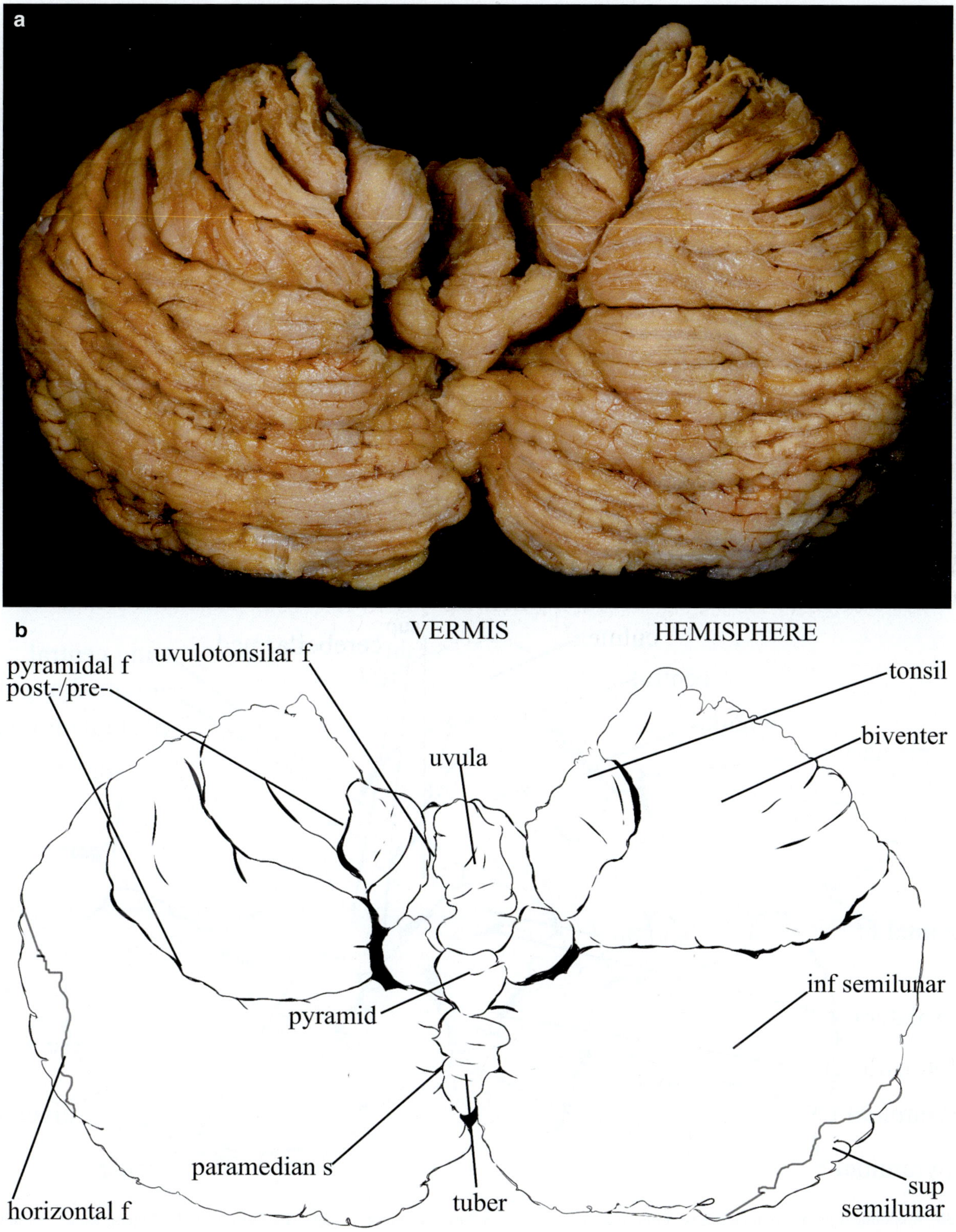

Fig. 8.64 Anatomical preparation and illustration of an isolated cerebellum. (**a**) Inferior view. (**b**) Illustration by Renata Kadri. *f* fascicle, *inf* inferior, *s* sulcus, *sup* superior (©Kadri 2023. All Rights Reserved)

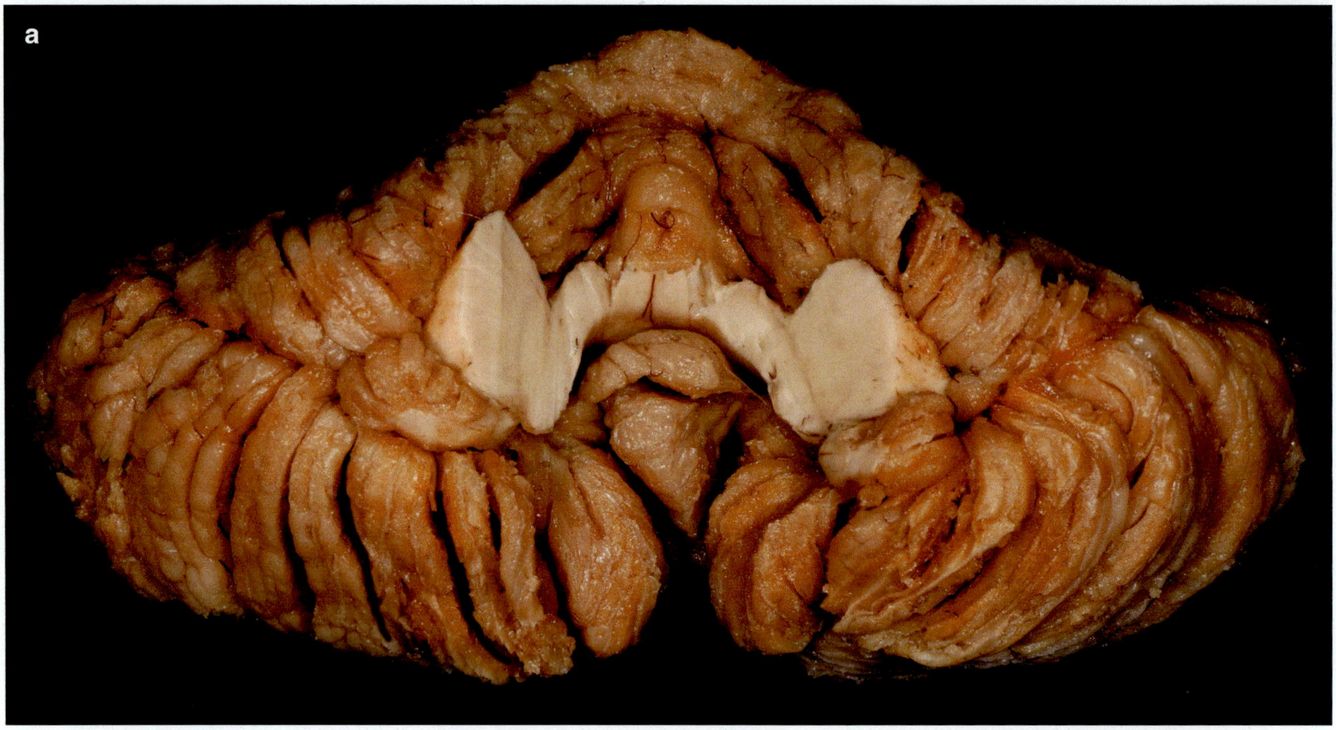

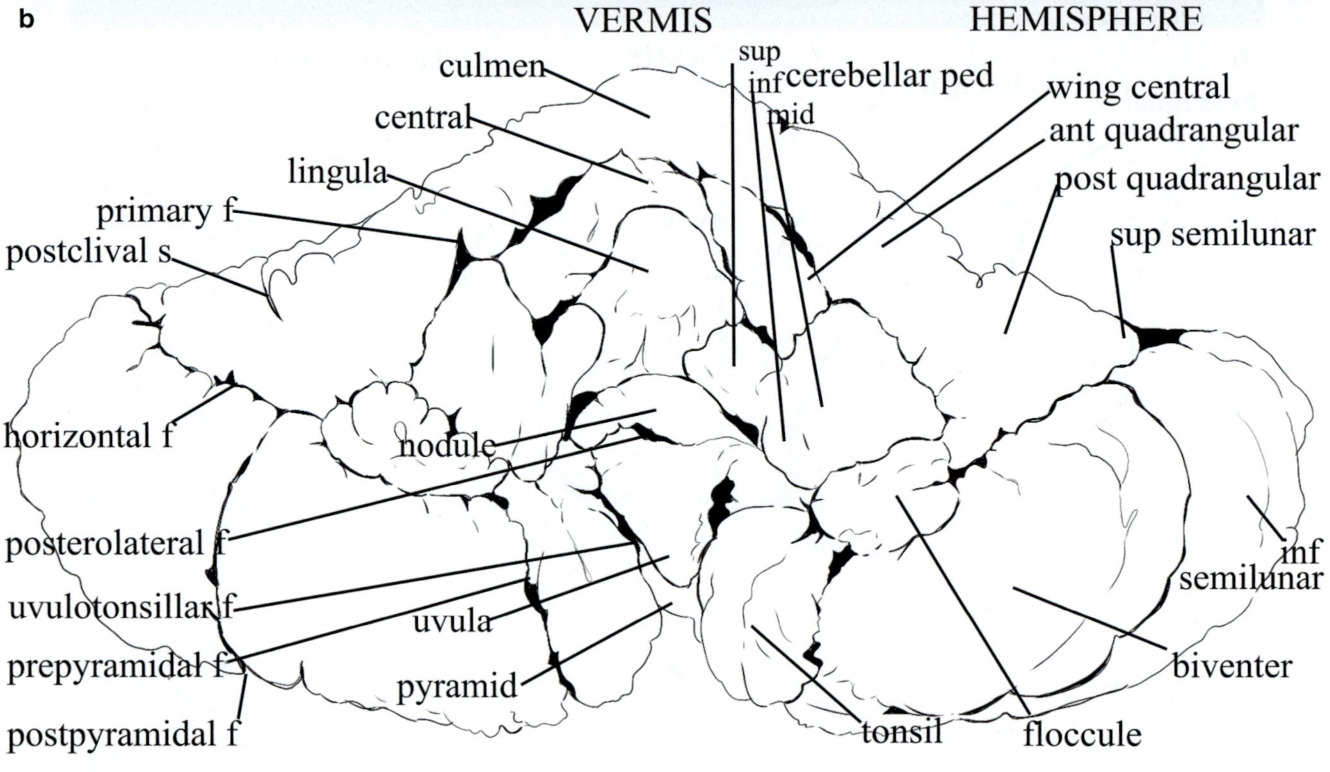

Fig. 8.65 Anatomical preparation and illustration of an isolated cerebellum. (**a**) Anterior view. (**b**) Illustration by Renata Kadri. *ant* anterior, *f* fascicle, *inf* inferior, *mid* middle, *ped* peduncle, *post* posterior, *s* sulcus, *sup* superior (©Kadri 2023. All Rights Reserved)

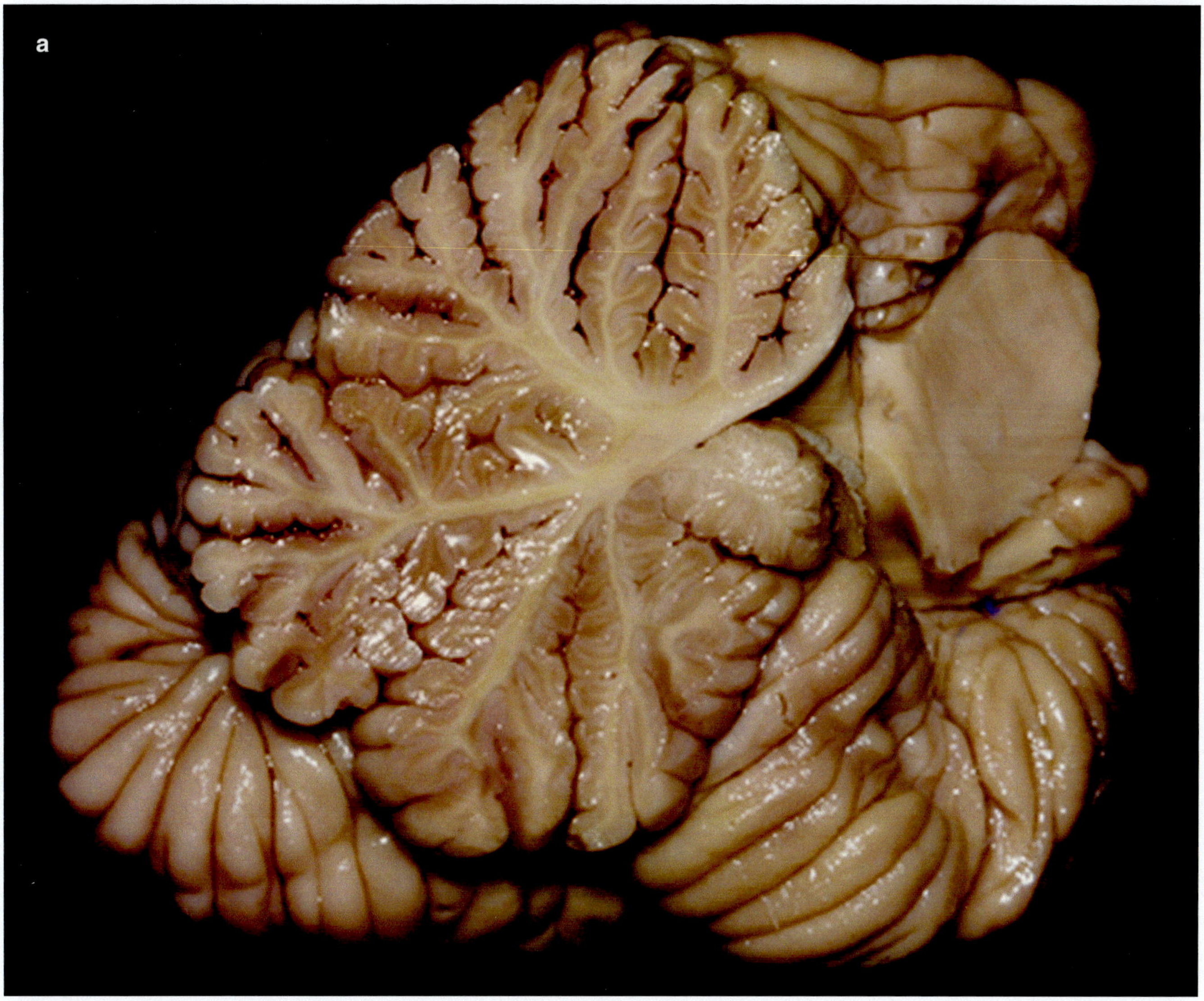

Fig. 8.66 Anatomical preparation and illustration of a cerebellum sectioned at the midline, exposing the lobules of the vermis cerebellar. (**a**) Medial view. (**b**) Illustration by Renata Kadri. *ant* anterior, *f* fascicle (©Kadri 2023. All Rights Reserved)

b

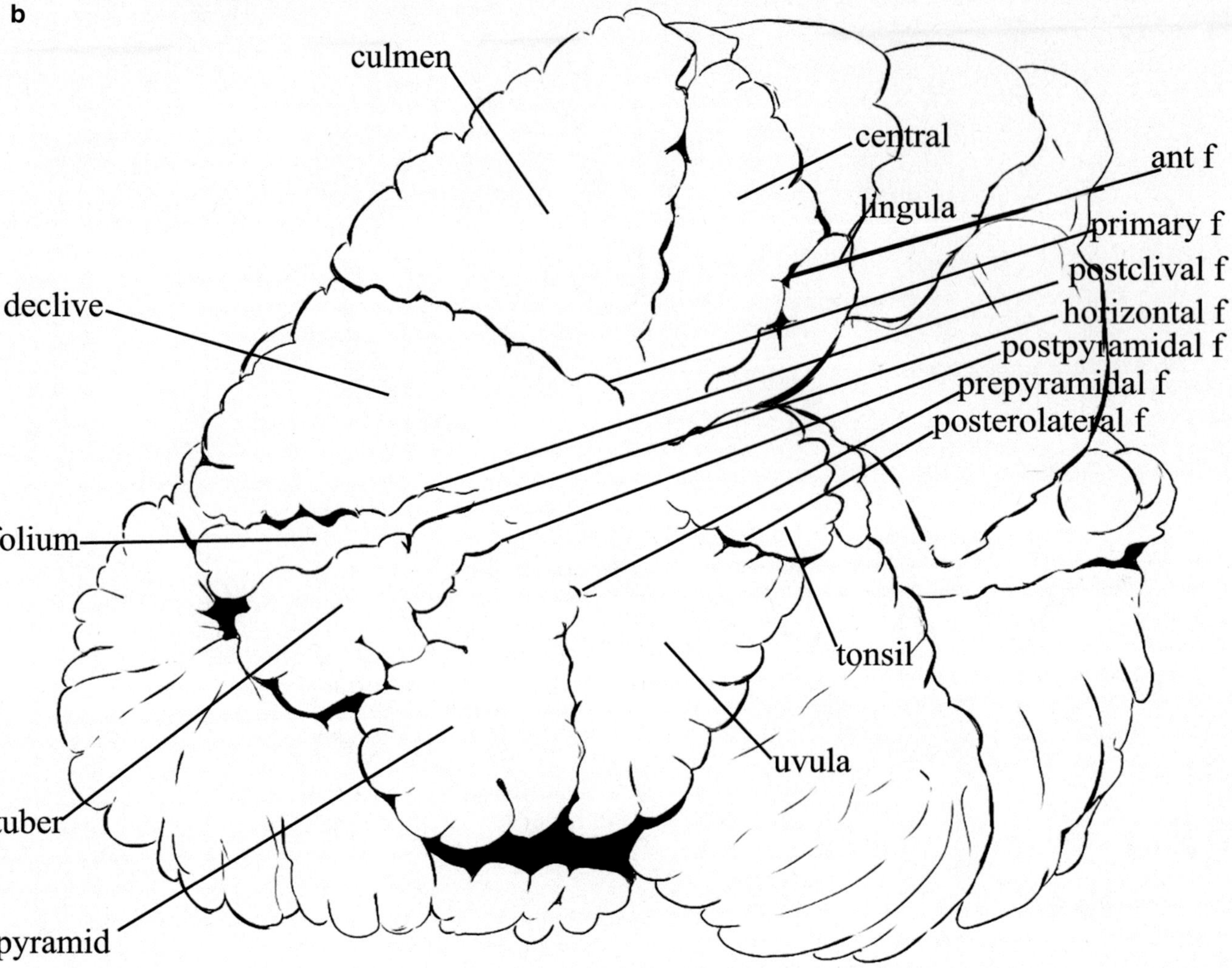

Fig. 8.66 (continued)

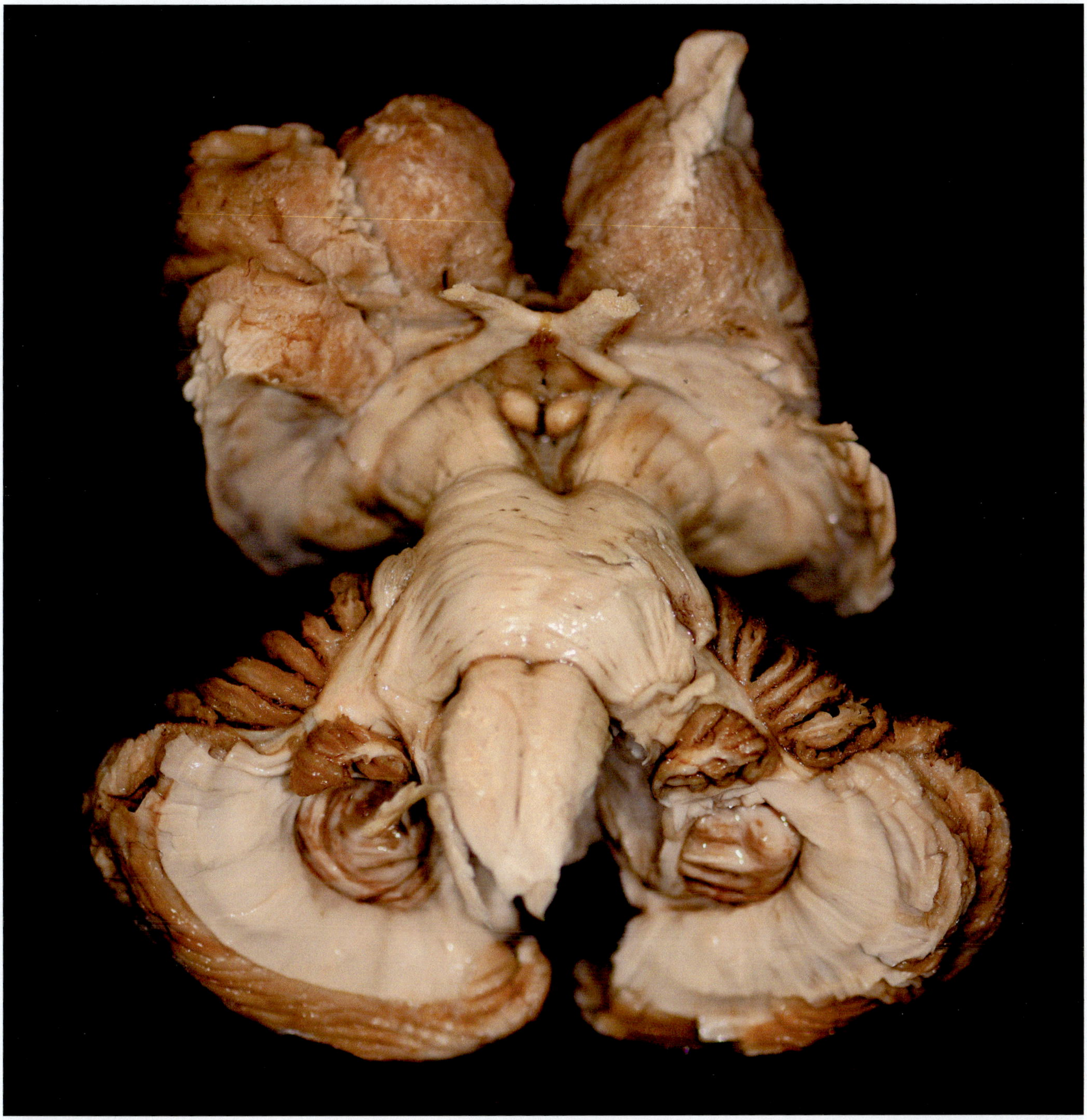

Fig. 8.67 Anatomical preparation of the deep nuclei of the cerebellum showing the inferior view of the dentate nuclei. *ant* anterior, *f* fascicle, *g* gyrus, *inf* inferior, *lat* lateral, *med* medial, *mid* middle, *n* nerve, *ped* peduncle, *s* sulcus, *sup* superior, *t* tract (©Kadri 2023. All Rights Reserved)

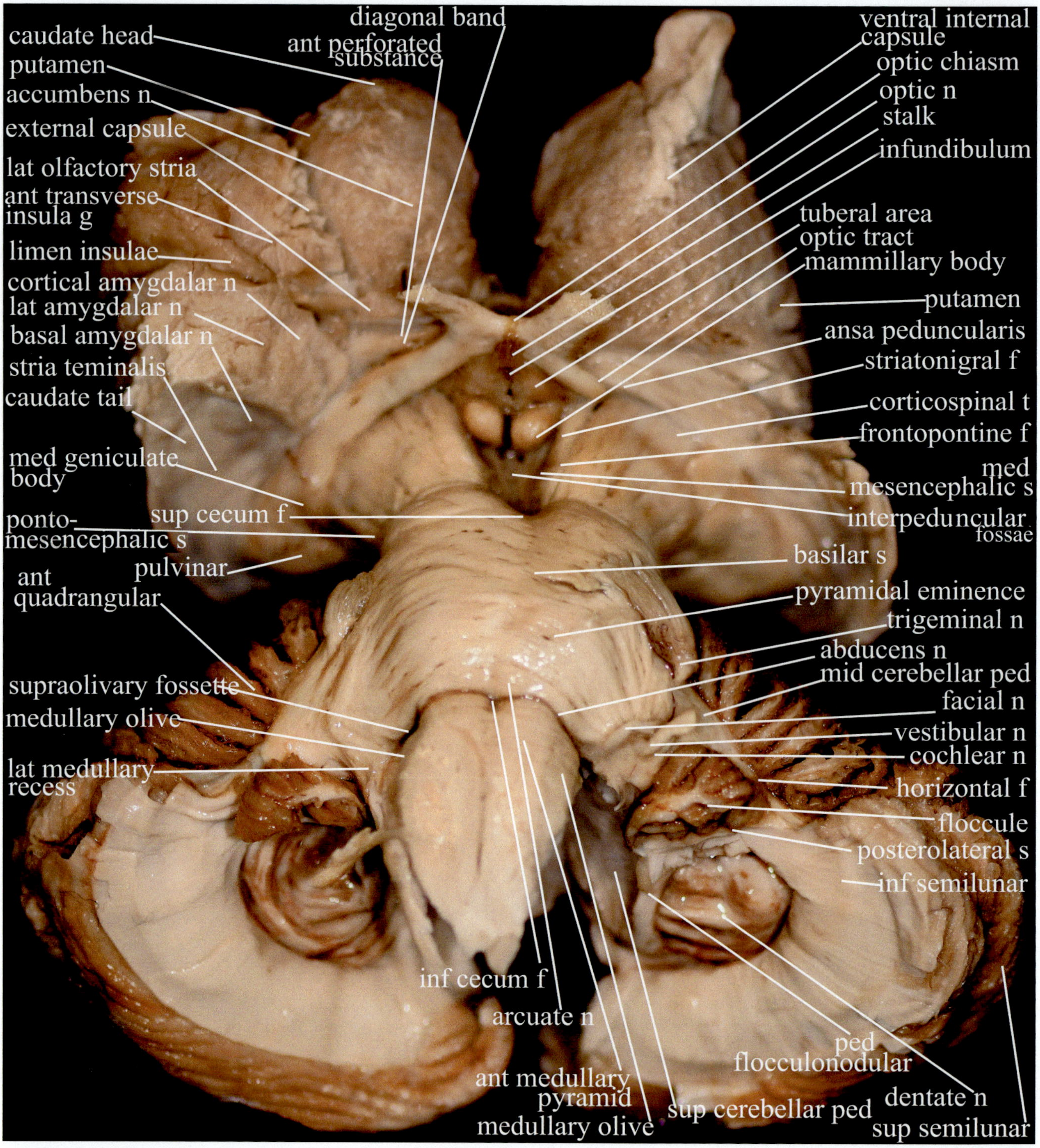

Fig. 8.67 (continued)

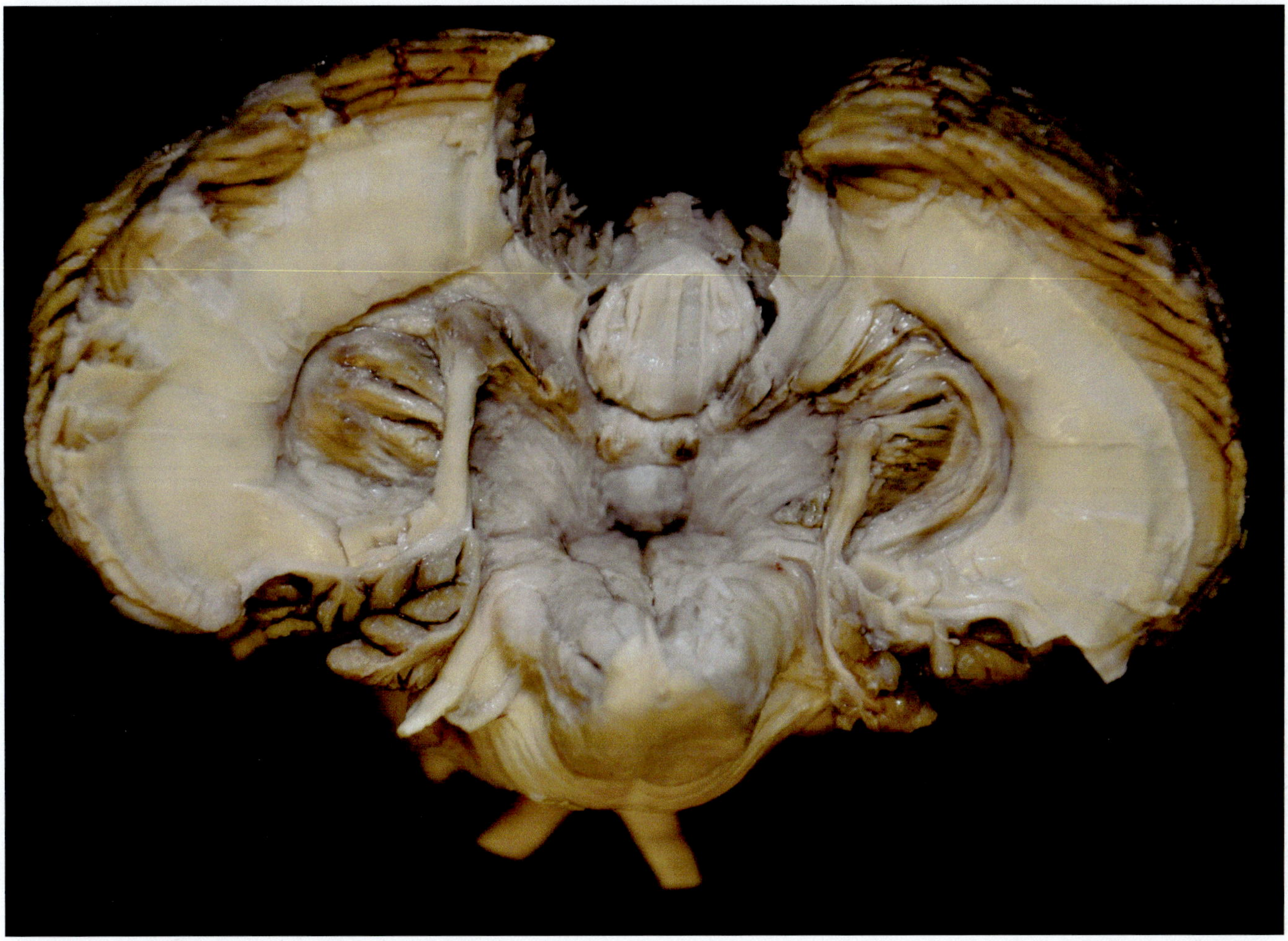

Fig. 8.68 Anatomical preparation of the deep nuclei of the cerebellum. The dentate, interpositum (emboliform and globose), and fastigial nuclei are the intrinsic gray matter of the cerebellum. *f* fascicle, *inf* inferior, *lat* lateral, *mid* middle, *n* nerve, *s* sulcus, *sup* superior

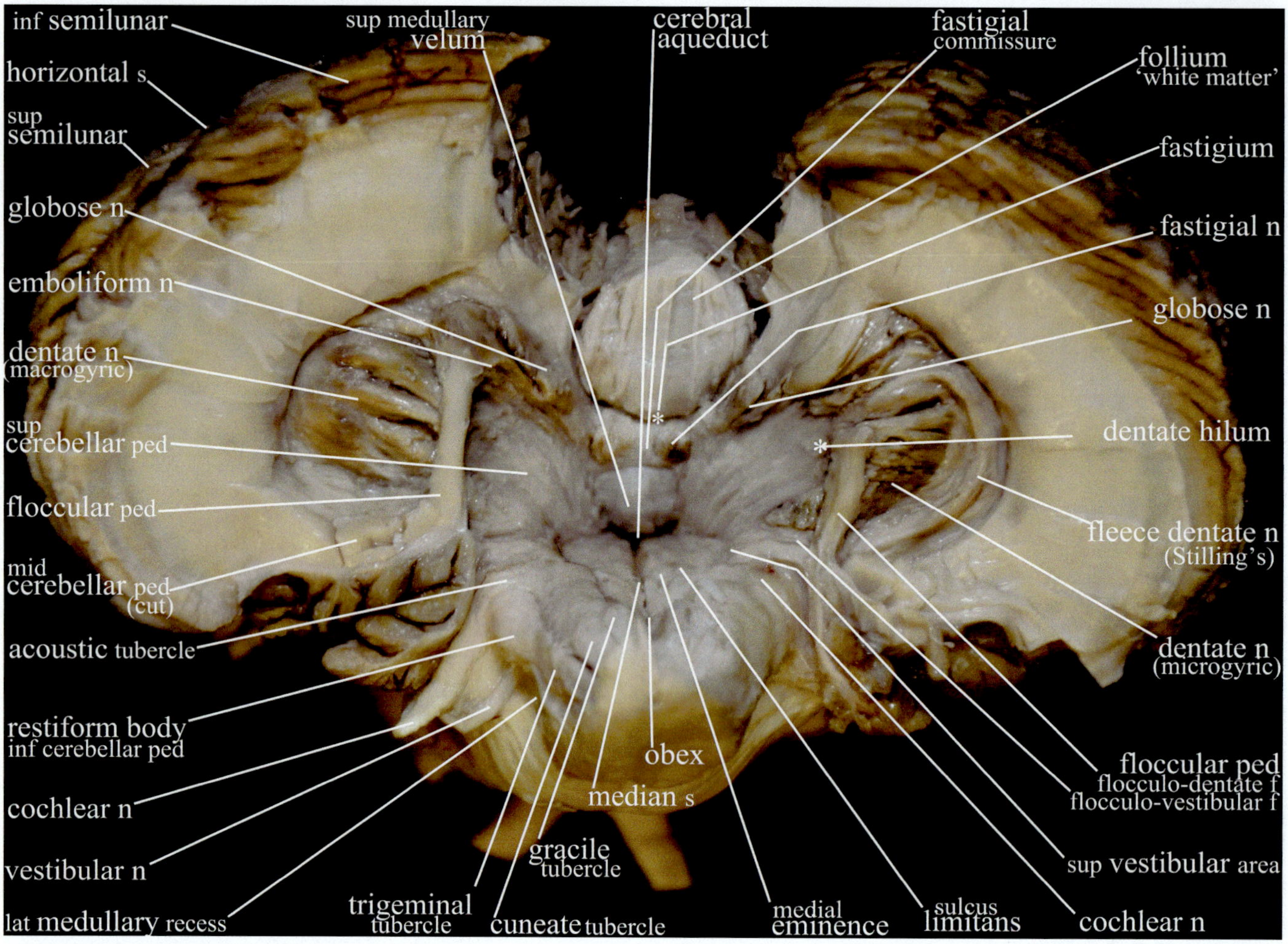

Fig. 8.68 (continued)

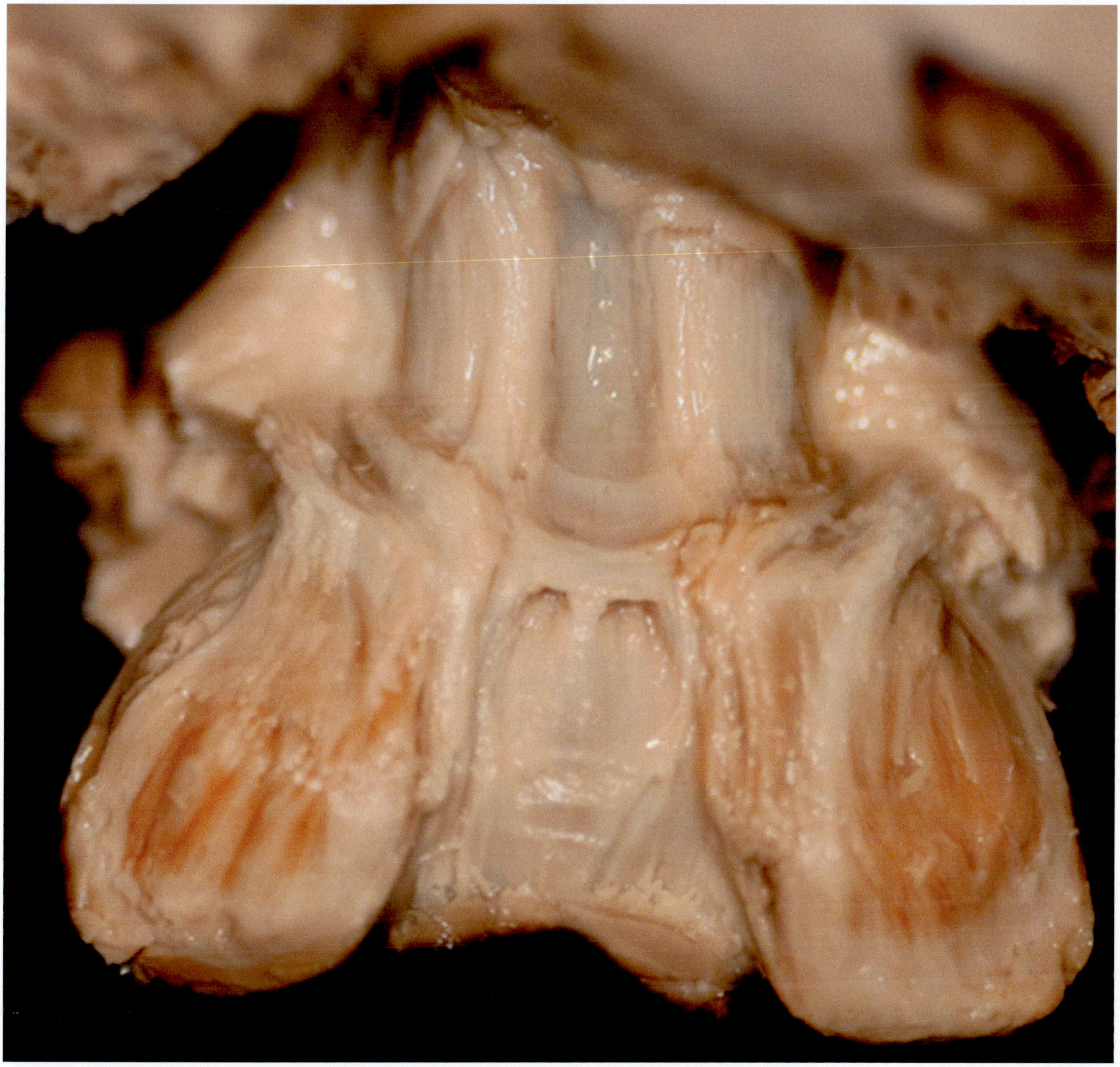

Fig. 8.69 Anatomical preparation of the deep cerebellar nuclei isolate and its connection via the superior and inferior cerebellar peduncles. *inf* inferior, *lat* lateral, *n* nerve, *ped* peduncle, *s* sulcus, *sup* superior, *t* tract (©Kadri 2023. All Rights Reserved)

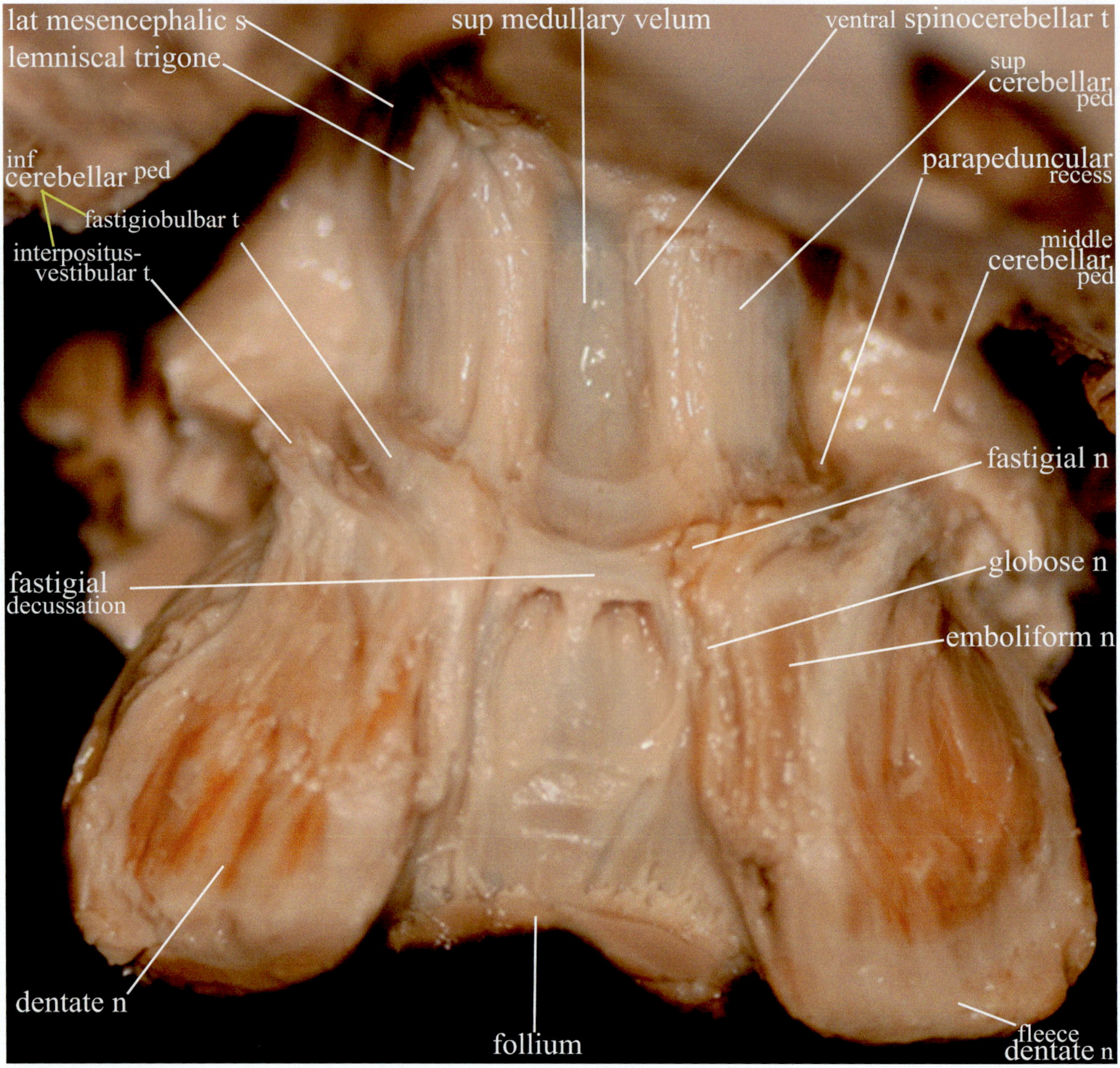

Fig. 8.69 (continued)

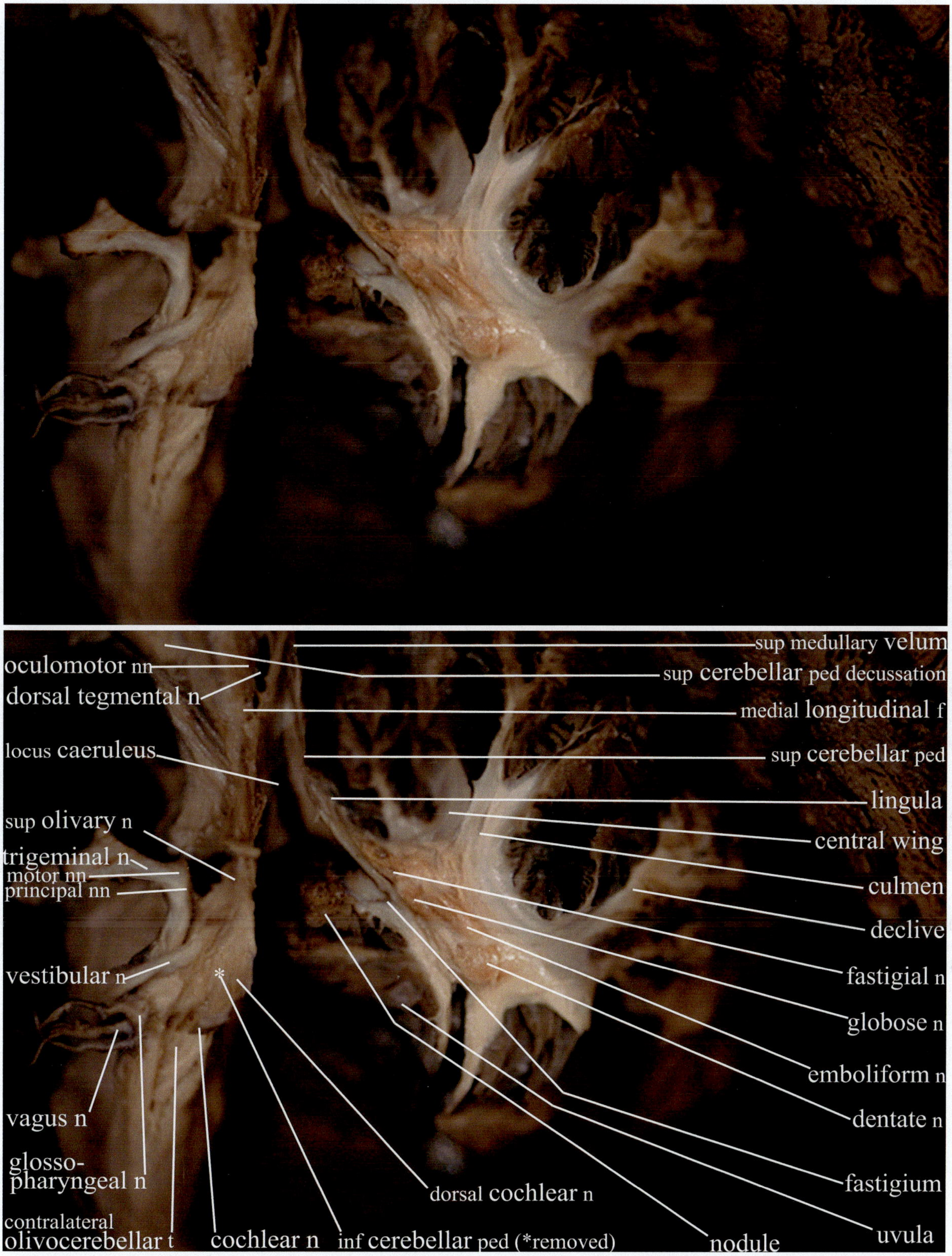

Fig. 8.70 Anatomical preparation of the deep cerebellar nuclei. The smallest and most medial is the fastigial nucleus. The largest and most lateral is the dentate nucleus. Interposed between the fastigial and dentate nuclei, and therefore known as the interpositus nuclei, are the emboliform (lateral) and globose (medial) nuclei. *f* fascicle, *inf* inferior, *n* nerve, *nn* nucleus, *ped* peduncle, *sup* superior, *t* tract (©Kadri 2023. All Rights Reserved)

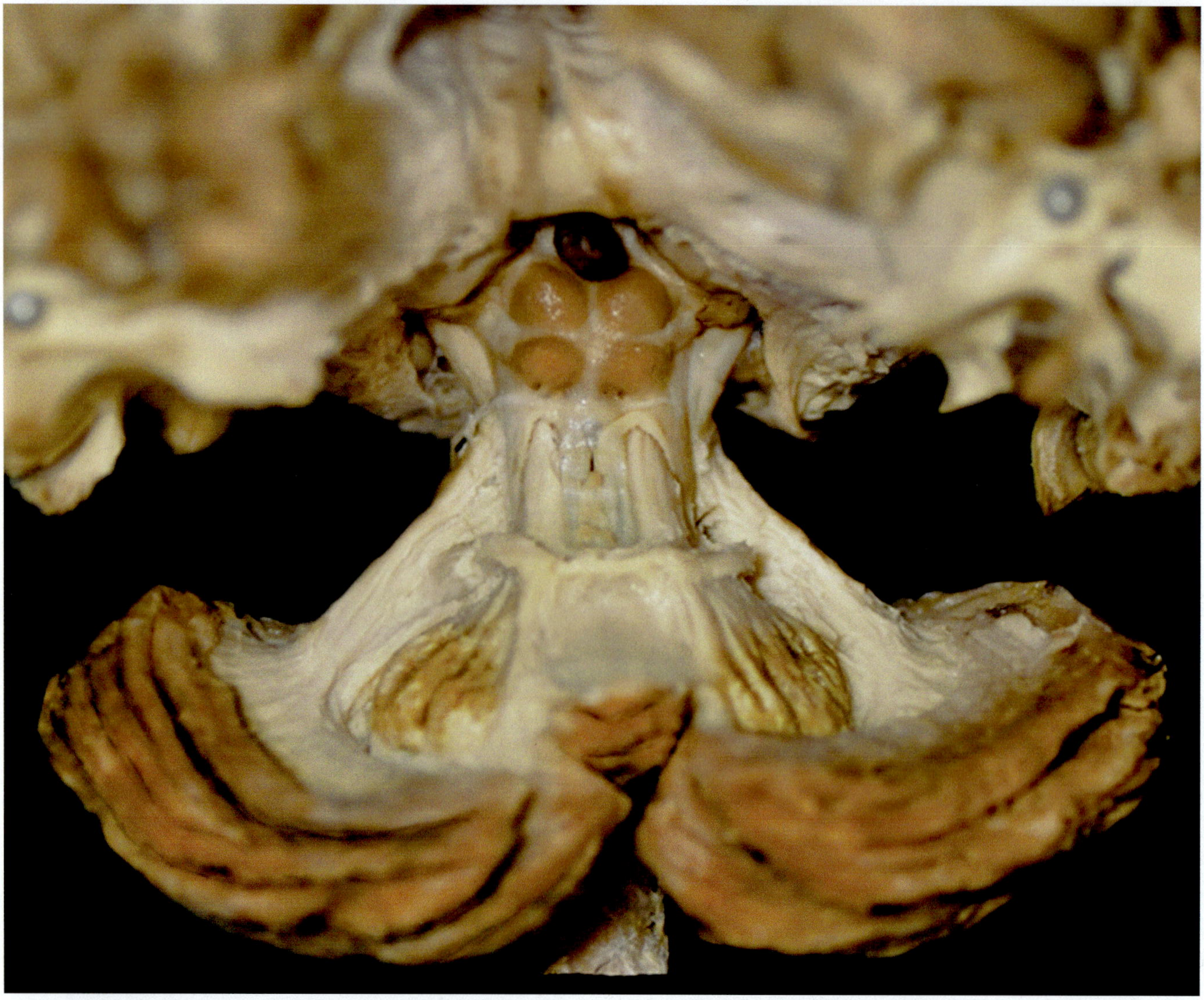

Fig. 8.71 Anatomical preparation of the deep cerebellar nuclei of the cerebellum, superior view. The anterior lobe and the posterior quadrangular and superior semilunar lobules of the posterior lobe of the cerebellar hemisphere have been removed. The vermis from the lingula to the declive has been removed to expose the folium. *f* fascicle, *g* gyrus, *inf* inferior, *lat* lateral, *med* medial, *n* nerve, *ped* peduncle, *s* sulcus, *sup* superior

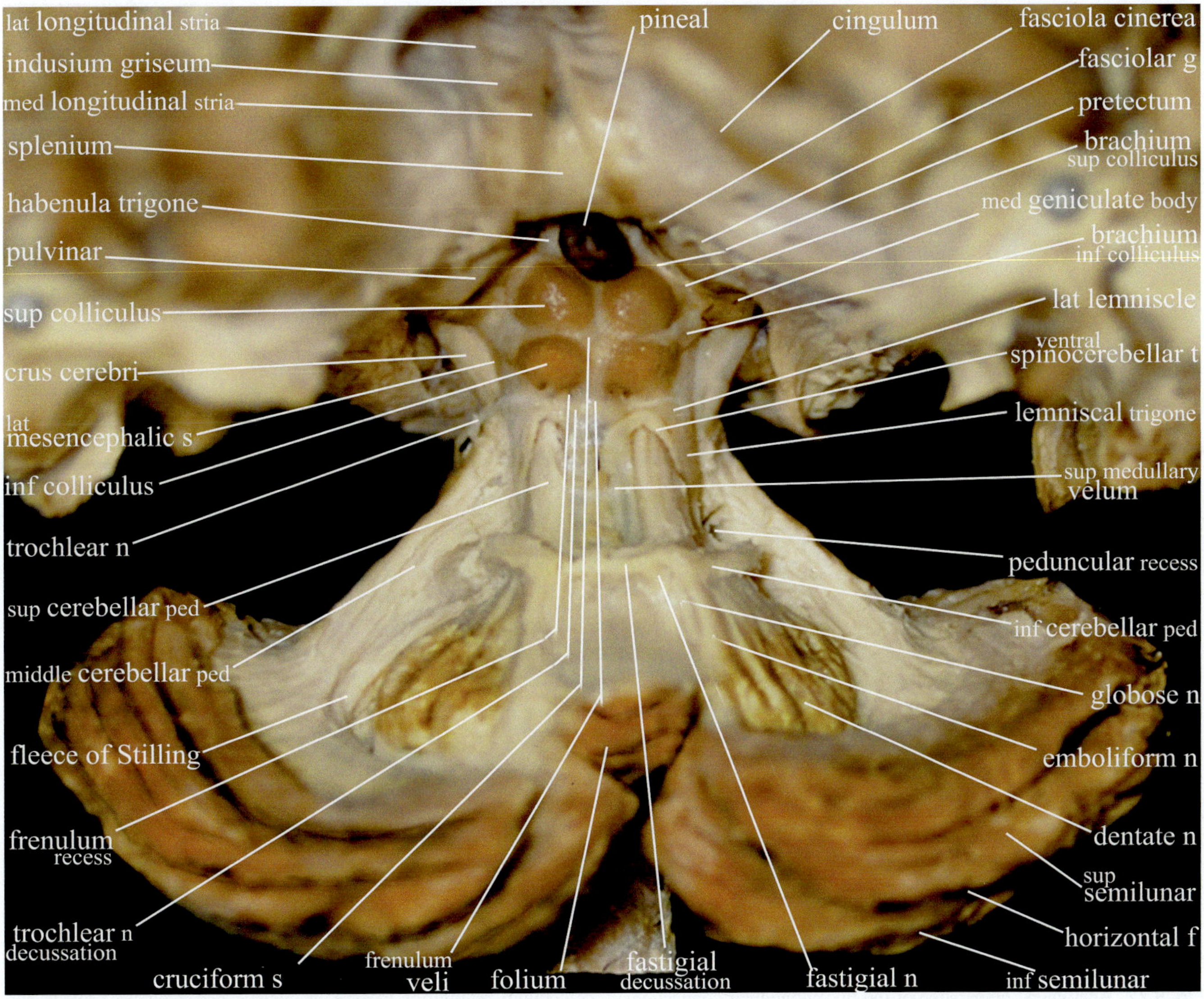

Fig. 8.71 (continued)

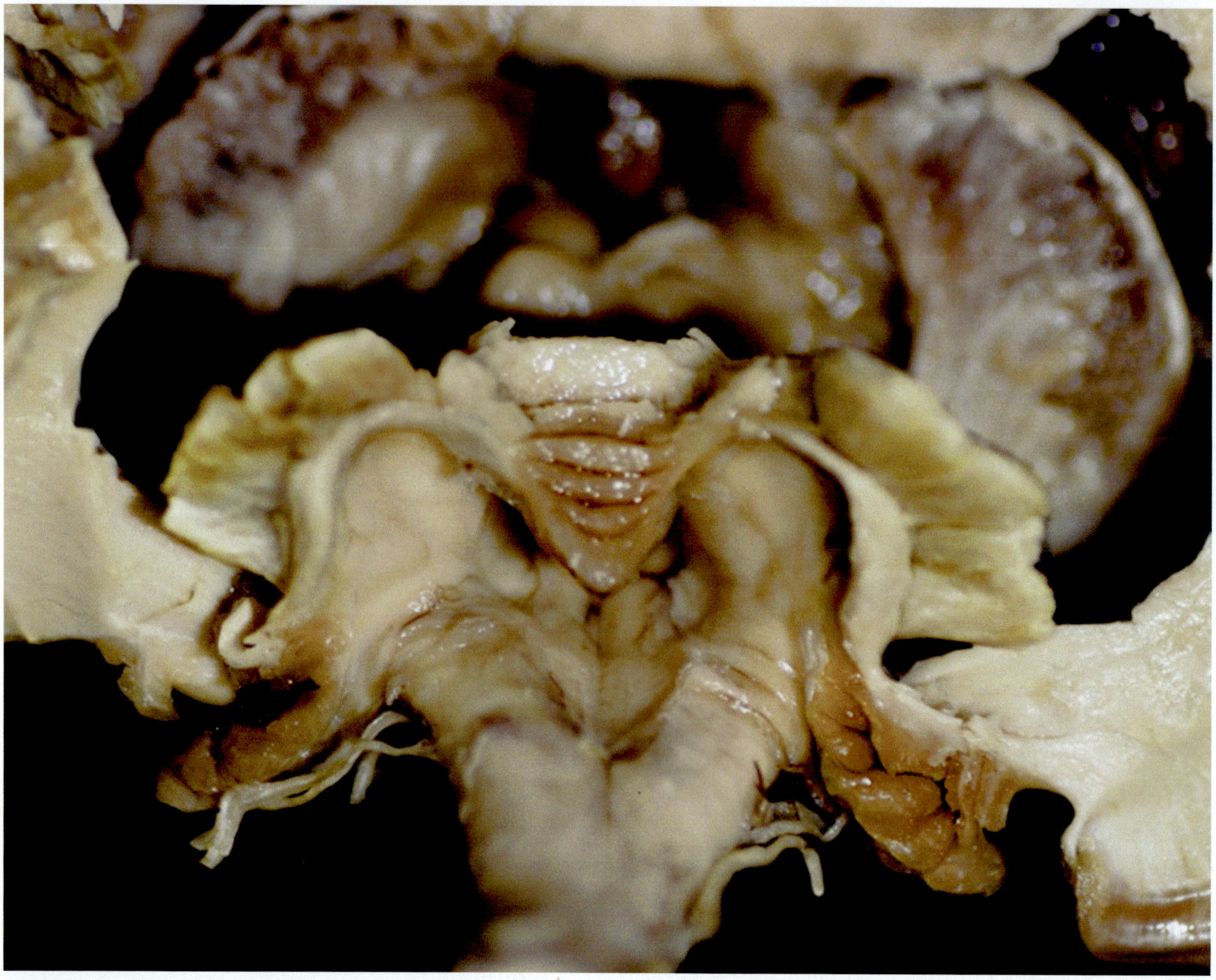

Fig. 8.72 Anatomical preparation of the flocculonodular connections and floccule vestibular connections via the floccular peduncle. The flocculonodular lobe is also known as the archicerebellum or vestibular cerebellum. *inf* inferior, *n* nerve, *nn* nucleus, *ped* peduncle, *s* sulcus, *sup* superior (©Kadri 2023. All Rights Reserved)

Fig. 8.72 (continued)

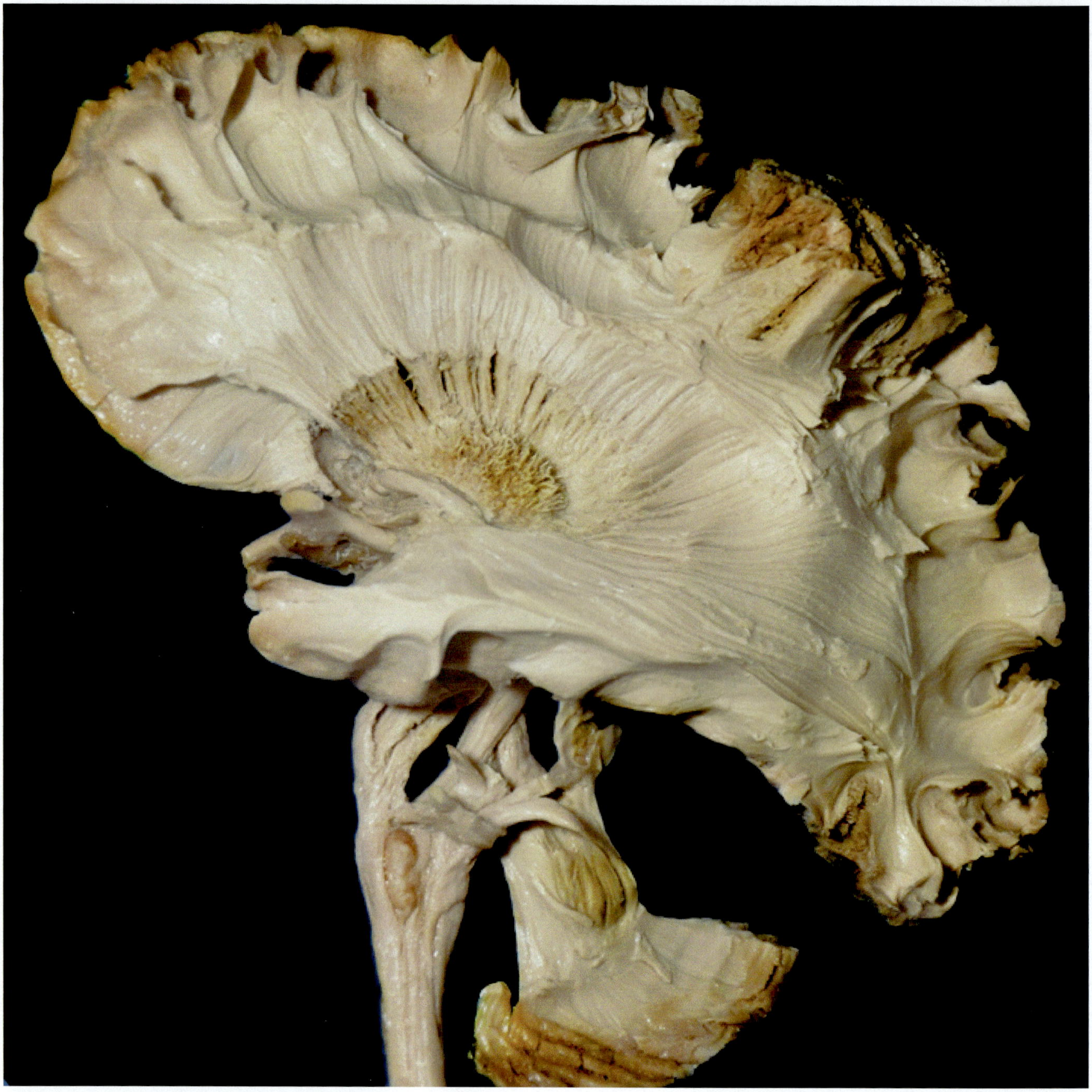

Fig. 8.73 Anatomical preparation of the cerebellar connections. The cerebellum is, in a broad sense, connected with the mesencephalon via the superior cerebellar peduncle, with the pons via the middle cerebellar peduncle, and with the medulla and the spinal cord via the inferior cerebellar peduncle. *ant* anterior, *f* fascicle, *inf* inferior, *n* nerve, *ped* peduncle, *sup* superior, *t* tract (©Kadri 2023. All Rights Reserved)

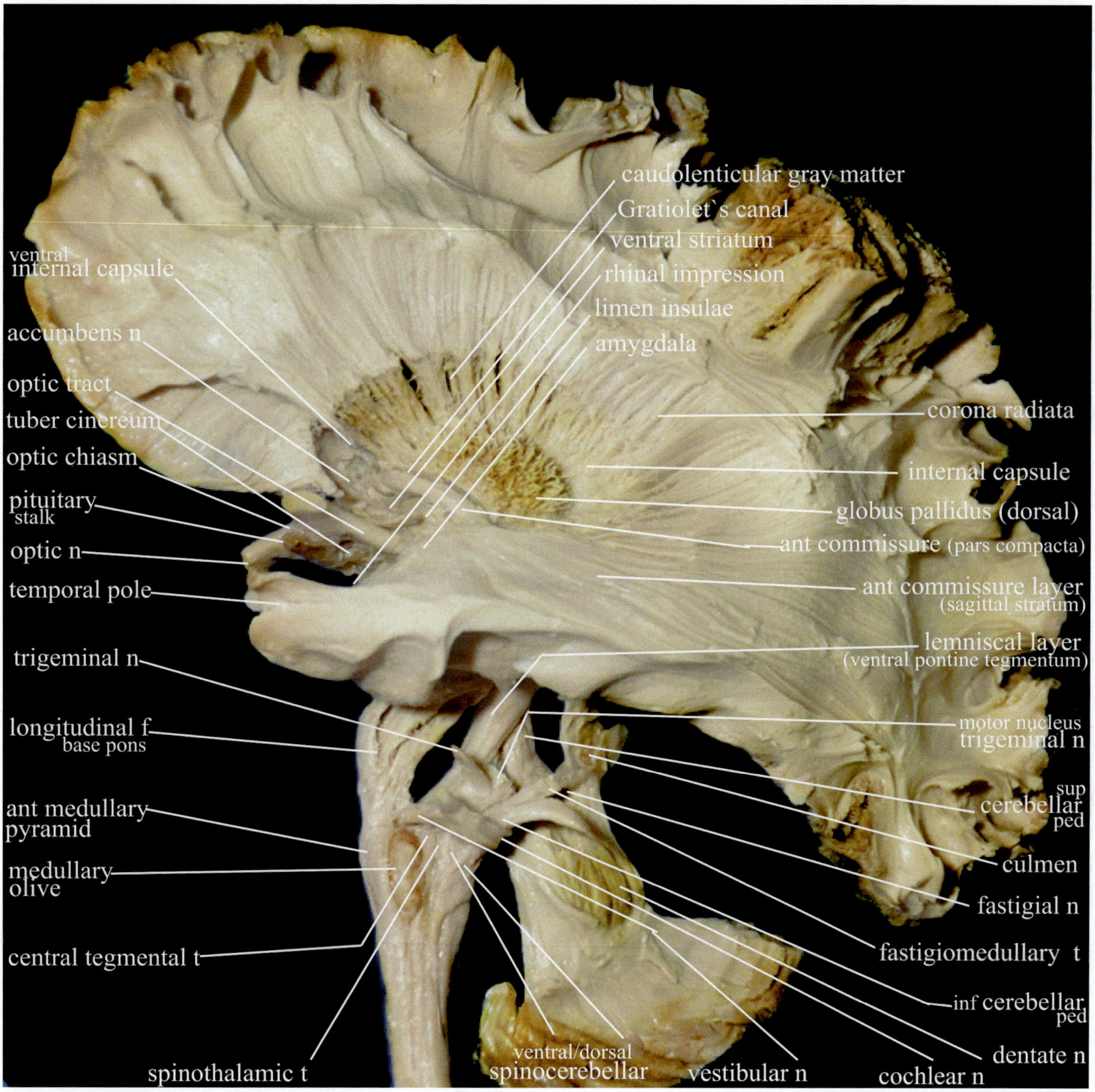

Fig. 8.73 (continued)

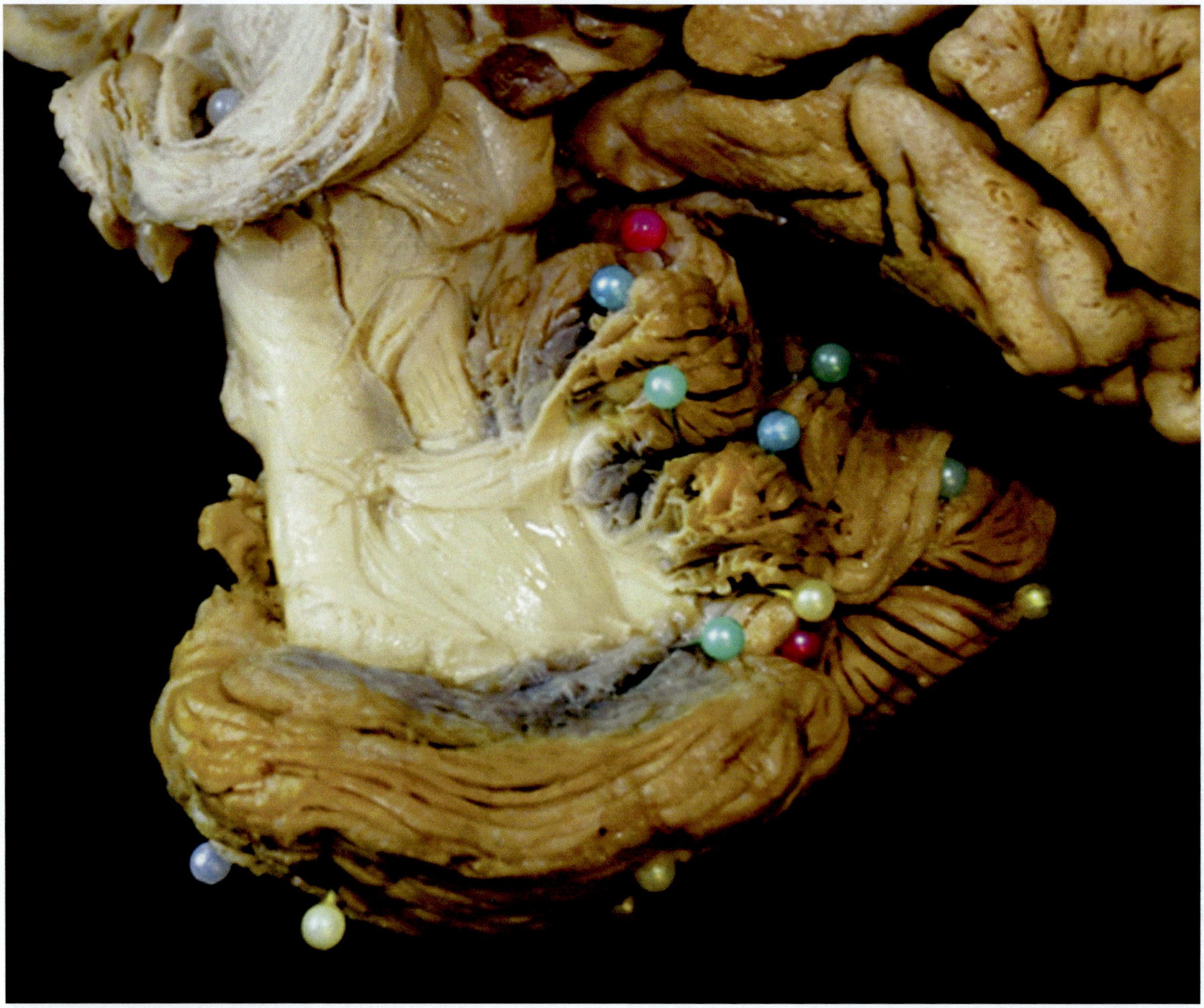

Fig. 8.74 Anatomical preparation of the fleece of white matter enveloping the dentate nucleus and the cerebellar connections. *ant* anterior, *f* fascicle, *g* gyrus, *inf* inferior, *lat* lateral, *med* medial, *mid* middle, *n* nerve, *post* posterior, *s* sulcus, *sup* superior, *t* tract (©Kadri 2023. All Rights Reserved)

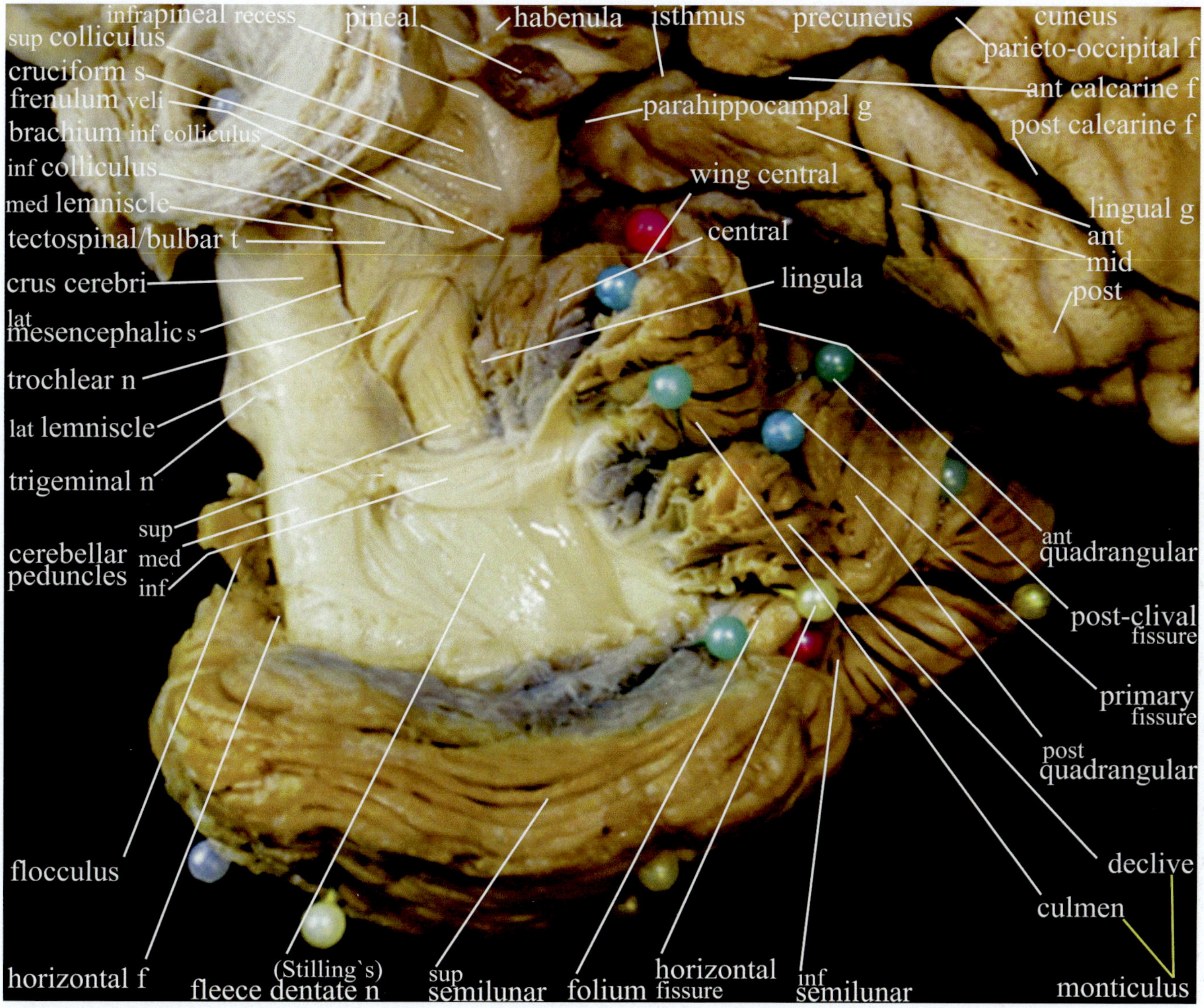

Fig. 8.74 (continued)

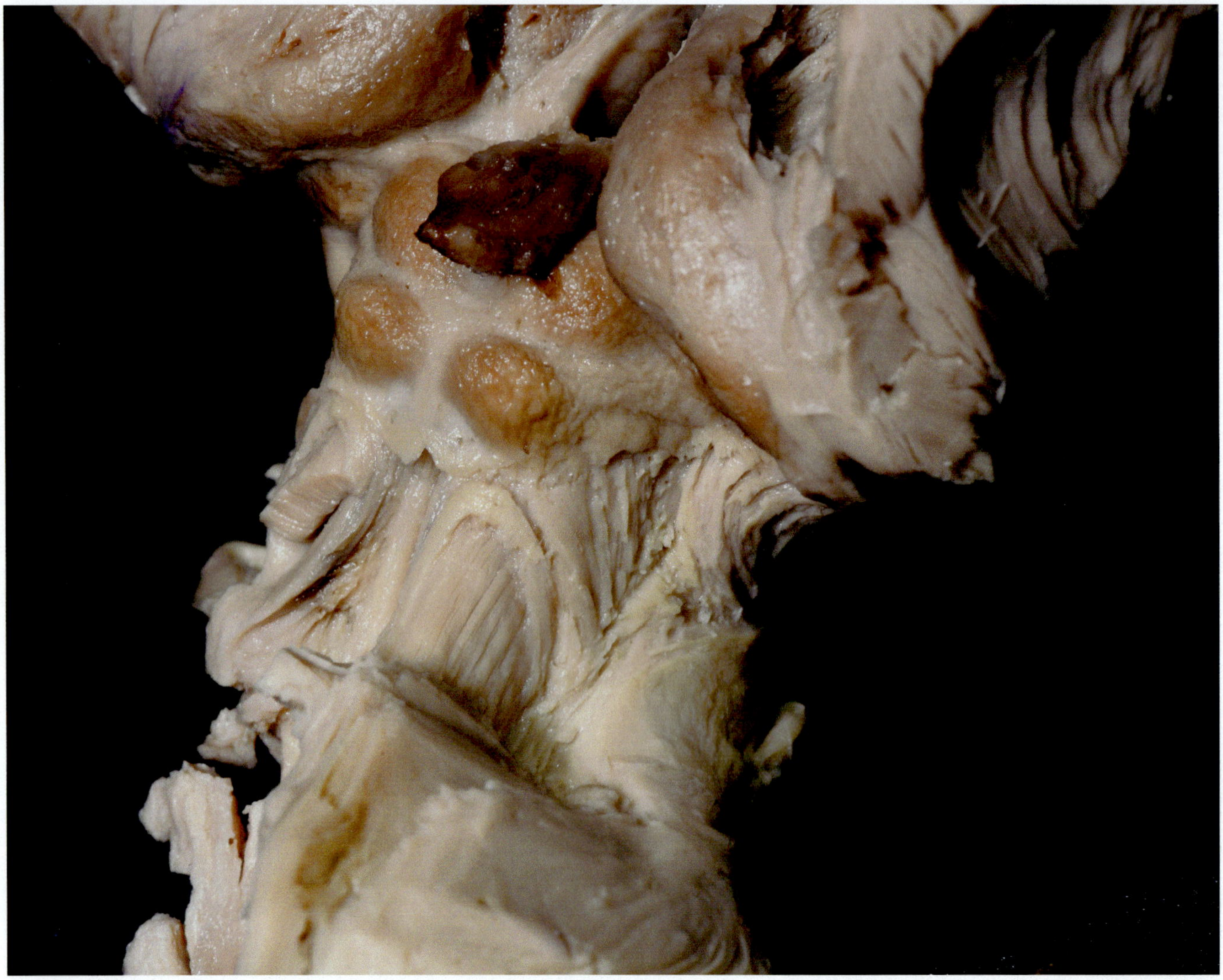

Fig. 8.75 Anatomical preparation of the superior cerebellar peduncle. *f* fascicle, *inf* inferior, *lat* lateral, *med* medial, *n* nerve, *nn* nucleus, *ped* peduncle, *s* sulcus, *sup* superior, *t* tract, *tt* tract (©Kadri 2023. All Rights Reserved)

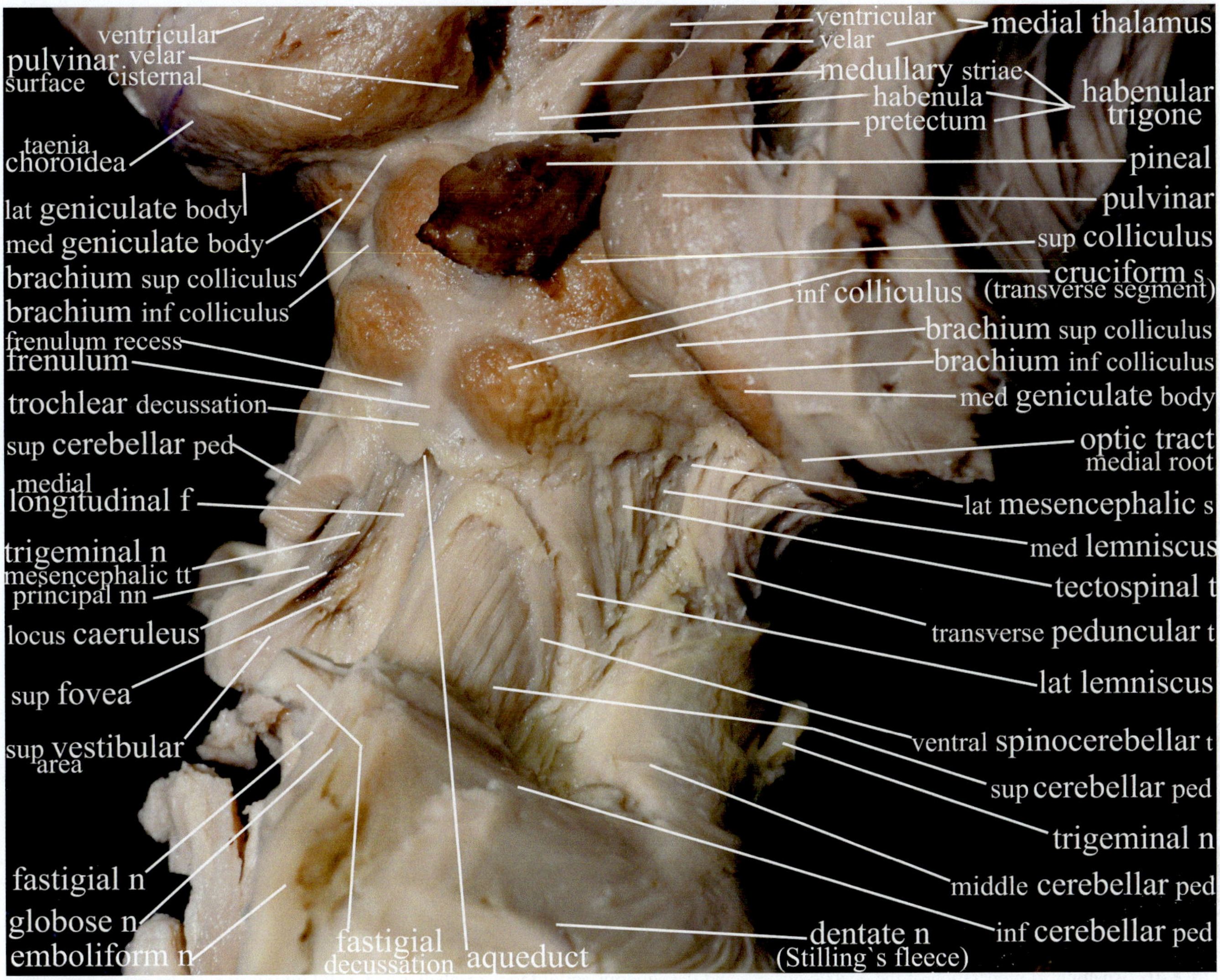

Fig. 8.75 (continued)

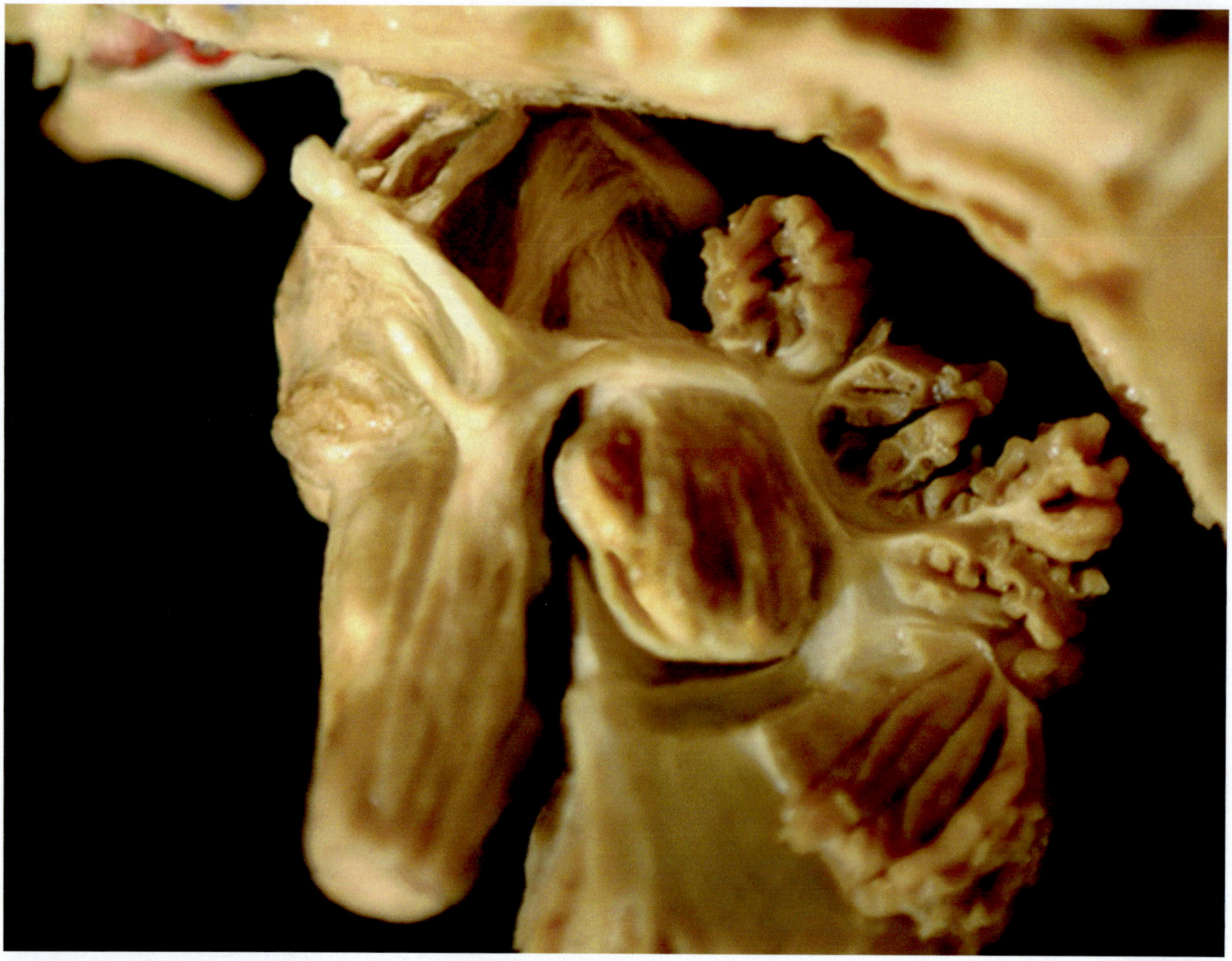

Fig. 8.76 Anatomical preparation of the dentate nucleus and the superior and inferior cerebellar peduncle. *inf* inferior, *lat* lateral, *n* nerve, *ped* peduncle, *sup* superior, *t* tract, *tt* tract (©Kadri 2023. All Rights Reserved)

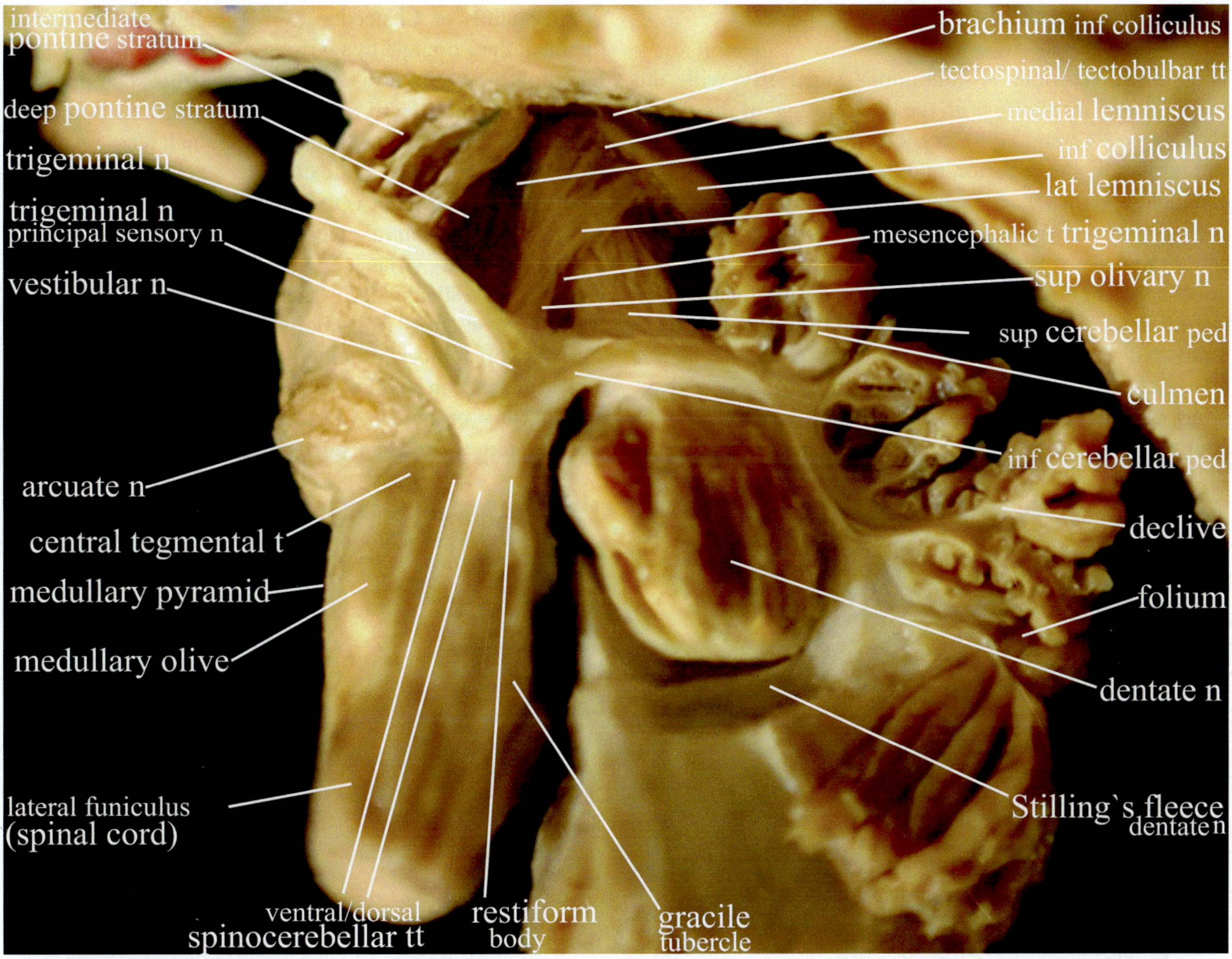

Fig. 8.76 (continued)

The Fourth Ventricle

The fourth ventricle is continuous superiorly with the aqueduct and inferiorly with the central canal of the spinal cord (Figs. 8.77 and 8.78). It is covered by the cerebellum and limited by the superior, middle, and inferior cerebellar peduncles. It has two surfaces, four borders, and four angles. The anterior surface is also known as the floor, while the posterior surface is also known as the roof. The floor has a rhomboid shape with a broad midline axis. A median sulcus divides it equally in two from its superior to inferior direction. A transverse line through the lateral recess divides it into a superior pontine triangle and an inferior medullary triangle. The pontine triangle is slightly larger [2].

Herophilus of Chalcedon (335 BC–280 BC) compared the medullary triangle to a writing pen; hence, it was named based on the Latin term *calamus scriptorius*. The tip is the most inferior portion of the triangle. At the transition of the tip to the central canal of the medulla, a small recess is named Arantius' ventricle. A delicate bridge between the gracile tubercles forms the obex, the transitional area of the fourth ventricle and the central medullary canal (Fig. 8.79) [2].

The medullary striae are at the base of the medullary triangle (Figs. 8.80 and 8.81). These striae are present as whitish transverse strips from the lateral recess to the midline. They vary in number (usually from three to five), dimension, and origin but can be totally absent. In the same manner, they can be very prominent or subtle in dimension. They can emerge from the median sulcus or from the sulcus limitans.

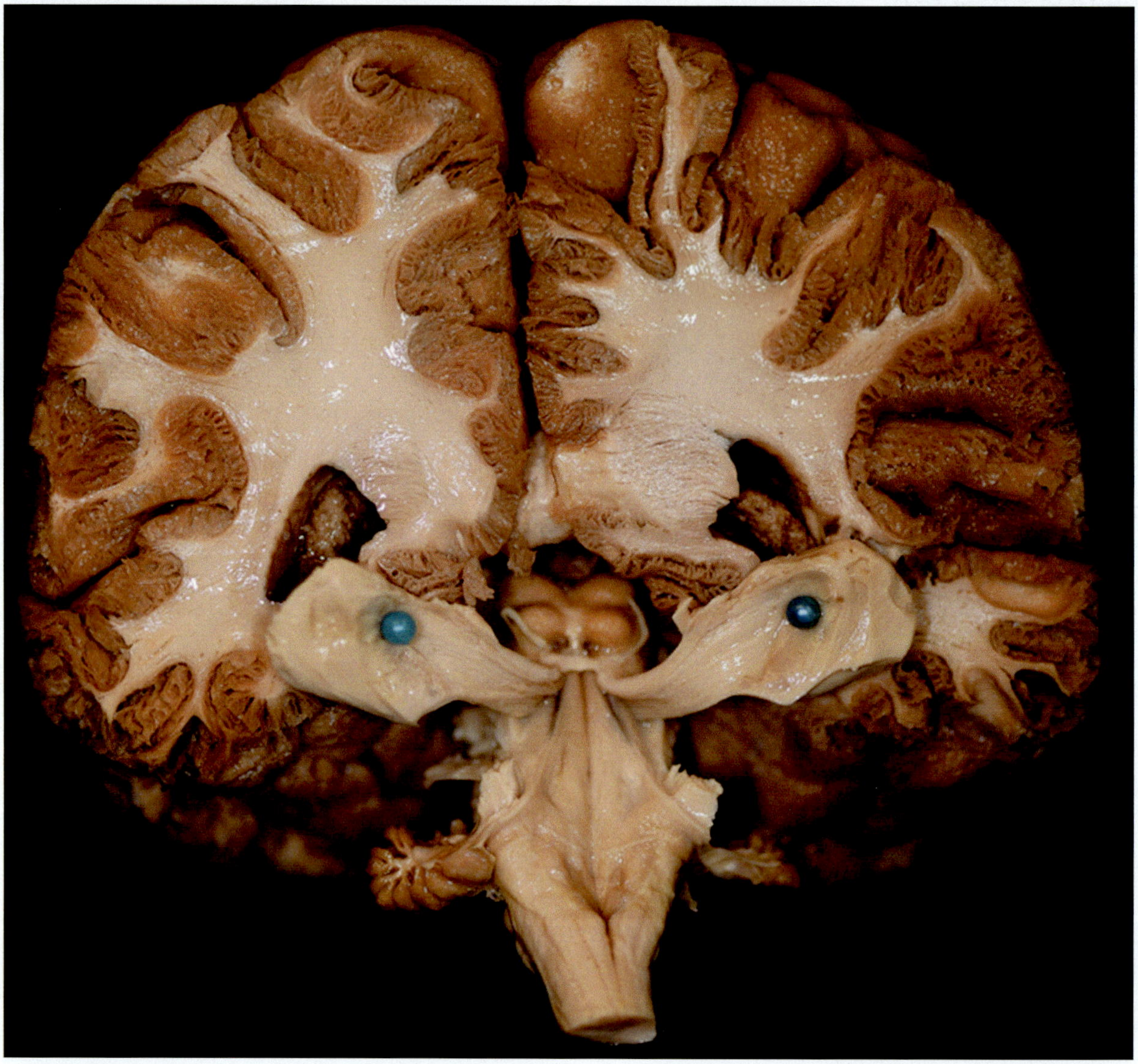

Fig. 8.77 Anatomical preparation of the floor of the fourth ventricle. The middle and inferior cerebellar peduncles have been removed, and the superior cerebellar peduncle and dentate nucleus were dislodged to expose the rhomboid fossa, also known as the floor of the fourth ventricle. *inf* inferior, *n* nerve, *nn* nucleus, *ped* peduncle, *post* posterior, *s* sulcus, *sup* superior, *t* tract, *tt* tract (©Kadri 2023. All Rights Reserved)

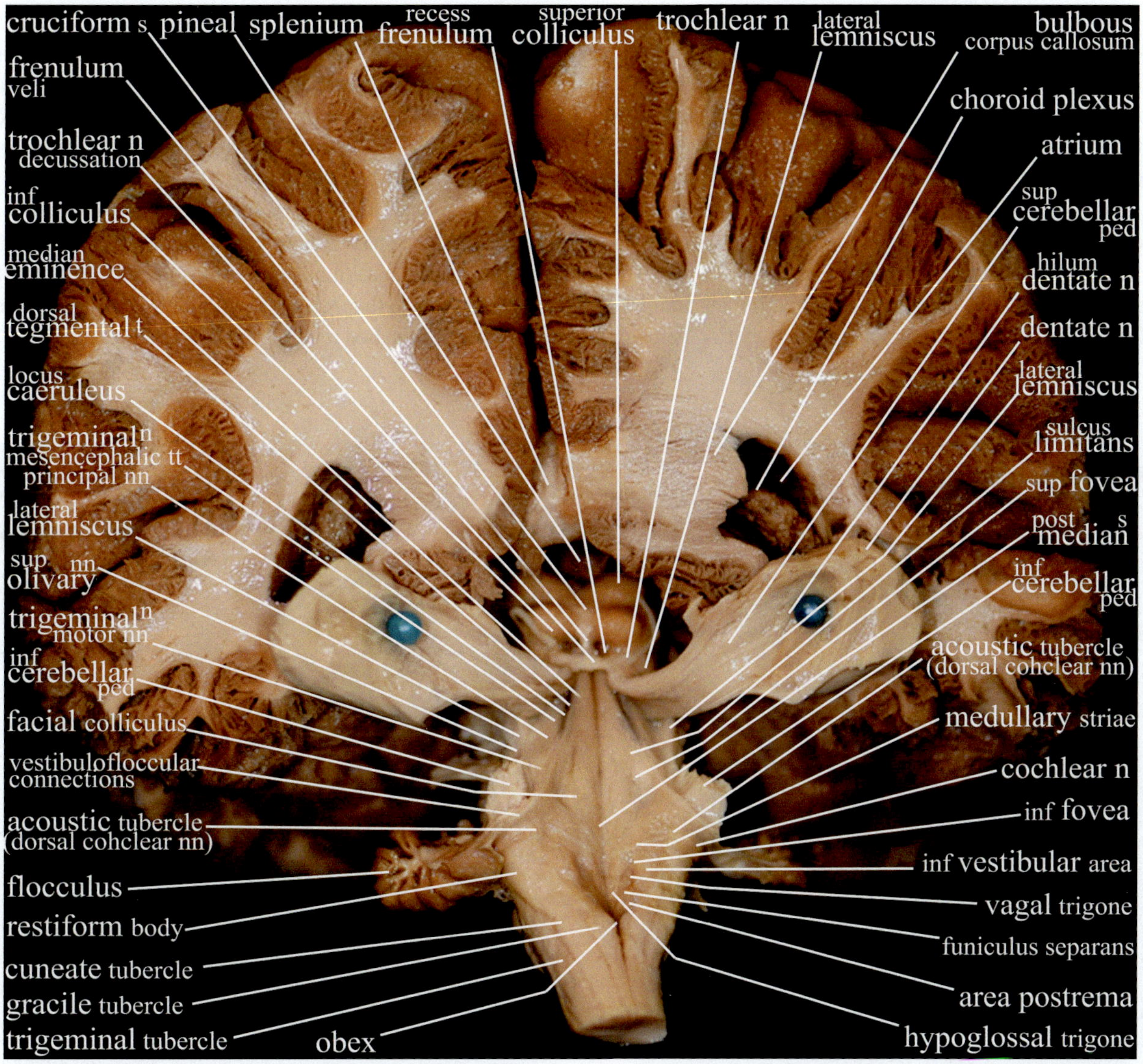

Fig. 8.76 (continued)

They may be directed laterally, turning around the restiform body (Figs. 8.82 and 8.83) to reach the lateral recess and arrive at the acoustic tubercle formed by the dorsal cochlear nucleus (Fig. 8.84).

The medullary floor is not homogenous, nor at the surface. It can be further divided into internal and external portions. At the first view, an internal, more whitish, zone and an external, more grayish, zone are encountered. The internal zone is the origin of the hypoglossal nerve and, because of its triangular shape, is known as the hypoglossal trigone (Figs. 8.85 and 8.86). The white external portion also has a triangular shape and contains the vestibular and cochlear nuclei. It is also called the acoustic area and acoustic trigone. The inferomedial portion of the acoustic triangle contains the main vestibular nucleus, also called the vestibular area. The superolateral portion is at the transition to the pontine triangle and harbors an eminence known as the acoustic tubercle or the cochlear area. The gray portion, or *area cinerea*, is located between the internal and external whitish areas and also has a triangular shape. However, instead of protruding like the white components, the gray part is a depression known as the inferior fovea, or vagal fovea, because of its relationship with the dorsal nuclei of the vagus (Figs. 8.87 and 8.88). This area is also known as the vagal trigone. A whitish strip known as the funiculus separans divides the vagal triangle from the area postrema. The area postrema may sometimes fuse with its contralateral counterpart and may form the coalescence interpostrema (Figs. 8.89 and 8.90).

The pontine triangle is slightly larger than the medullary one. On each side of the median sulcus, above the acoustic area, appears an eminence known as the facial colliculus. The facial colliculus is related to the abducens nerve nuclei (Fig. 8.91) [2]. An enlargement of the median sulcus above the facial colliculus is known as the median fovea. Between the facial colliculus and the vestibular area, another depression along the sulcus limitans is known as the superior fovea. The superior fovea is related to the motor nucleus of the trigeminal nerve; thus, it is also termed the trigeminal fovea. Above the superior fovea, an elongated pigmented area extends upward and is known as the locus coeruleus (Fig. 8.92).

The roof of the fourth ventricle has a superior and an inferior part. The superior part is formed by the superior cerebellar peduncles and the superior medullary velum, which is also known as the valve of Vieussens. The lingula of the vermis cerebellar is attached to the cisternal surface of the superior medullary velum. The inferior part is covered by the tectorial membrane, also known as the inferior medullary velum and valve of Tarin [2]. The inferior medullary velum has a triangular shape, the apex of which is inferior and attached laterally to the obex on both the sides. The inferior medullary velum is attached to the uvula in the midline and extends laterally, covered by the cerebellar tonsils, to reach the lateral recess. The inferior medullary velum is attached to the flocculonodular peduncle.

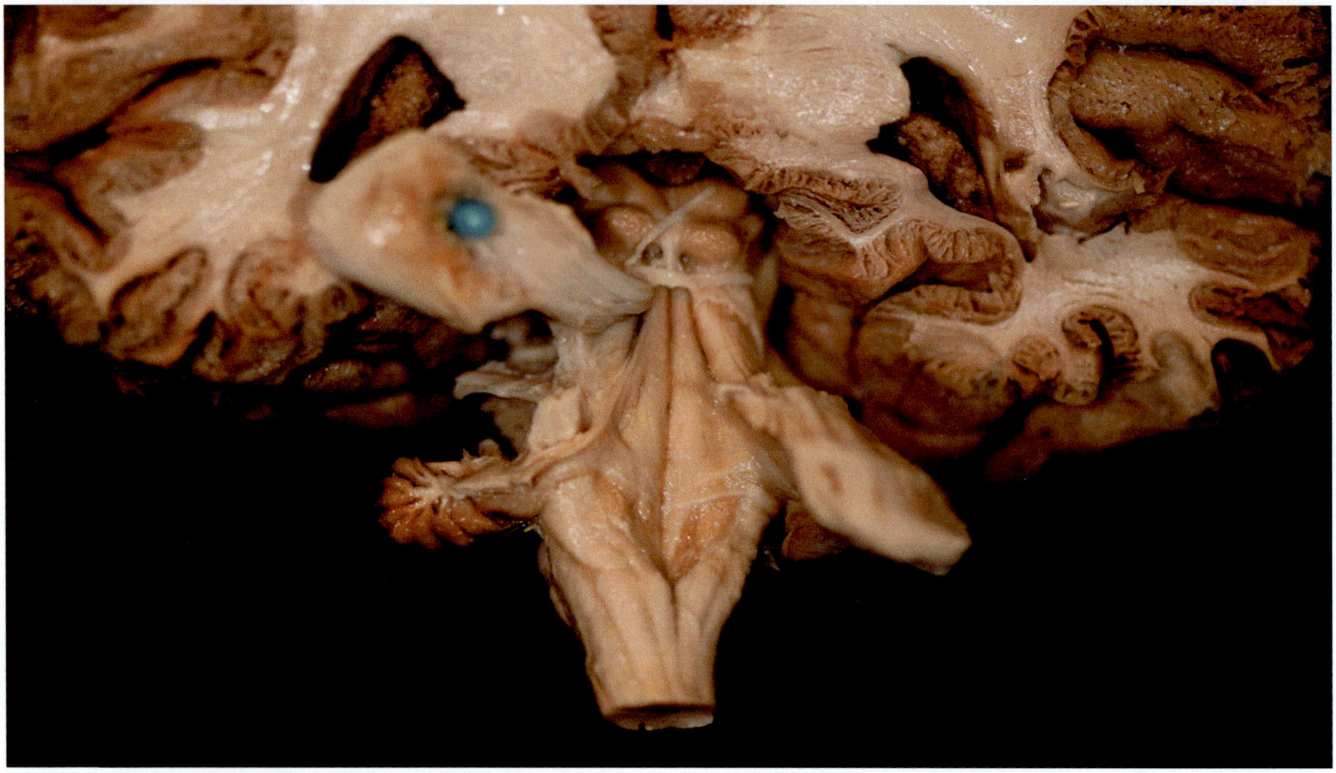

Fig. 8.78 Anatomical preparation of the floor of the fourth ventricle, closer view. *inf* inferior, *n* nerve, *nn* nucleus, *ped* peduncle, *s* sulcus, *sup* superior, *tt* tract (©Kadri 2023. All Rights Reserved)

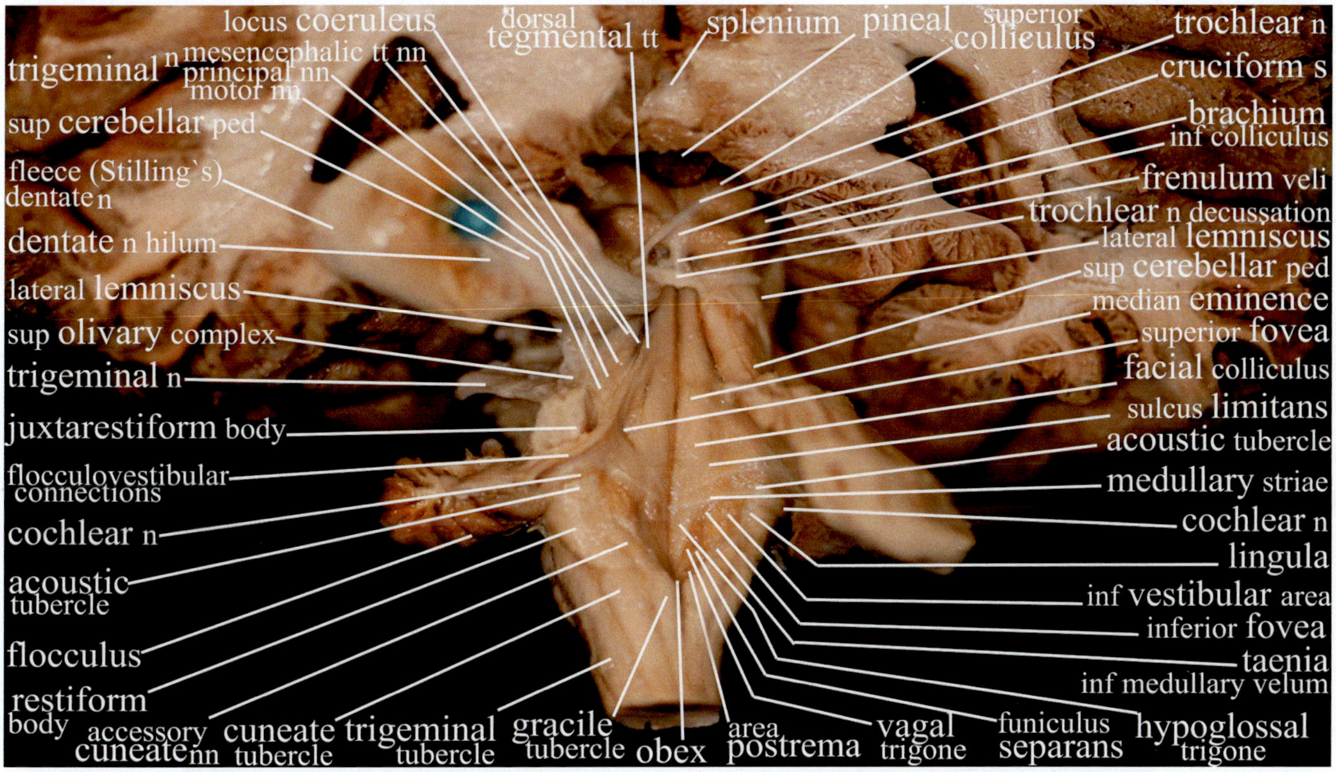

Fig. 8.78 (continued)

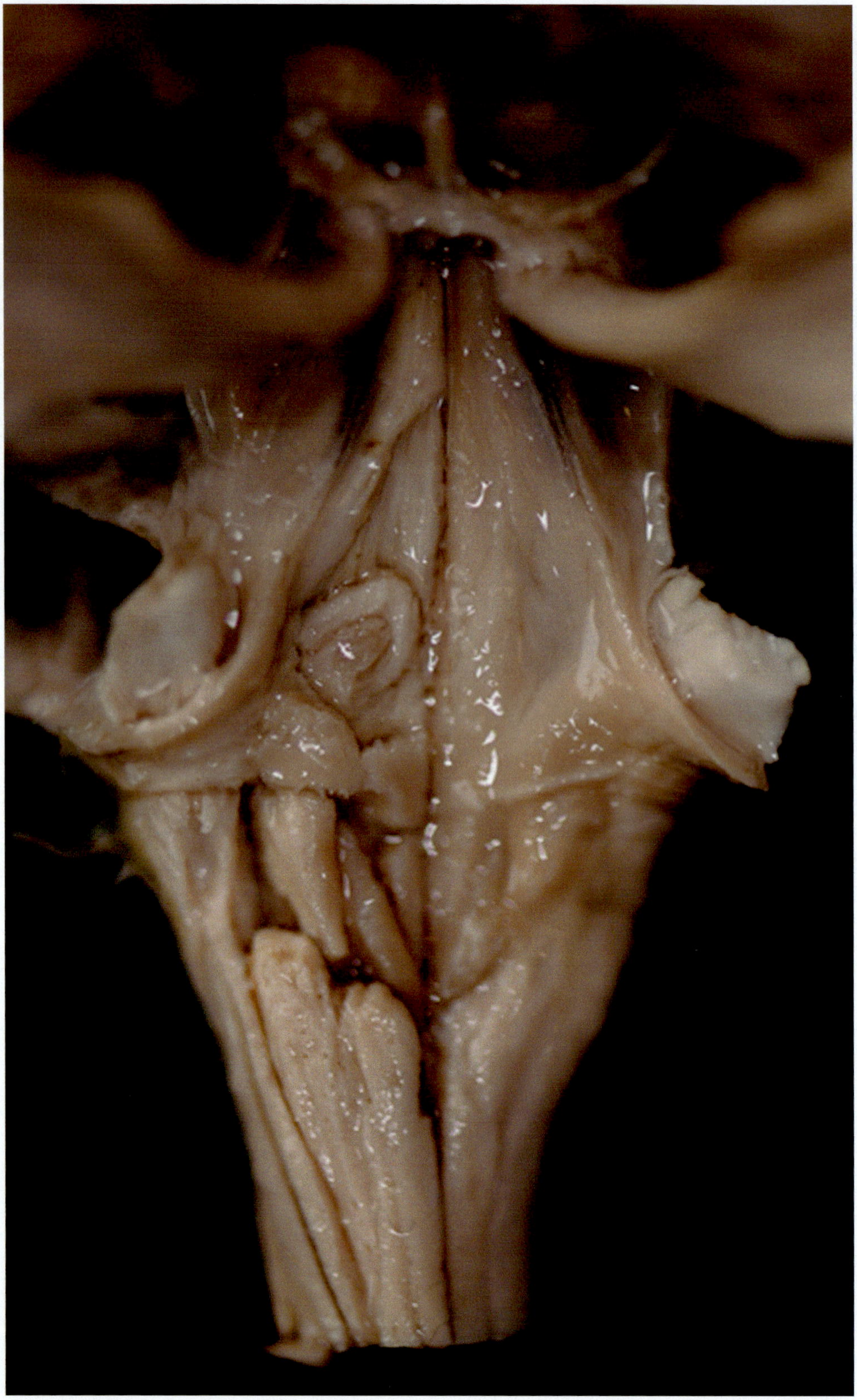

Fig. 8.79 Anatomical preparation of the floor of the fourth ventricle. The most superficial structures of the floor of the fourth ventricle were removed to depict the superficial layer of the pontine medullary tegmentum. *f* fasciculus, *inf* inferior, *lat* lateral, *med* medial, *n* nerve, *nn* nucleus, *ped* peduncle, *post* posterior, *s* sulcus, *sup* superior, *t* tract, *tt* tract (©Kadri 2023. All Rights Reserved)

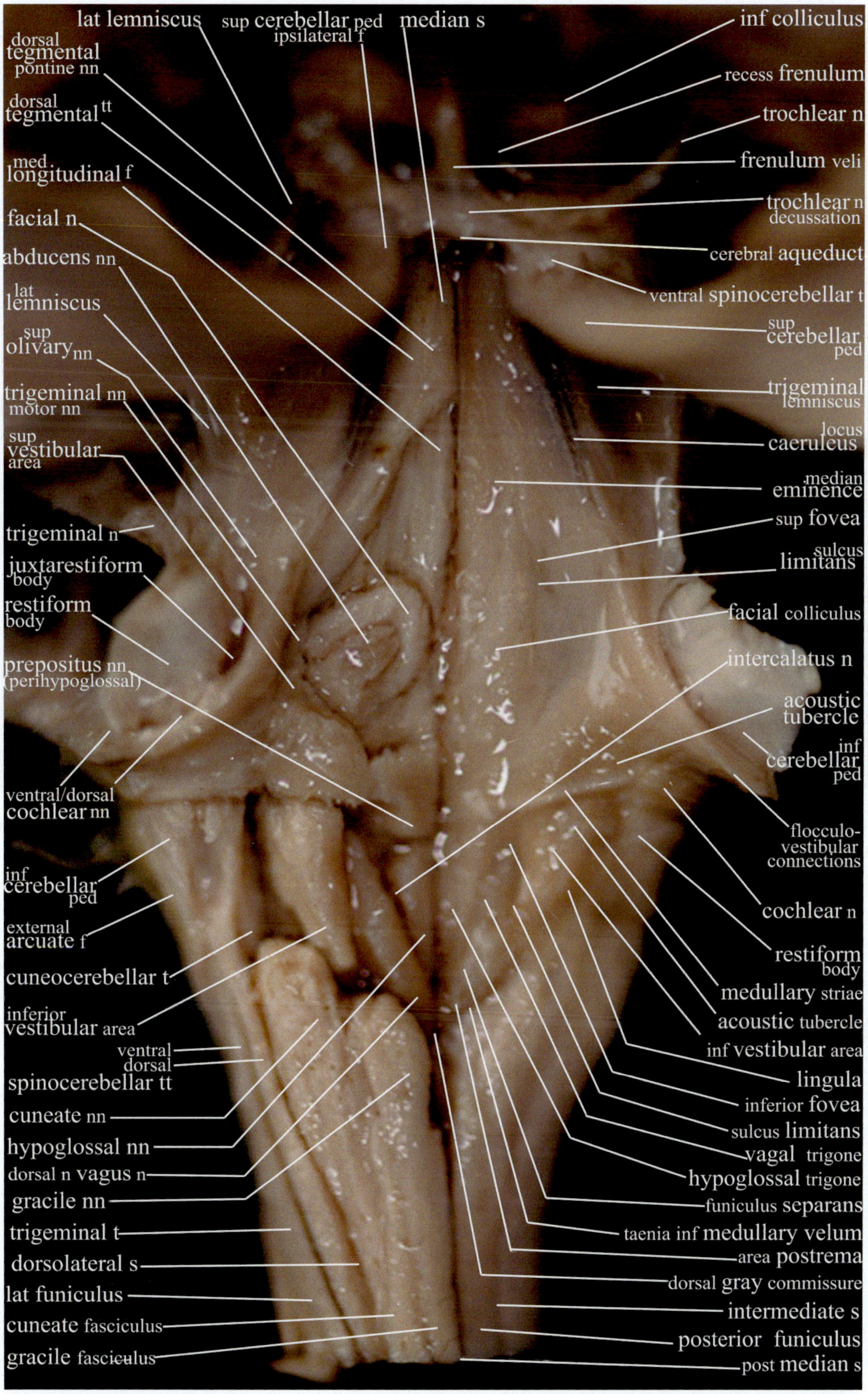

Fig. 8.79 (continued)

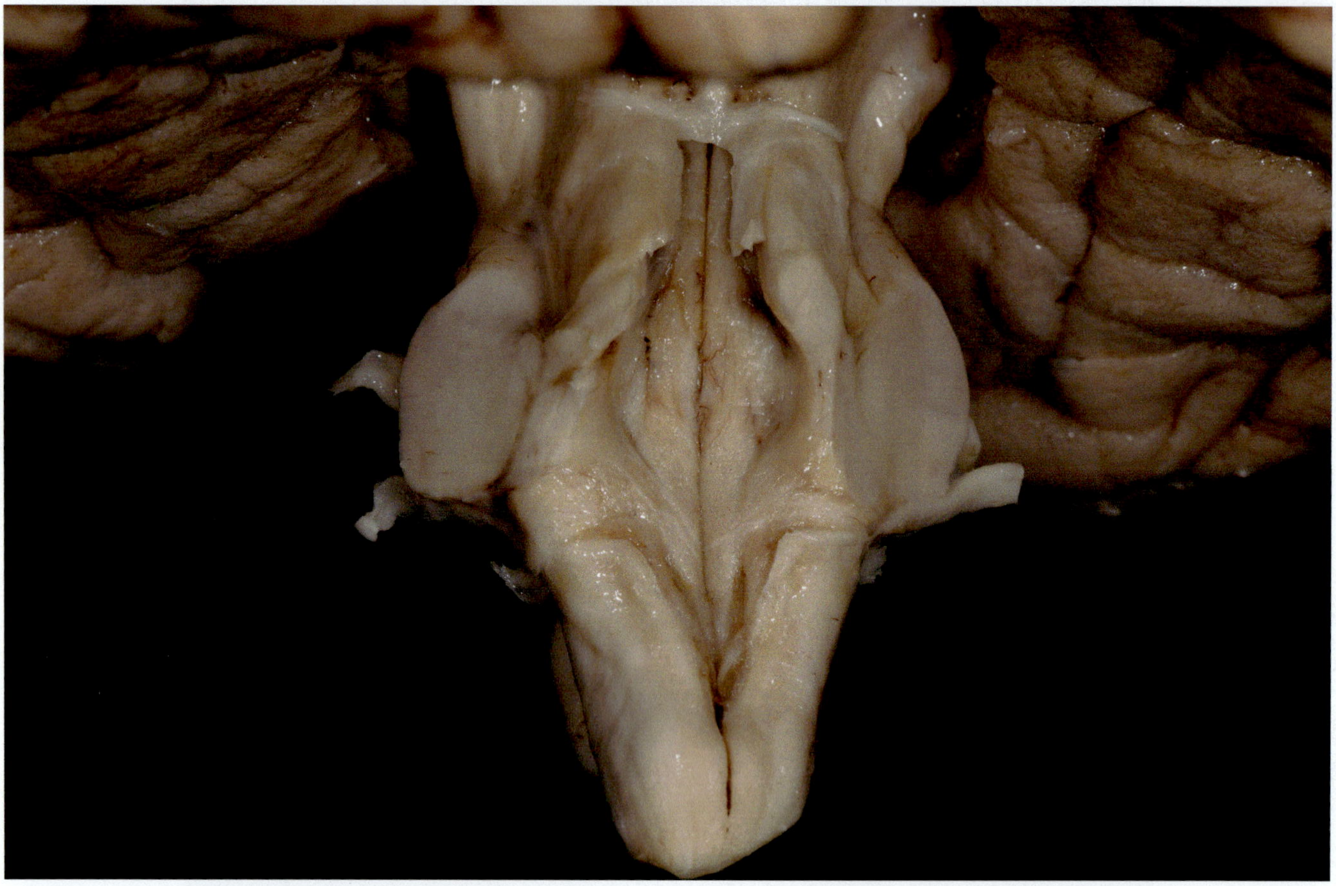

Fig. 8.80 Anatomical preparation of the floor of the fourth ventricle. *inf* inferior, *f* fasciculus, *lat* lateral, *n* nerve, *nn* nucleus, *ped* peduncle, *s* sulcus, *sup* superior, *t* tract (©Kadri 2023. All Rights Reserved)

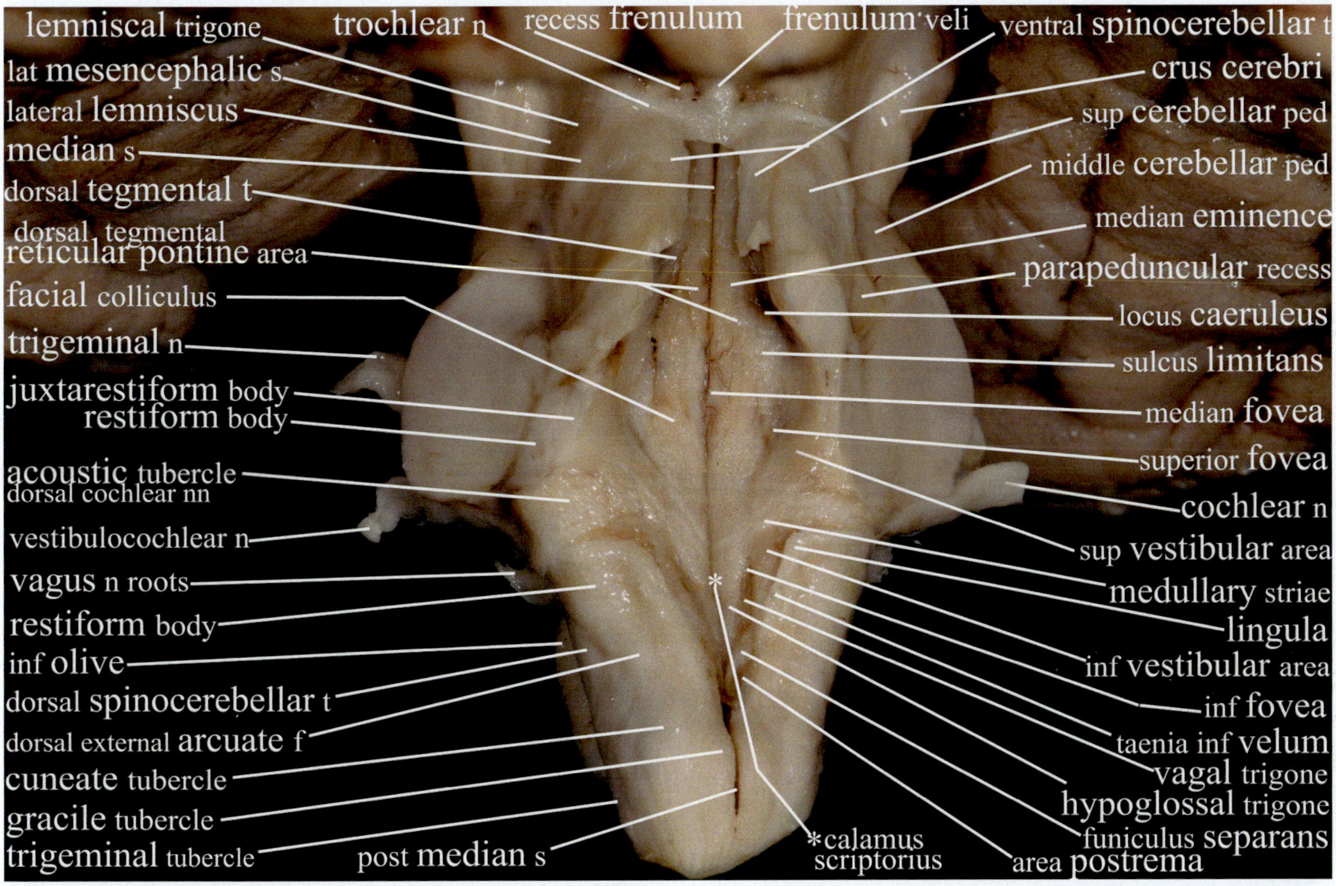

Fig. 8.80 (continued)

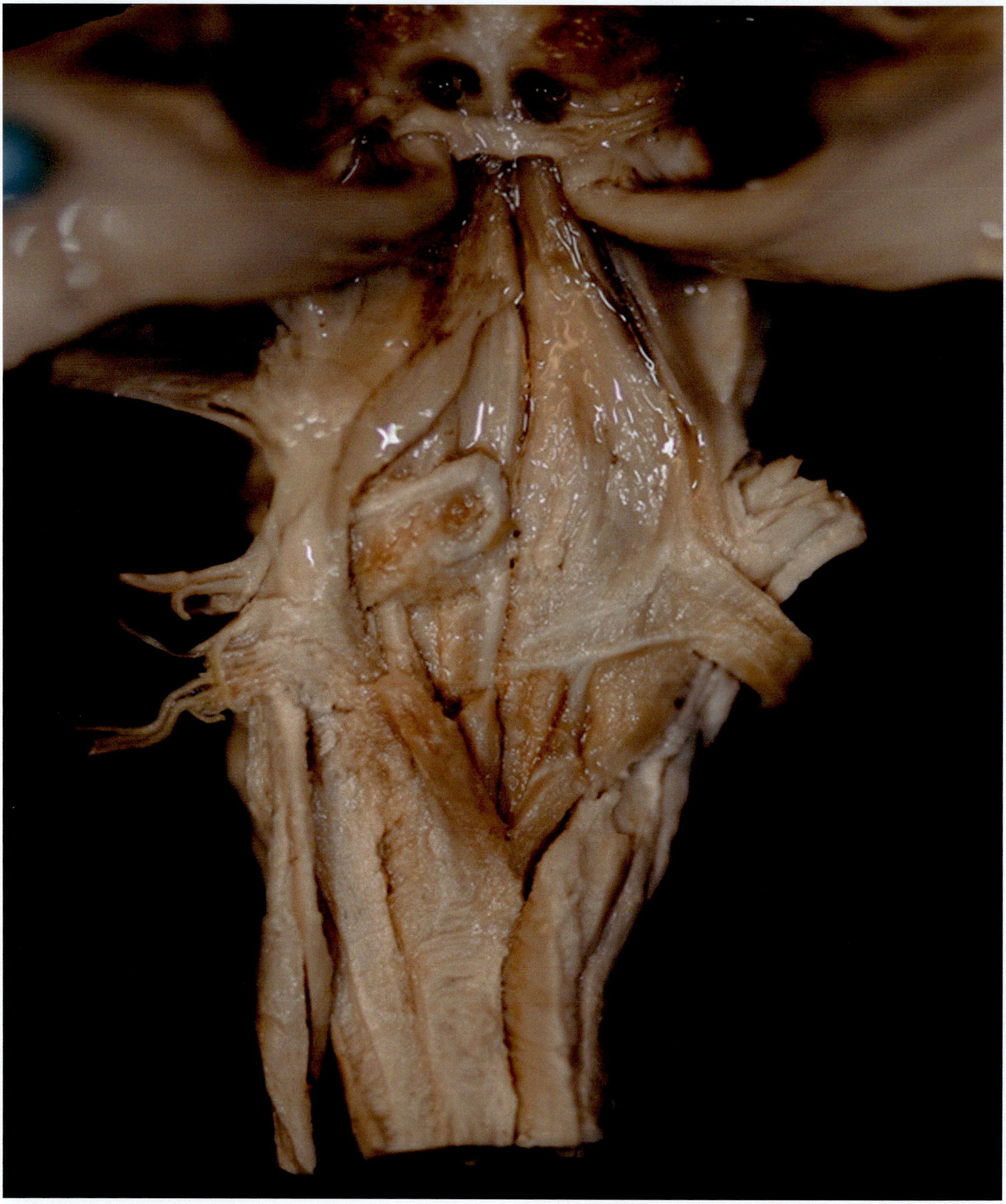

Fig. 8.81 Anatomical preparation of the floor of the fourth ventricle. The posterior funiculus and the gracile and cuneiform tubercles have been removed. *f* fasciculus, *inf* inferior, *lat* lateral, *n* nerve, *nn* nucleus, *ped* peduncle, *s* sulcus, *sup* superior, *t* tract, *tt* tract (©Kadri 2023. All Rights Reserved)

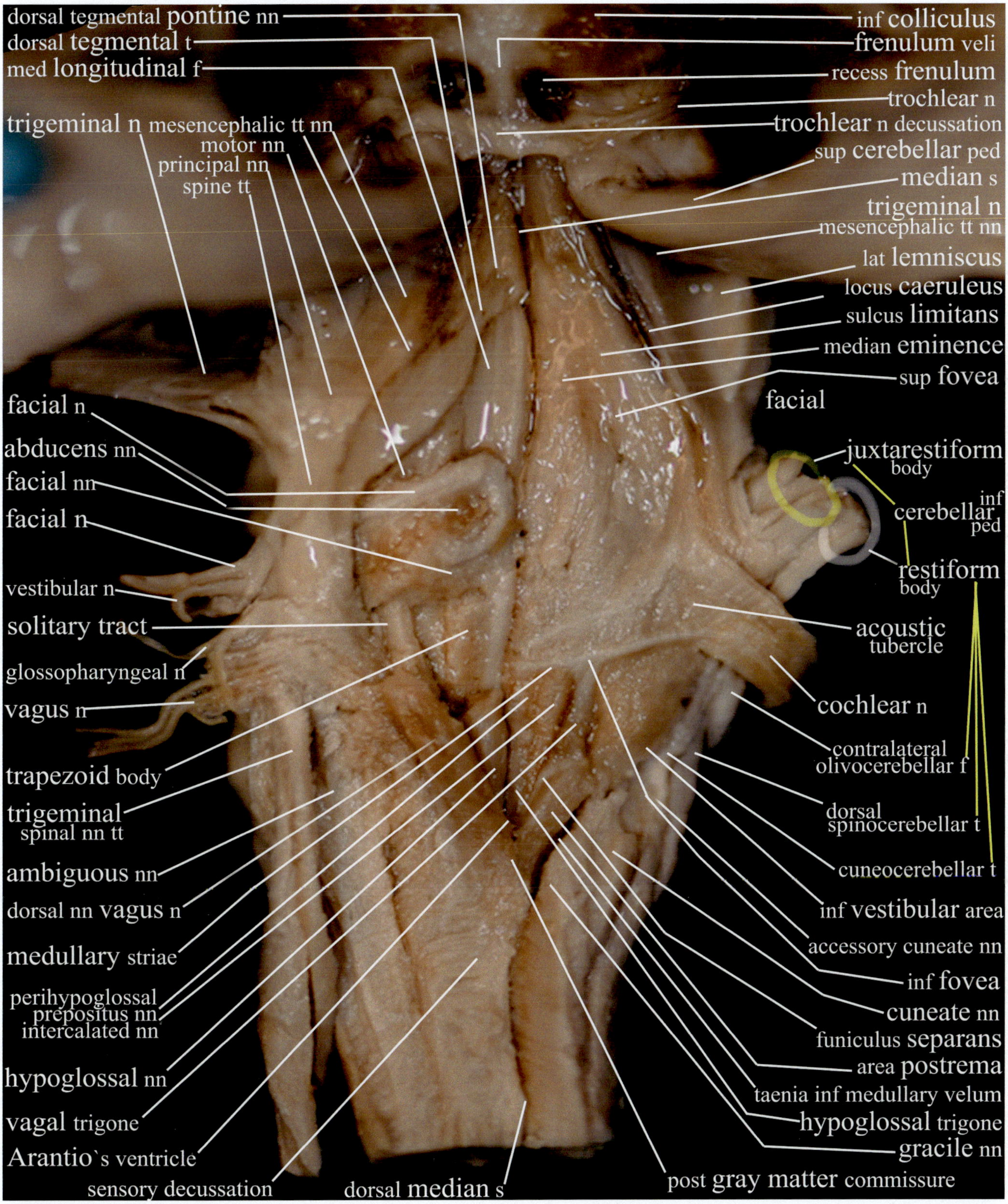

Fig. 8.81 (continued)

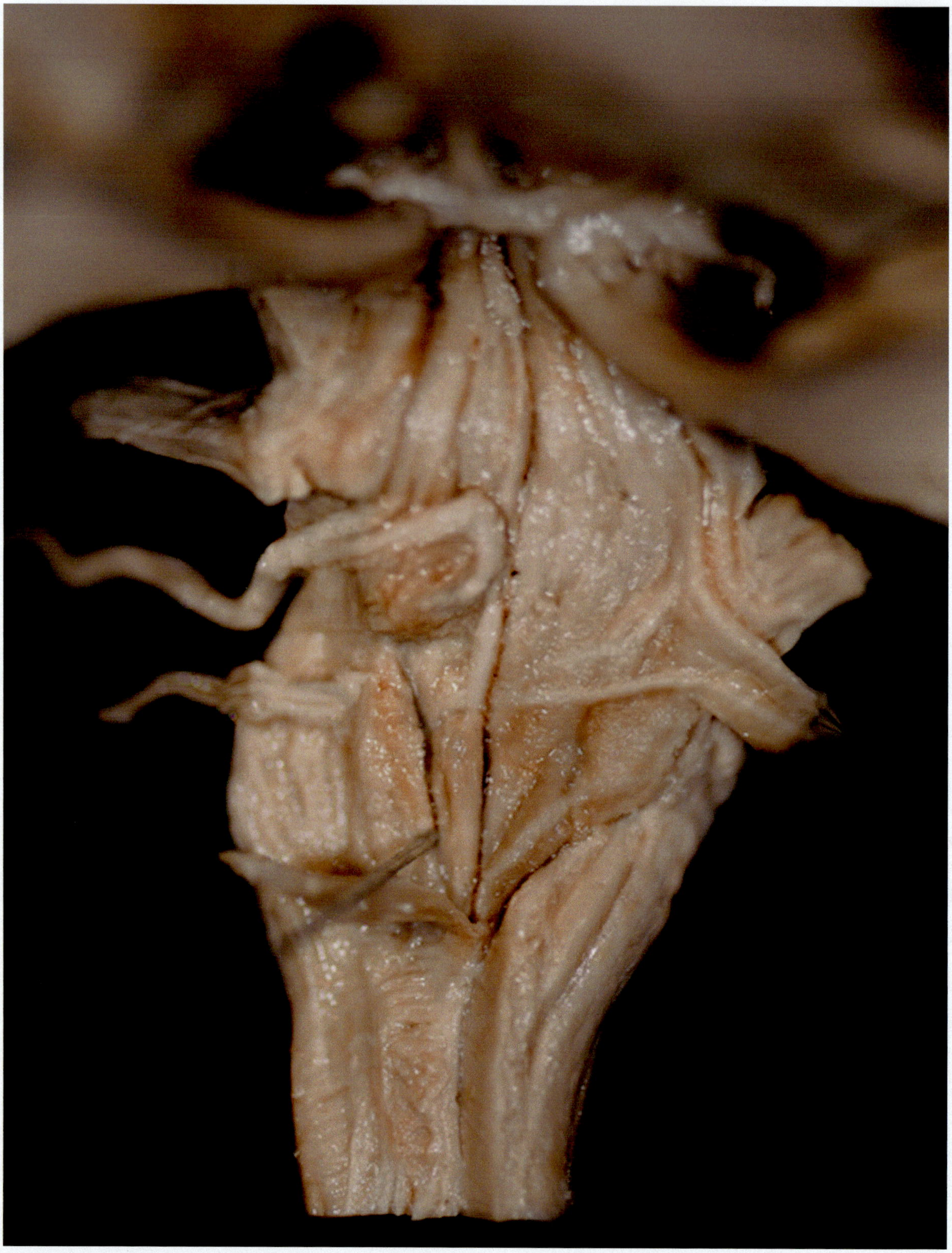

Fig. 8.82 Anatomical preparation of the floor of the fourth ventricle. The solitary tract, area postrema, and funiculus separans have been removed. *f* fasciculus, *inf* inferior, *lat* lateral, *n* nerve, *nn* nucleus, *ped* peduncle, *s* sulcus, *sup* superior, *t* tract, *tt* tract (©Kadri 2023. All Rights Reserved)

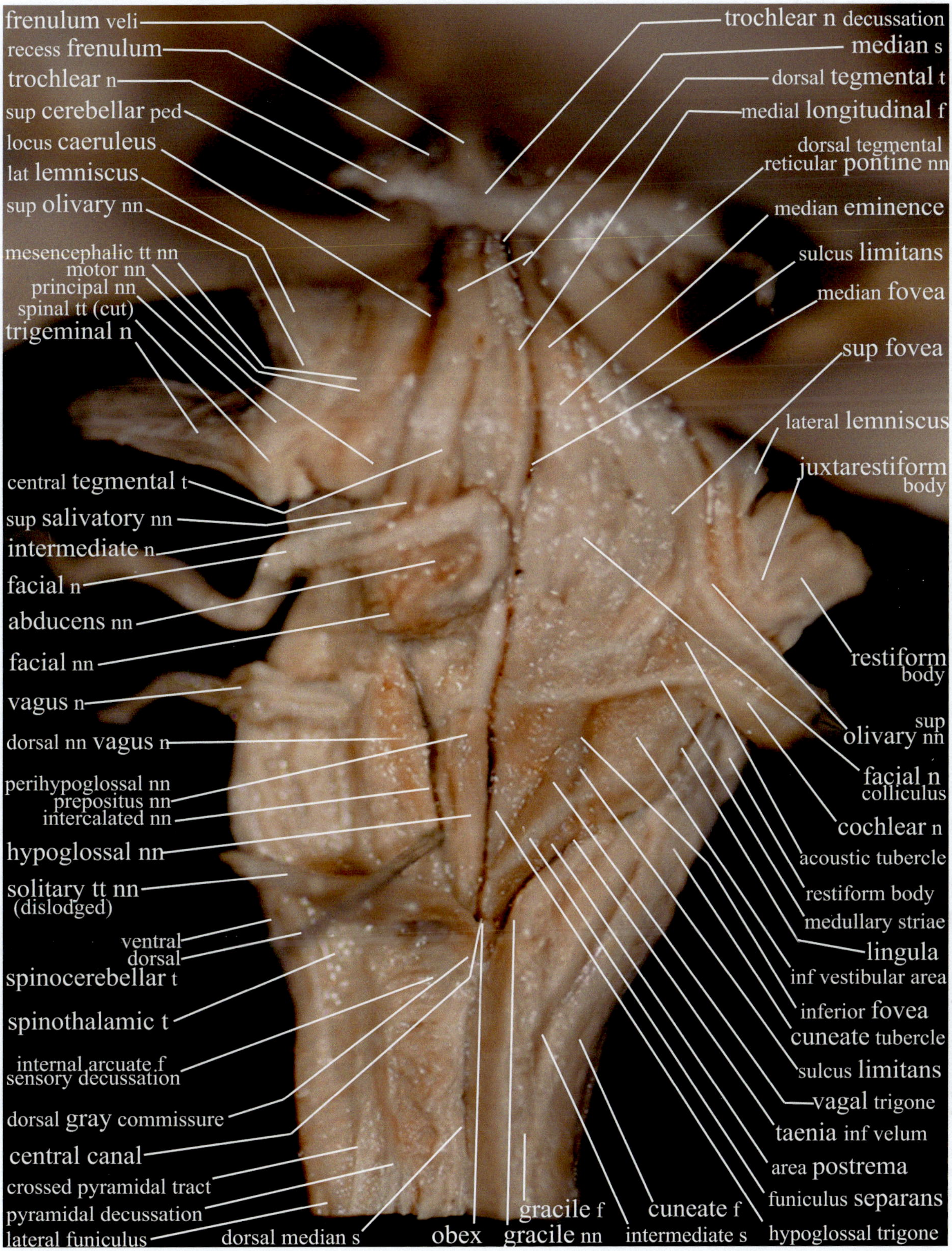

Fig. 8.82 (continued)

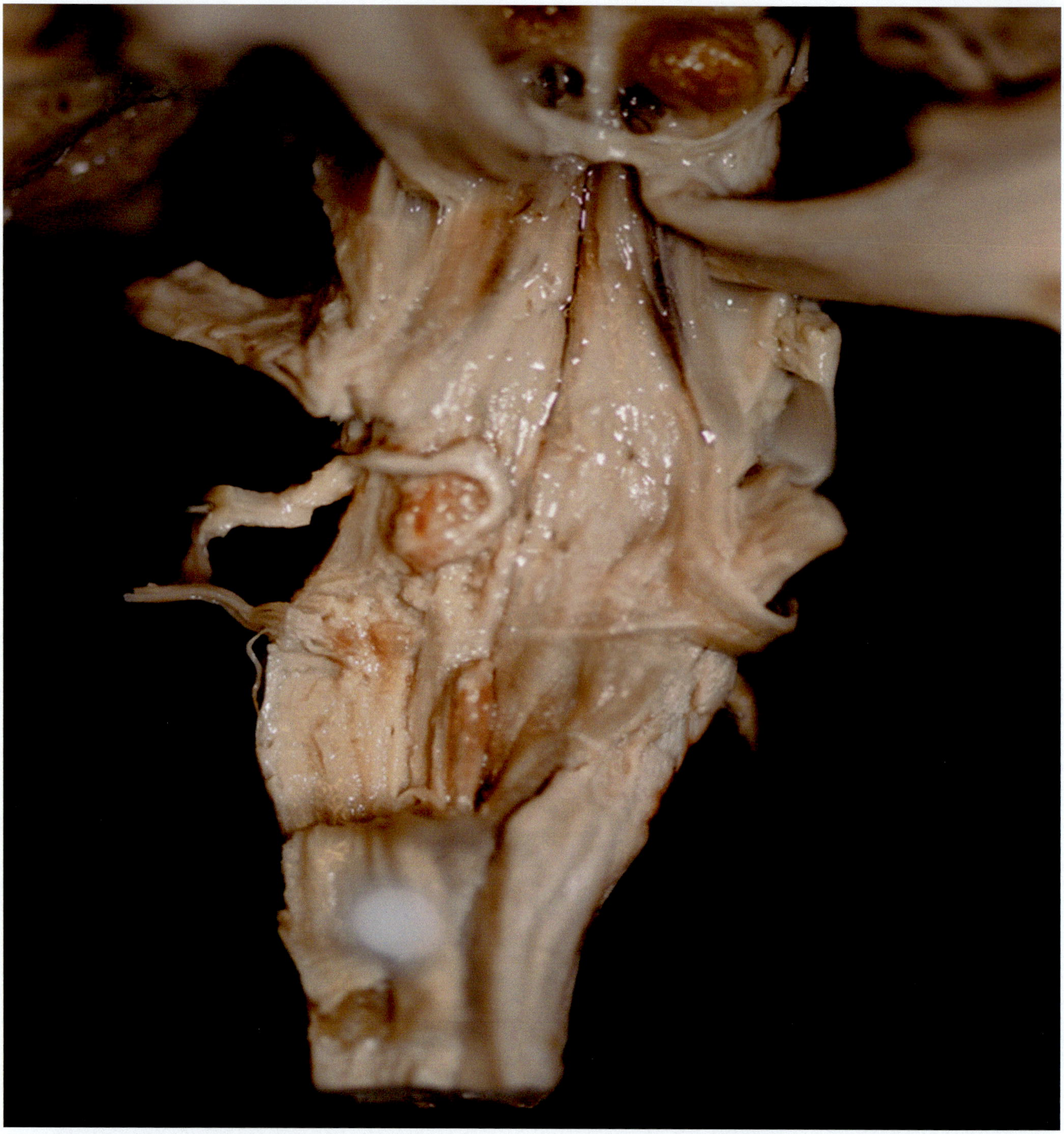

Fig. 8.83 Anatomical preparation of the floor of the fourth ventricle. The dorsal tegmentum was partially removed to expose the ventral layers of the tegmentum. *a* area, *f* fasciculus, *inf* inferior, *lat* lateral, *n* nerve, *nn* nucleus, *ped* peduncle, *s* sulcus, *sup* superior, *t* tract, *tt* tract (©Kadri 2023. All Rights Reserved)

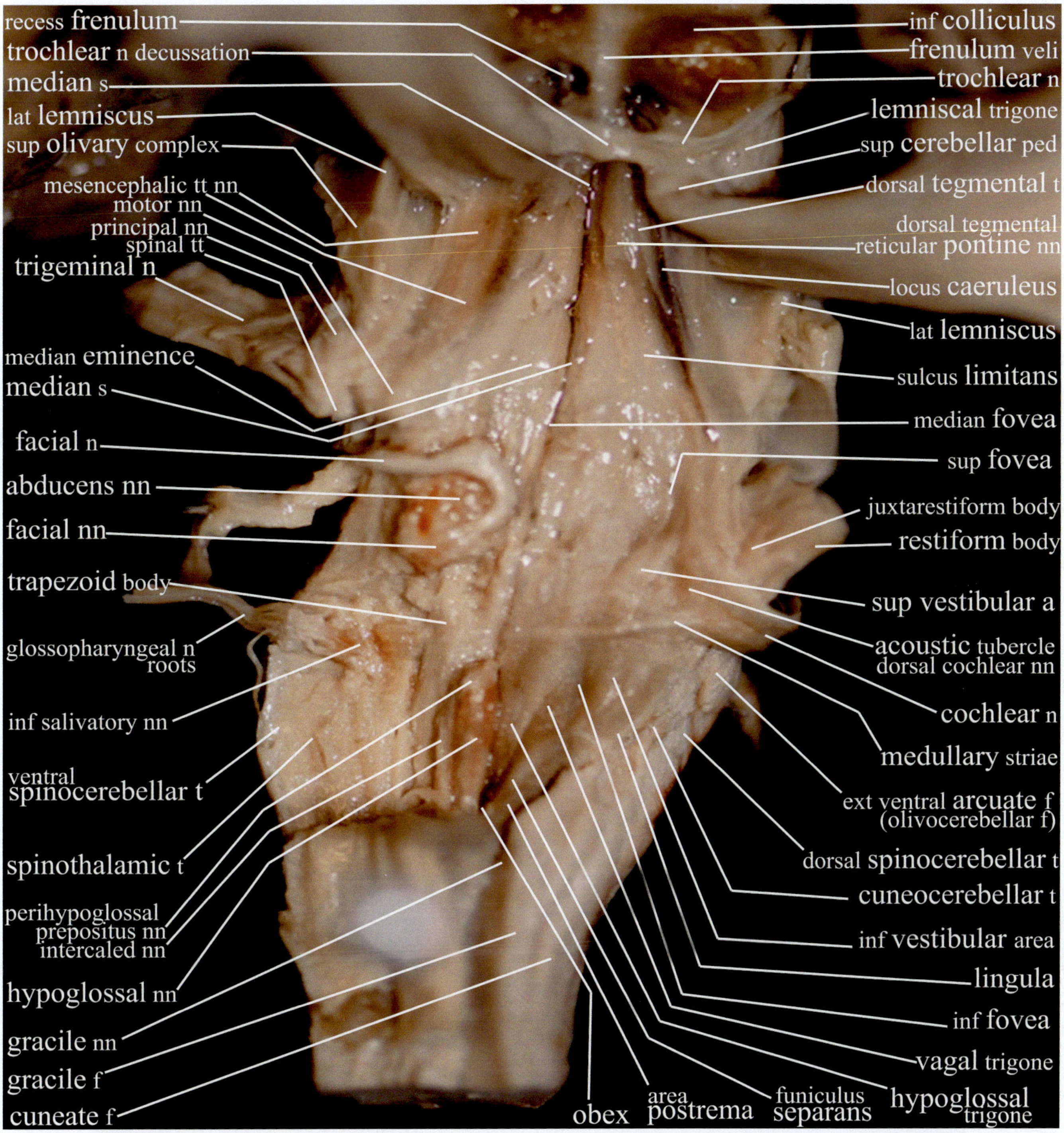

Fig. 8.83 (continued)

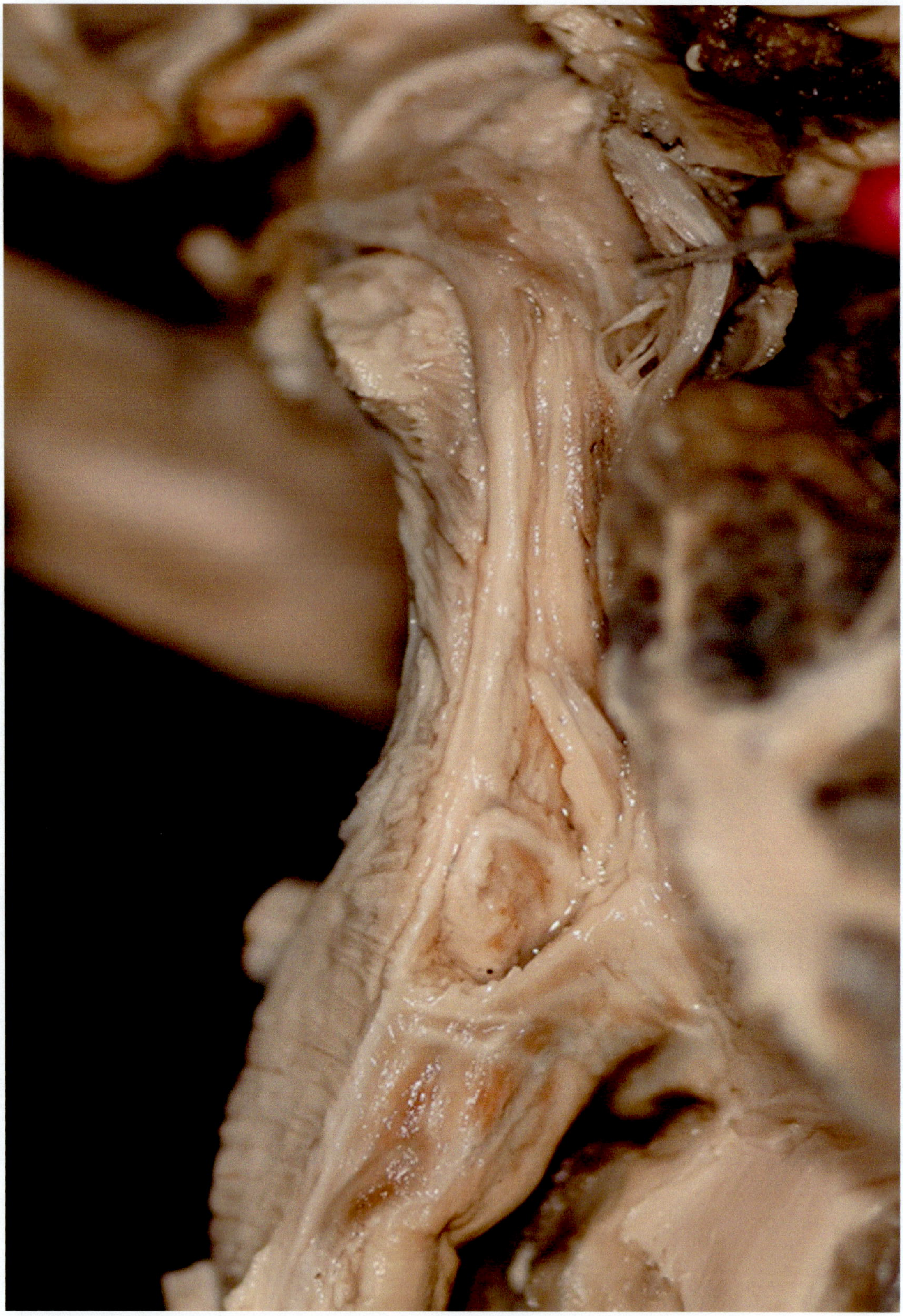

Fig. 8.84 Anatomical preparation of the floor of the fourth ventricle. *f* fasciculus, *inf* inferior, *lat* lateral, *n* nerve, *nn* nucleus, *ped* peduncle, *sup* superior, *t* tract (©Kadri 2023. All Rights Reserved)

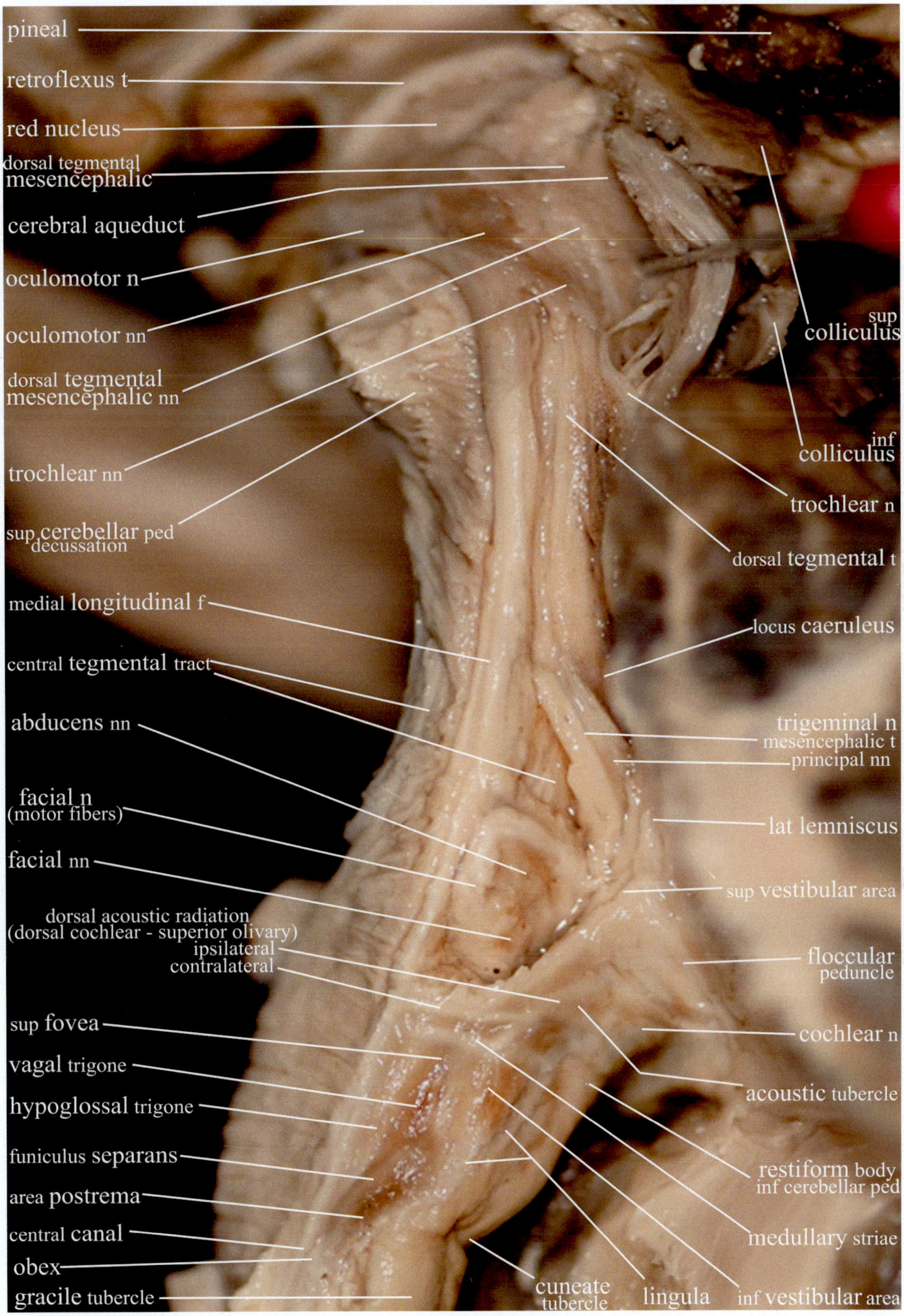

Fig. 8.84 (continued)

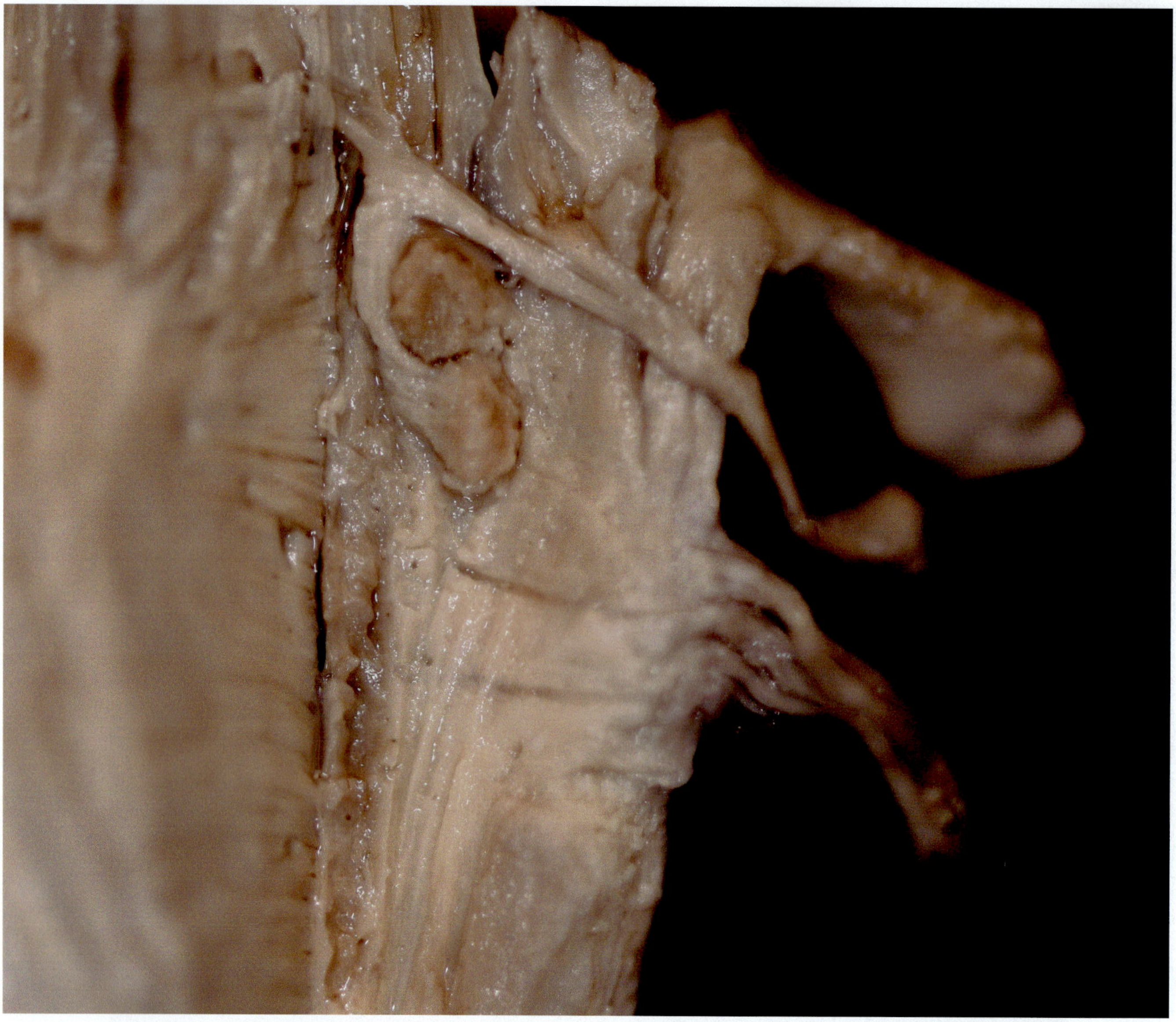

Fig. 8.85 Anatomical preparation of the motor column of the pons and medulla at the tegmentum. *n* nerve, *nn* nucleus, *sup* superior, *tt* tract (©Kadri 2023. All Rights Reserved)

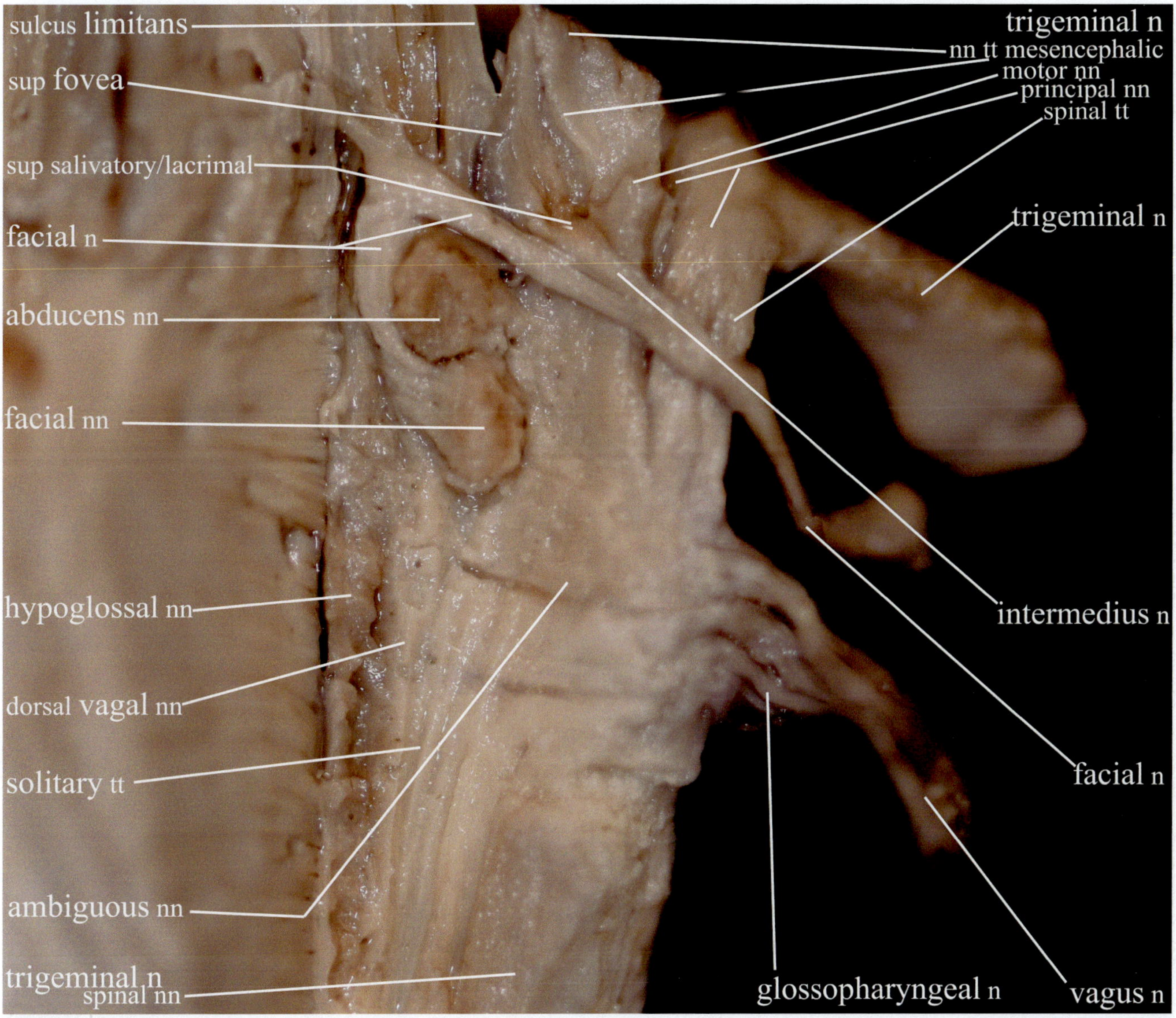

Fig. 8.85 (continued)

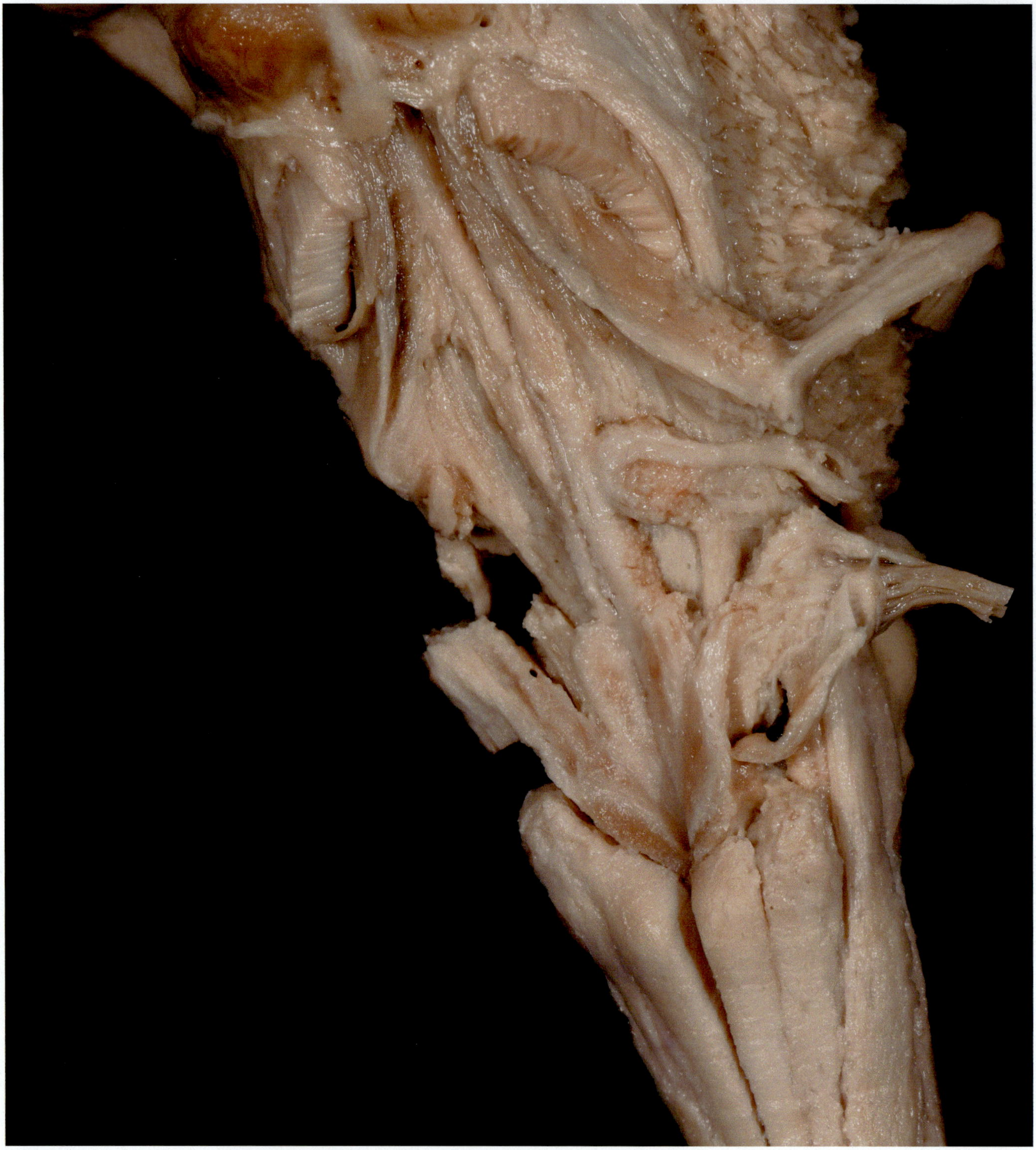

Fig. 8.86 Anatomical preparation of the floor of the fourth ventricle showing the connections of the lower cranial nerve roots. *f* fasciculus, *inf* inferior, *lat* lateral, *n* nerve, *nn* nucleus, *ped* peduncle, *s* sulcus, *sup* superior, *t* tract, *tt* tract (©Kadri 2023. All Rights Reserved)

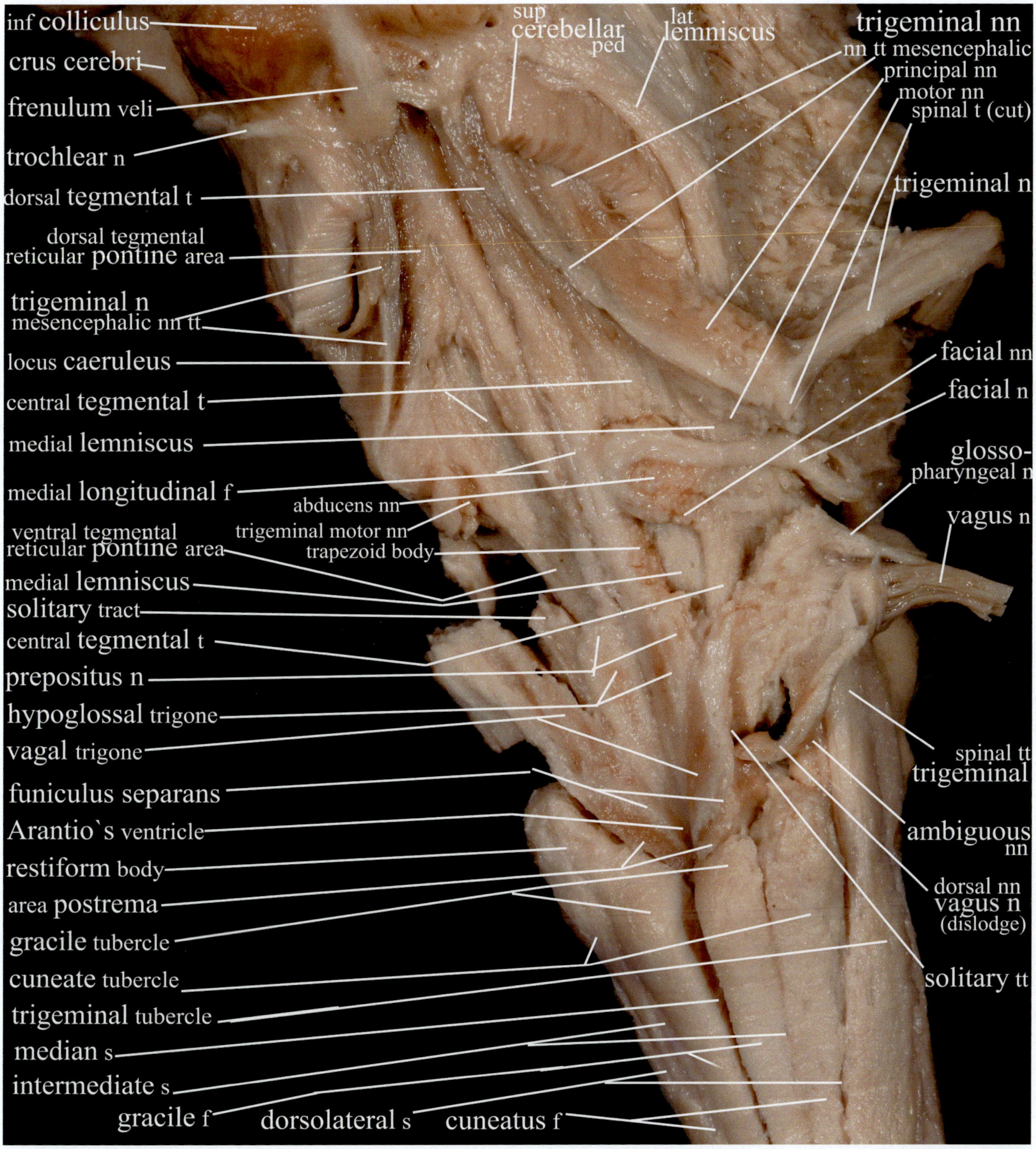

Fig. 8.86 (continued)

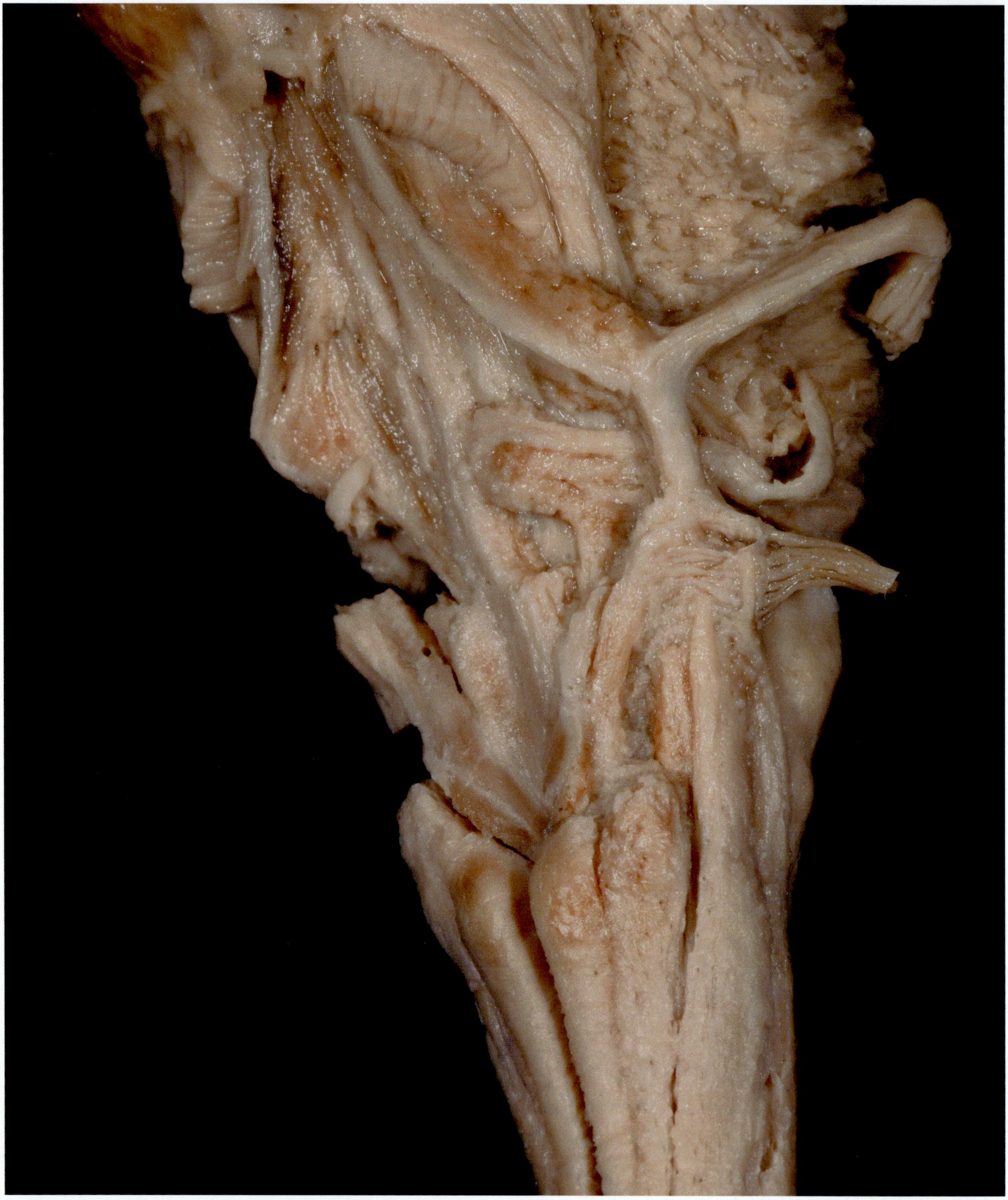

Fig. 8.87 Anatomical preparation of the floor of the fourth ventricle in the posterolateral view. The facial nerve traverses dorsolaterally to the medial lemniscus and ventromedially to the trigeminal nuclei complex in its tegmental route. *f* fasciculus, *inf* inferior, *lat* lateral, *med* medial, *n* nerve, *nn* nucleus, *s* sulcus, *sup* superior, *t* tract, *tt* tract (©Kadri 2023. All Rights Reserved)

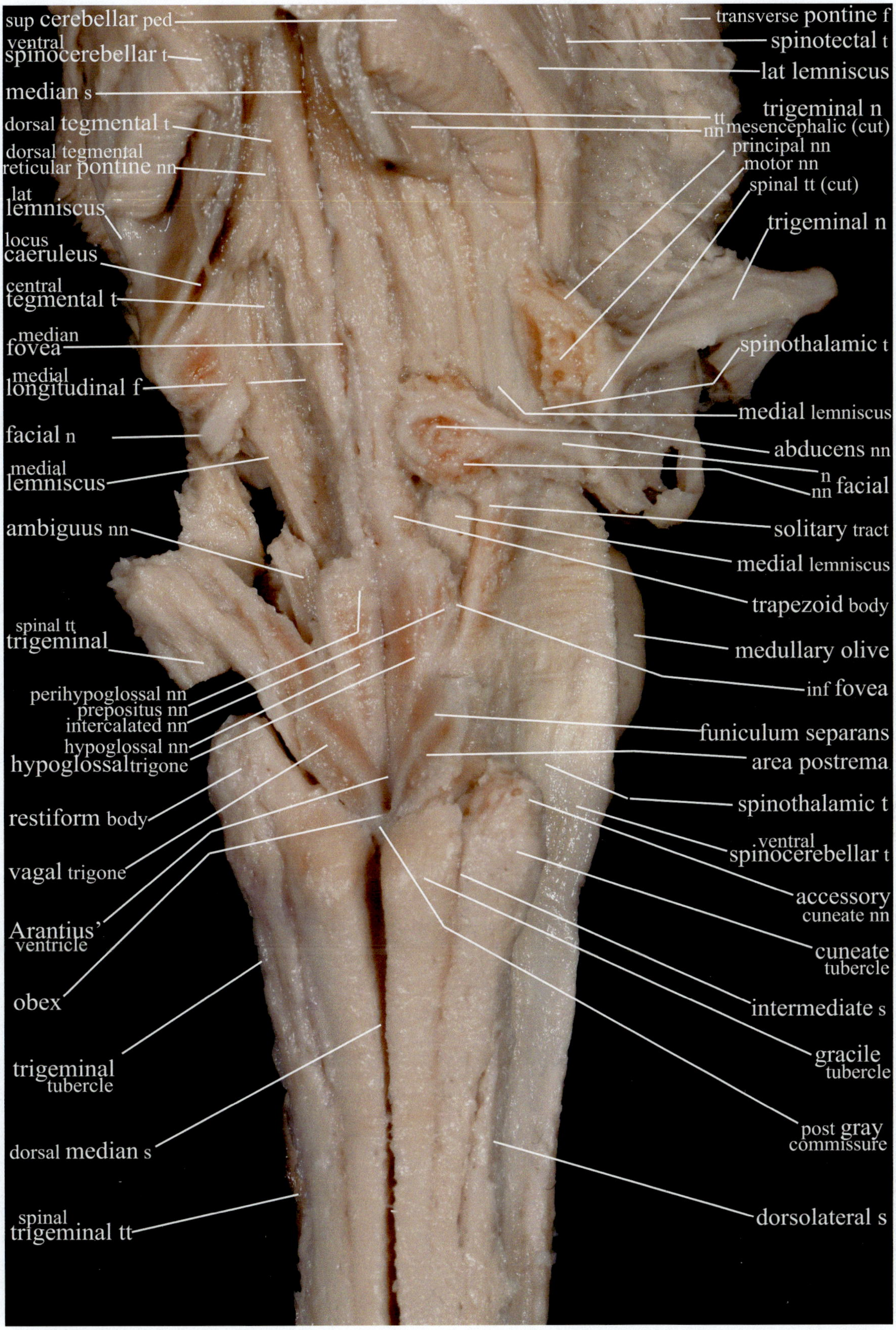

Fig. 8.89 (continued)

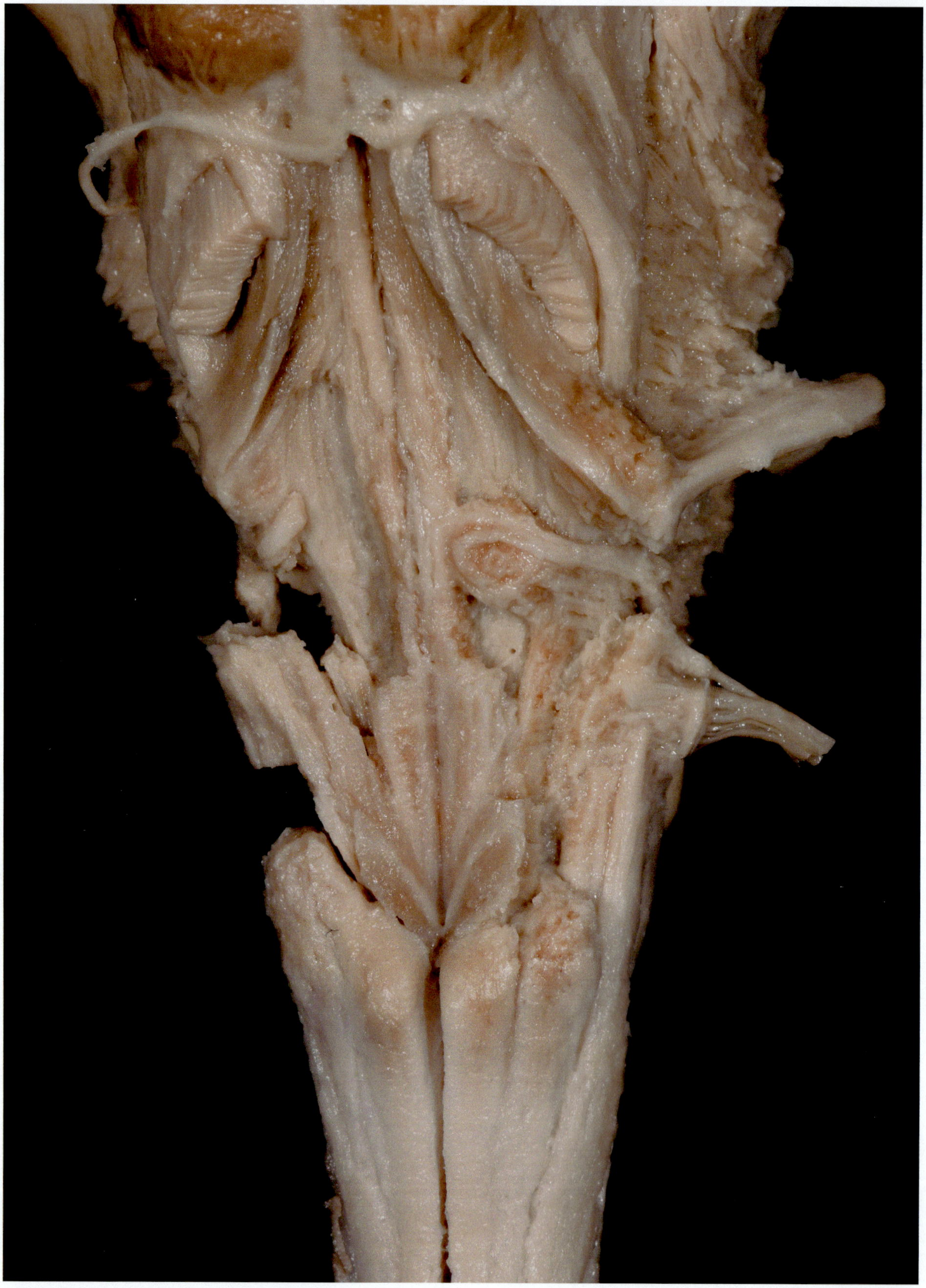

Fig. 8.90 Anatomical preparation of the floor of the fourth ventricle. The dissection of the lateral tegmental area exposed the connections of the nucleus ambiguus. *f* fasciculus, *inf* inferior, *lat* lateral, *med* medial, *n* nerve, *nn* nucleus, *ped* peduncle, *s* sulcus, *sup* superior, *t* tract, *tt* tract

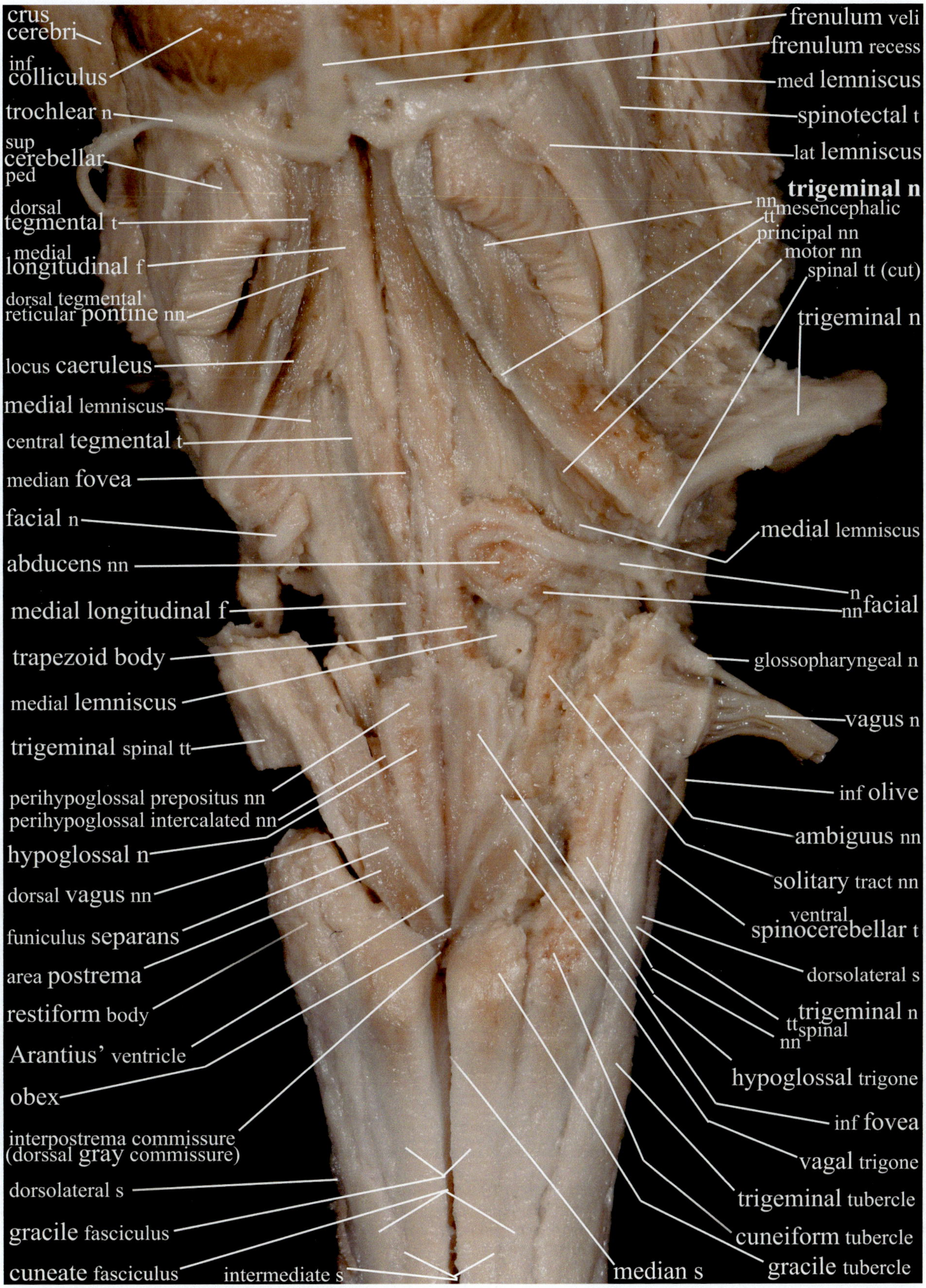

Fig. 8.90 (continued)

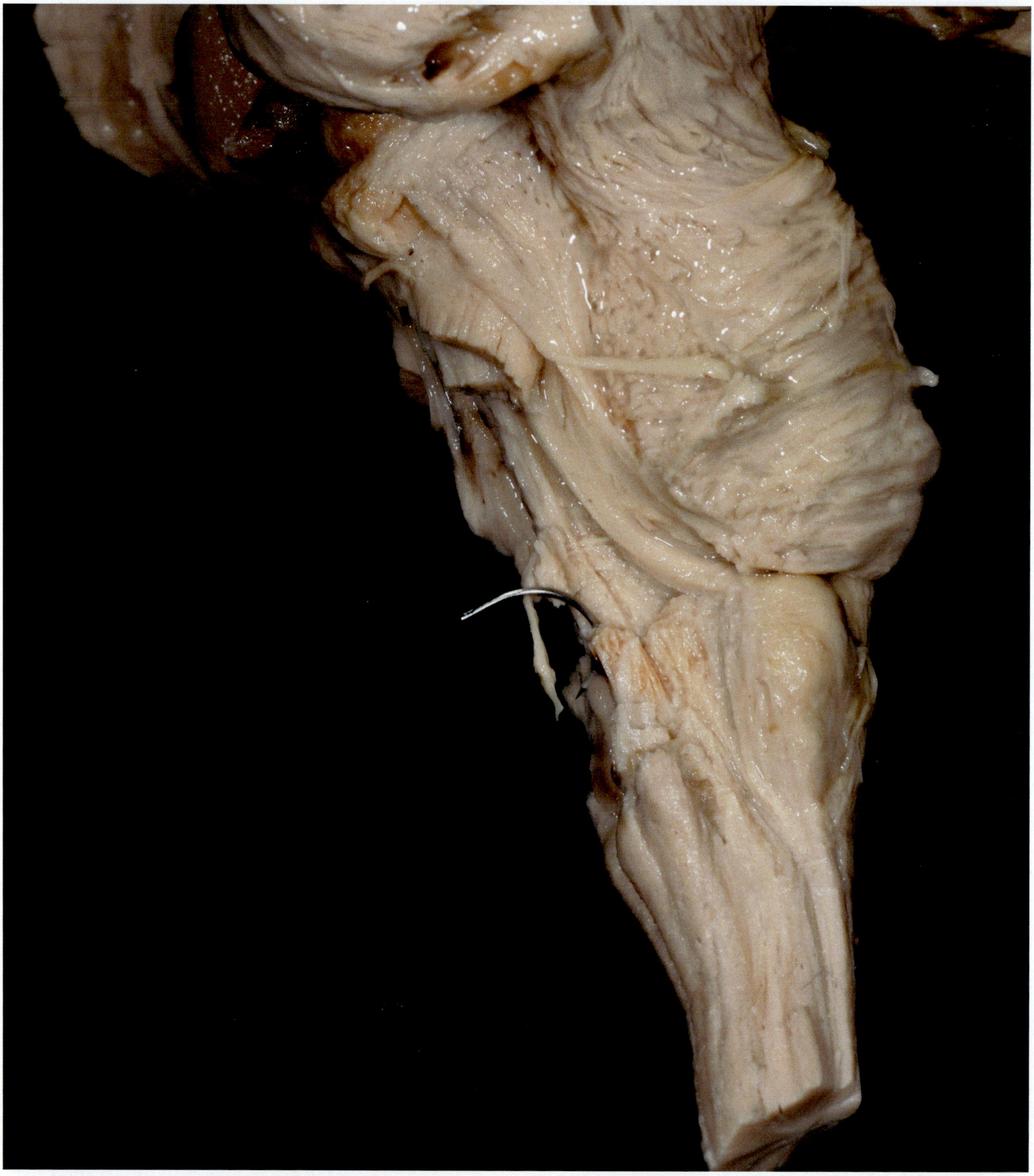

Fig. 8.91 Anatomical preparation of the tegmentum exposing the intra-axial roots of the sixth nerve. At the tegmentum, the roots of the sixth nerve extend from the abducens nucleus traversing the tegmentum ventrally between the central tegmental tract and the medial longitudinal fasciculus. *f* fasciculus, *inf* inferior, *lat* lateral, *n* nerve, *nn* nucleus, *ped* peduncle, *s* sulcus, *sup* superior, *t* tract, *tt* tract

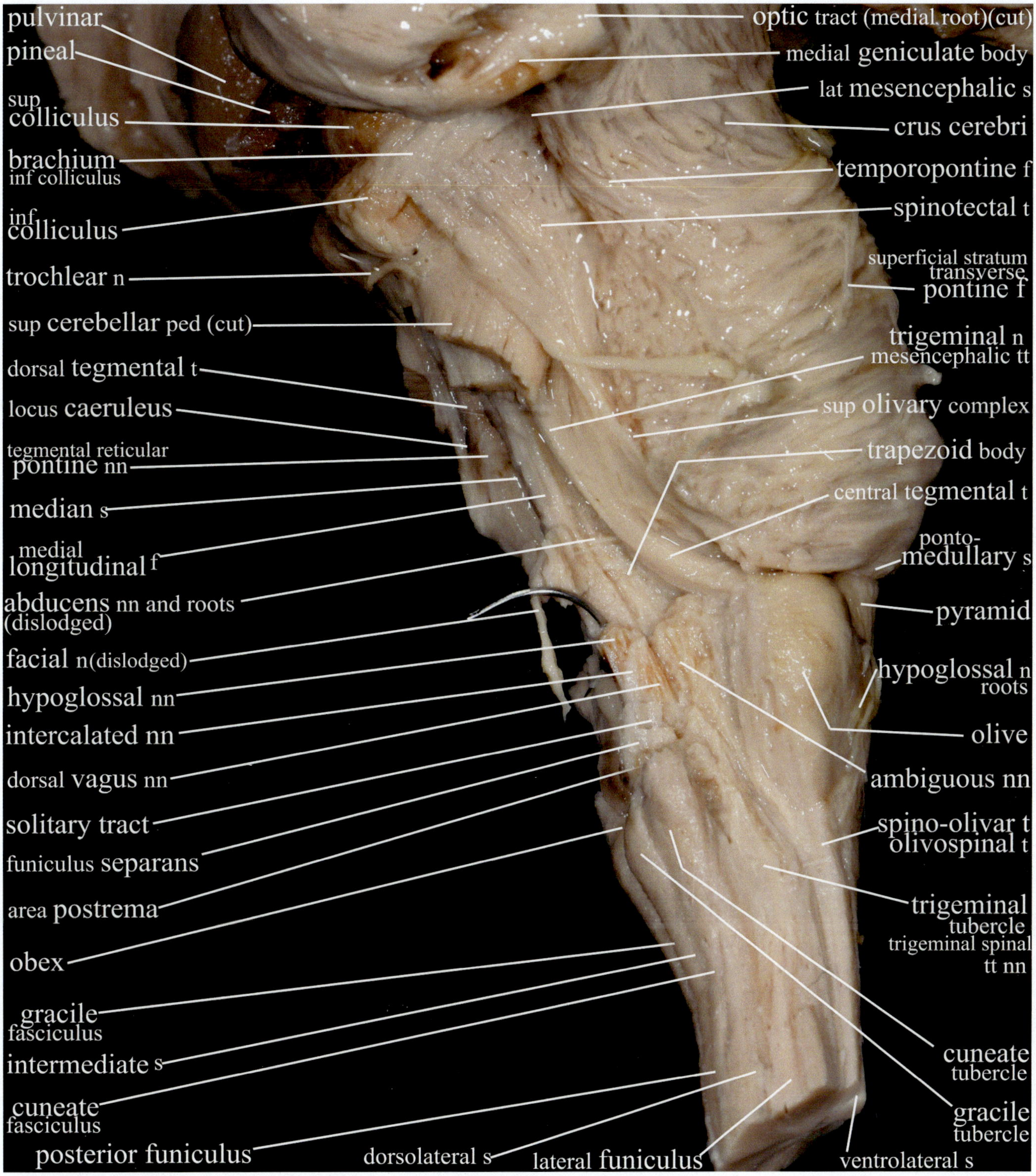

Fig. 8.91 (continued)

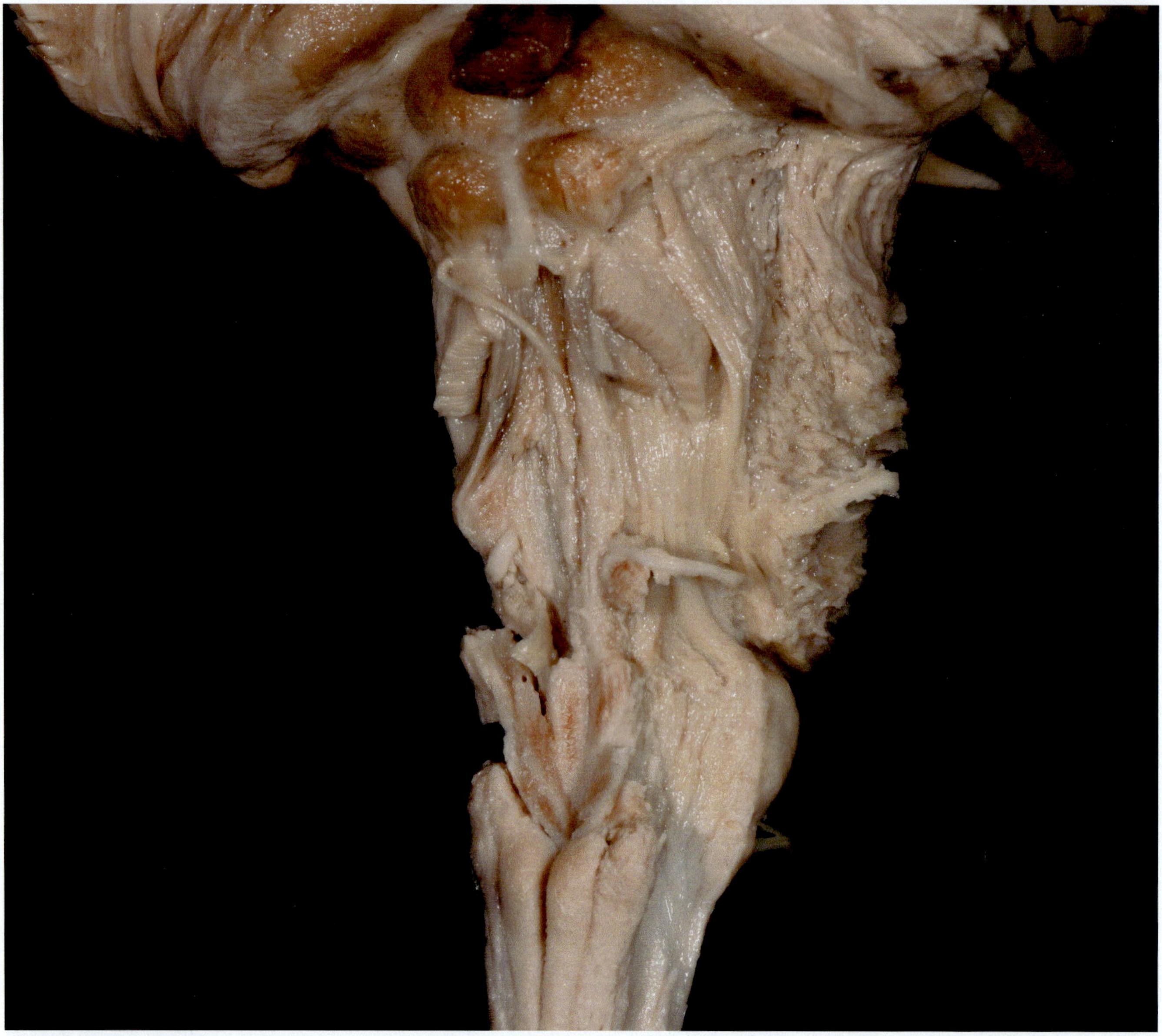

Fig. 8.92 Anatomical preparation of the floor of the fourth ventricle. The dorsal tegmentum has been partially removed to expose the longitudinal tracts of the ventral tegmentum. *f* fasciculus, *inf* inferior, *lat* lateral, *med* medial, *n* nerve, *nn* nucleus, *ped* peduncle, *s* sulcus, *sup* superior, *t* tract, *tt* tract (©Kadri 2023. All Rights Reserved)

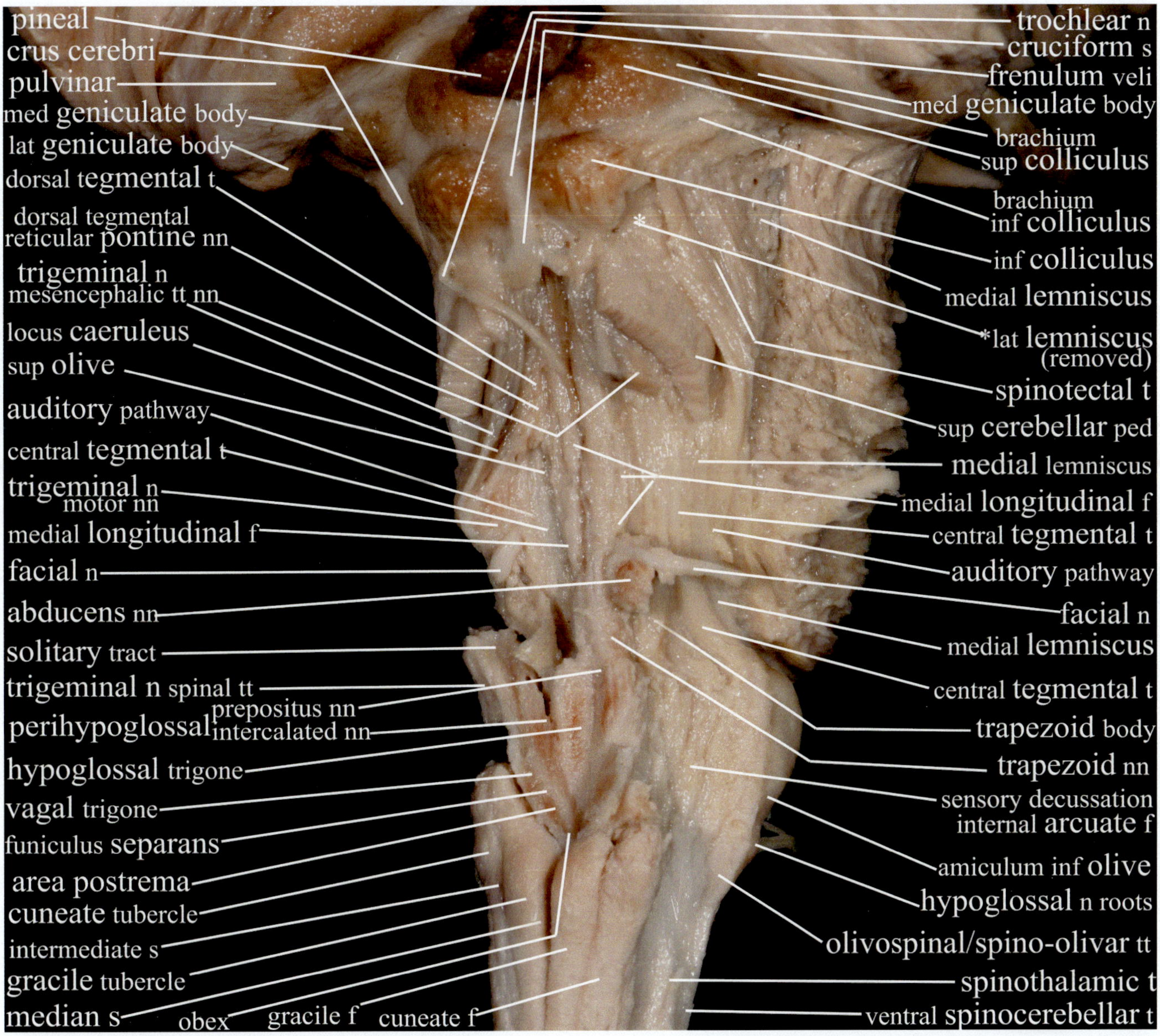

Fig. 8.92 (continued)

References

1. Nieuwenhuys R, Voogd J, van Huijzen C. The human central nervous system: a synopsis and atlas. 4th ed. Berlin: Springer; 2008.
2. Testut L, Latarjet A. Tratado de anatomia humana, vol. 2. 9th ed. Barcelona: Salvat Editores; 1966.
3. Federative Committee on Anatomical Terminology. Terminologia anatomica: international anatomical terminology. Stuttgart: Thieme; 1998.
4. Bochdalek VA. Anleitung zur praktischen Zergliederung des menschlichen Gehirnes, nebst einer anatomischen Beschreibung desselben; mit besonderer Rücksicht auf das kleine Gehirn. Prague: Druck und Papier von Gottlieb Haase Söhne; 1833.

Index

P. Kadri, *The Cartographic Atlas of the Brain*, https://doi.org/10.1007/978-3-031-38062-4